HEINEMANN PHYSICS 11

5TH EDITION

COORDINATING AUTHOR
Sam Trafford

AUTHORS
Liz Black
Amber Dommel
Norbert Dommel
Tracey Fisher
Kaitlin Jobson
Geoff Lewis
David Madden
Greg Moran
Gregory White

VCE UNITS 1 AND 2 • 2023–2027

Pearson Australia
(a division of Pearson Australia Group Pty Ltd)
459–471 Church Street
Level 1, Building B
Richmond, Victoria 3121
www.pearson.com.au

First published 2023 by Pearson Australia
2026 2025 2024 2023
10 9 8 7 6 5 4 3 2 1

VCE Heinemann Project Leads: Bryonie Scott, Misal Belvedere, Malcolm Parsons
Physics Content and Learning Specialists: Bryonie Scott, Zoe Hamilton
Senior Development Editor: Fiona Cooke
Development Editor: Lucy Bates
Schools Programme Manager: Michelle Thomas
Senior Production Editor: Laura Pietrobon
Series Designer: Anne Donald
Rights & Permissions Editor: Amirah Fatin Mohamed Sapi'ee
Publishing Services Analyst: Jit-Pin Chong
Design Analyst: Jennifer Johnston
Illustrators: QBS, DiacriTech
Editor: Marta Veroni

Proofreader: Marcia Bascombe
Indexer: Max McMaster

Printed in Malaysia (CTP-VVP)
National Library of Australia Cataloguing-in-Publication entry

A catalogue record for this book is available from the National Library of Australia

ISBN: 978 0 6557 0007 4

Pearson Australia Group Pty Ltd ABN 40 004 245 943

Disclaimer
The selection of internet addresses (URLs) provided for this book was valid at the time of publication and was chosen as being appropriate for use as a secondary education research tool. However, due to the dynamic nature of the internet, some addresses may have changed, may have ceased to exist since publication, or may inadvertently link to sites with content that could be considered offensive or inappropriate. While the authors and publisher regret any inconvenience this may cause readers, no responsibility for any such changes or unforeseeable errors can be accepted by either the authors or the publisher.

Indigenous Australians
We respectfully acknowledge the traditional custodians of the lands upon which the many schools throughout Australia are located. We acknowledge traditional Indigenous Knowledge systems are founded upon a logic that developed from Indigenous experience with the natural environment over thousands of years. This is in contrast to Western science, where the production of knowledge often takes place within specific disciplines. As a result, there are many cases where such disciplinary knowledge does not reflect Indigenous worldviews, and can be considered offensive to Indigenous peoples.

Some of the images used in *Heinemann Physics 11* 5th edition might have associations with deceased Indigenous Australians. Please be aware that these images might cause sadness or distress in Aboriginal or Torres Strait Islander communities.

Practical activities
All practical activities, including the illustrations, are provided as a guide only and the accuracy of such information cannot be guaranteed. Teachers must assess the appropriateness of an activity and take into account the experience of their students and facilities available. Additionally, all practical activities should be trialled before they are attempted with students and a risk assessment must be completed. All care should be taken and appropriate personal protective clothing and equipment should be worn when carrying out any practical activity. Although all practical activities have been written with safety in mind, Pearson Australia and the authors do not accept any responsibility for the information contained in or relating to the practical activities, and are not liable for loss and/or injury arising from or sustained as a result of conducting any of the practical activities described in this book.

HEINEMANN PHYSICS 11

5TH EDITION

Writing and development team

We are grateful to the following people for their time and expertise in contributing to the **Heinemann Physics 11** project.

Sam Trafford
Freelance science editor and communicator
Coordinating author, reviewer and digital question content lead

Doug Bail
Education consultant
Skills and Assessment author
SPARKlab developer

Stephanie Bernard
PhD candidate, Science communicator
Answer checker

Liz Black
Teacher
Author and digital question author

Reuben Bolt
Deputy Vice-Chancellor First Nations Leadership
Reviewer

Stephen Brown
Teacher
Answer checker

Jason Dicker
Retired teacher
Teacher support author and reviewer

Amber Dommel
Teacher and Science coordinator
Author

Norbert Dommel
Lecturer
Author

Tracey Fisher
Teacher
Author and digital question author

Richard Hecker
Tutor
Digital question author

Kaitlin Jobson
Teacher
Author and digital question author

Allan Knight
Science education consultant
Reviewer

Geoff Lewis
Retired teacher
Author

David Madden
Science education consultant
Author and digital question author

Svetlana Marchouba
Laboratory technician
Teacher support author

Veronika Miles
Tutor
Answer checker

Greg Moran
Teacher
Author and digital question author

Michael Ojaimi
Student
Answer checker

Gregory White
Lecturer
Author and digital question author

Maria Woodbury
Head of Mathematics and Science
Digital question author

Laurence Wooding
Laboratory technician
Reviewer and teacher support author

The Publisher wishes to thank and acknowledge Doug Bail, Keith Burrows, Robert Chapman and Carmel Fry for their contribution in creating the original works of the series and longstanding dedicated work with Pearson and Heinemann.

Contents

AREA OF STUDY 3

How can electricity be used to transfer energy?

Unit 2 How does physics help us to understand the world?

AREA OF STUDY 1

How is motion understood?

AREA OF STUDY 2

Heinemann Physics 11 5th edition includes a comprehensive set of resources to support Area of Study 2 via your Pearson Places bookshelf.

AREA OF STUDY 3

Heinemann Physics 11 5th edition includes a comprehensive set of resources to support Area of Study 3 via your Pearson Places bookshelf.

How to use this book

Heinemann Physics 11 5th edition

Heinemann Physics 11 5th edition has been written to the new VCE Physics Study Design 2023–2027. The book covers Units 1 and 2. Explore how to use this book below.

Chapter opener

Chapter opening pages link the study design to the chapter content. Key knowledge addressed in the chapter is clearly listed. To help you find where each outcome is covered in the chapter, the relevant section numbers are written in bold.

CHAPTER 08 Electrical physics

Every object around you is made up of charged particles. When these particles move relative to one another, we experience a phenomenon known as 'electricity'. This chapter looks at the fundamental concepts, such as current and voltage, that scientists have developed to explain electrical phenomena. This will provide the foundation for studying practical electric circuits in the following chapter.

Key knowledge

- apply concepts of charge (Q), electric current (I), potential difference (V), energy (E) and power (P), in electric circuits **8.1, 8.2, 8.3**
- analyse and evaluate different analogies used to describe electric current and potential difference **8.2, 8.3**
- investigate and analyse theoretically and practically electric circuits using the relationships: $I = \frac{Q}{t}, V = \frac{E}{Q}, P = \frac{E}{t} = VI$ **8.2, 8.3**
- justify the use of selected meters (ammeter, voltmeter, multimeter) in circuits **8.3**
- model resistance in series and parallel circuits using:
  - current versus potential difference (I–V) graphs **8.4**
  - resistance as the potential difference to current ratio, including R = constant for ohmic devices. **8.4**

VCE Physics Study Design extracts © VCAA (2022); reproduced by permission

Case study

Case studies place physics in an applied situation or relevant context. Text and artwork refer to the nature and practice of physics, applications of physics and associated issues, and the historical development of physics concepts and ideas.

CASE STUDY

Superconductors

Materials are classified as conductors, semiconductors or insulators when their electrical properties are taken into consideration. Scientists of the late 1800s also knew that the resistivity, hence resistance, of a material was temperature dependent and that this included some materials that were insulators at room temperature.

Further research on this led to the discovery of a new material classification of superconductors. Superconductivity occurs at very low temperatures at which the resistivity of a material is found to be zero and no interior magnetic field exists within it. Present research is trying to find materials that are superconductors at higher and near-room temperatures.

In 1908, Dutch physicist Kamerlingh Onnes (1853–1926) was the first to liquefy helium at 4.2 K (–269°C). Using this, he began investigating the resistance of pure metals at very low temperatures.

He immersed a solid wire of mercury into liquid helium and found that its resistance was indeed zero at 4.2 K. He called this the superconducting state. Theoretically, once an electric current was started in a loop maintained at such temperatures it would circulate indefinitely. For this work Onnes was awarded the 1913 Nobel Prize in Physics.

Other metals were soon found to become superconductors at extremely low temperatures; for example, aluminium at 1.2 K and lead at 7.9 K. However, not all metals exhibit superconductivity.

The mechanism by which a metal's electrical resistance drops to zero was not understood until 1957 when the American physicists John Bardeen, Leon Cooper and Robert Schrieffer published a paper that used quantum mechanics, a theory not known in Onnes's era, to explain the phenomenon. This is now known as the BCS theory, and for their work Bardeen, Cooper and Schrieffer were awarded the 1972 Nobel Prize in Physics.

Materials classified as high temperature (above 77 K) superconductors have also been found. These use liquid nitrogen instead of liquid helium as it is cheaper to produce, much easier to handle and more readily available. The ultimate goal, and the focus of current research, is to discover a material that will become superconducting at room temperature. Such a material would have many applications and would not require all the equipment and expense required to maintain it at low, or extremely low, temperatures.

The application of the extremely powerful and special magnetic properties of many superconducting materials has seen technological advancement in many areas. Some examples include transportation using magnetic levitation (Maglev trains, Figure 9.1.12), non-invasive medical diagnosis (MRI, Figure 9.1.13), cancer treatment using proton therapy, determining atomic structures of molecules (NMR), magnetic confinement fusion reactors, land-mine detectors, electricity generation and transmission, as beam steering and focusing magnets in particle accelerators (the Large Hadron Collider, Figure 9.1.14) and in the development of supercomputers.

FIGURE 9.1.12 Maglev train. This part of Japan's rail system uses superconductors to provide the magnetic levitation of the train and carriages.

FIGURE 9.1.13 Magnetic resonance imaging (MRI) devices use superconducting wire and electromagnets cooled with liquid helium to produce detailed images.

FIGURE 9.1.14 Two LHC magnets before connection. The blue cylinders contain the magnets and a liquid helium system that cools them so they become superconducting.

Case study: analysis

These case studies include real-world data that can be analysed and evaluated.

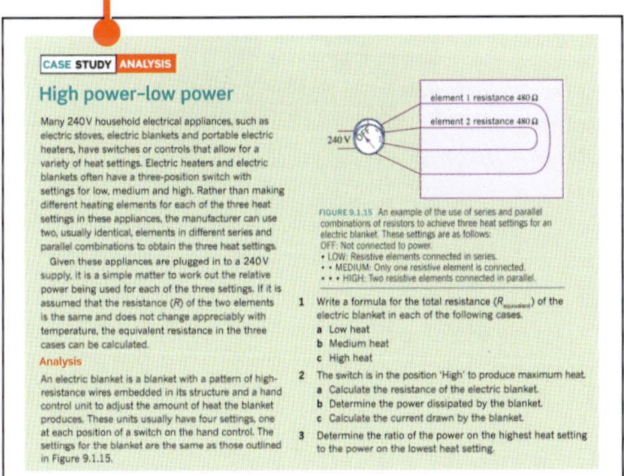

CASE STUDY ANALYSIS

High power–low power

Many 240 V household electrical appliances, such as electric stoves, electric blankets and portable electric heaters, have switches or controls that allow for a variety of heat settings. Electric heaters and electric blankets often have a three-position switch with settings for low, medium and high. Rather than making different heating elements for each of the three heat settings in these appliances, the manufacturer can use two, usually identical, elements in different series and parallel combinations to obtain the three heat settings.

Given these appliances are plugged in to a 240 V supply, it is a simple matter to work out the relative power being used for each of the three settings. If it is assumed that the resistance (R) of the two elements is the same and does not change appreciably with temperature, the equivalent resistance in the three cases can be calculated.

Analysis

An electric blanket is a blanket with a pattern of high-resistance wires embedded in its structure and a hand control unit to adjust the amount of heat the blanket produces. These units usually have four settings, one at each position of a switch on the hand control. The settings for the blanket are the same as those outlined in Figure 9.1.15.

FIGURE 9.1.15 An example of the use of series and parallel combinations of resistors to achieve three heat settings for an electric blanket. These settings are as follows:
OFF: Not connected to power.
• LOW: Resistive elements connected in series.
• • MEDIUM: Only one resistive element is connected.
• • • HIGH: Two resistive elements connected in parallel.

1 Write a formula for the total resistance ($R_{equivalent}$) of the electric blanket in each of the following cases.
   a Low heat
   b Medium heat
   c High heat
2 The switch is in the position 'High' to produce maximum heat.
   a Calculate the resistance of the electric blanket.
   b Determine the power dissipated by the blanket.
   c Calculate the current drawn by the blanket.
3 Determine the ratio of the power on the highest heat setting to the power on the lowest heat setting.

Highlight

Highlight boxes focus on important information such as key definitions and summary points.

Worked example

Worked examples are set out in steps that show thinking and working. Each Worked example is followed by a Worked example: Try yourself that allows students to test their understanding.

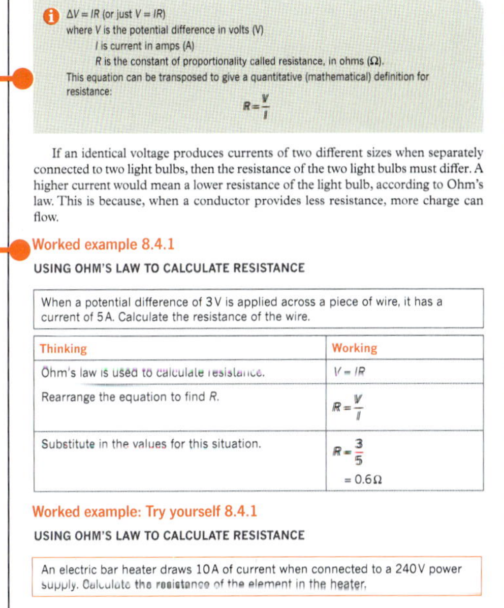

$\Delta V = IR$ (or just $V = IR$)
where V is the potential difference in volts (V)
I is current in amps (A)
R is the constant of proportionality called resistance, in ohms (Ω).
This equation can be transposed to give a quantitative (mathematical) definition for resistance:

$$R = \frac{V}{I}$$

If an identical voltage produces currents of two different sizes when separately connected to two light bulbs, then the resistance of the two light bulbs must differ. A higher current would mean a lower resistance of the light bulb, according to Ohm's law. This is because, when a conductor provides less resistance, more charge can flow.

Worked example 8.4.1

USING OHM'S LAW TO CALCULATE RESISTANCE

When a potential difference of 3 V is applied across a piece of wire, it has a current of 5 A. Calculate the resistance of the wire.

Thinking	Working
Ohm's law is used to calculate resistance.	$V = IR$
Rearrange the equation to find R.	$R = \frac{V}{I}$
Substitute in the values for this situation.	$R = \frac{3}{5}$ $= 0.6\,\Omega$

Worked example: Try yourself 8.4.1

USING OHM'S LAW TO CALCULATE RESISTANCE

An electric bar heater draws 10 A of current when connected to a 240 V power supply. Calculate the resistance of the element in the heater.

Icons

This icon is used to alert you to engage with auto-corrected questions through Pearson Places.

These icons indicate when it is the best time to engage with a worksheet (WS), a practical activity (PA) or exam questions (EQ) in the *Heinemann Physics 11 Skills and Assessment* book.

Chapter review

Each chapter concludes with a list of key terms and questions that test your understanding of the key knowledge covered in the chapter.

Section summary

Each section includes a summary to help you consolidate key points and concepts.

Section review

Each section concludes with questions that test your ability to recall, explain and apply key concepts.

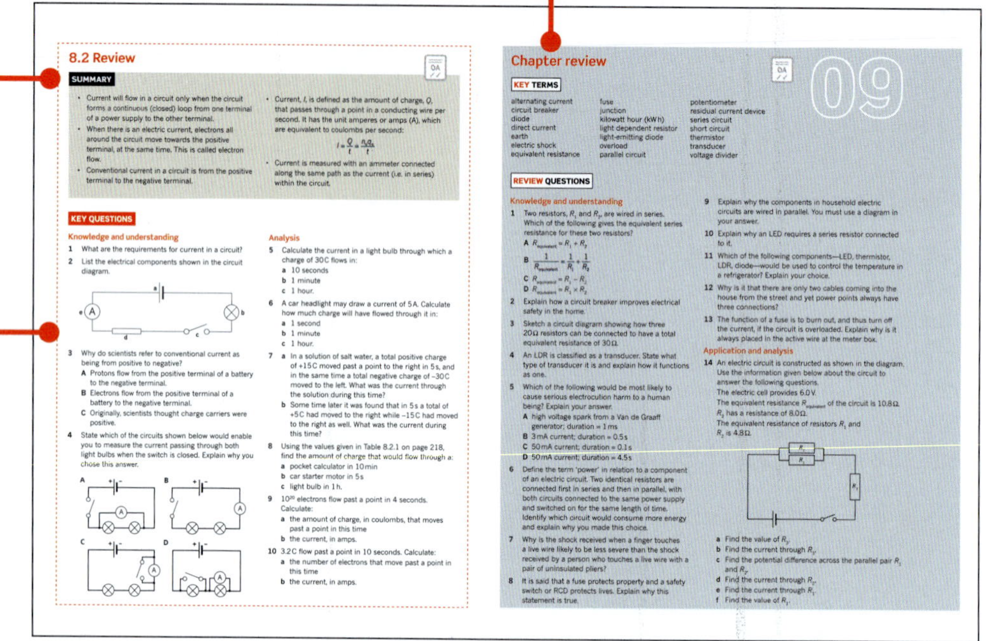

Area of Study review

Each area of study concludes with a comprehensive set of exam-style questions, including multiple choice and short answer, to support you in your exam preparation.

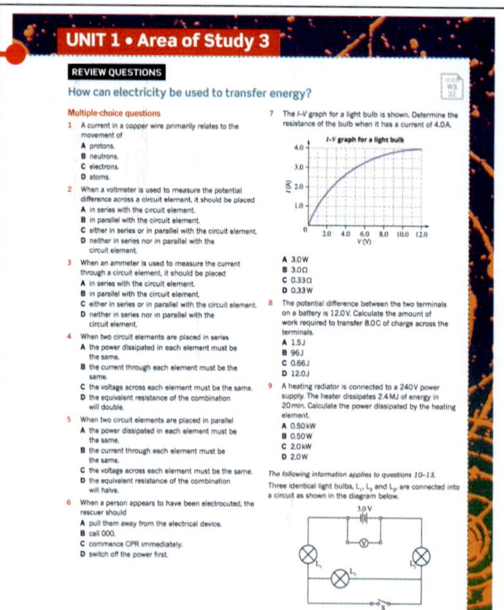

PhysicsFile

PhysicsFiles include interesting information and real-world examples.

Answers

Comprehensive answers for all section review, chapter review and Area of Study review questions are provided via the *Heinemann Physics 11 5th edition eBook + Assessment.*

Glossary

Key terms are shown in **bold** throughout and listed at the end of each chapter. A comprehensive glossary at the end of the book defines all key terms.

CHAPTER

01 Scientific investigation

Gaining a sound understanding of key science skills in a variety of contexts is essential preparation for undertaking scientific investigations and evaluating the research of others. The development of such skills is a core component of VCE Physics, and these skills apply across all units and areas of study.

Chapter 1 describes some of the most fundamental science skills. The chapter can be read as a whole or referred to as you work through other chapters. For example, it may provide you with a useful refresher on the scientific method of conducting investigations, or on what is typically included in a scientific report, at the time when you are to undertake an investigation as part of your studies.

Key science skills

Develop aims and questions, formulate hypotheses and make predictions

- identify, research and construct aims and questions for investigation **1.1**
- identify independent, dependent and controlled variables in experiments **1.1**
- formulate hypotheses to focus investigations **1.1**
- predict possible outcomes of investigations **1.1**

Plan and conduct investigations

- determine appropriate investigation methodology: case study; classification and identification; experiment; fieldwork; literature review; modelling; product, process or system development; simulation **1.1**
- design and conduct investigations: select and use methods appropriate to the selected investigation methodology, including consideration of equipment and procedures, taking into account potential sources of error and causes of uncertainty; determine the type and amount of qualitative and/or quantitative data to be generated or collated **1.2**
- work independently and collaboratively as appropriate and within identified research constraints, adapting or extending processes as required and recording such modifications in a logbook **1.2**

Comply with safety and ethical guidelines

- demonstrate safe laboratory practices when planning and conducting investigations by using risk assessments that are informed by safety data sheets (SDS), and accounting for risks **1.2**
- apply relevant occupational health and safety guidelines while undertaking practical investigations **1.2**
- demonstrate ethical conduct when undertaking and reporting investigations **1.2**

Generate, collate and record data

- systematically generate and record primary data, and collate secondary data, appropriate to the investigation, including use of databases and reputable online data sources **1.3**
- record and summarise both qualitative and quantitative data, including use of a logbook as an authentication of generated or collated data **1.2, 1.3**
- organise and present data in useful and meaningful ways, including tables and graphs **1.3**

Analyse and evaluate data and investigation methods

- process quantitative data using appropriate mathematical relationships and units **1.4**
- use appropriate numbers of significant figures in calculations **1.3**, **1.4**
- construct graphs that show the relationship between variables **1.4**
- extrapolate to determine graph intercepts of significance **1.4**
- construct linearised graphs and identify the significance of the gradient (using relationships relevant to the key knowledge outlined in the areas of study) **1.4**
- identify and analyse experimental data qualitatively, handling, where appropriate, concepts of: accuracy, precision, repeatability, reproducibility, resolution and validity of measurements; and errors (random and systematic) **1.2, 1.4**
- identify outliers, and contradictory, provisional or incomplete data **1.4**
- repeat experiments to evaluate the precision of data **1.4**
- evaluate investigation methods and possible causes of error and uncertainty, and suggest how precision can be improved, and how uncertainty can be reduced **1.4**, **1.5**

Construct evidence-based arguments and draw conclusions

- distinguish between opinion and evidence, and between scientific and non-scientific ideas **1.5**
- evaluate data to determine the degree to which the evidence supports the aim of the investigation, and make recommendations, as appropriate, for modifying or extending the investigation **1.5**
- evaluate data to determine the degree to which the evidence supports or refutes the initial prediction or hypothesis **1.5**
- use reasoning to construct scientific arguments, and to draw and justify conclusions consistent with evidence and relevant to the question under investigation **1.5**
- identify, describe and explain the limitations of conclusions, including identification of further evidence required **1.5**
- discuss the implications of research findings **1.5**

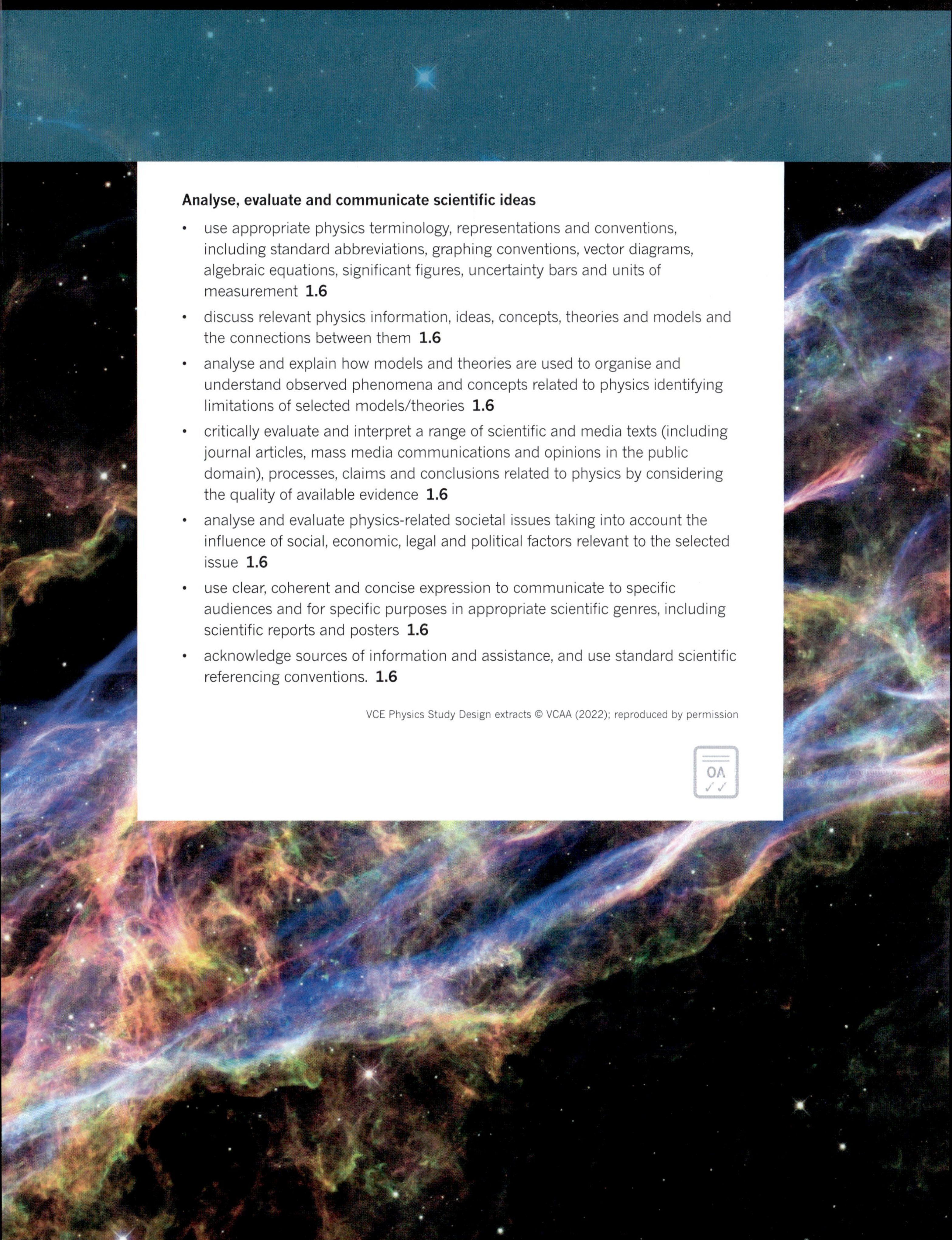

Analyse, evaluate and communicate scientific ideas

- use appropriate physics terminology, representations and conventions, including standard abbreviations, graphing conventions, vector diagrams, algebraic equations, significant figures, uncertainty bars and units of measurement **1.6**
- discuss relevant physics information, ideas, concepts, theories and models and the connections between them **1.6**
- analyse and explain how models and theories are used to organise and understand observed phenomena and concepts related to physics identifying limitations of selected models/theories **1.6**
- critically evaluate and interpret a range of scientific and media texts (including journal articles, mass media communications and opinions in the public domain), processes, claims and conclusions related to physics by considering the quality of available evidence **1.6**
- analyse and evaluate physics-related societal issues taking into account the influence of social, economic, legal and political factors relevant to the selected issue **1.6**
- use clear, coherent and concise expression to communicate to specific audiences and for specific purposes in appropriate scientific genres, including scientific reports and posters **1.6**
- acknowledge sources of information and assistance, and use standard scientific referencing conventions. **1.6**

1.1 Planning scientific investigations

FIGURE 1.1.1 Students conduct an experiment investigating the wave nature of light.

Physics is the study of motion, force and energy. It deals with the laws, theories and principles governing motion, force and energy across all imaginable scales throughout the universe. Its scope ranges from the interactions between particles that make up individual protons and neutrons inside every atomic nucleus to the motions of galaxies that comprise the universe.

As scientists, physicists extend their understanding using the scientific method, which involves investigations that are carefully designed, conducted and reported (Figure 1.1.1). Well-designed research is based on a sound knowledge of what is already understood about a subject, as well as careful preparation and observation.

Taking the time to carefully plan and design a scientific investigation before beginning will help you keep focused throughout. Preparation is essential. You should ensure that you understand the theory behind your investigation and prepare a detailed plan regarding the practical aspects of the investigation. This section is a guide to some of the key steps that should be taken when planning and designing a scientific investigation.

OBSERVATION

Scientific investigations start with careful **observation**. Observation involves using all your senses: sight, sound, smell, taste and touch. For example, an observation may involve seeing a change in colour in an object or noticing the smell from a chemical reaction. It also involves using instruments and laboratory techniques that may give you sharper and more detailed observations than those possible with just your senses, or perhaps may be safer to use (for example, using sensors to detect the radioactivity of a source).

The idea for a primary investigation of a complex problem arises from prior learning and observations that have raised further questions. How observations are interpreted depends on past experiences and knowledge. But to enquiring minds, observations will usually provoke further questions. Some examples are given below.

- How is the motion of an object modelled (e.g. why does a javelin travel farther than a cricket ball)?
- How does the motion of a student change using timed intervals of measured distance (e.g. student riding a bicycle)?
- What causes an object to accelerate (e.g. when an object is dropped from a certain height)?
- What modifications could be made to a circuit to reduce its power consumption (e.g. by lowering the resistance)?
- What applications are there of total internal reflection (e.g. fibre optic cables)?

Many of these questions cannot be answered by observation alone, but they can be answered through scientific investigation.

Many great discoveries have been made when a scientist is busy investigating some other problem. Good scientists have acute powers of observation and enquiring minds, and they make the most of these chance opportunities.

THE LOGBOOK

Careful and accurate record keeping is essential to good science. It is important that the methods we use in our investigations, our observations, analysis of data and conclusions are recorded. This ensures that the way our investigation was carried out, as well as our observations, analysis and conclusions can be shared with other scientists who can verify our work. This is an important part of the way scientific ideas are progressed.

One way of recording scientific investigations is with logbooks and throughout Units 1 and 2, and during your scientific investigations for Unit 2 Area of Study 2 and Unit 2 Area of Study 3, you must keep a logbook that includes every detail of your research.

The following checklist will help you remember what to include in your logbook:

- p your ideas when planning the research
- p clear protocols for each stage of your investigation (e.g. what standard procedures you will use and follow exactly each time)
- p instructions, noting exactly what needs to be recorded
- p tables ready for data entry
- p records of all materials, methods, experiments and raw data
- p all notes, sketches, photographs and results
- p records of any incidents or errors that may influence the results.

THE SCIENTIFIC METHOD

Before conducting an investigation, a scientist develops a clear and specific research question to explore. They state an aim that describes the purpose of their planned investigation. They then state a hypothesis; that is, a prediction based on scientific reasoning that can be tested experimentally. This is the basis of the **scientific method** (Figure 1.1.2).

- A **research question** describes the idea to be investigated. For example: What is the relationship between voltage and current?
- An **aim** is a statement describing in detail the purpose of the investigation. For example: The aim of the investigation is to investigate the relationship between voltage and the current in a circuit of constant resistance.
- A **hypothesis** is a prediction that proposes an answer to a research question. It is a prediction that can be tested by observation or experimentation. For example: For the 5 Ω resistor used in this investigation, the current will increase as the voltage is increased from 3 V to 12 V.

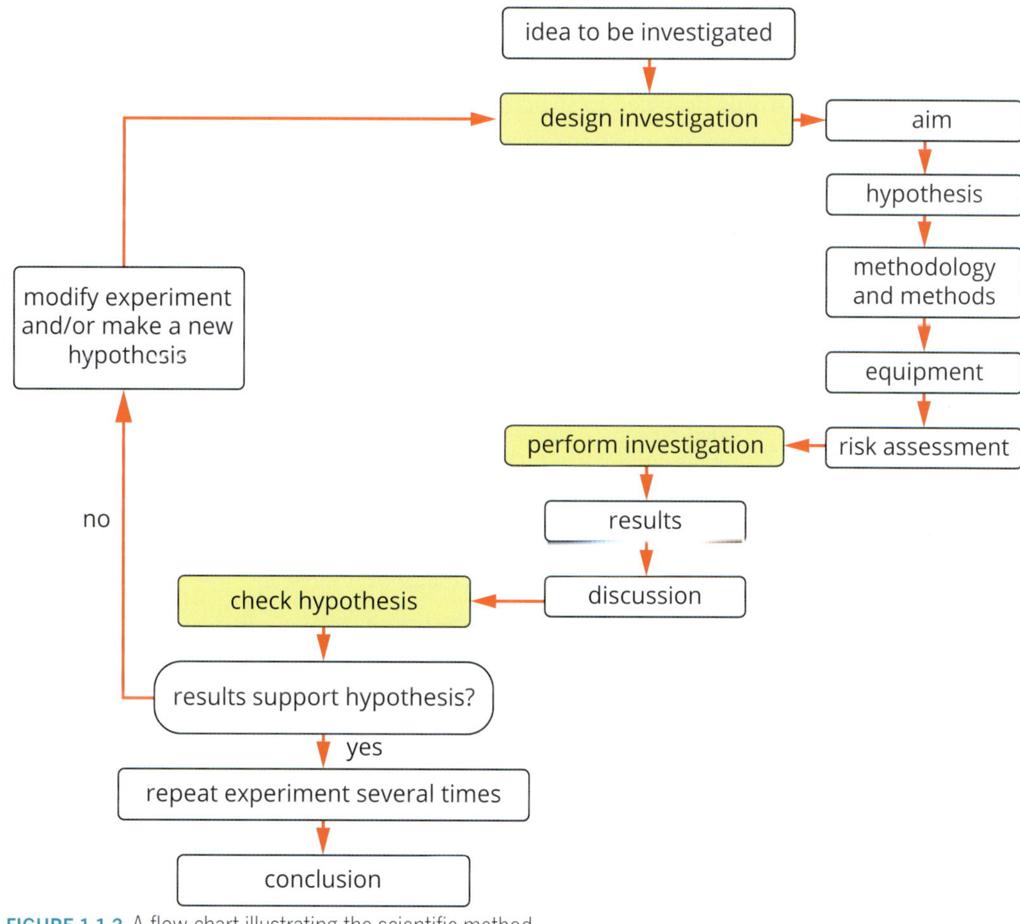

FIGURE 1.1.2 A flow chart illustrating the scientific method

CASE STUDY

The scientific method and gravitational waves

In the early part of the twentieth century, Einstein formulated the general theory of relativity. This theory predicted gravitational waves. Although hypothesised, scientists of the day were unsure how to detect them. Two American scientists, Kip Thorne from California Institute of Technology (Caltech) and Rainer Weiss from Massachusetts Institute of Technology (MIT), started to collaborate on possible ways to detect these waves. In 1984, Caltech and MIT set up a joint project called LIGO (Laser Interferometer Gravitational Wave Observatory).

There are over 1000 scientists working on the LIGO project. They are busy interpreting readings from their detection equipment, improving that equipment and, where necessary, modifying their hypotheses.

The LIGO team finally observed gravitational waves in 2015, almost 100 years after Einstein predicted them. It is the nature of scientific research and discovery that no scientist works alone. They work with others and build on the research done by their colleagues and predecessors.

To build on the discoveries by the LIGO team, the OzGrav program was started. OzGrav is an Australian Government funded project through the Australian Research Council (ARC), several Australian universities, and other collaborating organisations in Australia and overseas. OzGrav aims to understand the physics of black holes and curved spacetime, and to inspire future Australian scientists and engineers to continue this field of study.

Carefully designed scientific investigations are conducted to determine whether predictions are accurate or not. If the results of an investigation do not fall within an acceptable range, the hypothesis is rejected. If the predictions are found to be accurate, the hypothesis is supported. If, after many investigations, one hypothesis is supported by all the results obtained so far, then it is given the status of a theory or principle.

There is nothing mysterious about the scientific method. You might use the same process to find out how an unfamiliar machine works if you had no instructions. Careful observation is usually the first step.

● *You will now be able to answer key question 3.*

Formulating a research question

The research question at the centre of a scientific investigation directs the inquiry. Its primary purpose is to clearly set the boundaries of the investigation, specify the direction of the research, and guide all stages of inquiry, analysis, interpretation and evaluation. A research question should:

- clearly identify the topic of the experiment
- be specific enough to ensure a clear and unambiguous approach
- specify the scope or conditions of the inquiry
- propose to find trends, patterns or relationships between two measurable **variables**.

Relevant background research can help refine the question. The research could include:

- information about the variables to be explored
- correlations between these variables
- ideas for refining the question.

The structure of a research question

Table 1.1.1 provides examples of different types of research questions. You can use the guiding words provided to structure your research question.

If possible, research questions should make reference to the independent and the dependent variables that are to be explored. That is, the question should reference the variables the researcher will be manipulating (independent variables) and those that are expected to change in response (dependent variables). Each question should ask if, how or why the independent variable affects the dependent variable.

TABLE 1.1.1 Examples of how research questions can be constructed

Guiding word	Example research question
What	**What** difference does a change in resistance (independent variable) have on the current in a circuit (dependent variable)?
Will	**Will** a series circuit use more current than a parallel circuit?
How	**How** does the angle of an inclined plane (independent variable) affect the force of gravity on an object (dependent variable)?
Why	**Why** does the mass of an object affect the momentum?
Is/are	**Is** energy conserved in an inelastic collision?
Can	**Can** you charge a Perspex rod using a sheet of glass?
Do/does	**Does** the strength of Earth's gravitational field depend on the height above sea level?

Before formulating a research question, it is good practice to:

1. conduct a literature review of the topic to be investigated
2. become familiar with the relevant scientific concepts and key terms
3. write down questions or correlations as they arise
4. compile a list of possible ideas.

Avoid rejecting ideas that initially might seem impossible. Use these ideas to generate questions that are answerable.

Selecting and evaluating your question

When selecting a topic for your investigation, it is helpful to choose something that you already have some knowledge of and that you find interesting.

Ask yourself what the data you intend to collect might look like. If you cannot determine measurable variables, either pick another topic or modify your question to allow you to take measurements.

You should choose a question that can be explored within the conditions, time and equipment you have available to you.

Once a research question has been chosen, stop and evaluate it. Check if it requires further refinement or even further investigation before it is suitable as a basis for an achievable and worthwhile investigation. The checklist below will assist you in evaluating your research question.

- ☐ Relevance—make sure your research question is related to your chosen topic.
- ☐ Clarity and measurability—make sure your question can be framed as a clear hypothesis; otherwise it is going to be difficult to complete your research.
- ☐ Time frame—make sure your question can be answered within a reasonable period of time. This may not be possible if your question is too broad.
- ☐ Knowledge and skills—make sure you have the level of knowledge and laboratory, fieldwork and technical skills that will enable you to explore the question within the specified time frame. Keep the question simple and the outcomes of the research achievable.
- ☐ Practicality—make sure that the resources that will be needed, such as laboratory equipment and materials, are readily available. Keep things simple. Avoid investigations that require sophisticated or rare equipment. Instead, choose an investigation that requires only common laboratory equipment, such as timing devices, objects that could be used as projectiles and a tape measure.
- ☐ Safety and ethics—(discussed in more detail in Section 1.2) consider the safety and ethical issues associated with the research question you will be investigating. If there are issues, determine if these need to be addressed before you begin your investigation.
- ☐ Advice—seek advice from your teacher on your question. Their input may prove very useful, as their experience may lead you to consider aspects of the question that you have not thought about.

Finally, confirm the suitability of your research question with your teacher.

Defining the aim of the investigation

The aim sets out the purpose of the scientific investigation. It should directly mention the variables involved in the investigation and describe, in general terms, how each will be measured. An aim does not need to include the details of the proposed method.

An example for a laboratory experiment:

- Aim: The aim of the experiment is to investigate the relationship between force, mass and acceleration.
- In the first stage of this experiment, mass will be the independent variable (a number of different masses will be selected) and the force kept constant. The resulting acceleration (dependent variable) will be measured.
- In the second stage of the experiment, force will be the independent variable (a number of different forces will be tested) and the mass will be kept constant. The resulting acceleration (dependent variable) will be measured.
- Using the data collected from both stages of the experiment, the relationship between the three variables can be determined.

● *You will now be able to answer key question 1.*

PHYSICSFILE

Hypothetical neutrino

In 1934 Enrico Fermi proposed the nuclear electron hypothesis. He explained that the beta particles observed from radioactive processes were from the radioactive decay of a neutron. Theory suggested that a neutron would decay into a positively charged proton, a negatively charged electron, sometimes gamma radiation, and a then-theoretical particle (called the neutrino) to preserve energy and momentum considerations; experimental proof of the idea took another 20 years.

Formulating a hypothesis

Once the research question has been finalised and the aim of the research is clear, formulating a hypothesis is the next step. A hypothesis is the prediction that is to be tested by the evidence you will collect during your investigation. A hypothesis proposes a relationship between two or more variables. It should predict that a relationship exists or does not exist.

When writing the hypothesis, first identify the variables in your research question. There are different ways of constructing your hypothesis, but you may want to write the statement in the form of 'If … is done, then … will occur.' Ensure that the independent and dependent variables are included in this statement.

For example: If force (independent variable) is increased, then acceleration (dependent variable) will increase in proportion with the force.

A hypothesis does not need to include 'if' and 'then' in its wording. For example, the previous hypothesis could also be stated in the following way: As force increases, acceleration increases proportionally.

A good hypothesis should:

- be a statement, not a question
- be based on information contained in the research question and the aim
- be worded so that it can be tested in the experiment you are planning
- include an independent and a dependent variable
- include only variables that are measurable.

The hypothesis should also be falsifiable. This means that a negative outcome would disprove it. For example, the hypothesis that all apples are round cannot be proved beyond doubt, but it can be disproved—in other words, it is falsifiable. In fact, only one square apple is needed to disprove this hypothesis. Unfalsifiable hypotheses cannot be proved by science. These include hypotheses on ethical, moral and other subjective judgements. For instance, you could hypothesise that plagiarising your scientific report is wrong, but the results of this are a question of ethics not science.

● *You will now be able to answer key question 6.*

Variables

A good scientific hypothesis can be tested; that is, supported or refuted through investigation. To be a testable hypothesis, it must be possible to measure both what is changed and what will happen as a result. Thus a scientific investigation seeks to determine the relationship between variables.

There are three categories of variables:

- The **independent variable** is the variable that is controlled by the researcher (i.e. the one that is changed to see if there is an effect on the dependent variable).
- The **dependent variable** is the variable that may change in response to a change in the independent variable. This is the variable that will be observed.
- **Controlled variables** are the variables that must be kept constant during the investigation.

You should only test one variable at a time, otherwise you cannot be sure which variable has influenced a change in the dependent variable.

The three types of variables can be either qualitative or quantitative:

- **Qualitative variables** can be observed but not measured numerically. The data collected is known as **qualitative data**. They can only be sorted into groups or categories such as brightness, type of material of construction or type of device. There are two main types:
  - Nominal variables are those in which the order is not important; for example, the type of material or type of device.
  - Ordinal variables are those in which order is important and values are therefore ranked; for example, brightness (Figure 1.1.3).

FIGURE 1.1.3 When recording qualitative data, describe in detail how each variable will be defined. For example, if recording the brightness of light globes, pictures are a good way of clearly defining what each assigned term represents.

- **Quantitative variables** can be measured. The data collected is known as **quantitative data**. Length, area, weight, temperature and cost are examples of quantitative data. There are two main types:
  - **Discrete variables** can only take particular values; for example, the number of pins in a packet, the number of springs connected together or the energy levels in atoms.
  - **Continuous variables** allow for any numerical value within a given **range**; for example, the measurement of temperature, length, weight and frequency.

You will now be able to answer key questions 2, 4 and 5.

Methodology and methods

When planning your investigation you will need to decide on the methodology and methods.

The **methodology** section in a research plan is a brief description of the general approach taken to investigate the research question or hypothesis and the reasons why this approach is taken. Examples of scientific investigation methodologies are controlled experiments, fieldwork, literature reviews, modelling and simulation.

The **method** (also known as the procedure) is the set of specific steps that are to be taken to collect data during the investigation. The type of scientific investigation methodology and the methods selected will depend on the research question, the aim of the investigation and the equipment available to you.

For some investigations, setting up an experiment may require equipment that is not readily available to you in the school setting. This may mean you need to consider a computer simulation to model the outcomes of the investigation. Other approaches could include a literature review of other studies that considered a similar research question. The different approaches that you could use are outlined in Table 1.1.2.

TABLE 1.1.2 Scientific investigation methodologies

Type of methodology	Explanation	Example
case study	an investigation of a real or hypothetical situation, such as an activity, event, problem or behaviour, often involving analysis of data within a real-world context	determining the wavelength at which radiated power per unit area is maximum at any given temperature
classification and identification	arranging objects or events into manageable groups by identifying shared or similar features	classifying different stars in the sky
experiment	an experimental investigation that involves formulating a hypothesis and testing the effect of an independent variable on the dependent variable, while controlling all other variables in the experiment	investigating the effect of an object's mass on its momentum
fieldwork	collecting data outside of the laboratory, which could be initiated by the following stimuli: • an excursion • engagement with community experts	establishing a position to the path of the Moon in the night sky
literature review	a critical analysis of what has already been investigated and published, using secondary data from other people's investigations or from experimental research to explain events or propose new ideas or relationships	analysing of data looking at the impact of electromagnetic radiation on cells published in a variety of research papers to support, refute or develop new hypotheses
modelling	using models as representations of objects, systems or processes to aid understanding or make predictions	constructing models of the atom
product, process or system development	using scientific understanding and advances in technology to design a new tool, method or process to meet the demands or needs of society	developing a new cochlear implant to restore hearing
simulation	using mathematical models or simulations to test hypotheses, conduct virtual experiments or model the complexity of the complexity of systems or processes	performing a computer simulation of the spectrum of light emitted by a blackbody as a function of wavelength

PHYSICSFILE

Connecting the world

Have you ever wondered how the internet sends data around the world? Most people believe the world is connected using satellites that transfer information from device to device across countries separated by the oceans. But this is not the case. Data is transferred on fibre optic cables that lie across the vast ocean floors (see figure below). There are hundreds of underwater cables keeping the world connected.

The layout of these fibre optic cables can be complex. The likely impact on coral reefs, aquatic animals and other aquatic systems must be considered when cables are being laid. Underwater topography, volcanic activity and ocean currents also influence where the cables can be laid. Where the cables cross international borders, negotiations between countries may lead to routes that are less than optimal. And when something goes wrong in such a complex network, maintenance and repairs might need to be done in countries far away. So, the next time your internet connection is down, the solution may be far more complicated than simply resetting your modem.

Underwater cables form the backbone of the internet.

SOURCING INFORMATION

When you are sourcing information for your scientific investigation, consider whether the information is from primary or secondary sources. You should also consider the advantages and disadvantages of using such resources as books or the internet.

Primary and secondary sources

Primary sources of information are those created by a person directly involved in an investigation. An example of a primary source is a peer-reviewed scientific article. **Secondary sources** of information are syntheses, reviews or interpretations of primary sources. Examples of secondary sources are textbooks, newspaper articles and websites.

Secondary sources of information may have a bias, so you need to determine if they are reliable sources of information. You will learn about assessing the accuracy, reliability and validity of data in Section 1.2.

Table 1.1.3 compares primary and secondary sources.

TABLE 1.1.3 Summary of primary and secondary sources

Type of methodology	Primary sources	Secondary sources
Characteristics	• first-hand records of events or experiences • written at the time the event happened • original documents	• interpretation of primary sources • written by people who did not see or experience the event • use information from original documents but rework it
Examples	• results of experiments • articles in scientific journals or magazines • reports of scientific discoveries • photographs, specimens, maps and artefacts • interviews with experts • websites (if they meet the criteria above)	• textbooks • biographies • newspaper articles • magazine articles • radio and television documentaries • websites that interpret the scientific work of others • podcasts

ETHICAL CONSIDERATIONS

When you are planning a scientific investigation, identify all possible ethical considerations that might arise and consider how you could reduce or eliminate them. Asking questions such as those below may help you uncover ethical implications:

- How could this affect the wider society?
- Does one group benefit over another; for example, one individual, a group of individuals or a community? Is it fair?
- Who will have access to the data and results?
- Does it prevent anyone from meeting their basic needs?
- Could the investigation, though appearing to be ethically sound now, have ethical implications in the future?

SAFETY AND RISK ASSESSMENTS

When planning your investigation, you need to be aware of any risks or safety concerns in order to mitigate them. Always use safe procedures and common sense. For example, all equipment and instruments should be used at the back of the bench so that students walking by do not knock them or trip over them and cause an accident. Place a sign on the lab bench warning other students and staff not to touch the equipment.

You must follow the safety and **risk assessment** guidelines of your teacher and your school. Completing a risk assessment may require completing a form or completing an online process.

While conducting an investigation, it is important for your own safety and the safety of others that all potential risks are considered. Risk assessments are undertaken to identify, assess and control potential hazards. A risk assessment should be performed for any situation—in the laboratory or outside in the field. Always identify the risks and control them to keep everyone safe.

FIGURE 1.1.4 When planning an investigation, you need to identify, assess and control hazards.

For example, conduct voltage–current experiments only with low voltages (less than 6.0 V DC or 4 × 1.5 V batteries) coupled to resistors so that the currents in the circuits are of the order of milliamps. At all times avoid direct exposure to 240 V AC household voltages (Figure 1.1.4).

To identify risks think about:

- the activity that will be carried out
- where it will be carried out
- the equipment that will be used.

The following hierarchy of risk controls (Figure 1.1.5) is organised from most effective to least effective.

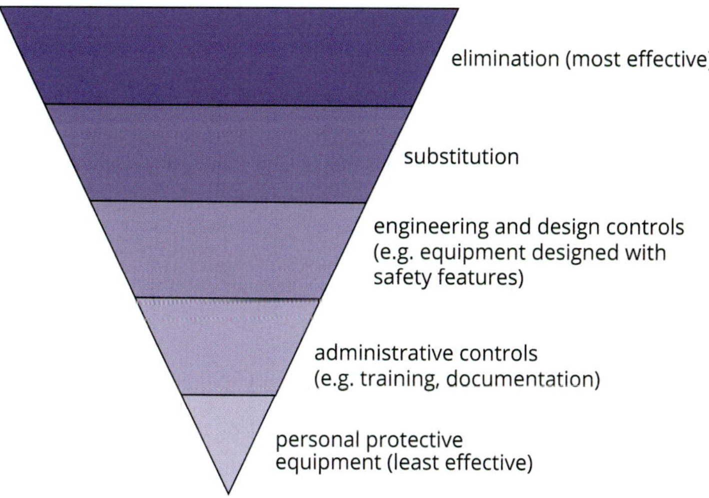

FIGURE 1.1.5 The hierarchy for risk control is shown in this pyramid, marked from bottom to top in order of increasing importance.

FIGURE 1.1.6 Examples of PPE shown are protective eyewear, lab coats and gloves.

Personal protective equipment

Everyone who works in a laboratory wears **personal protective equipment (PPE)** to help keep them safe. Consult your teacher or laboratory technician and safety data sheet (SDS) to see what PPE you are likely to need. Examples include:

- safety glasses
- lab coats
- shoes with covered tops
- disposable gloves.

Examples of PPE are shown in Figure 1.1.6.

Chemical codes

In January 2017, Australia adopted the Globally Harmonized System of Classification and Labelling of Chemicals (GHS). This system is used for labelling containers and in safety data sheets. The chemicals at school or at a hardware shop have a warning symbol on the label. These symbols—mandated by the GHS—are chemical codes that indicate the nature of the contents (Table 1.1.4). These chemical codes will need to be analysed when you are planning and conducting scientific investigations involving chemicals. You will perform a risk assessment in which these chemical codes will be provided. After analysing them, you may need to modify your experimental plan so that safety is improved.

TABLE 1.1.4 GHS symbols used as warnings on chemical labels

GHS symbol	Use	GHS symbol	Use	GHS symbol	Use
	flammable liquids, solids and gases; including self-heating and self-igniting substances		oxidising liquids, solids and gases, may cause or intensify fire		explosion, blast or projection hazard
	corrosive chemicals; may cause severe skin and eye damage and may be corrosive to metals		gases under pressure		fatal or toxic if swallowed, inhaled or in contact with skin
	low level toxicity; this includes respiratory, skin and eye irritation, skin sensitisers and chemicals harmful if swallowed, inhaled or in contact with skin		hazardous to aquatic life and the environment		chronic health hazards; this includes aspiratory and respiratory hazards, carcinogenicity, mutagenicity and reproductive toxicity

Safety data sheets

Every chemical substance has an associated document called a **safety data sheet (SDS)**. The SDS contains important safety, environmental and first aid information about the chemical, including how the chemical should be handled and stored (Figure 1.1.7). For example, if an SDS states the products of a reaction are toxic to the environment, you must pour your waste into a special container, not down the sink.

An SDS provides employers, workers, and health and safety representatives with the necessary information to safely manage the risk of exposure to a hazardous substance.

Safety Data Sheet
NITROGEN, REFRIGERATED LIQUID (N2)

Date of first issue: 30/07/2010 Revised date: 20/12/2016 Supersedes: 01/03/2013 Version: 6.0
SDS reference: AL613

Warning

SECTION 1: Identification of the substance/mixture and of the company/undertaking

1.1. Product identifier

Trade name	: Nitrogen (refrigerated)
SDS no	: AL613
Chemical description	: Nitrogen (refrigerated) CAS No : 7727-37-9 EC no : 231-783-9 EC index no : ---
Registration-No.	: Listed in Annex IV / V REACH, exempted from registration.
Chemical formula	: N2

1.2. Relevant identified uses of the substance or mixture and uses advised against

Relevant identified uses	: Industrial and professional. Perform risk assessment prior to use. Test gas/Calibration gas. Purge gas, diluting gas, inerting gas. Purging. Laboratory use. Use for manufacture of electronic/photovoltaic components. Shield gas for welding processes. Contact supplier for more information on uses.

FIGURE 1.1.7 An example of part of an SDS for liquid nitrogen showing the symbol for a compressed gas to alert the reader to potential hazards when using the substance. The SDS also includes measures to reduce the risk of harm.

Science outdoors

Sometimes investigations and experiments will be carried out outdoors. Working outdoors has its own set of potential risks and it is important to consider ways of eliminating or reducing those risks.

Table 1.1.5 gives examples of risks associated with working outdoors.

TABLE 1.1.5 Risks associated with working outdoors

Risks	Control measures
sunburn	wear sunscreen, a hat and sunglasses
hot or cold weather	wear clothing to protect against heat or cold
projectile launch	create barriers so that people know not to enter the area
trip hazards	minimise the use of cables (electrical, computer) and cover them with matting
landscape	be aware of tree roots, rocks, roads etc.

First aid measures

Minimising the risk of injury reduces the chance of requiring first aid assistance. However, it is still important to have someone with first aid training present during a practical investigation.

Always tell your teacher or laboratory technician if an injury or accident happens.

1.1 Review

OA ✓✓

SUMMARY

- A research question sets out what is being investigated.
- An aim is a statement describing in detail what will be investigated.
- A hypothesis is a prediction about the results of the investigation.
- Once a research question has been chosen, evaluate the question before proceeding.
- There are three main categories of variables.
  - The independent variable is the variable that is controlled by the researcher (i.e. the one that is changed to see if there is an effect on the dependent variable).
  - The dependent variable is the variable that may change in response to a change in the independent variable.
  - Controlled variables are all the variables that must be kept constant during the investigation.
- Qualitative data is descriptive and unmeasurable and uses descriptions or adjectives to record observations.
- Qualitative data can be characterised as either:
  - nominal, when the order of the data is not important
  - ordinal, when the order of the data is important.
- Quantitative data is empirically measurable and uses instruments to record observations.
- Quantitative data can be characterised as either:
  - discrete, when data can only be particular numerical values
  - continuous, when data is not restricted to particular numerical values.
- The methodology describes the general approach or style of the investigation.
- The method (or procedure) is the set of specific steps taken to collect data.
- Primary sources are those created by a person directly involved in the investigation.
- Secondary sources are syntheses, reviews or interpretations of primary sources.
- Identify and address any ethical issues while you plan your investigation.
- Risk assessments identify and assess hazards and propose controls to minimise their occurrence.
- GHS symbols identify the types of hazards associated with substances.

KEY QUESTIONS

Knowledge and understanding

1 What is the 'aim' of an experiment?

2 List the three categories of variables and describe each of them.

3 The following steps in the scientific method are out of order. Rewrite them in your notebook in the correct order.
- form a hypothesis
- collect the results
- plan the experiment and equipment
- draw conclusions
- question whether the results support the hypothesis
- state the research question to be investigated
- perform the experiment

4 A student is investigating whether the same ball rolls faster down one of two ramps of different slope (steepness). Identify the independent variable in this investigation.

5 Consider the hypothesis below. What are the dependent, independent and controlled variables?
Hypothesis: *As the voltage across a fixed resistor increases, the current through that resistor increases.*

6 Consider the following options (A–C):

A If the volume of a sound is increased, then the height of the wave will also increase.

B To investigate the effect of changing volume on the wave height.

C How does the wave height depend on the volume of a sound wave?

a Which of these (A–C) is an aim?

b Which of these (A–C) is a hypothesis?

c Which of these (A–C) is a research question?

7 For each of the following hypotheses, state the independent and dependent variables.

  a If the distance travelled is increased, the average speed will decrease if the time is kept the same.

  b If the angle of an inclined plane is increased, then the horizontal distance travelled will be decreased.

  c If the voltage applied to a circuit consisting of an ohmic resistor is increased, then the total current in the circuit will increase proportionally.

  d The intensity of light decreases the farther you move away from the light.

8 A physics student proposes the following research question: 'Is the current through a resistor proportional to the voltage across the resistor?'

  a Which of these (A–C) is the independent variable?

  b Which of these (A–C) is the dependent variable?

  c Which of these (A–C) is the controlled variable?

    A resistance value of the resistor

    B current through the resistor

    C voltage across the resistor

Analysis

9 Which is the most specific research question from the three options below? Explain your choice.

  A Will the current increase or decrease through a resistor when the voltage across the resistor is changed?

  B Will the current measured through a resistor (of fixed resistance value) increase when the voltage across the resistor is increased?

  C Will the current measured through a resistor (of fixed resistance value) change when the voltage across the resistor is increased?

1.2 Conducting investigations

Once you have written your research question, clarified the aim, stated the hypothesis and chosen a suitable methodology, you will need to begin to develop a method to conduct the investigation. In this section you will learn about designing and selecting methods to use in scientific investigations in the laboratory and in field work. You will be introduced to different techniques and understand how selecting appropriate equipment and methods will allow you to obtain accurate and precise measurements. Risks and safety precautions will also be discussed.

WRITING A METHOD

The method is a specific step-by-step procedure that you will follow when conducting your investigation. The method must be detailed enough so that someone else can conduct the investigation using the same steps, and should be recorded in your logbook. For example, the step 'Place a sheet of semiconductive paper flat, printed side up, on the lab bench,' ensures that whenever the method is followed, the paper will be placed the same way up. Number your steps sequentially, covering only one action per step.

You must also ensure that the method used enables the collection of data that is valid, repeatable and reproducible.

Validity

Validity refers to whether an experiment or investigation is in fact testing the set research question or hypothesis.

Factors influencing validity include:

- whether your experiment measures what it claims to measure; in other words, your experiment should test your research question or hypothesis
- the certainty that something observed in your experiment was the result of your experimental conditions and not some other cause that you did not consider; in other words, whether the independent variable influenced the dependent variable in the way you have concluded and was not influenced by other variables that should have been kept constant.

Make sure you have identified the independent variable and the dependent variable, the variables you will control in your experiment, and how you will control them. This information should be included in your method. In the example step discussed above, placing the paper printed side up is a controlled variable.

You should also be clear about what raw data you will collect (quantitative or qualitative). If necessary, re-read the relevant text about variables in Section 1.1 on pages 8–9.

Repeatability and reproducibility

Repeatability (sometimes called reliability) is the ability to obtain the same results if an investigation is repeated under the exact same set of conditions. Several steps can be taken to help improve the repeatability of an experiment, reducing the influence of natural variation, random error, uncalibrated instruments or instrument error, and the influence of unforeseen variables. These steps include:

- selecting the appropriate sample size to reduce natural variation, errors and uncertainty
- selecting the appropriate equipment to take the measurements you seek
- taking several measurements under the same conditions (repeat readings of each trial)
- specifying the materials and methods in detail.

Repetition can minimise random errors, but it will not minimise systematic errors. (Errors are discussed in the following sections.) Repeating the investigation and averaging the results will generate data that is more reliable. To ensure that all variables are being tested under the same conditions, modifications to your method may need to be considered before repeating the investigation. The goal is to ensure that every measurement can be repeated and the same result obtained (within a reasonable margin of experimental error, such as less than 5%).

Reproducibility is the ability to obtain the same results if an experiment is repeated under different conditions. Different conditions might include a different researcher conducting the experiment, the use of different equipment or instruments, or conducting the experiment at a different time or location. It is important to write a clear and detailed method so that the experiment can be reproduced successfully.

● *You will now be able to answer key questions 4 and 5.*

EQUIPMENT

When conducting your scientific investigation, it is important to choose the correct equipment so that your measurements are accurate (minimise error), and your results are reproducible and repeatable.

It is important to understand how to use the equipment correctly and how your choice of equipment will affect the accuracy and precision of the results you collect. You should ensure that the equipment is suitable for the measurement required. For example, a ticker timer (Figure 1.2.1(a)) and a motion detector (Figure 1.2.1(b)) would be suitable choices for measuring the velocity of an object. It is also important to ensure that the equipment is calibrated.

Calibrated equipment

Some equipment is sensitive to the conditions in which it operates, such as temperature and humidity. An example is the motion sensor. The accuracy of this equipment should be tested before each use to account for this. This is called 'calibration'. Before carrying out the investigation, make sure the equipment is properly calibrated and functioning correctly. For example, measure the temperature and, if necessary, apply a correction to the speed of sound to calibrate a motion sensor.

Correct use of equipment

Always use equipment properly. Complete whatever training in the use of the equipment is necessary and practise using the equipment before beginning your investigation. Improper use of equipment can result in inaccurate and imprecise data with large errors, which weakens the validity of the data.

● *You will now be able to answer key question 3.*

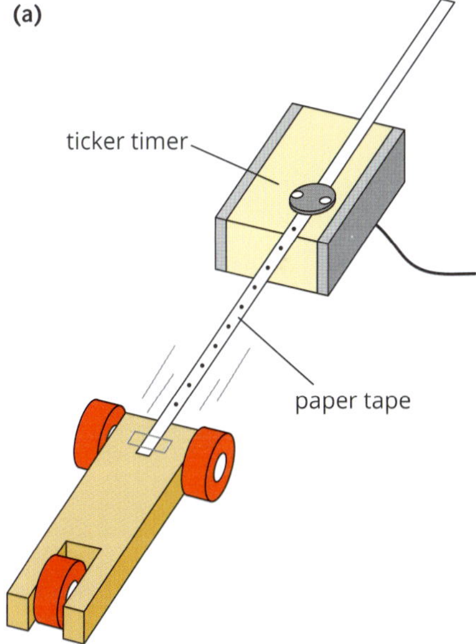

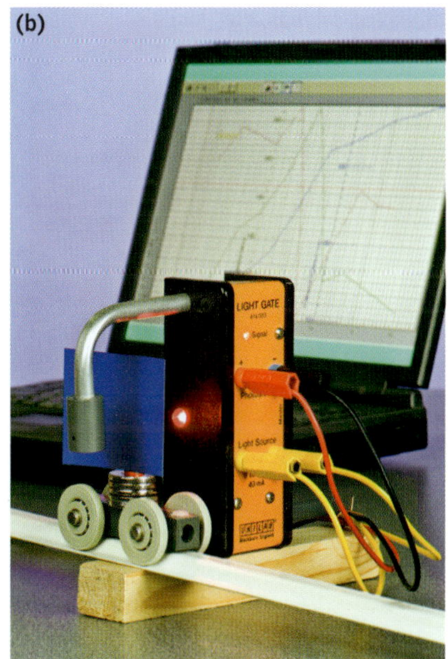

FIGURE 1.2.1 Some useful items of equipment in scientific investigations are (a) a ticker timer and (b) a motion sensor.

COLLECTING AND RECORDING DATA

For an investigation to be scientific, it must be objective and systematic. When working, keep asking questions. Is the work biased in any way? If changes are made, how will they affect the study? Will the investigation still be valid given the aim and hypothesis?

It is essential that you record the following information in your logbook during your investigation:

- all quantitative and qualitative data collected
- the methods used to collect the data
- any incident, feature or unexpected event that may have affected the quality or validity of the data.

The data recorded in a logbook is the **raw data**. Usually this data needs to be processed in some manner before it is presented. If an error occurs in the processing, or you decide to present the data in an alternative format, the recorded raw data will still be available for you to refer back to. How to collect and process your raw data will be covered in Section 1.3.

IDENTIFYING ERRORS

When an instrument is used to measure a physical quantity and obtain a numerical value, the aim is to determine the true value. The **true value** is the value, or range of values, that would be obtained if the variable could be measured perfectly. However, for a number of reasons the measured value is often not the true value.

Most practical investigations will have some errors in the data collected. Errors can occur for a variety of reasons. Being aware of potential errors helps you to avoid or minimise them. For an investigation to be valid, it is important to identify and record any errors.

There are three types of errors (Figure 1.2.2):

- systematic errors
- random errors
- mistakes.

Systematic errors

A **systematic error** is an error that is consistent and will occur again if the investigation is repeated in the same way.

Systematic errors are usually a result of instruments not being calibrated correctly, methods that are flawed, or environmental factors (e.g. electrical interference from other equipment or devices).

An example of a systematic error would be if a ruler mark indicating 5 cm from 0 cm was actually only 4.9 cm from 0 cm due to a manufacturing error or shrinkage of the wood. Another example would be if a researcher repeatedly used a piece of equipment incorrectly throughout the entire investigation.

Make sure you choose appropriate equipment that is in good working order before conducting your investigation. This will reduce the possibility of introducing systematic errors.

Random errors

Random errors occur in an unpredictable manner and are generally small. A random error could, for example, result from a researcher reading the same output value correctly one time and incorrectly another time. Another example would be if an instrument's readings fluctuated during periods of low battery power. To reduce the impact of random errors, make sure to build repetition into your method.

Mistakes

Mistakes are avoidable errors. Examples of mistakes include:

- misreading the numbers on a scale
- not identifying or labelling data points adequately.

A measurement that involves a mistake must be rejected and not included in any calculations or averaged with other measurements of the same quantity. Mistakes are often not referred to as errors because they are caused by the experimenter rather than the experiment or the experimental method. Mistakes are sometimes referred to as personal errors.

You will now be able to answer key question 2.

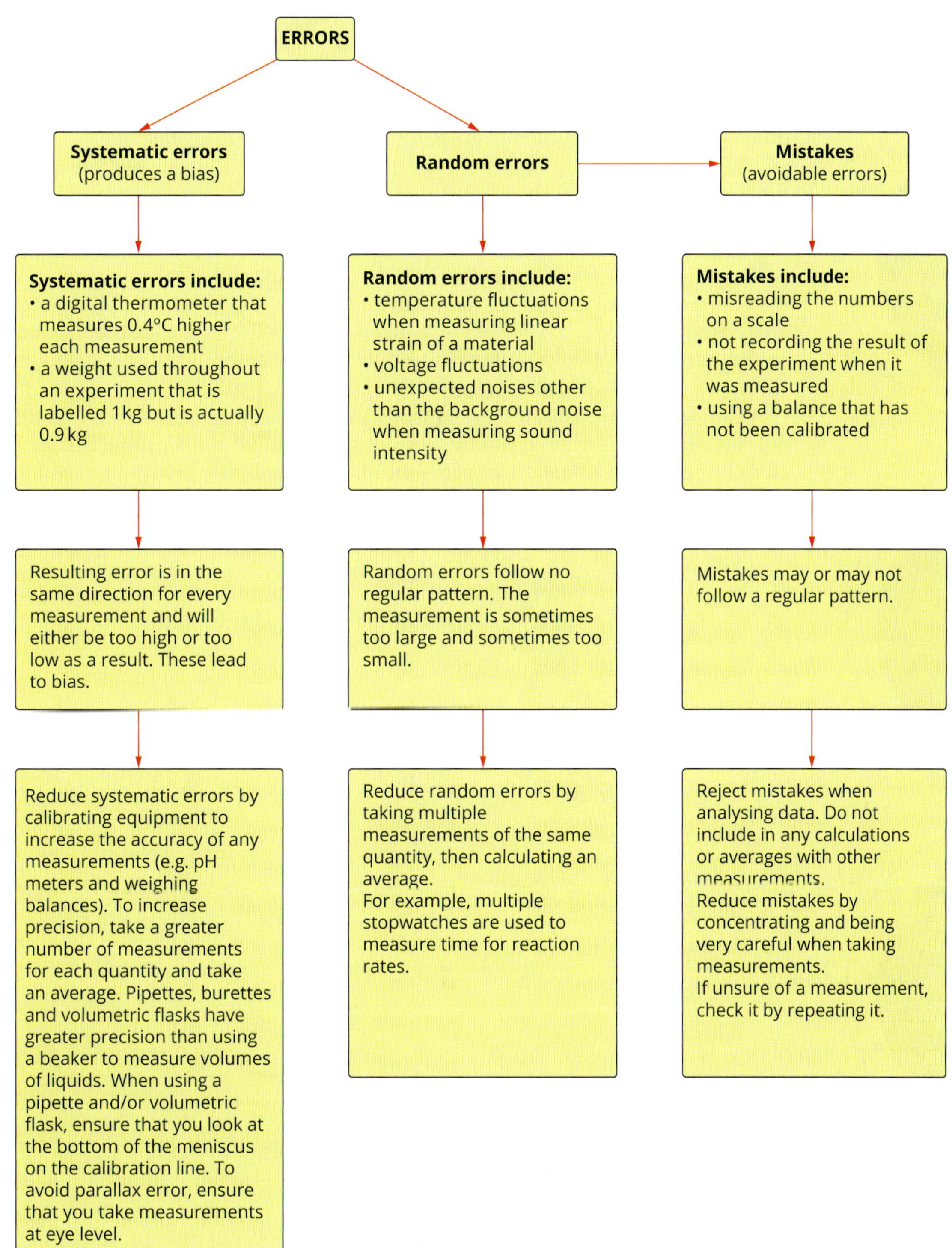

FIGURE 1.2.2 Types of error

(a)

(b)

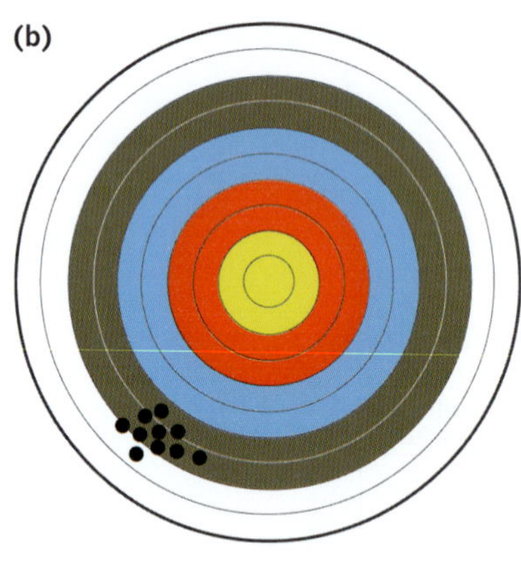

(c)

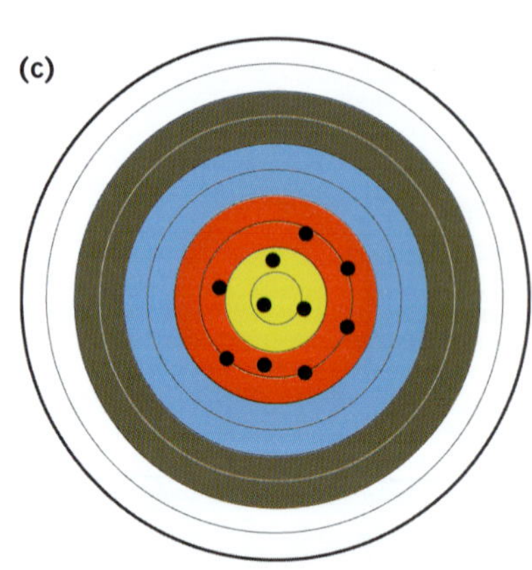

(d)

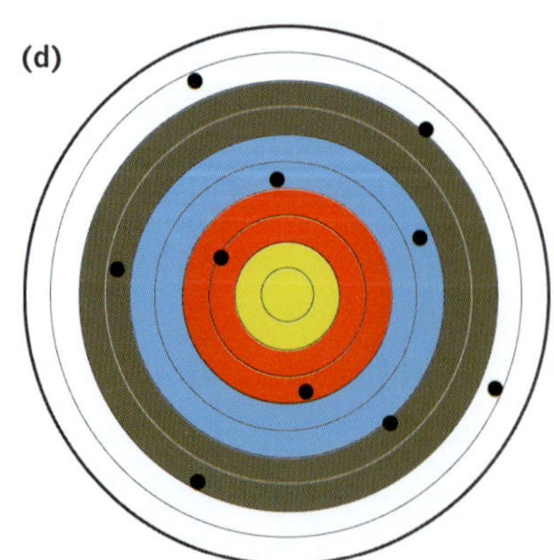

FIGURE 1.2.3 Examples of accuracy and precision: (a) both accurate and precise, (b) precise but not accurate, (c) accurate but not precise, and (d) neither accurate nor precise

Accuracy and precision

Two very important aspects of any measurement are accuracy and precision.

- **Accuracy** is the ability to obtain the true value of the variable being measured, including during repeated trials of the experiment. To obtain accurate results, you must minimise systematic errors.
- **Precision** is how closely a set of measurements agree with each other. A set of precise measurements will have values very close to the mean value of the measurements. Precision is different from accuracy in that it does not indicate how close the measurements are to the true value. To obtain precise results, you must minimise random errors.

Note that accuracy and precision are not the same thing.

Instruments are said to be accurate if they truly report the quantity being measured. For example, if a tape measure is correctly manufactured it can be used to measure lengths accurately to the nearest centimetre.

Instruments are said to be precise if they can differentiate between slightly different quantities. Precision refers to the fineness of the scale being used.

To understand more clearly the difference between accuracy and precision, think about firing arrows at an archery target (Figure 1.2.3). Accuracy is being able to hit the bullseye, whereas precision is being able to hit the same spot every time you shoot. If you hit the bullseye every time you shoot, you are both accurate and precise (Figure 1.2.3(a)). If you hit the same area of the target every time but not the bullseye, you are precise but not accurate (Figure 1.2.3(b)). If you hit the area around the bullseye each time but don't always hit the bullseye, you are accurate but not precise (Figure 1.2.3(c)). If you hit a different part of the target every time you shoot, you are neither accurate nor precise (Figure 1.2.3(d)).

Consider a metre rule, a tape measure and a measuring wheel used to mark out a sports field. All three will measure distance and all three can be accurate. However, the metre rule is the most precise of the three. This is because it measures to the nearest millimetre while the tape measure only to the nearest centimetre and the measuring wheel only to the nearest metre (Figure 1.2.4).

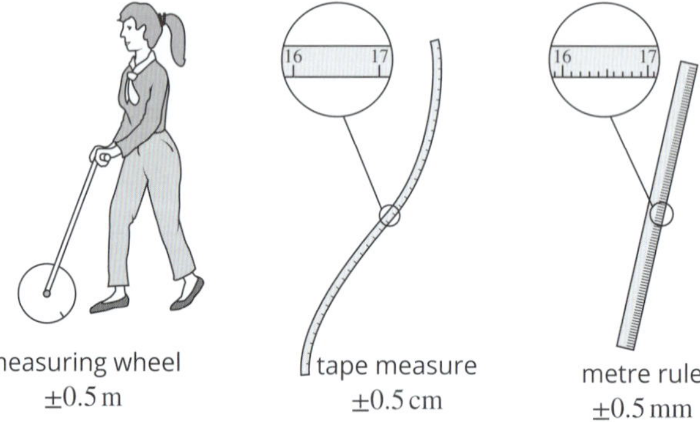

FIGURE 1.2.4 The measuring wheel has low precision and only measures to the nearest metre. It has an uncertainty of 0.5 m. The tape measure has more precision and has an uncertainty of 0.5 cm or 0.005 m. The metre rule has an uncertainty of 0.5 mm or 0.0005 m.

To see that the tape measure is a more precise instrument than the measuring wheel, suppose two distances of 2673 mm and 2691 mm are being measured with these two instruments. Each distance would be measured as 3 m, to the nearest metre, by the wheel. They would be measured by the tape measure as 2.67 m and 2.69 m to the nearest centimetre. The tape measure is more precise because it has a finer scale. You might also say that it has greater resolution. The measuring wheel has such low precision that it cannot be used to measure which of the two distances is greater or smaller. Measuring instruments with less precision give measurements that are less certain. The uncertainty in the measurement is due to the coarser scale. The measuring wheel gives less certain measurements than the tape measure even though both instruments may be equally accurate.

The **uncertainty** of a measurement is due to the limited precision of the instrument that does the measuring. All measurements have an amount of uncertainty. The uncertainty is generally one half of the finest scale division on the measuring instrument. This means that the actual measurement could be anywhere from half of the smallest graduation too big to half of the smallest graduation too small. The measuring wheel in the example above has an uncertainty of 0.5 m. The tape measure has an uncertainty of 0.5 cm. The metre ruler has an uncertainty of 0.5 mm.

Sometimes this uncertainty is referred to as error. It is not an error, or something that has gone wrong. All measuring instruments have limited precision.

When conducting experiments, check that the instruments to be used are sensitive enough. Build some testing into your investigation to confirm the accuracy and precision of those instruments.

● *You will now be able to answer key question 1.*

PHYSICSFILE

Uncertainty

The uncertainty in an instrument is half the smallest scale of division in the measurements offered by the instrument. This is written using a plus or minus symbol (±). This indicates that the true value of the measurement can be anywhere between the measured value less the uncertainty and the measured value plus the uncertainty. For example, a ruler may have gradations that are 1 mm apart. This makes the uncertainty in the ruler ±0.5 mm. If you were to measure the distance of a gap in a diffraction experiment using this ruler and found it to be 10 cm, then the true value is between 10 cm – 0.5 mm and 10 cm + 0.5 mm.

RECORDING INFORMATION FROM PRIMARY AND SECONDARY SOURCES

You may recall from Section 1.1 (page 10) that research results can come from secondary sources as well as primary sources. As you conduct your investigation, it is important to note any information you use that has come from secondary sources. This must be acknowledged in your final written report. Categorising the information and evidence you find while you are researching will make it easier to locate the information later when you are writing the report. For example, you could use the following categories for identifying the sources of your information:

- research methods—the steps and methods to conduct an experiment
- key findings—key information and facts related to the experiment
- research relevance—how relevant is the source of information to the experiment being conducted.

Record this information in your logbook.

MODIFYING THE METHOD

The method may need modifying as the investigation proceeds. The following actions will help you identify any issues and how to address them.

- Record everything in your logbook.
- Be prepared to make changes to your approach.
- Note any difficulties encountered and the ways they were overcome. What were the failures and successes? Every test carried out can contribute to an understanding of the investigation as a whole, no matter how much of a disaster it may first appear.
- Don't panic. Go over the theory again and talk to your teacher and other students. A different perspective can lead to a solution.

If the expected data is not obtained, don't worry. As long as it can be critically and objectively evaluated, the limitations of the investigation identified and further investigations proposed, your work will have been worthwhile.

1.2 Review

OA ✓✓

SUMMARY

- It is essential that during an investigation, the following are recorded in your logbook:
  - all quantitative and qualitative data collected
  - the methods used to collect the data
  - any incident, feature or unexpected event that may have affected the quality or validity of the data
  - all sources of information, both primary and secondary information.
- Validity refers to whether an experiment or investigation is in fact testing the set research question or hypothesis.
- Repeatability (or reliability) refers to the notion that if an experiment is repeated many times, the results obtained should be consistent.
- Repeatability is improved by:
  - replication (having multiple samples within an experiment)
  - repeat trials (repeating the experimental test).
- Accuracy refers to the ability to obtain the correct measurement.
- The precision of an instrument describes the smallest value it can measure.
- The uncertainty of a measurement due to the limited precision of the instrument is ± half of the finest scale division on the instrument.
- A systematic error is an error that is consistent. It will occur again if the investigation is repeated in the same way. Systematic errors are usually a result of instruments that are not calibrated correctly or methods that are flawed.
- Random errors are unpredictable and are generally insignificant. A random error could arise when a researcher misreads one of a number of measurements.
- Mistakes are avoidable. Do not include in your analysis of results any measurements that involve a mistake.

KEY QUESTIONS

Knowledge and understanding

1 Describe the difference between accuracy and precision.

2 Identify whether the following are systematic errors, random errors or mistakes.
   a A scale that should have measured a mass as 16.00 ± 0.03 g actually measured 15.96 ± 0.03 g.
   b Some reported measurements are above the true value and some below the true value.
   c A student misread the value on the ammeter while measuring the current in an investigation.
   d A student spills some water from the container while measuring the temperature increase.
   e The mass of a ball was taken three times with the following results: 78.97 g, 78.94 g and 78.92 g.
   f A 16.00 g mass measures 16.4 g on a scale.

3 State why it is important to choose appropriate equipment and instruments to conduct an experiment.

4 What is a method to ensure that an investigation is repeatable?

Analysis

5 You are conducting a practical investigation to find the acceleration due to gravity by dropping a ball from different heights and measuring the time it takes to fall to the ground. Write clear instructions for the method.

1.3 Data collection and quality

Measurements and observations made during a scientific investigation together form a picture of what occurred during the investigation. This is the raw data (Figure 1.3.1). The raw data needs to be analysed and then represented using tables, graphs, schematics or diagrams in accordance with correct mathematical and scientific conventions.

Choosing not to record certain measurements or observations in your raw data makes the experiment invalid, shows bias and is scientifically fraudulent. Unusual and unexpected measurements and observations may be due to valid relationships between variables that are unknown to science. This cannot be determined until the raw data is processed, analysed and interpreted.

In this section you will learn about recording quantitative data. You will also learn about the various factors that contribute to data quality, and the importance of controlled experiments in producing valid results.

FIGURE 1.3.1 All measurements and data collected during an experiment is the raw data.

RECORDING QUANTITATIVE DATA

The measurements or observations that you make during your investigation are your **primary data**. When planning your scientific investigation, you may have decided to also use **secondary data** (i.e. data you have not collected yourself).

The data you record in your logbook is raw data. This data needs to be processed or analysed before it can be presented. **Processed data** is raw data that has been organised, altered or analysed to produce meaningful information. If an error occurs in processing the data, or you decide to present the data in a different format, you will always have the recorded raw data to refer back to.

Raw data is unlikely to be used directly to validate a hypothesis. However, raw data is essential to the investigation, and plans for collecting the raw data should be made carefully. Consider the formulas or graphs that will be used to analyse the data at the end of the investigation. This will help to determine the type of raw data that needs to be collected.

For example, to calculate take-off velocity for a vertical jump, three sets of raw data will need to be collected using a force platform: the athlete's standing body mass, the ground reaction force and the time during the vertical jump. The data can then be processed to obtain the take-off impulse.

The data that you collect must relate directly to the variables in your experiment. The data collected and measured must be relevant to the proposed relationship set out in your research question and hypothesis. Also, it must be sufficient to provide accuracy and precision, otherwise the analysis and interpretation of the data will not be valid in relation to the research question or hypothesis.

Collecting sufficient data

You need to collect enough data to substantiate whether or not a relationship exists between the variables you are studying. This includes collecting an appropriate number of individual samples (also known as observational or collection points) and an appropriate number of replicates. It is also important to collect data around interesting points in your range, such as where a curve representing the data reaches a maximum or minimum.

Together, the number of individual samples and replicates determines the sample size. A sufficient sample size is essential for your interpretation to be considered supported.

Collecting relevant data

The variables you measure (i.e. the data collected) must be directly related to the independent–dependent variable relationship specified in your hypothesis. Additional variables can be measured that are indirectly related to your hypothesis, if your background research shows it could be beneficial in the analysis or interpretation of the relationship specified in your hypothesis.

● *You will now be able to answer key questions 1 and 2.*

MEASUREMENT AND UNITS

Every science needs a system of units in order to fully describe the measurements that are made. In physics, measurements are described using the International System of Units (known as SI).

Seven fundamental units are specified in the SI. They are the metre, kilogram, second, kelvin, ampere, mole and candela.

The majority of other units used in physics are a mathematical combination involving at least one of the seven fundamental units. These are called derived units.

Using unit symbols

The correct use of unit symbols removes ambiguity, as symbols are recognised internationally. The symbols for units are not abbreviations and should not be followed by a full stop unless they are at the end of a sentence.

The names and symbols for units are treated differently. Upper case letters are not used for the names of any physical quantities of units written in full. For example, we write newton for the unit of force, while we write Newton if referring to someone with that name. Upper case letters are only used for the symbols of the units that are named after people. For example, the unit of length is metre and its symbol is m. This is lower case because metre is not named after anyone. However, the unit for force is newton and its symbol is N; the unit for energy is joule and its symbol is J. N is upper case because the newton is named after Isaac Newton and J is upper case because the joule is named after James Joule (who is famous for his studies into energy conversion). The exception to this lower case–upper case rule is L for litre. Litre is not named after anyone, but we capitalise L because, in many fonts, a lower case 'l' looks very similar to the numeral 1, a similarity that could lead to ambiguity and confusion.

The product of units is shown by separating the symbol for each unit with a dot or a space. Most teachers prefer a space but a dot is equally correct. The division or ratio of two or more units can be shown in fraction form, by using a slash or by using negative indices.

Prefixes—such as the k in kg—should not be separated by a space.

Table 1.3.1 gives some examples of the correct symbols and format for SI derived units.

TABLE 1.3.1 Examples of the use of symbols for derived units

Majority preference	Also correct	Wrong
m s^{-2}	m.s^{-2} m/s^2	ms^{-2}
kW h	kW.h	kWh k Wh
kg m^{-3}	kg.m^{-3} kg/m^3	kgm^{-3}
μm		μ m
N m	N.m	Nm

Units take the plural form by adding an 's' when used with numbers greater than one. Never do this with the unit symbols. Hence, it is acceptable to write 'two newtons' but wrong to write '2 Ns'. It is also acceptable to say 'two newton'.

Numbers and symbols should not be mixed with words for units and numbers. For example, twenty metres and 20 m are correct but twenty m is incorrect.

Significant figures

Significant figures are the numbers that convey meaning and precision. The number of significant figures used depends on the scale of the instrument you are using. It is important to record data to the number of significant figures available from the instrument. Using either a greater or smaller number of significant figures can be misleading.

The following examples indicate how the number of significant figures is determined.

- Non-zero numbers are always significant: 15 has two significant figures; 3.5 has two significant figures.
- Trailing zeros to the right of the decimal point are significant: 3.50 has three significant figures.
- Leading zeros are not significant: 0.037 has two significant figures.
- Zeros between non-zero numbers are significant: 1401 has four significant figures.

Some numbers may have an ambiguous number of significant figures. For example, 100 could have three, two or one significant figure. In VCE physics, for simplicity, trailing zeros are significant: 100 is assumed to have three significant figures.

In VCE physics, for simplicity, trailing zeros are significant: 100 is assumed to have three significant figures.

The number of significant figures in the result of a calculation should never exceed the minimum number of significant figures in any component of the calculation. For example:

Calculate gravitational potential energy (E_g) using the formula $E_g = mg\Delta h$ when $g = 9.8\,\text{m}\,\text{s}^{-2}$, $m = 7.50\,\text{kg}$ and $h = 0.64\,\text{m}$.

The calculation is:

$$E_g = 9.8 \times 7.50 \times 0.64 = 47.04\,\text{J}$$

However, you should only quote the answer to the least number of significant figures in the component data. In this case, quote the answer to two significant figures: $E_g = 47\,\text{J}$.

Although digital scales can measure to many more than two figures, and calculators can give 12 figures, you should follow the significant-figures rule.

● *You will now be able to answer key questions 3 and 4.*

Scientific notation

For clarity, quantities are often written in scientific notation. A number between one and ten (but excluding ten) is written and then multiplied by an appropriate power of ten. Note that 'scientific notation', 'standard notation' and 'standard form' all have the same meaning.

Examples of some measurements rewritten in scientific notation are:

0.054 m becomes 5.4×10^{-2} m

245.7 J becomes 2.457×10^{2} J

2080 N becomes 2.080×10^{3} N.

You should routinely be using scientific notation to express very large or very small numbers. This also involves learning to use your calculator with scientific notation. Scientific and graphing calculators can be put into a mode in which all numbers are displayed in scientific notation. It is useful when doing calculations to use this mode rather than converting to scientific notation by counting digits on the calculator display. It is quite acceptable to write all numbers in scientific notation, although most people prefer not to use scientific notation when writing numbers between 0.1 and 1000.

An important reason for using scientific notation is that it removes ambiguity about the precision of some measurements. For example, a measurement recorded as 240 m could be a measurement to the nearest metre; that is, somewhere between 239.5 m and 240.5 m. It could also be a measurement to the nearest ten metres; that is, somewhere between 235 m and 245 m. Writing the measurement as 240 m does not indicate which level of precision is the case. If the measurement was taken to the nearest metre, it would be written in scientific notation as 2.40×10^{2} m. If it was taken only to the nearest ten metres, it would be written as 2.4×10^{2} m.

- *You will now be able to answer key questions 5 and 6.*

Prefixes and conversion factors

Conversion factors should be used carefully. You should be familiar with the prefixes and conversion factors in Table 1.3.2. The most common mistake made with conversion factors is multiplying rather than dividing. Some simple strategies can save you this problem. Note that the table gives all conversions as a multiplying factor.

TABLE 1.3.2 Prefixes and conversion factors

Multiplying factor	Index form	Prefix	Symbol
1 000 000 000 000	10^{12}	tera	T
1 000 000 000	10^{9}	giga	G
1 000 000	10^{6}	mega	M
1000	10^{3}	kilo	k
0.01	10^{-2}	centi	c
0.001	10^{-3}	milli	m
0.000 001	10^{-6}	micro	μ
0.000 000 001	10^{-9}	nano	n
0.000 000 000 001	10^{-12}	pico	p

It is important to give the symbol the correct case (upper or lower case). There is a big difference between 1 mm and 1 Mm.

There is no space between prefixes and unit symbols. For example, one-thousandth of an ampere is given the symbol mA. Writing it as m A is incorrect. The space would mean that the symbol is for a derived unit—a metre ampere.

1.3 Review

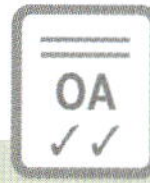

SUMMARY

- Record primary and secondary data in a logbook.
- Make sure you record your data with the correct units, to an appropriate number of significant figures.
- Collect enough data to substantiate whether or not a relationship exists between the variables you are studying.
- Make sure the data you collect is directly related to the independent–dependent variable relationship specified in your hypothesis.
- Physics uses the International System (SI) of units.
- There are seven fundamental units: metre, second, kilogram, kelvin, mole, ampere and candela. Most other units in physics are derived from these units.
- The SI prefixes are symbols that go before a unit and indicate multiplication of the unit by a power of 10.
- Units with a prefix (such as km or mA) should be converted into scientific notation before they can be used in a physics formula.
- Significant figures should be considered in calculations using your data. Quote the results of calculations to the least number of significant figures of your data.

KEY QUESTIONS

Knowledge and understanding

1 Describe why it is important to collect sufficient data.

2 Explain the difference between raw and processed data.

3 Explain why the use of significant figures is important.

Analysis

4 State how many significant figures are in each number.

a 4.56×10^{-6}

b 500

c 500.0

d 5.000

5 Convert the following to two significant figures and scientific notation.

a 6.24×10^{-4}

b 35 014

c 459

d 6.3000

6 Convert the following values to scientific notation.

a 132 kV

b 101.3 kPa

c 6 MJ

d 100 μA

1.4 Data analysis and presentation

A major problem with making a calculation from just one set of measurements is that a single incorrect measurement can significantly affect the result. Scientists like to collect a large amount of data and observe the trends in that data. This gives more precise measurements and allows scientists to recognise and eliminate problematic data.

Physicists commonly use tables and graphical techniques to analyse large sets of data. (Figure 1.4.1) Trends in data are often easier to observe in graphs than in tables of data. In this section, the basic graphing techniques physicists use will be outlined and a general method for fitting a mathematical relationship to a set of data will be explored.

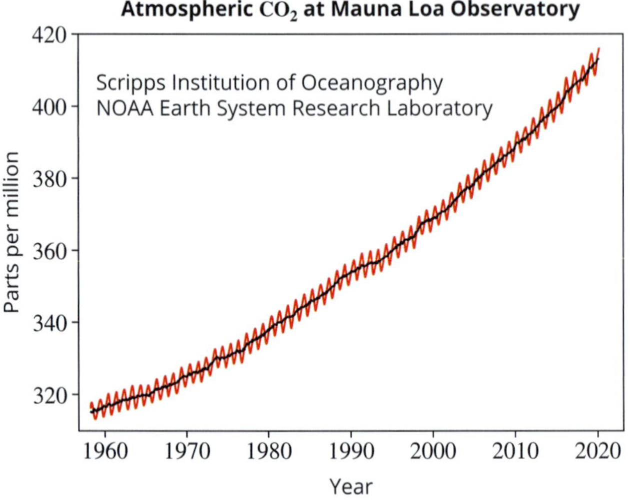

FIGURE 1.4.1 Graphical techniques are often used to find trends in data.

PRESENTING DATA

The raw data that has been obtained needs to be presented in a way that is clear, concise and accurate.

There are a number of ways of presenting data, including tables, graphs, flow charts and diagrams. The best way of visualising the data depends on its nature. To create the best possible presentation, try several formats before making a final decision.

Presenting raw and processed data in tables

Tables organise data into rows and columns and can vary in complexity according to the nature of the data. Tables can be used to organise raw data and processed data. They can also be used to summarise results.

The simplest form of a table is one with two columns. In a two-column table, the first column should contain the independent variable (the one being changed) and the second column should contain the dependent variable (the one that may change in response to a change in the independent variable).

Tables should have the following features:

- a descriptive title (preceded by the 'Table *n*' where *n* is the table number)
- column headings (including the units)
- scientific notation used in the column header for very small or very large numbers
- an indication of the precision of the data
- consistent use of significant figures
- figures that align on the decimal points
- the independent variable placed in the left column
- the dependent variable placed in the right column
- replicate measurements
- calculated averages as required.

The table in Figure 1.4.2 has been used to organise raw and processed data about the effect of current on voltage.

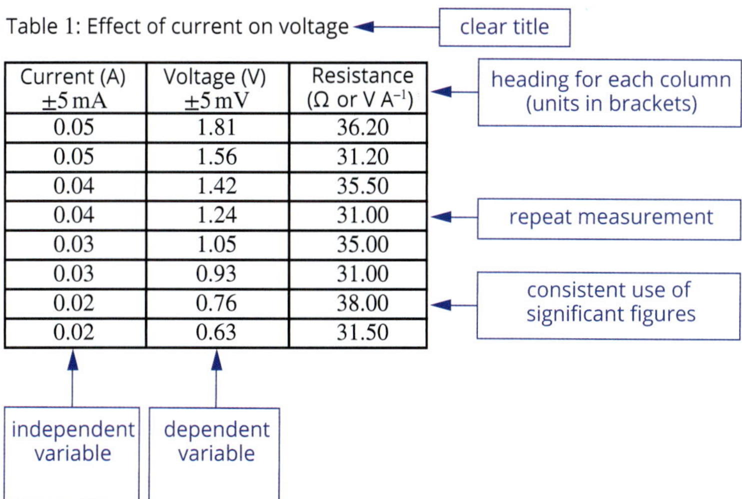

Table 1: Effect of current on voltage

Current (A) ± 5 mA	Voltage (V) ± 5 mV	Resistance (Ω or V A^{-1})
0.05	1.81	36.20
0.05	1.56	31.20
0.04	1.42	35.50
0.04	1.24	31.00
0.03	1.05	35.00
0.03	0.93	31.00
0.02	0.76	38.00
0.02	0.63	31.50

FIGURE 1.4.2 A simple table listing the raw data obtained in the first and second columns and processed data in the third column

Several statistical measures are used that help describe data accurately. This includes the mean, median, mode and uncertainty.

The **mean** is the average of the data, and can be obtained by adding all the measurements and dividing by the total number of data points

The **median** is the middle value of an ordered list of values (i.e. there are as many values less than the median as there are greater than it). For example, the median of the values 5, 5, 5, 8, 9, 10, 20 is 8. The median is preferred when the data range is spread, especially when the spread includes unusual results (also known as outliers). In this situation the mean is unreliable.

The **mode** is the value that appears most often in a data set. This measure is useful to describe qualitative or discrete data. For example, the mode of the values 5, 5, 5, 8, 9, 10, 20 is 5.

The measurement uncertainty, which for a single measurement is equal to ± one-half of the finest scale division on the measuring instrument. The measurement uncertainty is an indicator of the precision of the instrument (as discussed in Section 1.2, page 20).

- *You will now be able to answer key question 6.*

GRAPHICAL ANALYSIS OF DATA

There are several types of graphs that can be used, including line graphs, bar graphs and pie charts. The best one to use will depend on the nature of the data. Table 1.4.1 lists some suitable graphs for quantitative discrete data (data from measuring discrete variables) and continuous data (data from measuring continuous variables).

TABLE 1.4.1 Suitable graph types for quantitative data

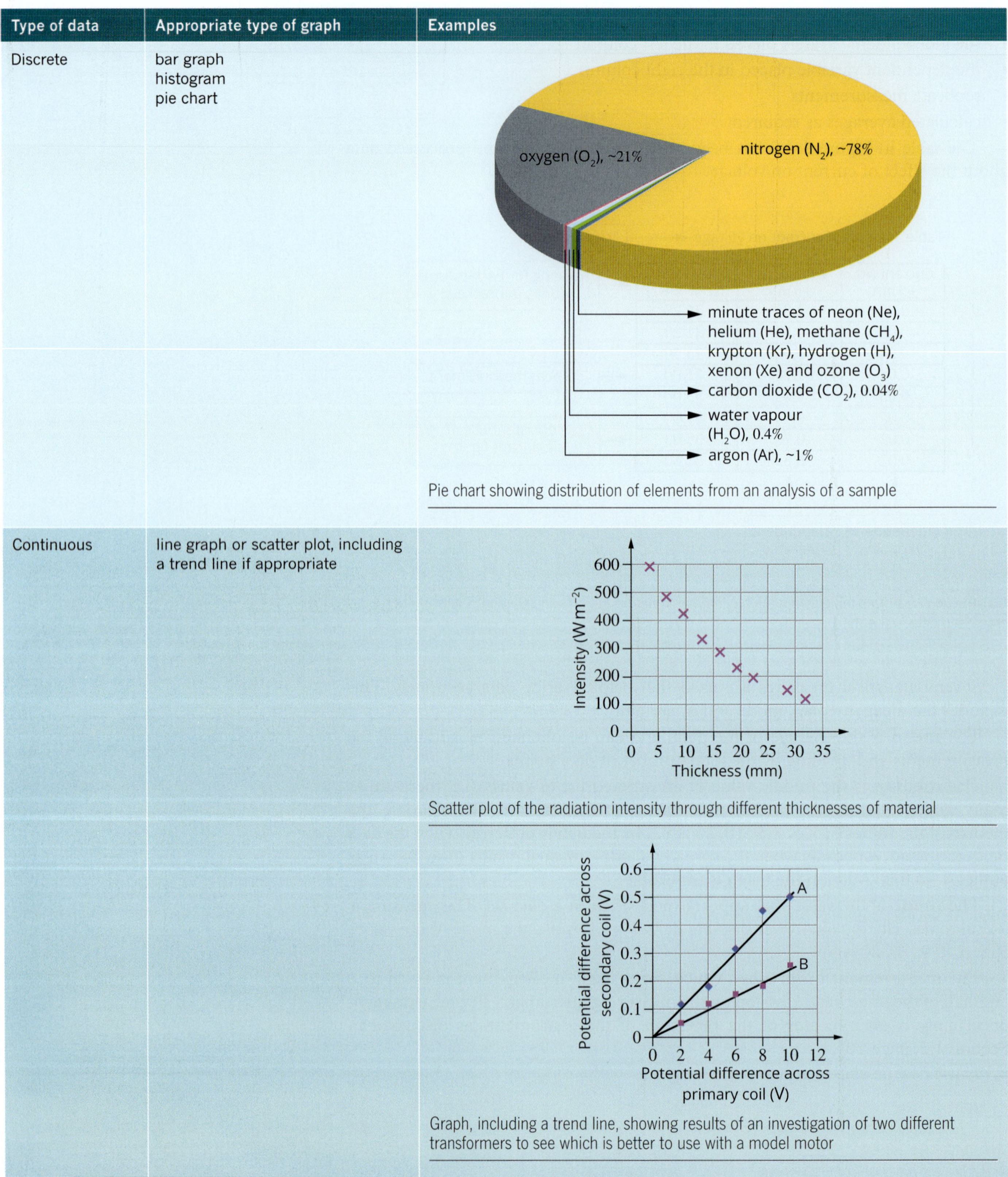

Type of data	Appropriate type of graph	Examples
Discrete	bar graph histogram pie chart	Pie chart showing distribution of elements from an analysis of a sample
Continuous	line graph or scatter plot, including a trend line if appropriate	Scatter plot of the radiation intensity through different thicknesses of material Graph, including a trend line, showing results of an investigation of two different transformers to see which is better to use with a model motor

General rules to follow when plotting a graph are listed below. Figure 1.4.3 illustrates these rules. (Note that these rules do not apply to all types of graph.)

- Keep the graph simple and uncluttered.
- Use a descriptive title.
- Represent the independent variable on the x-axis and the dependent variable on the y-axis.
- Make axes proportionate to the data.
- Clearly label axes with both the variable and the unit in which it is measured.
- Use scientific notation where appropriate.
- You can extrapolate (extend the trend line beyond the obtained data) to predict other values—for example, to determine where the graph intersects the axes. Take care, however, because the relationship between variables may not hold beyond the measured data.

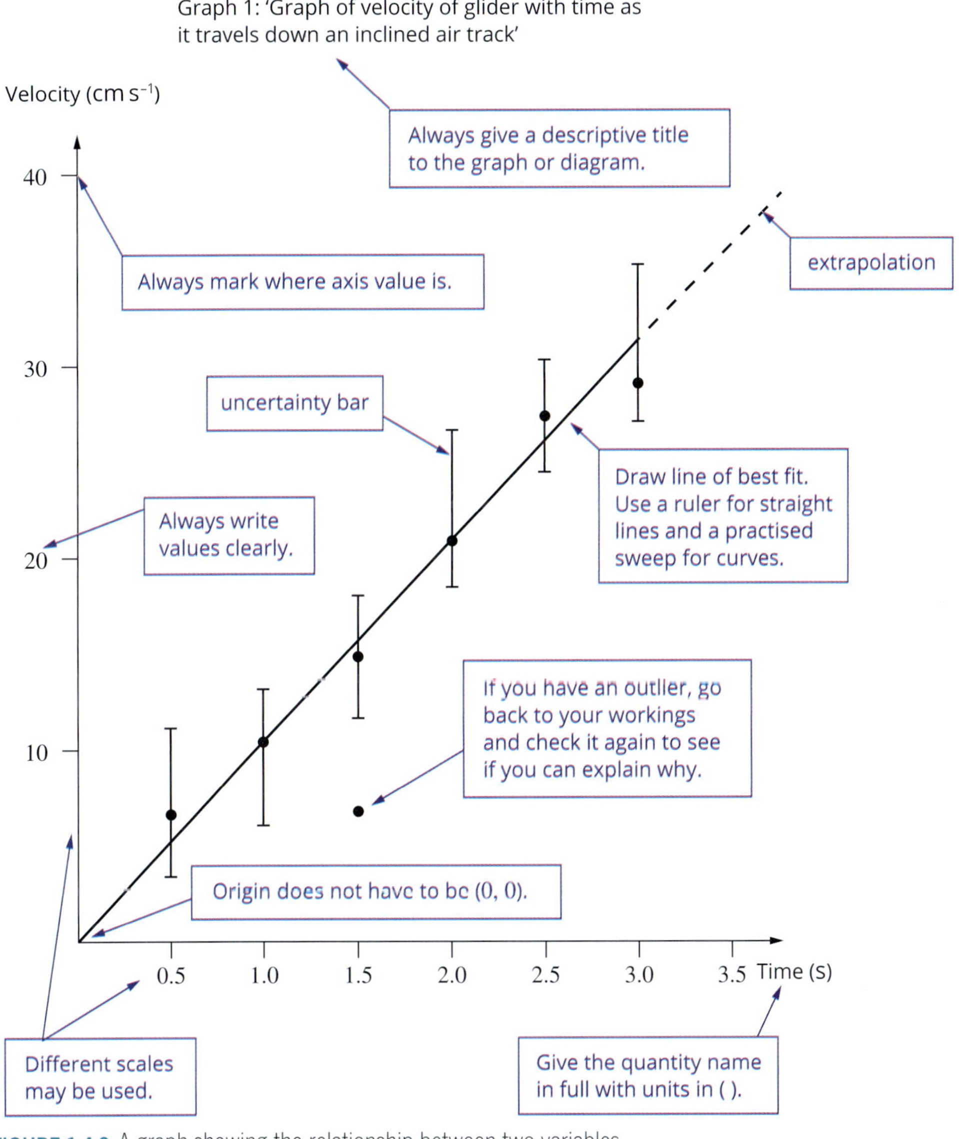

FIGURE 1.4.3 A graph showing the relationship between two variables

● *You will now be able to answer key question 4.*

Line graphs

Line graphs are a good way of representing continuous quantitative data. In a line graph, the values are plotted as a series of points on the graph. There are two ways of joining these points.

- A line can be ruled from each point to the next (Figure 1.4.4(a)). This may show the overall trend but it is not meant to predict the value of the points between the plotted data.
- The points can be joined with a single smooth line, straight or curved (Figure 1.4.4(b)). This creates a **line of best fit**, also known as a trend line. The line of best fit does not have to pass through every point but should go close to as many points as possible. It is used when there is an obvious relationship between the variables.

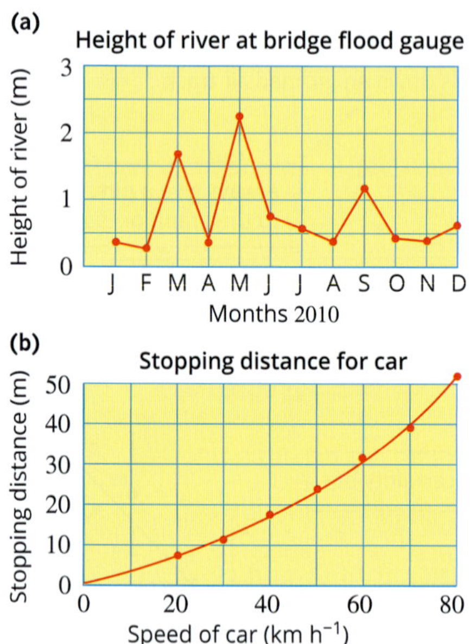

FIGURE 1.4.4 (a) Data in the graph is joined from point to point. (b) Data in the graph is joined with a line of best fit, which shows the general trend.

Outliers

Sometimes when the data is collected, there may be data points that do not fit with the trend and are clearly a mistake or a random error. These points are called **outliers**. An outlier is often caused by a mistake made in measuring or recording data, or from a random error that occurred during the investigation. If there is an outlier, include it on the graph, but ignore it when adding a line of best fit. In Figure 1.4.3 on page 31, the point (1.5, 6) is an outlier.)

You will now be able to answer key questions 1, 2 and 5.

Describing trends in line graphs

Graphs are drawn to show the relationship, or trend, between two variables (Figure 1.4.5).

- Variables that change in linear or direct proportion to each other produce a straight, sloping trend line (**linear relationship**) (Figure 1.4.5(a) and (b)).
- Variables that change exponentially in proportion to each other produce a curved trend line (Figure 1.4.5(c) and (d)). The inverse of an exponential function is a logarithmic function (Figure 1.4.5(g)).
- Variables that have a periodic relationship produce an oscillating relationship (Figure 1.4.5(e) and (f)).
- When there is no relationship between two variables, one variable does not change when the other variable changes.

(a) **Linear relationship**

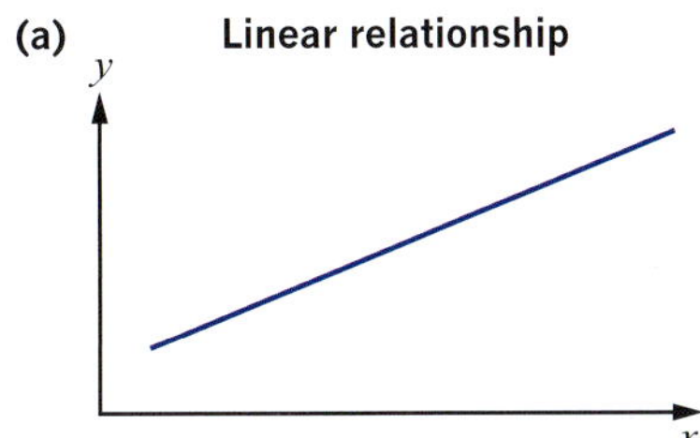

- Positive relationship—as x increases, y increases.

General equation:
$y = mx + c$
m = gradient
c = y-intercept

(b) **Linear relationship**

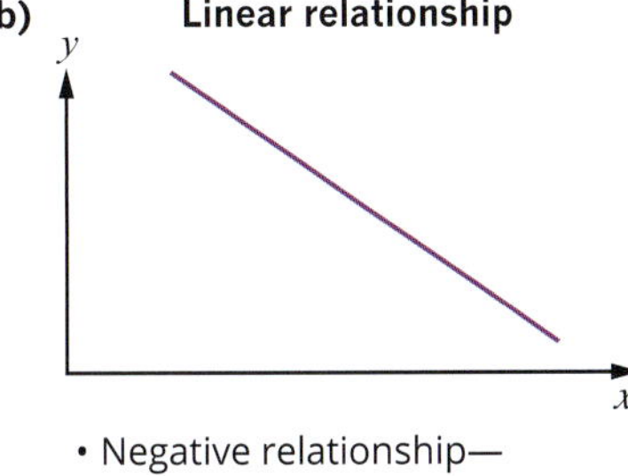

- Negative relationship—as x increases, y decreases.

General equation:
$y = mx + c$
m = gradient
c = y-intercept

(c) **Parabolic relationship**

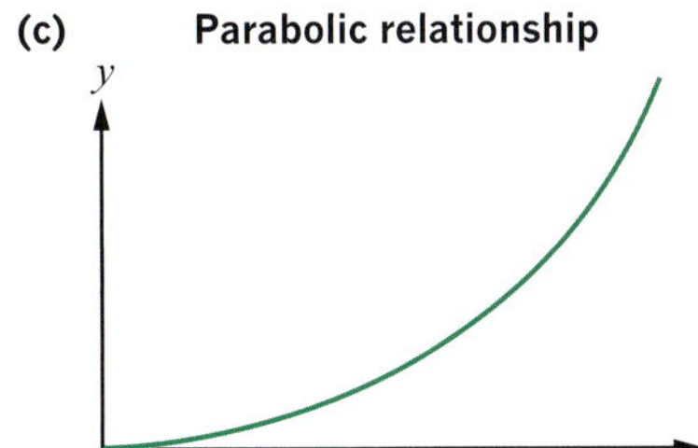

- As x increases, y increases slowly, then more rapidly.

General equation:
$y = kx^2$
k = constant

(d) **Exponential decay**

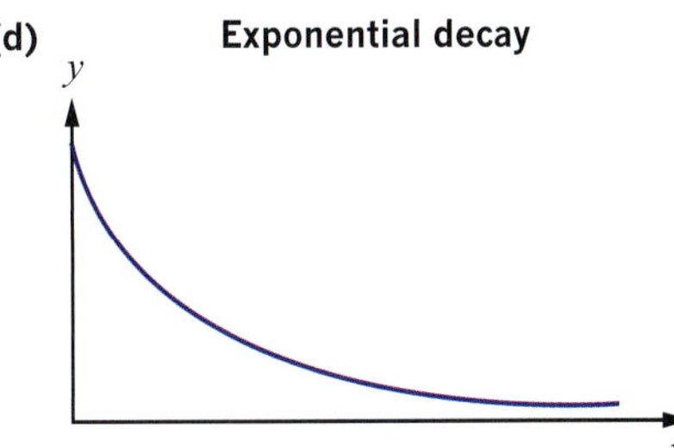

- As x increases, y decreases rapidly, then more slowly, until a minimum y-value is reached.

General equation:
$y = Ae^{-kx}$
k = constant
e = mathematical constant 2.718...

(e) **Sine**

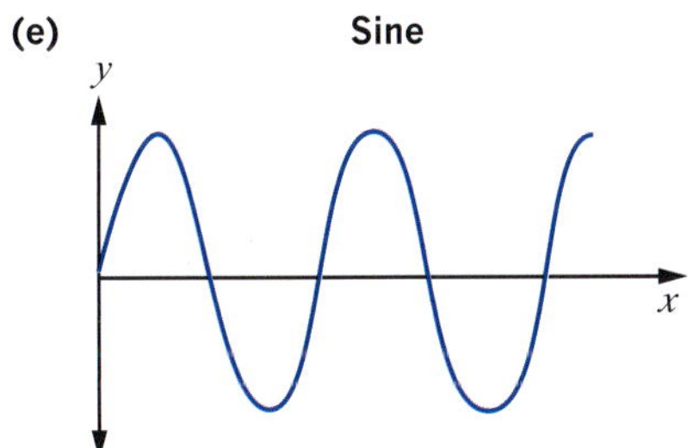

- Periodic relationship, oscillates between a maximum and minimum

General equation:
$y = A\sin(\omega x)$
A = amplitude
ω = angular frequency of the motion

(f) **Cosine**

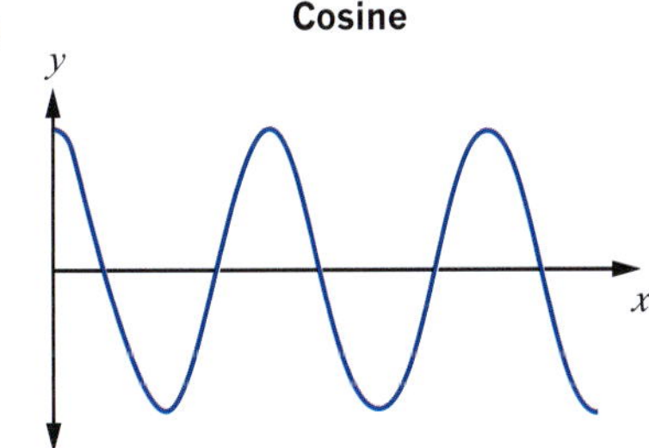

- Periodic relationship, oscillates between a maximum and minimum

General equation:
$y = A\cos(\omega x)$
A = amplitude
ω = angular frequency of the motion

(g) **Logarithmic relationship**

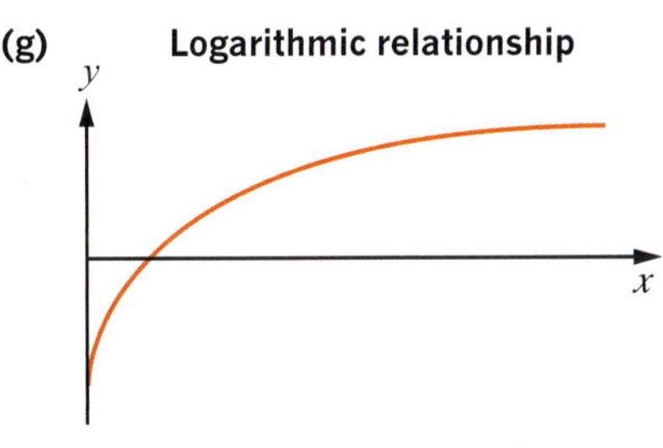

General equation:
$y = \log(x)$

(h) **No relationship**

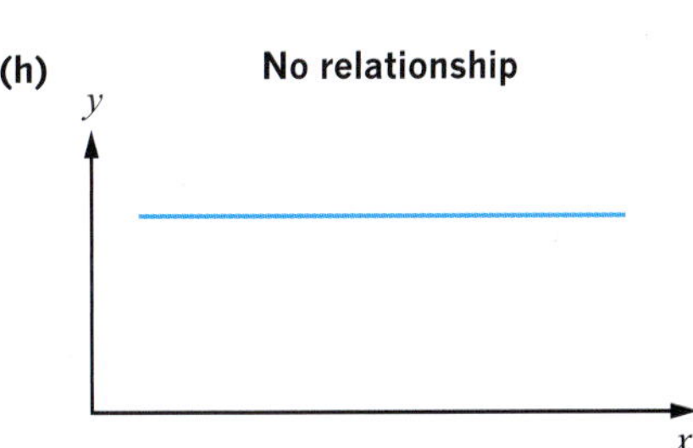

- As x increases, y remains the same.

FIGURE 1.4.5 Various relationships can exist between two variables.

Remember that your results may be unexpected and not match the type of graph you predicted. This does not make your investigation a failure. However, the findings you report must be related to the hypothesis, aims and method.

Linear relationships

Some relationships studied in physics are linear; that is, they can be represented by a straight line. Linear relationships and their graphs are fully specified with just two numbers: the gradient, m, and the vertical axis intercept, c. In general, linear relationships are written:

$$y = mx + c$$

The gradient, m, can be calculated from the coordinates of two points on the line:

$$m = \frac{\text{rise}}{\text{run}}$$
$$= \frac{y_2 - y_1}{x_2 - x_1}$$

where (x_1, y_1) and (x_2, y_2) are any two points on the line.

When analysing data from a linear relationship, you may need to find the equation for the line that best fits the data. This line of best fit is called a regression line. The entire process can be done on paper, but it may be more convenient to use a computer spreadsheet, calculator or some other computer-based application. The benefit of using a digital tool is that the straight line that is produced will be the very best fit for the data. If you plot a line of best fit by eye, there will be some subjectivity introduced.

> **i** Linear relationships are usually written as $y = mx + c$, but they can also be written as
> $y = c + mx$,
> where m is the gradient
> c is the y-intercept.

If you are plotting your graph manually on paper and fitting a regression line by eye, proceed as follows:

1 Plot each data point on clearly labelled, unbroken axes.
2 Label, but otherwise ignore, any suspect data points (outliers).
3 Draw by eye, the line of best fit for the points. The points should be evenly scattered either side of the line.
4 Locate the vertical axis intercept and record its value as c.
5 Choose any two points on the line of best fit and calculate the gradient. Do not use two of the original data points as this will not give you the gradient of the line of best fit. The gradient is the value m.
6 Write $y = mx + c$, replacing x and y with appropriate symbols, and use this equation for any further analysis.

If you are using a spreadsheet or calculator, proceed as follows:

1 Create a table of the raw data.
2 Plot a graph of the raw data.
3 Identify outliers and create another data table without them.
4 Plot a graph of the data without the outliers. Keep both graphs as you should not discard suspect data, but you can eliminate it from your analysis.
5 Plot the line of best fit—the regression line.
6 Compute the equation of the line of best fit that will give you values for m and c. (Many software programs and calculators will calculate these values for you.)
7 Write $y = mx + c$, replacing x and y with appropriate symbols, and substituting in the values of m and c that were calculated. Use this equation for any further analysis.

Don't forget that m and c have units. Omitting these is a common error.

Worked example 1.4.1

FINDING A LINEAR RELATIONSHIP FROM DATA

A group of students used a computer with an ultrasonic detector to obtain the speed–time data for a falling tennis ball. They wished to measure the acceleration of the ball as it fell. Their hypothesis is that the acceleration would be nearly constant and that the relevant relationship is $v = u + at$, where v is the speed of the ball at any given time, u is the speed when the measurements began, a is the acceleration of the ball and t is the time since the measurements began.

Their computer returned the following data:

Time (s)	Speed ($m\,s^{-1}$)
0.0	1.25
0.1	2.30
0.2	3.15
0.3	4.10
0.4	5.25
0.5	6.10
0.6	6.95

Using a spreadsheet or calculator, find their experimental value for acceleration.

Thinking	Working
Decide which axes each variable should be placed on.	The dependent variable is the speed, so it will go on the y-axis. The independent variable is the time, so it will go on the x-axis.
Graph the data as a scatter plot and generate the line of best fit through the points.	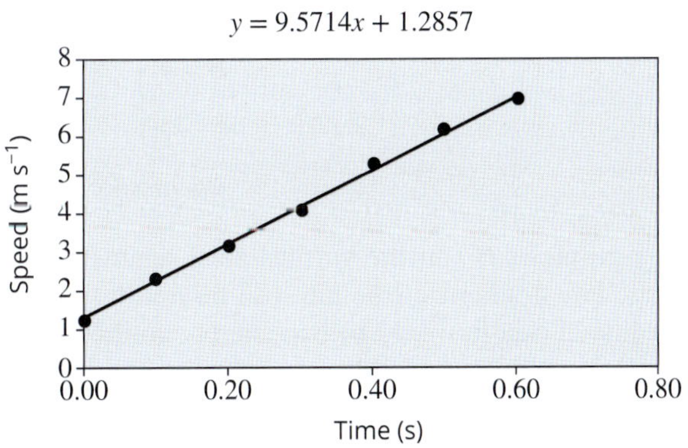 $y = 9.5714x + 1.2857$ Speed ($m\,s^{-1}$): 0, 1, 2, 3, 4, 5, 6, 7, 8 Time (s): 0.00, 0.20, 0.40, 0.60, 0.80
Using the functionality of the spreadsheet or calculator, find the equation for the line of best fit. Instead of y and x, you should use v and t.	$v = 9.5714t + 1.2857$
State the linear relationship in the form required. Express all numbers to two significant figures.	$v = 9.5714t + 1.2857$ $= 1.3 + 9.6t$ Note that linear relationships can be written in the form $y = mx + c$ or $y = c + mx$.
State the answer.	The acceleration is $9.6\,m\,s^{-2}$.

Worked example: Try yourself 1.4.1

FINDING A LINEAR RELATIONSHIP FROM DATA

A student conducted an experiment to calculate the acceleration due to gravity, *g*. The force due to gravity, *F*, is known to be equal to *mg*.

The downwards force was measured for a variety of different masses. It was measured using a spring scale with precision to one decimal place. The mass was measured with electronic scales to the nearest 10 g.

Mass (kg)	Force (N)
0.25	2.4
0.50	4.9
0.75	7.4
1.00	9.9
1.25	12.8

Using a spreadsheet or calculator, find the experimental value for the acceleration due to gravity, *g*, to two significant figures.

Non-linear relationships

Suppose you were examining the relationship between two quantities B and d and had good reason to believe that the relationship can be expressed as:

$$B = \frac{k}{d}$$

where k is some constant. This is similar to $y = \frac{1}{x}$. If you draw this graph using your calculator, you will see that this relationship is non-linear, therefore a graph of B against d will not be a straight line. However, the relationship can be restated by making the following substitutions:

$$B = k\frac{1}{d}$$

$$\uparrow \quad \uparrow\uparrow$$

$$y = m\,x + c$$

A graph of B (on the vertical axis) against $\frac{1}{d}$ (on the horizontal axis) will be linear. The gradient of the line will be k and the vertical intercept, c, will be zero. The line of best fit would be expected to go through the origin because, in this case, there is no constant added and so c is zero.

In this example, a graph of the raw data would simply show that as d increases, B decreases. It would be impossible to determine this relationship just by looking at a graph of the raw data.

Although a graph of raw data will not reveal the mathematical relationship between the variables, it can give some clues. The shape of the graph might suggest a possible relationship. Several relationships can be tried and then the best chosen. This does not *prove* any relationship, but it could provide strong evidence of a particular relationship.

When an experiment involves a non-linear relationship, the following procedure—called linearising the data—is followed.

1 Plot a graph of the original raw data.
2 Choose a possible relationship based on the shape of the initial graph and your knowledge of various graphical forms. Refer to Figure 1.4.5 on page 33.
3 Restate the relationship so that it mimics the form $y = mx + c$.
4 Make a new table of data using the linear relationship.
5 Follow the steps to find the line of best fit as outlined on page 34.

You might need to try several relationships to find the one that best fits the data.

Worked example 1.4.2

FINDING A NON-LINEAR RELATIONSHIP FROM DATA

A group of students were investigating the relationship between current and resistance for a new solid-state electronic device. They obtained the data shown in the table. The current was measured using an ammeter with precision to one decimal place and the resistance, to the nearest ohm (Ω), was measured using a multimeter.

Current, I (A)	Resistance, R (Ω)
1.5	22
1.7	39
2.2	46
2.6	70
3.1	110
3.4	145
3.9	212
4.2	236

According to the theory they had researched, the students believed that the relationship between I and R is

$$R = dI^3 + g$$

where d and g are constants.

By linearising, manipulating the data accordingly and graphing, find the experimental values for d and g. Use a spreadsheet or calculator to assist with finding the line of best fit.

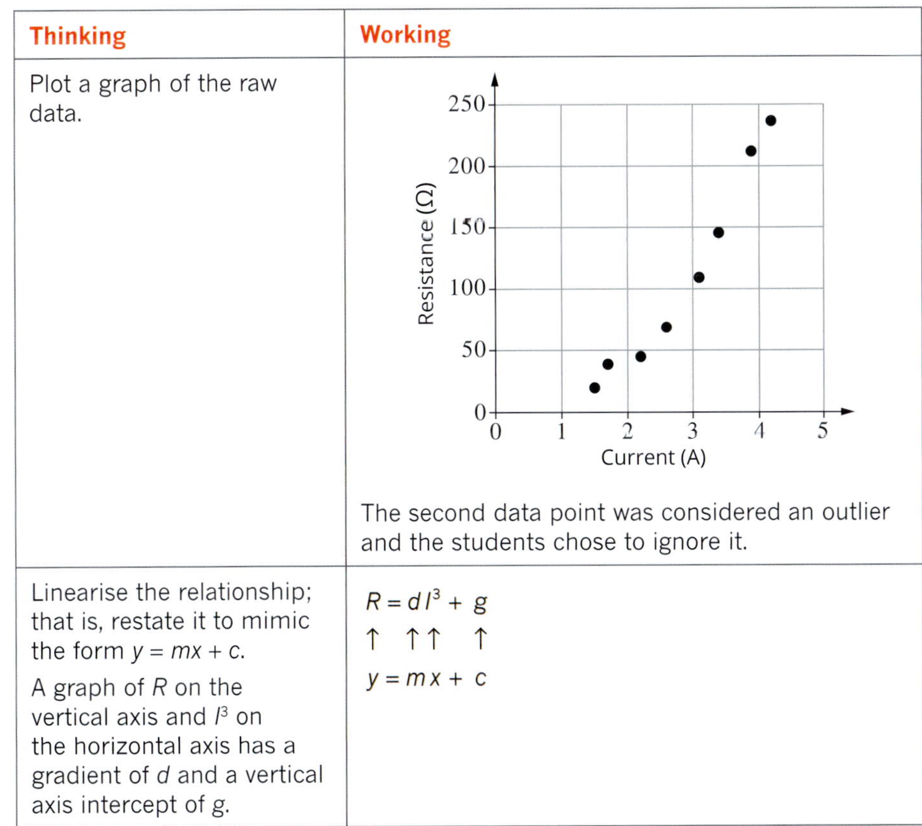

Thinking	Working
Plot a graph of the raw data.	[graph] The second data point was considered an outlier and the students chose to ignore it.
Linearise the relationship; that is, restate it to mimic the form $y = mx + c$. A graph of R on the vertical axis and I^3 on the horizontal axis has a gradient of d and a vertical axis intercept of g.	$R = d I^3 + g$ ↑ ↑↑ ↑ $y = m x + c$

<table>
<tr><td>Make a new table of data manipulated according to the linear relationship. The data is manipulated by finding the cube of each of the values for current; that is, I^3.</td><td>

Current, I^3 (A)	Resistance, R (Ω)
3.38	22
10.65	46
17.58	70
29.79	110
39.30	145
59.32	212
74.09	236

</td></tr>
<tr><td>Plot the graph using the manipulated data.</td><td> 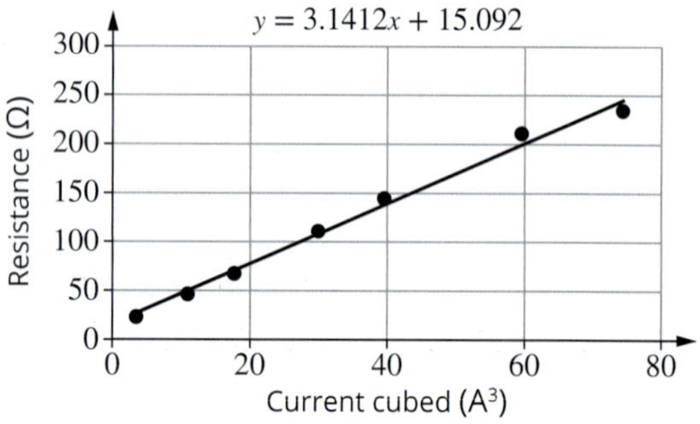

</td></tr>
<tr><td>Using the functionality of a spreadsheet or calculator, find the equation for the line of best fit.</td><td>$y = 3.1x + 15.1$</td></tr>
<tr><td>Restate the equation using I and R.</td><td>$R = 3.1I^3 + 15.1$</td></tr>
<tr><td>Write out the values for d and g. Remember to include the correct units.</td><td>$d = 3.1\,\Omega\,A^{-3}$
$g = 15.1\,\Omega$</td></tr>
</table>

Worked example: Try yourself 1.4.2

FINDING A NON-LINEAR RELATIONSHIP FROM DATA

A group of students were investigating the relationship between the distance from a source and the intensity of sound emanating from that source. They obtained the data shown in the table. Distance was measured using a metre ruler and intensity was measured using an app that displays the intensity to three decimal places.

Distance, r (m)	Intensity, I ($W\,m^{-2}$)
1	0.040
2	0.010
3	0.005
4	0.003
5	0.002

According to a theory they had researched, the students believed that the relationship between I and r is $I = \frac{P}{r^2}$, where P is a constant.

By appropriate manipulation and graphical techniques, find the students' experimental value for P. Use a spreadsheet or calculator to assist with finding the line of best fit.

You will now be able to answer key question 7.

EVALUATING THE QUALITY OF DATA

Raw data can be processed in numerous ways. Processing data is usually done to reveal any trends, patterns, uncertainties, mistakes, outliers and results of significance that may exist in the data, including revealing relationships between any variables (e.g. dependent and independent variables). During processing, trends, patterns, uncertainties, mistakes, outliers and results of significance could become obvious. It is important to discuss the limitations of your method of investigation and any effect these limitations may have had on the data collected. Specifically, you should look for anything that may have affected the validity, accuracy, precision or reliability of the data. Sources of errors and uncertainty must also be stated in the discussion section of the final report on your investigation.

When analysing data, it is important not to select processes that demonstrate only what you want to see. Bias will result from using analysis tools (e.g. statistics) inappropriately and this may lead to invalid conclusions. It might also be a case of academic fraud. Quality scientific analysis processes raw data as it is and is open to any result.

Bias

In Section 1.3 (page 23) we learnt about recording data during an experiment. It is important that bias is not introduced during the data-recording process. **Bias** is a form of systematic error resulting from the researcher's personal preferences or motivations. There are many types of bias, including:

- poor definitions of concepts or variables (e.g. classifying cricket pitch surfaces and their interaction with the ball according to resilience without defining 'slow' and 'fast')
- incorrect assumptions (e.g. that footwear type, model and manufacturer does not affect ground reaction forces and, as a result, failing to control for these variables during an investigation into slip risk on different indoor and outdoor surfaces)
- errors in the design or methodology of the investigation (e.g. taking more samples from one gender than another).

Bias may occur in any part of an investigation, including sampling and measurement.

Some biases cannot be eliminated, but they should be noted in the discussion.

● *You will now be able to answer key question 3.*

Analysing precision

Section 1.2 (page 16) highlighted the importance of designing an investigation that will minimise errors and ensure accuracy, precision and validity. Understanding uncertainty and precision is also vital in any analysis of data. In physics there is always variation in measurements. In your experiment, you should determine whether the variation is caused by systematic or random errors; in other words, how much variation in the collected data is due to instrumentation and how much is due to nature.

The precision and uncertainty of instruments must always be displayed as a range of data next to the results (measurement ± uncertainty). If calculations are performed with the results, then corresponding calculations must also be done with the uncertainties. When the total uncertainty is known, then it can be established whether variation in the data is due to the instrument or to the variables being tested.

If the measurements between trials fall within the uncertainty range of the instrument, then the variation in results could simply be due to the instrument. If the difference between the measured results is greater than the uncertainty range, then the variation in the results is not due to the instrument and must therefore be due to other variables.

It is important to understand the accuracy and precision of the instruments used in an investigation, as it affects the interpretation of the results. There are a few ways to analyse precision, including:

- measurement uncertainty, which displays the precision of the instrument and explains instrumental variation in the measured results
- range, which outlines the difference between the smallest and largest measurements
- tendency (e.g. mean), which is the potential variation in instrumental measurements due to the instrument's design or increments.

Analysing validity and theoretical relationships

Process the results and data to look for trends, patterns or differences. A common process for analysing data is to make statistical calculations to determine true values, uncertainties, errors and the significance of the measurements. Once the quality of the data is understood, then the validity can be analysed in relation to established theoretical concepts.

Analysis can also find anomalies and outliers in data that are not valid measurements. During the experiment, your record of observations may provide a reason for any outlier in your data. Based on this reason, you may be able to suggest improvements in the methodology that could eliminate outliers.

ESTIMATING THE UNCERTAINTY IN A RESULT

Uncertainty has been considered in this chapter in relation to instrument precision. You should be aware that scientists also routinely estimate uncertainty in calculations due to the variation in results. Uncertainty due to natural variation can be treated through statistical analysis of the data.

You will see a combination of the measurement uncertainty and the uncertainty due to variation indicated in processed data. The uncertainty will likely be shown in the column title of a table as ±uncertainty, as shown in Figure 1.4.2 on page 29. Uncertainty will also appear in graphical data with the use of uncertainty bars, as shown in Figure 1.4.3 on page 31, and in Figure 1.4.6. You are not expected to estimate uncertainty for the purposes of VCE Physics, but when you interpret results you should recognise that uncertainty indicates a limitation of the data.

Uncertainty bars

Uncertainty bars (often called error bars) indicate the absolute uncertainties in the independent values or dependent values. They are drawn as a horizontal line centred on the data point and with a length indicating the uncertainty of the independent variable, and as a vertical line centred on the data point with a length indicating the uncertainty of the dependent variable. The uncertainty bars can be viewed as forming an uncertainty rectangle, with the true measurement falling somewhere within that rectangle (Figure 1.4.6).

The smaller the uncertainty bars are for a given point, the more precise is the measurement.

You will now be able to answer key question 8.

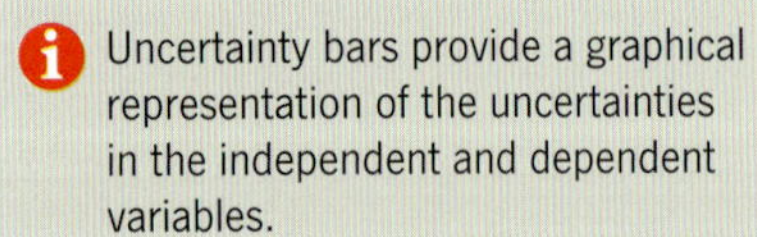

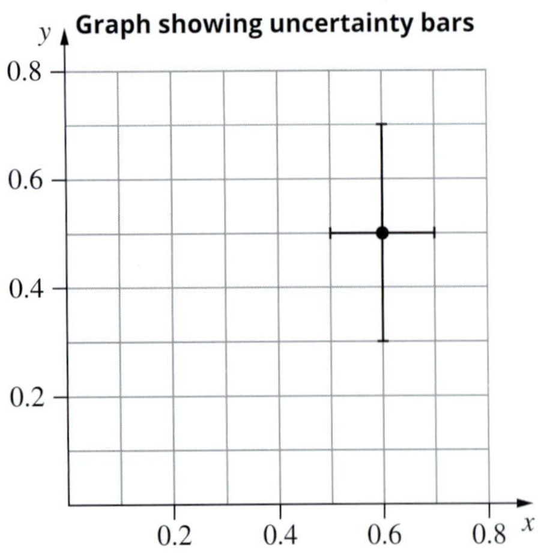

FIGURE 1.4.6 The horizontal uncertainty bar indicates an x-value of 0.6 ± 0.1. The vertical uncertainty bar indicates a y-value of 0.5 ± 0.2.

1.4 Review

OA ✓✓

SUMMARY

- Graphs can show a wealth of information relating two or more variables, including their uncertainties.
- Graphs can be used to reveal relationships between variables and spot outliers in data.
- Graphs in physics should display:
  - a title
  - labels, scales and units on each axis
  - plotted data
  - uncertainty bars in the *x*- and *y*-directions
  - line of best fit for linear and non-linear graphs.
- Commonly used graphs in physics are linear, parabolic (squared), sine and cosine, and exponential decay curves.
- An outlier is often caused by a mistake made in measuring or recording data, or from a random error in the measuring equipment. If there is an outlier, include it on the graph but ignore it when adding the line of best fit.
- A linear graph is easiest to analyse because it is straightforward to calculate the gradient and *y*-intercept, and these can be used in further processing the data.
- Any graph that is not linear can be linearised to produce a graph that has modified *x*- and *y*-axes.

KEY QUESTIONS

Knowledge and understanding

1 List a type of graph that can be used to represent discrete data and continuous data.

2 Identify information missing from the graph below.

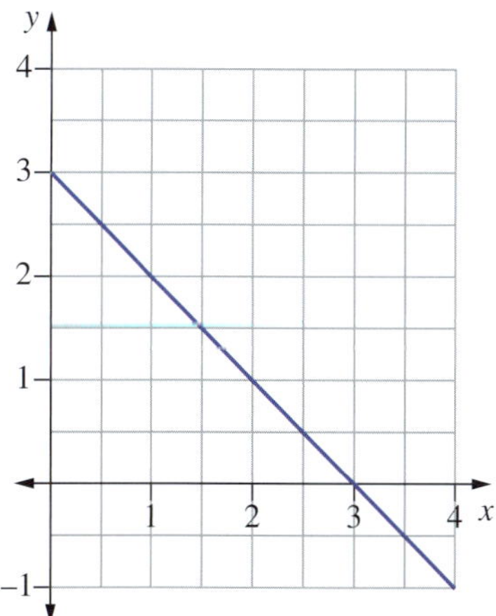

3 Describe what bias is in an investigation. Provide an example.

4 Plot the following data set, assigning each variable to the appropriate axis.

Current (A)	Voltage (V)
0.08	5.45
0.06	4.02
0.04	2.76
0.02	1.35
0.01	0.69

5 Explain what an outlier is and how outliers should be treated when analysing data. How would outliers be treated on a graph?

6 List two statistical measures used to describe data and how each measure is calculated.

Analysis

7 Using the data in question 4, obtain the line of best fit and determine the gradient, *m*, and the vertical axis intercept, *c*.

8 Isla and Sam performed an experiment to determine the resistance of an ohmic resistor, Ω. They measured the current, A, through the resistor when the voltage through the resistor was varied and obtained the following results.

Voltage, *V* (±0.2 V)	Current through the wire, *I* (±0.1 A)
0.0	0.0
1.8	0.5
3.4	0.9
5.6	1.4
7.8	2.0
9.8	2.5

The voltage and current through a resistor is given by $V = IR$.

a Draw a graph of the data in the table. Include a line of best fit and uncertainty bars.

b Determine the gradient of the line of best fit.

c Use your answer to part b to calculate the value of the strength of the resistor.

1.5 Conclusion and evaluation

Now that your chosen topic has been thoroughly researched, your investigation has been conducted and data collected and analysed, it is time to draw it all together. The final part of the investigation is to summarise your findings in an objective, clear and concise manner in a scientific report. Presenting your report and report formats will be discussed in Section 1.6.

EXPLAINING RESULTS IN THE DISCUSSION

The discussion is the part of a report on an investigation in which the method is explained and evaluated. It is also where the results are interpreted.

The key sections of the discussion are:

- an evaluation of the investigative method
- an analysis and evaluation of data
- an explanation of how the findings relate to established concepts in physics.

When writing the discussion section, consider the message to be conveyed and your expected audience. Statements need to be clear and concise. At the conclusion of the discussion, the audience must have a clear idea of the context, results and implications of your investigation.

Evaluating the investigative method

Your discussion should evaluate your investigative methodology and methods and identify any issues that could have affected the validity, reliability, accuracy or precision of the data. Any possible sources of error in your experiment should be stated. Remember that controls are essential to the reliability and validity of your investigation, so if you have overlooked or were unable to control a variable that should have been controlled, this may explain unexpected results.

The discussion should also make recommendations for modifying or extending the investigation. If there were sources of error in any of the methods or steps, provide suggestions for how this could be improved so that future researchers can benefit from your experience.

It is also important to acknowledge contradictions in data and information. Do any of the results not match the predictions? If so, is this a result of a limitation of the experimental design or methods? In your discussion, acknowledge these sorts of issues and make suggestions for further experiments to address them.

Some experimental findings may lead you to formulate new research questions and develop new hypotheses. An extension of the experiment may be to make an alteration that will enable further investigation. For example, if the effect of temperature has been investigated, further understanding of temperature could be determined by using a different temperature range in a modification of the original method.

Analysing and evaluating data

In the discussion section of your final report, the findings of the investigation need to be analysed and interpreted. A number of things need to be considered.

- State whether a pattern, trend or relationship was observed, including when relevant between the independent and dependent variables. Describe what kind of pattern it was and specify the conditions under which it was observed. Section 1.4 provided detailed guidance about how to analyse trends in data.
- Acknowledge and explain any discrepancies, deviations or anomalies in the data.
- Identify any limitations in the data you have collected. For example, you might think that a larger sample or further variations in the independent variable would have led to a more valid and reliable conclusion.

Relating your investigation to relevant physics concepts

To make your investigation more useful—and more interesting to other physicists or physics students—your discussion should explain how your investigation is related to established ideas, concepts, theories or models in physics. In particular, you should explain why you considered your hypothesis to be a worthwhile idea to explore.

For example, if studying the impact of temperature on the linear strain of a material (e.g. a rubber band), some of the information relevant to the domain of physics that might be included in the discussion could be:

- the functions of linear strain
- the factors known to affect linear strain
- existing knowledge on the role of temperature on linear strain
- the range of temperatures investigated and the reason they were chosen
- the materials studied and the reasons for this choice
- methods of measuring the linear strain of a material.

Framing your discussion

By relating your investigation to relevant concepts in physics, you will have created a framework for discussing whether the data you collected supports or refutes your hypothesis. Ask the following questions:

- Was the hypothesis supported?
- Has the research question been fully answered? (If not, give an explanation of why that is and suggest what could be done to either improve or complement the investigation.)
- Do the results contradict the hypothesis? If so, why? (The explanation must be plausible and must be based on the results and previous evidence.)

After identifying the major findings of your investigation, compare your results with existing relevant research and knowledge. Consider questions such as:

- How does the data fit with the literature?
- Does the data contradict the literature?
- Do the findings fill a gap in the literature?
- Do the findings lead to further questions?
- Can the findings be extended to other situations?

Be sure to discuss the broader implications of your findings. Ask such questions as:

- Do the findings contribute to our current knowledge of the topic?
- Do the findings suggest any practical applications?

Use the points in Figure 1.5.1 to help frame your discussion.

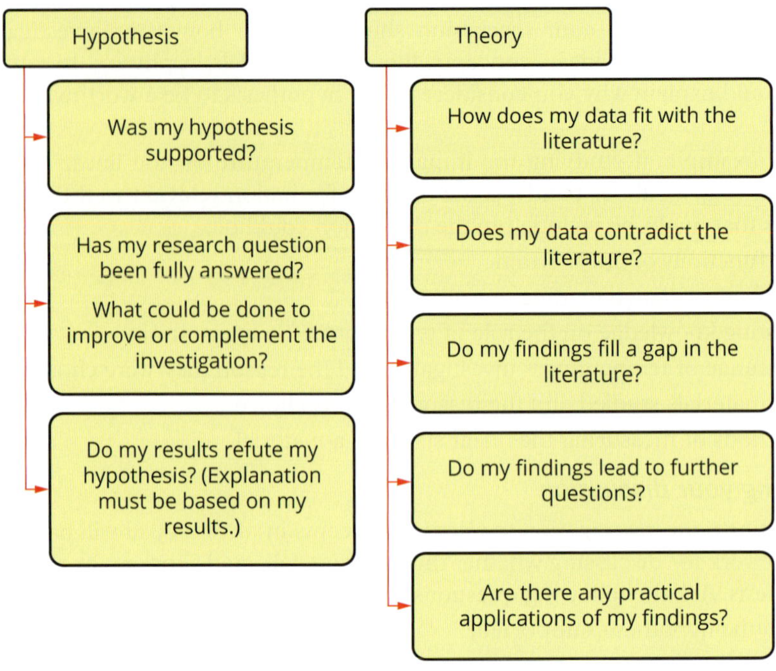

FIGURE 1.5.1 Points to help frame your discussion

● *You will now be able to answer key questions 1, 2 and 3.*

DRAWING EVIDENCE-BASED CONCLUSIONS

The **conclusion** to a scientific report or paper links the collected evidence to the hypothesis and provides a justified response to the research question.

Indicate whether the hypothesis was supported or refuted, and the evidence on which this is based (i.e. your results). Do not provide irrelevant information. Refer only to the specifics of the hypothesis and the research question and do not make generalisations.

The examples of poor and better conclusions in Table 1.5.1 may be of assistance.

TABLE 1.5.1 Examples of poor and better conclusions

Hypothesis/Research question	Poor conclusion	Better conclusion
An increase in temperature will cause an increase in linear deformation (change in length) before failure.	Linear deformation has value y_1 at temperature 1 and value y_2 at temperature 2.	An increase in temperature from 1 to 2 produced an increase in linear deformation in the rubber band.
Does temperature affect the maximum linear deformation the material can withstand?	The results show that temperature does affect the maximum deformation of a material.	Analysis of the results of the effect of an increase in temperature from 1 to 2 on the rubber band supports our current knowledge that an increase in temperature increases the maximum linear deformation.

● *You will now be able to answer key questions 4 and 5.*

1.5 Review

OA ✓✓

SUMMARY

- The discussion evaluates and explains the investigation methods and the results you obtained.
- Analyse and interpret your results in the discussion by:
  - identifying and describing any patterns, trends or relationships in the data. Specify the conditions under which they were observed
  - acknowledging and explaining any discrepancies, deviations or anomalies in the data
  - identifying any limitations in the data collected.
- Explain the hypothesis and investigation within the context of current thinking in physics.
- Use evidence from the data to conclude whether the hypothesis was supported or refuted.

KEY QUESTIONS

Knowledge and understanding

1. What is the purpose of a discussion of the results?
2. Explain why it is important for the discussion to be clear and concise.
3. What are the key sections of the discussion section of a scientific report?
4. What is the purpose of the conclusion?

Analysis

5. You conduct an investigation to test the following hypothesis: *If two objects are simultaneously dropped vertically from the same height, they will both land at the same time.*
   What is one conclusion you could draw if your results showed the following times for objects dropped from a height of 1 m?

Object	Time (s)
feather	3.5
tennis ball	0.65
bowling ball	0.3

1.6 Reporting investigations

FIGURE 1.6.1 Posters at a scientific conference are one way of presenting the findings of your research.

Scientists report their findings in a number of ways: as written peer-reviewed journal articles, on web pages, and at scientific conferences with short oral presentations or scientific posters (Figure 1.6.1). In this section you will learn how to present your findings effectively.

PRESENTATION FORMATS

There are numerous ways to present the results of a scientific investigation, each with varying emphases on visual and textual components. Table 1.6.1 provides some guidelines for different presentation formats.

TABLE 1.6.1 Main formats for presenting research work

Format	Characteristics	General guidelines for the presentation format
scientific poster	• concise visual display of information • suitable for presenting information to many people • summary of ideas	• title that attracts attention • large headings that stand out • subheadings of a smaller size • attractive presentation • combination of written material and visual material such as diagrams, photographs, tables, graphs • writing large enough to be read from a distance
written practical report	• presents clear and detailed information on a topic • suitable for providing detailed and more comprehensive background information	• appropriate written style for a report that provides an introduction, materials, methodology, methods, results, discussion and conclusion section • use subheadings to organise sections • text should be supported by tables, graphs, diagrams or photographs
oral communication with supporting slides and/or handouts	• easy-to-follow format • good for presenting to a large audience • supporting slides can be printed as notes and given to the audience • opportunity to answer questions from the audience	• brief oral descriptions • use clear visuals that complement what is spoken • minimal text on each slide • consistent format on all slides—background, colours and text • images, diagrams and graphs are clear and large
online presentation, e.g. website, blog	• accessible to a worldwide audience • easy to follow • easy to update with new information	• include hyperlinks to related information • include multimedia, such as video and audio, if appropriate • use the same format throughout—font, background, colours • use clear headings • list all hyperlinks on the main page • include your name, credentials and date of publication

● *You will now be able to answer key questions 1 and 3.*

STRUCTURING A WRITTEN REPORT

To write a scientific report, you need to follow some general conventions. Even though there are many ways to present a report, the report must broadly follow the structure set out below to meet the requirements of the VCE Physics Study Design. This section will provide a guide to writing an appropriate scientific report.

Headings are an essential feature of a scientific report. There is no single convention for headings in scientific report writing and what is required is often specific to a particular journal. Figure 1.6.2 lists headings that are commonly used in scientific reports and describes the information that is usually provided under each heading. Sections can be broken down further into subsections, as shown. Although some subheadings may be suitable for more than one section, a scientific report will only use each heading and subheading once. It is best to ask your teacher about which headings are preferred to meet the requirements of the VCE Physics Study Design.

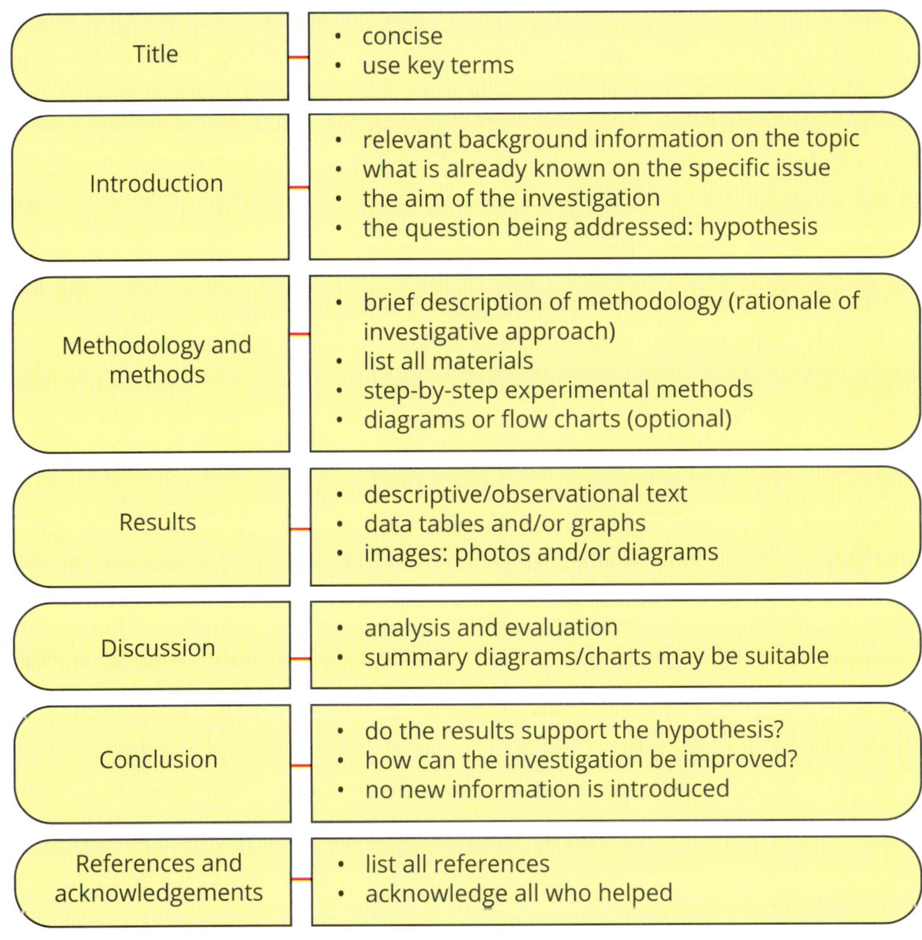

FIGURE 1.6.2 Elements of a scientific report or presentation

Title

The title should give a clear idea of what the report is about, without being too long. It should include key terms that tell the reader what your study is about.

Introduction

The introduction sets the context of your report. It should outline relevant physics ideas, concepts, theories and models, and how they relate to your specific research question and hypothesis. It introduces the key terms, the specific question to be addressed, and states your hypothesis and aim. Any references used in the introduction should be correctly cited. This section should also identify the independent, dependent and controlled variables (these are discussed in Section 1.1).

Methodology and methods

The methodology and methods section outlines the reason why you adopted your particular investigative approach and describes in detail all the steps that were undertaken during the investigation. It also includes a list of the materials used. For a scientific report use stepwise lists, diagrams of specific methods, and/or flow charts of the overall experimental design.

There should be enough detail in this section for someone else to be able to replicate your experiment; therefore, your method needs to be in the correct sequence and include how you observed, measured and recorded your results.

Results

The results section is a record of your observations. It is where you present your data using graphs, diagrams, tables or photographs. In Section 1.4 you learnt tips on using graphs and tables appropriately.

In general, tables provide more detailed data than graphs. However, it is easier to observe trends and patterns in graphs, making them a very useful tool for presenting evidence. Pie charts illustrate percentages well, while scatterplots illustrate relationships between variables. Bar charts are best used for qualitative data and discrete quantitative data. Scatterplots are best used for continuous quantitative data.

Discussion

In the discussion section you interpret your results, and discuss how they relate to your initial research question and hypothesis, and to the research of others. See Section 1.5 for more information about how to do this. You should link your discussion to the key concepts discussed in the introduction.

It is also important to evaluate the methods used and the impact of any errors on the results and on conclusions that can be drawn.

Conclusion

Your conclusion should include a carefully considered response to the evidence (i.e. your data) that supports or refutes your hypothesis. It should provide a carefully considered response to your research question based on your results and discussion. You should clearly state whether your hypothesis was supported or not. Draw your conclusions by identifying trends, patterns and relationships in the data.

It is important to recognise the limitations of your data and the limitations of the scientific method. Be careful not to overstate your conclusion. Your results will support or refute the hypothesis. They will not prove that something is true, as you can only ever provide evidence that indicates the probability of something being true.

Do not provide irrelevant information, or introduce new information, in your conclusion. Refer to the specifics of your hypothesis and research question, and do not make generalisations.

The conclusion section should be a short, succinct paragraph.

References

All the scientific papers and other sources that are mentioned in the report are to be listed at the end of your report. Cite the source of any information you obtained from all sources in the text of your report whenever it is used and referred to, and provide a list of references at the end of your report. This demonstrates that you are aware of previous work in the area and allows readers to locate sources of information if they want to study them further.

Acknowledgements

It is good practice to acknowledge anyone who helped you during your investigation; for example, your teacher or lab tech, or anyone else who provided you with guidance. This only needs to be a sentence or two: 'Thanks to my teacher for useful discussions about the feasibility of my experiment, to our lab tech for help setting up the motion sensor correctly and to my fellow students for repeatedly running along the sports track so I could time them'.

- *You will now be able to answer key question 5.*

WRITING FOR SCIENCE

A scientific report is written for a scientific audience, so it is important to ensure that the report uses appropriate scientific language and follows the expected conventions. This language and its conventions are different from everyday English writing.

Scientific reports should be written using:

- past tense—the experiment was conducted in the past, so the report should be in the past tense
- either third-person (passive) voice, or first-person (active) voice—ask your teacher if they have a preference and make sure to keep voice consistent throughout your report (Table 1.6.2)
- scientific language—the terms used are specific to concepts, models and theories
- objective, unbiased language—avoid subjective and emotional or persuasive writing (Table 1.6.3)
- concise language—avoid unnecessary repetition and express ideas succinctly. Scientific language allows more details, knowledge and understanding to be communicated in fewer words. Use short sentences (Table 1.6.4).

TABLE 1.6.2 Examples of first-person and third-person writing

First person	Third person
I first tied a rubber stopper of known mass onto one end of a piece of fishing line, and a brass cradle of 200 g to the other end.	First, a rubber stopper of known mass was tied to one end of a piece of fishing line. A brass cradle of mass 200 g was tied to the other end.
After the current was switched on, I found that ...	After the current was turned on, the results showed ...
My colleagues and I found ...	Researchers found ...

TABLE 1.6.3 Persuasive writing versus scientific writing styles

Persuasive writing examples	Scientific writing equivalent examples
Use of biased and subjective language: • The results are extremely bad, atrocious, wonderful etc. • This is terrible because ...	Use of unbiased and objective language: • The results showed ... • The implications of these results suggest ... • The results imply ...
Use of exaggeration: • The object weighed a colossal amount, like an elephant. • Safety crisis ...	Use of non-emotive language: • The object weighed 256 kg. • Safety issue ...
Use of everyday or colloquial language: • The experiment didn't work because we didn't know what we were doing. • We did not obtain the results we expected because our we were not sure how to use the equipment, which led to significant errors in measurement. • The results don't ... • The researchers had a sneaking suspicion that ...	Use of formal language: • Further research is needed to fully determine why the results of the experiment were not as expected. • The results do not ... • The researchers predicted (or hypothesised or theorised) that ...

TABLE 1.6.4 Examples of wordy and concise language

Wordy language	Concise language
Due to the fact that ...	Because ...
Anog and Walsh (2022) undertook an investigation into ...	Anog and Walsh (2022) investigated ...
It is possible that the cause could be ...	The cause may be ...
End result ...	Result ...
In the event that ...	If ...
Shorter in length ...	Shorter ...

Scientific language must be used without error so that the reader understands the meaning of the information easily. For the report to be concise, there should not be any repetition. The report must remain within the required word limit. Being precise and concise will help you stay within the specified word limit.

Paragraphs in scientific writing

In a scientific report each paragraph should explain only one topic. The first sentence of a paragraph (called the topic sentence) introduces the topic. The sentences that follow the topic sentence provide details about the topic, with the final sentence concluding the discussion of the one topic that the paragraph is about.

Each sentence in a paragraph should refer to only one subject (two only if necessary), and each sentence should flow on to the next. Readers should be able to see how each sentence relates to the previous one.

● *You will now be able to answer key questions 2 and 6.*

EDITING YOUR REPORT

Editing your report is an important part of the process. After editing your report, save new drafts with a different file name and always back up your files in another location. Once you have completed a draft, it is good practice to read your work a day or two after you have completed it.

When reading your own work, do not read it as you intend it to be read by others. Instead, carefully read your work, following the punctuation, grammar and spelling as it appears on the page. This is more easily achieved if you read the report aloud. When editing, look for content that:

- is ambiguous or unclear
- is repetitive
- is awkwardly phrased
- is too lengthy
- is not relevant to your research question
- is poorly structured
- lacks evidence
- lacks a reference (if it is another researcher's work)
- contains spelling mistakes.

ACKNOWLEDGING SOURCES

The source of all quotations used in your report must be listed in the references section. You should also acknowledge the ideas of others that have been instrumental in forming the idea for your research or the interpretation you have made of your results. References and acknowledgements also give credibility to your study and allow the audience to locate information sources for further study.

Plagiarism is using other people's work without acknowledging them as the author or creator. To avoid plagiarism, include a reference every time you report the work of others, placing it at the end of a sentence or following a diagram. If you use a direct quotation from a source, enclose it in quotation marks.

Referencing

Each time you write about the findings of other people or organisations, you need to provide an in-text citation and the full details of the source in a reference list. Numerous referencing systems are in use, but the American Psychological Association system is common in the sciences. Check with your teacher as to which system you should use for your report.

In most referencing systems, sources in a reference list are listed in alphabetical order (by the author's last name or the organisation's name). Compile your references in a separate document as you conduct your practical investigation. This will save you time later.

Your Unit 2 Area of Study 3 practical investigation report does not require a bibliography. A reference list is sufficient. A bibliography is a list of all the sources used during your research that helped you develop an understanding of your research topic even if such a source is not cited in your final report. A reference list only lists those sources that you cite.

The following examples show the use of in-text citation and the corresponding reference list entry for an article in a journal in APA (seventh edition) style. The format varies slightly according to the type of document or source you are referencing. Useful guides can be found online including many university websites.

Source of information and example of reference in text	Format for listing references and example of a reference as written in the reference list
Research article or review article in a scientific journal A single atom of the rare-earth metal holmium has been made into the world's smallest, stable magnet. This was then used to make an atomic hard drive, in which each holmium atom stored one bit of information (Natterer et al., 2017).	Author, initials. (year). Title of article. Journal title, volume number (issue number), page numbers. Digital object identifier (doi) or URL Natterer, F., Yang, K., Paul, W., Willke, P., Choi, T., Greber, T., Heinrich, A., & Lutz, C. (2017), Reading and writing single-atom magnets. *Nature, 543*, 226–228.
Book Hawking (1988) discusses the theories of general relativity and quantum mechanics and seeks to describe a unifying theory that combines these.	Author, initials. (year). Title of book (edition, if not first). Publisher Hawking, S., (1988). *A brief history of time: From the big bang to black holes*. Bantam Books.
Online article or page A layered crystal (created with hafnium oxide and zirconium oxide) reduces the required voltage by around 30% (Perfetto, 2017).	Author, initials/name of organisation. (year). Title of web page or web document. URL Perfetto, I., (2022, April 1). New crystal could help transistors run on less power. Cosmosmagazine.com. https://cosmosmagazine.com/technology/computing/crystal-for-transistors/

● *You will now be able to answer key question 4.*

STRUCTURING A POSTER

Scientific posters are used at conferences to grab people's attention and quickly summarise your research. A poster should follow much the same format as that of a scientific report outlined earlier in this section. A poster is meant to be more succinct and direct in its approach to presenting your findings. You should pick the information that you want to present carefully so that the impact of your investigation is best communicated. For instance, in analysing your data you may have produced both a table and a graph. In the poster, it may be best to simply show the graph, which will more clearly represent any trends in your data.

A scientific poster should be understood by both technical and non-technical audiences. Make sure to keep the language concise and as free of jargon as possible.

Figure 1.6.3 is an example of a poster from a student's investigation of the research question: 'Is the period of a pendulum affected by the mass or length of the pendulum?' You can follow a similar structure if you choose to report on your investigation for Unit 2 Area of Study 3 in poster format.

Is the period of a pendulum affected by the mass or length of the pendulum?

Yarran Jones
Heinemann Physics College
Unit 3–4 Physics 2022

Introduction

A simple pendulum consists of a mass or 'bob' attached to a pivot point. As the bob is lifted and released, the pendulum sweeps back and forth. One full movement, from left to right and back again, is a 'period' (Figure 1).

When a pendulum is lifted, it acquires gravitational potential energy (E_g). When the pendulum is released, the bob's mass causes it to swing back and forth. Its E_g is transformed into kinetic energy (E_k). When the pendulum passes through its equilibrium, the E_g it has 'lost' has been transformed into E_k. As it continues to swing and move upwards on the next arc, the E_k is transformed back into E_g.

(Museum of Science and Industry, Chicago, n.d.)

AIM To determine if the mass or length of a pendulum affects its period.

HYPOTHESIS The period of a pendulum is independent of its mass and depends on the square root of its length.

Figure 1 Period

Methodology and methods

METHODOLOGY

Controlled experiment

METHODS

Set up

1. Fold A4 paper into a 'ramp' approximately 11 cm × 2 cm with a V-notch in one end.
2. Mark a line at 45° on vertical desk edge.
3. Extend ramp over desk edge, with V lined up with line on desk edge.
4. Tie a washer to a 0.45 m piece of string.
5. Measure from the bottom of the washer along the string. Put coloured dots at 0.10 m, 0.15 m, 0.20 m, 0.25 m, 0.30 m and 0.35 m.
6. Make two more pendulums of two and three washers each and mark their lengths.

Testing the pendulums

1. Hang pendulum over ramp with the 0.30 m mark on the string with one washer attached.
2. Pull washer so string lines up with 45° line.
3. Release pendulum, start timing. Stop after three complete swings.
4. Repeat twice more, calculate averages.

Part A: Change the mass

1. Change mass of the pendulum by adding a washer.
2. Repeat test procedure, record three times for this pendulum.
3. Change to three-washer pendulum and repeat.

Part B: Change the length

1. Return to one washer on the pendulum.
2. Change length of the pendulum to 0.35 m by adjusting the string length.
3. Use different lengths for one-washer pendulum (0.10 m, 0.15 m, 0.20 m and 0.25 m).
4. Calculate averages.

Communication statement

The period of a pendulum is independent of its mass. The period does depend on the square root of its length (i.e. $T \propto \sqrt{L}$).

Results

Table 1 Effect of mass on period

No. of washers	Length (m)	Release angle (°)	Time for three swings (s)				Period T (s)
			Trial 1	Trial 2	Trial 3	Average	
1	0.30	45	3.39	3.36	3.30	3.35	1.12
2	0.30	45	3.24	3.49	3.42	3.38	1.13
3	0.30	45	3.32	3.31	3.36	3.33	1.11

Table 2 Effect of length on period

No. of washers	Length (m)	Release angle (°)	Time for three swings (s)				Period T (s)
			Trial 1	Trial 2	Trial 3	Average	
1	0						0
1	0.10	45	1.60	1.69	1.63	1.64	0.55
1	0.15	45	2.27	2.26	2.15	2.23	0.74
1	0.20	45	2.69	2.76	2.68	2.71	0.90
1	0.25	45	3.02	3.00	3.01	3.01	1.00
1	0.30	45	3.30	3.31	3.43	3.35	1.12
1	0.5	45	3.62	3.60	3.61	3.61	1.20

Discussion

EVALUATION

Graph 1 Period squared v length

Table 1 shows no significant variation in the three periods when the mass changed. The period of a pendulum is not, therefore, dependent on the mass.

According to the formula $T = 2\pi\sqrt{\frac{L}{g}}$ (Bunn, 1990), the relationship between T and L is a square root relationship. This is evident in Table 2 and confirmed by the linear relationship of Graph 1.

LIMITATIONS

- measurements of length limited by use of a ruler
- error bars not included as graph is too small to show up precisely
- V supported string and held it away from desk edge by a few mm to reduce friction while pendulum swinging.

IMPROVEMENTS

- better method of removing friction at pivot point
- more accurate measure of length of the pendulum
- measure masses of washers.

Conclusion

The hypothesis was supported by the results showing that the period of a pendulum is independent of the mass and dependent on the square root of the length.

Further experiments could be conducted:

- with a larger range of masses to further confirm the independence of the period on mass
- to determine if the angle of release (other than 45°) has an effect on the period
- to determine if different types of strings have an effect on the period
- to determine the acceleration due to gravity.

References and acknowledgements

1. Bunn, D.J. (1990). *Physics for a modern world.* (p. 237). John Wiley & Sons Australia.
2. Museum of Science and Industry, Chicago. (n.d.). *Back and Forth*. https://www.msichicago.org/fileadmin/assets/educators/learning_labs/documents/Back_and_Forth.pdf

Acknowledgements: with thanks to lab technicians Jarrah and Ivan and my lab partner Michelle Chan.

FIGURE 1.6.3 Example of a scientific poster

1.6 Review

OA ✓✓

SUMMARY

- A scientific report generally includes the following headings:
  - Title
  - Introduction
  - Methodology and methods
  - Results
  - Discussion
  - Conclusion
  - References
  - Acknowledgements
- Your scientific report must meet the requirements of the VCE Physics Study Design.
- A scientific poster:
  - follows a similar structure to that of a scientific report
  - should be more succinct than a report
  - should be written for both technical and non-technical audiences.

KEY QUESTIONS

Knowledge and understanding

1 What format for presenting research would be appropriate when addressing a large group? Explain your answer.

2 Which one or more of the following statements is written in the third person?

A The results of the study concluded ...

B I added 3 mL of solution to the graduated cylinder ...

C Samples were analysed using ...

D We repeated the experiment three times ...

3 What reporting format is best suited for presenting the details of an experiment you conducted, including the equipment, methodology, results and a discussion of the results?

4 Describe what plagiarism is.

5 Explain the purpose of a clear introduction and conclusion.

Analysis

6 A scientist conducted an experiment to test the following hypothesis: *Adding more resistance to a circuit decreases the current through the circuit.* The discussion section of the scientist's report included comments to support the accuracy and precision of the investigation. Determine whether the following sentences indicate precision or accuracy.

a The current through the circuit was measured using a calibrated digital ammeter.

b The digital ammeter measured the current to the nearest μA, whereas an analog ammeter would only measure to the nearest mA.

Chapter review

01

KEY TERMS

accuracy
aim
bias
conclusion
continuous variable
controlled variable
dependent variable
discrete variable
hypothesis
independent variable
line of best fit
linear relationship
mean
median
method
methodology
mistake
mode
observation
outliers
personal protective equipment (PPE)
precision
primary data
primary source
processed data
qualitative data
qualitative variable
quantitative data
quantitative variable
random errors
range
raw data
repeatability
reproducibility
research question
risk assessment
safety data sheet (SDS)
scientific method
secondary data
secondary source
significant figures
systematic error
true value
uncertainty
uncertainty bars
validity
variable

UNIT 1 How is energy useful to society?

To achieve the outcomes in Unit 1, you will draw on key knowledge outlined in each area of study and the related key science skills on pages 11 and 12 of the study design. The key science skills are discussed in Chapter 1 of this book.

AREA OF STUDY 1

How are light and heat explained?

Outcome 1: On completion of this unit the student should be able to model, investigate and evaluate the wave-like nature of light, thermal energy and the emission and absorption of light by matter.

AREA OF STUDY 2

How is energy from the nucleus utilised?

Outcome 2: On completion of this unit the student should be able to explain, apply and evaluate nuclear radiation, radioactive decay and nuclear energy.

AREA OF STUDY 3

How can electricity be used to transfer energy?

Outcome 3: On completion of this unit the student should be able to investigate and apply a basic DC circuit model to simple battery-operated devices and household electrical systems, apply mathematical models to analyse circuits, and describe the safe and effective use of electricity by individuals and the community.

CHAPTER

02 Waves and electromagnetic radiation

Have you ever watched ocean waves heading towards the shore? For many people their first thought when encountering a topic called 'waves' is to picture a water wave moving across the surface of an ocean. The wave may be created by some kind of disturbance, such as the action of wind on water or a boat as it moves through the water.

In fact, waves are everywhere. Sound, visible light, radio waves, waves in the string of an instrument, the wave of a hand, the 'Mexican wave' at a stadium and the recently discovered gravitational waves—all are waves or wave-like phenomena. Understanding the physics of waves provides a broad base upon which to build your understanding of the physical world. A knowledge of waves gives an introduction to the concepts that describe the nature of light.

Key knowledge

- identify all electromagnetic waves as transverse waves travelling at the same speed, c, in a vacuum as distinct from mechanical waves that require a medium to propagate **2.1, 2.3**
- identify the amplitude, wavelength, period and frequency of waves **2.2**
- calculate the wavelength, frequency, period and speed of travel of waves using: $\lambda = \frac{v}{f} = vT$ **2.2, 2.3**
- explain the wavelength of a wave as a result of the velocity (determined by the medium through which it travels) and the frequency (determined by the source) **2.2**
- describe electromagnetic radiation emitted from the Sun as mainly ultraviolet, visible and infrared **2.3**
- compare the wavelength and frequencies of different regions of the electromagnetic spectrum, including radio, microwave, infrared, visible, ultraviolet, x-ray and gamma, and compare the different uses each has in society. **2.3**

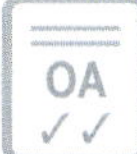

2.1 Longitudinal and transverse waves

Throw a stone into a pool or lake, and you will see circular waves form and move outwards from the source as ripples. Stretch a cord out on a table and wriggle one end back and forth across the table surface and another type of wave can be observed. Water waves, sound waves and waves in strings are all examples of **mechanical waves**. These waves require a **medium** (a physical substance) to **transmit** (carry or transfer) energy: water waves use water molecules, sound waves use air and the wave on a string uses the string (Figure 2.1.1).

FIGURE 2.1.1 In this tin can phone, sound waves vibrate the string. The vibrating string transfers the sound between the children.

FIGURE 2.1.2. Light travels from the Sun through the vacuum of space and does not need a medium.

Electromagnetic waves, which include visible light, do not require a medium to transfer energy. Thus, light from the Sun can transmit across the vacuum of space (Figure 2.1.2).

MECHANICAL WAVES

Watch a piece of driftwood, a leaf, or even a surfer resting in the water as a smooth wave goes past. The object moves up and down but doesn't move forwards with the wave. The movement of the object on the water reveals how the particles in the water move as the wave passes; that is, the particles in the water move up and down from an average position.

Any wave that needs a medium (such as water) through which to travel is called a mechanical wave. Mechanical waves can move over very large distances, but the particles of the medium only have very limited movement.

Mechanical waves transfer energy from one place to another through a medium. The particles of the matter vibrate back and forth or up and down about an average position, which transfers the energy from one place to another. For example, energy is given to an ocean wave by the action of the wind far out at sea. The energy is transported by waves to the shore, but (except in the case of a tsunami event) most of the ocean water itself does not travel to the shore.

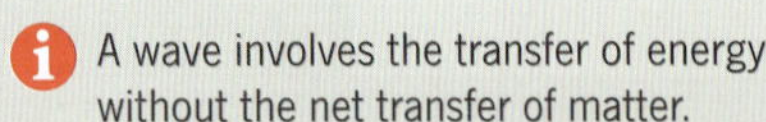

A wave involves the transfer of energy without the net transfer of matter.

Pulses versus periodic waves

A single wave **pulse** can be formed by giving a slinky spring or a rope a single up-and-down motion, as shown in Figure 2.1.3(a). As the hand pulls upwards, the adjacent parts of the slinky will also feel an upwards force and begin to move upwards. The source of the wave energy is the movement of the hand.

If the up-and-down motion is repeated, each successive section of the slinky will move up and down, moving the wave forwards along the slinky, as shown in Figure 2.1.3(b). Connections between each loop of the slinky cause the wave to travel away from the source, carrying with it the energy from the source.

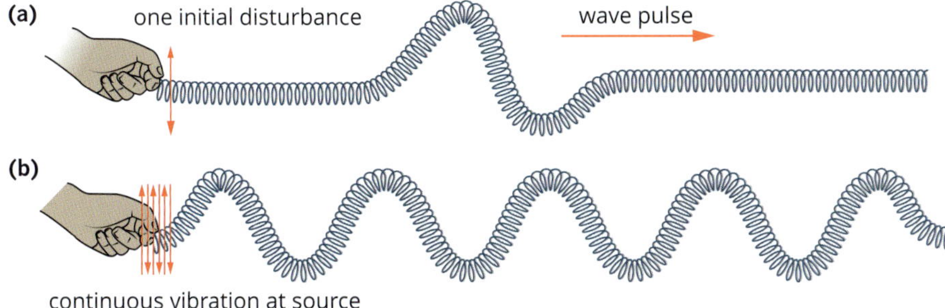

FIGURE 2.1.3 (a) A single wave pulse can be sent along a slinky by a single up and down motion. (b) A continuous or periodic wave is created by a regular, repeated movement of the hand.

In a continuous or periodic wave, continuous vibration of the source, such as that shown in Figure 2.1.3(b), will cause the particles within the medium to **oscillate** (move about their average position in a regular, repetitive or periodic pattern). The source of any mechanical wave is this repeated motion or vibration. The energy from the vibration moves through the medium and constitutes a mechanical wave.

Transverse waves

When waves travel on water, or through a rope, spring or string, the particles within the medium vibrate up and down in a direction perpendicular, or **transverse**, to the direction of motion of the wave energy, as can be seen from the position of the cork in Figure 2.1.4. Such a wave is called a transverse wave. When the particles are displaced upwards from the average position, or resting position, they reach a maximum positive displacement called a **crest**. Particles below the average position fall to a maximum negative position called a **trough**.

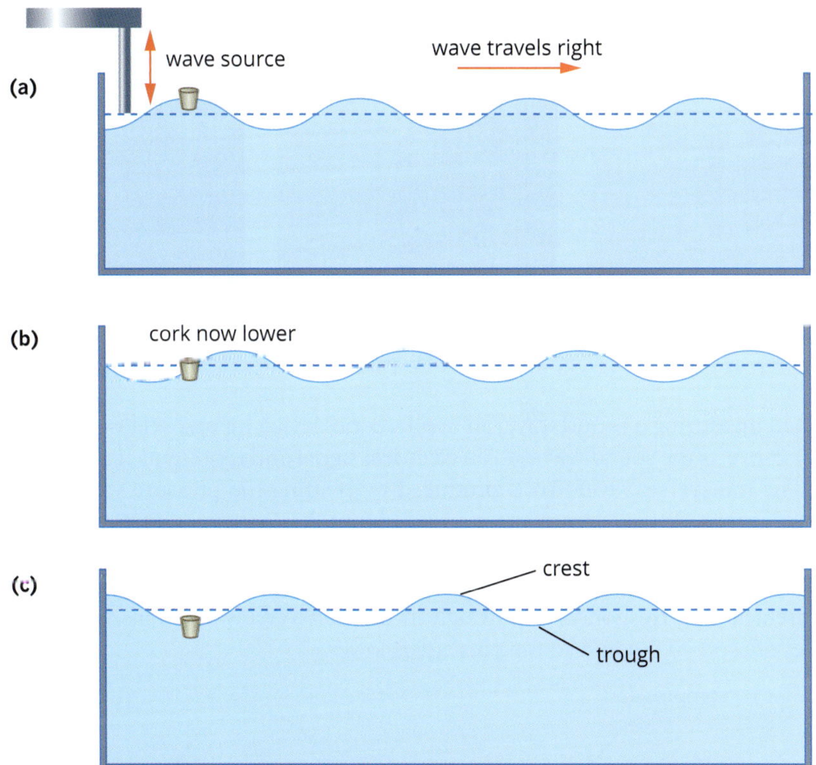

FIGURE 2.1.4 A continuous water wave moves to the right. As it does so, the up-and-down displacement of the particles transverse to the wave motion can be monitored using a cork. The cork simply moves up and down as the wave passes through it.

Longitudinal waves

In a **longitudinal** mechanical wave, the vibration of the particles within the medium is in the same direction, or parallel to, the direction of the energy flow of the wave. You can demonstrate this type of wave with a slinky by moving your hand backwards and forwards in a line parallel to the length of the slinky, as shown in Figure 2.1.5.

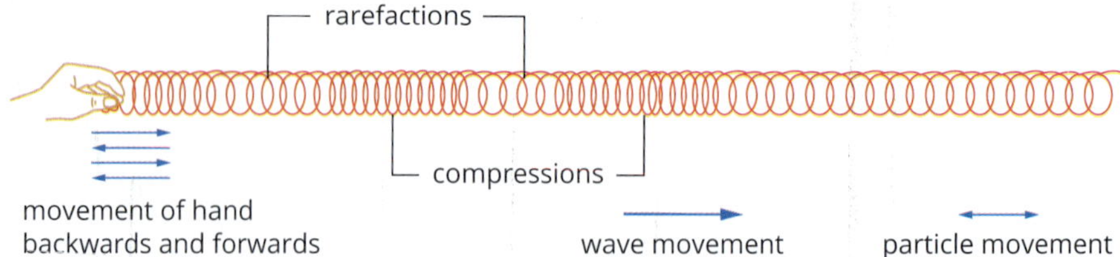

FIGURE 2.1.5 When the direction of the vibrations of the medium and the direction of travel of the wave energy are parallel, a longitudinal wave is created. This can be demonstrated with a slinky.

As you move your hand, a series of compressed and expanded areas form along the slinky (Figure 2.1.5). **Compressions** are those areas where the coils of the slinky come together. Expansions are regions where the coils are spread apart. Areas of expansion are termed **rarefactions**. The compressions and rarefactions in a longitudinal wave correspond to the crests and troughs of a transverse wave.

An important example of a longitudinal wave is a sound wave. As the cone of a loudspeaker vibrates, the layer of air next to it is alternately pushed away and drawn back, creating a series of compressions and rarefactions in the air (Figure 2.1.6). This vibration is transmitted through the air as a sound wave. As in transverse waves, the individual molecules vibrate over a very small distance while the wave itself can carry energy over very long distances. If the vibration was from a single point, then the waves would tend to spread out spherically.

FIGURE 2.1.6 The motion of a flame in front of a loudspeaker is clear evidence of the continuous movement of air backwards and forwards as the loudspeaker creates a sound wave.

When measuring a sound wave, an oscilloscope device (or an oscilloscope app on a phone) converts the sound waves to an electrical signal and represents it as a transverse wave. The transverse waveform is produced by plotting the pressure variation in the medium against distance from the source. Figure 2.1.7 shows that the sound wave compression corresponds to a peak or crest in the transverse wave representation. This is because the compression is an area of high pressure—the particles are close together. The rarefaction corresponds to a trough in the transverse wave. This is because the particles are spread out and so the pressure is lower.

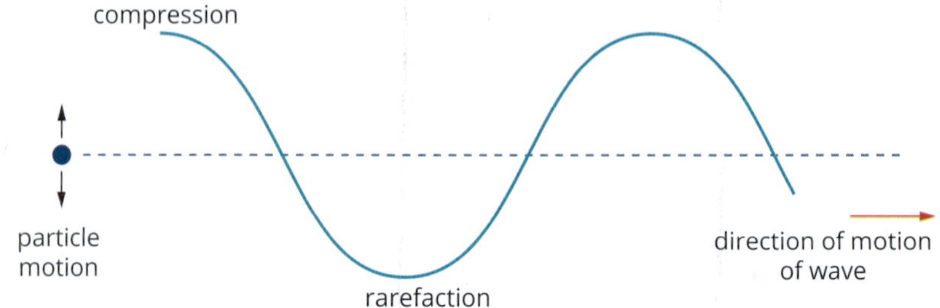

FIGURE 2.1.7 The compression of a longitudinal wave coincides with the peak of the transverse wave representation, while the rarefaction coincides with the trough.

PHYSICSFILE

Water waves

Water waves are often classified as transverse waves, but this is an approximation. In practical situations, transverse and longitudinal waves don't always occur in isolation. The breaking of waves on a beach produces complex wave forms that are a combination of transverse and longitudinal waves (see below).

If you looked carefully at a cork bobbing about in gentle water waves, you would notice that it doesn't move straight up and down but that it has a more elliptical motion. It moves up and down, and very slightly forwards and backwards as each wave passes. However, since this second aspect of the motion is so subtle, in most circumstances it is adequate to treat water waves as if they were purely transverse waves.

Although this surfer rides forwards on the wave, the water itself only moves in an elliptical motion as the wave passes.

2.1 Review

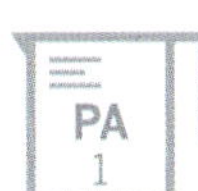

SUMMARY

- Vibrating objects transfer energy through waves, travelling outwards from the source.
- A wave may be a single pulse, or it may be continuous or periodic (successive crests and troughs or compressions and rarefactions).
- A wave only transfers energy from one point to another. There is no net transfer of matter or material.
- Mechanical waves require a medium to transmit energy. Waves on water or on a string, and sound waves in air are examples of mechanical waves.
- Mechanical waves can be either transverse or longitudinal.
  - In a transverse wave, the oscillations are perpendicular to the direction in which the wave energy is travelling. A wave in a string is an example of a transverse wave.
  - In a longitudinal wave, the oscillations are parallel to the direction the wave energy is travelling. Sound is an example of a longitudinal wave.
- Electromagnetic radiation includes visible light and does not require a medium to transmit energy.

KEY QUESTIONS

Knowledge and understanding

1 Describe the motion of particles within a medium and the transmission of energy as a mechanical wave passes through the medium.

2 Which of the following are examples of mechanical waves?
light, sound, ripples on a pond, vibrations in a rope

3 Classify the waves described below as either longitudinal or transverse.
- **a** sound waves
- **b** a vibrating violin string
- **c** slinky moved with an upwards pulse
- **d** slinky pushed forwards and backwards

4 For the wave shown below, describe the direction of energy transfer of the sound between the tuning fork and point X. Justify your answer.

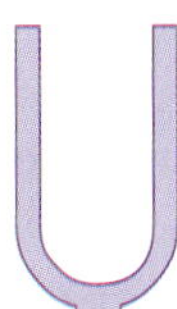

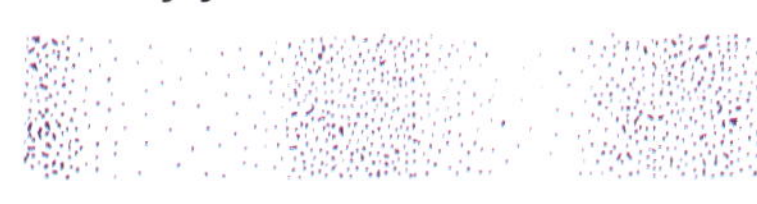

Analysis

5 A mechanical arm moves to produce a pulse that transfers energy to a piece of string. The pulse travels from the left to the right of the string, as shown below. The dots represent the particles on the string.
Describe the movement of particles A and B for one complete oscillation, as the particle moves to the right.

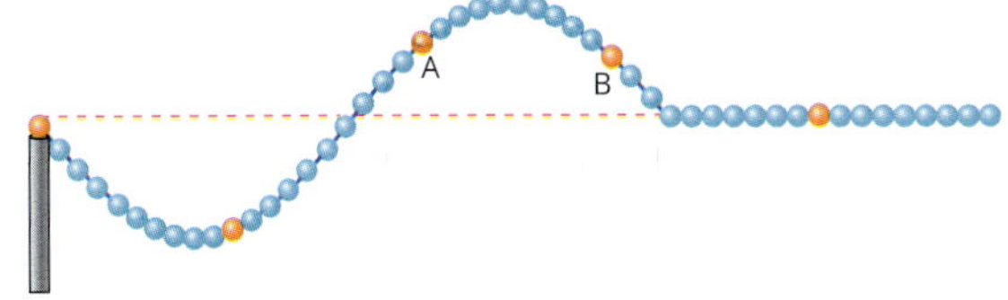

6 The diagrams below shows dots representing the average displacement of air particles at one moment in time as a sound wave travels to the right.

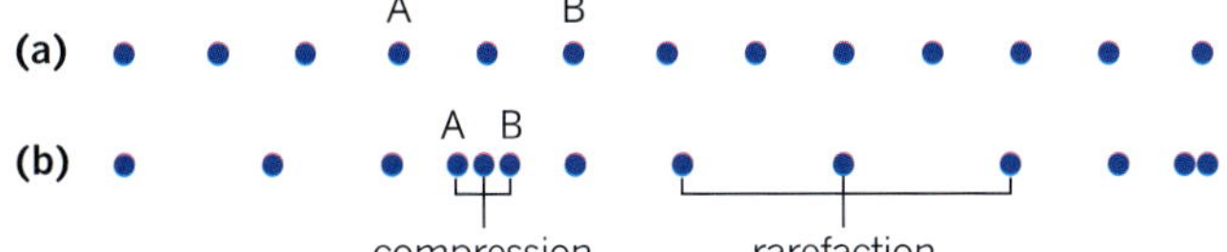

Initially the particles, including A and B, are equally spaced as shown in (a). A wave passes through, forming compressions and rarefactions as shown in (b). Describe how particles A and B have moved from their initial positions to form the compression.

7 Compare similarities and differences between the properties of longitudinal and transverse waves and give an example of each.

8 Why can't sound waves travel through the vacuum of space?

9 Compare the similarities and differences between a wave on a guitar string and light.

2.2 Measuring waves

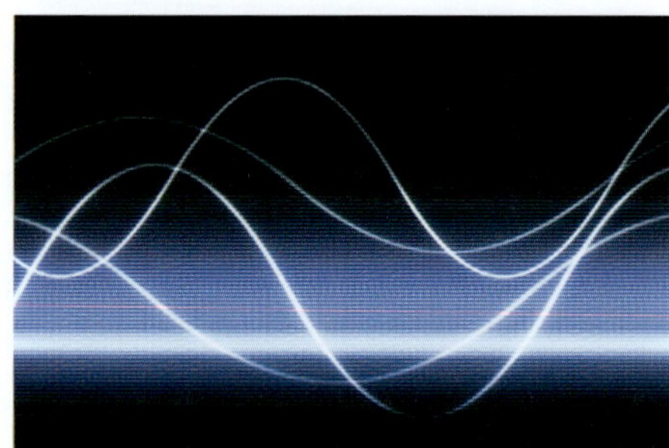

FIGURE 2.2.1 Waves can have different wavelengths, amplitudes, frequencies, periods and speeds, which can all be represented on a graph.

The features of a mechanical wave can be represented using a graph. In this section you will explore how the displacement of particles within the wave can be represented using graphs. From these graphs several key features of a wave can be identified:

- amplitude
- wavelength
- frequency
- period
- speed.

Waves of different amplitudes and wavelengths can be seen in Figure 2.2.1.

DISPLACEMENT–DISTANCE GRAPHS

The displacement–distance graph in Figure 2.2.2 shows the displacement of all particles along the length of a transverse wave at a particular point in time.

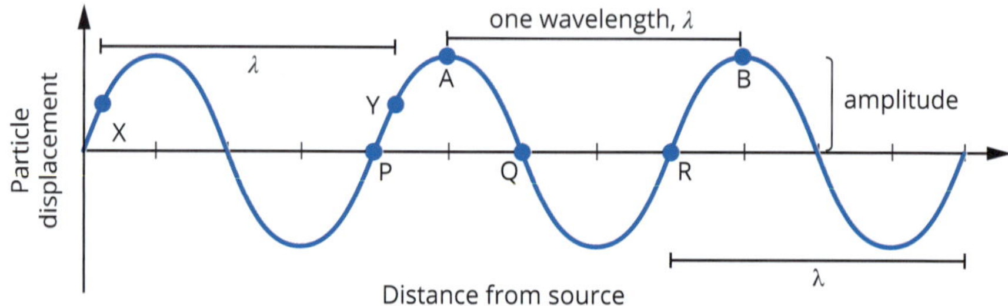

FIGURE 2.2.2 A sine wave represents the particle displacements along a wave.

Have a look back at Figure 2.1.3(b) on page 61 of a continuous wave in a slinky. This 'snapshot' in time shows the particles moving up and down sinusoidally about a central rest position. As a wave passes a given point, the particle at that point will go through a complete cycle before returning to its starting point. The wave spread along the length of the slinky has the shape of a sine or cosine function, which you will recognise from mathematics. A displacement–distance graph shows the position (displacement) of the particles at any moment in time along the slinky about a central position.

From a displacement–distance graph, the amplitude and wavelength of a wave are easily recognisable.

- The **amplitude** of a wave is the maximum displacement of a particle from the average or rest position. That is, the amplitude is distance from the middle of a wave to the top of a crest or to the bottom of a trough. The total distance a particle will move through in one cycle is twice the amplitude.
- The **wavelength** of a wave is the distance between any two successive points in phase (e.g. points A and B or X and Y in Figure 2.2.2). Wavelength is denoted by the Greek letter λ (lambda) and is measured in metres. Two particles on the wave are said to be in phase if they have the same displacement from the average position and are moving in the same direction. Points P and R in Figure 2.2.2 are two such particles that are in phase, as are points A and B and X and Y, but not P and Q.
- The **frequency**, f, is the number of complete cycles that pass a given point per second and is measured in hertz (Hz). By drawing a series of displacement–distance graphs at various times, you can see the motion of the wave. By comparing the changes in these graphs, the travelling speed and direction of the wave can be found, as well as the direction of motion of the vibrating particles.

Worked example 2.2.1

DISPLACEMENT–DISTANCE GRAPH

The displacement–distance graph below shows a snapshot of a transverse wave as it travels along a spring towards the right. Use the graph to determine the amplitude and the wavelength of this wave.

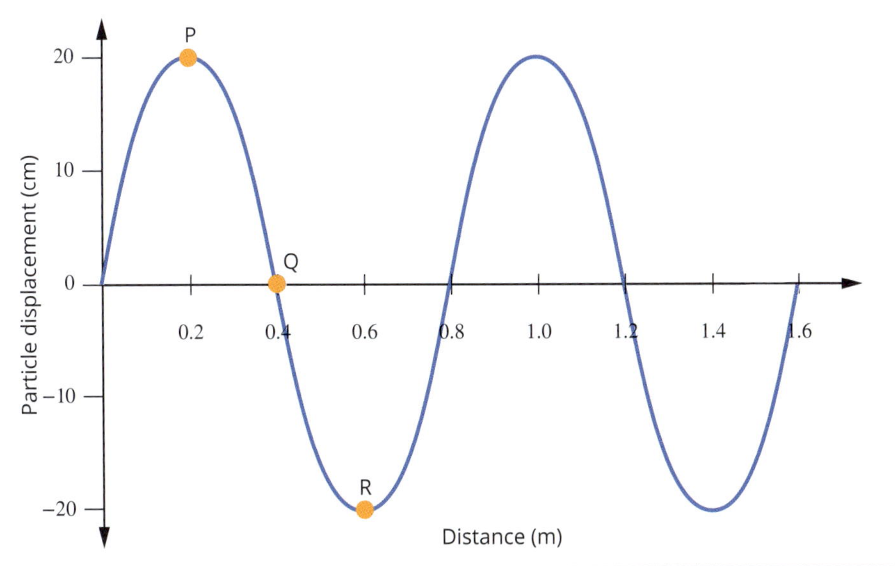

Thinking	Working
Amplitude on a displacement–distance graph is the distance from the average position to a crest (P) or a trough (R). Read the displacement of a crest or a trough from the vertical axis. Convert to SI units where necessary.	Amplitude is 20 cm = 0.2 m.
Wavelength is the distance for one complete cycle. Any two consecutive points in phase and at the same position on the wave could be used.	The first cycle runs from the origin through P, Q, and R to intersect the horizontal axis at 0.8 m. This intersection is the wavelength. Wavelength λ is 0.8 m.

Worked example: Try yourself 2.2.1

DISPLACEMENT–DISTANCE GRAPH

The displacement–distance graph below shows a snapshot of a transverse wave as it travels along a spring towards the right. Use the graph to determine the wavelength and the amplitude of this wave.

Particle displacement (cm)

2, 1, 0, −1, −2

0.1, 0.2, 0.3, 0.4, 0.5, 0.6, 0.7, 0.8

Distance (m)

The displacement–time graph looks very similar to a displacement–distance graph of a transverse wave, so be careful to check the horizontal axis label.

DISPLACEMENT-TIME GRAPHS

A displacement–time graph, such as the one shown in Figure 2.2.3, tracks the position of one point over time as the wave moves through that point.

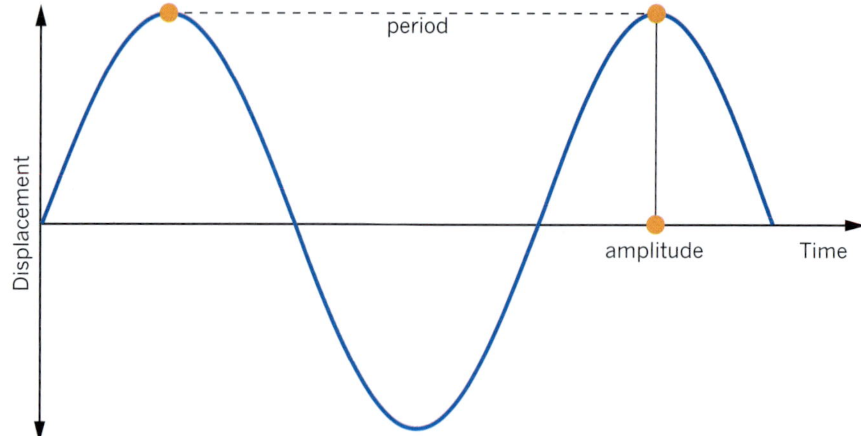

FIGURE 2.2.3 The graph of displacement versus time from the source of a transverse wave shows the movement of a single point on a wave over time as the wave passes through that point.

Crests and troughs are shown in the same way in both graphs. The amplitude is still the maximum displacement from the average or rest position of either a crest or a trough, but the distance between two successive points in phase in a displacement–time graph represents the period of the wave, T, measured in seconds.

The **period** is the time it takes for any point on the wave to go through one complete cycle (e.g. from crest to successive crest). The period of a wave is inversely related to its frequency.

$$T = \frac{1}{f}$$

where T is the period of the wave (s)

f is the frequency of the wave (Hz).

The amplitude and period of a wave, and the direction of motion of a particular particle, can be determined from a displacement–time graph.

Worked example 2.2.2

DISPLACEMENT–TIME GRAPHS

The displacement–time graph below shows the motion of a single part of a rope (point P) as a wave passes by travelling to the right. Use the graph to find the amplitude, period and frequency of the wave.

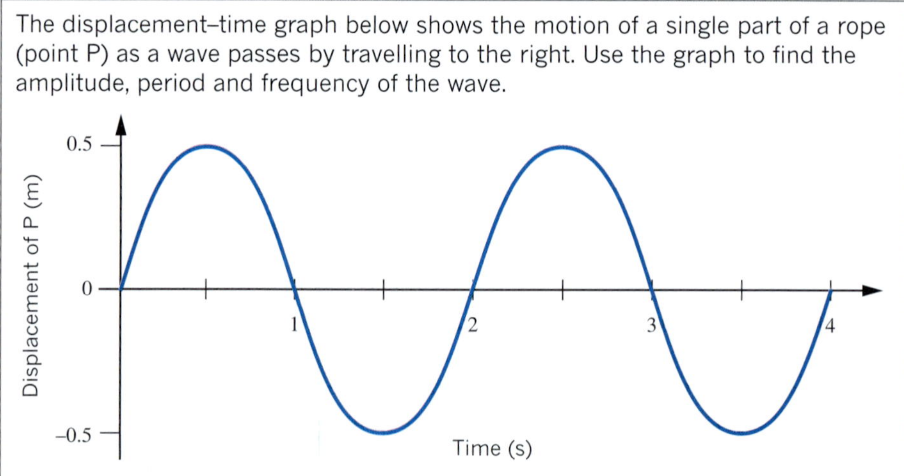

Thinking	Working
The amplitude on a displacement–time graph is the displacement from the average position to a crest or trough. Note the displacement of successive crests and/or troughs on the wave and carefully note units on the vertical axis.	Maximum displacement is 0.5 m. Therefore amplitude is 0.5 m.
Period is the time it takes to complete one cycle and can be identified on a displacement–time graph as the time between two successive points on the graph that are in phase. Identify two points on the graph at the same position in the wave cycle, e.g. the origin and $t = 2$ s. Confirm by checking two other points, e.g. two crests or two troughs.	Period T is 2 s.
Frequency can be calculated using $f = \frac{1}{T}$, measured in hertz (Hz).	$f = \frac{1}{T} = \frac{1}{2} = 0.5$ The frequency is 0.5 Hz.

Worked example: Try yourself 2.2.2

DISPLACEMENT–TIME GRAPHS

The displacement–time graph below shows the motion of a single part of a rope as a wave passes travelling to the right. Use the graph to find the amplitude, period and frequency of the wave.

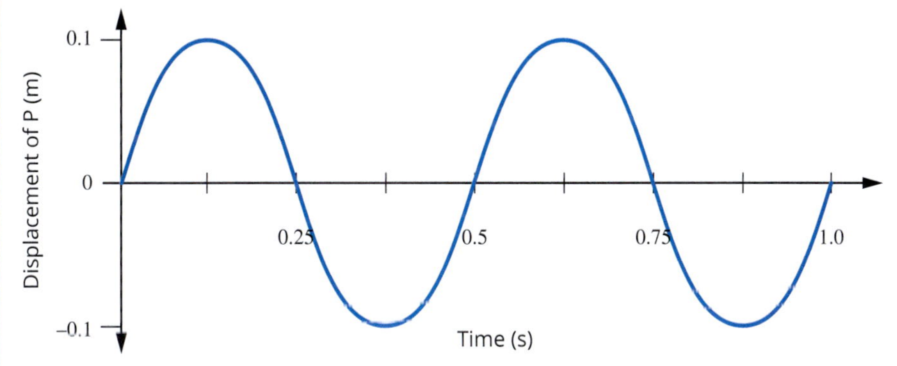

THE WAVE EQUATION

Although the speed of a wave can vary, there is a relationship between the speed of a wave and other significant wave characteristics.

In general, the speed (v) of an object is given by:

$$v = \frac{\text{distance travelled}}{\text{time taken}} = \frac{d}{\Delta t}$$

For a wave, the distance between any two successive points in phase is one wavelength ($d = \lambda$). This occurs in the time of one period ($t = T$). Therefore, the equation becomes:

$$v = \frac{\lambda}{T}$$

As $f = \frac{1}{T}$, we can substitute $T = \frac{1}{f}$ into the expression for v.
This gives

$$v = \lambda f$$

Rearrange this expression to make wavelength the subject, and you can see that wavelength depends on both the speed of the wave and the frequency.

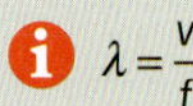

$$\lambda = \frac{v}{f}$$

where λ is the wavelength (m)
v is the speed (m s^{-1})
f is the frequency (Hz).

This is known as the wave equation and applies to both longitudinal and transverse mechanical waves.

Worked example 2.2.3

THE WAVE EQUATION

A longitudinal wave has a wavelength of 2.00 m and a speed of 340 m s^{-1}. What is the frequency, f, of the wave?

Thinking	Working
The wave equation states that $\lambda = \frac{v}{f}$. Both v and λ are known, so the frequency, f, can be found. Rewrite the wave equation in terms of f.	$\lambda = \frac{v}{f}$ $f = \frac{v}{\lambda}$
Substitute the known values and solve.	$f = \frac{v}{\lambda}$ $= \frac{340}{2.00} = 170$ The frequency is 170 Hz.

Worked example: Try yourself 2.2.3

THE WAVE EQUATION

A transverse wave has a wavelength of 4.0×10^{-7} m and a speed of 3.0×10^{8} m s^{-1}. What is the frequency, f, of the wave?

Worked example 2.2.4

THE WAVE EQUATION

A longitudinal wave has a wavelength of 2.00 m and a speed of 340 m s^{-1}.
What is the period, T, of the wave?

Thinking	Working
Rewrite the wave equation in terms of T.	$\lambda = \frac{v}{f}$ and $f = \frac{1}{T}$ Substitute $f = \frac{v}{\lambda}$ into $T = \frac{1}{f}$. $T = \frac{1}{\frac{v}{\lambda}}$ $T = \frac{\lambda}{v}$
Substitute the known values and solve.	$T = \frac{\lambda}{v}$ $= \frac{2.00}{340}$ $= 5.90 \times 10^{-3}$ Period T is 5.90×10^{-3} s.

Worked example: Try yourself 2.2.4

THE WAVE EQUATION

A transverse wave has a wavelength of 4.0×10^{-7} m and a speed of 3.0×10^{8} m s^{-1}.
What is the period, T, of the wave?

CASE STUDY ANALYSIS

Seismic waves and the composition of Earth

On 28 December 1989, an earthquake devastated the region in and around Newcastle, New South Wales. The earthquake was rated 5.6 on the Richter Scale. It was not the most powerful earthquake recorded in Australia; however, it did cause the most damage. Contributing factors were that the epicentre was close to the city centre, it occurred at a shallow depth, soft sediments in the ground amplified the vibrations, and buildings were not designed to adequately withstand earthquakes.

Three main types of seismic waves are produced in an earthquake: two types of body waves (P- and S-waves) and surface waves. These waves can have different wavelengths. The difference in speed between the S and P waves can give a measure of the location of the epicentre, the source of the waves.

Body waves travel through Earth. The primary (P) waves are longitudinal waves (Figure 2.2.4(a) on page 70) and they travel through both liquids, such as molten rocks in Earth's mantle, and solids, such as rocks that comprise most of Earth's crust. They have speeds between 1.5 and 8.0 km s^{-1}, with a typical speed of about 6.0 km s^{-1}. The blue grid shows the compressions and rarefactions as the wave oscillates in the direction of the motion.

The secondary (S) waves are transverse waves (Figure 2.2.4(b) on page 70). They do not travel through liquids and their speed is slower than that of P-waves. The blue grid shows the displacement of the wave perpendicular to the direction of travel. The difference in speed between these two waves allows scientists to determine the location of the epicentre of the earthquake.

The third type of wave, the surface wave (Figure 2.2.4(c) on page 70), has a rolling motion and travels along Earth's surface. The blue grid shows it is a transverse wave, but the perpendicular displacement is only at the surface. This type of wave typically causes the most damage.

continued over page

CASE STUDY ANALYSIS *continued*

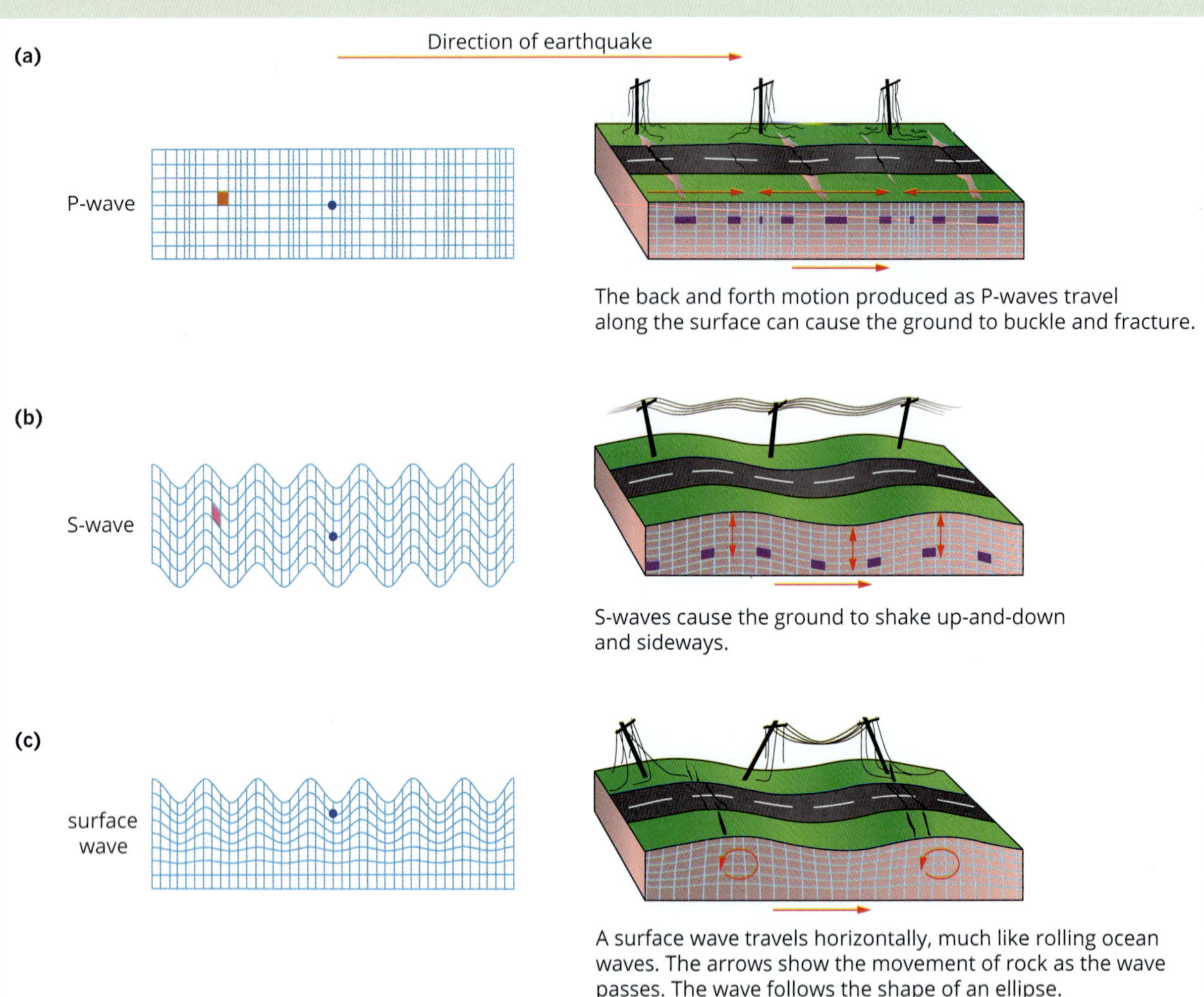

FIGURE 2.2.4 The three different wave types have different effects on Earth's crust.

Analysis

Scientists can measure the seismic waves using seismometers. From the information collected, they can determine the composition of Earth.

1. If a large earthquake occurred in Melbourne, the P-waves would travel through the centre of Earth and be detected on the other side of the world in England. However, the S-waves would not be detected. From the information above, determine the likely state of the material comprising the centre of Earth.
2. An average P-wave has a speed of $6.0\,km\,s^{-1}$. If it has a period of 0.20 seconds, calculate the frequency and the wavelength of the P-wave.

The difference in arrival time at a seismometer between a P-wave (t_P) and an S-wave (t_S) is given by $\Delta t = t_P - t_S$. Each wave travels the same distance.

3. Derive an expression for Δt in terms of the distance travelled, the speed of the P-wave (v_P) and the speed of the S-wave (v_S).
4. Use your answer to question 3 to calculate the distance from the seismometer to the epicentre of the earthquake, if $v_S = 3.45\,km\,s^{-1}$ and $v_P = 8.00\,km\,s^{-1}$, and the difference in the time for arrival of the waves is 9.00 seconds.

2.2 Review

OA ✓✓

SUMMARY

- Waves can be represented by displacement–distance graphs and displacement–time graphs.
- From a displacement–time graph, you can determine amplitude, frequency and period.
- The period of a wave has an inverse relationship to the frequency, according to the relationship:

$$T = \frac{1}{f}$$

- The speed of a wave can be calculated using the wave equation:

$$\lambda = \frac{v}{f}$$

KEY QUESTIONS

Knowledge and understanding

1 From the displacement–distance graph below, give the correct term or letters for the following:

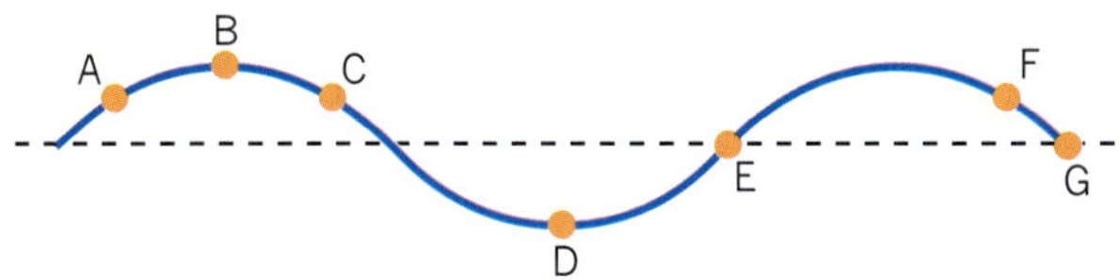

a two points on the wave that are in phase
b the name for the distance between these two points
c two particles with maximum displacement from their rest position
d the term for this maximum displacement.

2 Use the graph below to determine the wavelength and the amplitude of this wave.

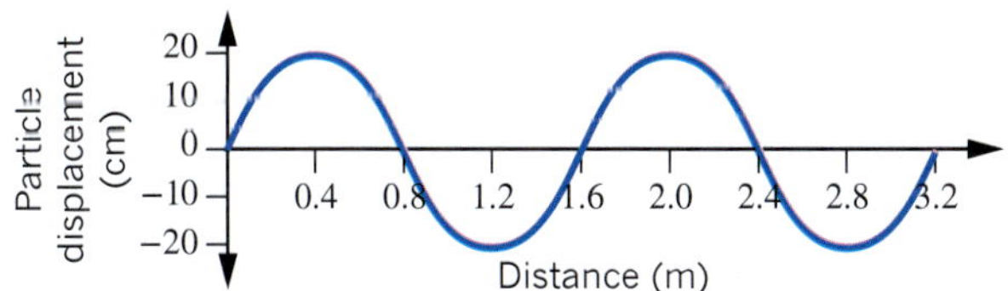

3 This is the displacement–time graph for a particle.

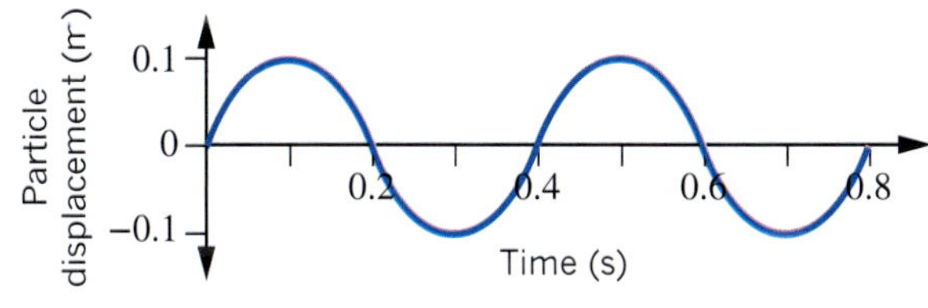

a Determine the period of the wave.
b Calculate frequency of the wave.

4 Calculate the period of a wave with frequency 2×10^5 Hz.

5 Five wavelengths of a wave pass a point each second. The amplitude is 0.3 m and the distance between successive crests of the waves is 1.3 m. What is the speed of the wave?

Analysis

6 Consider the displacement–distance graph below.

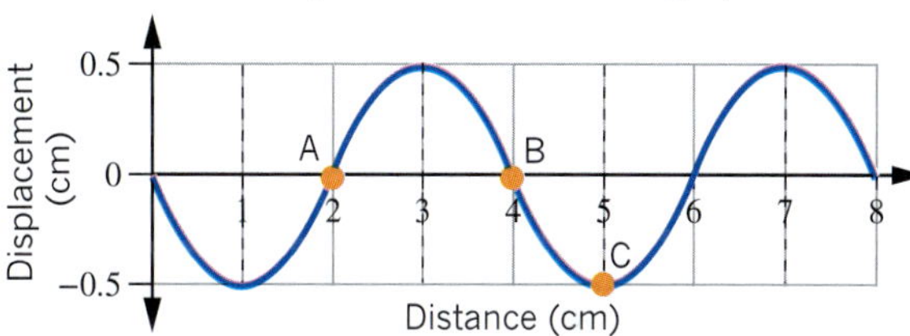

a State the wavelength and amplitude of the wave.
b If the wave moves through one wavelength in 2 s, what is the speed of the wave?
c If the wave is moving to the right, which of the particles is moving down?

7 Five complete waves pass a point in 8.0 s. The amplitude of the wave is 0.70 m and distance between successive troughs is 1.20 m. Calculate the speed of the wave.

2.3 The electromagnetic spectrum

Light, like all electromagnetic radiation (EMR), is a transverse wave that does not need a medium in order to travel from its source. As will be explored more fully in Year 12, light consists of two transverse waves perpendicular to each other: one is an electric field wave and the other is the magnetic field wave, as shown in Figure 2.3.1.

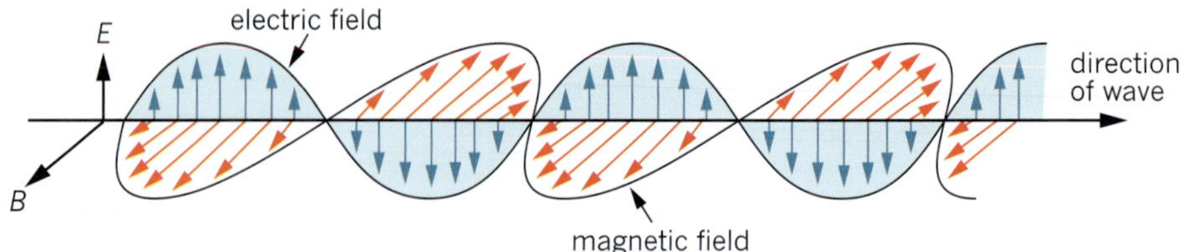

FIGURE 2.3.1 The electric field (*E*) and magnetic field (*B*) in electromagnetic radiation are perpendicular to each other and both are perpendicular to the direction of propagation of the radiation.

Our eyes are receptive to the visible spectrum. The wavelengths of all the different colours of visible light fall between 390 nm (violet) and 780 nm (red), Naturally, physicists were bound to inquire about other wavelengths of electromagnetic radiation. It is now understood that the visible spectrum is just one small part of a much broader set of possible wavelengths known as the **electromagnetic spectrum** (Figure 2.3.2).

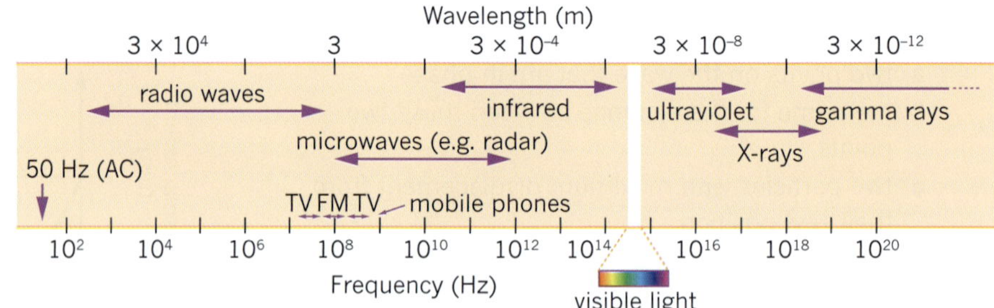

FIGURE 2.3.2 The electromagnetic spectrum

THE WAVE EQUATION AND ELECTROMAGNETIC RADIATION

The wave equation, introduced in Section 2.2, also applies to EMR. However, the speed of light is always constant in a vacuum regardless of the speed of the source or the observer, so it is given its own constant, *c*.

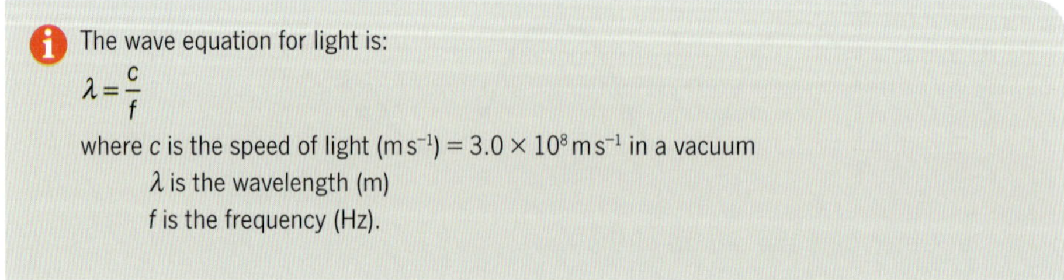

The wave equation for light is:

$$\lambda = \frac{c}{f}$$

where c is the speed of light (m s^{-1}) = $3.0 \times 10^8\ \text{m s}^{-1}$ in a vacuum

λ is the wavelength (m)

f is the frequency (Hz).

You will note that Worked example 2.3.1 and Worked example: Try yourself 2.3.1 are examples of an electromagnetic wave travelling through a vacuum or air. The speed of light in air is not significantly different from its speed in a vacuum. However, the speed of light in a medium such as glass is lower than in a vacuum. This is discussed further in Chapter 3.

Worked example 2.3.1

THE WAVE EQUATION AND ELECTROMAGNETIC RADIATION

A laser of blue light travelling through a vacuum has a frequency of 6.7×10^{14} Hz. What is the wavelength, λ, of the light?

Thinking	Working
State your variables and the wave equation.	$f = 6.7 \times 10^{14}$ Hz $v = c = 3.0 \times 10^{8}$ m s^{-1} $\lambda = ?$ $\lambda = \frac{v}{f}$ $\lambda = \frac{c}{f}$
Substitute the known values and solve.	$\lambda = \frac{c}{f}$ $= \frac{3.0 \times 10^{8}}{6.7 \times 10^{14}}$ $= 4.5 \times 10^{-7}$ m

Worked example: Try yourself 2.3.1

THE WAVE EQUATION AND ELECTROMAGNETIC RADIATION

A beam of red light travelling through air has a frequency of 4.3×10^{14} Hz. What is the wavelength, λ, of the light?

CASE STUDY

Why do we see the wavelengths we do?

Earth's atmosphere blocks many types of EMR (Figure 2.3.3). The highest level of the atmosphere, the ionosphere, contains charged particles and effectively blocks the high-energy ionising EMR (gamma rays, X-rays). Lower down, molecular ozone, O_3, and nitrogen, N_2, absorb and block about 70% of the ultraviolet EMR. Visible light is transmitted well, as it is not energetic enough to be absorbed. The atmosphere becomes increasingly opaque in the infrared and microwave bands, due mainly to absorption by water vapour.

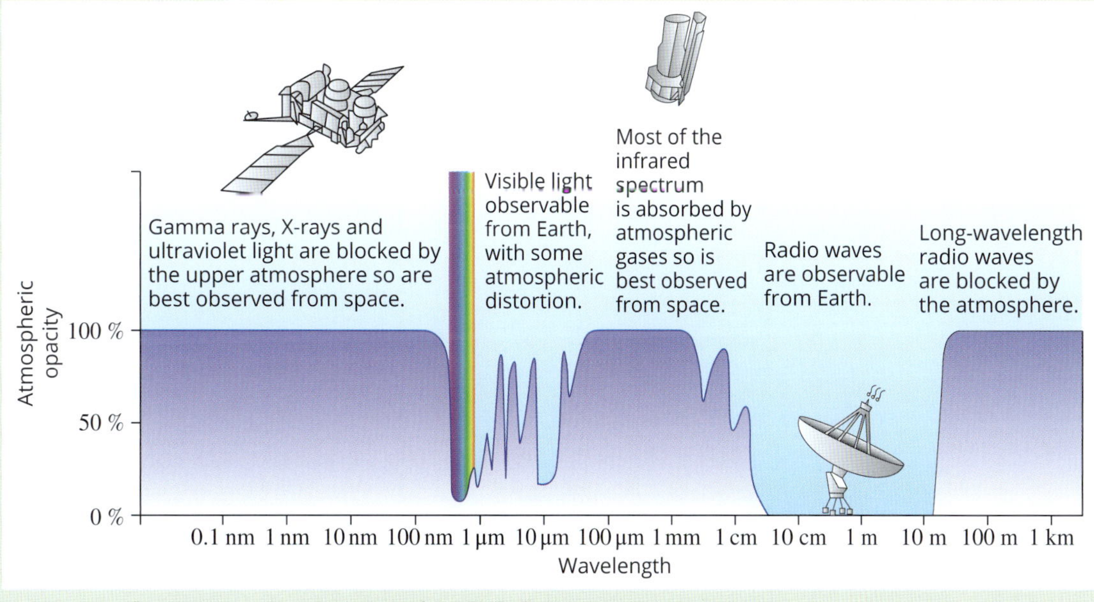

FIGURE 2.3.3 Depending on the wavelength of the EMR, Earth's atmosphere is transparent, translucent or opaque.

continued over page

CASE STUDY *continued*

At still lower energies, the atmosphere becomes transparent again to shorter wavelength radio waves, until the lowest energy longer wavelength EMR cannot penetrate the atmosphere.

During our evolution, our eyes have developed photoreceptors ('cones') that respond to the visible spectrum. It is not the only option, however. Dogs have two types of photoreceptors, green and blue, which enable them to see blue, green and yellow. Humans have three types, which are sensitive to red, green and blue, and allow us to see colours derived from red, such as orange and purple, which are invisible to dogs. Honeybees also have three types of photoreceptors, but the evolution of bees led to their photoreceptors being sensitive to ultraviolet, blue and green, which makes the pollen of flowers stand out more strongly. Butterflies have five types, and the mantis shrimp (Figure 2.3.4) has sixteen types of photoreceptor. We see an entire rainbow with just three photoreceptors. What must a mantis shrimp see?

FIGURE 2.3.4 The magnificent mantis shrimp

TYPES OF ELECTROMAGNETIC RADIATION

Changing the frequency and wavelength of the waves changes the properties of the EMR, and so the electromagnetic spectrum is divided into 'bands' according to its properties and how the particular types of EMR are used. The shorter the wavelength of the electromagnetic wave, the greater its penetrating power. This means that waves with extremely short wavelengths, such as X-rays, can pass through some materials (e.g. skin), revealing the structures inside (e.g. bone).

Long wavelength waves, such as AM radio waves, have such low penetrating power that they cannot even escape Earth's atmosphere, and can be used to 'bounce' radio signals around to the other side of the world. Table 2.3.1 compares the characteristics of different waves in the electromagnetic spectrum.

TABLE 2.3.1 Comparison of the different waves in the electromagnetic spectrum

Type of wave	Typical wavelength (m)	Typical frequency (Hz)	Comparable object	Effect on matter
AM radio wave	100	3×10^6	sports oval	causes movement of free electrons in a conductor
FM radio or TV wave	3	1×10^8	small car	causes movement of free electrons in a conductor
microwaves	0.03	1×10^{10}	50c coin	causes molecular rotation
infrared	10^{-5}	3×10^{13}	white blood cell	makes chemical bonds vibrate
visible light	10^{-7}	3×10^{15}	small cell	affects electronic states in atoms or molecules
ultraviolet	10^{-8}	3×10^{16}	large molecule	affects electronic states in atoms or molecules
X-ray	10^{-10}	3×10^{18}	atom	excites electrons in atomic orbitals
gamma ray	10^{-15}	3×10^{23}	atomic nucleus	causes disintegration of atomic nuclei

Our Sun emits electromagnetic radiation mainly in the infrared, visible and ultraviolet bands. Some high-energy radiations such as X-rays and gamma rays are also emitted, but we are protected from these by Earth's magnetic field and atmosphere.

Radio waves

One of the most revolutionary applications of electromagnetic radiation is the use of radio waves to transmit information from one point to another over long distances. Radio waves are the longest type of electromagnetic radiation, with wavelengths ranging from 1 mm to hundreds of kilometres, as shown in Figure 2.3.2 on page 72. The principle of radio transmission is relatively simple, and neatly illustrates the nature of electromagnetic waves.

The radio transmitter converts the signal (e.g. radio announcer's voice, music or stream of data) into an alternating current. When this alternating current flows in the transmission antenna, the electrons in the antenna oscillate backwards and forwards. This oscillation of charges in the antenna produces a corresponding electromagnetic wave that radiates outwards in all directions from the antenna.

When the radio wave hits the antenna of a radio receiver, the electrons in the receiver's antenna start to oscillate in exactly the same way as in the transmitting antenna. The radio receiver then reverses the process of the transmitter, converting the alternating current from the reception antenna back into the original signal, as seen in Figure 2.3.5.

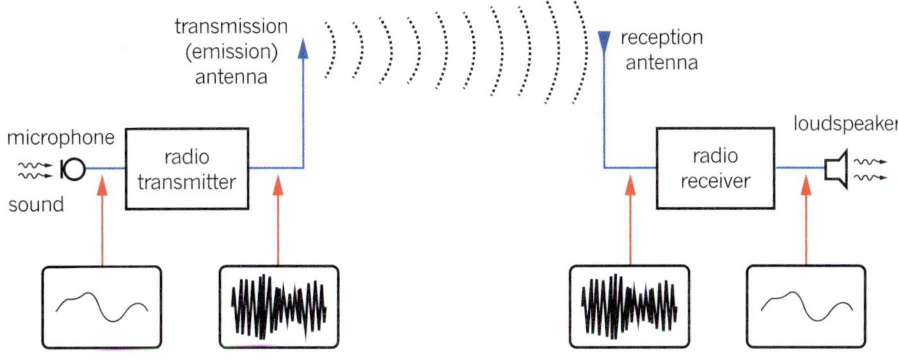

FIGURE 2.3.5 A typical radio transmission system

Microwaves

Microwaves have wavelengths between those of radio waves and visible light, as shown in Figure 2.3.2 on page 72. The most familiar example of microwaves is the microwave oven, used in heating and cooking food. A microwave oven is 'tuned' to produce a particular frequency of electromagnetic radiation: 2.45 GHz (2.45×10^9 Hz). This is the resonant or natural vibration frequency of water molecules. The energy from the microwaves is transferred to the water molecules, causing the water molecules to vibrate more strongly, thus heating up the food.

Microwaves are also particularly useful in personal communication devices such as mobile phones, and for wireless internet transmission (WiFi), something we are incredibly reliant on and take for granted, as well as many other applications. Microwaves have shorter wavelengths and therefore greater penetrating power than radio waves, and so can be produced by devices with short antennas.

CASE STUDY

Australia invents WiFi

In 1990 there were no wireless devices. If people attempted to send complex signals at radio frequencies wirelessly across a room, the signals would reflect, interfere with each other and cause reverberations (delayed echoes). The solution to this problem came from a small team of Australian scientists, mathematicians and engineers at the CSIRO Department of Radiophysics. The team was led by physicist and radio-astronomer John O'Sullivan and included Terry Percival, Diet Ostry, Graham Daniels and John Deane.

Using a solution from Dr O'Sullivan's work in radio-astronomy, the team developed the Fast Fourier Transform (FFT) chip. A complex wave can be modelled as the superposition of individual waves. In WiFi, the original signal undergoes an FFT process through a computer chip, and is transmitted on a carrier signal to the receiver. There, the wave undergoes a reverse FFT and other signal processing, which results in the original waveform or signal.

Two common bands are used for the WiFi carrier signal, depending on the amount of data being sent: 2.4 GHz and 5 GHz. The two frequencies are split into multiple channels so as to prevent high traffic and interference.

FIGURE 2.3.6 A common symbol for wireless communication (WiFi)

FIGURE 2.3.7 The coals of a fire emit red light as well as infrared radiation, which you experience as heat.

Infrared

The infrared section of the electromagnetic spectrum lies between microwaves and visible light (Figure 2.3.2 on page 72). Infrared waves are longer than the red waves of the visible spectrum, hence their name.

Infrared waves become useful because they are emitted by objects, to varying degrees, due to their temperature. The warmth that you feel standing next to an electric bar heater or a fire is due to infrared radiation (Figure 2.3.7). The radiant heat Earth receives from the Sun is transmitted in the form of infrared waves; life on Earth would not be possible without this important form of electromagnetic radiation.

Carbon dioxide is an important greenhouse gas that **absorbs** (takes in) and re-emits infrared radiation. This cycle is part of an important energy balance that keeps Earth warm enough for life.

PHYSICSFILE

Night vision

Infrared radiation can be detected by sensors in night-vision goggles and cameras, and can be used to form images at night. For example, researchers can record the movement of many native Australian animals that are mostly active at night. Infrared radiation is also used in your television remote control.

Ultraviolet light

As the name suggests, ultraviolet (UV) waves have wavelengths that are shorter than those of violet light (Figure 2.3.2 on page 72), and therefore cannot be detected by the human eye. The shorter wavelengths means that UV rays have a stronger penetrating power than visible light. In fact, UV rays can actually penetrate human skin and overexposure can cause skin cancers. It should be noted that some exposure to UV radiation is essential for the production of vitamin D, which helps absorb calcium and potassium from food.

UV radiation is divided into three bands: UVA, UVB and UVC. As shown in Table 2.3.2, UVA is not blocked by the atmosphere, while only 10% of UVB reaches Earth. Both bands can be blocked by a good sunscreen. UVC is blocked by the ozone layer. As exposure to UVC increases the risk of cancer to 10 000 times more than for UVA and UVB, scientists became concerned in the 1980s when a depletion in the ozone layer over the poles was measured. This was caused by chlorofluorocarbons, used in refrigeration. International efforts to reduce their use and use alternatives has resulted in a reduction in the size of the ozone hole.

TABLE 2.3.2 The UV radiation band is divided into three wavelength ranges according to how much reaches Earth.

UV band	Wavelength range	Penetrating power
UVA	315–400 nm	not blocked by Earth's atmosphere
UVB	280–315 nm	10% reaches Earth
UVC	100–280 nm	blocked by the ozone layer

Scientists can make use of UV light to take images. Figure 2.3.8 is a UV image of the surface of the Sun taken after a solar flare has occurred. The image has been recoloured so that it highlights areas of different temperature. Here, areas that are coloured white are the hottest. Images like this help scientists learn about the temperatures of very hot objects. Taking an image of the Sun using visible light would not allow this same distinction.

FIGURE 2.3.8 Recoloured UV image of the surface of the Sun. The white areas reveal the hottest parts.

X-rays and gamma rays

X-rays and gamma rays have much shorter wavelengths than visible light (Figure 2.3.2 on page 72). This means that these forms of electromagnetic radiation have very high penetrating powers. For example, some X-rays can pass through different types of human tissues, which means that they are very useful in medical imaging (Figure 2.3.9).

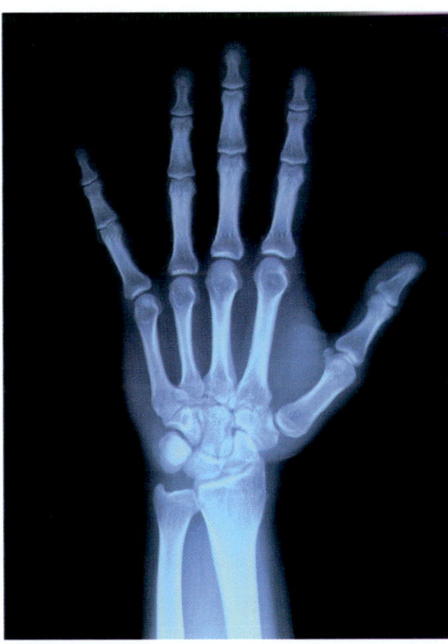

FIGURE 2.3.9 This X-ray image of a hand can be formed because X-rays can pass through human tissues.

Unfortunately, this useful penetrating property of X-rays comes with inherent dangers. As X-rays pass through a human cell, they can do damage to the tissue, sometimes killing the cells or damaging the DNA in the cell nucleus, leading to harmful cancers. For this reason, a person's exposure to X-rays has to be carefully monitored to avoid harmful side effects.

Similarly, exposure to gamma rays can be very dangerous to human beings. The main natural sources of gamma radiation exposure are the Sun and radioactive isotopes. Fortunately, Earth's atmosphere protects us from most of the Sun's harmful gamma rays, and radioactive isotopes are not commonly found in sufficient quantities to produce harmful doses of radiation.

2.3 Review

SUMMARY

- Light is a form of electromagnetic radiation.
- Electromagnetic waves are transverse waves made up of mutually perpendicular, oscillating electric and magnetic fields.
- Electromagnetic waves can travel through a vacuum. As they do not require a medium to travel through, they are not mechanical waves.
- Electromagnetic radiation travels through a vacuum at approximately $c = 3.0 \times 10^{8}\,\text{m s}^{-1}$.
- The wave equation $\lambda = \frac{c}{f}$ can be used to calculate the frequency and wavelength of electromagnetic waves.
- Electromagnetic radiation can be used for a variety of purposes depending on the frequency of the waves.
- The electromagnetic spectrum consists of radio waves, microwaves, infrared waves, visible light, ultraviolet light, X-rays and gamma rays.

KEY QUESTIONS

Knowledge and understanding

1 Outline the key difference between a mechanical wave and a light wave.

2 Arrange the types of electromagnetic radiation below in order of increasing wavelength.
FM radio waves / visible light / infrared radiation / X-rays / microwaves

3 What type of electromagnetic radiation would have a wavelength of 200 nm?
- **A** radio waves
- **B** microwaves
- **C** visible light
- **D** ultraviolet light

4 Give the form of electromagnetic radiation used or emitted in the following applications.
- **a** TV remote control
- **b** mobile phone
- **c** TV signal
- **d** the torch on your phone
- **e** astronomy
- **f** imaging a broken bone

Analysis

5 Calculate the frequencies of the following wavelengths of light.
- **a** red of wavelength 656 nm
- **b** yellow of wavelength 589 nm
- **c** blue of wavelength 486 nm
- **d** violet of wavelength 397 nm

6 Calculate the wavelength (in nm) of light with a frequency of 6.0×10^{14} Hz.

7 Calculate the wavelength of a UHF (ultra-high frequency) television signal with a frequency of 7.0×10^{7} Hz.

8 Calculate the frequency of an X-ray with a wavelength of 200 pm ($1\,\text{pm} = 1 \times 10^{-12}\,\text{m}$).

Chapter review

KEY TERMS

absorb
amplitude
compression
crest
electromagnetic spectrum
electromagnetic wave
frequency
longitudinal
mechanical wave
medium
oscillate
period
pulse
rarefaction
transmit
transverse
trough
wavelength

REVIEW QUESTIONS

Knowledge and understanding

1 Imagine that you watch from above as a stone is dropped into water. Describe the movement of the particles on the surface of the water.

2 State whether the following statements are true or false. Rewrite the false statements to make them true.
- **a** Longitudinal waves occur when particles of the medium vibrate in the opposite direction to the direction of the wave.
- **b** Transverse waves are created when the direction of vibration of the particles is at right angles to the direction of the wave.
- **c** A longitudinal wave is able to travel through air.
- **d** The vibrating string of a guitar is an example of a transverse wave.

3 A sound wave is emitted from a speaker and heard by Lee who is 50 m from the speaker. Lee made a number of statements once he heard the sound. Which one or more of the following statements made by Lee would be correct? Explain your answers.
- **A** Hearing a sound wave tells me that air particles have travelled from the speaker to me.
- **B** Air particles carried energy with them as they travelled from the speaker to me.
- **C** Energy has been transferred from the speaker to me.
- **D** Energy has been transferred from the speaker to me by the oscillation of air particles.

4 State whether the following statements are true or false. Rewrite the false statements to make them true.
- **a** The frequency of a wave is inversely proportional to its wavelength.
- **b** The period of a wave is inversely proportional to its wavelength.
- **c** The amplitude of a wave is not related to its speed.
- **d** Only the wavelength of a wave determines its speed.

5 If you decreased the wavelength of the sound made by a loudspeaker, what effect would this have on the frequency of the sound waves? The speed of sound in air is constant, for a constant temperature.

6 What form of electromagnetic radiation is used in the following applications?
- **a** night-vision goggles
- **b** medical imaging

7 Using ideas about the movement of particles in air, explain how you know sound waves only carry energy and not matter from one place to another.

Application and analysis

8 The graph below shows a wave moving to the right at a moment in time. In which directions are the particles U and V moving?

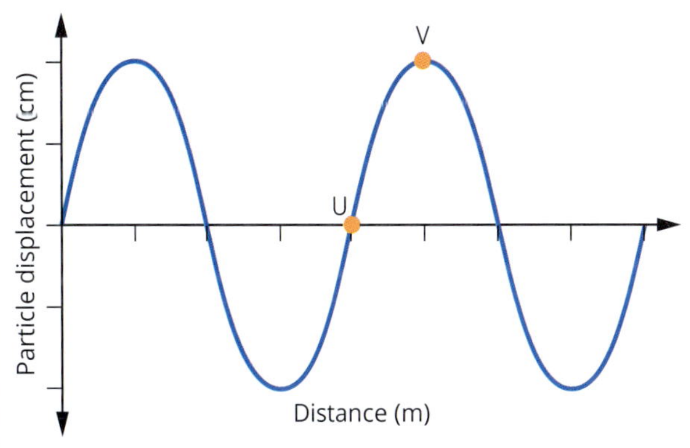

continued over page

9 The displacement–time graph below shows the variation of displacement of a specific point on a wave with time. Identify which of the following wave characteristics can be determined from this type of graph: amplitude, frequency, period, wavelength, wave speed. Clearly state the values of these characteristics.

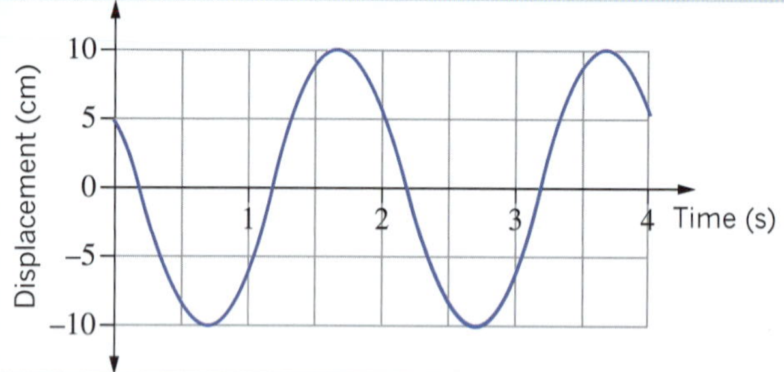

10 The displacement–distance graph below indicates the disturbance of any point on the rope at a specific moment in time. Identify which of the following wave characteristics can be determined from this type of graph: amplitude, frequency, period, wavelength, wave speed. Clearly state the values of these characteristics.

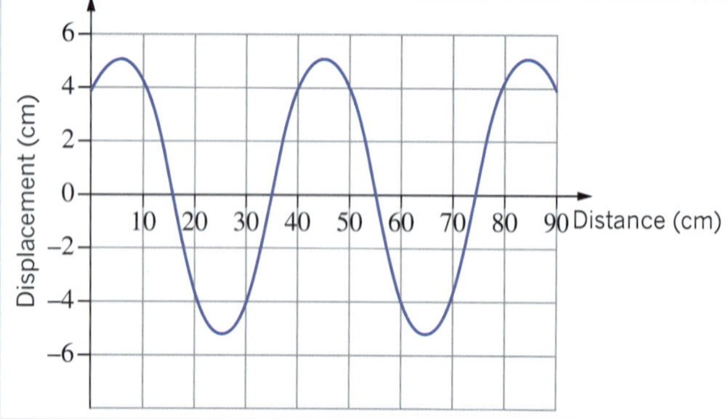

11 The source of waves in a ripple tank vibrates at a frequency of 10.0 Hz. If the wave crests formed are 30.0 mm apart, what is the speed of the waves (in $m s^{-1}$) in the tank?

12 A submarine's sonar sends out a signal with a frequency of 32 kHz. If the wave travels at 1400 $m s^{-1}$ in seawater, what is the wavelength of the signal?

13 Assuming the speed of sound in water is 1500 $m s^{-1}$, what would be the wavelength of a sound of frequency 300 Hz?

14 Blue light (6.00×10^{14} Hz) has a wavelength of 375 nm in water. Calculate the speed of blue light in water.

15 An AM radio station has a frequency of 612 kHz. If the speed of light is 3.00×10^{8} $m s^{-1}$, calculate the wavelength of these waves to the nearest metre.

16 Many WiFi routers have a 2.4 GHz band with a range from 2.40 GHz to 2.50 GHz, and a 5 GHz band that ranges from 5.180 GHz to 5.825 GHz. Calculate the range of wavelengths in each band.

17 Compare the microwave oven frequency with the WiFi 2.4 GHz band. Does this interfere with your WiFi router signal? Explain why? You may need to do some research.

18 Concerns have been raised that microwave radiation from mobile phone usage could cause cancers by damaging cells in a similar way to ionisation caused by X-rays and UV radiation. Given your knowledge of wavelengths and that the size of the human cell is 100 μm, how could you respond to this?

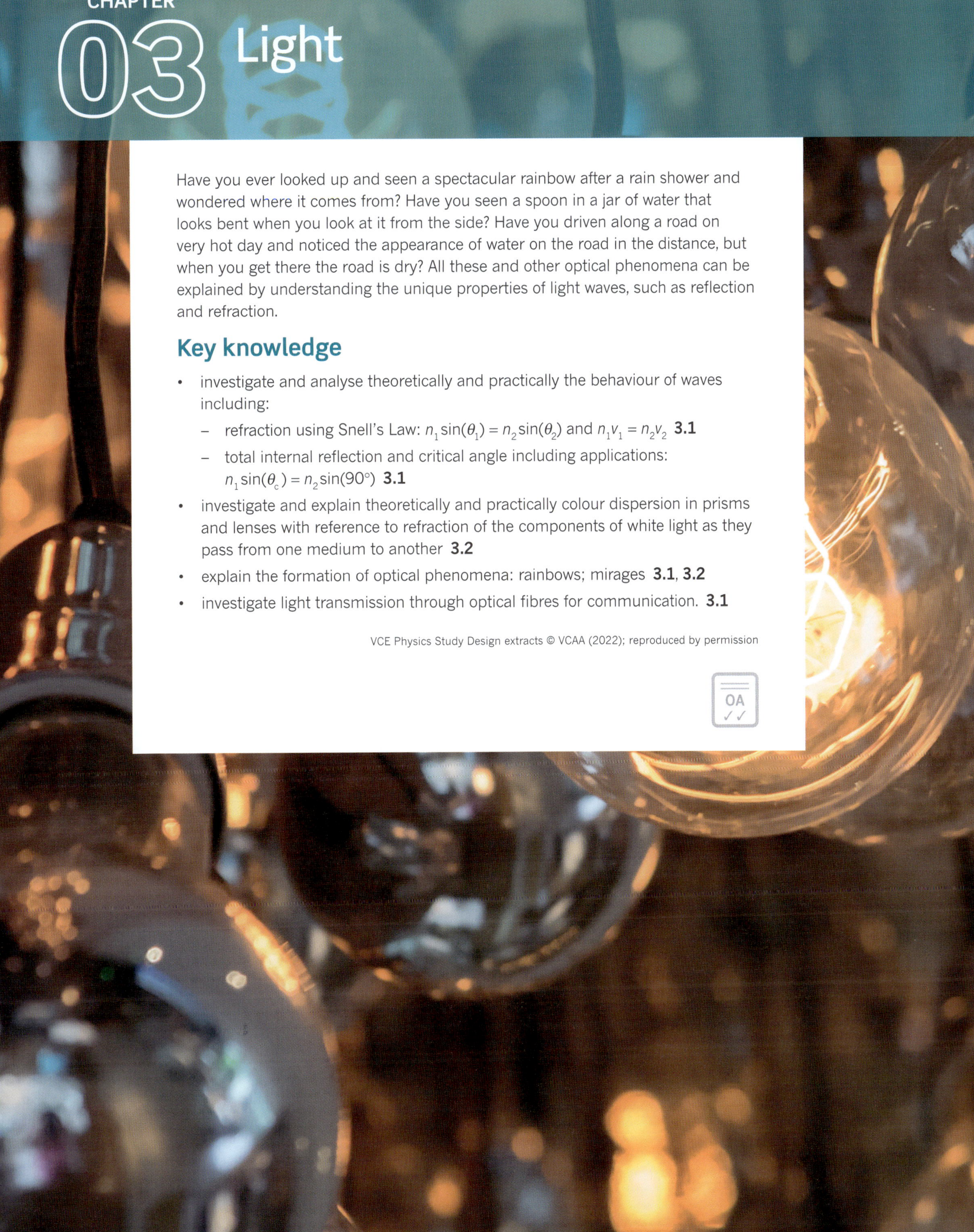

CHAPTER

03 Light

Have you ever looked up and seen a spectacular rainbow after a rain shower and wondered where it comes from? Have you seen a spoon in a jar of water that looks bent when you look at it from the side? Have you driven along a road on very hot day and noticed the appearance of water on the road in the distance, but when you get there the road is dry? All these and other optical phenomena can be explained by understanding the unique properties of light waves, such as reflection and refraction.

Key knowledge

- investigate and analyse theoretically and practically the behaviour of waves including:
  - refraction using Snell's Law: $n_1 \sin(\theta_1) = n_2 \sin(\theta_2)$ and $n_1 v_1 = n_2 v_2$ **3.1**
  - total internal reflection and critical angle including applications: $n_1 \sin(\theta_c) = n_2 \sin(90°)$ **3.1**
- investigate and explain theoretically and practically colour dispersion in prisms and lenses with reference to refraction of the components of white light as they pass from one medium to another **3.2**
- explain the formation of optical phenomena: rainbows; mirages **3.1**, **3.2**
- investigate light transmission through optical fibres for communication. **3.1**

OA

3.1 Reflection and refraction

If you drop a stone into water, water waves will ripple out in a circular fashion, as shown in Figure 3.1.1. The crests of the wave appears as wavefronts that ripple out in two dimensions.

FIGURE 3.1.1 If a stone is dropped into still water the waves will ripple out in a circular fashion.

WAVEFRONTS

All two- and three-dimensional waves, such as water waves, travel as **wavefronts**. A wavefront is a continuous line (or surface) that includes all the points reached by a wave at the same instant. When drawing wavefronts (see red curves in Figure 3.1.2), it is common to show the crests of the waves. When close to the source, wavefronts can show considerable curvature (Figure 3.1.2(a)) or may even be spherical when generated in three dimensions. For a wave that has travelled a long distance from its source, the wavefront is nearly straight and is called a **plane wave**. A plane wave is shown in Figure 3.1.2(b). Plane waves in water can also be generated by a long, flat source in a ripple tank.

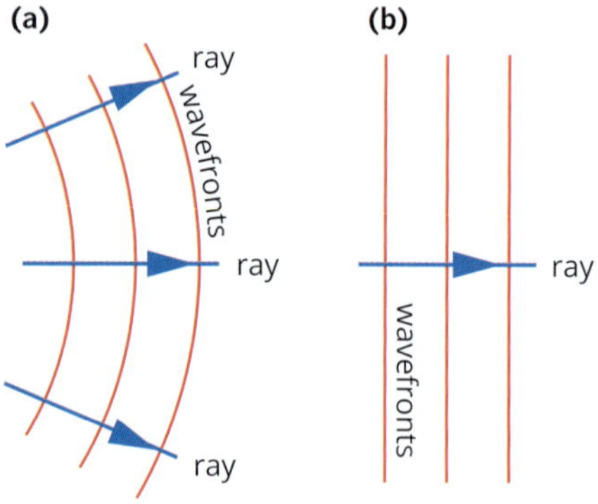

FIGURE 3.1.2 The crests of waves are drawn as wavefronts, shown in red. Rays can be used to illustrate the direction of motion of a wave and are drawn perpendicular to the wavefront of a two- or three-dimensional wave; (a) illustrates circular waves near a point source while (b) shows plane waves.

The direction of motion of any wavefront can be represented by a line drawn perpendicular to the wavefront and in the direction the wave is moving (see blue arrows in Figure 3.1.2). This is called a **ray**.

The wavefront from a light wave can also be drawn in this way.

HUYGENS' PRINCIPLE

The theoretical basis for wave propagation in two dimensions was first explained by the Dutch scientist Christiaan Huygens. Huygens' principle states that each point on a wavefront can be considered as a source of secondary wavelets (i.e. small waves).

Consider the plane wave shown in Figure 3.1.3. Each point on the initial wavefront can be treated as if it is a point source producing circular waves, some of which are shown in green. After one period, these circular waves will have advanced by a distance equal to one wavelength. Huygens proved mathematically that when the amplitudes of each of the individual circular waves are added, the result is another plane wave as shown by the new wavefront.

This process is repeated at the new wavefront, causing the wave to propagate in the direction shown.

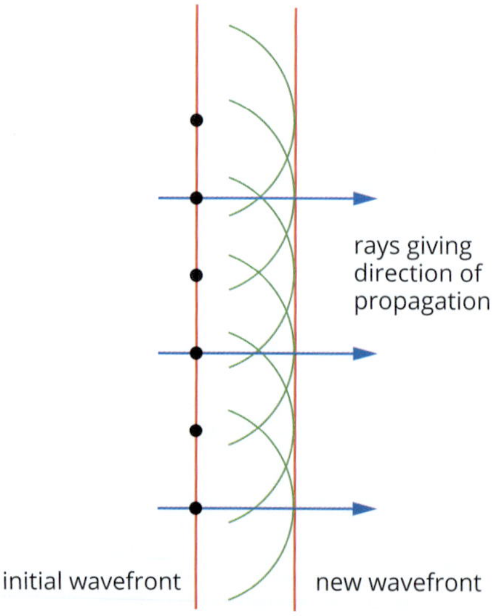

FIGURE 3.1.3 Each point on the wavefront of a plane wave can be considered as a source of secondary wavelets. These wavelets combine to produce a new plane wavefront.

Circular waves are propagated in a similar way, as shown in Figure 3.1.4.

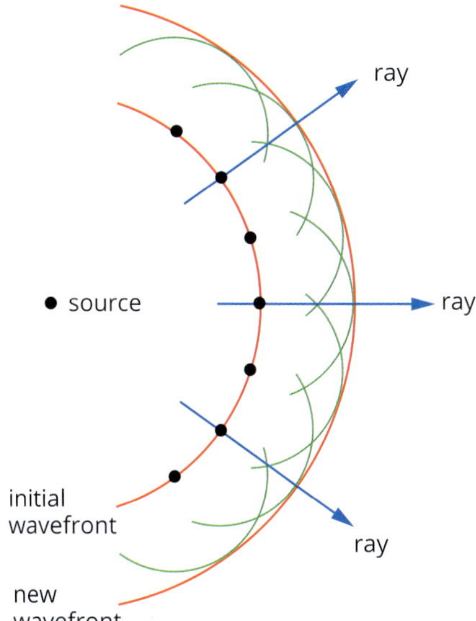

FIGURE 3.1.4 Each point on the wavefront of a circular wave can be considered as a source of secondary wavelets. These wavelets combine to produce a new circular wavefront.

Worked example 3.1.1

APPLYING HUYGENS' PRINCIPLE

On the plane wave shown moving from left to right below, sketch some of the secondary wavelets on the outer wavefront and draw the appearance of the new wave formed after one period.

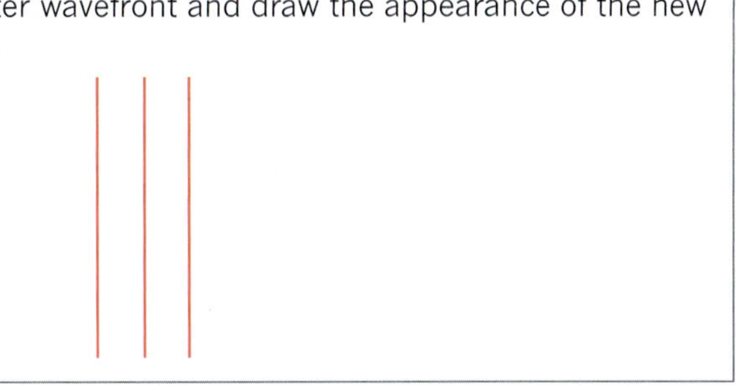

Thinking	Working
Sketch a number of secondary wavelets on the advancing wavefront. The radius of each secondary wavelet will be the same as the distance between the existing wavefronts.	
Sketch the new wavefront, by drawing a line joining the peak of each secondary wavelet.	

Worked example: Try yourself 3.1.1

APPLYING HUYGENS' PRINCIPLE

On the circular waves shown below, sketch some of the secondary wavelets on the outer wavefront and draw the appearance of the new wave formed after one period.

REFLECTED WAVEFRONTS

By using rays to illustrate the path of a wavefront reflecting from a surface, it can be shown that for a two- or three-dimensional wave, the angle from the normal at which the wave strikes a surface will equal the angle from the normal to the reflected wave. The **normal** is an imaginary line at 90° (i.e. perpendicular) to the surface.

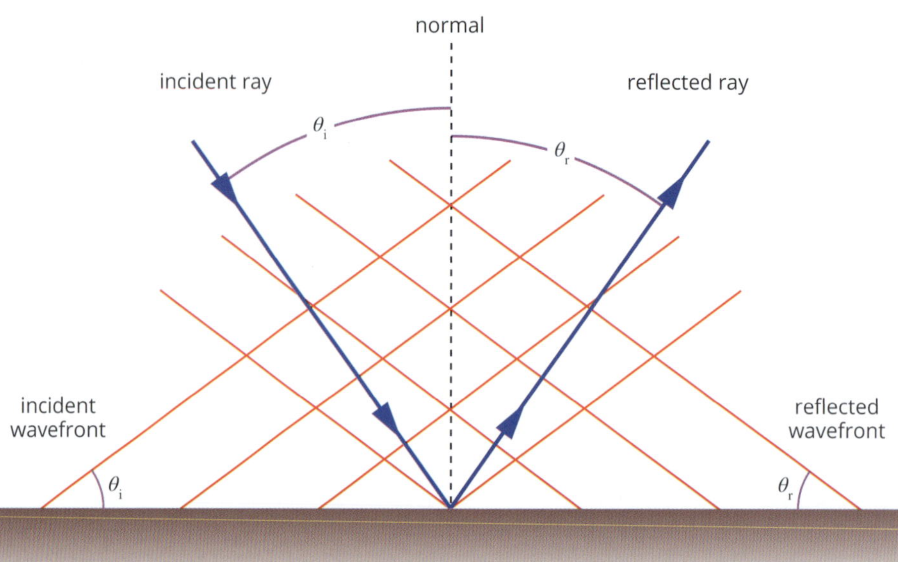

FIGURE 3.1.5 The law of reflection. The angle between the direction of the incident wave and the normal (θ_i) is the same as the angle between the normal and the reflected wave (θ_r).

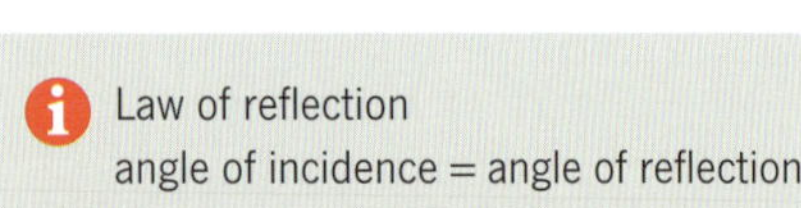

Law of reflection

angle of incidence = angle of reflection

$\theta_i = \theta_r$

These angles of the incident and reflected waves from the normal are labelled θ_i and θ_r, respectively, in Figure 3.1.5. This is known as the law of reflection. The law of reflection states that the **angle of reflection**, measured from the normal, equals the **angle of incidence** measured from the normal; that is, $\theta_i = \theta_r$.

The law of reflection is true for any surface whether it is straight, curved or irregular. For all surfaces, including curved or irregular surfaces, the normal is drawn perpendicular to the surface at the point of contact of the incident ray or rays.

When wavefronts meet an irregular, rough surface, the resulting reflection can be spread over a broad area. This is because each point on the surface may reflect the portion of the wavefront reaching it in a different direction, as seen in Figure 3.1.6. This is referred to as **diffuse** (spread out) reflection.

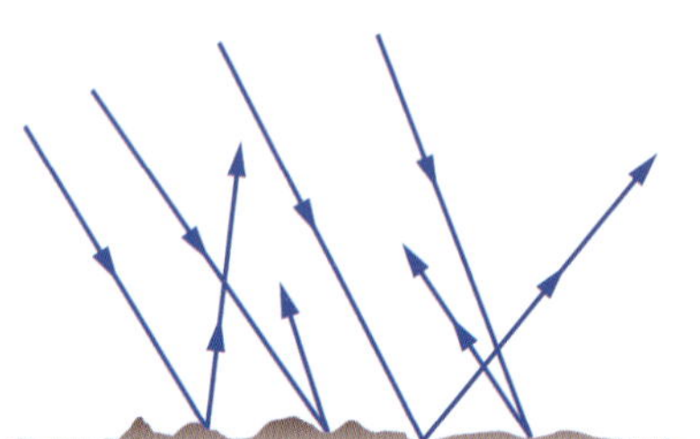

FIGURE 3.1.6 Reflection from an irregular surface. Each incident ray may be reflected in a different direction, depending upon how rough or irregular the reflecting surface is. The resulting wave will be diffuse (spread out).

When you walk on the beach at the height of summer, there is often a strong glare, a result of diffuse reflection from the sand.

REFRACTION

Refraction is a change in the direction of light caused by changes in its speed. Changes in the speed of light occur when light passes from one medium (substance) into another. In Figure 3.1.7, the light changes direction as it enters the glass prism, and then again when it leaves the glass prism and re-enters the air.

FIGURE 3.1.7 Light refracts as it moves from the air into the glass, causing a change in direction. The light refracts again as it goes from glass to air.

Consider Figure 3.1.8, in which light waves are moving from an incident medium where they have high speed, v_1, into a transmitting medium in which they have a lower speed, v_2. For the same time interval, Δt, in which the wave travels a distance $v_1\Delta t$ (B–D) in the incident medium, it travels a shorter distance $v_2\Delta t$ (A–C) in the transmitting medium. In order to do this, the wavefronts must change direction or 'refract' as shown.

Light waves behave in a similar way when they move from a medium such as air into water. The direction of the refraction depends on whether the waves speed up or slow down when they move into the new medium. In Figure 3.1.9, the light waves slow down as they move from air into glass, so the wavelength decreases and the direction of propagation of the wave is refracted towards the normal.

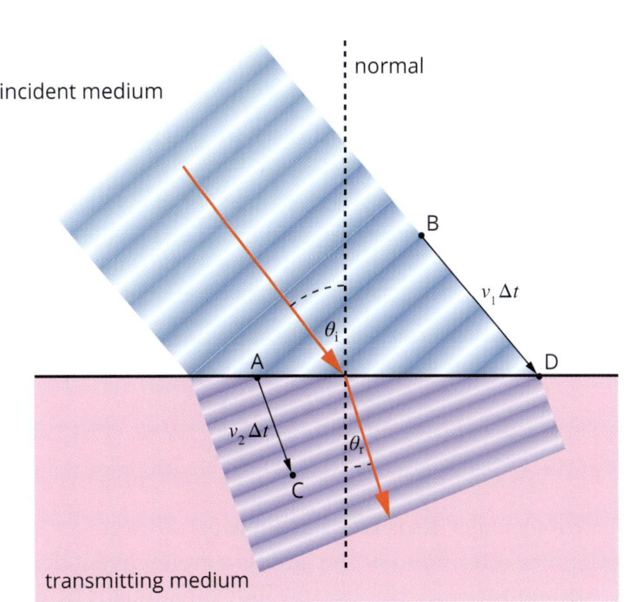

FIGURE 3.1.8 Wave refraction occurs because the distance A–C travelled by the wave in the transmitting medium is shorter than the distance B–D that it travels in the same time in the incident medium.

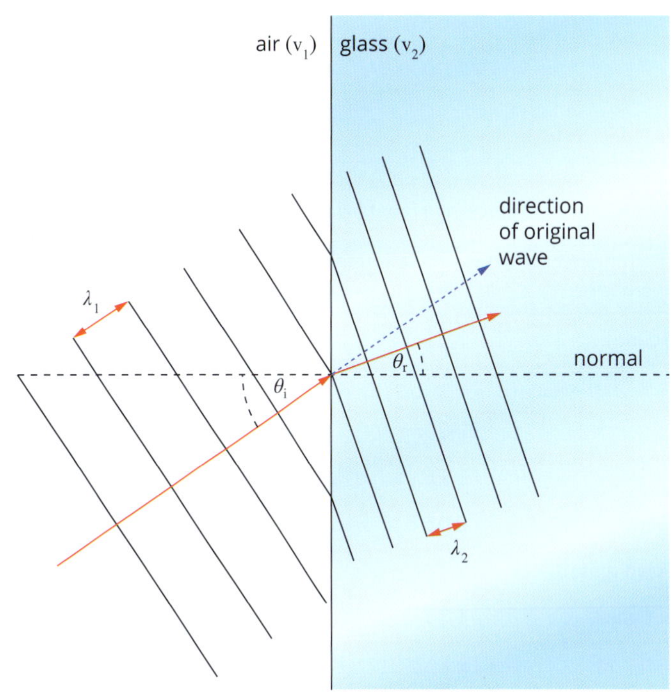

FIGURE 3.1.9 Light waves refract towards the normal when they slow down, $v_2 < v_1$.

The angle of incidence, θ_i, which is defined as the angle between the direction of propagation and the normal, is greater than the angle of refraction, θ_r.

Conversely, when a light wave moves from glass, in which it has low speed, into air, in which it travels more quickly, it is refracted away from the normal, as shown in Figure 3.1.10. In other words, the angle of incidence, θ_i, is less than the angle of refraction, θ_r.

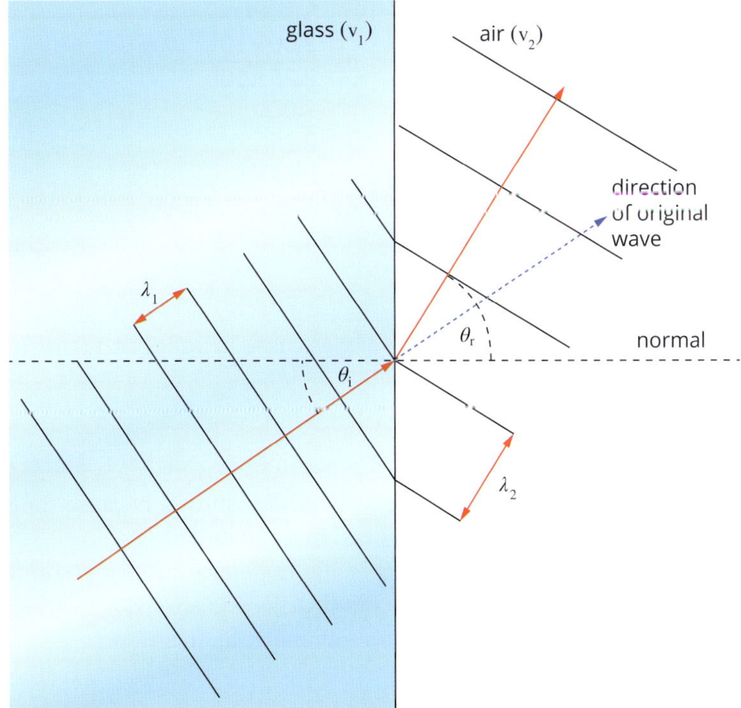

FIGURE 3.1.10 Light waves refract away from the normal when they speed up, $v_2 > v_1$.

Note that when a wave changes its speed, its wavelength also changes correspondingly, but its frequency does not change, as the number of waves per second remains the same.

Refractive index

The amount of refraction that occurs depends on how much the speed of light changes as light moves from one medium to another—when light slows down greatly, it will undergo significant refraction.

The speed of light in a range of different materials is shown in Table 3.1.1.

TABLE 3.1.1 The speed of light in various materials, correct to three significant figures

Material	Speed of light ($\times 10^8$ m s^{-1})
vacuum	3.00
air	3.00
ice	2.29
water	2.25
quartz	2.05
crown glass	1.97
flint glass	1.85
diamond	1.24

Scientists find it convenient to describe the change in speed of a wave using a property called the **refractive index**. The refractive index of a material, n, is defined as the ratio of the speed of light in a vacuum, c, to the speed of light in the medium, v.

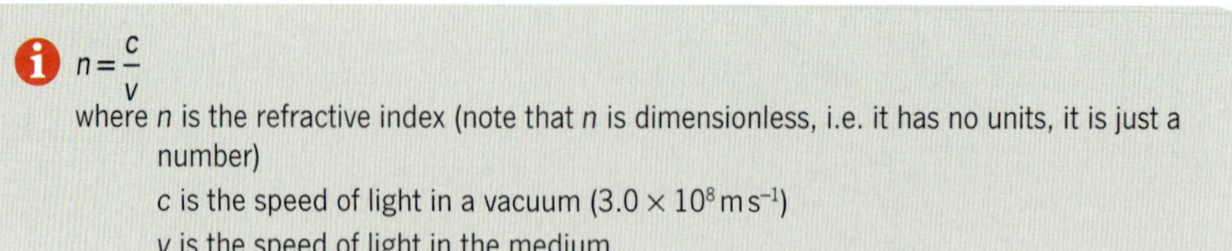

$n = \frac{c}{v}$

where n is the refractive index (note that n is dimensionless, i.e. it has no units, it is just a number)

c is the speed of light in a vacuum (3.0×10^8 m s^{-1})

v is the speed of light in the medium.

The refractive index for various materials is given in Table 3.1.2.

TABLE 3.1.2 Refractive indices of various materials

Material	Refractive index, n
vacuum	1.00
air	1.00
ice	1.31
water	1.33
quartz	1.46
crown glass	1.52
flint glass	1.62
diamond	2.42

This quantity is also sometimes referred to as the 'absolute' refractive index of the material, to distinguish it from the 'relative' refractive index that might be used when a light wave moves from one medium to another, such as from water to glass.

Worked example 3.1.2

CALCULATING REFRACTIVE INDEX

The speed of light in water is 2.25×10^8 m s^{-1}. Given that the speed of light in a vacuum is 3.00×10^8 m s^{-1}, calculate the refractive index of water.

Thinking	Working
Recall the definition of refractive index.	$n = \frac{c}{v}$
Substitute the appropriate values into the formula and solve.	$n = \frac{3.00 \times 10^8}{2.25 \times 10^8}$ $= \frac{3.00}{2.25}$ $= 1.33$

Worked example: Try yourself 3.1.2

CALCULATING REFRACTIVE INDEX

The speed of light in crown glass (a type of glass used in optics) is 1.97×10^8 m s^{-1}. Given that the speed of light in a vacuum is 3.00×10^8 m s^{-1}, calculate the refractive index of crown glass.

By definition, the refractive index of a vacuum is exactly 1, since $n = \frac{c}{c} = 1$. Similarly, the refractive index of air is effectively equal to 1, because the speed of light in air is practically the same as its speed in a vacuum.

The definition of refractive index allows you to determine changes in the speed of light as it moves from one medium to another.

$n = \frac{c}{v}$, therefore $c = nv$. This applies for any material, therefore:

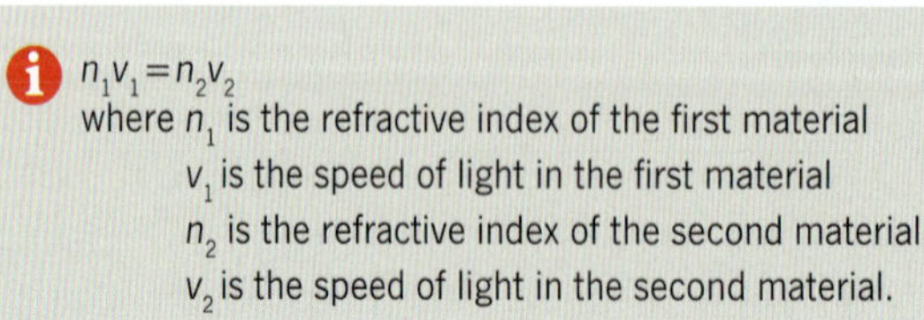

$n_1 v_1 = n_2 v_2$

where n_1 is the refractive index of the first material

v_1 is the speed of light in the first material

n_2 is the refractive index of the second material

v_2 is the speed of light in the second material.

Worked example 3.1.3

SPEED OF LIGHT CHANGES

A wave of light travels from crown glass (n = 1.52), in which it has a speed of $1.97 \times 10^8\,\text{m s}^{-1}$, into water ($n$ = 1.33). Calculate the speed of light in water.

Thinking	Working
Recall the formula.	$n_1v_1 = n_2v_2$
Substitute the appropriate values into the formula and solve.	$1.52 \times 1.97 \times 10^8 = 1.33 \times v_2$ $\frac{1.52 \times 1.97 \times 10^8}{1.33} = v_2$ $v_2 = 2.25 \times 10^8\,\text{m s}^{-1}$

Worked example: Try yourself 3.1.3

SPEED OF LIGHT CHANGES

A light wave travels from water (n = 1.33), in which it has a speed of $2.25 \times 10^8\,\text{m s}^{-1}$, into glass ($n$ = 1.85). Calculate the speed of light in glass.

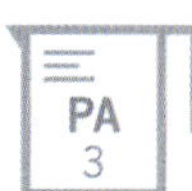

Snell's law

The refractive indices can also be used to determine how much a light wave will refract as it moves from one medium to another. Consider the situation shown in Figure 3.1.11, in which light refracts as it moves from air into water. To simplify the diagram, only the ray is shown and not the wavefronts.

In 1621, the Dutch mathematician Willebrord Snell described the geometry of this situation with a formula that is now known as **Snell's law**.

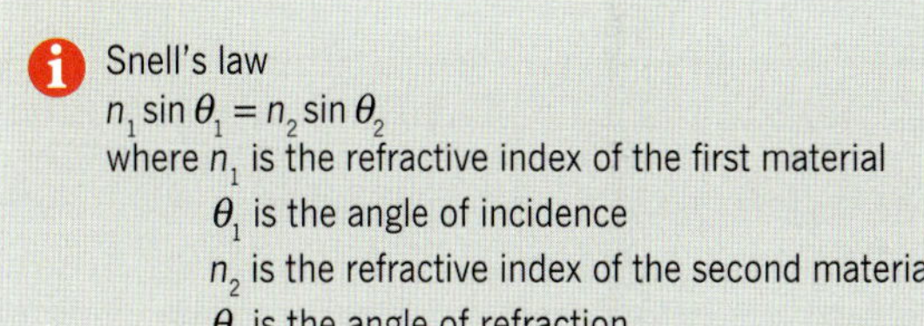

Snell's law

$n_1 \sin\theta_1 = n_2 \sin\theta_2$

where n_1 is the refractive index of the first material

θ_1 is the angle of incidence

n_2 is the refractive index of the second material

θ_2 is the angle of refraction.

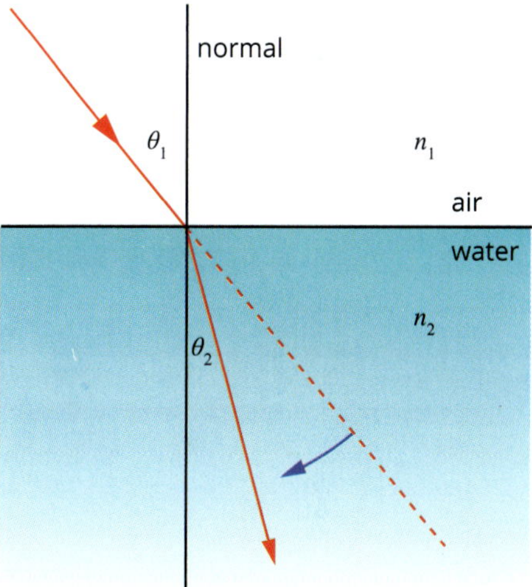

FIGURE 3.1.11 Light refracts as it moves from air into water.

Worked example 3.1.4

USING SNELL'S LAW

A light wave in air strikes the surface of a pool of water (n = 1.33) at angle of 30° to the normal. Calculate the angle of refraction of the light in water.

Thinking	Working
Recall Snell's law.	$n_1 \sin\theta_1 = n_2 \sin\theta_2$
Recall the refractive index of air.	$n_1 = 1.00$
Substitute the appropriate values into the formula to find a value for $\sin\theta_2$.	$1.00 \times \sin 30° = 1.33 \times \sin\theta_2$ $\sin\theta_2 = \frac{1.00 \times \sin 30°}{1.33}$ $= 0.3759$
Calculate the angle of refraction.	$\theta_2 = \sin^{-1} 0.3759$ $= 22.1°$

Worked example: Try yourself 3.1.4

USING SNELL'S LAW

A light wave in air strikes a piece of flint glass ($n = 1.62$) at angle of incidence of 50° to the normal. Calculate the angle of refraction of the light in the glass.

Total internal reflection

When light passes from a medium with low refractive index into one with higher refractive index, it is refracted towards the normal. Conversely, as shown in Figure 3.1.12, when light passes from a medium with a high refractive index to one with a lower refractive index, it is refracted away from the normal (Figure 3.1.12(a)). In this case, as the angle of incidence increases, the angle of refraction gets closer to 90° (Figure 3.1.12(b)). Eventually, at an angle of incidence known as the **critical angle**, the angle of refraction becomes 90° and the light is refracted along the interface between the two mediums (Figure 3.1.12(c)). If the angle of incidence is increased beyond this value, the light ray does not undergo refraction; instead, it is reflected back into the original medium, as if it was striking a perfect mirror (Figure 3.1.12(d)). This phenomenon is known as **total internal reflection** and is used in fibre-optic cables, as shown in Figure 3.1.13. The working of a fibre-optic cable is discussed below.

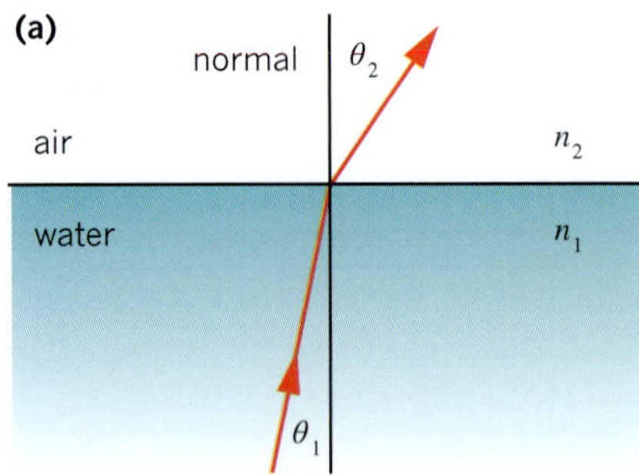

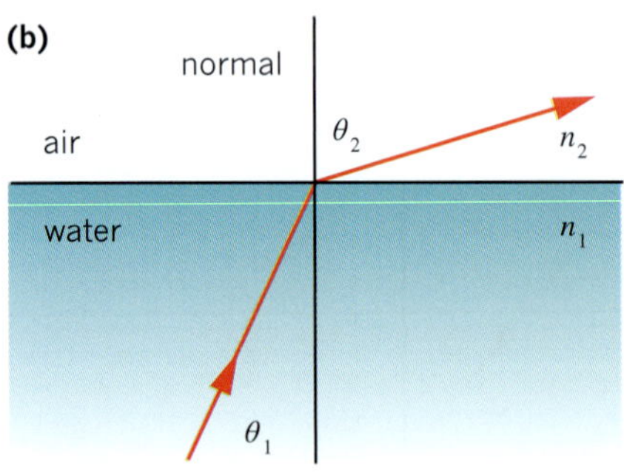

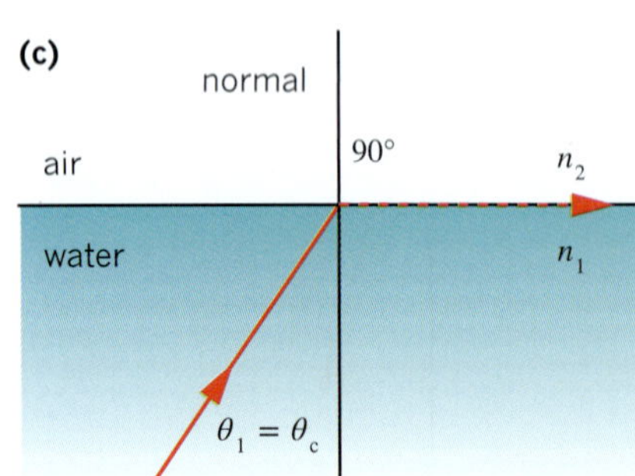

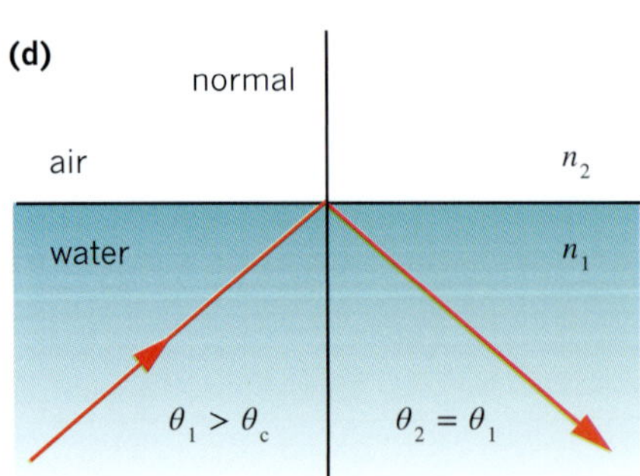

FIGURE 3.1.12 Light refracts as it moves from water into air as shown in diagrams (a) and (b). In diagram (c), the angle of refraction is exactly 90° to the normal and in (d) the light is undergoing total internal reflection.

As the angle of refraction for the critical angle is 90°, the critical angle is defined by the formula:

$n_1 \sin\theta_c = n_2 \sin 90°$

$\sin 90° = 1$, therefore $n_1 \sin\theta_c = n_2$

Therefore:

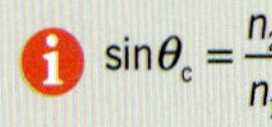

$$\sin\theta_c = \frac{n_2}{n_1}$$

Worked example 3.1.5

CALCULATING CRITICAL ANGLE

Calculate the critical angle for light passing from water into air.

Thinking	Working
Recall the equation for critical angle.	$\sin\theta_c = \frac{n_2}{n_1}$
Substitute the refractive indices of water and air into the formula. (Unless otherwise stated, assume that the second medium is air with $n_2 = 1$.)	$\sin\theta_c = \frac{1.00}{1.33}$ $= 0.7519$
Solve for θ_c.	$\theta_c = \sin^{-1} 0.7519$ $= 48.8°$

Worked example: Try yourself 3.1.5

CALCULATING CRITICAL ANGLE

Calculate the critical angle for light passing from diamond into air.

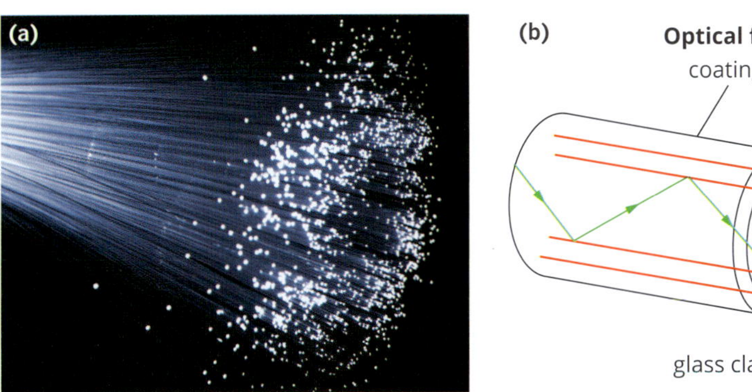

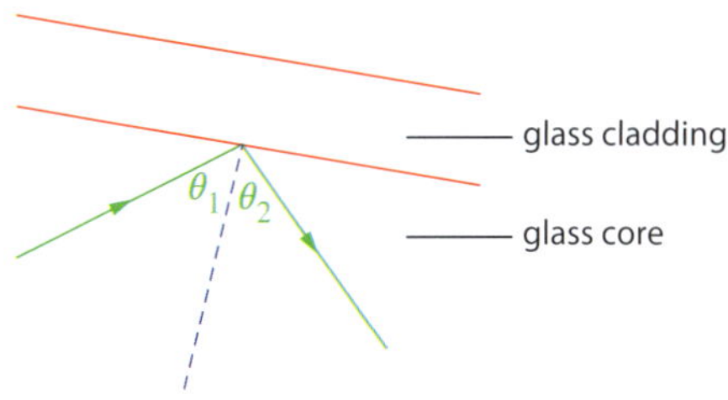

FIGURE 3.1.13 (a) Optical fibres transmit light using total internal reflection. (b) The outer glass cladding is made of glass with a slightly lower refractive index (n_2) than the glass core (n_1). (c) The angle of incidence of the incoming light is shown as θ_1 and the angle of reflection is shown as θ_2.

Fibre-optic cables and total internal reflection

The National Broadband Network (NBN) is integral to the functioning of modern society. A key part of this is the fibre-optic network used to send optical signals to your home, school and work. The information is turned into a light wave signal and sent down the fibre-optic cable using a semiconductor laser diode. Fibre-optic cables can also be used to send light down a cable for decorative or lighting effects, as shown in Figure 3.1.13(a).

A fibre-optic cable consists of an inner glass core with refractive index n_1, surrounded by an outer glass cladding with refractive index n_2, as shown in Figure 3.1.13(b). The refractive index of the cladding is less than that of the core, ($n_2 < n_1$). At angles greater than the critical angle, total internal reflection occurs from the cladding and light is propagated down the core. Some refraction can occur at the interface between the core and the cladding, so the intensity of the signal gradually reduces. To counteract this, the fibre-optic cable is connected to a semiconductor detector, the signal is electrically amplified and then converted into a light wave by another semiconductor laser diode and sent down another length of fibre-optic cable. The fibre-optic cable is protected by an outer plastic coating as shown.

Optical effects due to refraction

Some everyday optical phenomena can be explained using the principles of refraction.

Apparent position of objects under water

When you look at a fish in the water, where you see the fish is not the real position of the fish. Indigenous Australian fishing techniques allow for this and include aiming lower than where the fish appears to be and using a spear that has a number of points evenly spread along the shaft, which increases the probability of striking a fish at its real depth.

The **apparent depth** (D_a) and the **real depth** (D_r) of the fish is shown in Figure 3.1.14. This difference between real depth and apparent depth is due to the refraction of the light rays travelling up from the fish through the water to the water–air boundary, and into the air above to the observer. The change in medium results in a change in velocity of the light waves. The light waves travel faster in air, making the light ray bend, or refract, at the water–air boundary. In this case, the light ray speeds up and bends away from the imaginary normal line. The light rays are refracted at the water–air boundary before they enter the observer's eyes, so the fish appears at the apparent depth (D_a) and not the real depth (D_r).

PHYSICSFILE

Refractive index of diamonds

Diamond has a very high refractive index; therefore, it has a small critical angle. This means that a light wave that enters a diamond will often bounce around inside the diamond many times before leaving the diamond. A jeweller can cut a diamond to take advantage of this property; this causes the diamond to 'sparkle' (see below), as it appears to reflect more light than is falling on it.

The refractive properties of diamonds mean they appear to sparkle.

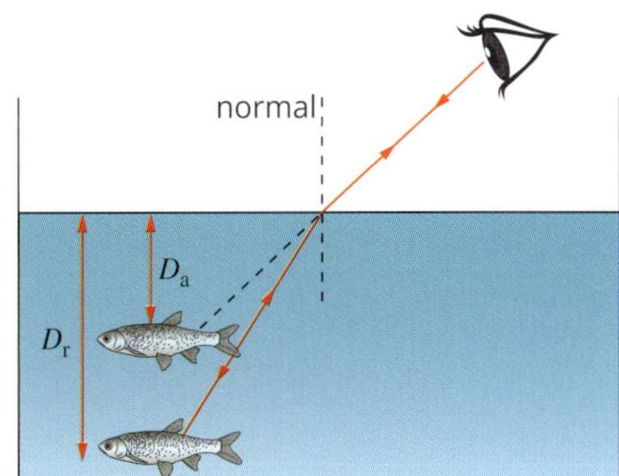

FIGURE 3.1.14 The apparent depth, D_a, of a fish compared to the real depth, D_r, as seen from above the air water interface

Early sunrise and late sunset

At sunset, you are seeing the Sun when it is already below the horizon. Also, when the Sun is near the horizon during sunset, it appears to be more oval-shaped than circular, as shown in Figure 3.1.15(a). Notice that the shape of the Sun appears to distort as it approaches the horizon.

During sunrise and sunset, light waves from the Sun travel a greater distance through the atmosphere than at midday, when the Sun is directly overhead. The atmosphere consists of layers due to the air thinning with increasing altitude. This means that the density of air decreases with increasing altitude, or in other words, the refractive index of air at the surface of Earth is higher than that of air in the upper atmosphere. This layered structure of the atmosphere continuously refracts the light rays until they reach the observer's eye. As light travels through the atmosphere it slows down and is continually refracted towards the normal. This leads to the light wave travelling in a curved path as shown in Figure 3.1.15(b). To the observer, the Sun appears to be in a higher position than it actually is. Due to this effect, we can see the Sun even if it is actually below our horizon. The length of a day appears to be about 4 minutes longer than it actually is due to the refraction of sunlight.

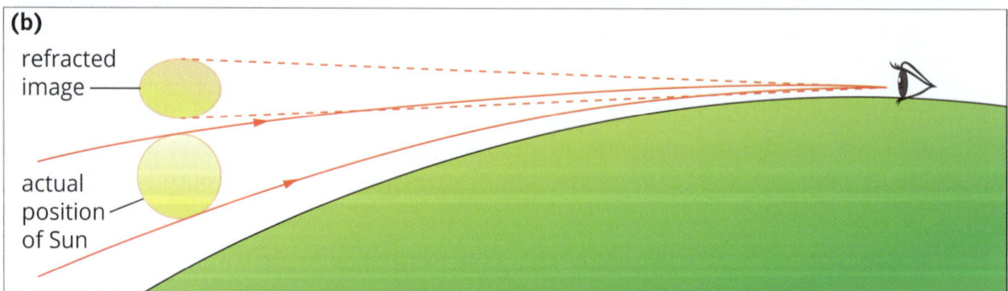

FIGURE 3.1.15 (a) The shape of the Sun appears flatter as it approaches the horizon. It looks less circular and more oval-shaped. (b) At sunset and sunrise, the observed position of the Sun is a refracted image and appears higher than the actual position of the Sun.

Refraction of light by the atmospheric layers also makes the Sun appear flattened or distorted. At sunset and sunrise, the lower part of the Sun is closer to the horizon and light from it is refracted more than light from the top part of the Sun. The observed effect is that the bottom of the Sun is lifted up more than the top. The Sun appears more oval in shape. The circular Sun is the object and the flatter, oval Sun is the image seen by the observer.

Mirages

You are in a car that is travelling down a road on a very hot day. Looking ahead you see the illusion of water on the road similar to that in Figure 3.1.16(a). When you get there, however, the road is completely dry. This effect is known as a mirage and occurs due to refraction effects in the atmosphere.

On a very hot day the atmosphere heats up, leading to the formation of hotter (less dense) layers of air rising above colder (denser) layers of air, which sink down. The variation in temperature and density produces a variation in the refractive index of the air, which effectively curves the direction of the light. This can result in light from the sky being refracted upwards towards you in the oncoming car, as shown in Figure 3.1.16(b), giving the appearance of water. Under certain conditions you can also see a refracted image from an object in front of you, as shown by the red ute.

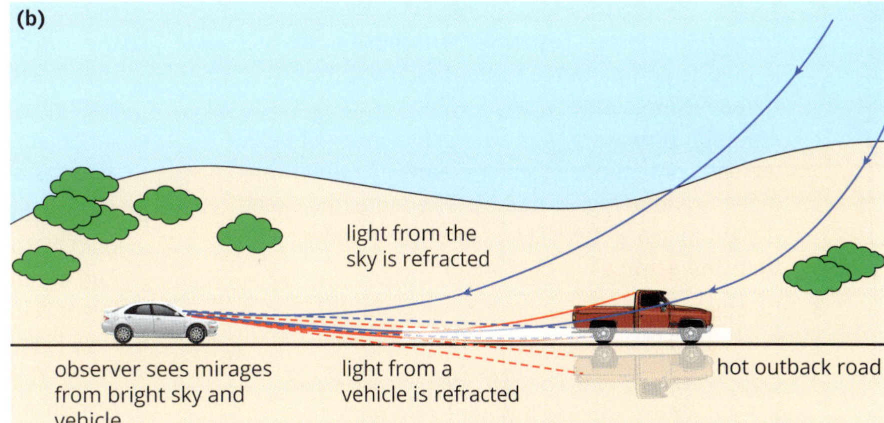

FIGURE 3.1.16 (a) On a hot day a mirage gives the illusion of water on the road. (b) Variations in the atmospheric refractive index lead to refraction of blue light from the sky. In addition, refraction effects can lead to the illusion of the occupants of the blue car seeing an inverted image of the red car.

3.1 Review

OA ✓✓

SUMMARY

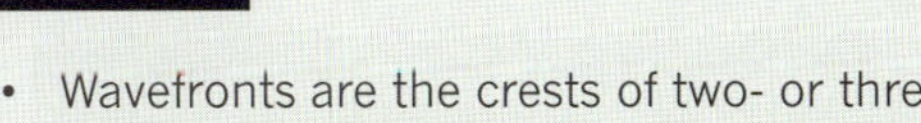

- Wavefronts are the crests of two- or three-dimensional waves.
- Huygens' principle states that each point on a wavefront can be considered as a source of secondary wavelets. These wavelets combine to produce a new wavefront.
- In reflection, the angle of incidence of the wave relative to the normal is equal to the angle of reflection relative to the normal.
- Refraction is the change in the direction of light that occurs when light moves from one medium to another due to a change in the speed of the light waves.
- The refractive index, n, of a material is given by the formula $n = \frac{c}{v}$, where c is the speed of light in a vacuum and v is the speed of light in the material.
- When light moves from one material to another, the change in speed can be calculated using: $n_1 v_1 = n_2 v_2$

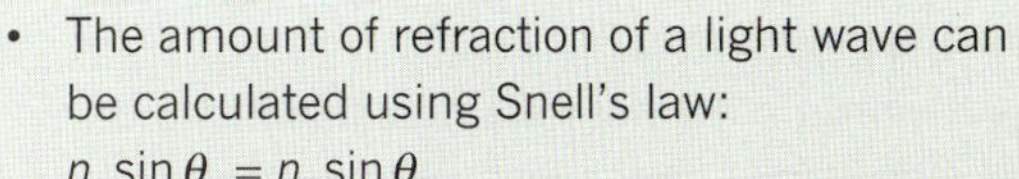

- The amount of refraction of a light wave can be calculated using Snell's law: $n_1 \sin\theta_1 = n_2 \sin\theta_2$
- The critical angle of a material denotes the angle of incidence when the angle of refraction is 90°. It can be calculated using $n_1 \sin\theta_c = n_2 \sin 90°$ or $\sin\theta_c = \frac{n_2}{n_1}$.
- Total internal reflection occurs when the angle of incidence is greater than the critical angle.
- Light is propagated through the glass inner core of fibre-optic cables. Total internal reflection occurs between the inner core glass and the outer cladding glass.
- Optical effects in which the apparent position of an object does not match its true position occur due to changes in refractive index between the source of the light wave and the observer.

continued over page

3.1 Review *continued*

KEY QUESTIONS

Knowledge and understanding

1 Copy the diagram below and use Huygens' principle to draw a new wavefront of the plane wave after one period.

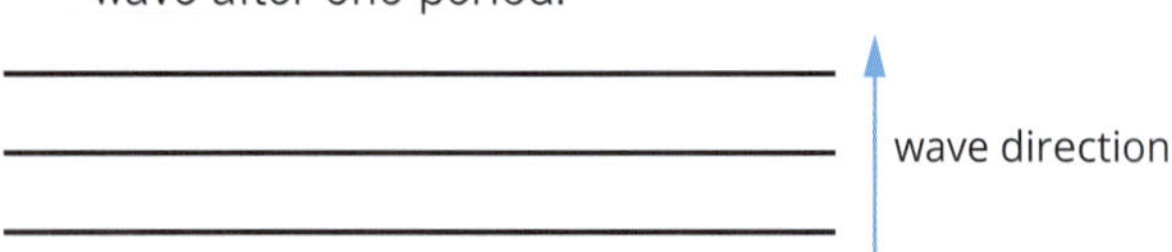

2 Choose the correct response from those given in bold to complete the sentences about the refractive indices of types of water.
Although pure water has a refractive index of 1.33, the salt content of sea water means its refractive index is a little higher at 1.38. Therefore, the speed of light in sea water will be **faster than/slower than/the same as** in pure water.

3 Calculate the speed of light in sea water that has a refractive index of 1.38.

4 Light travels at of $2.25 \times 10^8 \, \text{m s}^{-1}$ in water and $2.29 \times 10^8 \, \text{m s}^{-1}$ in ice. If water has a refractive index of 1.33, use this information to calculate the refractive index of ice.

5 Light travels from water ($n = 1.33$) into glass ($n = 1.60$). The incident angle is 44°. Calculate the angle of refraction.

6 Wavefronts of light initially travelling in air are incident and parallel to an air–glass boundary (the angle of incidence of the light ray is 0°). Identify which one or more of the following statements is true regarding the wavefront and ray travelling through the glass.

- **A** The wavefronts will not refract or bend as the wave slows down in the glass medium.
- **B** The wavefronts will not refract or bend as the wave speeds up in the glass medium.
- **C** The wavefronts will refract or bend as the wave slows down in the glass medium.
- **D** The wavefronts will refract or bend as the wave speeds up in the glass medium.

7 Assess whether these statements are true or false regarding different types of waves and the angles of incidence and refraction. Rewrite the false statements to make them true.

- **a** When light rays refract away from the normal line at a boundary between two different media, the light wave is travelling faster in this new medium.
- **b** The setting and rising Sun appears flatter than the afternoon Sun because light from the higher part of the Sun refracts more than that from the lower part of the Sun. The observed effect is that the top of the Sun is lowered more than the bottom of the Sun.
- **c** An object in water appears lower than it actually is.

8 **a** Choose the correct response from those given in bold to complete the sentence.
The fibre-optic glass cladding has a **higher/lower** refractive index than the glass core.

b How is light propagated down the fibre-optic cable? (Explain your answer incorporating your response from part a).

Analysis

9 **a** The cladding of a fibre-optic cable has a refractive index of 1.348. The critical angle is 62.3°. Determine the refractive index of the core.

b Determine the speed of light in the core.

c Determine the range of angles, relative to the core–cladding interface, that will be reflected.

10 For which of the following situations can total internal reflection occur?

	Incident medium	Refracting medium
a	air ($n = 1.00$)	glass ($n = 1.55$)
b	glass ($n = 1.55$)	air ($n = 1.00$)
c	glass ($n = 1.55$)	water ($n = 1.33$)
d	glass ($n = 1.55$)	glass ($n = 1.58$)

3.2 Dispersion and polarisation

A rainbow is often seen when the Sun appears after a rain shower. The rainbow illustrates that visible light is made up of a spectrum of colours. In Section 2.3 on page 72, light is described as a wave, and its properties are explored in Section 3.1. Further examples of the wave behaviour of light are dispersion, which is seen in rainbows, and polarisation.

FIGURE 3.2.1 When white light enters a prism, it is split into its component wavelengths or colours.

DISPERSION

When white light passes through a triangular glass prism (as shown in Figure 3.2.1), it undergoes **dispersion**. This spreading out into its component colours is a result of refraction.

Each different colour of light has a different wavelength (Table 3.2.1). White light is a mixture of light waves of many different wavelengths.

As discussed in the previous section and shown in Figure 3.1.8 on page 85, when light travels from air into a medium such as glass, the wavelength decreases as the waves bunch up. However, each colour travels at a different speed, so that, in effect, a medium has a different refractive index for each wavelength of light. Figure 3.2.2 shows the wavelength dependence of the refractive index for crown glass, acrylic and silica. The refractive index for each material decreases with wavelength. Given that $n = \frac{c}{v}$, the velocity of a wave in these materials increases with wavelength.

For light of longer wavelengths (such as red light), a medium has a lower refractive index, and light travels the fastest. Therefore it will refract at a smaller angle.

For light of shorter wavelengths (such as violet light), a medium has the highest refractive index, and light travels the slowest. Therefore it will refract the most.

This leads to the components of visible light being refracted at a range of angles, leading to the rainbow effect seen in Figure 3.2.1. Each wavelength of light is incident at a different angle on the opposite face of the prism, which further exaggerates the dispersion effect.

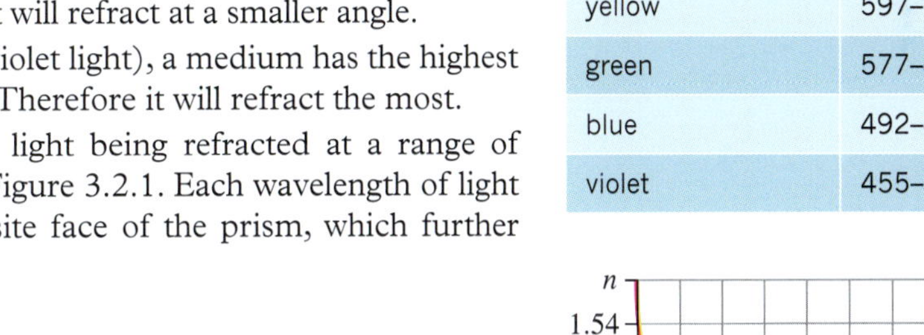

TABLE 3.2.1 Approximate wavelength ranges for the colours in the visible spectrum in air. 1 nm = 10^{-9} m

Colour	Wavelength (nm)
red	780–622
orange	622–597
yellow	597–577
green	577–492
blue	492–455
violet	455–390

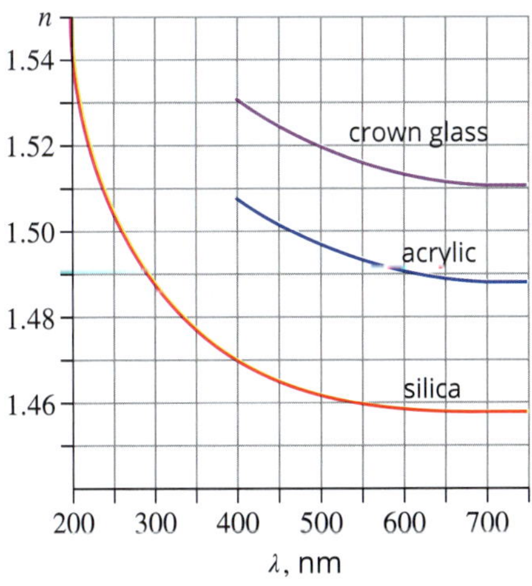

FIGURE 3.2.2 The refractive index of a material varies with wavelength. For glass, the refractive index decreases as wavelength increases.

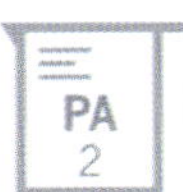

PHYSICSFILE

Where does colour come from?

In the seventeenth century, many people believed that white light was 'stained' by its interaction with earthly materials. Isaac Newton very neatly disproved this with a simple experiment using two prisms—one to split light into its component colours and the other to turn it back into white light (see right). This showed that the various colours were intrinsic components of white light since, if colour was a result of 'staining', the second prism should have added more colour rather than less.

Newton's double prism experiment showed that white light is made up of its component colours.

Newton was the first to identify the colours of the spectrum—red, orange, yellow, green, blue, indigo and violet. He chose seven colours by inventing the colour 'indigo', because seven was considered a sacred number.

You can see how white light is formed by the combination of other colours by using a ×10 lens (a microscope objective lens works well) to look at the white part of a computer screen. You will see the red, blue and green pixels that are used to generate the white light.

Worked example 3.2.1

CALCULATING RANGE OF ANGLES FOR DISPERSION

White light is incident on the surface of a triangular prism made of crown glass at an angle of 38° to the normal. Use the graph in Figure 3.2.2 on page 93 to determine the range of refracted angles for visible light entering the prism. Assume a wavelength range of 400 nm to 700 nm.

Thinking	Working	
Recall Snell's law.	$n_1 \sin\theta_1 = n_2 \sin\theta_2$	
Identify the variables for air.	$n_1 = 1$ for air, $\theta_1 = 38°$	
Use the graph to determine the refractive index for crown glass at 400 nm and 700 nm.	Reading from the graph: At 400 nm: $n_2 = 1.531$ At 700 nm: $n_2 = 1.512$	
Substitute the refractive index of crown glass at 400 nm and 700 nm into Snell's law to determine the range of wavelengths.	For 400 nm: $\sin\theta_2 = \frac{n_1 \sin\theta_1}{n_2}$ $= \frac{1.00 \times \sin 38°}{1.531}$ $= 0.402$ $\theta_2 = 23.7°$	For 700 nm: $\sin\theta_2 = \frac{n_1 \sin\theta_1}{n_2}$ $= \frac{1.00 \times \sin 38°}{1.512}$ $= 0.407$ $\theta_2 = 24.0°$
State the range.	The range of angles within the prism varies from 23.7° to 24.0°.	

Worked example: Try yourself 3.2.1

CALCULATING RANGE OF ANGLES FOR DISPERSION

White light is incident on the surface of a triangular prism made of acrylic at an angle of 70° to the normal. Use the graph in Figure 3.2.2 on page 93 to determine the range of angles for visible light entering the prism. Assume a wavelength range of 400 nm to 700 nm.

Optical effects due to dispersion

Some everyday optical phenomena can be explained using the principles of dispersion.

Colour dispersion in lenses

As each colour of light effectively has a different refractive index in glass, light passing through a glass lens always undergoes some dispersion. This means that coloured images formed by optical instruments such as microscopes and telescopes can suffer from a type of distortion known as chromatic aberration (Figure 3.2.3).

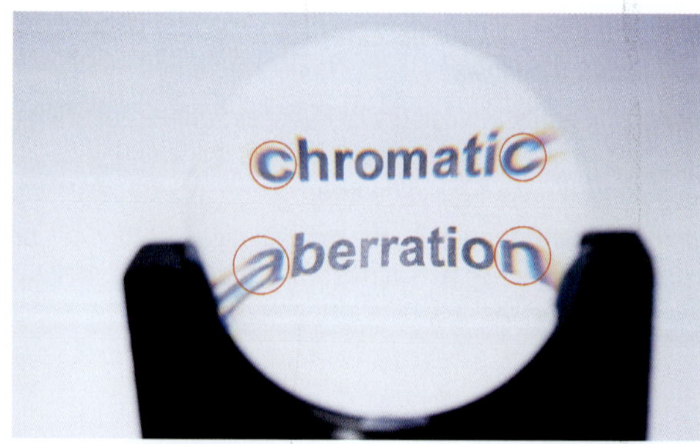

FIGURE 3.2.3 Chromatic aberration causes the coloured fringes that can be seen in the circled regions in this image.

Scientists have developed a number of techniques to deal with this problem, including:

- using lenses with very long focal lengths
- using 'achromatic' lenses—compound lenses that are made of different types of glass with different refractive properties
- taking separate images using coloured filters and then combining these images to form a single multi-coloured image.

The formation of rainbows

Rainbows are spectacular optical phenomena that occur after rainfall or on showery days. They are quite often seen as single rainbows, but they can sometimes form the double image shown in Figure 3.2.4(a).

The next time you see a rainbow, have a look at the direction of the Sun relative to the position of the rainbow. The Sun will generally be behind you. As discussed, light from the Sun that enters Earth's atmosphere consists of a range of wavelengths from violet-blue through to red. After rain, light waves can enter a raindrop. The refractive index of water is higher than the refractive index of air, so refraction occurs as the light enters the raindrop. The refractive index of water, like that for the prism, varies with wavelength; thus, dispersion effects occur such that the angle of refraction is higher for red light than for blue light (Figure 3.2.4(b)). Total internal reflection occurs at the back of the raindrop, then the light is refracted again as it exits the raindrop. This leads to an angular spread of colour from blue through to red, which gives the rainbow its circular shape. We don't see a full circle as Earth gets in the way.

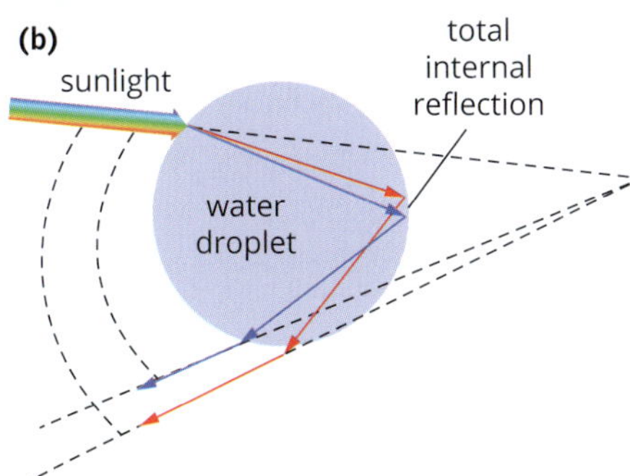

FIGURE 3.2.4 (a) A double rainbow is formed over Phillip Island, Victoria. (b) The formation of a rainbow is due to a combination of reflection and dispersion in raindrops.

POLARISATION

One of the most convincing pieces of evidence for the wave nature of light is the phenomenon of polarisation.

Light is a transverse wave (as discussed in Chapter 2), which means the wave is vibrating perpendicular to the direction of propagation. Light produced by some sources, such as a light globe or the Sun, is unpolarised and can be thought of as a collection of waves, each vibrating in a different plane but still perpendicular to the direction of travel, as shown in Figure 3.2.5.

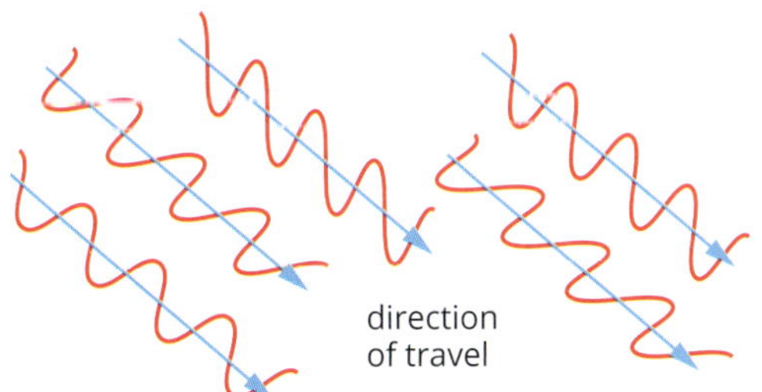

FIGURE 3.2.5 Unpolarised light waves consist of a collection of waves that vibrate perpendicular to the direction of travel but in different planes. Each wave has a different plane of polarisation.

Polarisation occurs when a transverse wave is allowed to vibrate in only one plane. This can be done by using a polarising filter. For example, the light wave in Figure 3.2.6 is already vertically polarised—the wave oscillations occur in the vertical plane only. This means that this wave is unaffected by a polarising filter that is orientated in the vertical plane.

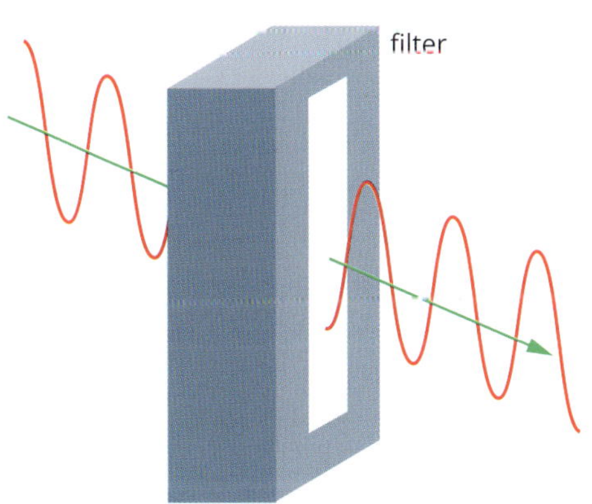

FIGURE 3.2.6 A vertically polarised wave can pass through a vertically orientated polarising filter.

The wave in Figure 3.2.7 is horizontally polarised. It is completely blocked by the vertical polarising filter.

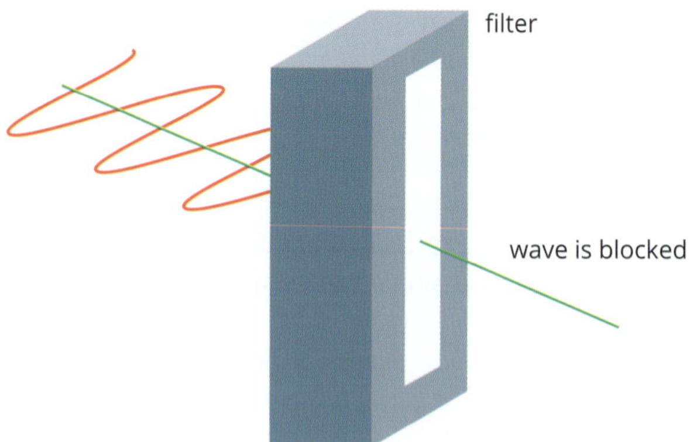

FIGURE 3.2.7 A horizontally polarised wave cannot pass through a vertically orientated polarising filter.

In Figure 3.2.8, the incoming wave is polarised at 45° to the horizontal and vertical planes. The horizontal component of this wave is blocked by the vertical filter, so the ongoing wave is vertically polarised and has a smaller amplitude than the original wave.

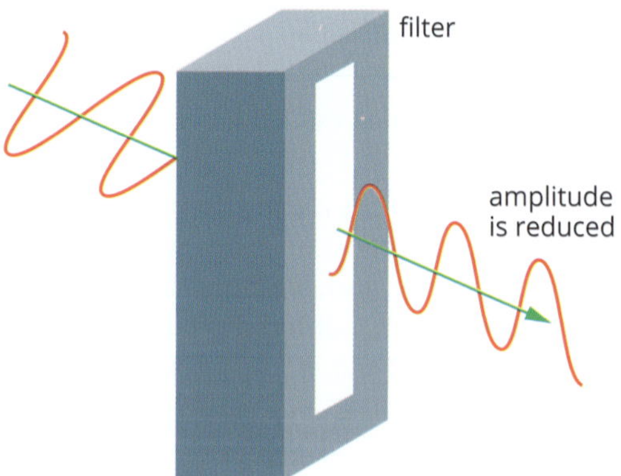

FIGURE 3.2.8 A diagonally polarised wave has its horizontal component blocked by the vertically orientated polarising filter. A vertically polarised wave of reduced amplitude passes through it.

Certain materials can act as polarising filters for light. These materials only transmit the waves or components of waves that are polarised in a particular direction and absorb the rest. Polarising sunglasses work by absorbing the light polarised parallel to a surface, thus reducing glare. Photographers use polarising filters to reduce the glare in photographs or to achieve specific effects (Figure 3.2.9).

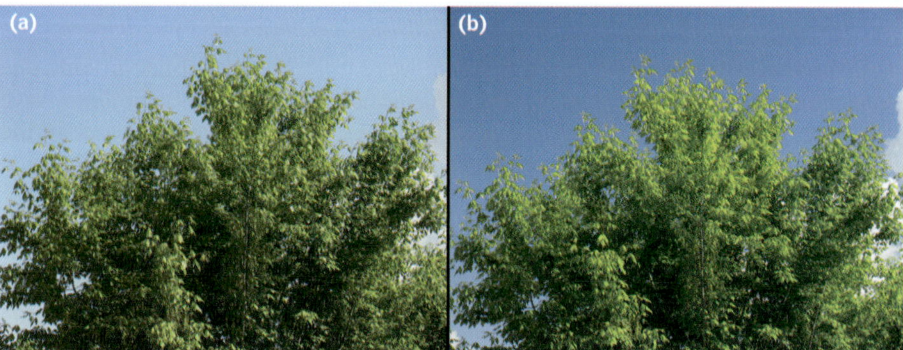

FIGURE 3.2.9 These are photographs taken of the same tree, one (a) without a polarising filter and (b) with a polarising filter.

PHYSICSFILE

Polarising sunglasses

Light that is reflected from a surface, such as water, snow or sand, is partially polarised in a direction parallel to the surface from which it reflects (see right). The polarising plane of polarising sunglasses is selected to absorb this reflected light. This makes polarising sunglasses particularly effective for people involved in outdoor activities such as boating, fishing or skiing.

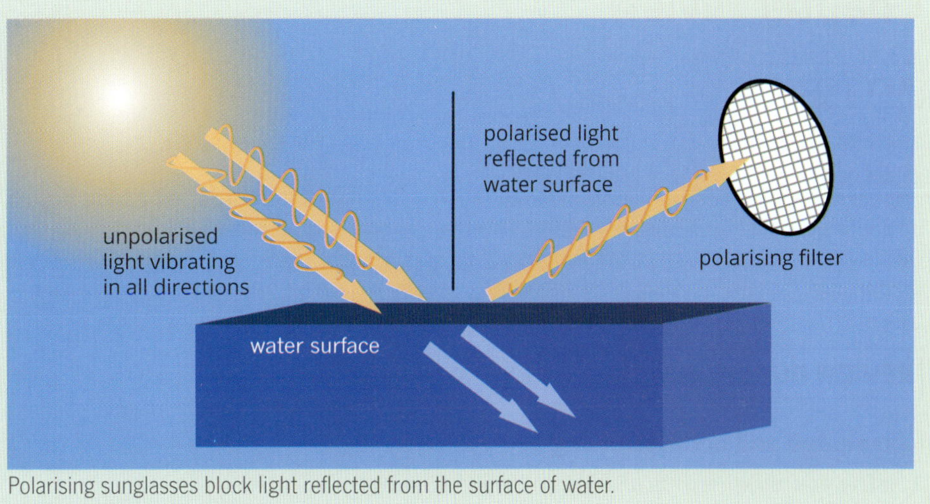

Polarising sunglasses block light reflected from the surface of water.

3.2 Review

OA

SUMMARY

- Different colours of light have different wavelengths.
- Although all light travels at $3.00 \times 10^{8}\,m\,s^{-1}$ in a vacuum, red light travels faster than blue light in a medium. Dispersion occurs because when white light is incident on a medium, such as water or glass, the angle of refraction of blue light is greater than that of red light.
- Rainbows from water droplets or prisms, and coloured fringes on images from lenses are all due to dispersion.
- Light waves emitted from an unpolarised source can oscillate in multiple planes perpendicular to the direction of propagation.
- Polarisation occurs when light travels through a polarising filter and is only allowed to vibrate in one plane.

KEY QUESTIONS

Knowledge and understanding

1 A rainbow occurs because of dispersion effects in water. State whether these statements are true or false. Rewrite the false statements to make them true.
  - a Light travels through a water droplet at a higher velocity than in air, which leads to refraction effects in water.
  - b Red light travels at slower speeds than blue light, which leads to a greater angle of refraction of red light.

2 What is dispersion in lenses and why does it occur?

3 What direction do polarisers on polarising sunglasses need to be to block out glare?

Analysis

4 Calculate the angle of refraction of green light of wavelength 550 nm after it enters a rectangular prism of crown glass at an angle of 35°. Use the graph in Figure 3.2.2 on page 93.

5 A white light wave enters an acrylic prism at an angle of 60°.
  - a Explain whether dispersion will occur and why.
  - b If dispersion will occur, calculate the angle of refraction of red light at 700 nm and blue light at 400 nm.

Chapter review

KEY TERMS

angle of incidence
angle of reflection
apparent depth
critical angle
diffuse
dispersion
normal
plane wave
polarisation
ray
real depth
refraction
refractive index
Snell's law
total internal reflection
wavefront

REVIEW QUESTIONS

Knowledge and understanding

1 The figure represents a situation involving the refraction of light. Identify the correct label for each letter from the choices provided:
boundary between media, reflected wave, incident wave, normal, refracted wave

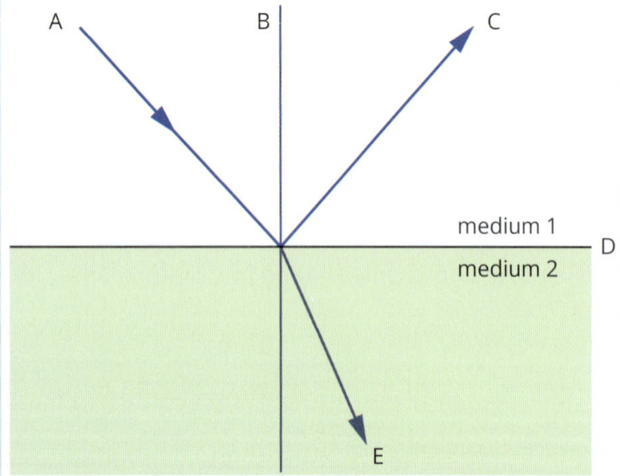

2 Why can chromatic aberration occur in basic lenses?

3 On a very hot day it can look as if there is water on the road. Explain why.

4 Explain briefly why snowboarders and sailors are likely to wear polarising sunglasses.

5 Choose the correct responses from those given in bold to complete the following sentence about refraction.
As light travels from quartz ($n = 1.46$) to water ($n = 1.33$), its speed **increases/decreases**, which causes it to refract **away from/towards** the normal.

6 Red light (4.5×10^{14} Hz) has a wavelength of 500 nm in water. Calculate the speed of red light in water.

7 A light wave travelling in air strikes a glass boundary at an angle such that the angle between the direction of the light wave and the glass boundary is 90°.

a Explain what happens to the light wave as it passes into the glass. Explain whether the wave refracts.

b Determine whether the frequency of the light ray changes in the glass medium.

8 A person is looking down at a fish below the surface of the water. Select the most correct statement regarding the apparent position and the real position of the fish.

A The real position and the apparent position are identical, as the reflected light from the water surface and the incident light make the same angle with the water's surface.

B The apparent position of the fish would be lower and closer to the person than the real position.

C The real position would be lower in the water than the apparent position.

D The apparent position would be lower and further away than the real position of the fish.

Application and analysis

9 The refractive index of a material is 1.20. Calculate the speed of light in the material.

10 A diver shines a torch up from under the water at an angle of incidence of 32.0°. The light enters the glass of a glass-bottom boat. If the refractive index of water is 1.33 and that of the glass is 1.52, what is the angle of refraction within the glass?

11 The speed of light in air is 3.00×10^8 m s^{-1}. Light strikes an air–perspex boundary at an angle of incidence of 43.0° and its angle of refraction is 28.5°.
Calculate the speed of light in perspex.

12 A light wave, represented by the ray of light, travels from air, through a layer of glass and then into water as shown. Calculate angles *a*, *b* and *c*.

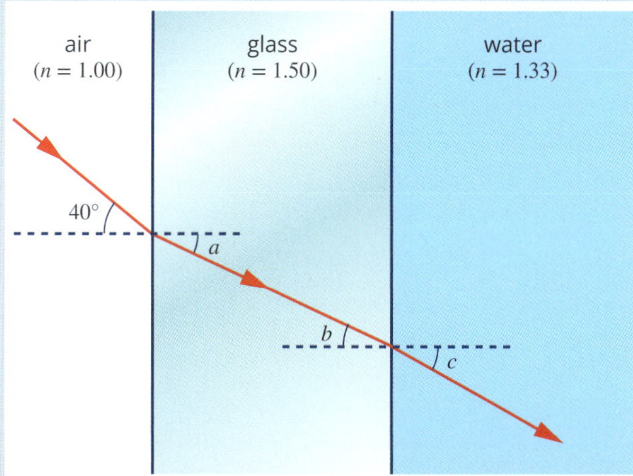

13 A light wave exiting a glass block strikes the inside wall of the glass block and makes an angle of 58.0° with the glass–air boundary. The index of refraction of the glass is 1.52.

a Calculate the angle of incidence.

b Calculate the angle of refraction of the transmitted ray (assuming n_{air} = 1.00).

c Determine the angle of deviation (angle between the direction of the incident wave and the refracted wave).

d Calculate the speed of light in the glass.

14 Calculate the critical angle for light travelling between the following media.

	Incident medium	Refracting medium
a	ice (n = 1.31)	air (n = 1.00)
b	salt (n = 1.54)	air (n = 1.00)
c	cubic zirconia (n = 2.16)	air (n = 1.00)

15 A narrow beam of white light enters a crown glass prism with an angle of incidence of 30.0°. In the prism, the different colours of light are slowed to varying degrees. The refractive index for red light in crown glass is 1.50 and for violet light the refractive index is 1.53.

a Calculate the angle of refraction for the red light.

b Calculate the angle of refraction for the violet light.

c Determine the angle through which the spectrum is dispersed.

d Calculate the speed of the violet light in the crown glass. Use $c = 3.00 \times 10^8\,\text{m s}^{-1}$.

16 When a light wave refracts, the difference between the angle of incidence and angle of refraction is known as the *angle of deviation*. Sort the following boundaries between media in order of increasing angle of deviation.

A water (n = 1.33) to diamond (n = 2.42)

B water (n = 1.33) to air (n = 1.00)

C air (n = 1.00) to diamond (n = 2.42)

D glass (n = 1.50) to air (n = 1.00)

17 Two students find a piece of an unknown glass in the laboratory and want to determine its refractive index. They design an experiment in which they vary the angle of incidence and measure the refracted angle. The results obtained by the students are tabulated below.

Angle of incidence, θ_1 (°)	Angle of refraction, θ_2 (°)
0	0
10	4
20	10
30	17
40	25
50	27
60	32
70	33
80	35

a Plot a suitable graph and draw a line of best fit.

b Use the line of best fit to determine the refractive index of the material.

18 a A particular fibre optic cable has a core with refractive index n_1 = 1.557 and cladding with refractive index n_2 = 1.343. Calculate the speed of light in the core and in the cladding.

b Calculate the critical angle at the interface between the core and the cladding.

c State the range of angles at which total internal reflection will occur. State this angle relative to the surface of the core–cladding interface. This is shown as θ_2 in Figure 3.1.13(c) on page 89.

d A light wave travelling down the core hits the wall of the core at an angle of 15° (relative to the wall of the core). Does total internal reflection or refraction occur?

e Explain why the fibre-optic light wave signal loses intensity.

OA ✓✓

CHAPTER

04 Thermal energy

Thermal energy is part of our everyday experience—for example, we can see how some substances transfer heat better than others. A metal pan heats up quickly on a stove, yet the water in it takes much longer to reach boiling point. The air gap in double glazing acts as insulation to retain heat in your home, yet you can feel warmth from the Sun that has travelled across the near-vacuum of space.

In this chapter, you will learn about thermodynamic principles and thermal energy, including the difference between temperature and heat, and the different ways in which heat energy is transferred. You will also learn about the effects of heat transfer, including what happens to the energy of a substance when it is heated or changes state.

Key knowledge

- convert between Celsius and kelvin scales **4.1**
- describe how an increase in temperature corresponds to an increase in thermal energy (kinetic and potential energy of the atoms) of a system:
  - distinguish between conduction, convection and radiation with reference to heat transfers within and between systems **4.4**
  - explain why cooling results from evaporation using a simple kinetic energy model **4.2**
- investigate and analyse theoretically and practically the energy required to:
  - raise the temperature of a substance: $Q = mc\Delta T$ **4.2**
  - change the state of a substance: $Q = mL$. **4.3**

4.1 Heat and temperature

In the sixteenth century, Sir Francis Bacon, an English essayist and philosopher, proposed the radical idea that heat is motion. He went on to write that heat is the rapid vibration of tiny particles within every substance. At the time, his ideas were dismissed because the nature of particles wasn't fully understood. An opposing theory at the time was that heat was related to the movement of a fluid called 'caloric' that filled the spaces within a substance.

Today, it is understood that all matter is made up of small particles (atoms or molecules). Using this knowledge, it is possible to look more closely at what happens during heating processes.

KINETIC PARTICLE MODEL

This section starts by looking at the **kinetic particle model**, which states that the small particles (atoms or molecules) that make up all matter have kinetic energy. This means that all particles are in constant motion, even in extremely cold solids. It was thought centuries ago that if a material was continually made cooler, there would be a point at which the particles would eventually stop moving. This coldest possible temperature is called **absolute zero** and will be discussed later in this section.

Recall that a model is a representation that describes or explains the workings of an object, system or idea.

These are the assumptions behind the kinetic particle model:

- All matter is made up of many very small particles (atoms or molecules).
- The particles are in constant motion.
- No kinetic energy is lost or gained overall during collisions between particles.
- There are forces of attraction and repulsion between the particles in a material.
- The distances between particles in a gas are large compared with the size of the particles.

The kinetic particle model applies to the states (or phases) of matter: solids, liquids and gases.

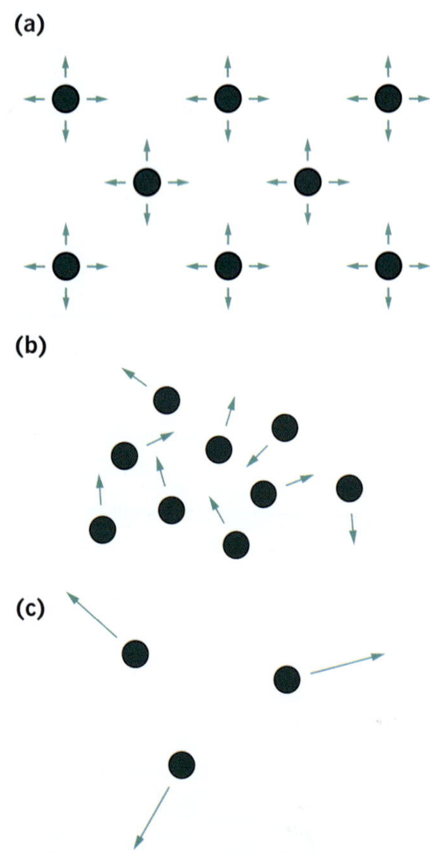

FIGURE 4.1.1 (a) Molecules in a solid have low kinetic energy and vibrate around average positions within a regular arrangement. (b) The particles in a liquid have more kinetic energy than those in a solid. They move more freely and will take the shape of the container. (c) Gas molecules are free to move in any direction.

Solids

Within a solid, the particles must be exerting attractive forces or bonds on each other for the matter to hold together in its fixed shape. There must also be repulsive forces, without which the attractive forces would cause the solid to collapse. In a solid, the attractive and repulsive forces hold these particles in more or less fixed positions, usually in a regular arrangement or lattice (Figure 4.1.1(a)). But the particles in a solid are not completely still; they vibrate around average positions. The forces on individual particles are sometimes predominantly attractive and sometimes repulsive, depending on their exact position relative to neighbouring particles.

Liquids

Within a liquid, there is still a balance of attractive and repulsive forces. Compared with a solid, the particles in a liquid have more freedom to move around each other and will therefore take the shape of the container (Figure 4.1.1(b)). Generally, the liquid takes up a slightly greater volume than it would in the solid state. Particles collide but remain attracted to each other, so the liquid remains within a fixed volume but with no fixed shape.

Gases

In a gas, particles are in constant, random motion, colliding with each other and the walls of the container. The particles move rapidly in every direction, quickly filling the volume of any container and occasionally colliding with each other (Figure 4.1.1(c)). A gas has no fixed volume. The particle speeds are high enough that, when the particles collide, the attractive forces are not strong enough to keep the particles close together. The repulsive forces cause the particles to separate and move off in other directions.

PHYSICSFILE

Energy

Energy is a very important concept in the study of the physical world, and is a focus in all areas of scientific study. Later chapters investigate energy in more detail.

Energy is a measure of an object's ability to do work. For example, raising the temperature of an object or lifting an object is referred to as 'doing work'. Work is measured in joules. The symbol for joule is J. The figure to the right shows the number of joules available from some energy sources.

Kinetic energy is the energy of movement. It is equal to the amount of work needed to bring an object from rest to its present speed or to return it to rest. Potential energy is stored energy. There are many forms of potential energy—for example, gravitational, nuclear, spring and chemical. Chemical potential energy is associated with the bonds between the particles within a substance. An increase in the potential energy of particles in a substance results in movement of the particles from their equilibrium positions.

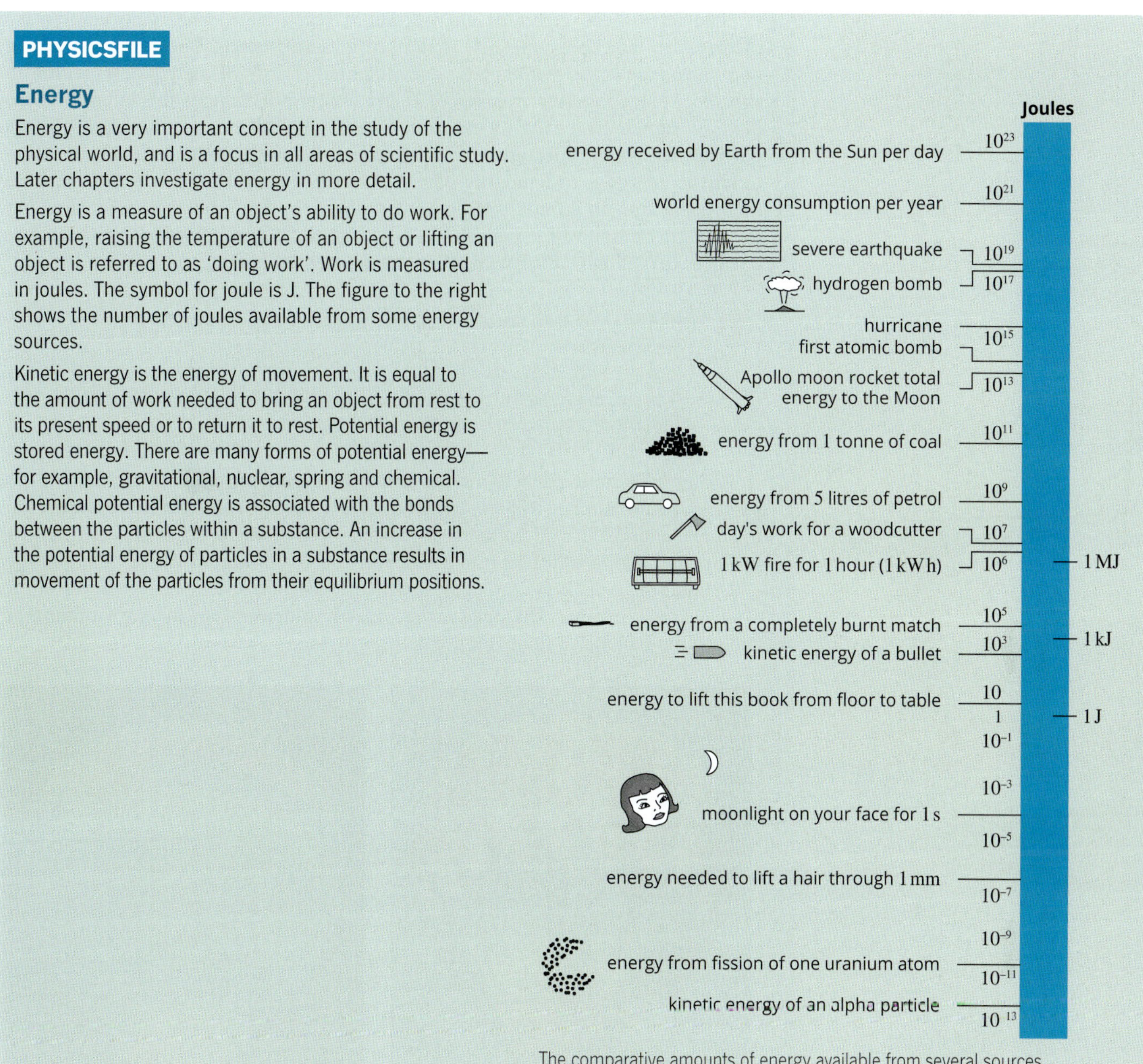

The comparative amounts of energy available from several sources

Plasmas

Plasma exists when matter is heated to very high temperatures and electrons are freed from atoms (ionisation). A gas that is ionised and has an equal number of positive ions and negative electrons is called a plasma. A plasma shares some properties with a gas (for example, they both have no fixed volume), but its charge gives it unique properties. Most of the matter in the universe is plasma.

Internal energy and temperature

The kinetic particle model can be used to explain the idea of heat as a transfer of energy. **Heat** (measured in joules) is the transfer of **thermal energy** from a hotter body to a colder one. Heating is observed by the change in **temperature**, the change of state or the expansion of a substance.

When a solid substance is 'heated', the particles within the material gain either **kinetic energy** (move faster) or **potential energy** (move away from their equilibrium positions).

The term 'heat' refers to energy that is being transferred (moved), so it is incorrect to talk about the heat *contained* in a substance. The **internal energy** of a substance is the total kinetic and potential energy of the particles within the substance. Heating (the transfer of thermal energy) changes the internal energy of a substance by changing the kinetic energy and/or potential energy of the particles within the substance. The movement of the particles in a substance due to kinetic energy is ordered: the particles move back and forth and we can model their behaviour. In comparison, the internal energy of a system is associated with the chaotic motion of the particles: it relates to the behaviour of a large number of particles that all have their own kinetic and potential energy.

- Heating is a process that always transfers thermal energy from a hotter substance to a colder substance.
- Heat is measured in joules (J).
- Temperature is related to the average kinetic energy of the particles in the substance. The faster the particles move, the higher the temperature of the substance.

Using the kinetic particle model, an increase in the total internal energy of the particles in a substance will result in an increase in temperature if there is a net gain in kinetic energy. Hot air balloons are an example of this process in action. The air in a hot air balloon is heated by a gas burner to a maximum of 120°C. The nitrogen (78%) and oxygen (21%) molecules in the hot air gain kinetic energy and so move a lot faster. The air in the balloon becomes less dense than the surrounding air, causing the balloon to float (Figure 4.1.2).

FIGURE 4.1.2 (a) Nitrogen and oxygen molecules gain energy when the air is heated, lowering the density of the air and causing the hot air balloon to rise off the ground. (b) A thermal image shows the temperature of the air inside the balloon from pink and red (hottest) to blue (coolest).

Sometimes heating results only in the change of state or the expansion of an object, and not a change in temperature. In such cases, the total internal energy of the particles has still increased, but only the potential energy has increased; the kinetic energy has not changed.

For example, particles in a solid that is being heated will continue to be mostly held in place, due to the relatively strong interparticle forces. For the substance to change state from solid to liquid, it must receive enough energy to separate the particles from each other and disrupt the regular arrangement of the solid. During this 'phase change' process, the energy is used to overcome the strong interparticle forces, but not to change the overall speed of the particles. In this situation, the temperature does not change. This will be discussed in more detail in Section 4.3 'Latent heat'.

MEASURING TEMPERATURE

Only four centuries ago, there were no thermometers and people described heating effects by vague terms such as 'hot', 'cold' and 'lukewarm'. In about 1593, Italian inventor Galileo Galilei made one of the first thermometers. His 'thermoscope' was not particularly accurate, as it did not take into account changes in air pressure, but it did suggest some basic principles for determining a suitable scale of measurement. His work suggested that there be two fixed points: the hottest day of summer and the coldest day of winter. A scale such as this is referred to as an arbitrary scale; zero is not its lowest point.

Celsius and Fahrenheit scales

Two of the better-known arbitrary temperature scales are the Fahrenheit and Celsius scales. Gabriel Fahrenheit of Germany invented the first mercury thermometer in 1714. Most countries use the Celsius scale to measure temperature; the Fahrenheit scale is used in the United States of America.

- 0 degrees Celsius (0°C) is the freezing point of water at standard atmospheric pressure.
- 100°C is the boiling point of water at standard atmospheric pressure.

Kelvin scale

Absolute scales are different from arbitrary scales. For a scale to be regarded as 'absolute', it should have no negative values—zero must be its lowest value. It must have fixed points that are reproducible. The triple point of water provides one such fixed point. This is the point at which the combination of temperature and air pressure allows all three states of water to coexist. For water, the triple point is only slightly above the standard freezing point (0.01°C) and provides a unique and repeatable temperature with which to adjust the Celsius scale.

The absolute or **kelvin** temperature scale is based on absolute zero and the triple point of water. See Figure 4.1.3 for a comparison of the kelvin and Celsius scales.

- The freezing point of water (0°C) is equivalent to 273.15 K (kelvin). This is often approximated to 273 K.
- The size of each unit, 1°C or 1 K, is the same.
- The word 'degrees' or the degree symbol are not used with the kelvin scale.
- To convert a temperature from degrees Celsius to kelvin, add 273.15.
- To convert a temperature from kelvin to degrees Celsius, subtract 273.15.

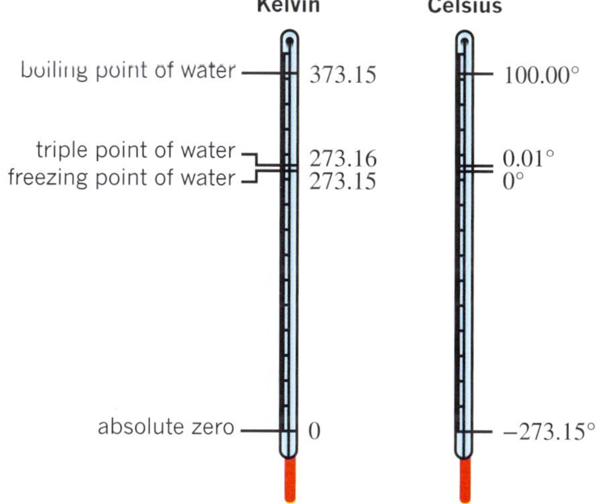

FIGURE 4.1.3 Comparison of the kelvin and Celsius scales. Note that there are no negative values on the kelvin scale.

PHYSICSFILE

Unit conventions in physics

The unit for energy, the joule, is named after James Joule in recognition of his work. When a unit is named after a person, its symbol is usually a capital letter, but the unit name is always lower case; for example, joule (J), newton (N), kelvin (K).

Exceptions are degrees Celsius (°C) and degrees Fahrenheit (°F), which also include a degree symbol.

Units not named after people usually have both the symbol and the name in lower case; for example, metre (m), second (s).

PHYSICSFILE

Close to absolute zero

As temperatures get close to absolute zero, atoms start to behave in strange ways. Since 1699, when the French physicist Guillaume Amontons first proposed the idea of an absolute lowest temperature, physicists have theorised about the effects of such a temperature and how it could be achieved. The laws of physics dictate that absolute zero itself can be approached but not reached. Very low temperature research is being carried out in the Cold Atom Laboratory on the International Space Station, where temperatures as low as 1 picokelvin ($1\,pK = 10^{-12}\,K$) can be reached. At this temperature, all elementary particles merge into a single state called a Bose–Einstein condensate, losing their separate properties and behaving as a single 'super atom'. This state was first proposed by Albert Einstein and Satyendra Nath Bose a century ago.

Absolute zero

Experiments indicate that there is a limit to how cold things can get. The kinetic theory suggests that if a given quantity of an ideal gas is cooled, its volume decreases. The volume can be plotted against temperature and results in a straight-line graph, as shown in Figure 4.1.4.

- Absolute zero = 0 K = −273.15°C.
- All molecular motion ceases at absolute zero. This is the coldest temperature possible.

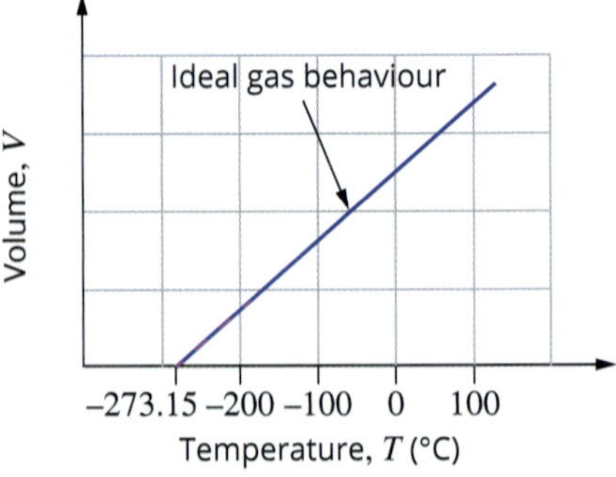

FIGURE 4.1.4 Gases have smaller volumes as they cool. This relationship is linear. Extrapolating (extending) the line to where the volume is zero gives a theoretical value of absolute zero.

THE LAWS OF THERMODYNAMICS

The topic of thermal physics involves the phenomena associated with energy transfer between objects at different temperatures. Since the nineteenth century, scientists have developed four laws for this subject. The first two will be studied in this section.

The zeroth law of thermodynamics

If objects A and B are each in thermal equilibrium with object C, then objects A and B are in thermal equilibrium with each other.

This is the zeroth law of thermodynamics.

The **zeroth law of thermodynamics** relates to **thermal equilibrium** and **thermal contact** and allows temperature to be defined. It was discovered after the first, second and third laws of thermodynamics, but was considered so important that it was decided to place it first.

If two objects are in thermal contact, energy can flow between them. For example, if an ice cube is placed in a copper pan, the ice molecules are in thermal contact with the copper atoms. Assuming that the copper is warmer than the ice, thermal energy will flow from the copper to the ice.

Thermal equilibrium occurs when two objects in thermal contact stop having a flow of energy between them. If frozen peas are placed in a container of warm water, energy is transferred from the water to the peas. The peas gain energy and warm up. The water loses energy and cools down. Eventually the transfer of energy between the water and the peas will stop. This point is called the thermal equilibrium, and the peas and water will be at the same temperature.

Two objects in thermal equilibrium with each other must be at the same temperature.

The first law of thermodynamics

The **first law of thermodynamics** states that energy simply changes from one form to another and the total internal energy of a system is constant. The internal energy of the system can be changed by heating or cooling, or by **work** being done on or by the system.

Any change in the internal energy (ΔU) of a system is equal to the energy added by heating ($+Q$) or removed by cooling ($-Q$), minus the work done on ($-W$) or by ($+W$) the system.

$\Delta U = Q - W$

The internal energy (U) of a system is defined as the total kinetic and potential energy of the system. As the average kinetic energy of a system is related to its temperature and the potential energy of the system is related to the state, then a change in the internal energy of a system means that either the temperature changes or the state changes.

If heat (Q) is added to the system, then the internal energy (U) rises by either increasing the temperature or changing state from solid to liquid or liquid to gas. Similarly, if work (W) is done on a system, then the internal energy rises and the system will once again increase in temperature or will change state by melting or boiling. When heat is added to a system or work is done on a system, ΔU is positive.

If heat (Q) is removed from the system, then the internal energy (U) decreases by either decreasing temperature or changing state from liquid to solid or gas to liquid. Similarly, if work (W) is done by the system, then the internal energy decreases and the system will once again decrease in temperature or change state by condensing or solidifying. When heat is removed from a system or work is done by a system, ΔU is negative.

If heat is added to the system and work is done by the system, then whether the internal energy increases or decreases will depend on the magnitude of the energy into the system compared to the magnitude of the energy out of the system.

The equation for the first law of thermodynamics can be rearranged to solve for the total heat (Q) or the work done (W).

$Q = \Delta U + W$

or

$W = Q - \Delta U$

Worked example 4.1.1

CALCULATING THE CHANGE IN INTERNAL ENERGY

A 1 L beaker of water has 25 kJ of work done on it. It also loses 30 kJ of thermal energy to the surroundings. Calculate the change in the internal energy of the water.

Thinking	Working
Heat is removed from the system, so Q is negative. Work is done on the system, so W is negative.	$\Delta U - Q - W$ $= -30 - (-25)$
Note that the units are kJ, so express the final answer in kJ.	$\Delta U = -5$ kJ

Worked example: Try yourself 4.1.1

CALCULATING THE CHANGE IN INTERNAL ENERGY

A student places a heating element and a paddle-wheel apparatus in an insulated container of water. She calculates that the heater transfers 2530 J of thermal energy to the water and that the paddle does 240.0 J of work on the water. Calculate the change in internal energy of the water.

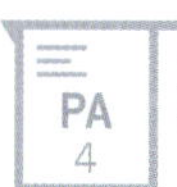

4.1 Review

OA ✓✓

SUMMARY

- The kinetic particle model proposes that all matter is made of atoms or molecules (particles) that are in constant motion.
- In solids, the attractive and repulsive forces hold the particles in more or less fixed positions, usually in a regular arrangement or lattice. These particles are not completely still—they vibrate about average positions.
- In liquids, there is still a balance of attractive and repulsive forces between particles but the particles have more freedom to move around. Liquids maintain a fixed volume.
- In gases, the particle speeds are high enough that, when particles collide, the attractive forces are not strong enough to keep them close together. The repulsive forces cause the particles to move off in other directions.
- Internal energy refers to the total kinetic and potential energy of the particles within a substance.
- Temperature is related to the average kinetic energy of the particles in a substance.
- Heating is a process that always transfers thermal energy from a hotter substance to a colder substance.
- Temperatures can be measured in degrees Celsius (°C) or in kelvin (K).
- Absolute zero is called simply 'zero kelvin' (0 K) and it is equal to −273.15°C.
- The size of each unit, 1°C or 1 K, is the same.
- To convert from Celsius to kelvin, add 273.15; to convert from kelvin to Celsius, subtract 273.15.
- The zeroth law of thermodynamics states that if objects A and B are each in thermal equilibrium with object C, then objects A and B are in thermal equilibrium with each other. A, B and C must be at the same temperature.
- The first law of thermodynamics states that energy simply changes from one form to another and the total energy in a system is constant.
- Any change in the internal energy (ΔU) of a system is equal to the energy added by heating ($+Q$) or removed by cooling ($-Q$), minus the work done on ($-W$) or by ($+W$) the system: $\Delta U = Q - W$.

KEY QUESTIONS

Knowledge and understanding

1 Which of the following is true of a solid?
  - **A** Particles are moving around freely.
  - **B** Particles are not moving.
  - **C** Particles are vibrating in constant motion.
  - **D** A solid is not made up of particles.

2 An uncooked chicken is placed into an oven that has been preheated to 180°C. The chicken is left in the oven for an hour, after which time it has a temperature of 180°C.
  - **a** Which of the following statements describe what happens as soon as the chicken is placed in the oven? (More than one answer is possible.)
    - **A** Thermal energy flows from the chicken into the hot air.
    - **B** The chicken and the air in the oven are in thermal equilibrium.
    - **C** Thermal energy flows from the hot air into the chicken.
    - **D** The chicken and the air in the oven are not in thermal equilibrium.
  - **b** Which of the following statements best describes what is happening at the end of the hour before the chicken is removed from the oven?
    - **A** Thermal energy is flowing from the chicken into the hot air.
    - **B** The chicken and the air in the oven are in thermal equilibrium.
    - **C** Thermal energy is flowing from the hot air into the chicken.
    - **D** The chicken and the air in the oven are not in thermal equilibrium

3 Which of the following temperature(s) cannot possibly exist? (More than one answer is possible.)
  - **A** 1 000 000°C
  - **B** −50°C
  - **C** −50 K
  - **D** −300°C

4 Covert the following temperatures:
  - **a** 30°C into kelvin
  - **b** 375 K into degrees Celsius

5 Tank A is filled with hydrogen gas at 0°C and tank B is filled with hydrogen gas at 300 K. Describe the difference between the average kinetic energy of the hydrogen particles in the two tanks.

Analysis

6 A hot block of iron is placed onto one side of a balance. The block pushes the balance down, doing 50 kJ of work on it. The block of iron also transfers 20 kJ of heat energy to the air and balance. Calculate the change in energy (in kJ) of the iron block.

7 A chef vigorously stirs a pot of cold water and does 150 J of work on the water. The water also gains 80.0 J of thermal energy from the surroundings. Calculate the change in energy of the water.

8 A scientist very carefully does mechanical work on a container of liquid sodium. The liquid sodium loses 350 J of energy to its surroundings but gains 250 J of energy overall. How much work did the scientist do?

4.2 Specific heat capacity

A small amount of water in a kettle will experience a greater change in temperature than a larger amount of water if heated for the same time. A large heater warms a room faster than a small one. A metal object left in the sunshine gets hotter faster than a wooden object.

These simple observations suggest that the mass, the amount of energy transferred and the material of the object influence any change of temperature.

CHANGING TEMPERATURE

The temperature of a substance is a measure of the average kinetic energy of the particles inside the substance. To increase the temperature of the substance, the kinetic energy of its particles must increase. This happens when heat is transferred to that substance. The amount by which the temperature increases depends on a number of factors.

The greater the mass of a substance, the greater the energy required to change the kinetic energy of all the particles. So, the heat required to raise the temperature by a given amount is proportional to the mass of the substance:

$$\Delta Q \propto m$$

where ΔQ is the heat energy transferred in joules (J)
m is the mass of material being heated in kilograms (kg).

The more heat that is transferred to a substance, the more the temperature of that substance increases. The amount of energy transferred is therefore proportional to the change in temperature:

$$\Delta Q \propto \Delta T$$

where ΔT is the change in temperature in °C or K.

Heating experiments using different materials will confirm that these relationships hold true regardless of the material being heated. However, heating the same masses of different materials will show that the amount of energy required to heat a given mass of a material through a particular temperature change also depends on the nature of the material being heated. For example, a volume of water requires more energy to change its temperature by a given amount than the same volume of methylated spirits. For some materials, temperature change occurs more easily than for others. This is called the **specific heat capacity** of the material.

> The specific heat capacity of a material, c, is the amount of energy that must be transferred to change the temperature of 1 kg of the material by 1°C or 1 K.

Combining these observations, the amount of energy added to or removed from the substance is proportional to the change in its temperature, its mass and its specific heat capacity.

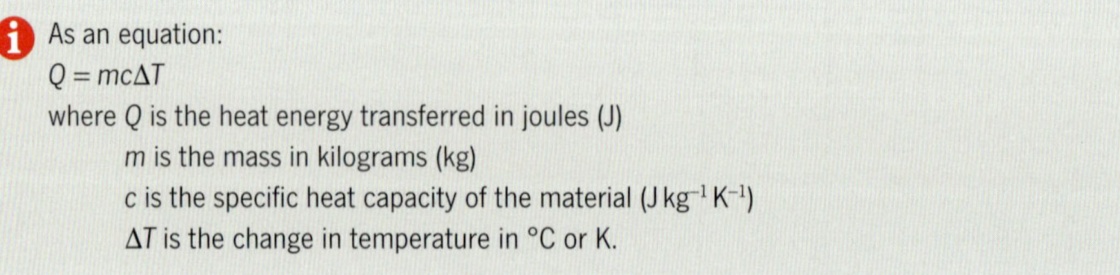

As an equation:
$Q = mc\Delta T$
where Q is the heat energy transferred in joules (J)
m is the mass in kilograms (kg)
c is the specific heat capacity of the material (J kg^{-1} K^{-1})
ΔT is the change in temperature in °C or K.

This is the case as long as the material does not change state; the specific heat capacity of a material changes when the material changes state.

Table 4.2.1 lists the specific heat capacities for some common materials. You can see that it also lists the average value for the human body, which takes into account the various materials within the body and the percentage that each material contributes to the body's total mass.

PHYSICSFILE

The mass of water

Since water is a familiar material, many of the examples in this section use it as the liquid being heated. One kilogram of pure water has a volume of one litre at 4°C.

TABLE 4.2.1 Approximate specific heat capacities of common substances

Material	c (J kg^{-1}K^{-1})
human body	3500
methylated spirits	2500
air	1000
aluminium	900
glass	840
iron	440
copper	390
brass	370
lead	130
mercury	140
ice (water)	2100
liquid water	4200
steam (water)	2000

Worked example 4.2.1

CALCULATIONS USING SPECIFIC HEAT CAPACITY

A hot water tank contains 135 L of water. Initially the water is at 20.0°C. Calculate the amount of energy that must be transferred to the water to raise the temperature to 70.0°C.

Thinking	Working
Calculate the mass of water. 1 L of water = 1 kg	Volume = 135 L So mass of water = 135 kg
ΔT = final temperature − initial temperature	$\Delta T = 70.0 - 20.0 = 50.0$°C
From Table 4.2.1, $c_{water} = 4200\ \text{J kg}^{-1}\text{K}^{-1}$. Use the equation $Q = mc\Delta T$.	$Q = mc\Delta T$ $= 135 \times 4200 \times 50.0$ $= 28\,350\,000$ J $= 2.84 \times 10^7$ J

Worked example: Try yourself 4.2.1

CALCULATIONS USING SPECIFIC HEAT CAPACITY

A bath contains 75.0 L of water. Initially the water is at 50.0°C. Calculate the amount of energy that must be transferred from the water to cool the bath to 30.0°C.

Worked example 4.2.2

COMPARING SPECIFIC HEAT CAPACITIES

Different states of matter of the same substance have different specific heat capacities. What is the ratio of the specific heat capacity of liquid aluminium ($c_{liquid} = 1180\ \text{J kg}^{-1}\text{K}^{-1}$) to that of solid aluminium ($c_{solid} = 900\ \text{J kg}^{-1}\text{K}^{-1}$)?

Thinking	Working
State the specific heat capacities given in the question.	$c_{liquid} = 1180\ \text{J kg}^{-1}\text{K}^{-1}$ $c_{solid} = 900\ \text{J kg}^{-1}\text{K}^{-1}$
Divide the specific heat capacity of liquid aluminium by the specific heat capacity of solid aluminium.	Ratio $= \frac{c_{liquid}}{c_{solid}}$ $= \frac{1180}{900}$
Note that ratios have no units because the unit of each quantity is the same and cancels out.	Ratio = 1.3

Worked example: Try yourself 4.2.2

COMPARING SPECIFIC HEAT CAPACITIES

What is the ratio of the specific heat capacity of liquid water to that of steam? Refer to Table 4.2.1 for the specific heat capacities of water in different states.

PHYSICSFILE

Specific heat capacity of water

One of the notable values in Table 4.2.1 of specific heat capacities is the high value for water. It is 10 times, or an order of magnitude, higher than those of most metals listed. The specific heat capacity of water is higher than those of most common materials. As a result, water makes a very useful cooling and heat storage agent and is used in applications such as generator cooling towers and car-engine radiators.

Life on Earth also depends on the specific heat capacity of water. About 70% of Earth's surface is covered by water, and this water can absorb large quantities of thermal energy without great changes in temperature. Oceans both heat up and cool down more slowly than the land areas next to them. This helps to maintain a relatively stable range of temperatures for life on Earth.

Scientists are monitoring the temperatures of the oceans, both at the surface and in the deep ocean, as part of research on global warming. They use a variety of methods to measure the temperature, including a fleet of robots floating at different depths known as Argo floats. For hard-to-reach places in Antarctica, they have fitted seals with instruments that measure temperature and salinity.

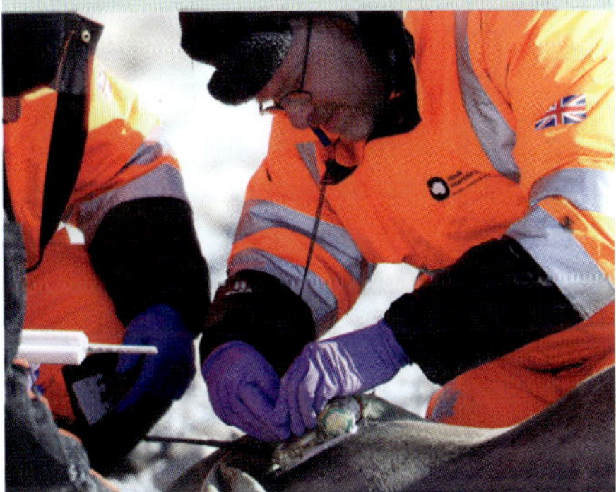

A seal is fitted with a tracking device to measure water temperatures and salinity.

4.2 Review

SUMMARY

- When heat is transferred to or from a system or object, the temperature change depends upon the amount of energy transferred, the mass of the material(s) and the specific heat capacity of the material(s): $Q = mc\Delta T$

  where Q is the heat energy transferred in joules (J)

  m is the mass of material being heated in kilograms (kg)

  c is the specific heat capacity of the material ($\text{J kg}^{-1}\,\text{K}^{-1}$)

  ΔT is the change in temperature (°C or K).

- A substance will have different specific heat capacities at different states (solid, liquid, gas).

KEY QUESTIONS

Knowledge and understanding

1 Which one or more of the following statements about specific heat capacity is true?

A All materials have the same specific heat capacity when in solid form.

B The specific heat capacity of the liquid form of a material is different from that of its solid and gas forms.

C Good conductors of heat generally have high specific heat capacities.

D Specific heat capacity is independent of temperature.

2 Equal masses of aluminium and copper are heated through the same temperature range. Using the values of c from Table 4.2.1 on page 110, which material requires the more energy to achieve this result?

3 Which has more thermal energy: 10 kg of iron at 20°C or 10 kg of aluminium at 20°C?

Analysis

4 To make a cup of tea, Sam heats one cup (250 mL) of water from room temperature at 25.0°C to 100°C. How much energy is transferred to the water to achieve this temperature change?

5 For a 1 kg block of aluminium, x J of energy are needed to raise the temperature by 10°C. How much energy, in J, is needed to raise the temperature by 20°C?

6 Equal amounts of energy are absorbed by equal masses of aluminium and water. What is the ratio of the temperature rise of the aluminium to that of water?

7 A cup holds 250 mL of water at 20°C. What temperature does the water reach after 10.5 kJ of heat energy is transferred to the water?

8 A block of iron is left to cool for a short time. When 13.2 kJ of energy has been transferred away from the block of iron, its temperature has decreased by 30°C. What is the mass of the block of iron?

4.3 Latent heat

If water is heated, its temperature will rise. If enough energy is transferred to the water, eventually the water will boil. The water changes state (from liquid to gas). The **latent heat** is the energy released or absorbed during a change of state. 'Latent' means 'hidden or unseen', so latent heat is the 'hidden' energy that has to be added or removed from a material in order for the material to change state.

While a substance is changing state, its temperature remains constant. The energy used in, say, melting ice into water is hidden in the sense that the temperature doesn't rise while the change of state is occurring.

ENERGY AND CHANGE OF STATE

Look at the heating curve for water shown in Figure 4.3.1. This graph shows how the temperature of water changes as energy is added at a constant rate. Although the rate at which the energy is added is constant, the increase in temperature is not always constant. There are sections of increasing temperature, and sections where the temperature remains unchanged (the horizontal sections) while the material changes state. Temperature remains constant during the change in state from ice to liquid water and again from liquid water to steam.

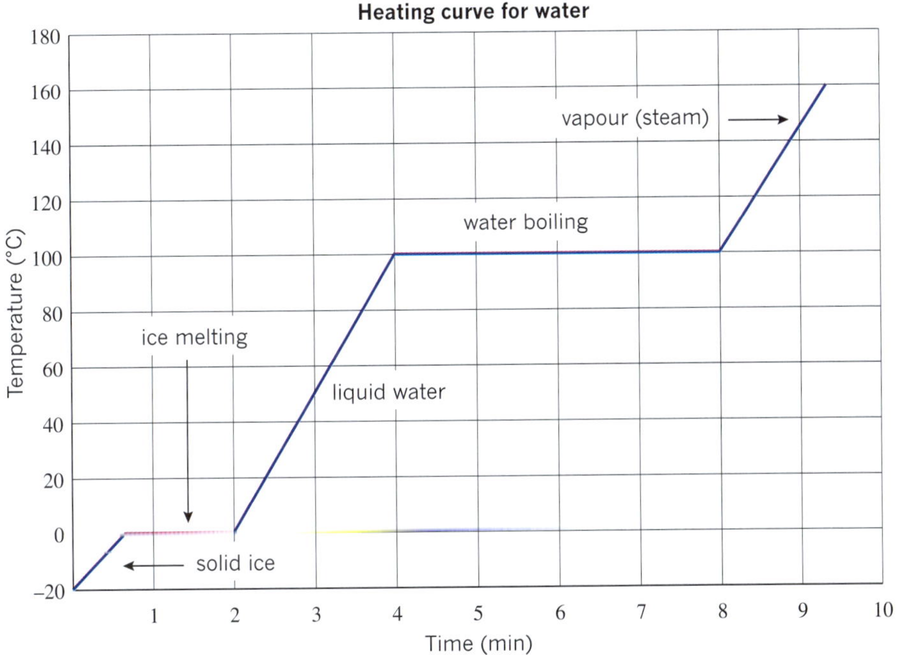

FIGURE 4.3.1 A heating curve for water

Calculating latent heat

The latent heat is calculated using the equation:

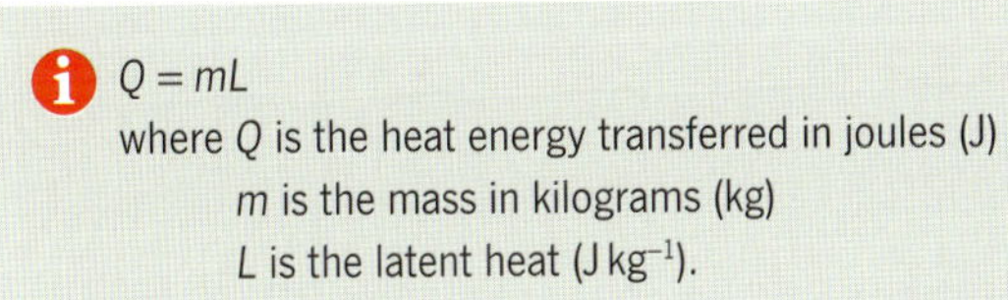

$Q = mL$

where Q is the heat energy transferred in joules (J)

m is the mass in kilograms (kg)

L is the latent heat (J kg^{-1}).

Latent heat of fusion (melting)

As thermal energy is transferred to a solid, the temperature of the solid increases. The particles within the solid gain internal energy (as kinetic energy and some potential energy) as their speed of vibration increases. At the point where the solid begins to melt, the particles move further apart, overcoming the attraction to their neighbouring particles that is holding them in place. At this point, instead of raising the temperature, the extra energy increases the potential energy of the particles, reducing the interparticle or intermolecular forces. No change in temperature occurs as all the extra energy supplied is used in reducing these forces between particles.

The amount of energy required to melt a solid is exactly the same as the amount of potential energy released when the liquid re-forms into a solid. It is called the **latent heat of fusion**.

The amount of energy required will depend on the particular substance.

For a given mass of a substance:

heat energy transferred = mass of substance × specific latent heat of fusion.

$$Q = mL_{fusion}$$

where Q is the heat energy transferred in joules (J)

m is the mass in kilograms (kg)

L_{fusion} is the latent heat of fusion in $J\,kg^{-1}$.

It takes almost 80 times as much energy to turn 1 kg of ice into water (with no temperature change) as it does to raise the temperature of 1 kg of water by 1°C. It takes a lot more energy to overcome the large intermolecular forces within the ice than it does to simply add kinetic energy in raising the temperature.

The latent heats of fusion for some common materials are shown in Table 4.3.1.

TABLE 4.3.1 The latent heats of fusion for some common materials

Substance	Melting point (°C)	L_{fusion} ($J\,kg^{-1}$)
water	0	3.34×10^5
oxygen	−219	0.139×10^5
lead	327	0.230×10^5
ethanol	−130	1.08×10^5
silver	961	1.05×10^5

Worked example 4.3.1

LATENT HEAT OF FUSION

How much energy must be removed from 2.50 L of water at 0°C to produce a block of ice at 0°C?

Thinking	Working
Calculate the mass of water involved.	1 L of water = 1 kg, so 2.50 L = 2.50 kg
Cooling from liquid to solid at the same temperature involves the latent heat of fusion, as the energy is removed from the water. Use Table 4.3.1 to find the latent heat of fusion for water.	$L_{fusion} = 3.34 \times 10^5\,J\,kg^{-1}$
Use the equation $Q = mL_{fusion}$.	$Q = mL_{fusion}$ $= 2.50 \times 3.34 \times 10^5$ $= 8.35 \times 10^5\,J$

Worked example: Try yourself 4.3.1

LATENT HEAT OF FUSION

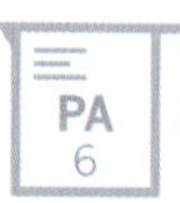

How much energy must be removed from 5.50 kg of liquid lead at 327°C to produce a block of solid lead at 327°C?

Latent heat of vaporisation (boiling)

It takes much more energy to convert a liquid to a gas than it does to convert a solid to a liquid. This is because to convert a liquid to a gas requires the intermolecular bonds to be broken. During the change of state, the energy supplied is used solely in overcoming intermolecular bonds. The temperature will not rise until all of the material in the liquid state has been converted to a gas, assuming that the liquid is evenly heated. For example, when liquid water is heated to boiling point, a large amount of energy is required to change its state from liquid to steam (gas). The temperature will remain at 100°C until all of the water has turned into steam. Once the water has been completely converted to steam, then the temperature can start to rise again.

The amount of energy required to change a liquid to a gas is exactly the same as the potential energy released when the gas returns to a liquid. It is called the **latent heat of vaporisation**.

The amount of energy required will depend on the particular substance.

For a given mass of a substance:

heat energy transferred = mass of substance × specific latent heat of vaporisation

$$Q = mL_{\text{vapour}}$$

where Q is the heat energy transferred in joules (J)

m is the mass in kilograms (kg)

L_{vapour} is the latent heat of vaporisation (J kg^{-1}).

Note that, in just about every case, the latent heat of vaporisation of a substance will be different from the latent heat of fusion for that substance. Some values for latent heat of vaporisation are listed in Table 4.3.2.

In many instances, it is necessary to consider the energy required to heat a substance and also change its state. Problems like this are solved by considering the rise in temperature separately from the change of state.

TABLE 4.3.2 The latent heat of vaporisation of some common materials

Substance	Boiling point (°C)	L_{vapour} (J kg^{-1})
water	100	22.5×10^5
oxygen	−183	2.1×10^5
lead	1750	8.7×10^5
ethanol	78	8.7×10^5
silver	2162	23.9×10^5

Worked example 4.3.2

CHANGE IN TEMPERATURE AND STATE

50.0 mL of water is heated from a room temperature of 20.0°C to its boiling point at 100°C. It is boiled at this temperature until it is completely evaporated. How much energy in total was required to raise the temperature and boil the water?

Thinking	Working
Calculate the mass of water involved.	50.0 mL of water = 0.0500 kg
Find the specific heat capacity of water (see Table 4.2.1 on page 110).	$c = 4200\ \text{J kg}^{-1}\ \text{K}^{-1}$
Use the equation $Q = mc\Delta T$ to calculate the heat energy required to change the temperature of water from 20.0°C to 100.0°C.	$Q = mc\Delta T$ $= 0.0500 \times 4200 \times (100 - 20.0)$ $= 16\,800\ \text{J}$
Find the latent heat of vaporisation of water from Table 4.3.2.	$L_{\text{vapour}} = 22.5 \times 10^5\ \text{J kg}^{-1}$
Use the equation $Q = mL_{\text{vapour}}$ to calculate the latent heat required to boil the water.	$Q = mL_{\text{vapour}}$ $= 0.0500 \times 22.5 \times 10^5$ $= 112\,500\ \text{J}$
Find the total energy required to raise the temperature and change the state of the water.	Total $Q = 16\,800 + 112\,500$ $= 129\,300\ \text{J}$ $= 1.29 \times 10^5\ \text{J}$

Worked example: Try yourself 4.3.2

CHANGE IN TEMPERATURE AND STATE

3.00 L of water is heated from a fridge temperature of 4.00°C to its boiling point at 100°C. It is boiled at this temperature until it is completely evaporated. How much energy in total was required to raise the temperature and boil the water?

CASE STUDY ANALYSIS

Extinguishing fire

Water is often used by fire fighters to extinguish a fire. As seen in Section 4.2 (page 110) and Table 4.3.2 (page 115), the specific heat capacity and latent heat of vaporisation of water are both very high, meaning that water can absorb vast amounts of thermal energy before it evaporates. This is due to the molecular structure of water. This characteristic of water makes it very useful for extinguishing fires. By pouring water onto a fire, energy is transferred away from the fire to heat the water. Then, even more (in fact much more) heat is transferred away from the fire to convert the water into steam.

Water isn't suitable to use on every type of fire, however. If you search online for 'water on oil fire', you'll see the dramatic consequences of using water on an oil fire. This is because the temperature of burning oil is much higher than the boiling point of water. The water would rapidly change state and expand, propelling oil over a large area and making the fire flare dangerously.

Carbon dioxide is often used in fire extinguishers like the ones you might see in your school. These are suitable for putting out flammable liquid and electrical fires. The carbon dioxide in the extinguisher is a mixture of liquid and gaseous states under very high pressure. When it is sprayed on the fire, the carbon dioxide displaces the oxygen and thus smothers the fire. It also expands rapidly as it leaves the extinguisher, changing completely to a gas, and in the process taking thermal energy from the fire as it changes state.

Analysis

To understand how water is used to extinguish a fire, analyse the energy absorbed by 1.00 kg of water after it leaves the nozzle of the firefighter's hose.

1. The water leaves the hose at approximately 10.0°C and enters the room full of hot fire gases at 600°C. The water nearly instantly heats to 100°C. Calculate the energy absorbed by the mass of water.
2. When the water reaches 100°C it is still cooler than the surrounding gases. Which law of thermodynamics determines the direction of heat transfer? For this situation, describe the direction of heat transfer.
3. The mass of water turns to water vapour. Calculate the energy transfer that occurs to convert the 1.00 kg of liquid water at 100°C to water vapour at 100°C.
4. The water vapour disperses through the fire gases. The temperature of the water vapour will continue to rise until thermal equilibrium is reached. Assume the fire gases cool through this process to 300°C. Calculate the energy absorbed by the water vapour as it is heated from 100°C to 300°C.
5. Calculate the total energy the 1.00 kg mass of water has absorbed since leaving the firefighter's hose.

EVAPORATION AND COOLING

If you spill some water on the floor and then come back in a couple of hours, the water will probably be gone. It will have evaporated. It has changed from a liquid into a vapour at room temperature in a process called **evaporation**. The reason for this is that the water particles, if they have sufficient energy, are able to escape through the surface of the liquid into the air. Over time, no liquid remains.

Evaporation is more noticeable with **volatile** liquids such as methylated spirits, mineral turpentine, perfume and liquid paper. The surface bonds are weaker in these liquids and they evaporate rapidly. This is why you should never leave the lid off bottles of these liquids. They are often stored in narrow-necked bottles for this reason.

The rate of evaporation of a liquid depends on:

- the volatility of the liquid—more-volatile liquids evaporate faster
- the surface area—greater evaporation occurs when greater surface areas are exposed to the air
- the temperature—hotter liquids evaporate faster
- the humidity—less evaporation occurs in more humid conditions
- air movement—if a breeze is blowing over the liquid's surface, evaporation is more rapid.

Whenever evaporation occurs, higher-energy particles escape the surface of the liquid, leaving the lower-energy particles behind, as is shown in Figure 4.3.2. As a result, the average kinetic energy of the particles remaining in the liquid drops and the temperature decreases. Humans use this cooling principle when sweating to stay cool. When rubbing alcohol is dabbed on your arm before an injection, the cooling of the volatile liquid numbs your skin.

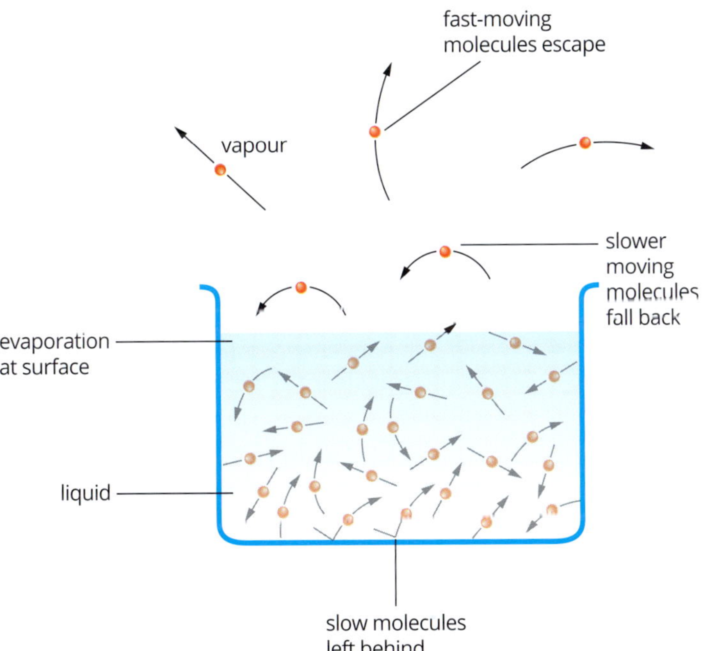

FIGURE 4.3.2 Fast-moving molecules with high kinetic energy can escape the liquid, leaving molecules with lower kinetic energy behind.

4.3 Review

OA ✓✓

SUMMARY

- When a solid material changes state, energy is needed to separate the particles by overcoming the attractive forces between the particles.
- Latent heat is the energy required to change the state of 1 kg of material at a constant temperature.
- In general, for any mass of material the energy required (or released) is $Q = mL$

  where Q is the energy transferred in joules (J)

  m is the mass in kilograms (kg)

  L is the latent heat ($J\,kg^{-1}$).
- The latent heat of fusion, L_{fusion}, is the energy required to change 1 kg of a material between the solid and liquid states.
- The latent heat of vaporisation, L_{vapour}, is the energy required to change 1 kg of a material between the liquid and gaseous states.
- The latent heat of fusion of a material will be different from (and usually less than) the latent heat of vaporisation for that material.
- Evaporation is the process in which a liquid turns into gas at room temperature. The temperature of the liquid falls as this occurs.
- The rate of evaporation depends on the volatility, temperature and surface area of the liquid and the presence of a breeze.

KEY QUESTIONS

Refer to the values in Table 4.3.1 on page 114 and Table 4.3.2 on page 115. You may also need to refer to Table 4.2.1 on page 110.

Knowledge and understanding

1 The graph below represents the heating curve for mercury, a metal that is a liquid at normal room temperature. Thermal energy is added to 10 g of solid mercury, initially at a temperature of –39.0°C, until all of the mercury has evaporated.

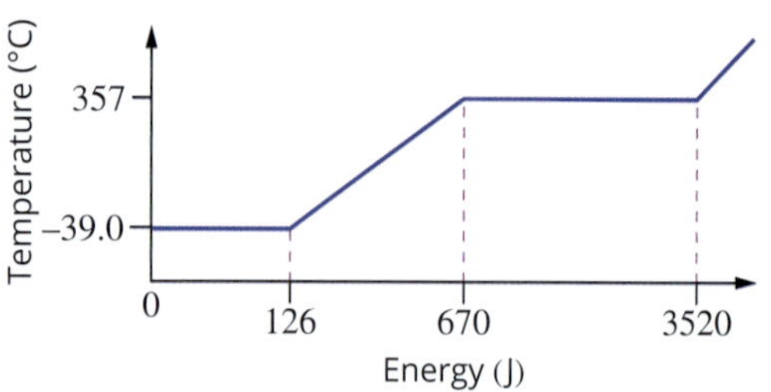

a Why does the temperature remain constant during the first part of the graph?

b What is the melting point of mercury, in degrees Celsius?

c What is the boiling point of mercury, in degrees Celsius?

d From the graph, calculate the latent heat of fusion of mercury.

e From the graph, calculate the latent heat of vaporisation of mercury.

2 Which of the following explains why hot water in a spa pool evaporates more rapidly than cold water?

A Hot water molecules have less energy than cold water molecules.

B Hot water molecules have more energy than cold water molecules.

C Hot water forms water vapour and bubbles to the surface.

D Hot water dissolves into the pool material more rapidly.

3 Describe the ideal kind of weather for drying clothes on the line, and explain why this is the case.

Analysis

4 How much heat energy must be transferred away from 100 g of steam at 100°C to change it completely to a liquid?

5 How many kJ of energy are required to melt exactly 100 g of ice initially at –4.00°C? Assume no loss of energy to the surroundings.

6 Use the kinetic particle model to differentiate between water ice, liquid water and water vapour.

4.4 Conduction, convection and radiation

In your home in winter, the room might feel warm from hydronic-heating radiators, you might feel a cool breeze from a gap under the front door, or have cold feet from walking on a tiled floor. These are all types of heat transfer.

In Section 4.1 you learnt that if two objects are at different temperatures and are in thermal contact (that is, they can exchange energy via heat processes), then thermal energy will transfer from the hotter object to the cooler object. This is the zeroth law of thermodynamics.

There are three possible means by which heat can be transferred:

- conduction
- convection
- radiation.

CONDUCTION

Figure 4.4.1 shows how, by preventing the chick's thermal contact with the cold ice, this adult penguin is able to protect its vulnerable offspring.

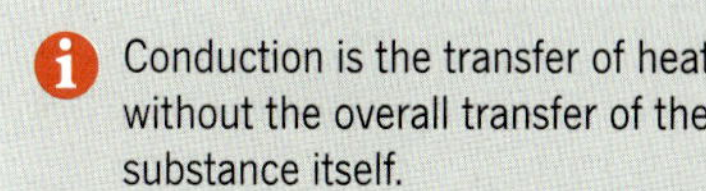

Conduction is the process by which heat is transferred from one place to another without the net movement of particles (atoms or molecules). Conduction can occur within a material or between materials that are in thermal contact. For example, if one end of a steel rod is placed in a fire, heat will travel along the rod so that the far end of the rod will also heat up; or if a person holds an ice cube, then heat will travel from their hand to the ice.

FIGURE 4.4.1 Emperor penguin chicks avoid heat loss through conduction by sitting on the adult's feet. In this way they avoid contact with the ice.

Although all materials will conduct heat to some extent, this process is most significant in solids. It is important in liquids but plays a lesser role in the movement of energy in gases.

Materials that conduct heat readily are referred to as good **conductors**. Materials that are poor conductors of heat are referred to as **insulators**. An example of a good conductor and a good insulator can be seen in Figure 4.4.2.

In secondary physics, the terms 'conductor' and 'insulator' are used in the context of both electricity and heating processes. What makes a material a good conductor of heat doesn't necessarily make it a good conductor of electricity. The two types of conduction are related, but it's important not to confuse the two processes. A material's ability to conduct heat depends on how conduction occurs within the material.

Conduction can happen in two ways:

- energy transfer through molecular or atomic collisions
- energy transfer by free electrons.

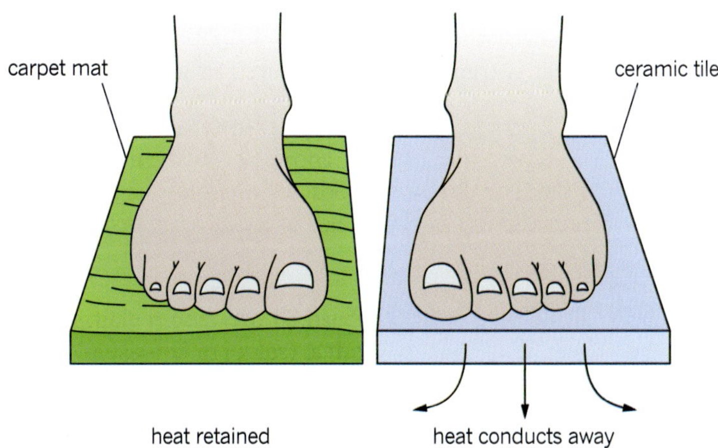

FIGURE 4.4.2 Ceramic floor tiles are good conductors of heat. They conduct heat away from the foot readily and so your feet feel cold on tiles. The carpet mat is a thermal insulator. Thermal energy from the foot is not transferred away as quickly and so your feet don't feel as cold.

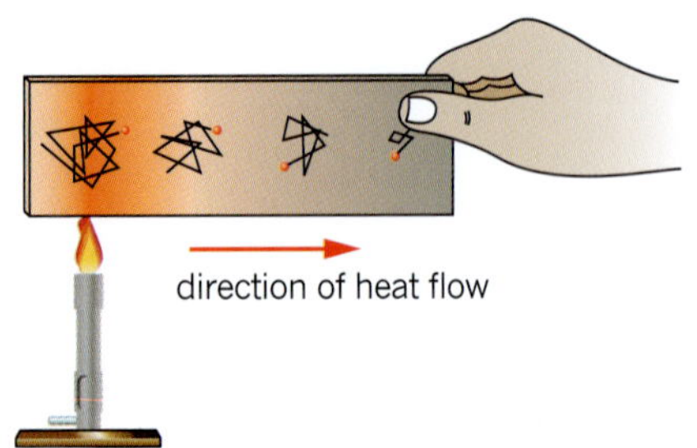

FIGURE 4.4.3 Thermal energy is passed on by collisions between adjacent particles.

Thermal transfer by collision

The kinetic particle model explains that particles in a solid substance are constantly vibrating within the material structure and so interact with neighbouring particles. If one part of the material is heated, then the particles in that region will vibrate more rapidly. Interactions with neighbouring particles will pass on this kinetic energy throughout the system via the bonds between the particles (Figure 4.4.3).

The process can be quite slow since the mass of the particles is relatively large and the vibrational velocities are fairly low. Materials for which this method of conduction is the only means of heat transfer are likely to be poor conductors of heat or even thermal insulators. Materials such as glass, wood and paper are poor conductors of heat.

Thermal transfer by free electrons

Some materials, particularly metals, have electrons that are not directly involved in any one particular chemical bond. Therefore, these electrons are free to move throughout the lattice of positive ions.

For example, if a metal is heated, then not only will the positive ions within the metal gain extra energy but so will these free electrons. As an electron's mass is considerably less than that of the positive ions, even a small energy gain will result in a very large gain in velocity. Consequently, these free electrons provide a means by which heat can be quickly transferred throughout the whole of the material. It is therefore no surprise that metals, which are good electrical conductors because of these free electrons, are also good thermal conductors.

Thermal conductivity

Thermal conductivity describes the ability of a material to conduct heat. It is temperature dependent and is measured in watts per metre per kelvin ($W\,m^{-1}\,K^{-1}$). Table 4.4.1 highlights the difference in conductivity between metals and other substances.

TABLE 4.4.1 Thermal conductivities of some common materials

Material	Conductivity ($W\,m^{-1}\,K^{-1}$)
silver	420
copper	380
aluminium	240
steel	60
ice	2.2
brick, glass	≈ 1
concrete	≈ 1 (depending on composition)
water	0.6
human tissue	0.2
rubber	≈ 0.5
wood	0.15
polystyrene	0.08
paper	0.06
fibreglass	0.04
air	0.025

Factors affecting thermal conduction

The rate at which heat will be transferred through a system depends on the:

- nature of the material. The larger a material's thermal conductivity, the more rapidly it will conduct heat energy.
- temperature difference between the two objects. A greater temperature difference will result in a faster rate of energy transfer.
- thickness of the material. Thicker materials require a greater number of collisions between particles or movement of electrons to transfer energy from one side to the other.
- surface area. Increasing the surface area relative to the volume of a system increases the number of particles involved in the transfer process, increasing the rate of conduction.

The rate at which heat is transferred is measured in joules per second ($J\,s^{-1}$), or watts (W).

PHYSICSFILE

Igloos

It seems strange that an igloo can keep a person warm when ice is so cold. Igloos are constructed from compressed snow, which contains many air pockets. The air in these pockets is a poor conductor of heat, which means that heat inside the igloo is not easily transferred away. The body heat of the occupants, as well as that of their small heat source, is trapped inside the igloo and keeps them warm.

An igloo

CONVECTION

Although liquids and gases are generally not very good conductors of thermal energy, heat can be transferred quite quickly through liquids and gases by **convection**. Convection is the transfer of thermal energy within a fluid (a liquid or a gas) by the movement of hot areas from one place to another. Unlike other forms of heat transfer such as conduction and radiation, convection involves the mass movement of particles within a system over a distance that can be quite considerable.

Convection is the transfer of heat within a fluid (liquid or gas).

PHYSICSFILE

Fluid

In everyday life, the word 'fluid' is often used interchangeably with 'liquid'. However, in physics, it means something that can flow, such as a liquid or a gas.

Heating by convection

As a fluid is heated, the particles within it gain kinetic energy and push apart due to the increased vibration of the particles. This causes the density of the heated fluid to decrease and the heated fluid rises. Colder fluid, with slower-moving particles, is more dense and heavier and hence falls, moving in to take the place of the warmer fluid. A convection current forms when there is warm fluid rising and cool fluid falling. This action can be seen in Figure 4.4.4. Upwellings in oceans, wind and weather patterns are at least partially due to convection on a very large scale.

It is difficult to quantify the thermal energy transferred via convection but some estimates can be made. The rate at which convection will occur is affected by:

- the temperature difference between the heat source and the convective fluid
- the surface area exposed to the convective fluid.

In a container, the effectiveness of convection to transfer heat depends on the placement of the source of heat. For example, the heating element in a kettle is always found near the bottom of the kettle. From this position, convection currents form throughout the water to heat it more effectively (Figure 4.4.5(a)). If the heating element was placed near the top of the kettle, convection currents would form only near the top. This is because the hotter water is less dense than the cooler water below and would remain near the top. Convection currents would not form throughout the water (Figure 4.4.5(b)).

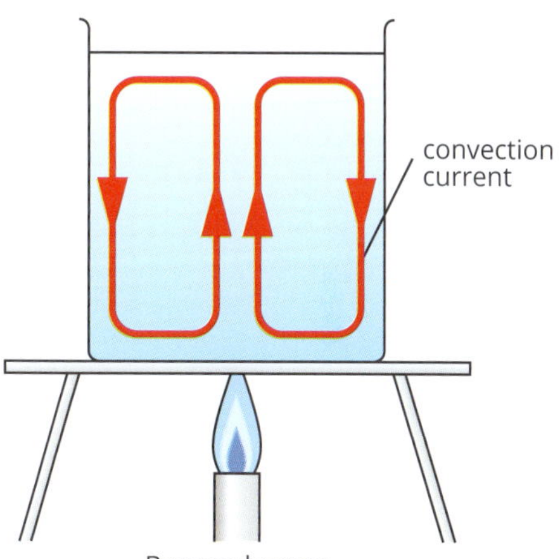

FIGURE 4.4.4 When a liquid or gas is heated, it becomes hotter and less dense and so will rise. The colder, denser fluid will fall. As this fluid heats up, it in turn will rise, creating a convection current.

(a)

(b)

FIGURE 4.4.5 (a) By placing the heating element at the bottom of the kettle, the water near the bottom is heated and rises, forming convection currents throughout the entire depth of the water. (b) If the heating element is placed near the top of the kettle, the convection currents form near the top and heat transfer is slower.

There are two main causes of convection:

- forced convection; for example, ducted heating in which air is heated and then blown into a room
- natural convection, such as that illustrated in Figure 4.4.4, in which a fluid rises as it is heated.

A dramatic example of natural convection is the thunderhead clouds of summer storms (as seen in Figure 4.4.6), which form when hot, humid air from natural convection currents is carried rapidly upwards into the cooler upper atmosphere.

FIGURE 4.4.6 The thunderheads of summer storms are a very visible indication of natural convection in action.

CASE STUDY

Wind chill

Convection is the main means of heat transfer that leads to the 'wind chill' factor. In cool still air, our body warms a thin layer of relatively still air near our skin, which acts as a partial insulator. However, in windy weather, the wind blows this warm air away. When this happens, cooler air comes in closer contact with the skin and heat loss increases. It feels as if the 'effective' temperature of the surrounding air has decreased. The warm air near the skin can be blown away by the wind, or it can be removed through motion such as when a bike rider travels fast.

The chilling effect is even more dramatic when moisture is next to the skin, which results in evaporative cooling.

In contrast to cotton, modern fibres used in athletic clothing do not absorb large quantities of water. The water that is absorbed is carried away from the skin to the outer layers of the clothing to evaporate. Bushwalkers look for clothing that dries rapidly after rain and which carries moisture from the perspiration of heavy exertion away from the skin.

The wind chill factor can be used in hot climates to cool down. A room with a fan circulating the air past your body will feel cooler than a room with still air at the same temperature.

Heat is transferred from one place to another without the movement of particles by electromagnetic radiation.

RADIATION

Both convection and conduction involve the transfer of heat through matter. However, life on Earth depends upon the transfer of energy from the Sun through the near-vacuum of space. If heat could only be transferred by the action of particles, then the Sun's energy would never reach Earth. **Radiation** is a means of transfer of heat without the movement of matter.

In this context, the term 'radiation' is a shortened form of the term 'electromagnetic radiation', which includes visible, ultraviolet and infrared light. Together with other forms of light, these make up the electromagnetic spectrum. You learnt about the electromagnetic spectrum in Section 2.3 on page 72.

Electromagnetic radiation travels at the speed of light. When electromagnetic radiation (light) hits an object, it will be partially **reflected**, partially transmitted and partially absorbed. The absorbed radiation transfers thermal energy to the absorbing object and causes a rise in temperature. When you hold a marshmallow by an open fire, you are using radiation to toast the marshmallow, as shown in Figure 4.4.7.

FIGURE 4.4.7 Heat transfer from the flame to the marshmallow is an example of radiation.

You will learn more about heat transfer by electromagnetic radiation in Chapter 5.

PHYSICSFILE

Paragliders

Paragliders fly by sitting in a harness suspended beneath a fabric wing. They gain altitude by catching thermals. Thermals are columns of rising hot air—convection currents—created by dark regions on the ground that have been heated up by the Sun.

Paragliders can gain altitude by finding a thermal. These are areas of rising hot air created by hot regions on the ground. These paragliders are flying near Bright, Victoria.

4.4 Review

SUMMARY

Heat is transferred by conduction, convection or radiation.

- Conduction is the process of heat transfer within a material or between materials without the overall transfer of the substance itself.
- All materials conduct heat to a greater or lesser degree. Materials that readily conduct heat are called good thermal conductors. Materials that conduct heat poorly are called thermal insulators.
- Whether a material is a good conductor depends on the method of conduction.
  - Heat transfer by molecular collisions alone occurs in poor to very poor conductors.
  - Heat transfer by molecular collisions and free electrons occurs in good to very good conductors.
- The rate of conduction depends on the temperature difference between two materials, the thickness of the material, the surface area and the nature of the material.
- Convection is the transfer of heat within a fluid (liquid or gas).
- Convection involves the mass movement of particles within a system over a distance.
- A convection current forms when there is warm fluid rising and cool fluid falling.
- Radiation transfers thermal energy from one place to another by means of electromagnetic waves.
- When electromagnetic radiation falls on an object, it will be partially reflected, partially transmitted and partially absorbed.

KEY QUESTIONS

Knowledge and understanding

1. a Why are metals better conductors of heat than wood?
   b List the properties of a material that affect its ability to conduct heat.
   c Stainless steel saucepans are often manufactured with a copper base. What is the most likely reason for this?
2. a Through what states of matter can convection occur?
   b On a hot day, the top layer of water in a swimming pool can heat up while the lower, deeper parts of the water can remain quite cold.
   Explain, using the concept of convection, why this happens.
3. Light is shone on an object.
   a List three interactions that can occur between the light and the object.
   b Which of the interactions from part a are associated with a rise in temperature?
4. Why is it impossible for heat to travel from the Sun to Earth by conduction or convection?

Analysis

5. On a cold day, the plastic or rubber handles of a bicycle feel much warmer than the metal surfaces. Explain this in terms of the thermal conductivity of each material.
6. Convection is referred to as a method of heat transfer through fluids. Evaluate whether it is possible for solids to pass on their heat energy by convection.

Chapter review

OA ✓✓

04

KEY TERMS

absolute zero
conduction
conductor
convection
evaporation
first law of thermodynamics
heat
insulator
internal energy
kelvin
kinetic energy
kinetic particle model
latent heat
latent heat of fusion
latent heat of vaporisation
potential energy
radiation
reflect
specific heat capacity
temperature
thermal contact
thermal energy
thermal equilibrium
volatile
work
zeroth law of thermodynamics

REVIEW QUESTIONS

Knowledge and understanding

1 How does temperature differ from heat?

2 Convert:
  a 5°C to kelvin
  b 200 K to °C.

3 A tank of pure helium is cooled to its freezing point of −272.2°C. Describe the energy of the helium particles at this temperature.

4 Sort the following temperatures from coldest to hottest:
  freezing point of water
  100 K
  absolute zero
  −180°C
  10 K

5 The specific heat capacity of copper is approximately three times that of lead. A ball of copper and a ball of lead of equal mass, both at 50°C, are dropped into a thermally insulated jar that contains a mass of water, equal to that of the balls, at 20°C. Thermal equilibrium is eventually reached.
  a Describe the energy of each of the metal balls before they are dropped into the water.
  b Describe the final temperatures of each of the metal balls.

6 Two cubes, one copper and one silver, have the same mass and are initially at 60°C. A student places them both in a cold-water bath until they are cooled to 20°C. He argues that they have both had the same amount of energy removed because they had the same initial and final temperatures. Is he correct? Explain your answer.

7 Describe the type(s) of heat transfer involved in each situation.
  a Oceans move thermal energy over large distances.
  b Heat is transferred around Earth's atmosphere.
  c The Sun's energy reaches Earth.

8 A solid substance is heated but its temperature does not change. Explain what is occurring.

9 Which possesses the greater internal energy—1 kg of water boiling at 100°C or 1 kg of steam at 100°C? Explain why.

10 If 4.0 kJ of energy is required to raise the temperature of 1.0 kg of paraffin by 2.0°C, how much energy (in kJ) is required to raise the temperature of 5.0 kg of paraffin by 1.0°C?

11 A vacuum flask has a tight-fitting stopper at the top. Its walls are made up of an inner and outer layer, which are shiny and are separated by a layer of air. Describe how this design makes a vacuum flask good at keeping liquids inside hot.

12 Solar hot water panels are placed on a roof, and can be used to supply hot water for a house to reduce electricity and gas consumption.
  a Describe the energy transfer and transformations involved in a solar hot water panel.
  b Why does the warm water move to the top of the solar panel?

Application and analysis

13 Calculate the energy required to raise the temperature of a 1.50 kg block of copper from 24.0°C to 80.0°C.

14 150 mL of water is heated from 10.0°C to 50.0°C. What amount of energy is required for this temperature change to occur?

15 A 1.00 kg metal object requires 8.50×10^3 J of heat to raise its temperature from 25.0°C to 50.0°C. What is the specific heat capacity of the metal in $J\,kg^{-1}\,K^{-1}$? Give your answer to the nearest whole number.

16 Shasha fills a 250 mL ice tray with water at 25.0°C and places it in the freezer to make ice cubes. How many joules of energy need to be removed from the water for it to freeze?

17 An insulated container holding 4.55 kg of ice at 0.00°C has 2.65 MJ of work done on it, while a heater provides 14 600 J of heat to the ice. If the latent heat of fusion of ice is $3.34 \times 10^5\,J\,kg^{-1}$, calculate the final temperature of the water. Assume that the increase in internal energy is first due to an increase in the potential energy and then an increase in the kinetic energy.

18 Describe an experiment to compare how two different solids conduct heat at different rates.

CHAPTER

05 Thermal energy, electromagnetic radiation and Earth's climate

Without the effects of greenhouse gases in our atmosphere, life on Earth as we know it would not exist. Our atmosphere keeps Earth at a steady temperature, which has allowed life to evolve. But human activity has changed this delicate balance. Due to increasing levels of carbon dioxide in the atmosphere, Earth is getting warmer. These higher temperatures have impacts across the world. In Australia, their effects include an increase in the severity of bushfires and storms, more severe and frequent droughts, and more frequent bleaching of coral reefs.

This chapter explains further concepts of heat and heating processes, and builds on this knowledge to describe what happens to the energy from the Sun when it reaches Earth and how heat energy is moved around Earth. The physics behind human-induced climate change is explored.

Key knowledge

- calculate the peak wavelength of the radiated electromagnetic radiation using Wien's Law: $\lambda_{max}T = \text{constant}$ **5.1**
- compare the total energy across the electromagnetic spectrum emitted by objects at different temperatures **5.1**
- apply concepts of energy transfer, energy transformation, temperature change and change of state to climate change and global warming. **5.2**

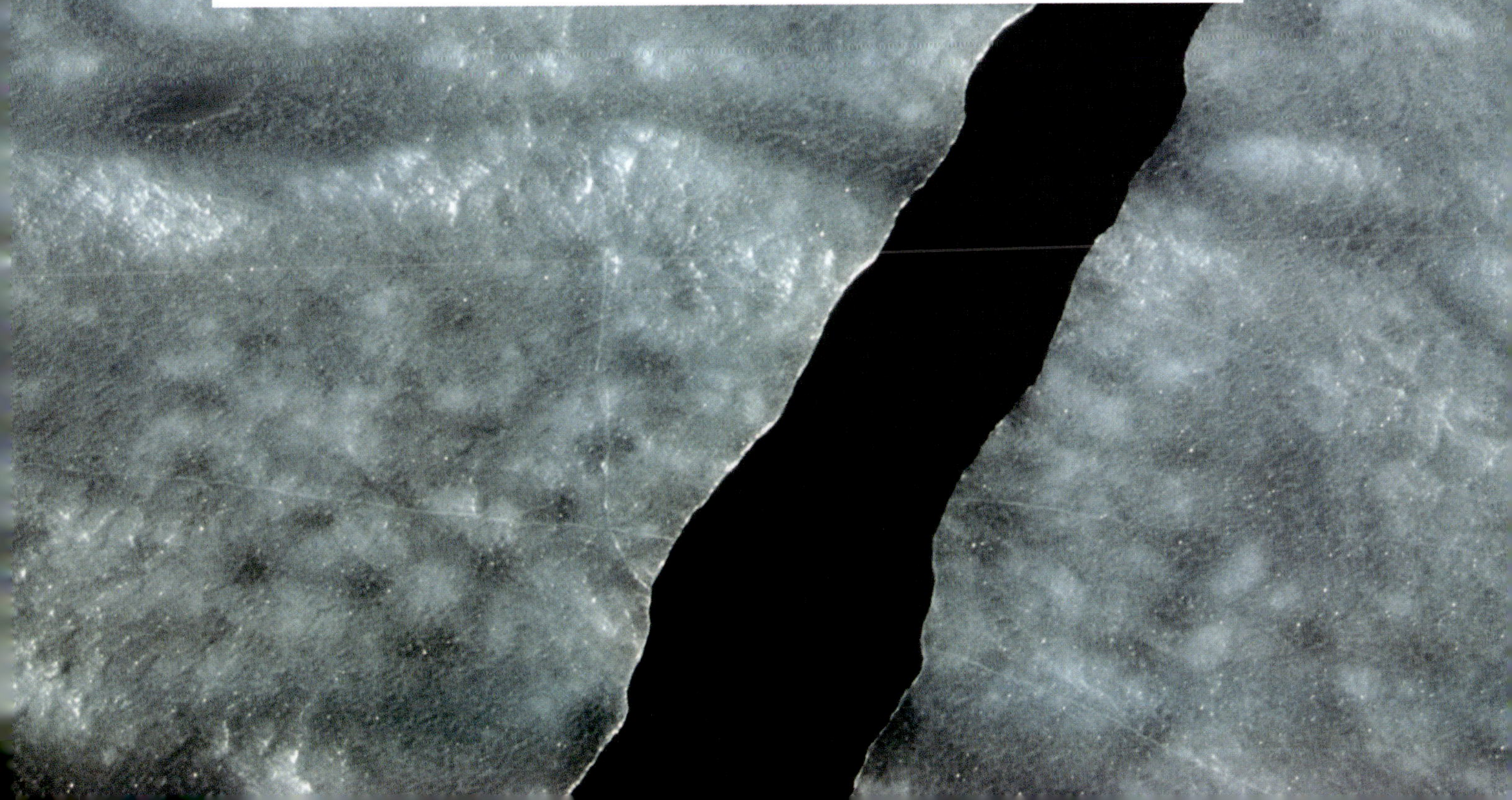

5.1 Wien's law and black-body radiation

As you saw in Chapter 4, heat energy can be transferred from one place to another without the movement of particles by electromagnetic radiation. It is quite easy to picture radiation that is emitted from hot objects—we feel warmer when we absorb the heat that is emitted from a fire or from the Sun, and we see the light emitted from an incandescent light bulb. However, all objects with temperatures above absolute zero (0 K or −273.15°C) emit and absorb electromagnetic radiation.

Solids, liquids and dense gases emit electromagnetic radiation in a continuous spectrum. However, the amount of energy they emit is not the same for all wavelengths—they emit a maximum amount of energy at a particular wavelength. This wavelength depends on the internal energy (that is, temperature) of the object. The higher the temperature of the object, the shorter the wavelength (and therefore the higher the frequency) of the radiation emitted. This can be seen in Figure 5.1.1.

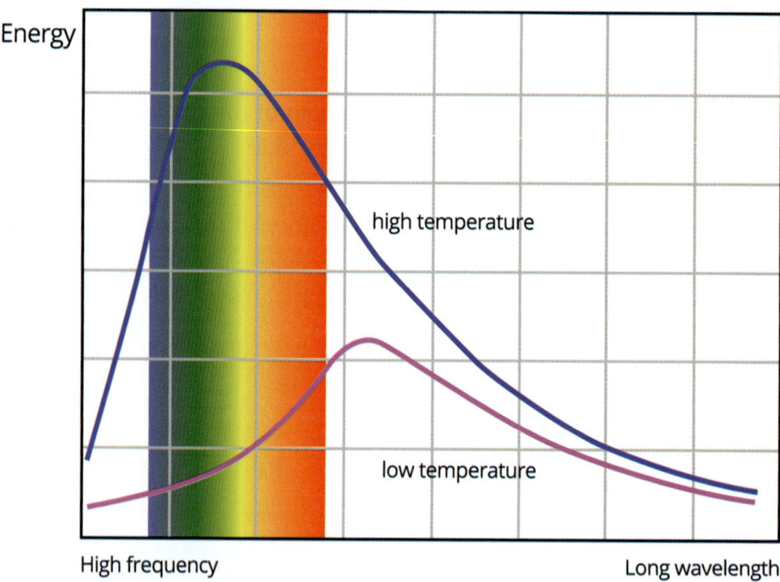

FIGURE 5.1.1 An object emits radiation over a range of frequencies. At a low temperature, it will emit small amounts of radiation and this radiation is of longer wavelengths. As the temperature of the object increases, the total radiant energy emitted increases and more short-wavelength radiation is emitted.

A cool object emits radiation at wavelengths longer than the visible range. For example, the human body emits radiation in the infrared range, which our eyes can't detect but which can be detected by an infrared camera. Hotter objects emit radiation in the range of visible, ultraviolet and shorter wavelengths of the electromagnetic spectrum. For example, as a poker heats up in a fire, it becomes red and then yellow as it emits radiation at shorter wavelengths. The Sun emits radiation mainly in the ultraviolet to infrared range, including visible light.

PHYSICSFILE

Our eyes and the Sun

It is no coincidence that the human eye is very good at detecting visible light. Human eyes have evolved to be most receptive to wavelengths of light that correspond to the highest intensity of light produced by our Sun. This light is within what is known as the visible range. If the Sun had a lower surface temperature, of say around 3000 K, it is highly probable that human eyes would be adapted to the infrared range.

WIEN'S LAW

Wilhelm Wien, a German physicist, formulated laws that describe the properties of heat radiation. Wien discovered that the **peak wavelength** at which an object emits the maximum intensity of radiation is dependent on its surface temperature. Wien's displacement law, more commonly known just as **Wien's law**, can be used to determine the peak wavelength for an object at a particular surface temperature.

Wien's law states that

$\lambda_{max} T = \text{constant}$

where λ_{max} is the peak wavelength of the emitted radiation in metres (m)

T is the surface temperature of the object in kelvin (K).

That is, no matter what the surface temperature of an object, the product of the temperature and the wavelength at which the peak intensity of the emitted radiation occurs is a constant and is equal to about 2.898×10^{-3} m K.

The graph in Figure 5.1.2 shows the continuous spectrum emitted by any solid, liquid or even a dense gas at particular temperatures. The wavelength corresponding to the highest intensity for the 12 000 K curve is in the ultraviolet range. That is, an object at a temperature of around 12 000 K emits most of its energy with wavelengths in the ultraviolet range.

The 6000 K curve in the graph in Figure 5.1.2 corresponds to the surface temperature of our Sun. The intensity maximum corresponds to a peak wavelength at about 500 nm, which is within the visible band of the electromagnetic spectrum.

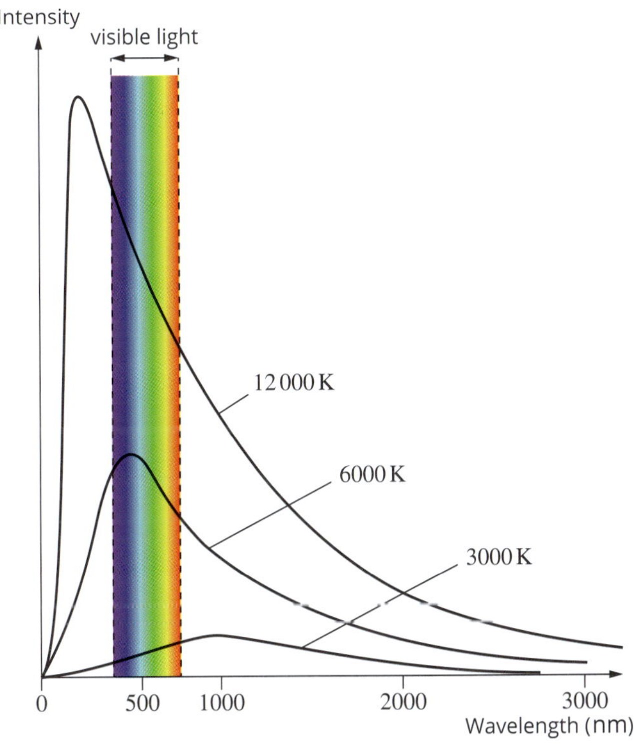

FIGURE 5.1.2 The spectrum of wavelengths emitted for an idealised black body at different temperatures. The radiation approximates the surface temperature of many real objects.

BLACK-BODY RADIATION

Wien's work on the relationship between temperature and wavelength of the radiation emitted by an object was based initially on a theoretical object called a **black body**. A black body is a hypothetical object that is a perfect absorber or emitter of radiation—it completely absorbs all the electromagnetic radiation incident on it regardless of the wavelength of the radiation and, therefore, does not reflect any radiation. Its radiation is known as **black-body radiation**. A black body does not necessarily have to be black, and black objects are not necessarily black bodies.

PHYSICSFILE

Wilhelm Wien

Wilhelm Wien (1864–1928), a German physicist, was awarded the 1911 Nobel Prize in Physics for making a significant contribution to the thermodynamics of radiation. In 1893 he had discovered Wien's displacement law, which paved the way for Planck's quantum theory of radiation. This led to the development of quantum theory, the theoretical basis of modern physics, which explains the nature and behaviour of matter and energy. So Wien's law was a very significant discovery indeed!

Wilhelm Wien

The radiation emitted by many objects, such as the Sun, can be approximated as the radiation emitted by a black body at the same temperature. The spectrum emitted by a hot solid, liquid or dense gas is continuous but has a peak intensity at a wavelength inversely proportional to the surface temperature. This relationship is more simply stated by rearranging Wien's law:

$$\lambda_{max} \propto \frac{1}{T} \text{ and } T \propto \frac{1}{\lambda_{max}}$$

Wien's law makes it possible to determine the approximate temperature of stars, assuming that they emit radiation similar to that emitted by a black body. During astronomic observations, it was discovered that stars at different temperatures have peaks in the graph of emissive power at different wavelengths. When the wavelength that corresponds to the peak of the power emitted by a star is known, the temperature of the star can be found by applying Wien's law. This is shown in Worked example 5.1.1.

Worked example 5.1.1

THE TEMPERATURE AT A STAR'S SURFACE

The Sun emits a continuous electromagnetic spectrum with a peak wavelength of approximately 500.0 nm. Based on this wavelength, estimate the surface temperature of the Sun.

Thinking	Working
Express the peak wavelength in metres.	$\lambda_{max} = 500.0\,\text{nm} = 500.0 \times 10^{-9}\,\text{m}$
Rearrange Wien's law to solve for T.	$\lambda_{max} T = 2.898 \times 10^{-3}\,\text{m K}$ $T = \frac{2.898 \times 10^{-3}}{\lambda_{max}}$
Substitute the value for λ_{max} and solve for T.	$T = \frac{2.898 \times 10^{-3}}{500.0 \times 10^{-9}}$ $= 5796\,\text{K}$

Worked example: Try yourself 5.1.1

THE TEMPERATURE AT A STAR'S SURFACE

WS 6

A newly discovered star is observed to have a peak emitted radiation wavelength of approximately 90 nm. Based on this wavelength, estimate the surface temperature of this star.

EMITTED AND ABSORBED ENERGY

As stated above, all objects above absolute zero both absorb and emit thermal energy by radiation. However, they do not all emit energy at the same rate. A number of factors affect the total rate of energy emission by radiation across the electromagnetic spectrum: surface area, temperature, and surface colour and texture. An object's effectiveness at emitting energy is called its **emissivity** and is given the symbol e. Emissivity is defined as the fraction of energy that is emitted relative to that emitted by a thermally black surface (a black body). A black body is a perfect emitter of heat energy and has an emissivity value of 1. A perfect reflector of thermal energy has an emissivity value of 0. Good emitters are also good absorbers.

As an example, matte black surfaces will emit radiant energy at a greater rate than shiny, white surfaces. Matte black surfaces have a value of e close to 1, whereas shiny surfaces have an e value close to 0. This means that a roughened, dark surface will heat up faster than a shiny, light one. It will also cool down faster, since it will radiate energy just as efficiently as it absorbs it. Car radiators are painted black for this reason—to increase the emission of thermal energy collected from the car engine.

Table 5.1.1 shows the emissivity of some common materials.

TABLE 5.1.1 The emissivity of some common materials

Material	Emissivity
anodised aluminium	0.77
brick	0.81–0.86
concrete	0.92
powdered charcoal	0.96
glass	0.92
ice	0.97
black paper	0.90
black rubber stopper	0.97
human skin	0.98
snow	0.80
water	0.95

5.1 Review

OA ✓✓

SUMMARY

- Any object whose temperature is greater than absolute zero emits thermal energy by radiation.
- The rate of emission or absorption of radiant heat will depend upon the:
  - temperature difference between the object and the surrounding environment
  - surface area and surface characteristics of the object
  - wavelength of the radiation.
- The peak wavelength, at which an object will emit the maximum intensity of radiation, is dependent on the object's surface temperature and is given by Wien's law: $\lambda_{max}T = 2.898 \times 10^{-3}\,\text{m K}$.
- A black body is a perfect emitter and absorber of heat energy—none of the radiation incident on it is reflected.
- The emissivity of an object gives an idea of how effective it is at emitting radiation. A black body has an emissivity value of 1. A perfect reflector of thermal energy has an emissivity value of 0. Good emitters are also good absorbers.

KEY QUESTIONS

Knowledge and understanding

1 All objects with a temperature above absolute zero emit radiant thermal energy. As an object's temperature increases, what happens to the wavelength and frequency of the emitted radiation?

2 The emissivity, e, of a surface depends on the colour, characteristics (matte or shiny, for example) and temperature of the surface. Which of the following statements about the value of e for a particular surface is true?

A A matte black surface will have a value of e close to 0.

B A shiny, white surface will have a value of e close to 1.

C A shiny, white surface will have a value of e close to 0.

D None of the above.

3 Three identical, sealed beakers are filled with near-boiling water. One beaker is painted matte black, one is dull white and the third is glossy white.

a Which beaker will cool fastest, and which beaker will cool slowest?

b When the beakers have cooled to room temperature, they are placed in strong sunlight. Which will warm fastest, and which will warm slowest?

4 Computer chips generate a lot of thermal energy that must be dispersed for the computer to function efficiently. Devices called heat sinks are used to help this process. What would you predict the heat sinks to be made of?

Analysis

5 A star has a surface temperature of 9000 K. What is the peak wavelength of the energy being emitted by this star?

6 A star emits a continuous electromagnetic spectrum with a peak wavelength of approximately 800 nm. What is the surface temperature of this star?

7 The element of an electric heater is just seen to glow a dull red. This colour corresponds to the lower end of the visible spectrum at approximately 700 nm. What temperature, in kelvin, is the element of the heater?

5.2 Radiation and the enhanced greenhouse effect

FIGURE 5.2.1 Energy from the Sun enables life on Earth to exist.

Without the energy of the Sun, ecosystems like that shown in Figure 5.2.1 could not exist. All life on Earth depends upon the transfer of radiant energy from the Sun through space. However, that alone is not enough to keep Earth warm enough for life to exist. Earth's atmosphere acts as a greenhouse, trapping some of the Sun's energy to keep our planet at a constant temperature. Without this **greenhouse effect**, Earth would be a very cold place.

The composition of Earth's atmosphere determines how much energy is reflected, absorbed and re-emitted by Earth. In recent decades, evidence has shown that human activity has increased the levels of specific gases in the atmosphere, in particular carbon dioxide. As the levels of these gases rise, more thermal energy is being retained than previously, resulting in a rise in global temperatures. This has implications for all life on Earth.

In this section, you will apply the concepts of radiation, convection and conduction to the theory of the greenhouse effect and how thermal energy is moved around Earth.

HEATING EARTH

The overall temperature of Earth is determined by the total thermal energy received from the Sun and the amount that is lost back to space. Any change in that balance will lead to a warming or cooling of Earth as a whole.

Energy gain—radiant energy from the Sun

Most of the thermal energy received by Earth is short-wave radiant energy from the Sun. Most of this is within or close to the visible spectrum.

Of the incoming radiant energy from the Sun, about 47% is eventually absorbed by Earth's surface. As shown in Figure 5.2.2, of the remaining energy:

- about 23% of the energy reaching Earth is absorbed by the atmosphere, predominantly by the ozone layer, but also by water vapour and **greenhouse gases** such as carbon monoxide and carbon dioxide
- about 26% is reflected back towards outer space by clouds and the atmosphere
- about 4% is reflected by Earth's surface.

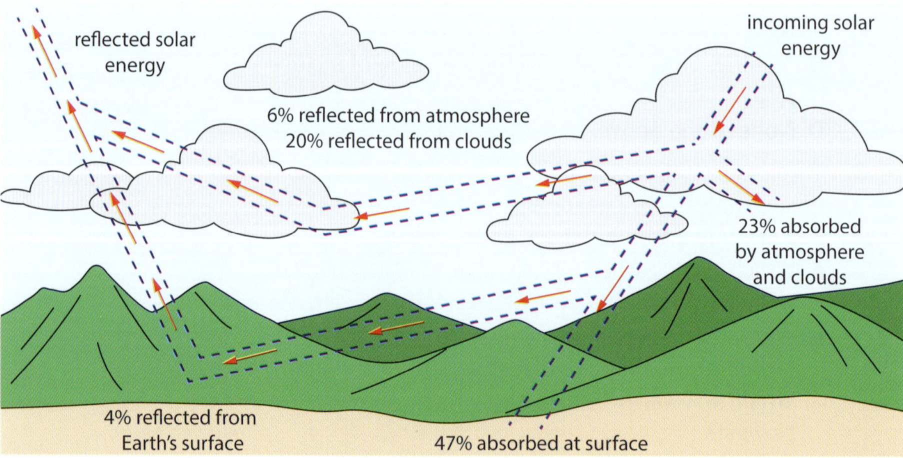

FIGURE 5.2.2 Of the radiant energy reaching Earth from the Sun, only 47% is eventually absorbed by Earth's surface. The remainder is reflected back into space or absorbed by the atmosphere.

The radiant energy that is absorbed by Earth's surface transforms to thermal energy, increasing the temperature of the absorbing surfaces and of the air in contact with those surfaces. Over the longer term, very little of this energy is retained. Almost all is re-radiated by the surface as long-wavelength radiant energy. The peak wavelength of the re-radiated electromagnetic radiation can be approximated using Wien's law, which was introduced in Section 5.1. This is shown in the worked example below.

Worked example 5.2.1

RE-RADIATED ENERGY FROM EARTH

Earth's average surface temperature is 289 K. What is the peak wavelength of the re-radiated electromagnetic radiation?

Thinking	Working
State Wien's law.	$\lambda_{max}T = 2.898 \times 10^{-3}\,\text{m K}$
Rearrange Wien's law to express it in terms of λ_{max}.	$\lambda_{max} = \dfrac{2.898 \times 10^{-3}}{T}$
Substitute the value for T and solve for λ_{max}.	$\lambda_{max} = \dfrac{2.898 \times 10^{-3}}{289}$ $= 1.00 \times 10^{-5}\,\text{m} = 10.0\,\mu\text{m}$

Worked example: Try yourself 5.2.1

RE-RADIATED ENERGY FROM EARTH

Earth's average surface temperature at the equator is 300 K. What is the peak wavelength of the re-radiated electromagnetic radiation from this portion of Earth's surface?

Energy retention—greenhouse gases

With Earth's surface re-radiating heat from the Sun back out towards space, you may wonder how Earth remains warm. Earth stays warm because the greenhouse gases in the atmosphere absorb some of this energy and re-radiate it back down towards Earth's surface. Only about 12% is lost directly out to space. This energy retention has resulted in a long-term energy balance and relatively stable temperatures, allowing life to evolve.

The long-wavelength infrared radiation emitted by Earth is readily absorbed by the greenhouse gases in the atmosphere. This creates what is called the greenhouse effect and the relatively consistent day and night temperatures required by the life forms that have evolved on Earth.

This cycle is shown in Figure 5.2.3.

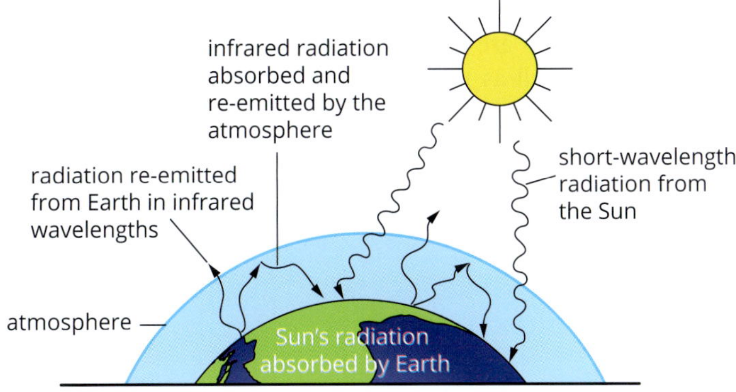

FIGURE 5.2.3 The Sun's energy that has been absorbed by Earth is re-radiated away from Earth as infrared radiation. A significant amount of this is absorbed by the greenhouse gases in the atmosphere and re-radiated back to Earth's surface.

PHYSICSFILE

Venus greenhouse effect

Many scientists believe that Venus used to have an environment similar to that on Earth, with lower temperatures and even liquid water on its surface. Now, the carbon dioxide atmosphere on Venus is 92 times denser than Earth's atmosphere at the surface. The average surface temperature on Venus is 462°C. It is thought that Venus experienced a 'runaway' greenhouse effect.

On Venus, approximately 10% of the Sun's energy reaches the surface and heats it up. The surface radiates the infrared energy back towards space. As you can see in the image below, this radiant energy used to continue out to space, but it is now retained by the dense atmosphere, which has caused the planet to heat up.

The greenhouse effect on Venus. The changed atmosphere, which now consists mainly of carbon dioxide, blocks most of the radiated heat energy from leaving the atmosphere.

ENHANCED GREENHOUSE EFFECT

In recent decades, evidence has shown that the atmosphere is absorbing and retaining more of the long-wavelength infrared radiation emitted from Earth's surface. This **enhanced greenhouse effect** is driving the changes in climate we are starting to experience.

Earth's atmosphere is mostly made up of nitrogen, oxygen and argon. These gases are almost totally unaffected by infrared radiation and so have no effect on the absorption or re-emission of infrared wavelengths. However, some gases present in the lower atmosphere in very small proportions are responsible for most of the absorption of these longer wavelengths. The most significant of these have increased in concentration since the mid-eighteenth century, coinciding with the industrialisation of modern society. These greenhouse gases can absorb and emit long-wavelength infrared radiation.

The most abundant greenhouse gases in Earth's atmosphere are, in order of abundance:

- water vapour (H_2O)
- carbon dioxide (CO_2)
- methane (CH_4)
- nitrous oxide (N_2O)
- ozone (O_3)
- chlorofluorocarbons (CFCs).

The enhanced greenhouse effect is caused by the combined effects of these greenhouse gases. The extent to which each gas contributes depends on the chemical characteristics of the gas, on its percentage abundance and on some indirect effects, such as water vapour turning to ice. For example, although methane absorbs 72 times as much thermal energy as carbon dioxide, it is present in much smaller concentrations, so overall it doesn't contribute as much to the enhanced greenhouse effect. Table 5.2.1 ranks greenhouse gases based on their overall contribution to the enhanced greenhouse effect.

TABLE 5.2.1 Greenhouse gas contribution to the enhanced greenhouse effect

Compound	Formula	Contribution (%)
water vapour/clouds	H_2O	36–72%
carbon dioxide	CO_2	9–26%
methane	CH_4	4–9%
ozone	O_3	3–7%
nitrous oxide	N_2O	1.5%
chlorofluorocarbons	CFCs	0.1%

Other than CFCs, which are entirely human produced, most greenhouse gases have natural sources as well as those from human activity. The sources of some greenhouse gases are summarised in Table 5.2.2.

TABLE 5.2.2 Sources of greenhouse gases

Gas	Natural sources	Human sources
carbon dioxide, CO_2	respiration volcanic eruptions	burning fossil fuels deforestation land-use changes
methane, CH_4	digestion in animals	decomposition of wastes in landfills agriculture and especially rice cultivation energy use domestication of livestock
nitrous oxide, N_2O	soils under natural vegetation and the oceans	fertiliser use burning fossil fuels nitric acid production biomass burning
chlorofluorocarbons (CFCs)	none	industrial processes refrigerants (such as those used in air-conditioning) a variety of consumer products

Greenhouse gas levels

By comparing current levels of greenhouse gases with those found trapped in air in Antarctic ice cores, scientists can develop a good idea of the changes in greenhouse gas levels over long periods of time. Before the Industrial Revolution (which started in the middle of the 1700s), concentrations of these gases were relatively constant. In the modern industrial era, human activities have increased the proportions of greenhouse gases in the atmosphere. This has mainly occurred as a result of the burning of **fossil fuels** (natural fuels such as coal or gas, formed in the geological past from the remains of living organisms) and large-scale land clearing. The graph in Figure 5.2.4 clearly shows how significant the rise in greenhouse gases has been in modern times.

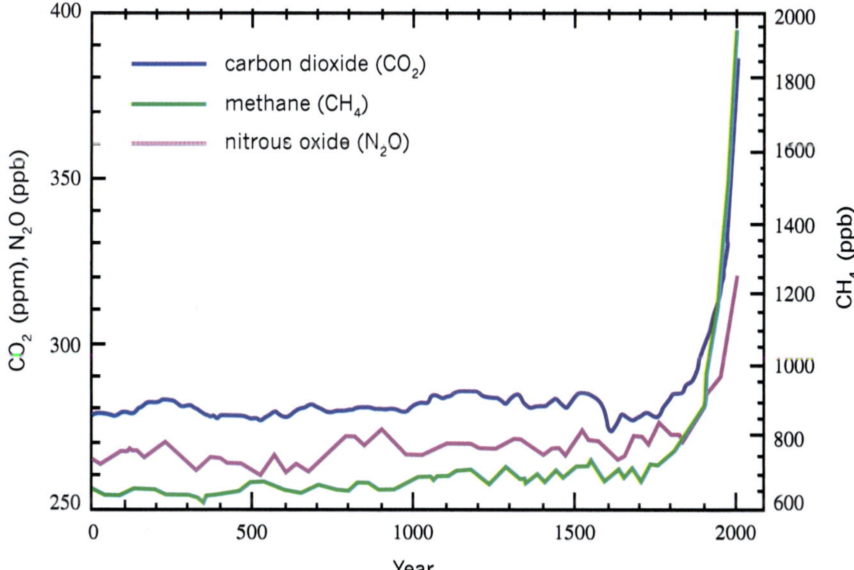

FIGURE 5.2.4 This graph shows that atmospheric carbon dioxide, methane and nitrous oxide have increased since the Industrial Revolution. Note that two different units are used on the left-hand *y*-axis: parts per million (ppm) and parts per billion (ppb).

For example, based on data from ice-core samples, carbon dioxide was found to have remained between 180 and 300 parts per million (ppm) for at least 400 000 years, up until the Industrial Revolution. Figure 5.2.5 shows the levels of carbon dioxide (CO_2) in the atmosphere. The graph shows that in 2020, carbon dioxide levels passed 400 ppm. With increased levels of carbon dioxide comes an increase in the absorption of infrared radiation and its re-radiation back to Earth's surface rather than reflection into space, causing Earth to heat up.

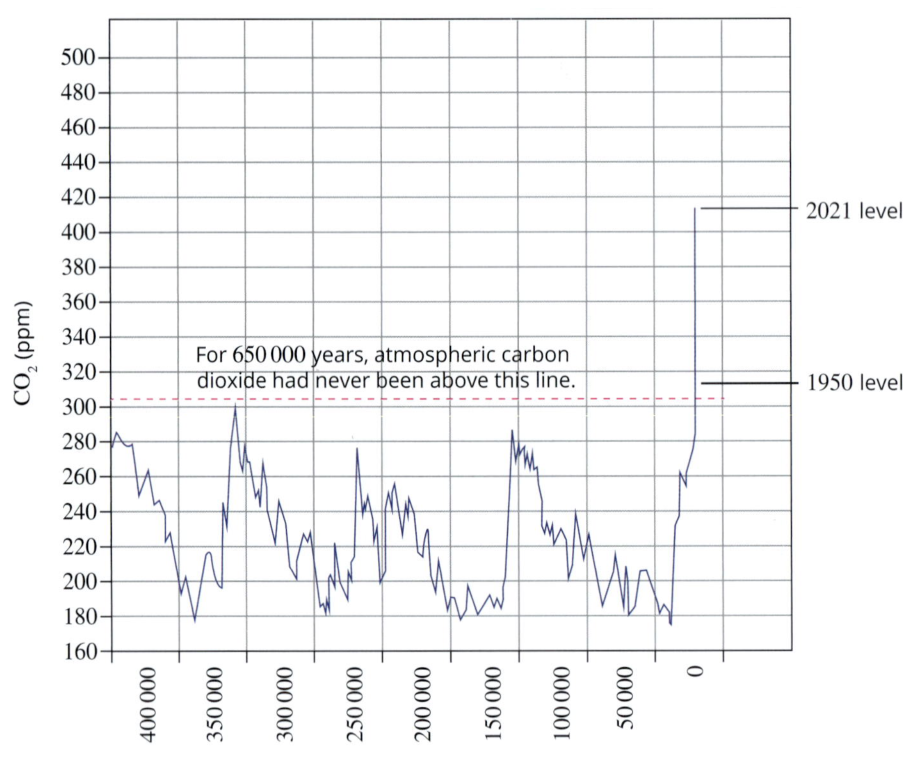

FIGURE 5.2.5 Carbon dioxide levels over the last 400 000 years, based on ice-core samples. The recent sudden rise coincides with the beginning of the industrial age.

Carbon dioxide levels are not the only gas levels that have increased, as shown in Table 5.2.3.

TABLE 5.2.3 Changes in greenhouse gas concentrations over the industrial era

Greenhouse gas	Pre-industrial-age concentration	Current concentration
carbon dioxide, CO_2	280 ppm (parts per million)	412.5 ppm
methane, CH_4	722 ppb (parts per billion)	1800 ppb
nitrous oxide, N_2O	270 ppb	325 ppb
ozone, O_3	237 ppb	337 ppb

CASE STUDY

Effects of a changing climate

As levels of carbon dioxide in the atmosphere rise, the enhanced greenhouse effect increases and Earth gets warmer and warmer. The consequences for the environment can be devastating. Currently, Earth's average temperature (land and ocean) is about 15°C, which is an increase of about 1.1°C since pre-industrial times. That may not sound much, but it represents a significant amount of energy that has been retained by our planet. As Earth keeps getting warmer, it is impacting the entire planet: the oceans, weather patterns and all living things. If Earth becomes too hot, it is likely to cause significant changes that will prevent Earth from sustaining our current way of life. We will face more frequent and severe extreme weather events including floods and droughts, and more frequent and more severe heatwaves. This will reduce Earth's capacity to grow enough food. For example, across the world, the area that is classified as drylands, which includes arid, semi-arid and dry sub-humid climates, has increased over the last 60 years. These drylands are prone to land degradation, and do not support large crops. They are currently home to 2.7 billion people, a number that is set to increase as the population increases and the area classified as drylands increases.

There is evidence that the predicted changes are occurring already. Globally, 2020 and 2016 are tied as the hottest years on record; nine of the 10 hottest years on record have occurred since 2010. In Australia, 2019 was the hottest and driest year on record, and nine of our ten hottest years have occurred since 2005. It is likely these records will be broken in the coming years. Australia is vulnerable to the effects of climate change in many ways, including more frequent and severe bushfires and storms, more frequent marine heatwaves, which lead to coral bleaching, and increases in the frequency of heatwaves, which affect humans, wildlife and crops.

HUMAN ACTIVITIES AND ENERGY RE-RADIATED BY EARTH

Very little of the radiant energy that reaches Earth's surface from the Sun is retained. However, the rate at which energy is re-radiated by Earth's surface depends on the surface material. Recall from Section 5.1 that the properties of materials affect how readily they will absorb and re-radiate thermal energy.

Energy retention: Building

The building of cities has affected the greenhouse effect in a number of ways.

- The materials used to build cities re-radiate thermal energy into the atmosphere at a higher rate than uncleared land.
- Increased concentrations of greenhouse gases around cities can also act as urban heat traps. This leads to increased localised temperatures and more thermal energy in cities than in rural areas.

Studies by the bureaus of meteorology around the world have found that the centre of a modern city may be several degrees warmer than the surrounding suburbs and countryside. For example, the Melbourne Central Business District (CBD) may be as much as 7°C warmer than outlying suburbs. This can be seen in the thermal image in Figure 5.2.6. Although that may sound great in the depths of winter, in mid-summer it becomes a cause of greater heat-related mortality (deaths), morbidity (illness) and damage to infrastructure, and it can be a factor in increased energy use as people rely on air-conditioning to stay cool.

Much of this additional heat energy comes from dark surfaces, such as bitumen roads, thermal energy waste by vehicles and poorly insulated buildings. The effect is compounded by more carbon dioxide in the air from car exhausts, building heating and so on. Dark surfaces absorb heat during the day and release energy overnight, keeping night-time temperatures in cities 2°C or so above those of surrounding suburbs.

FIGURE 5.2.6 In this thermal image of Melbourne, you can see the bluer park areas (such as Flemington racecourse, top left, or the MCG, bottom right) are cooler than roads, which crisscross the image in red.

Energy retention: Land clearing

Since European arrival, about 90% of native vegetation in the eastern temperate zone has been removed as a result of human activity, for building or crops, and across Australia about 44% of forests have been cleared, according to government reports. This has slowed in recent decades, largely because there is less forest left—about 39% of forest was cleared before 1972. The map in Figure 5.2.7 shows the areas most affected by clearing.

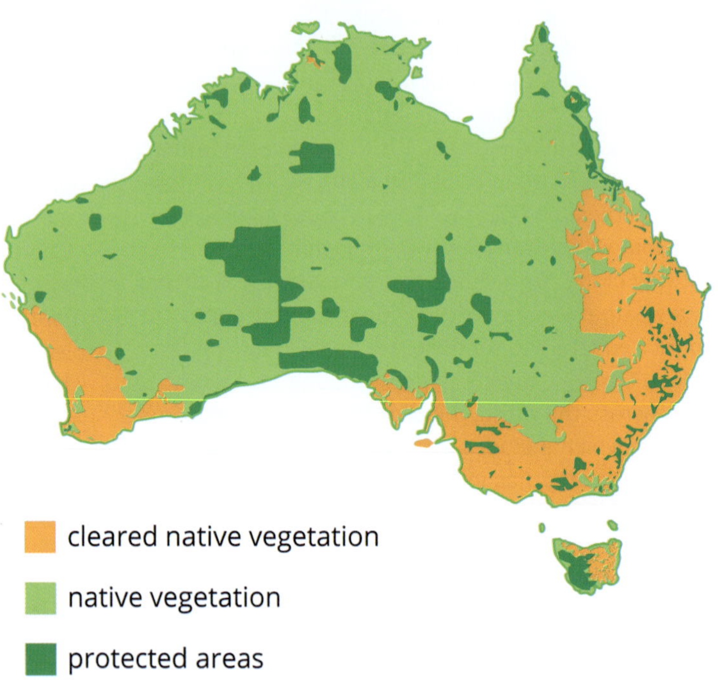

FIGURE 5.2.7 Land cleared in Australia since European arrival

Land clearing on a large scale changes the surface characteristics of the land and therefore the amount of thermal energy absorbed or reflected. In Australia, land clearing contributes approximately 12% of Australia's total emissions. Clearing vegetation causes:

- an increase in the number of dry days
- an increase in days over 35°C
- a decrease in daily rainfall intensity and cumulative rainfall on rainy days
- an increase in the duration of droughts.

Clearing native vegetation contributes to higher temperatures and decreased rainfall by reducing both shade and humidity.

Energy retention: Melting ice

Earth's surface is covered in many different types of snow and ice, including sea ice, glaciers and polar icecaps. This ice is important for our climate. Recall from Section 5.1 that bright, white surfaces reflect radiation much more than matte, dark surfaces. The bright surface of ice and snow reflects radiation from the Sun, which passes through the atmosphere back into space. Only 10% of radiation from the Sun is absorbed as heat by ice, whereas 94% of the Sun's radiation is absorbed by water. However, increasing temperatures are causing this ice to melt. In turn, this causes more heat to be absorbed, which causes more ice to melt.

Sea ice in the Arctic is at a minimum in September after the northern hemisphere summer. The extent of Arctic sea ice in September 2020 was the lowest on record. It is estimated that by 2035 there may no longer be ice in the Arctic in summer. Since 1992, the giant ice sheets that cover Greenland and Antarctica have each lost more than 100 billion metric tons of ice each year on average.

HOW HEAT MOVES AROUND EARTH

Scientists build models that can predict the effects the enhanced greenhouse effect is having, or will have, on the climate. This requires an understanding of the processes by which thermal energy is moved around Earth.

The main mechanisms for moving heat around Earth are conduction, convection and radiation. Evaporation also moves heat energy around Earth's atmosphere and is sometimes considered a fourth process. More correctly though, evaporation is a product of heat transfer due to conduction, convection and radiation. (These processes are described in more detail in Chapter 4.)

Heat flow inside Earth

Although most of the thermal energy needed to support life on Earth comes from the Sun, a small proportion comes from **geothermal energy** (the internal heat of Earth itself).

At Earth's core, it is estimated that the temperature is the same or higher than that on the surface of the Sun, at about 7000 K.

Heat flow inside Earth is mainly through convection. Heat from the **mantle** (between the crust and the outer core) moves towards the surface, as shown in Figure 5.2.8. Heat flows constantly from within Earth to Earth's surface at a rate estimated to be around 47 terawatts (4.7×10^{13} W). That sounds like an enormous amount, but it is an average of 0.087 W m^{-2} or just 0.03% of the total radiant energy from the Sun that is absorbed by Earth.

When the hot magma reaches the surface, it transfers heat to the crust through conduction and then sinks back to the centre of Earth. Where the crust is thinnest, conduction is highest, as shown in the thermal image in Figure 5.2.9.

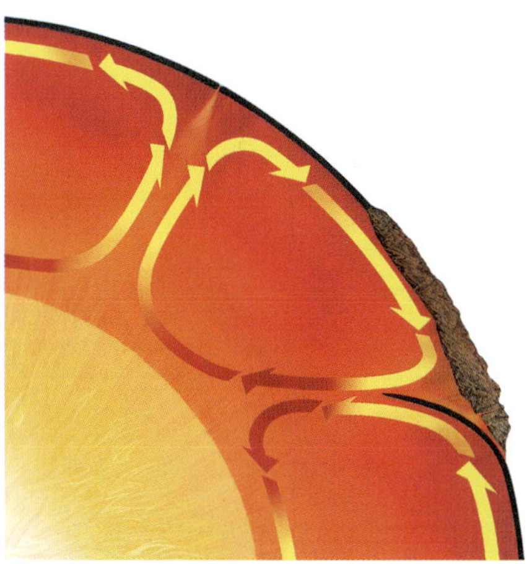

FIGURE 5.2.8 Mantle convection in the modern Earth. Hot plumes reach the outer layer of the upper mantle and fan out before finally sinking as cooler magma, having transferred thermal energy through Earth's crust by conduction.

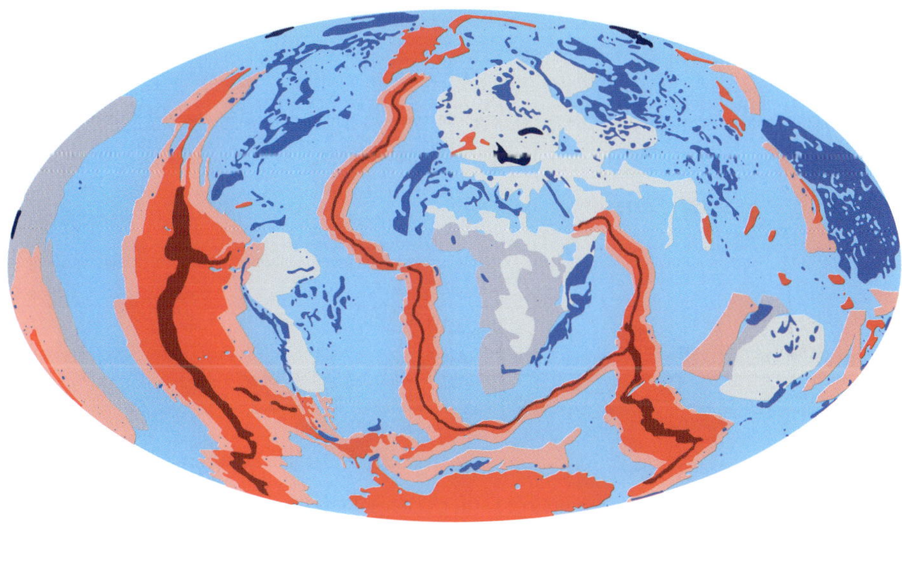

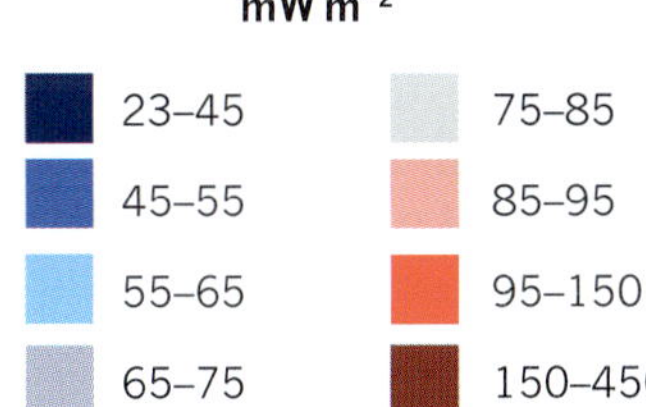

FIGURE 5.2.9 The flow of heat by conduction from Earth's interior to the surface. The highest heat exchanges are observed at the locations of mid-ocean fault lines, where the crust is thinnest.

PHYSICSFILE

Modelling complex behaviour

For decades, scientists have been making evidence-based predictions about our climate. Much research has gone into predicting how increasing greenhouse gas levels will affect global temperatures, and what effects these rising temperatures will have, such as making sea levels rise and causing extreme weather events and stronger hurricanes. With so many different factors affecting our climate and our weather, this is a complex task that relies on sophisticated mathematical modelling.

In 2021, Syukuro Manabe, Klaus Hasselmann and Giorgio Parisi were awarded the Nobel Prize in Physics for their work on modelling complex systems, including our climate. Manabe developed some of the first climate models in the 1960s. These models linked increased levels of carbon dioxide with increased temperatures and predicted global warming. Hasselmann developed models that connect our weather with our climate, and his work has shown that our climate models are accurate despite the chaotic behaviour of weather. The mathematical models of all three physicists have also been used in fields such as machine learning, biology and neuroscience.

Heat flow in the oceans

Ocean currents are mass movers of thermal energy over very large distances through the process of convection. The large-scale circulation of water and thermal energy via ocean currents is called the Great Ocean Conveyor Belt. It occurs because of variations in both water temperature and salinity (salt concentration).

Although the direction of ocean currents is not due to convection, warm water in the ocean will rise and cold water will sink through normal convection. Tropical regions of Earth's oceans receive a large amount of radiant energy from the Sun. This warmed water travels via currents towards the polar regions. The water cools at the poles, sinks and begins its journey back to the equator. Currents carrying warm, less-dense water move on the surface in one direction while cold, salty water moves in the opposite direction.

The major surface currents of Earth's oceans are caused by the wind and are influenced by land masses. The prevailing winds of Earth's atmosphere push the surface water along until it reaches land, at which point the currents will divert along the coasts of the land masses. In the major ocean basins, surface currents form circular patterns that are influenced by Earth's rotation. Currents flow clockwise in the northern hemisphere and anticlockwise in the southern hemisphere.

The Gulf Stream is part of the Great Ocean Conveyor Belt and is shown in the centre of Figure 5.2.10. The Gulf Stream carries the warm, salty water up along the east coast of the Americas, then towards Europe. It makes the climate of Western Europe much warmer than that of other regions at the same latitudes. At colder northern latitudes, the water becomes so dense that it sinks to the sea floor and travels south (shown in blue).

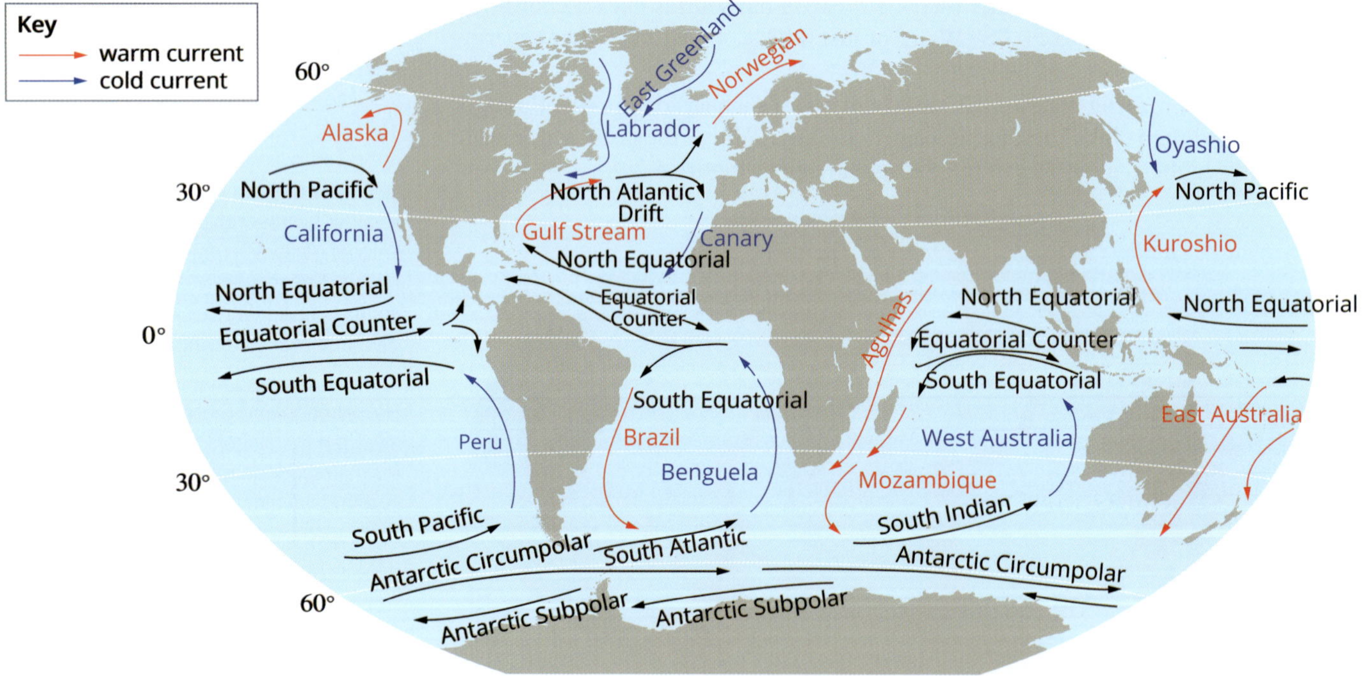

FIGURE 5.2.10 The major surface currents of Earth's oceans. Warm currents are shown in red, cold currents in blue.

Circulation via this process is extremely slow, taking about 1600 years to complete one cycle. Although slow, this system of ocean currents plays a major role in determining the climate of many of the regions of Earth.

Earth's present system of ocean currents is one of delicate balance. Minor changes can upset the balance of temperature differentials and prevailing winds. Scientists modelling Earth's climate have become concerned at the significant melting of Greenland's ice caps. There is also concern about large-scale melting of ice in the Antarctic. This produces large volumes of low-density fresh water that could, for example, prevent the surface currents in the north from sinking and returning as a southwards deep current. In turn, this would cause oceans and regions closer to the equator to warm while northern Europe would get colder.

Heat flow in the atmosphere

Earth's atmosphere is an essential part of our climate system. It is able to absorb and store thermal energy, so it acts as a heat sink and has a major impact on our climate. By transferring thermal energy quickly around Earth, our atmosphere regulates the temperature of Earth and keeps it stable.

Energy is transferred within the atmosphere by:

- radiation. Radiant energy largely comes from the Sun as short-wavelength radiation and, to a lesser degree, as long-wavelength reflection and emission from Earth's surface. Greenhouse gases retain radiant energy within the atmosphere.
- conduction. This happens only in the very low levels of Earth's atmosphere, as air is a very poor conductor of thermal energy.
- convection. It is the major process by which thermal energy is moved around Earth's atmosphere.

At the coast there is often a temperature difference between the land and the sea. The water in the sea hardly changes temperature between night and day due to its high specific heat capacity (covered in Section 4.2), but the land can become much hotter through the day. The air in contact with the land becomes hotter than the air in contact with the sea. As the air over the land is heated, it rises and is replaced by cooler, denser air from over the sea. Figure 5.2.11 shows how the cycle works.

This moving air creates a sea breeze and is experienced in most coastal areas in Australia during summer. This makes the coastal climate more pleasant on a hot day than in inland regions.

At night the process is reversed, as shown in Figure 5.2.12. The land and the air above it cools more quickly. This cooler, denser air moves out over the ocean, displacing the now relatively lighter and warmer air, and a land breeze is created.

On a global scale, the radiant energy from the Sun at the equator heats the air and it becomes less dense. Cooler, denser air moves in and forces the warm air higher into the atmosphere. This creates an area of low pressure. Once the warm air is high in the atmosphere, it spreads out towards the poles, where it cools down. The cooler air sinks back to Earth's surface creating areas of high pressure. These circular currents are created purely by convection and are one of the main transport mechanisms of thermal energy in the atmosphere.

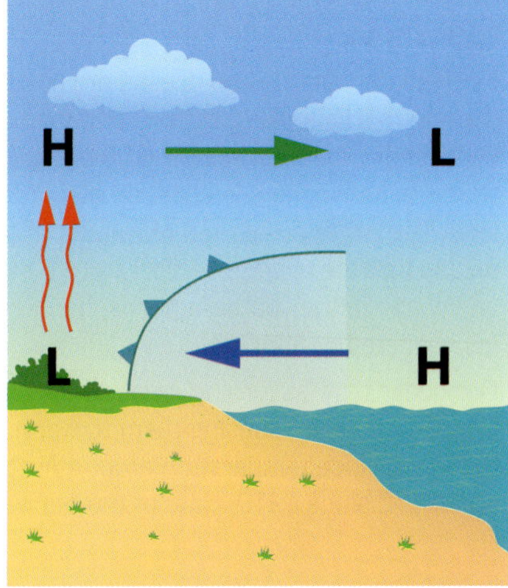

FIGURE 5.2.11 Sea breezes are created by convection currents caused by the temperature difference between air over the sea and air over the land. In this diagram, L represents regions of relatively low pressure, while H shows regions of higher pressure.

FIGURE 5.2.12 The convection currents created at night are the opposite of those during the day. At night, the land cools more quickly, denser air moves over the ocean and a land breeze occurs.

PHYSICSFILE

Earth's oceans as a temperature buffer

Water has a significantly higher specific heat capacity than air, as discussed in Chapter 4. As such, the world's oceans provide a significant temperature buffer and climate-stabilising effect, absorbing thermal energy when the oceans are colder than the atmosphere and releasing it when they are warmer than the atmosphere. This is evident in differences in seasonal weather and temperature between the northern and southern hemispheres.

Earth's land masses and oceans are not distributed evenly. You can see this by looking at the world map on the right. The northern hemisphere is approximately 61% ocean and 39% land. The southern hemisphere is approximately 81% ocean and just 19% land. The larger proportion of water in the southern hemisphere means that the average temperature variation between summer and winter in the southern hemisphere is 7.3°C. In the northern hemisphere the temperature variation is up to 14.3°C.

The difference in proportion of land and sea between the southern and northern hemispheres is immediately apparent when you look at (composite) satellite images.

Our oceans are absorbing much of the extra energy that is trapped on Earth, causing them to warm. According to CSIRO, the sea surface temperature around Australia has warmed by about 1°C since 1910. Nine of the ten warmest years on record have occurred since 2010. An increase of 1°C might not sound like much, but it has already changed the geographic distribution of some species of marine life. Species that cannot move, such as coral reefs, are dying. Also, as the water warms it increases in volume, which, combined with the melting of the polar ice caps and glaciers, is causing the sea levels to rise.

CASE STUDY

Australia's climate

The modelling of the influence of the enhanced greenhouse effect on Australia's climate is affected by two other major climate phenomena:

- the Southern Oscillation
- the Indian Ocean Dipole.

Southern Oscillation

The Southern Oscillation is a sequence of changes to the way the atmosphere and water circulate across the Pacific Ocean. Changes to the Southern Oscillation have significant effects on the climate of the countries across the tropical regions of the Pacific Ocean, including Australia.

At one extreme of the Southern Oscillation is an El Niño event. This event causes drier conditions in eastern Australia, often leading to droughts. On the other side of the Pacific, South America experiences warmer, wetter conditions. The El Niño event is illustrated in Figure 5.2.13.

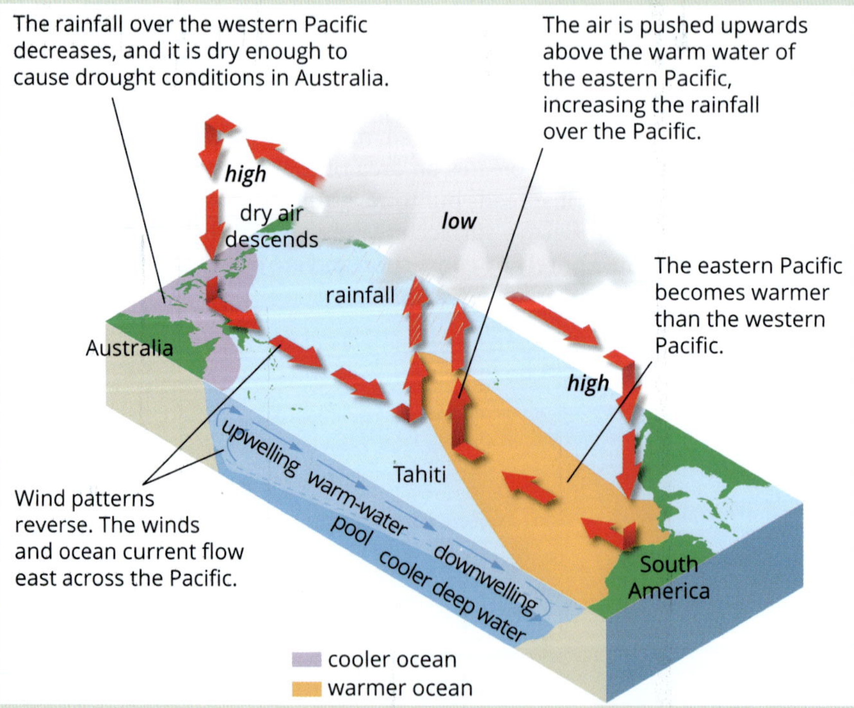

FIGURE 5.2.13 Conditions for an El Niño event. While South America may experience wetter conditions, large areas of Australia will experience hotter, drier conditions.

At the other extreme is a La Niña event, shown in Figure 5.2.14. During this event north-eastern Australia, Malaysia, the Philippines and Indonesia experience wetter conditions.

Climate scientists are predicting that the enhanced greenhouse effect will amplify El Niño and La Niña events, so the effects will be increased, they will occur more often and they will affect a wider area.

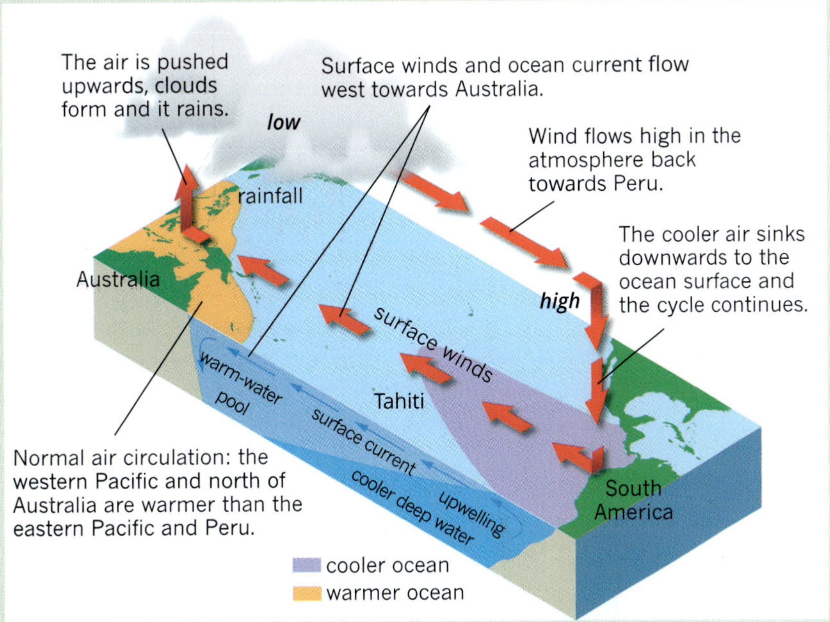

FIGURE 5.2.14 Conditions for a La Niña event. South America will be drier while eastern Australia will experience wetter conditions and stronger cyclones in northern regions.

Indian Ocean Dipole

The Indian Ocean Dipole is a cycle of change in the water temperature between the eastern and western areas of the Indian Ocean that borders Australia's western coast. Its effects are not as strong as the Southern Oscillation. Cool surface waters in the eastern Indian Ocean near Western Australia mean cooler, drier air. Hence there is less rainfall, particularly in central and southern Australia, as shown in Figure 5.2.15(a).

When warm waters are near Australia, moist air circulates over north-western Australia, and southern regions can expect more rainfall. This can be seen in Figure 5.2.15(b). The size of the difference between the sea temperatures and how long this difference lasts will affect the length of dry periods.

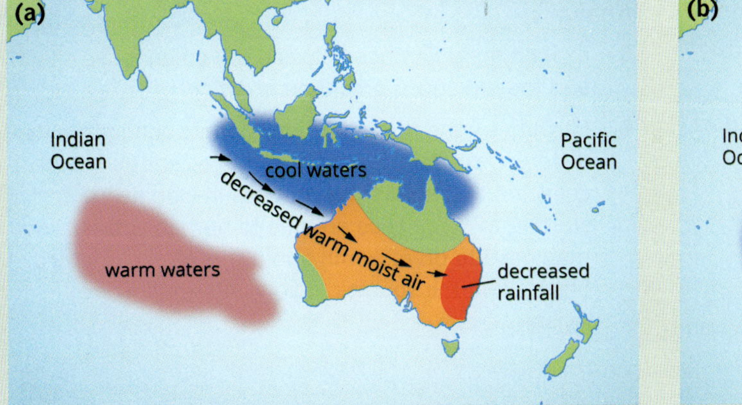

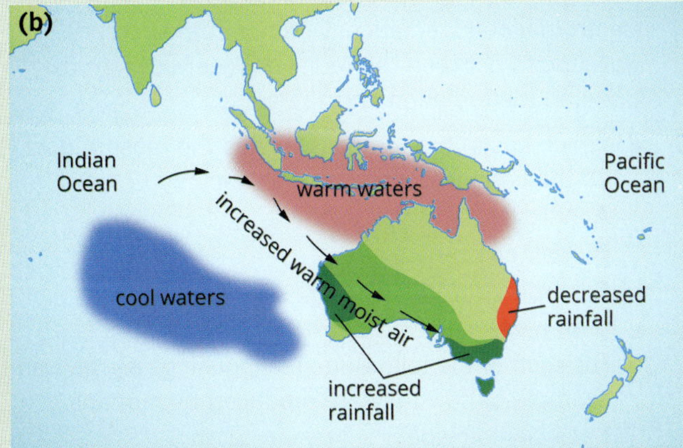

FIGURE 5.2.15 (a) Cooler water off the coast of Australia causes lower rainfall in central and southern Australia. (b) Warmer waters to the north cause increased rainfall.

5.2 Review

OA ✓✓

SUMMARY

- The overall temperature of Earth is determined by the total thermal energy received, largely from the Sun, and the amount that is lost back to space, outside Earth's atmosphere. Any change in that balance will lead to a warming or cooling of Earth as a whole.
- Only about half of the incoming radiant energy from the Sun is eventually absorbed by Earth's surface. This is because about 23% is absorbed by the atmosphere. Of the remainder, clouds reflect about 26% back towards outer space and Earth's surface reflects about 4%.
- The energy that reaches Earth's surface heats up the surface and is then partially radiated back out into space as longer-wavelength radiation.
- Specific gases present in very small proportions in the lower atmosphere absorb the long-wavelength radiation from Earth and re-radiate it back to Earth's surface.
- Evidence suggests that the atmosphere is absorbing and retaining more of the long-wavelength infrared radiation from Earth's surface, in an enhanced greenhouse effect. This is due to increased concentrations of greenhouse gases due to human activities since the mid-1700s.
- The rate at which the energy is reflected or absorbed and re-radiated by Earth depends on the surface material. Cleared land and built-up areas absorb and re-emit larger amounts of heat than uncleared land. Snow and ice reflect large amounts of radiation.
- A small proportion of the thermal energy heating Earth comes from Earth itself.
- Movement of thermal energy through Earth occurs by conduction at the surface through the crust and convection through the mantle deep within Earth.
- Ocean currents are mass movers of thermal energy over very large distances. The large-scale ocean circulation of water is called the Great Ocean Conveyor Belt.
- Convection is the major process by which thermal energy is moved around Earth's atmosphere.

KEY QUESTIONS

Knowledge and understanding

1. What is the main source of thermal energy heating Earth?
2. Radiant energy from the Sun reaches Earth largely as electromagnetic radiation. Describe the difference in wavelength between the radiation received from the Sun and the radiation Earth re-emits.
3. **a** Which gas in the atmosphere has the most impact on the enhanced greenhouse effect?
   A nitrogen
   B methane
   C carbon dioxide
   D oxygen
   b For each of the following greenhouse gases, give an example of how they are produced by humans.
   i carbon dioxide
   ii methane
   iii CFCs
   iv nitrous oxide
4. Describe the major process by which thermal energy moves in:
   a Earth's mantle
   b Earth's atmosphere.

Analysis

5. Explain the difference between the greenhouse effect and the enhanced greenhouse effect.
6. Earth has gone through many climate changes in its history. Describe the evidence that scientists have collected that shows the changes in climate we are starting to experience are due to human activity. You may need to do research about this on the internet. Ensure that the sources of your information are reliable.

Chapter review

KEY TERMS

black body
black-body radiation
emissivity
enhanced greenhouse effect
fossil fuel
geothermal energy
greenhouse effect
greenhouse gas
mantle
peak wavelength
Wien's law

REVIEW QUESTIONS

Knowledge and understanding

1 What does the enhanced greenhouse effect refer to?

2 How do high-density urban areas and land clearing contribute to the enhanced greenhouse effect?

3 Select the correct statement about the total radiant thermal energy re-radiated by Earth.
  - **A** About 12% is directly lost to space.
  - **B** About 50% is directly lost to space.
  - **C** 100% is absorbed by the atmosphere.
  - **D** 100% is reflected back to Earth by the atmosphere.

4 Complete the paragraph below by choosing the correct response from the choices given in brackets.
Radiant energy from the Sun reaches Earth largely as electromagnetic radiation in or near the [**infrared/visible/ultraviolet/radio**] wavelengths. Earth reflects radiant energy as [**shorter/same/longer**] wavelength, [**infrared/visible/ultraviolet/radio**] radiation.

5 Why does carbon dioxide have such a big impact on the enhanced greenhouse effect?

6 State the type(s) of heat transfer involved in each situation.
  - **a** Oceans move thermal energy over large distances.
  - **b** Heat is transferred around Earth's atmosphere.
  - **c** The Sun's energy reaches Earth.
  - **d** Heat is transferred around the surface and centre of Earth.

7 Describe how sea breezes form.

Application and analysis

8 Thermal imaging technology can be used to locate people lost in the Australian bush. How can thermal imaging technology 'see' people when the naked eye cannot?

9 A light globe is labelled as 'cool daylight' and another as 'warm white'. The 'warm white' globe appears more yellow than the 'cool daylight' globe. Based on this observation, which of the statements below is correct?
  - **A** The apparent temperature of the 'warm white' globe is higher than the one marked 'cool daylight'.
  - **B** The apparent temperature of the 'warm white' globe is lower than the one marked 'cool daylight'.
  - **C** Both globes have the same apparent temperature.
  - **D** The apparent temperature of the globes can't be determined from the information available.

10 A black solar water heater has an emissivity of 0.98. What does this imply about the heater's ability to
  - **a** absorb radiant thermal energy?
  - **b** emit radiant thermal energy?

11 Earth's centre is estimated to be at a temperature of 7000 K. Using Wien's law, determine the peak wavelength of the radiant energy emitted at this temperature.

12 The floor of the Sustainability Learning Centre in Tasmania is heated by the Sun's radiation to a comfortable 30°C. The floor re-emits this thermal energy as radiation, heating the room. What would be the peak wavelength of the re-emitted radiation?

13 The quartz element of particular radiant heater glows orange at a peak wavelength of 650 nm. Based on this wavelength, what temperature, in kelvin, is the element of the heater?

14 The surface of a particular star is at a temperature of 9300 K. With what region of the electromagnetic spectrum does the peak wavelength of the radiant thermal energy for this temperature coincide?

continued over page

15 The James Webb Space Telescope is designed to detect infrared light, in contrast to the Hubble Space Telescope, which detects near-infrared, visible and ultraviolet light. Infrared radiation is not as affected by dust and gas in space, allowing astronomers to see objects that are dimmer and further away.

a Using your knowledge of Earth's atmosphere, explain why infrared astronomy is difficult to carry out from Earth.

b Would you expect the objects the James Webb Space Telescope will view to be hotter or colder than those the Hubble can view? Explain your answer.

c Astronomers must also allow for red shift when making calculations about the universe. Radiation from objects that are moving away from us appears to have a longer wavelength than it did at its source—it's shifted to the red end of the spectrum. If a student calculated the temperature of a distant star moving away from Earth using data from the telescope and Wien's law, would that star be hotter or colder than suggested by their calculation?

OA
✓✓

REVIEW QUESTIONS

WS 8

How are light and heat explained?

Multiple-choice questions

The following information relates to questions 1 and 2.

The diagram shows the displacement of the air molecules in a sound wave from their mean positions as a function of distance from the source, at a particular time. The wave is travelling to the right at 340 m s^{-1}.

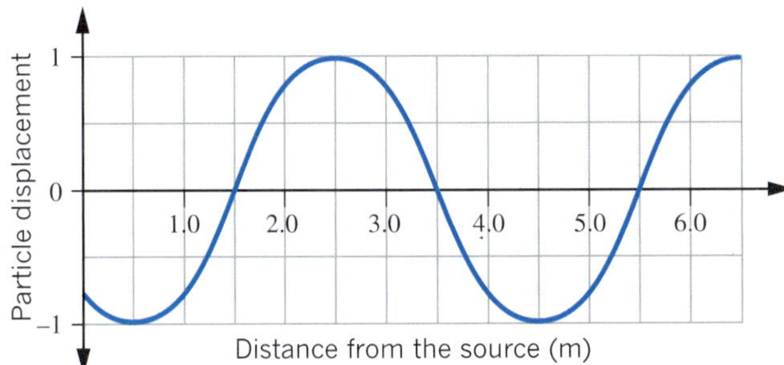

1 What is the wavelength of the sound wave?
- **A** 1.0 m
- **B** 2.0 m
- **C** 4.0 m
- **D** 5.0 m

2 Which arrow below describes the direction of transfer of acoustic energy by this wave?
- **A** →
- **B** ←
- **C** ↑
- **D** ↓
- **E** No energy is transferred.

3 Which of the following properties of sound is independent of the source producing the sound?
- **A** frequency
- **B** amplitude
- **C** wavelength
- **D** speed

4 What wavelengths of the electromagnetic spectrum does the Sun mainly emit?
- **A** microwaves, infrared, visible
- **B** infrared, visible, ultraviolet
- **C** visible, ultraviolet, X-rays
- **D** ultraviolet, X-rays, gamma

5 Light is travelling through air (n_{air} = 1.00) and is incident on a glass block with refractive index of n_{glass} = 1.52. Which statement is true about the velocity of the incident light travelling through these mediums?
- **A** $v_{air} > v_{glass}$
- **B** $v_{air} < v_{glass}$
- **C** $v_{air} = v_{glass}$
- **D** Not enough information is provided.

6 In which of the following scenarios will total internal reflection **not** occur?
- **A** The incident light source is red visible light.
- **B** The incident light is travelling from air into water.
- **C** The angle of incidence is greater than the critical angle.
- **D** The incident light is travelling from a higher refractive index material to a lower refractive index material.

7 Three metals A, B and C are placed so that all three are in thermal contact with one another. The only flow of heat that occurs is from A to B and from C to B. What can you say about the relative temperatures of metals A and C?
- **A** A is at a higher temperature than C.
- **B** A is at a lower temperature than C.
- **C** A is at the same temperature as C.
- **D** There is insufficient information to compare the temperatures of A and C.

8 A chemical engineer is doing a gas law calculation and understands that she needs to use temperature in kelvin. Her thermometer in the reactor vessel reads the temperature as 1550°C.

What temperature in kelvin would this be?
- **A** 1277 K
- **B** 1550 K
- **C** 1823 K
- **D** 2732 K

9 In the constellation Orion, Rigel is blue-white and Betelgeuse is reddish. Based on this observation alone, what conclusion can be made?
- **A** Rigel is further away than Betelgeuse.
- **B** Rigel is closer than Betelgeuse.
- **C** Rigel is cooler than Betelgeuse.
- **D** Rigel is hotter than Betelgeuse.

10 If boiling water changes to steam at the same temperature, which of the following has/have changed for the water molecules in the steam? More than one answer may be correct.
- **A** average kinetic energy of the particles
- **B** average potential energy of the particles
- **C** total internal energy of the particles
- **D** the total number of particles

11 Which of the following examples supports the statement that it takes a larger amount of thermal energy to melt ice than to warm air?

A Moisture forms on the outside of a glass of cold water.
B Ice cubes in a freezer can be colder than 0°C.
C Glaciers and ice flows last throughout summer in some areas of New Zealand.
D Snow storms can occur at low altitudes in winter during extreme weather conditions.

12 A bucket is filled with equal amounts of hot and cold water. The hot water is originally at 80°C and the cold water at 10°C. The temperature of the final mixture will be approximately:

A 10°C
B 45°C
C 70°C
D 90°C

Short-answer questions

13 Define what is meant by a mechanical wave.

14 Compare transverse and longitudinal waves.

15 Determine the wavelength and amplitude of the wave depicted in the following graph.

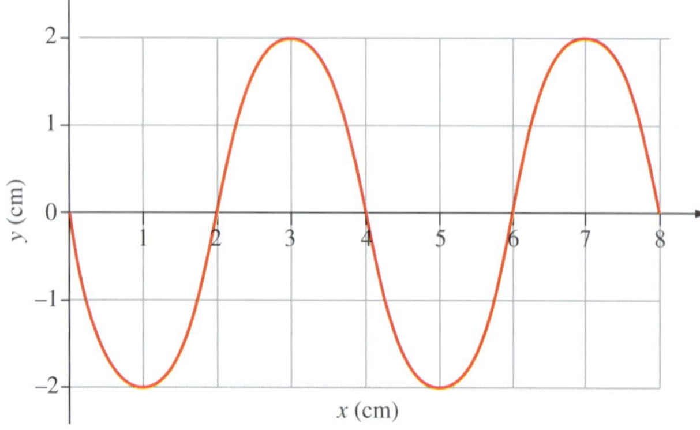

The following diagram relates to question 16.

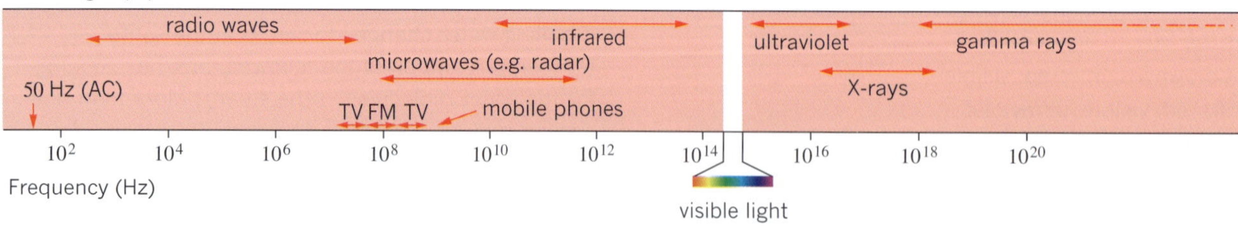

16 A source emits electromagnetic radiation with a frequency of 10^{16} Hz.

a What is the speed of these waves travelling in air?
b What is the wavelength of these waves?
c What specific type of electromagnetic radiation are these waves?
d Explain one helpful use of electromagnetic radiation of this wavelength.

The following information relates to questions 17–19.

Light travelling in air (n = 1.0) enters a second medium of refractive index 2.4 at an angle of incidence of 30°.

17 What is the angle of refraction?

18 Calculate the speed of light in medium 2.

19 The light source has moved and is now travelling from the second medium into air. What is the angle of incidence required for total internal reflection to occur in the second medium (the critical angle)?

The following information relates to questions 20 and 21.

Two students were given the task by their teacher to demonstrate the dispersion of white light in the school science laboratory.

20 What equipment would they need and how would they demonstrate dispersion?

21 What visible result will the students achieve if they perform the experiment correctly?

The following information relates to questions 22 to 24.

When a copper rod at room temperature (25°C) is placed in a 1500°C furnace, heat flows into the rod.

22 Describe the flow in terms of the first law of thermodynamics.

23 Explain what happens to the kinetic energy of the metal atoms and relate this to the temperature of the metal rod.

24 The copper rod begins to glow bright red and is removed and placed on a steel plate in a cooling chamber filled with nitrogen gas. Explain the different means by which the copper rod can lose heat.

25 Hot water tanks are often located outside the home and the pipe carrying the hot water away from the tank will be run along an outer wall for part of the journey to the tap inside the home. This pipe is usually wrapped in a foam sleeve about 1 cm thick. The pipe carrying the cool water to the hot water tank is not covered in a foam sleeve. Explain, using thermal energy principles, this difference.

26 A student attempts to identify a metal by measuring its specific heat capacity. A 100 g block of the metal is heated to 75.0°C and then transferred to a 70.0 g copper calorimeter containing 200 g of water at 20.0°C. Before the metal sample is transferred, the copper calorimeter and the water are at thermal equilibrium. When they return to thermal equilibrium the final temperature of the system is 25.0°C.

Using the table below, what metal is the student probably testing?

Material	Specific heat capacity (J kg^{-1} K^{-1})
human body	3500
methylated spirits	250
air	1000
aluminium	900
glass	840
iron	440
copper	390
brass	370
lead	130
mercury	140
ice	2100
liquid water	4200
steam	2000

The following information applies to questions 27–30.

An entrepreneur is considering opening an ice cream parlour that makes ice cream using liquid nitrogen. The liquid nitrogen serves the dual purpose of freezing the water in the ice cream and aerating it at the same time. He intends to make ice cream from a cream and sugar mix that is 70% water. He assumes that only the water needs to freeze to set the ice cream.

He collects the following data:

Boiling temperature of liquid nitrogen at atmospheric pressure: 77.0 K

Specific heat capacity for nitrogen gas: 1.34 kJ kg^{-1} K^{-1}

Heat of vaporisation of liquid nitrogen: 199 kJ kg^{-1}

Specific heat capacity of cream and sugar mix: 3.80 kJ kg^{-1} K^{-1}

Heat of fusion of water: 334 kJ kg^{-1}

27 How much heat is required to vaporise 1.00 kg of liquid nitrogen at 77.0 K?

28 How much heat does 1.00 kg nitrogen gas absorb as it heats from 77.0 K to 0°C?

29 How much heat must be removed to cool 200 g of refrigerated cream and sugar mix at 8.00°C down to make ice cream at 0°C?

30 What mass of liquid nitrogen does the entrepreneur need per 200 g ice cream portion if he takes the sugar and cream mix from the fridge at 8.00°C and freezes it at 0°C?

31 A 50.0 g minted coin is at 250.0°C when it is dropped into 500 g of water in a beaker at 20.0°C. If the final temperature of the system is 22.5°C, calculate the specific heat capacity of the metal. (Specific heat capacity of water: 4200 J kg^{-1} K^{-1})

32 Wien's law can be written as $l_{max}T = 2.898 \times 10^{-3}$ m K, where l is in m and the temperature is in K.

a Calculate the peak radiation wavelength for the Sun at 5778 K and the range of the electromagnetic spectrum this wavelength emits.

b Calculate the temperature of a piece of steel glowing red hot in a furnace that radiates at 4.14 μm.

33 Explain using your knowledge of thermal energy and electromagnetic radiation the following observations about the planets Mercury and Venus.

a Mercury is the planet closest to the Sun. It has essentially no atmosphere. The surface temperature of the side of the planet facing the Sun is typically around 350–400°C while the surface of the side away from the Sun is typically around −160°C.

b Venus is the second closest planet to the Sun and it has a dense atmosphere of approximately 96% carbon dioxide. The surface temperature of both the Sun-facing and non-facing sides of the planet is typically around 450°C.

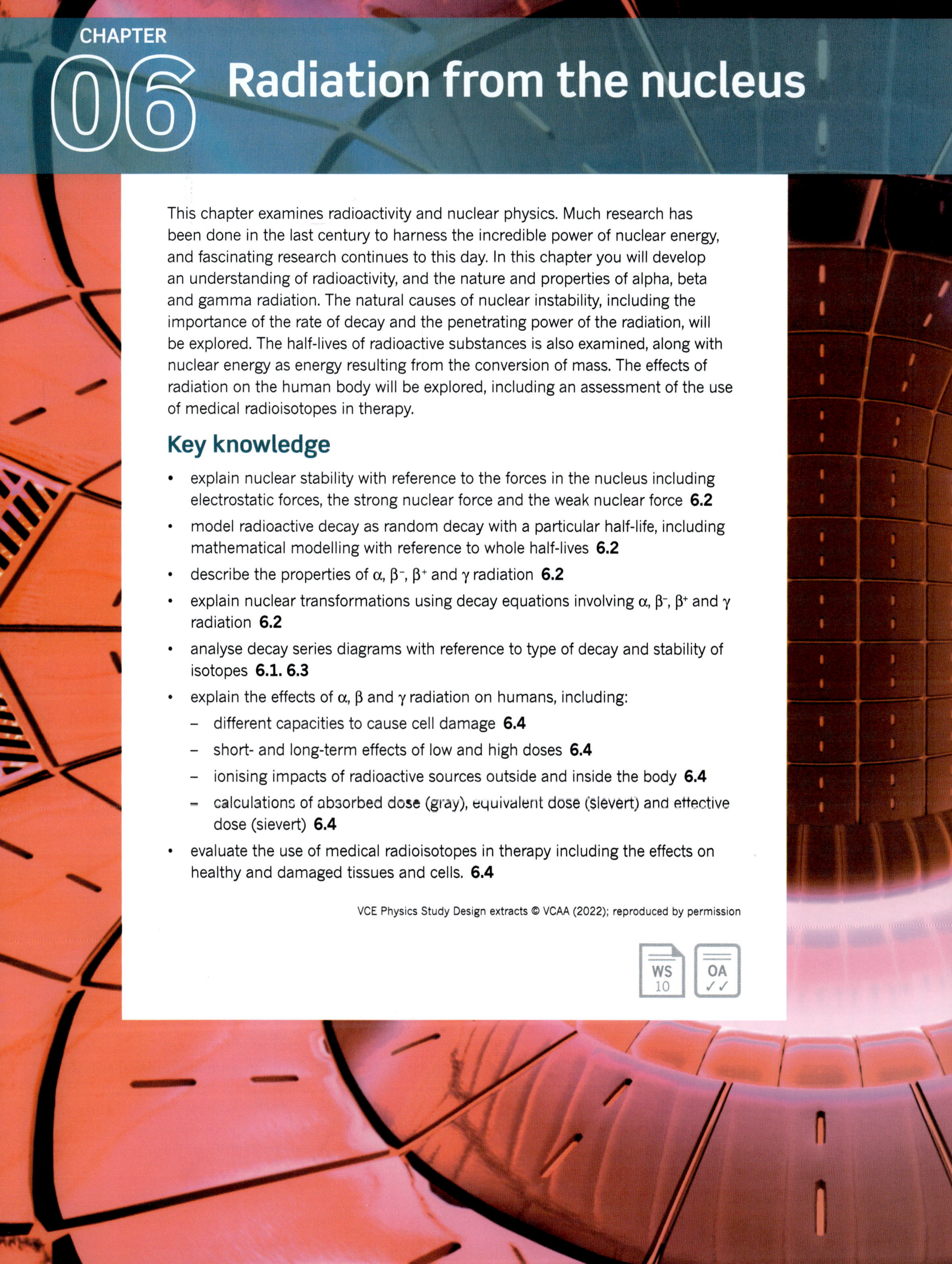

CHAPTER

06 Radiation from the nucleus

This chapter examines radioactivity and nuclear physics. Much research has been done in the last century to harness the incredible power of nuclear energy, and fascinating research continues to this day. In this chapter you will develop an understanding of radioactivity, and the nature and properties of alpha, beta and gamma radiation. The natural causes of nuclear instability, including the importance of the rate of decay and the penetrating power of the radiation, will be explored. The half-lives of radioactive substances is also examined, along with nuclear energy as energy resulting from the conversion of mass. The effects of radiation on the human body will be explored, including an assessment of the use of medical radioisotopes in therapy.

Key knowledge

- explain nuclear stability with reference to the forces in the nucleus including electrostatic forces, the strong nuclear force and the weak nuclear force **6.2**
- model radioactive decay as random decay with a particular half-life, including mathematical modelling with reference to whole half-lives **6.2**
- describe the properties of α, β^-, β^+ and γ radiation **6.2**
- explain nuclear transformations using decay equations involving α, β^-, β^+ and γ radiation **6.2**
- analyse decay series diagrams with reference to type of decay and stability of isotopes **6.1. 6.3**
- explain the effects of α, β and γ radiation on humans, including:
  - different capacities to cause cell damage **6.4**
  - short- and long-term effects of low and high doses **6.4**
  - ionising impacts of radioactive sources outside and inside the body **6.4**
  - calculations of absorbed dose (gray), equivalent dose (sievert) and effective dose (sievert) **6.4**
- evaluate the use of medical radioisotopes in therapy including the effects on healthy and damaged tissues and cells. **6.4**

WS 10

OA ✓✓

6.1 Atoms, isotopes and radioisotopes

Many people mistakenly think that they never come into contact with radioactive materials or the radiation that these materials produce. However, Earth is a radioactive planet and it is impossible to avoid exposure to radioactivity. Our senses cannot detect the radiation from radioactive atoms. High-energy radiation in higher than normal doses can be damaging to living tissue. Radiation and radioactive elements can also be used in a variety of applications that are beneficial. These radioactive atoms, or radioisotopes, will be discussed in this section.

ATOMS

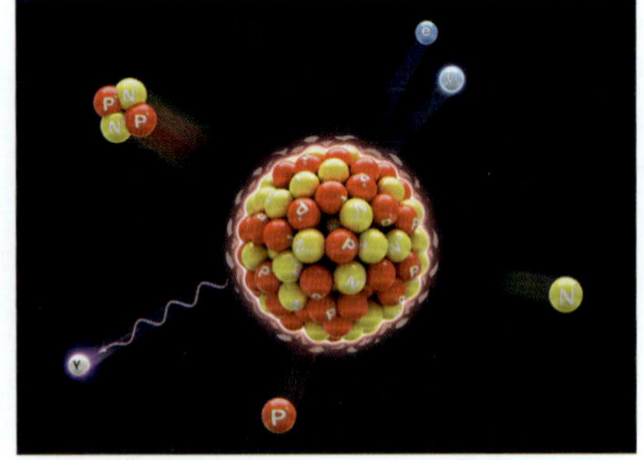

FIGURE 6.1.1 Radiation is spontaneously emitted from a radioactive nucleus.

If an atom is **radioactive**, it will spontaneously emit **radiation** from its nucleus. Figure 6.1.1 shows this radiation emitted in the form of particles and electromagnetic energy (light).

To understand radiation and radioactivity, it is necessary to know about the structure of the atom. The central part of an atom, the **nucleus**, consists of particles known as **protons** and **neutrons**. These particles are almost identical in mass and size, and collectively are called **nucleons**.

The nucleons have very different electrical properties. Protons have a positive charge. Neutrons are electrically **neutral**, so they have no charge. The nucleus contains nearly all of the atom's mass.

Most of the atom is empty space occupied only by negatively charged particles called **electrons**. These are much smaller and lighter than protons or neutrons. The nucleus of an atom occupies about 10^{-12} of the volume of the atom, yet it contains more than 99% of its mass. Figure 6.1.2 shows the structure of a typical atom.

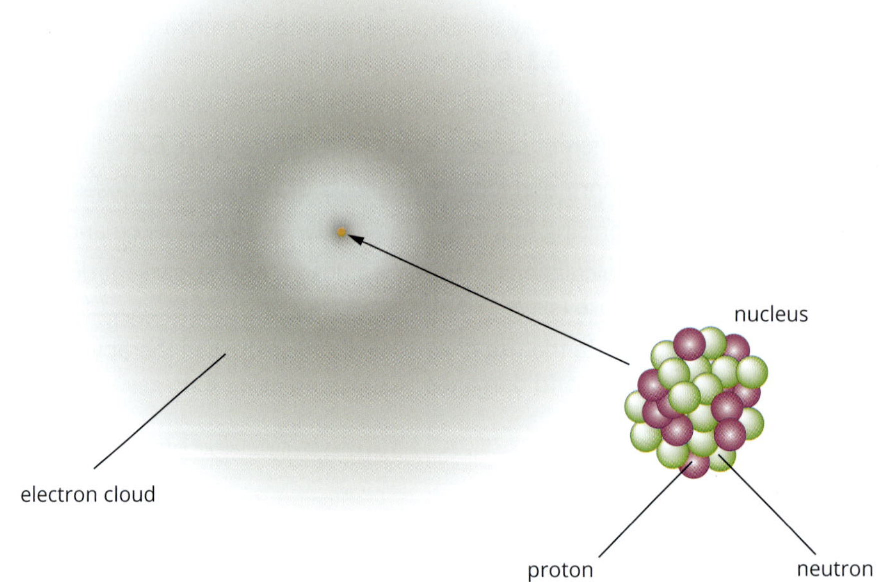

FIGURE 6.1.2 The typical structure of an atom. Atoms are mostly empty space. (Note, this atom is not drawn to scale.)

A particular atom can be identified by using an atomic symbol that has the format shown in Figure 6.1.3.

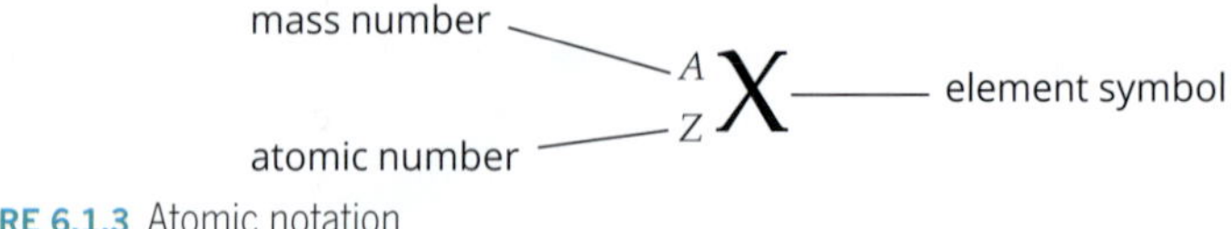

FIGURE 6.1.3 Atomic notation

The **mass number** (A) is the total number of protons and neutrons in the nucleus.

The **atomic number** (Z) is the number of protons in the nucleus.

Atoms with the same number of protons belong to the same element. For example, if an atom has six protons in its nucleus (i.e. $Z = 6$), then the atom must be carbon. The number of neutrons does not affect which element the atom is, but it does affect the mass of the atom. Figure 6.1.4 shows how the size of the nucleus depends on the mass number. The more protons and neutrons there are in a nucleus, the heavier and larger it is.

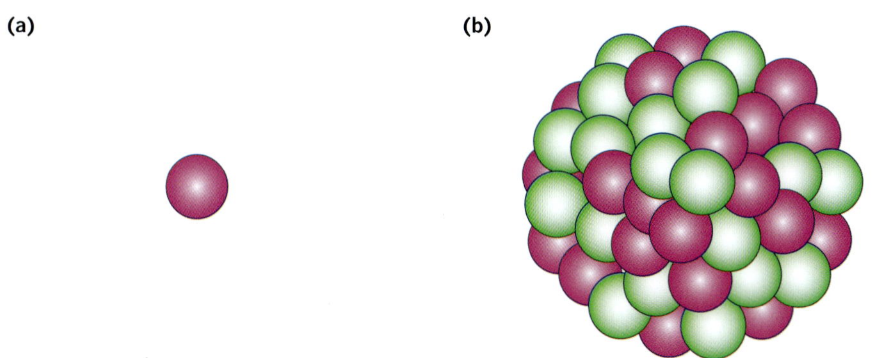

FIGURE 6.1.4 (a) and (b) are both nuclei; however, the nucleus of a hydrogen atom (a) is a very different size from that of a uranium atom (b). (Note, the nuclei are not drawn to scale.)

In an electrically neutral atom, the number of electrons is equal to the number of protons. For example, any neutral atom of uranium ($Z = 92$) has 92 protons in the nucleus and 92 electrons in the electron cloud.

ISOTOPES

All atoms of a particular element will have the same number of protons, but may have a different number of neutrons. For example, lithium exists naturally in two different forms. One form has three protons and three neutrons, the other has three protons and four neutrons. These different forms of lithium are called **isotopes** of lithium and are illustrated in Figure 6.1.5.

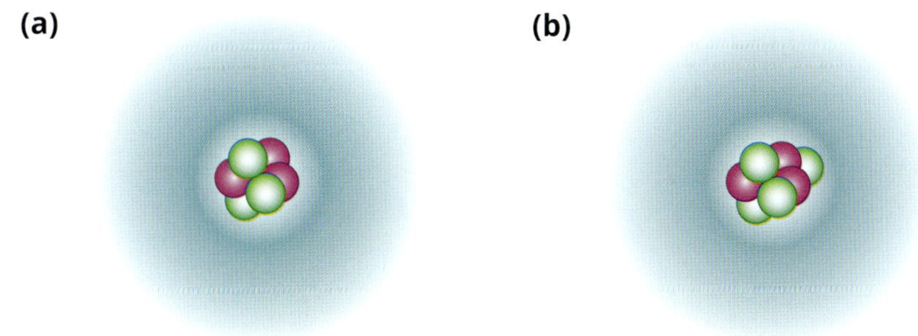

FIGURE 6.1.5 Two different isotopes of lithium: (a) ${}^{6}_{3}\text{Li}$ and (b) ${}^{7}_{3}\text{Li}$

Isotopes are atoms that have the same number of protons but different numbers of neutrons. Isotopes have the same chemical properties but different physical properties such as density and volume.

The term **nuclide** is used when referring to a particular nucleus. For example, lithium-6 is a nuclide that has three protons and three neutrons.

There are three isotopes of hydrogen: the nuclide with one proton is called hydrogen, the nuclide with one proton and one neutron is called deuterium, and the nuclide with one proton and two neutrons is called tritium.

PHYSICSFILE

Neutron stars

In the universe there are objects whose density is almost equal to that of nuclear matter. These are called neutron stars. They are gigantic balls with radii of ten or more kilometres, and are made only of neutrons—something like a gigantic atomic nucleus. A one-litre carton filled with this type of matter would weigh a thousand times more than the largest Egyptian pyramid (which weighs approximately 5 750 000 tonnes).

PHYSICSFILE

Heavy water

A compound of oxygen and deuterium has identical chemical properties to ordinary water. However, the molecular mass of ordinary water is about 18 (16 + 1 + 1), while the molecular mass of water containing deuterium is 20 (16 + 2 + 2). Thus, water that contains deuterium has a higher density (by about 11%) and is commonly known as 'heavy water'.

Worked example 6.1.1

WORKING WITH ISOTOPES

Consider the isotope of molybdenum, $^{95}_{42}\text{Mo}$. Work out the number of protons, nucleons and neutrons in this isotope.

Thinking	Working
The lower number is the atomic number. This gives the number of protons.	atomic number = 42 This nuclide has 42 protons.
The upper number is the mass number. This gives the number of particles in the nucleus, i.e. the number of nucleons.	mass number = 95 This nuclide has 95 nucleons.
Subtract the atomic number from the mass number to find the number of neutrons.	This isotope has 95 – 42 = 53 neutrons.

Worked example: Try yourself 6.1.1

WORKING WITH ISOTOPES

Consider the isotope of thorium, $^{230}_{90}\text{Th}$. Work out the number of protons, nucleons and neutrons in this isotope.

RADIOISOTOPES

Most of the atoms that make up the world around us are stable. Their nuclei have not altered in the billions of years since they were formed. These atoms will stay unchanged in the future. There are about 270 stable isotopes in nature. Tin ($Z = 50$) has ten stable isotopes, while aluminium ($Z = 13$) has just one.

There are also many naturally occurring isotopes that are unstable. An unstable nucleus may spontaneously become more stable by emitting a particle and so change into a different element or isotope. Unstable atoms are radioactive. An individual radioactive isotope is known as a **radioisotope** (sometimes called a radionuclide). Carbon has two stable isotopes: carbon-12 and carbon-13. Carbon also has one naturally occurring isotope that is unstable: carbon-14. The nucleus of a carbon-14 atom may spontaneously decay into a different substance, emitting high-energy particles that can be harmful. A known radioactive substance is identified by the radiation warning symbol shown in Figure 6.1.6.

FIGURE 6.1.6 This symbol is used to label and identify a radioactive source.

Figure 6.1.7 shows that every isotope of every element with an atomic mass equal to or greater than that of bismuth ($Z = 83$) is radioactive. (Bismuth was long thought to be stable, but in 2003 it was found to be very weakly radioactive with a half-life of over 20 billion billion years.) The first 92 elements are naturally occurring.

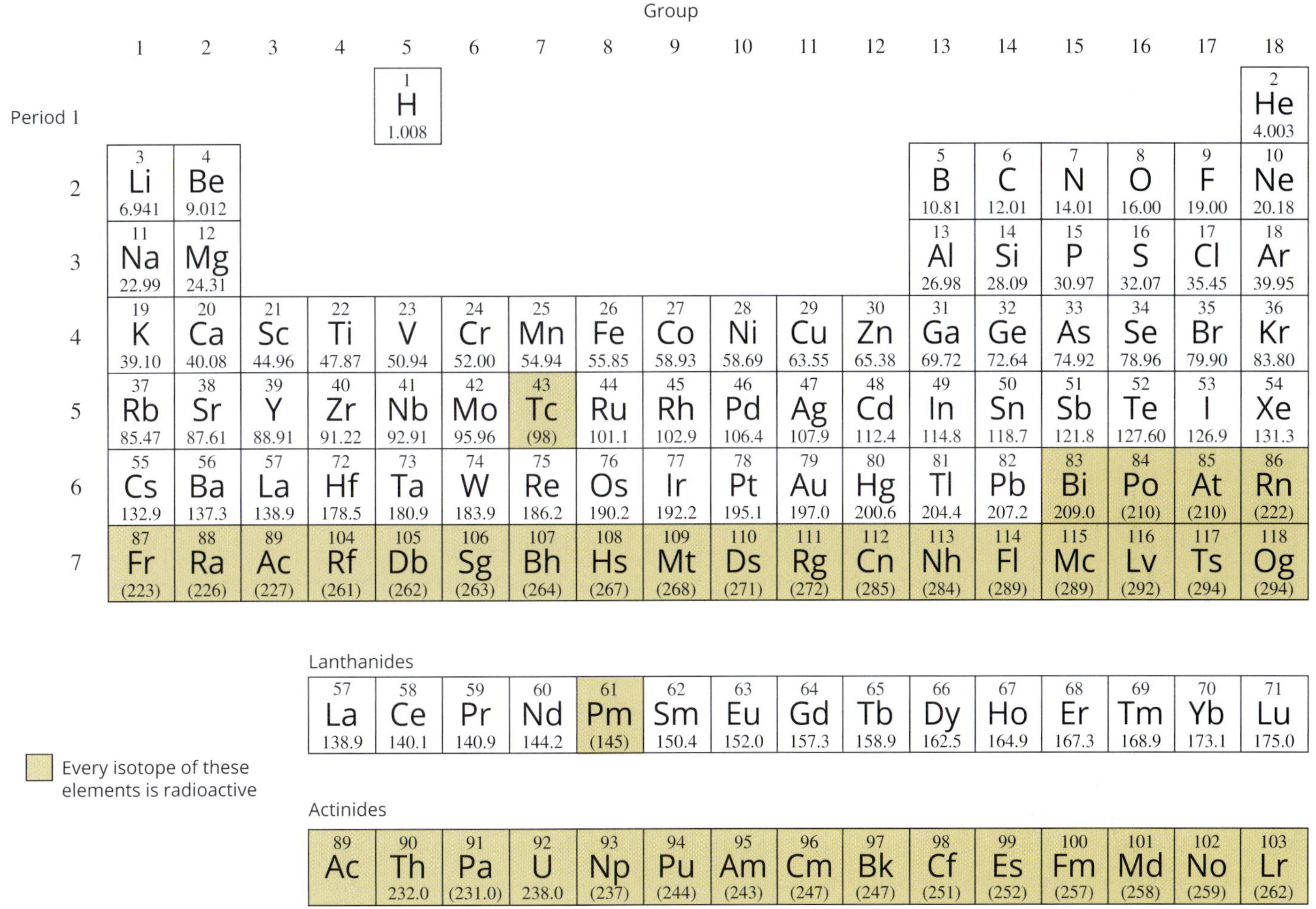

FIGURE 6.1.7 The periodic table of the elements

Most of the elements found on Earth have naturally occurring radioisotopes; there are about 200 of these natural radioisotopes. During the twentieth century, an enormous number of radioisotopes were also artificially produced. Most of the radioisotopes used in industry, medicine and for scientific research are artificially produced. Artificial radioisotopes are produced in nuclear reactors or particle accelerators.

PHYSICSFILE

Oganesson

The element with the highest atomic number and highest atomic mass so far discovered is element 118, oganesson (Og). Three atoms of this element were made in a particle accelerator in 2006 when calcium-48 nuclei were bombarded with californium-249 nuclei. The 20 protons of calcium combined with the 98 protons of californium to make just one or two atoms of Og. Og is very unstable and decays very rapidly, with a half-life of less than 1 ms.

6.1 Review

SUMMARY

- The nucleus of an atom consists of positively charged protons and neutral neutrons. Collectively, protons and neutrons are known as nucleons. Negatively charged electrons surround the nucleus.
- The nucleus of the atom is extremely small but contains most of the atom's mass.
- The atomic number, *Z*, is the number of protons in the nucleus. The mass number, *A*, is the number of nucleons in the nucleus; it is the combined number of protons and neutrons. Elements are represented as $^{A}_{Z}X$.
- Isotopes of an element have the same number of protons but different numbers of neutrons. Isotopes of an element are chemically identical to each other, but have different physical properties.
- An unstable isotope—a radioisotope—may spontaneously decay by emitting a particle from the nucleus.

KEY QUESTIONS

Knowledge and understanding

1. What is the collective term used for protons and neutrons?
2. How is the number of electrons in a neutral atom determined?
3. Explain the meaning of the term 'isotope'.

Analysis

4. What is the difference between a stable isotope and a radioisotope?
5. Can a natural isotope be radioactive? If so, give an example of such an isotope.

6.2 Radioactivity

At about the turn of the twentieth century, scientists such as Marie Curie were investigating the newly discovered radioactive substances polonium and radium. Ernest Rutherford (Figure 6.2.1) and Paul Villard found that there were three different types of emission from these mysterious substances. They named them alpha, beta and gamma radiation.

Further experiments showed that the alpha and beta emissions were actually particles expelled from the nucleus. Gamma radiation was found to be high-energy electromagnetic radiation (light) also expelled from the nucleus. The term **radioactive decay** refers to the process that emits these particles and electromagnetic radiation from a nucleus.

The nature of these radiations will be discussed in this section.

FIGURE 6.2.1 Ernest Rutherford

In any nuclear reaction, including radioactive decay, atomic number and mass number are conserved. Energy is released during these decays.

ALPHA (α) DECAY

When a heavy unstable nucleus undergoes radioactive decay, it may eject an **alpha particle**. This is a positively charged particle that consists of two protons and two neutrons. An alpha particle, symbol α, is identical to a helium nucleus and can also be written as $^{4}_{2}\text{He}$.

Uranium-238 is radioactive and may decay by emitting an alpha particle from its nucleus. Figure 6.2.2 shows the unstable nucleus of uranium-238 ejecting an alpha particle. This can be represented in a nuclear equation that shows the changes occurring in the nuclei. Electrons are not considered in these equations, only the nucleons. The equation for the alpha decay of uranium-238 is:

$$^{238}_{92}\text{U} \rightarrow {}^{234}_{90}\text{Th} + {}^{4}_{2}\text{He}$$

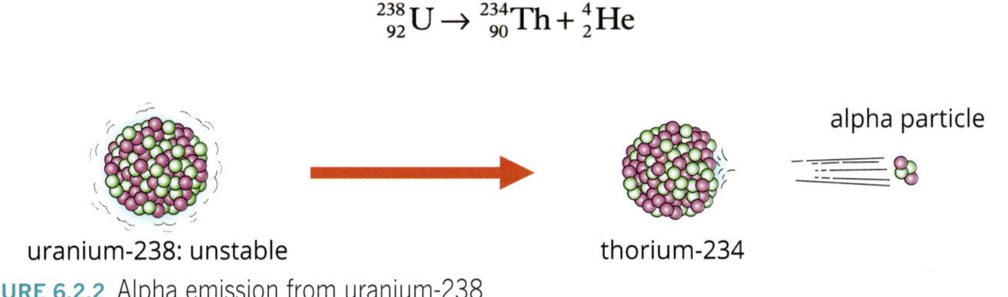

FIGURE 6.2.2 Alpha emission from uranium-238

The **parent nucleus** $^{238}_{92}\text{U}$ has spontaneously emitted an alpha particle (α) and has changed into a completely different element, $^{234}_{90}\text{Th}$. Thorium-234 is called the **daughter nucleus**. Energy is also emitted, mostly in the form of kinetic energy carried by the fast-moving alpha particle.

When an atom changes into a different element, it is said to undergo a **nuclear transmutation**. In nuclear transmutations, electric charge is conserved. This results in the conservation of atomic number (i.e. the number of protons). The sums of atomic numbers on both sides of a nuclear equation must be equal. In the uranium decay equation, the atomic number is conserved: 92 = 90 + 2. The mass number is also conserved: 238 = 234 + 4.

BETA (β) DECAY

Many radioactive materials emit **beta particles**. There are two different types of beta particles: beta-minus (β^-) and beta-plus (β^+).

The **weak nuclear force** is responsible for both types of beta decay. This allows the parent nucleus to become more stable by rearranging the number of protons and neutrons into a more energy-favourable ratio.

PHYSICSFILE

Radioactive lamps

The wicks or mantles used in old-style camping lamps are slightly radioactive (see figure below). They contain a radioisotope of thorium, an alpha-particle emitter. They have not been banned from sale so far because they contain only small amounts of the radioisotope and can be used safely by taking simple precautions such as washing hands and avoiding inhalation or ingestion.

However, a scientist from the Australian National University in Canberra has called for these mantles to be banned because they tend to crumble and turn to dust as they age. If this dust were inhaled, alpha particles could settle in someone's lung tissue, possibly causing cancers to form.

An old radioactive gas-light mantle

Beta-minus (β^-)

This type of beta decay occurs when an electron is emitted from the nucleus of a radioactive atom, rather than from the electron cloud. This type of beta particle can be written as ${}_{-1}^{0}\beta$.

The atomic number of −1 indicates that the beta particle (the electron) has a single negative charge. The mass number of zero indicates that its mass is far less than that of a proton or a neutron.

Typically, beta-minus decay occurs if a nucleus has too many neutrons to be stable. A neutron spontaneously changes into a proton, a beta-minus particle (β^-, an electron), and an uncharged massless antimatter particle called an **antineutrino** ($\overline{\nu}$). This makes the nucleus more stable.

An example of an isotope that undergoes beta-minus decay is carbon-14. The other isotopes of carbon (carbon-12 and carbon-13) are both stable. Carbon-14 is unstable. It has too many neutrons and undergoes a beta-minus decay to become stable. In this process, one of the neutrons changes into a proton. Nitrogen-14 is then formed and energy is released. The beta-minus decay of carbon-14 is shown in Figure 6.2.3.

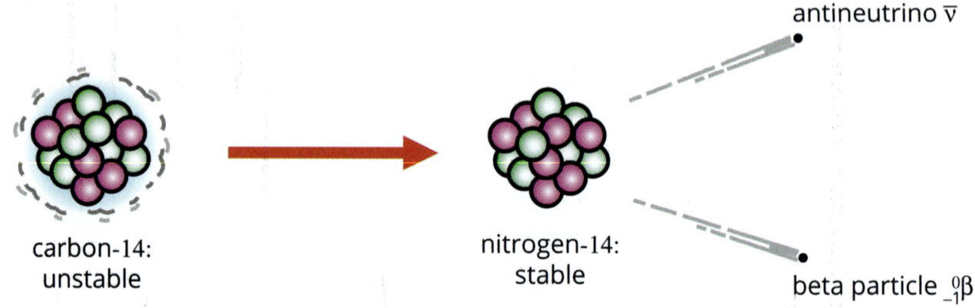

FIGURE 6.2.3 The beta-minus decay of carbon-14

The nuclear equation for this decay is:

$${}_{6}^{14}C \rightarrow {}_{7}^{14}N + {}_{-1}^{0}\beta + \overline{\nu}$$

The transformation taking place inside the nucleus is:

$${}_{0}^{1}n \rightarrow {}_{1}^{1}p + {}_{-1}^{0}\beta + \overline{\nu}$$

Notice that, in all these equations, the atomic and mass numbers are conserved. The antineutrino has no charge and has so little mass that both its atomic and mass numbers are zero.

Beta-plus (β^+)

A different form of beta decay occurs when a nucleus has too many protons. In this case, a proton may spontaneously change into a neutron and emit a neutrino (ν) and a positively charged beta particle. This process is known as β^+ (beta-positive) decay. The positively charged beta particle is called a **positron**.

Positrons (${}_{+1}^{0}\beta$) have the same properties as electrons, but their electrical charge is positive rather than negative.

GAMMA (γ) DECAY

After a radioisotope has emitted an alpha or beta particle, the daughter nucleus usually has excess energy. The protons and neutrons in the daughter nucleus then rearrange slightly and offload this excess energy by releasing a gamma ray, ${}_{0}^{0}\gamma$.

Gamma rays are high-energy electromagnetic radiation and so have no mass, are uncharged and travel at the speed of light ($3.0 \times 10^8 \, m \, s^{-1}$).

A common example of a gamma ray emitter is iodine-131. It decays by beta and gamma emission to form xenon-131, as shown in Figure 6.2.4.

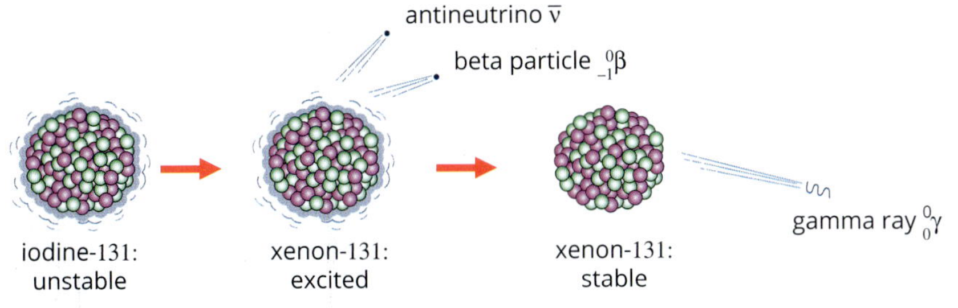

FIGURE 6.2.4 The gamma and beta decay of iodine-131

The equations for this decay are:

$$^{131}_{53}\text{I} \rightarrow {}^{131}_{54}\text{Xe}^{*} + {}^{0}_{-1}\beta + \bar{\nu}$$

$$^{131}_{54}\text{Xe}^{*} \rightarrow {}^{131}_{54}\text{Xe} + {}^{0}_{0}\gamma$$

The asterisk indicates that the nuclide is in an excited state.

Gamma rays carry no charge and have no mass, so they have no effect when balancing the atomic or mass numbers in a nuclear equation.

Worked example 6.2.1

RADIOACTIVE DECAY

Strontium-90 decays by radioactive emission to form yttrium-90.
The equation is:

$$^{90}_{38}\text{Sr} \rightarrow {}^{90}_{39}\text{Y} + \text{X}$$

Determine the atomic number and mass number for X and identify the type of radiation being emitted.

Thinking	Working
Balance the mass numbers.	The mass numbers of 90 are already balanced. The mass number of X is zero.
Balance the atomic numbers.	$38 = 39 + a$ $a = 38 - 39 = -1$ atomic number = −1
Use the atomic number and mass number to identify X.	X has an atomic number of −1 and a mass number of zero. X is a beta-minus particle, ${}^{0}_{-1}\beta$.

Worked example: Try yourself 6.2.1

RADIOACTIVE DECAY

Polonium-218 decays by emitting an alpha particle and a gamma ray. The nuclear equation is:

$$^{218}_{84}\text{Po} \rightarrow \text{X} + {}^{4}_{2}\alpha + \gamma$$

Determine the atomic number and mass number for X, then use the periodic table on page 155 to identify the element.

WHY RADIOACTIVE NUCLEI ARE UNSTABLE

In Figure 6.2.5 the stable nuclides that exist in nature are indicated by purple squares. The radioisotopes that are alpha and beta emitters can be identified by the black alpha and plus and minus symbols. Most of these radioisotopes also emit gamma radiation.

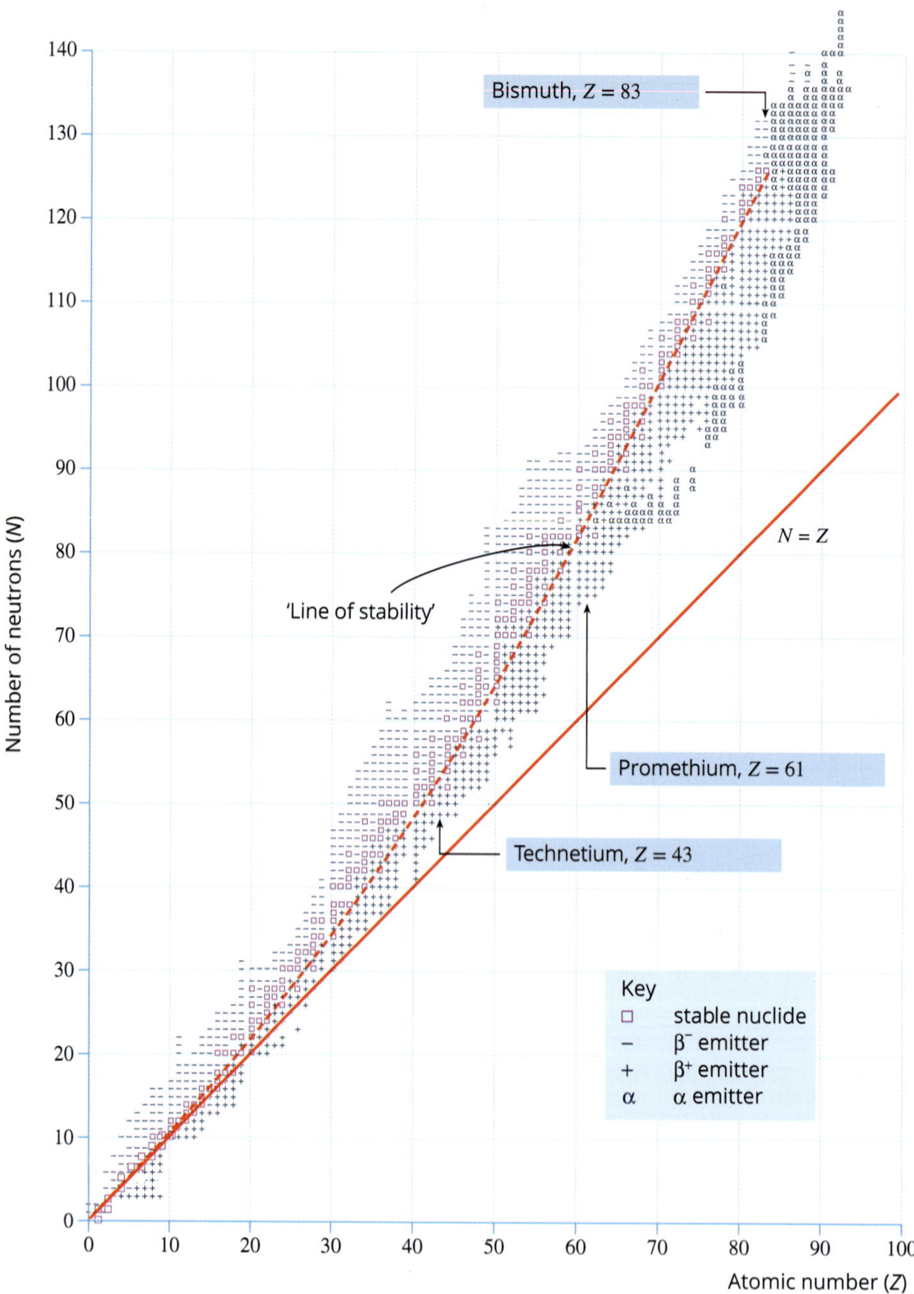

FIGURE 6.2.5 The chart shows stable and radioactive isotopes, plotted according to their number of protons (atomic number) and number of neutrons.

Electrostatic force describes the force of attraction or repulsion between charged particles. Strong nuclear force describes the force of attraction that holds the nucleus together.

Within the nucleus, protons are in close proximity to other protons. This should seem odd since protons exert strong electrostatic forces of repulsion over each other. **Electrostatic forces** act between charged particles and can act over relatively large distances. In the nucleus, this means that each proton strongly repels every other proton so this force is trying to make the nucleus break apart. Neutrons are uncharged so are unaffected by electrostatic forces.

To stop the nucleus breaking apart, a force known as the **strong nuclear force** is also acting. The strong nuclear force is a force of attraction that acts between every nucleon regardless of whether or not they are charged. This force acts like 'nuclear glue'. Neutrons are attracted to nearby neutrons and protons. Protons are also attracted to nearby neutrons and protons. However, this force only acts over relatively short distances (up to 1 fm or 10^{-15} m) so for nucleons on the opposite sides of a large nucleus, this force is not significant.

In a stable nucleus, there is a delicate balance between the repulsive electrostatic force and the attractive strong nuclear force. For example, bismuth-209 has 83 protons and 126 neutrons. It was long thought to be the heaviest stable isotope, although we now know that it is very weakly radioactive with a half-life of over 20 billion billion (2×10^{19}) years. Here, the electrostatic repulsion of the protons is balanced by the strong attractive nuclear forces between the nucleons to make the nucleus stable. Compare this with bismuth-211, which has a half-life of just over 2 minutes. Its two extra neutrons upset the balance between forces. The nucleus of ^{211}Bi is unstable and ejects an alpha particle to become more stable.

From Figure 6.2.5, it is evident that there is a 'line of stability' (indicated by the curved red dashed line on the graph) along which the stable nuclei tend to cluster. Nuclei away from this line are unstable.

For small nuclei with atomic numbers up to about 20, the ratio of neutrons to protons in stable nuclei is close to one. However, as the nuclei become bigger, this ratio increases for stable nuclei. Zirconium ($Z = 40$) has a neutron-to-proton ratio of about 1.25, while for mercury ($Z = 80$) the ratio is close to 1.5. This indicates that for higher numbers of protons, nuclei must have even more neutrons to remain stable. These neutrons act to dilute the repelling forces that exist between the extra protons.

Elements with more protons than bismuth ($Z = 83$) simply have too many repulsive charges in the nucleus. Additional neutrons are unable to stabilise these nuclei. All of these elements are unstable and radioactive. Figure 6.2.6 illustrates stable and unstable nuclei.

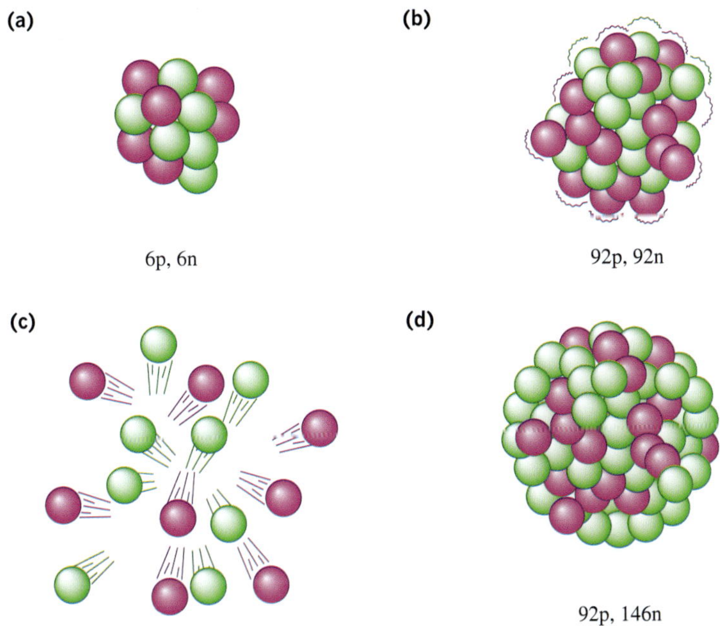

FIGURE 6.2.6 Stable and unstable nuclei. (a) A small nucleus such as carbon-12 is stable. This is because the electrostatic force of repulsion that acts between the protons is overcome by the strong nuclear force of attraction. (b) and (c) A large nucleus with equal numbers of protons and neutrons cannot exist. The electrostatic forces of repulsion between the protons would overcome the strong nuclear forces. (d) Additional neutrons increase the stability of large nuclei. The extra neutrons increase the influence of the strong nuclear force and act like a 'nuclear glue', holding the nucleus together.

PROPERTIES OF ALPHA, BETA AND GAMMA RADIATION

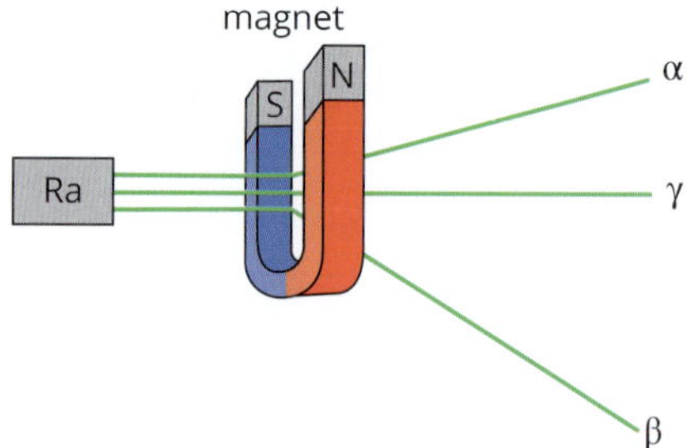

FIGURE 6.2.7 Applying a magnetic field shows that there are three different types of emissions from a radium source.

In the early experiments with radioactivity, emissions from a sample of radium were directed through a magnetic field, as shown in Figure 6.2.7. The emissions followed three distinct paths, which suggested that there were three different forms of radiation being emitted. The emissions each had different charges, masses and speeds.

Alpha (α) particles

Alpha particles consist of two protons and two neutrons. This means that they are relatively heavy and slow moving. Alpha particles are emitted from the nucleus at speeds of up to 20 000 km s^{-1} (2.0×10^7 m s^{-1}), just less than 10% of the speed of light, *c*.

Alpha particles have a double positive charge. This, combined with their relatively slow speed, makes them very easy to stop—they have a poor **penetrating ability**. They only travel a few centimetres in air before losing their energy, and will be completely absorbed by thin card or a human hand (Figure 6.2.8).

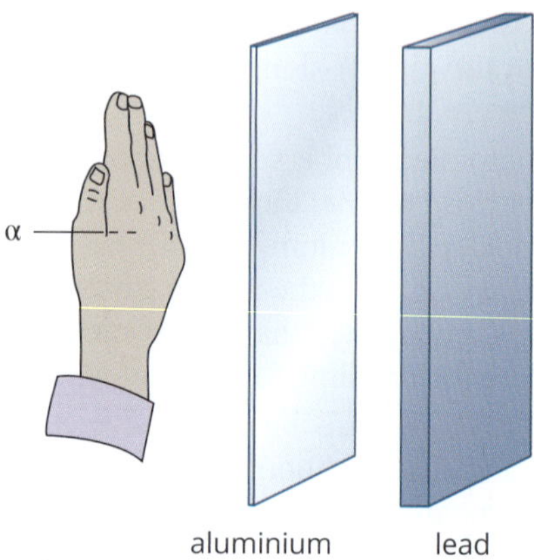

FIGURE 6.2.8 The penetrating ability of α radiation

An example of an isotope that emits α radiation (or undergoes α decay) is the isotope of americium $^{241}_{95}$Am. Americium can be found in ionisation smoke detectors (Figure 6.2.9).

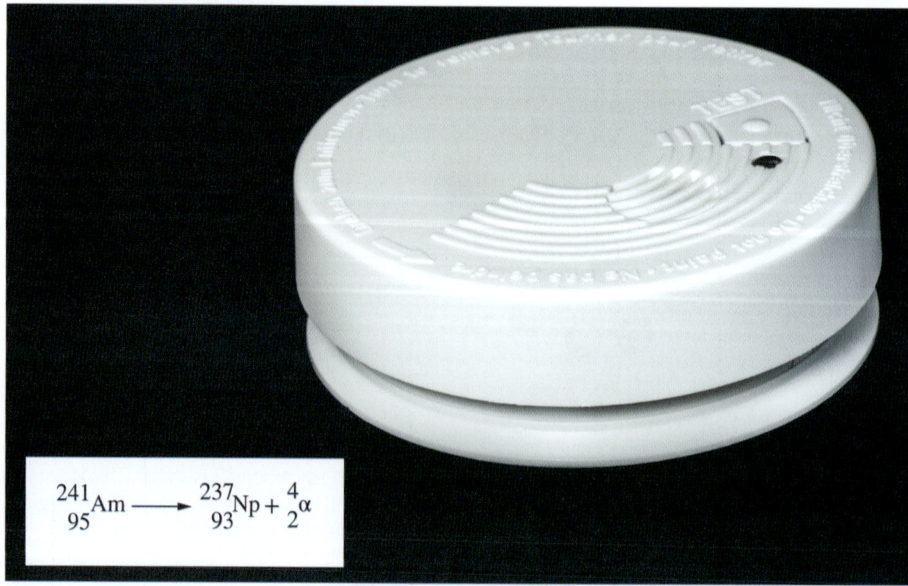

FIGURE 6.2.9 A domestic smoke detector contains a radioactive alpha emitter.

CASE STUDY

How radiation is detected

Our bodies cannot detect alpha, beta or gamma radiation. Therefore, a number of devices have been developed to detect and measure radiation.

A common detector is the Geiger counter. These are used:

- by geologists searching for radioactive minerals such as uranium
- to monitor radiation levels in mines
- to measure the level of radiation after a nuclear accident, such as the accident at Fukushima, Japan, in 2011
- to check the safety of nuclear reactors
- to monitor radiation levels in hospitals and factories.

A Geiger counter consists of a Geiger–Muller tube filled with argon gas, as shown in Figure 6.2.10.

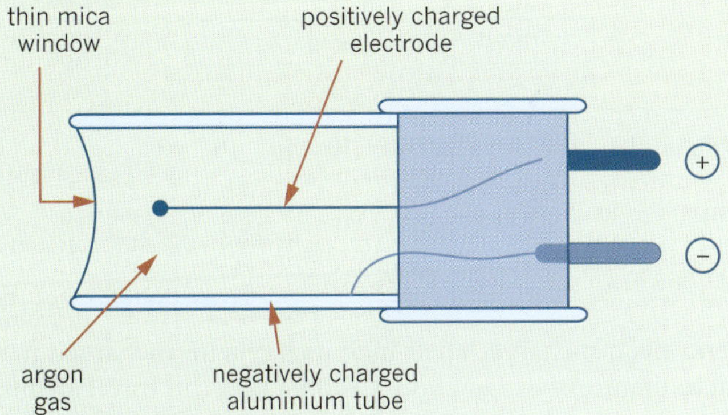

FIGURE 6.2.10 A schematic diagram of Geiger counter used for detecting ionising radiation

An electrical potential difference of about 400V is maintained between the positively charged central electrode and the negatively charged aluminium tube. When radiation enters the tube through the thin mica window, the argon gas becomes ionised and releases electrons. These electrons are attracted towards the central electrode and ionise more argon atoms along the way. For an instant, the gas between the electrodes becomes ionised enough to conduct a pulse of current between the electrodes. This pulse is registered as a count. The counter is often connected to a small loudspeaker so that the count is heard as a 'click' (Figure 6.2.11).

People who work in occupations that involve ongoing exposure to levels of ionising radiation usually pin a small radiation-monitoring device to their clothing. This is usually a thermoluminescent dosimeter, as pictured in Figure 6.2.12. These are used by personnel in nuclear power plants, radiotherapy departments at hospitals, airport security gates and uranium mines.

Thermoluminescent dosimeters contain a disc of lithium fluoride encased in plastic. Lithium fluoride can detect beta and gamma radiation, as well as X-rays and neutrons. They are a cheap and reliable method for measuring radiation doses.

FIGURE 6.2.11 A Russian scientist uses a Geiger counter to measure radiation levels.

FIGURE 6.2.12 Thermoluminescent dosimeters are used by doctors, radiologists and scientists who work with radiation to monitor their exposure levels.

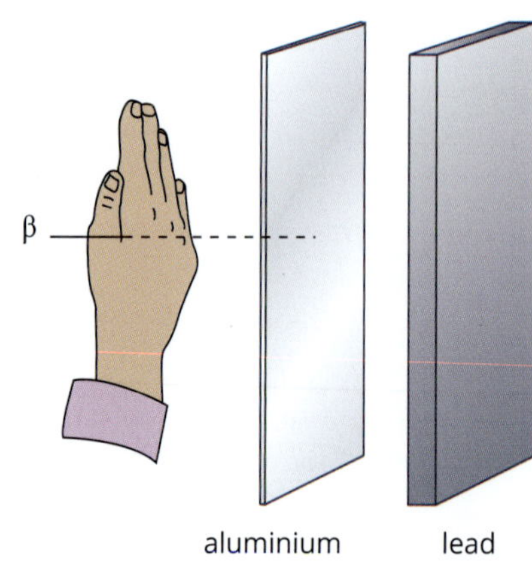

FIGURE 6.2.13 The penetrating ability of β radiation

Beta (β) particles

Beta particles are fast-moving electrons (β^-) or positrons (β^+).

Beta particles are much lighter than alpha particles. As a result, they leave the nucleus with far higher speeds—up to 90% of the speed of light, *c*.

Beta particles are more penetrating than α particles. They are faster and have a smaller charge than α particles. Beta particles will travel a few metres through air and through a human hand. Typically, a sheet of aluminium about 1 mm thick will stop them (Figure 6.2.13).

Gamma (γ) rays

Gamma rays are electromagnetic radiation with a very high frequency. Figure 6.2.14 shows where γ rays lie along the electromagnetic spectrum. They have no rest mass and travel at the speed of light: $3.0 \times 10^8\,\mathrm{m\,s^{-1}}$ or $300\,000\,\mathrm{km\,s^{-1}}$.

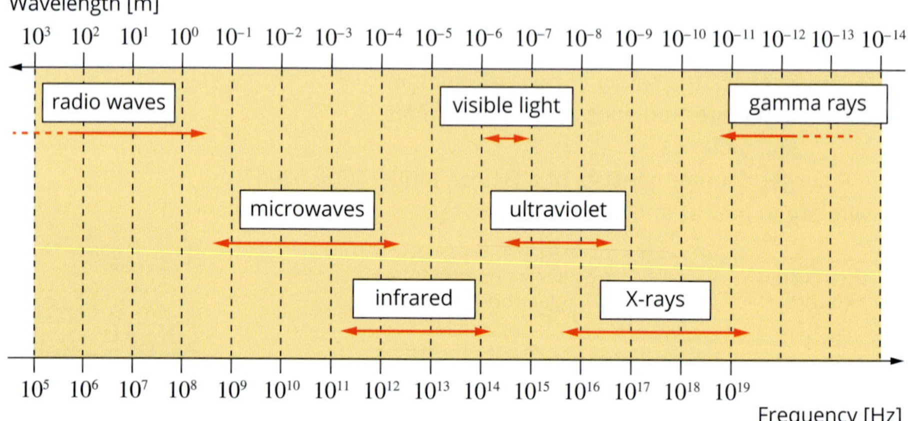

FIGURE 6.2.14 The electromagnetic spectrum contains many different types of radiation that differ in their wavelength and frequency. Gamma rays have very high frequencies and very short wavelengths, making them very energetic and highly penetrating.

Gamma rays have no electric charge. Their high energy and uncharged nature make them a very penetrating form of radiation. Gamma rays can travel an almost unlimited distance through air and even through a human hand, an aluminium sheet and a few centimetres of lead (Figure 6.2.15). Even a metre of concrete would not completely absorb a beam of γ rays.

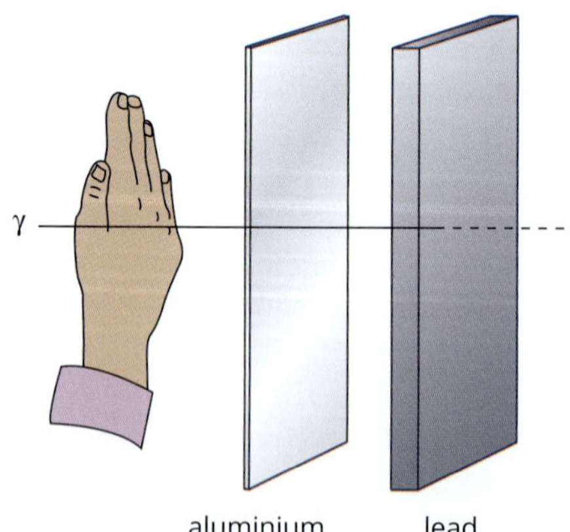

FIGURE 6.2.15 The penetrating ability of γ radiation

CASE STUDY

Monitoring the thickness of sheet metal

Beta particles can be used to monitor the thickness of rolled sheets of metal and plastic during manufacture (Figure 6.2.16). A β particle source is placed under the newly rolled sheet and a detector is placed on the other side. If the sheet being made is too thick, fewer β particles will penetrate and the detector count will fall. This information is instantaneously fed back to the rollers and the pressure is increased until the correct reading is achieved and hence the right thickness of metal is attained.

Alpha particles or γ rays would not be appropriate for this task. Alpha particles have a very poor penetrating ability, so none would pass through the metal. Gamma rays usually have a high penetrating ability and so a thin metal sheet would not stop them. In addition, workers would need to be shielded from γ radiation.

The penetrating properties of β particles make them ideal for this job. The thickness of photographic film and plastic sheets is also monitored in this way.

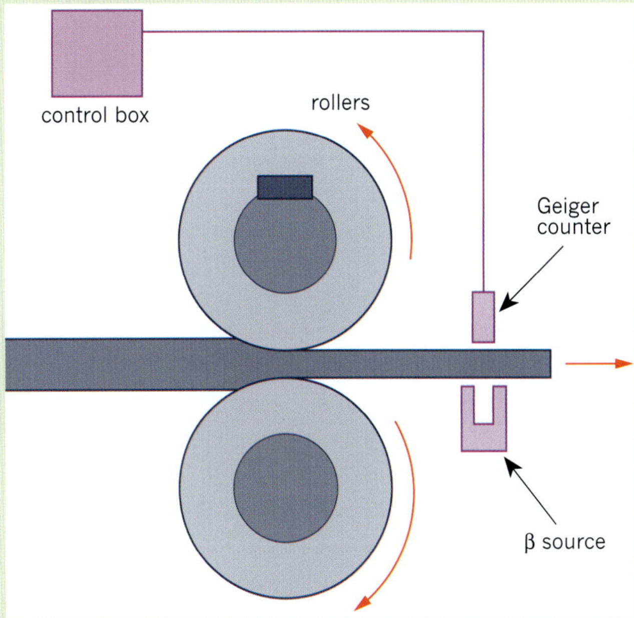

FIGURE 6.2.16 Beta emitters are used to monitor metal-sheet thickness.

Comparing α, β and γ radiation

The properties of alpha (α), beta (β) and gamma (γ) radiation are summarised in Table 6.2.1.

TABLE 6.2.1 A comparison of the properties of alpha, beta and gamma radiation

Property	α particle	β particle	γ ray
mass	heavy	light	none
speed	up to 20 000 km s^{-1} or about 10% of the speed of light	about 90% of the speed of light	the speed of light
charge	+2	−1 or +1	0
range in air	a few centimetres	1 or 2 m	many metres
penetration in matter	~10^{-2} mm	a few mm	high

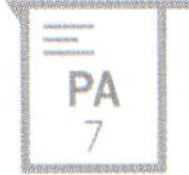

6.2 Review

OA ✓✓

SUMMARY

- In any nuclear reaction, both the atomic and mass numbers are conserved.
- The strong nuclear force is a short-range but powerful force of attraction that acts on all nucleons (protons and neutrons) to hold the nucleus together. It opposes electrostatic repulsion between protons and results in nuclear stability.
- The weak nuclear force acts in the nucleus to cause radioactive decay. It allows a nucleus to become more stable by rearranging the number of protons and neutrons into a more energy-favourable ratio via beta decay.
- Radioactive isotopes may decay by emitting alpha, beta or gamma radiation from their nuclei.
- Alpha particles (α):
  - consist of two protons and two neutrons and thus have a double positive charge
  - are identical to helium nuclei and can be written as $^{4}_{2}He$
  - are emitted from the nuclei of some radioisotopes at around 10% of the speed of light
  - are relatively heavy and have poor penetrating power.
- Beta particles (β):
  - emanate from the nucleus at up to 90% of the speed of light
  - are much lighter than α particles and have moderate penetrating ability
  - have a single negative charge (β^-) or single positive charge (β^+).
- A beta-minus particle (β^- or $^{0}_{-1}\beta$) is an electron that has been emitted from the nucleus of a radioactive atom as a result of a neutron transmutating into a proton. An antineutrino is always emitted with a β^- particle.
- A beta-plus particle (β^+) or positron is a positively charged electron ($^{0}_{+1}\beta$) that has been emitted from the nucleus of a radioactive atom as a result of a proton transmutating into a neutron. A neutrino is always emitted with a β^+ particle.
- Gamma rays (γ):
  - are high-energy electromagnetic radiation that is emitted from the nuclei of radioactive atoms
  - usually accompany an alpha or a beta emission
  - travel at the speed of light and have no charge
  - have high penetrating power.

KEY QUESTIONS

Knowledge and understanding

1 What type of decay occurs when a nucleus has too many protons?

2 What is the nature (or identity) of alpha, beta and gamma radiation?

3 For the unknown nuclides X and Y in each of these decay equations, determine the atomic number and mass number, and use the periodic table to identify the unknown elements.

a $^{235}_{92}U \rightarrow \alpha + X + \gamma$

b $^{228}_{88}Ra \rightarrow Y + \beta^- + \gamma$

4 Carbon-14 decays by beta-minus emission to form nitrogen-14. The equation for this is:

$$^{14}_{6}C \rightarrow ^{14}_{7}N + ^{0}_{-1}\beta + \bar{\nu}$$

a How many protons and neutrons does the nitrogen atom have?

b What nucleon on the left side of the equation has transformed into what particle(s) on the right side of the equation?

5 Where in the atom does each type of radiation originate?

a alpha

b beta

c gamma

6 What is the strong nuclear force?

7 Briefly explain why alpha particles have a very poor penetrating ability.

Analysis

8 A radiation oncologist inserts a radioactive wire into a breast cancer to destroy the cancerous cells close to the wire. Should this wire be primarily an alpha, beta or gamma emitter? Explain your reasoning.

6.3 Half-life and decay series

Different radioisotopes will emit radiation and decay at very different rates. A **Geiger counter** held close to a small sample of polonium-218 will initially detect a very high level of radiation. Half an hour later, this count rate will have dropped to almost zero.

Compare this with a similar sample of uranium-235. A Geiger counter held close to the uranium will show a low count rate. However, as time passes, the count rate does not seem to change. If you came back decades later, the count level would still be the same. Figure 6.3.1 compares the activity of both of these radioisotopes.

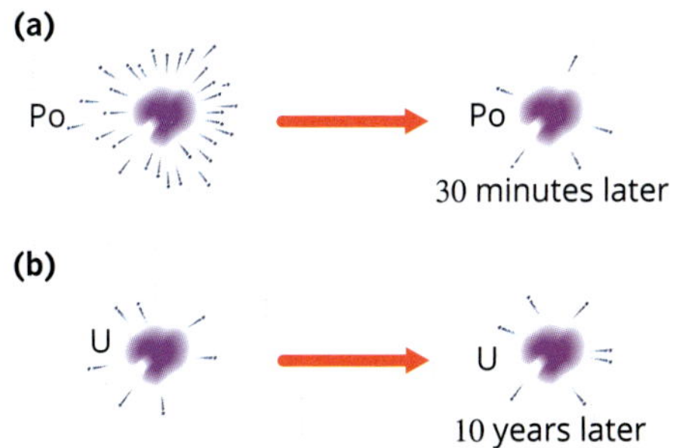

FIGURE 6.3.1 The activity of (a) polonium-218 and (b) uranium-235

The half-life ($t_{1/2}$) of a radioisotope is the time that it takes for half of the nuclei of the sample of radioisotope to decay.

The **half-life** of a radioisotope describes how long it takes for half of the atoms in a given mass to decay. The count rate is the **activity** of the sample. These ideas will be studied in this section.

HALF-LIFE

All radioisotopes are unstable but some are more unstable than others. In the previous example, polonium-218 is more unstable than uranium-235. One way of determining this instability is by measuring the half-life ($t_{1/2}$) of the radioisotope.

The half-life of polonium-218 is 3 minutes. Consider a sample of polonium that initially contains 100 million undecayed polonium-218 nuclei, as shown in Figure 6.3.2. Over the first 3 minutes about half of these will have decayed, leaving about 50 million polonium-218 nuclei. Over the next 3 minutes half of these 50 million polonium-218 nuclei will decay, leaving approximately 25 million of the original radioactive nuclei. The process continues as time passes.

The number of nuclei remaining after a particular number of half-lives can be found mathematically using:

$$N = N_0\left(\frac{1}{2}\right)^n$$

where N is the number of radioactive nuclei remaining

N_0 is the initial number of radioactive nuclei

n is the number of half-lives elapsed.

The number of half-lives in a period of time can be found using:

$$n = \frac{T}{t_{1/2}}$$

where n is the number of half-lives elapsed

T is the period of time that the radioactive nuclei has decayed

$t_{1/2}$ is the half-life of the radioactive nuclei.

Key: ● = 1 million ^{218}Po nuclei

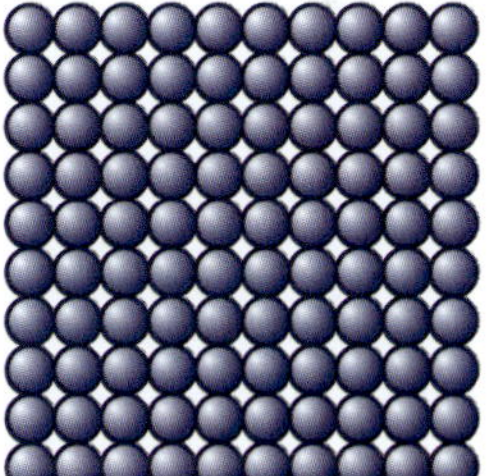

Initially:
100 million ^{218}Po nuclei

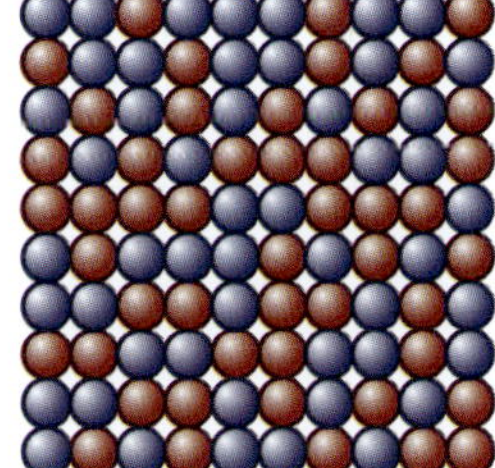

After 3 minutes:
~ 50 million ^{218}Po nuclei

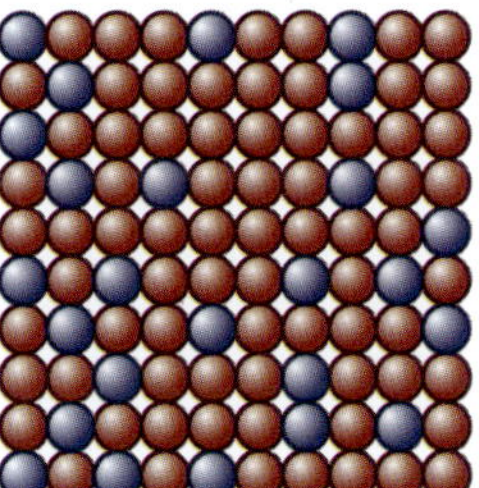

After 6 minutes:
~ 25 million ^{218}Po nuclei

FIGURE 6.3.2 The decay of polonium-218 over two half-lives. Only one quarter (25%) of the original radioisotope remains after two half-lives.

As time passes, a smaller and smaller proportion of the original radioisotope remains in the sample. The graph in Figure 6.3.3, known as a decay curve, shows this.

Even a very small radioactive sample will contain billions of atoms. It is important to know that although the behaviour of such a large sample of nuclei can be predicted mathematically, the behaviour of one particular nucleus cannot. It has a 50% chance of decaying in each half-life. Also, the half-life of a radioisotope is constant and cannot be changed by chemical reactions, heat and so on. Half-life is solely determined by the instability of the nuclei of the radioisotope.

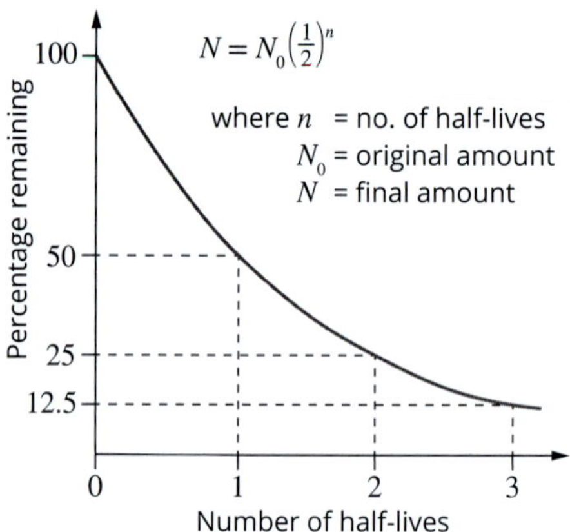

FIGURE 6.3.3 A decay curve for a radioisotope

Worked example 6.3.1

HALF-LIFE

A sample of the radioisotope thorium-234 contains 8.0×10^{12} nuclei. The half-life of thorium-234 is 24 days. How many thorium-234 nuclei will remain in the sample after 120 days?

Thinking	Working
Calculate how many half-lives 120 days corresponds to.	$n = \frac{T}{t_{1/2}}$ $= \frac{120}{24}$ = 5 half-lives
Substitute $N_0 = 8.0 \times 10^{12}$ and $n = 5$ into the equation. Calculate the number of nuclei remaining.	$N = N_0\left(\frac{1}{2}\right)^n$ $= 8.0 \times 10^{12} \times \left(\frac{1}{2}\right)^5$ $= 2.5 \times 10^{11}$ nuclei

Worked example: Try yourself 6.3.1

HALF-LIFE

A sample of the radioisotope sodium-24 contains 4.0×10^{10} nuclei. The half-life of sodium-24 is 15 hours. How many sodium-24 atoms will remain in the sample after 150 hours?

ACTIVITY

A Geiger counter can be used to record the number of radioactive decays occurring in a sample each second. It measures the activity of the sample.

Over time, the activity of a sample of a radioisotope will decrease. More and more of the radioactive nuclei will have decayed and at some point it will no longer emit radiation.

If a sample of polonium-218 ($t_{1/2}$ = 3 minutes) has an initial activity of 2000 Bq, then after one half-life its activity will be 1000 Bq. After a further 3 minutes, the activity of the sample will have reduced to 500 Bq, and after a further 3 minutes it will be 250 Bq and so on.

Uranium-235 has a half-life of 700 000 years. Its activity will remain virtually constant for decades; it will certainly not change over 3 minutes.

The same half-life equation can be used to calculate the final activity of a radioactive sample after a number of half-lives.

Activity is measured in becquerels, Bq.
1 Bq = 1 disintegration per second.

The activity of the nuclei remaining after a number of half-lives can be found mathematically using:

$$A = A_0\left(\frac{1}{2}\right)^n$$

where A is the activity of radioactive nuclei remaining
A_0 is the initial activity of radioactive nuclei
n is the number of half-lives elapsed.

COMMON RADIOISOTOPES AND THEIR APPLICATIONS

The half-lives of some common radioisotopes are shown in Table 6.3.1. The half-life of a radioisotope will determine what it is used for. For example, the most commonly used medical tracer, technecium-99, has a short half-life of just 6 hours. The short half-life means that radioactivity does not remain in the body any longer than necessary. On the other hand, the radioisotope used in a smoke detector, americium-241, is chosen because of its long half-life, 461 years. The smoke detector can continue to function for a very long time, as long as the battery is replaced each year.

TABLE 6.3.1 Some common radioisotopes and their half-lives

Isotope	Emission	Half-life	Application
Natural			
polonium-214	α	0.000 16 s	nothing at this time
strontium-90	β	28.8 years	cancer therapy
radium-226	α	1630 years	once used in luminous paints
carbon-14	β	5730 years	carbon dating of fossils
uranium-235	α	700 000 years	nuclear fuel, rock dating
uranium-238	α	4.5 billion years	nuclear fuel, rock dating
thorium-232	α	14 billion years	fossil dating, nuclear fuel
Artificial			
technetium-99m	β	6 h	medical tracer
sodium-24	β	15 h	medical tracer
iodine-131	β	8 days	medical tracer
phosphorus-32	β	14.3 days	medical tracer
cobalt-60	β	5.3 years	radiation therapy
americium-241	α	460 years	smoke detectors
plutonium-239	α	24 000 years	nuclear fuel, rock dating

DECAY SERIES

Generally, when a radioisotope decays, its daughter nucleus is not completely stable and is itself radioactive. This daughter nucleus will then undergo further decay. Eventually a stable isotope is reached and the sequence ends. This is known as a **decay series**. An example of a decay series is shown in Figure 6.3.4. This particular series shows the decay of uranium-238 (shown at the top of the chart) into lead-206 (shown at the bottom of the chart).

Earth is 4.5 billion years old—old enough to have only four naturally occurring decay series that remain active:

- the uranium series in which uranium-238 eventually becomes lead-206
- the actinium series in which actinium-235 eventually becomes lead-207
- the thorium series in which thorium-232 eventually becomes lead-208
- the neptunium series in which neptunium-237 eventually becomes bismuth-209. (Since neptunium-237 has a relatively short half-life, it is no longer present in Earth's crust, but the rest of its decay series is still continuing.)

Geologists analyse the proportions of the radioactive elements in a sample of rock to gain a reasonable estimate of the age of the rock. This technique is known as rock dating.

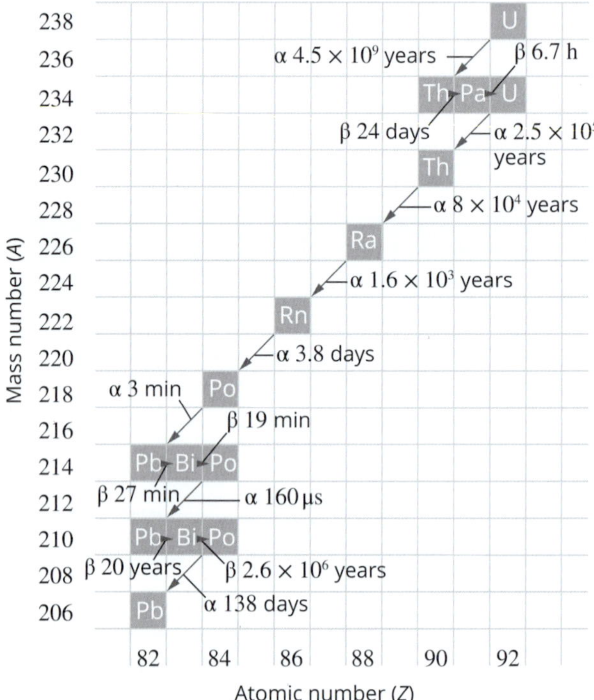

FIGURE 6.3.4 The uranium decay series. The half-lives and emissions are indicated on each of the decays as radioactive uranium-238 is gradually transformed into stable lead-206. Mining companies find significant quantities of lead at uranium mines.

CASE STUDY **ANALYSIS**

Radiocarbon dating

Carbon dating is a technique used by archaeologists to determine the ages of fossils and ancient objects that were made from plant matter. This method involves measuring and comparing the proportion of two isotopes of carbon, ^{12}C and ^{14}C, in the specimen.

Carbon-12 is a stable isotope but carbon-14 is radioactive and only exists in trace amounts in nature. Carbon-12 atoms are about 1000000000000 (10^{12}) times more common than carbon-14 atoms.

Carbon-14 has a half-life of 5730 years and decays by β^- emission to nitrogen-14. Its decay equation is:

$$^{14}_{6}C \rightarrow {}^{14}_{7}N + \beta^- + \bar{\nu}$$

Both carbon-12 and carbon-14 can combine with other atoms in the environment. For example, they both combine with oxygen to form carbon dioxide. Plants and animals take in carbon-based molecules from the air and food. This means that all living organisms contain the same percentage of carbon-14. In the environment, the production of carbon-14 is matched by its decay and so the proportion of carbon-14 atoms to carbon-12 remains constant.

After an organism dies, the amount of carbon-14 it contains will decrease as these atoms decay to form nitrogen-14 and are not replaced from the environment. The number of atoms of carbon-12 does not change as carbon-12 is a stable atom. So, over time, the proportion of carbon-14 to carbon-12 atoms decreases.

The proportion of carbon-14 to carbon-12 in a dead organism can be compared with that found in living organisms and the approximate age of the specimen can be determined from the half-life of carbon-14.

Radiocarbon dating is an important aid to anthropologists who are interested in finding out about the migration patterns of early peoples—including Indigenous Australians. This technique is very powerful since it can be applied to the remains of ancient campfires.

Radiocarbon dating has been used to date the remains of Indigenous Australians found at Lake Mungo (New South Wales) in the Willandra Lakes Region. Mungo Lady was discovered in 1969, the remains estimated to be between 40000 and 42000 years old. Mungo Man was discovered in 1974, and the remains estimated to be about 40000 years old.

Radiocarbon dating is accurate and reliable for samples up to about 60000 years old. Carbon dating cannot be used to date dinosaur bones as they are more than 60 million years old, but it can be used to determine the age of more recently extinct mammoth fossils, such as that shown in Figure 6.3.5.

FIGURE 6.3.5 A fossilised mammoth analysed by carbon dating

Analysis

The count rate from a 1 g sample of carbon that has been extracted from an ancient wooden spear is 10 Bq. A 1 g sample of carbon from a living piece of wood has a count rate of 40 Bq. Assume that this was also the initial count rate of the spear.

1 What is the approximate age of the ancient wooden spear?

2 If the activity of a 1 g sample of carbon extracted from the ancient wooden spear is 5 Bq, what would be the approximate age of the spear?

3 Suppose a fossil found near the ancient wooden spear had 20 g of ^{14}C at its time of death. Approximately how much ^{14}C would remain after 22920 years?

6.3 Review

OA ✓✓

SUMMARY

- The rate of decay of a radioisotope is measured by its half-life, $t_{1/2}$. This is the time that it takes for half of the radioisotope to decay.
- The activity of a sample indicates the number of emissions per second. Activity is measured in becquerels (Bq), where 1 Bq = 1 emission per second.
- The number of atoms of a radioisotope will decrease over time. Over one half-life, the number of atoms of a radioisotope will halve.
- The half-life equation can be used to calculate the number (N) or activity (A) of a radioisotope remaining after a number of half-lives (n) has passed:

$$N = N_0\left(\frac{1}{2}\right)^n, \ A = A_0\left(\frac{1}{2}\right)^n$$

- When a radioisotope decays, its daughter nucleus is usually itself radioactive. This daughter will then decay to a grand-daughter nucleus, which may also be radioactive, and so on. This is called a decay series.

KEY QUESTIONS

Knowledge and understanding

1 What is meant by the 'activity' of a radioisotope?

2 The activity of a radioisotope changes from 6000 Bq to 375 Bq over a period of 60 minutes. Calculate the half-life of this radioisotope.

3 A Geiger counter is used to measure the radioactive emissions from a certain radioisotope. The activity of the sample is shown in the graph.

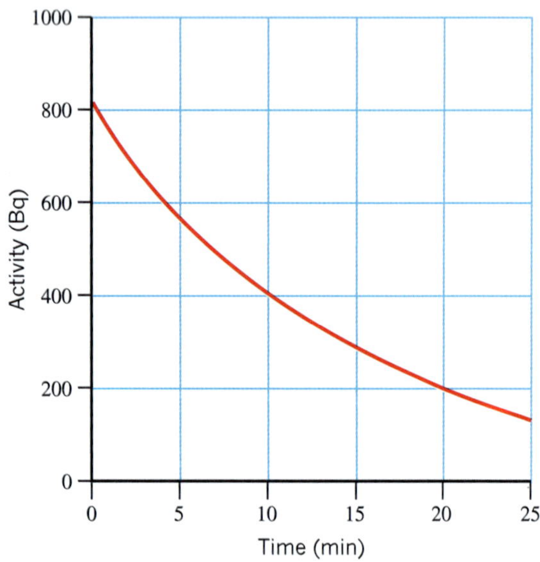

a What is the half-life of the radioisotope according to the graph?

b What would be the activity of the sample after 40 minutes have elapsed?

Analysis

4 A hospital in Alice Springs needs 12 μg of the radioisotope technetium-99m. The specimen has to be ordered from Sydney. The half-life of technetium-99m is 6 hours and the delivery takes 24 hours. How much must be produced in Sydney to satisfy the Alice Springs order?

5 According to Figure 6.3.4 on page 170, what type of decay does lead-210 undergo and what is its half-life?

6.4 Radiation and the human body

Ionising radiation from alpha, beta and gamma sources is harmful to humans and other living things. People who work with radiation in fields such as medicine, mining, nuclear power plants and industry must be able to closely monitor their radiation dose (see Figure 6.4.1). Similarly, radiologists who administer courses of radiation treatment to cancer patients need to be able to measure the radiation dose that they are applying. Radiation can also have positive impacts depending on how it is utilised; for example, in the treatment of cancer. In this section, radiation doses, their effect on the human body and how they are measured will be explained.

FIGURE 6.4.1 A dosimeter used to monitor gamma radiation exposure

IONISING RADIATION

Some types of radiation, such as radio waves, are harmless. Other types, however, are dangerous to humans. Known as ionising radiation, these harmful types of radiation interact with atoms. They have enough energy to remove outer-shell electrons and create ions. Alpha particles, beta particles and gamma rays are all ionising. So too is electromagnetic radiation with a frequency above 2×10^{16} Hz. So, X-rays and ultraviolet-B radiation are also ionising. When ionising radiation interacts with human tissue, the ions it produces are harmful and can lead to the development of cancers and tumours. Of the three types of radiation, alpha has the greatest ionising power, followed by beta and then gamma rays.

Lower energy electromagnetic radiation, such as radio waves, microwaves, infrared light, visible light and ultraviolet-A radiation, is non-ionising. Exposure to this radiation, even in significant amounts, has no serious consequences. Non-ionising radiation does not have enough energy to change the chemistry of the atoms and molecules that make up body cells.

EFFECTS OF RADIATION ON LIVING ORGANISMS

If possible, exposure to ionising radiation should be avoided. When alpha, beta or gamma radiation or an X-ray pass through a body cell, it may turn molecules in the cell into an ion pair. For example, if the radiation ionises a water molecule, then a hydrogen ion and a hydroxide ion can be formed. This is shown in Figure 6.4.2. These ions are highly reactive and can damage the DNA that forms the chromosomes in the nucleus of the cell. This may cause the cell either to die or to divide and reproduce at an abnormally rapid rate. When the latter occurs, a cancerous tumour may form.

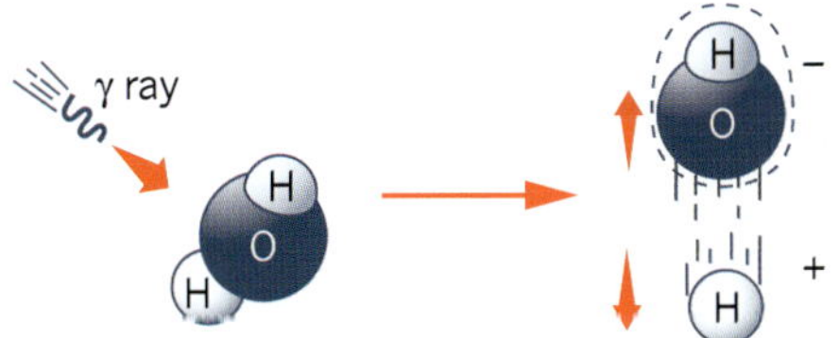

FIGURE 6.4.2 Ionising radiation has enough energy to break the bonds within a water molecule and create a pair of ions.

The effects of a dose of ionising radiation can be divided into two groups: the somatic effects and the genetic effects.

Somatic effects

Somatic (physical) effects arise when ordinary body cells are damaged; they depend on the size of the dose. Very high doses lead to almost immediate symptoms; lower doses could lead to symptoms developing many years later, such as an increased chance of developing cancer. Table 6.4.1 outlines the somatic effects of radiation doses. Note the units for the whole body dose of radiation used in Table 6.4.1 is the sievert (Sv). These units will be explained on page 175.

TABLE 6.4.1 The somatic effects of radiation doses (Sv)

Whole body dose (Sv)	Symptom
<1	non-fatal only minor symptoms such as nausea white blood cell level drops
2	death unlikely radiation sickness, i.e. nausea, vomiting and diarrhoea skin rashes hair loss bone-marrow damage
4	50% likelihood of death within 2 months severe radiation sickness high probability of leukaemia and tumours
8	almost certain death within 1 or 2 weeks acute radiation sickness—convulsions, lethargy

Genetic effects

When cells in the reproductive organs (ovaries or testes) are damaged, the body suffers **genetic** effects of ionising radiation. Cells in the reproductive organs develop into ova and sperm, so if the DNA in the chromosomes of these cells is damaged, this genetic change could be passed on to a developing embryo. The DNA changes in these damaged cells are known as **mutations**.

There are many different ways in which genetic effects can show up in future generations. Possible effects include poor limb development, cleft lip and palate and other birth anomalies. Genetic effects may surface in the next generation or lie dormant for several generations. In other words, if a person suffers damage to their reproductive cells, their children may seem to be unaffected but their grandchildren may be genetically affected.

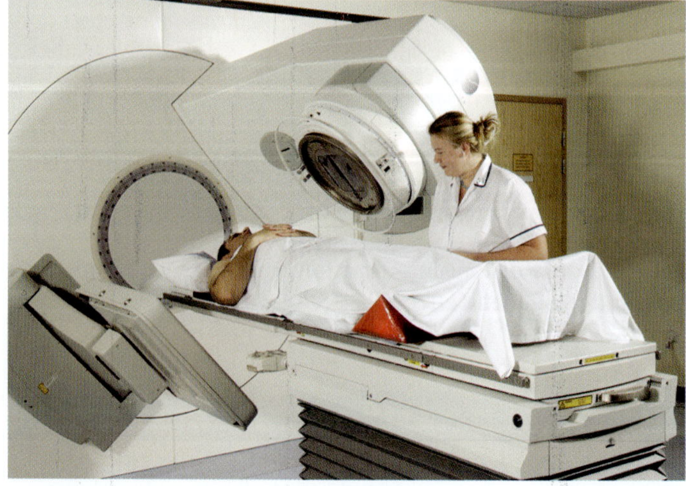

FIGURE 6.4.3 A cancer patient about to receive radiation therapy. Medical staff would take precautions to protect the patient's reproductive organs before exposure to ionising radiation.

For these reasons, when a patient is undergoing radiotherapy, it is most important that their reproductive organs are well shielded from the radiation (Figure 6.4.3). These organs are among the most radiosensitive organs (i.e. easily damaged by radiation) in the body. Patients who may wish to have children in future may have eggs removed from ovaries or sperm samples taken and frozen for later use.

MEASURING RADIATION EXPOSURE

Because of the damage radiation can do to the human body, it is important to be able to accurately measure the dose a person receives. This can either be as a relatively simple absorbed dose (the amount of energy per kilogram of body mass) or, more usefully, by taking into account the type of radiation and the parts of the body it affected, as well as the amount received.

Absorbed dose

Exposure to high-energy radiation such as alpha, beta or gamma radiation is harmful to living tissue. The energy of the radiation acts to break apart molecules and ionise atoms in the body's cells. The severity of this exposure depends on the amount of radiation energy that has been absorbed and the mass of tissue involved.

The radiation energy absorbed per kilogram of tissue is called the **absorbed dose**.

$$\text{Absorbed dose} = \frac{\text{energy absorbed by tissue}}{\text{mass of tissue}}$$

The absorbed dose is measured in joules per kilogram (J kg^{-1}) or grays (Gy): $1\,\text{Gy} = 1\,\text{J kg}^{-1}$.

Equivalent dose

The absorbed dose is not widely used when measuring the radiation dose. That is because it does not take into account the type of radiation involved. **Equivalent dose**, which does take the type of radiation involved into account, is the most common way of measuring radiation doses.

Alpha particles are the most ionising form of radiation. This is because of their relatively low speed, high charge and large mass. Alpha particles interact with, and ionise, almost every atom that lies in their path. This means that an absorbed dose of alpha radiation is about 20 times more damaging to human tissue than an equal absorbed dose of beta or gamma radiation.

The biological impact of different types of radiation is given a weighting called the **quality factor (QF)**. A list of quality factors is shown in Table 6.4.2.

Gamma rays and X-rays have a relatively low ionising power. They have no charge and move at the speed of light, so they fly straight past most atoms and interact only occasionally as they pass through a material or substance. Gamma rays and X-rays would cause only slight damage to living cells. Beta-minus particles (electrons) are considered to be as damaging as gamma rays and X-rays. This is reflected in their low QF of 1, as shown in Table 6.4.2.

Neutrons are a damaging form of radiation and have a quality factor of 10. Neutrons are produced as part of chain reactions in nuclear reactors, making accidental exposure to radioactive material very dangerous to workers who are present.

The equivalent dose (ED) takes into account the absorbed dose (AD) and the type of radiation. This gives a more accurate picture of the actual effect of the radiation on a person.

$$\text{Equivalent dose} = \text{absorbed dose} \times \text{quality factor}$$

$$\text{ED} = \text{AD} \times \text{QF}$$

The equivalent dose is measured in sieverts (Sv).

An absorbed dose of just 0.05 Gy of alpha radiation is equally as damaging to a person as an absorbed dose of 1.0 Gy of beta radiation. Although α particles have less energy than the β particles, each α particle does far more damage. In each case, the equivalent dose is 1 Sv, and 1 Sv of any radiation causes the same amount of damage.

TABLE 6.4.2 Quality factors for different types of radiation

Radiation	Quality factor
α particles	20
neutrons* (10 keV)	10
β particles	1
γ rays	1
X-rays	1

*Radiation from neutrons is only found around nuclear reactors and neutron bomb explosions.

Background radiation

It is important to note that 1 Sv is a massive dose of radiation. It would not be fatal, but it would certainly lead to a severe case of radiation sickness. In Australia, the average annual background radiation dose is about 2.0 mSv, or 2000 μSv. Using Table 6.4.3, which outlines annual radiation doses in Australia, it is possible to estimate your dose over the past year.

TABLE 6.4.3 Radiation doses from different sources

Radiation source	Average annual dose (μSv)	Local or temporary variations
cosmic radiation	300	plus 200 μSv for each round-the-world flight plus 20 μSv for each 10° of latitude (Melbourne is at 37.8° latitude) plus 150 μSv if you live 1000 m above sea level
rocks, air and water	1350	plus 1350 μSv if you live underground plus 1350 μSv if your house is made of granite minus 140 μSv if you live in a weatherboard house
radioactive food and drink	350	plus 1000 μSv if you have eaten food affected by the Fukushima fallout
manufactured radiation	60	plus 60 μSv if you live near a coal-fired power station plus 30 μSv from nuclear testing in the Pacific
medical exposure	variable	plus 30 μSv for a chest X-ray plus 300 μSv for a pelvic X-ray plus 5000 μSv for a CT scan plus 40 000 000 μSv for a course of radiotherapy using cobalt-60

PHYSICSFILE

Share of the different sources of radiation

The average equivalent of the dose absorbed by a person in a year is equal to a few millisieverts.

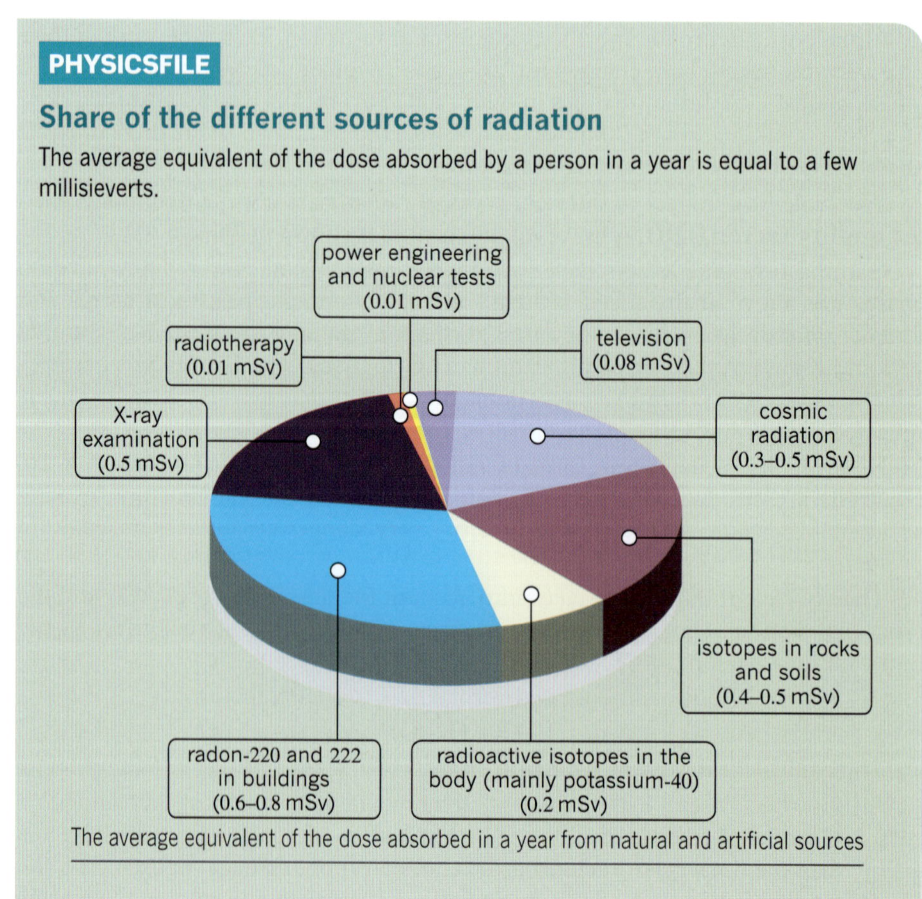

The average equivalent of the dose absorbed in a year from natural and artificial sources

Worked example 6.4.1

TREATING TUMOURS

A 10 g cancer tumour absorbs 2.5×10^{-3} J of energy from an applied radiation source. Calculate the equivalent dose if the source is an alpha emitter.

Thinking	Working
Convert the mass of the tumour from grams to kilograms.	$10\,g = \frac{10}{1000} = 0.010\,kg$
Calculate the absorbed dose using the formula: $\text{absorbed dose} = \frac{\text{energy absorbed}}{\text{mass of tissue}}$	$\text{absorbed dose} = \frac{\text{energy absorbed}}{\text{mass of tissue}}$ $= \frac{2.5 \times 10^{-3}}{0.010}$ $= 0.25\,Gy$
Calculate the equivalent dose using the formula: $ED = AD \times QF$ From Table 6.4.2 on page 175, QF = 20 for alpha particles.	$ED = AD \times QF$ $= 0.25 \times 20$ $= 5.0\,Sv$

Worked example: Try yourself 6.4.1

TREATING TUMOURS

A 25 g cancer tumour absorbs 5.0×10^{-3} J of energy from an applied radiation source. Calculate the equivalent dose if the source is an alpha emitter.

Effective dose

The different organs of the body have different sensitivities to radiation doses. For example, if a person's lung was exposed to a dose of 10 mSv, it would be more than twice as likely that cancers could develop than if the same 10 mSv dose was delivered to the liver. The weightings assigned by the International Commission of Radiological Protection (ICRP) to the various organs are shown in Table 6.4.4.

Effective dose takes into account the sensitivity of the organ to ionising radiation. Effective dose is found by calculating the sum of the equivalent doses multiplied by the **weighting factor**, W, for each organ affected. In physics, the symbol sigma (Σ) is used to represent the 'sum of'.

$$\text{Effective dose} = \Sigma(\text{equivalent dose} \times W)$$

Effective dose is measured in sieverts (Sv).

TABLE 6.4.4 The ICRP weighting values

Body part	Weighting (W)
ovaries/testes	0.20
bone marrow	0.12
colon	0.12
lung	0.12
stomach	0.12
bladder	0.05
breast	0.05
liver	0.05
oesophagus	0.05
thyroid	0.05
rest of body	0.07
total	1.00

Worked example 6.4.2

CALCULATING EFFECTIVE DOSE

A patient undergoing radiotherapy receives the following exposures: ovaries 35 mSv, bladder 30 mSv, colon 45 mSv.

Calculate the effective dose the patient receives.

Thinking	Working
Recall the equation for effective dose. The units of sievert (Sv) mean that the values given in the question are equivalent doses.	Effective dose = Σ(equivalent dose × W)
Calculate the effective dose. Remember to convert mSv to Sv.	Effective dose $= (35 \times 10^{-3} \times 0.20) + (30 \times 10^{-3} \times 0.05) + (45 \times 10^{-3} \times 0.12)$ = 13.9 = 14 mSv

Worked example: Try yourself 6.4.2

CALCULATING EFFECTIVE DOSE

A patient undergoing radiotherapy receives the following exposures: colon 50 mSv, liver 10 mSv.

Calculate the effective dose the patient receives.

CASE STUDY

Iodine-131 and the thyroid

On 11 March 2011, a catastrophic earthquake and tsunami hit Japan, killing tens of thousands of people and severely damaging the nuclear power station at Fukushima. Radioactive materials, including caesium-137 and iodine-131, escaped into the surrounding environment. These have half-lives of 30 years and 8 days respectively.

Your body needs iodine for the healthy functioning of the thyroid gland, which maintains proper metabolism. Foods rich in iodine include seafood, vegetables and salt. However, the body cannot tell the difference between normal iodine and radioactive iodine. To prevent the evacuees from absorbing radioactive iodine into their thyroid glands, they were issued with iodine tablets. Taking an iodine tablet each day ensured that the thyroid gland was saturated with iodine and so any radioactive iodine ingested by eating contaminated food would not be taken into the body and deposited in the thyroid (Figure 6.4.4).

Many victims of the Chernobyl nuclear disaster in 1986 died of thyroid cancer years after the accident. They ingested radioactive iodine and this accumulated in the thyroid gland, eventually leading to cancer. They had not been issued iodine tablets.

FIGURE 6.4.4 Children in Japan receiving iodine tablets after the Fukushima nuclear accident

RADIATION IN THERAPY

'Cancer' is a general term that actually incorporates different diseases. The term 'tumour' describes an abnormal mass of tissue due to the increased growth of cells. These can be either benign, containing only normal cells, or malignant, in which growth of cancerous cells takes place.

Malignant tumours can grow in just about any part of the body and invade the surrounding healthy tissue. Moreover, some cancer cells may break away and be carried by the bloodstream to then settle in other parts of the body, spreading the cancer (Figure 6.4.5).

This section focuses on the use of radioisotopes in the diagnosis and treatment of cancer, but this is just one practical area of research in this field. The use of radiation in medicine is inclusive of various types of ionising radiation. If you are interested in learning more about the uses of radiation in medicine, you may want to explore the Option in Unit 2 Area of Study 2: How is radiation used to maintain human health?

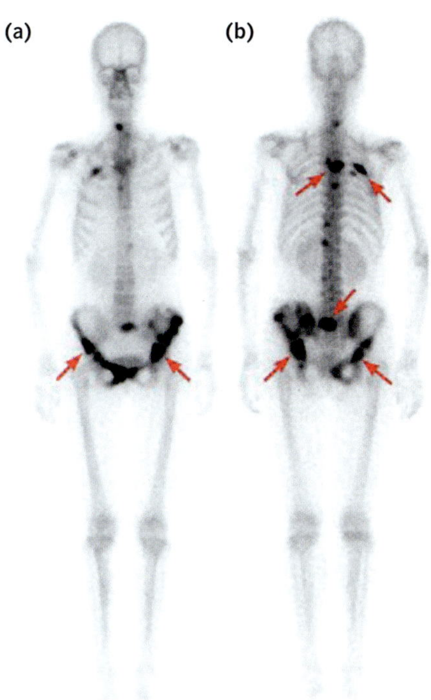

FIGURE 6.4.5 A bone scan using technetium-99m of a patient with lung cancer. The skeletal metastases (cancers) take up the radioactive isotope differently from the rest of the body, and appear as black dots spread out across the entire skeleton: (a) anterior view; (b) posterior view.

Diagnosis

Medical imaging, or radiology, is used in the diagnosis of different diseases. A variety of imaging techniques have been developed for this purpose. These include X-rays, computed tomography (CT), gamma (γ) radiation scans, magnetic resonance imaging (MRI), positron emission tomography (PET) scans and single photon emission computed tomography (SPECT). Some of these use ionising radiation to create images of the body, but in some modern clinical practices radioisotopes can be used to collect information about a person's physiological state. They can be used to investigate the functioning of specific organs, monitor certain biological processes (such as thyroid or liver function), or determine the advancement stage of a specific disease, such as cancer.

In such imaging techniques, radioisotopes called **radioactive tracers** are attached to specific biomolecules or drugs, thus creating **radiopharmaceuticals**. These are then administered to a patient either orally or by injection (see Figure 6.4.6).

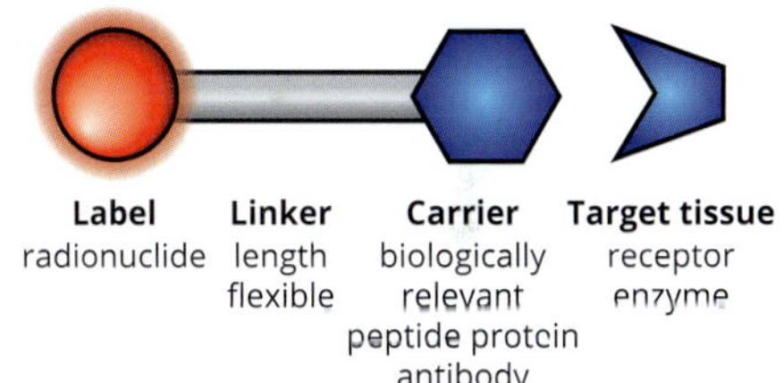

FIGURE 6.4.6 Illustration of a radiopharmaceutical, showing the radioactive label and the biomolecule or drug to which is attached

The specific radioisotope used in the generation of a radiopharmaceutical depends on the tissue or organ that is suspected of malfunctioning or of growing abnormal cells. The body naturally distributes different molecules to different organs and this is used to target specific organs in the body. For example, iodine is normally sent to the thyroid gland by the liver. So if a radiopharmaceutical containing radioactive iodine is taken, most of this iodine will end up in the thyroid gland.

When the tracer has reached the target organ, a radiation scan is taken with a gamma camera. An unusual pattern on the scan indicates a possible health problem.

For a radioisotope to be used for diagnostic imaging, it must:

- have a short half-life (hours or days) that is appropriate for the time taken for the diagnostic procedure. Radioactive materials are considered to be relatively safe after about 10 half-lives have passed
- emit only gamma radiation of an energy that can be detected by a gamma camera
- not emit alpha or beta radiation, because these particles would be trapped in the patient's tissues and they would not be detected externally
- have a high enough activity to give useful information without being toxic to the patient or react with drugs administered at the same time.

PET scans

Positron emission tomography (PET) scanning is a medical imaging technique that gives an assessment of how an organ is working, rather than providing information about its structure. The technique is very sensitive for detecting the early stages of an illness, and can detect malfunctions even without any structural changes in the tissue.

Emitted positrons (beta-positive particles) travel for a short distance from their site of origin and lose energy as they pass through tissue. When most of their kinetic energy has been lost, they undergo a process called annihilation. This happens when the positrons react with electrons from the immediate tissue area, and results in the emission of two very-high-energy photons in the form of gamma rays. These gamma rays are recorded by a gamma camera and are reconstructed into images by a computer.

The basic principle of PET scanning relies on the selection of a biologically relevant molecule that is selectively taken up by a specific organ or tissue under examination (such as glucose, estradiol and methionine). Then, this molecule is labelled with a positron-emitting radiotracer, such as ^{18}F, ^{15}O, ^{82}Rb or ^{64}Cu.

For example, for PET scans of the brain, the radioactive tracer ^{18}F is attached to molecules of D-glucose, creating the radiopharmaceutical fludeoxyglucose (^{18}F-FDG). This arrangement is chosen because the brain uses high levels of glucose in its metabolism.

Other applications of brain imaging with ^{18}F-FDG include that shown in Figure 6.4.7, in which an FDG-PET scan is used to predict learning and memory problems that might arise with Alzheimer's disease.

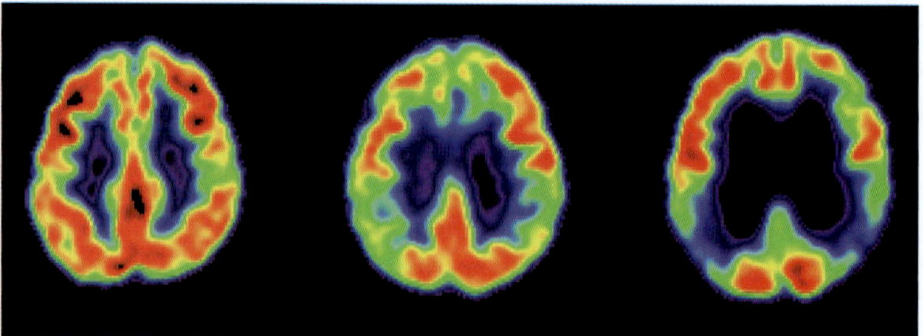

FIGURE 6.4.7 FDG-PET images showing reduced glucose metabolism in selected brain regions in patients with mild cognitive impairment and Alzheimer's disease. Areas with lowered uptake of FDG, and therefore reduced glucose metabolism, are shown in black in the images.

SPECT scans

A single photon emission computed tomography (SPECT) scan is a more-advanced type of nuclear imaging that is also used for diagnostic purposes. A radioactive tracer and a special camera are used to create 3D images of specific organs in the body. For example, SPECT can give information about blood flow to tissues and chemical reactions taking place in specific locations in the body.

In a similar manner to PET scans, SPECT uses radioactive tracers attached to specific biomolecules and a scanner to record data and construct 3D images. For the procedure to take place, a small amount of a gamma-emitting radiopharmaceutical with a short half-life (hours or days) is injected into a patient's vein. Shortly after, a gamma camera is used to generate detailed images of areas inside the body where the radiopharmaceutical has been taken up by the cells.

Radioisotopes such as technetium-99m, fluorine-18 and iodine-123 are commonly used in SPECT scans. They have half-lives of 6 hours, 110 minutes and 2.8 days respectively.

Treatment

Radiation therapy, or radiotherapy, uses ionising radiation as a cancer treatment by controlling and/or killing malignant cells. There are a variety of ways in which ionising radiation from radioisotopes can be used to treat diseases, especially cancer. The radiation can be applied internally or externally to a patient's body.

Cobalt-60 external beam therapy

Cobalt-60 machines were first developed in the 1950s. A beam of gamma rays from a cobalt-60 source is directed from an external source through a patient into the tumour site.

- Cobalt-60 has a half-life of 5.3 years, so the source lasts for a long time before it needs replacing.
- The source cannot be switched off, and it needs to be shielded using lead so that the gamma rays do not irradiate everyone in the region.

Cobalt-60 machines have now been replaced by linear accelerators (linacs) in many countries, including Australia.

Tomotherapy

Tomotherapy is applied externally and utilises a combination of medical imaging and radiotherapy. A three-dimensional image of the tumour is taken using computed tomography (CT) scanning.

In CT scans, an image of a slice of the body is created using soft X-rays. To understand the usefulness of CT scans, imagine a large sausage. By taking a set of thin slices of the sausage you could see exactly what it is made of.

If you saw anything unexpected in your slice (or cross-section), you could determine exactly where in the sausage it was originally located. An example of the image produced is shown in Figure 6.4.8. Although this technique provides much more detailed images for doctors, the doses of radiation received by the patient are much higher than for a standard X-ray radiograph.

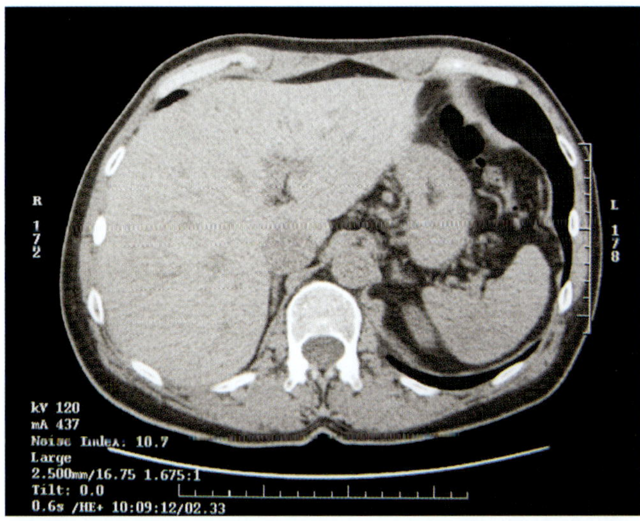

FIGURE 6.4.8 A cross-sectional view from a CT scan

Once the three-dimensional image of the tumour is created by taking multiple 'slices' or images of the area with the CT scan, ionising radiation is then applied to the tumour. This radiation is applied using a linac beam. As the shape and position of the tumour is found using the CT scan, the dose of the applied radiation from the linac can be higher and still spare the normal surrounding tissue.

PHYSICSFILE

How technetium is produced

Technetium-99m is the most widely used radioisotope in nuclear medicine. It is used for diagnosing and treating cancers. However, this radioisotope decays relatively quickly and so usually needs to be produced close to where it will be used. Technetium-99m is produced in small nuclear generators that are located in hospitals around the country. In this process, the radioisotope molybdenum-99, obtained from the Australian Nuclear Science and Technology Organisation in Lucas Heights, New South Wales, is used as the parent nuclide. Molybdenum-99 decays by beta emission to form a relatively stable (or metastable) isotope of technetium, technetium-99m, as shown:

$${}^{99}_{42}\text{Mo} \rightarrow {}^{99m}_{43}\text{Tc} + {}^{0}_{-1}\beta + \nu$$

Technetium-99m is flushed from the generator using a saline solution. The radioisotope is then diluted and attached to an appropriate chemical compound before being administered to the patient as a tracer. Technetium-99m is purely a gamma emitter. This makes it very useful as a diagnostic tool for locating and treating cancer. Its decay equation is:

$${}^{99m}_{43}\text{Tc} \rightarrow {}^{99}_{43}\text{Tc} + \gamma$$

The Gamma Knife

Despite its name, the Gamma Knife isn't a real knife but a machine that delivers a focused, high dose of gamma radiation to its target, made possible by the 200 or so cobalt-60 sources contained in it. The overlapping beams destroy the tumour while the individual beams do minimal damage to healthy tissue. The first Gamma Knife in Australia was installed in 2010 at the Macquarie University Hospital in Sydney. A Gamma Knife is now commonly used as an external source for treating brain tumours.

Chemotherapy

Radiopharmaceuticals may also be used in **chemotherapy**, in which the radiation acts internally. In this case, the half-life of the radioisotope is an important consideration. Recall that radioactive materials take around 10 half-lives to become effectively non-radioactive. Therefore, a half-life of 5 seconds, for example, would not be suitable because the radioisotope would have almost completely decayed in less than a minute and not have a chance to do its job. A half-life of 5 years, on the other hand, would mean that the patient would be continually exposed to radiation for decades, which could cause more harm than good.

For a radioisotope to be used for this purpose, it must:

- have a short half-life (hours or days) that is appropriate for the time taken for the therapeutic procedure
- emit alpha or beta radiation, because these particles would be trapped in the patient's tissues and would destroy the cells in the tumour
- not emit too much gamma radiation because its high penetrating ability would mean that healthy cells and bystanders would be irradiated; however, some gamma radiation allows a gamma camera to monitor the tumour.

Brachytherapy

Brachytherapy, also known as Radioactive Seed Implantation Therapy, is an advanced form of cancer treatment. It uses small wires or 'seeds' made of radioactive elements that are temporarily or permanently implanted in the body at a site near or within the tumour growth area. From there they radiate ionising energy that kills the abnormal cells. Dozens of these seeds can be surgically placed around the tumour site, as shown in Figure 6.4.9. This method is most commonly used for tumours in the breast and prostate gland. There are two modes of delivery of this treatment:

- low-dose rate brachytherapy: Seeds are applied permanently and emit a low dose of radiation.
- high-dose rate brachytherapy: Seeds are applied temporarily with high radiation potency.

Some of the radioisotopes now used in brachytherapy include iridium-192, iodine-125, cobalt-60 and caesium-137.

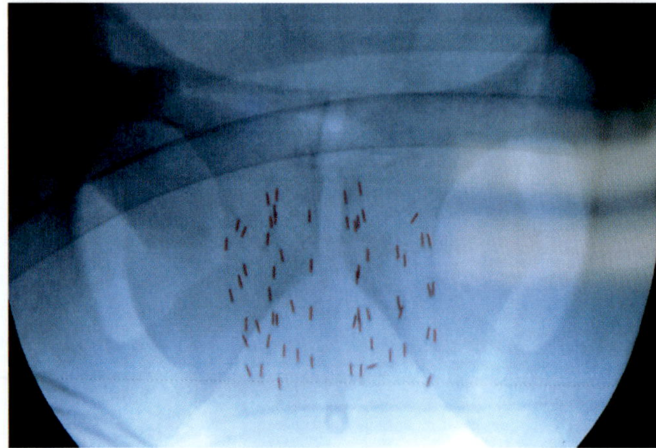

FIGURE 6.4.9 Numerous iodine-125 (I-125) seeds can be seen in this X-ray of a prostate cancer patient. I-125 is a gamma emitter with a half-life of 59.4 days.

6.4 Review

OA ✓✓

SUMMARY

- Alpha, beta, gamma and high-energy electromagnetic radiation are ionising and are harmful to humans.
- Exposure to some ionising radiation is a natural part of human existence. However, unnecessary exposure to ionising radiation can be dangerous and should be avoided.
- The equivalent dose (ED) gives a measure of the biological damage that a dose of radiation causes. ED = AD × QF and is measured in sieverts (Sv).
- The effective dose is found by calculating the sum of the dose equivalents multiplied by the weighting factor, W, for each organ affected.
- Effective dose = Σ(equivalent dose × W) and is measured in sieverts (Sv).
- When ionising radiation passes through human tissue, it may ionise atoms and molecules in the body's cells, which can lead to the development of cancerous cells.
- Exposure to ionising radiation can lead to both somatic and genetic effects.
- Depending on the radiation dose, somatic effects can vary from feelings of nausea to severe illness and even death.
- If a person's reproductive cells are damaged by radiation, genetic anomalies may arise in future generations.
- A radioisotope can also be called a radionuclide or radioactive tracer. All terms refer to a chemical element that has an unstable nucleus and emits radioactivity during its decay to form a stable element.
- A radiopharmaceutical is a drug that has a radioisotope chemically bonded to its structure. The type of radioisotope used depends on the target tissue and the specific application. These drugs can be used for the diagnosis and treatment of disease.
- Radiopharmaceuticals can also be used as part of chemotherapy treatments, in which they deliver radiation from inside the body to kill the cancer cells at the site where they occur.

KEY QUESTIONS

Knowledge and understanding

1 What is a somatic effect of exposure to ionising radiation? Give three examples.

2 What type of radiation is emitted from the radioisotopes used in PET scans? Why is this?

3 A cancer tumour of mass 150 g is exposed to 0.30 J of radiation energy. What is the absorbed dose in grays?

4 What are two characteristics that a radioisotope should have for it to be suitable for use as a tracer?

5 Which of the following is the most damaging radiation dose: 200 μGy of gamma radiation, 20 μGy of alpha radiation, 50 μGy of beta radiation or 30 μGy of neutron radiation? Explain your answer.

6 During a course of radiotherapy, a cancer patient may be exposed to 40 Sv of radiation. Why is this massive dose not fatal for the patient?

7 Refer to Table 6.4.4 of W values on page 177. Calculate the effective dose for a patient whose organs received the following exposures during a course of radiotherapy: ovaries and bladder 35 mSv each, colon 50 mSv.

Analysing

8 A woman is exposed to a large whole-body dose of radiation and was later found to be anaemic (to have low red blood cell count). Is this a genetic or somatic effect? Explain your answer.

9 Most forms of diagnostic imaging involve the patient being exposed to ionising radiation. If ionising radiation is dangerous, why is it used in diagnostic procedures?

Chapter review

KEY TERMS

absorbed dose
activity
alpha particle
antineutrino
atomic number
beta particle
chemotherapy
daughter nucleus
decay series
effective dose
electron
electrostatic force
equivalent dose
gamma ray
Geiger counter
genetic
half-life
isotope
mass number
mutation
neutral
neutron
nuclear transmutation
nucleon
nucleus
nuclide
parent nucleus
penetrating ability
positron
proton
quality factor (QF)
radiation
radioactive
radioactive decay
radioactive tracer
radioisotope
radiopharmaceutical
somatic
strong nuclear force
weak nuclear force
weighting factor

REVIEW QUESTIONS

Knowledge and understanding

1 How many protons and neutrons are in the $^{45}_{20}Ca$ nuclide?

2 Use the periodic table in Figure 6.1.7 on page 155 to determine the number of protons, neutrons and nucleons in cobalt-60.

3 What type of radiation does potassium-48 (atomic number 19) emit? Use Figure 6.2.5 on page 160 to answer this question.

4 Some nuclei can be made unstable by firing neutrons into them. The neutron is captured and the nucleus becomes unstable. The nuclear equation when the stable isotope boron-10 transmutates by neutron capture into a different element, X, by emitting alpha particles is:

$$^{10}_{5}B + ^{1}_{0}n \rightarrow X + ^{4}_{2}He$$

Identify the unknown element, X, and its mass and atomic numbers.

5 Identify the unknown particles X and Y in the following nuclear transmutations.

a $^{14}_{7}N + \alpha \rightarrow ^{17}_{8}O + X$

b $^{27}_{13}Al + Y \rightarrow ^{27}_{12}Mg + ^{1}_{1}H$

c $^{2}_{1}H + ^{3}_{1}H \rightarrow ^{a}_{b}X + ^{1}_{0}n$

6 The half-life of uranium-236 is 23.4 million years. Explain how a single individual nucleus of uranium-236 can spontaneously decay at any time.

7 Positron emission tomography (PET) scans are used in the diagnosis of cancer.

a What will a patient be injected with prior to undergoing a PET scan?

b How do positrons interact with a patient's body in order to produce an image during a PET scan?

8 Which of the following is the most damaging dose of radiation: 1 Gy of alpha, 1 Gy of beta or 1 Gy of gamma radiation? Explain your answer.

9 An oncology patient receives a radiation dose of 1 Sv. Which of the following is the most damaging dose: 1 Sv of alpha, 1 Sv of beta or 1 Sv of gamma? Explain your answer.

10 An airline pilot of mass 90 kg absorbs a gamma radiation dose of 300 mGy during a return flight to New York. Calculate the equivalent dose the pilot received in mSv.

11 What is the most widely used radioactive medical tracer and how is it produced?

12 Health workers who deal with radiation to treat cancer often have to wear a lead vest to protect their vital organs from exposure. Which type(s) of radiation is the lead apron shielding them from?

Application and analysis

13 Iodine-131 has a half-life of 8 days. A sample of the radioisotope initially contains 2.4×10^{12} iodine-131 nuclei. Determine how many iodine-131 nuclei remain after 24 days.

14 According to the decay series diagram shown, what type of decay does polonium-218 undergo and what is its half-life?

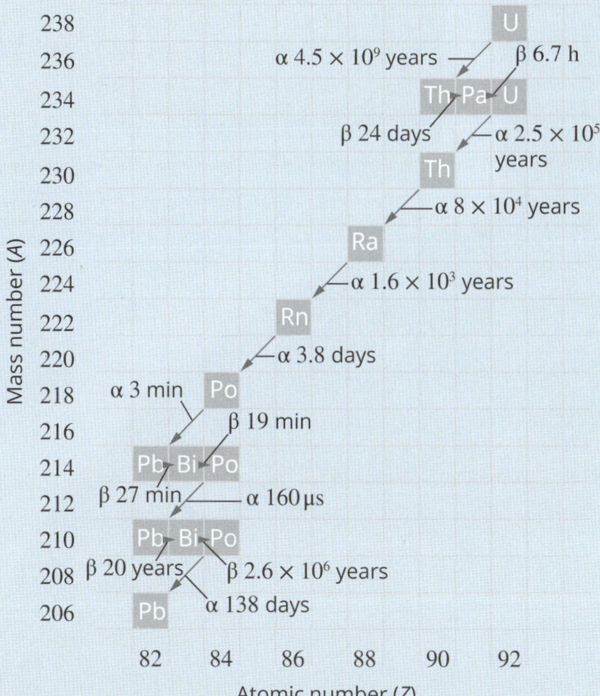

15 A stable isotope of neon has 10 protons and 10 neutrons in each nucleus. Every proton is repelling all the other protons. Why is the nucleus stable?

16 In a major incident in a nuclear reactor, a 75 kg employee received a full-body absorbed radiation dose of 5.0 Gy of gamma rays.

a Calculate the amount of energy that was absorbed during this exposure. Give your answer to two significant figures.

b Calculate the equivalent dose for this person. Give your answer to two significant figures.

17 A worker in an X-ray clinic takes an average of ten X-ray radiographs each working day and receives an annual radiation equivalent dose of 7900 μSv.

a Calculate the dose (to one significant figure), in μSv, that the worker receives from taking each X-ray radiograph. (Assume they work for 5 days per week for 45 weeks a year.)

b How many times greater than the normal background radiation dose is the worker's annual dose?

18 A radioisotope can be used as a tracer in the diagnosis of certain conditions in the human body. For this particular use, the radioisotope should ideally be a gamma emitter. Why is this?

19 Max performed an experiment to find the half-life of a sample of solid, crystal radioactive iodine-131 in a laboratory. The sample was weighed at the start of the observations and was found to be 436 g. Max then weighed the remaining sample at various times and these results were recorded in the table below.

Time after first observation	Remaining mass of sample (± 1 g)
0	436
1 minute	436
1 hour	424
12 hours	411
1 day	395
2 days	360
5 days	291
10 days	166
15 days	131
20 days	77

a Draw a graph of the data above, showing how the mass of the iodine-131 changes over time. Include a line of best fit on your graph.

b Use the graph to determine the half-life of iodine-131.

c Hypothesise why Max recorded the mass of the sample at irregular times after the first observation until 5 days into the experiment.

d Iodine-131 decays via beta-minus emission into gaseous xenon-131. Devise two safety precautions Max had to take when he made mass measurements.

e Max realises that the mass of the sample is probably more than the actual mass of the remaining iodine-131. Draw a conclusion as to why he thought this. Suggest an experiment he could carry out to make sure that the measured mass is the actual mass of the remaining iodine-131 and nothing else.

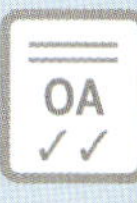

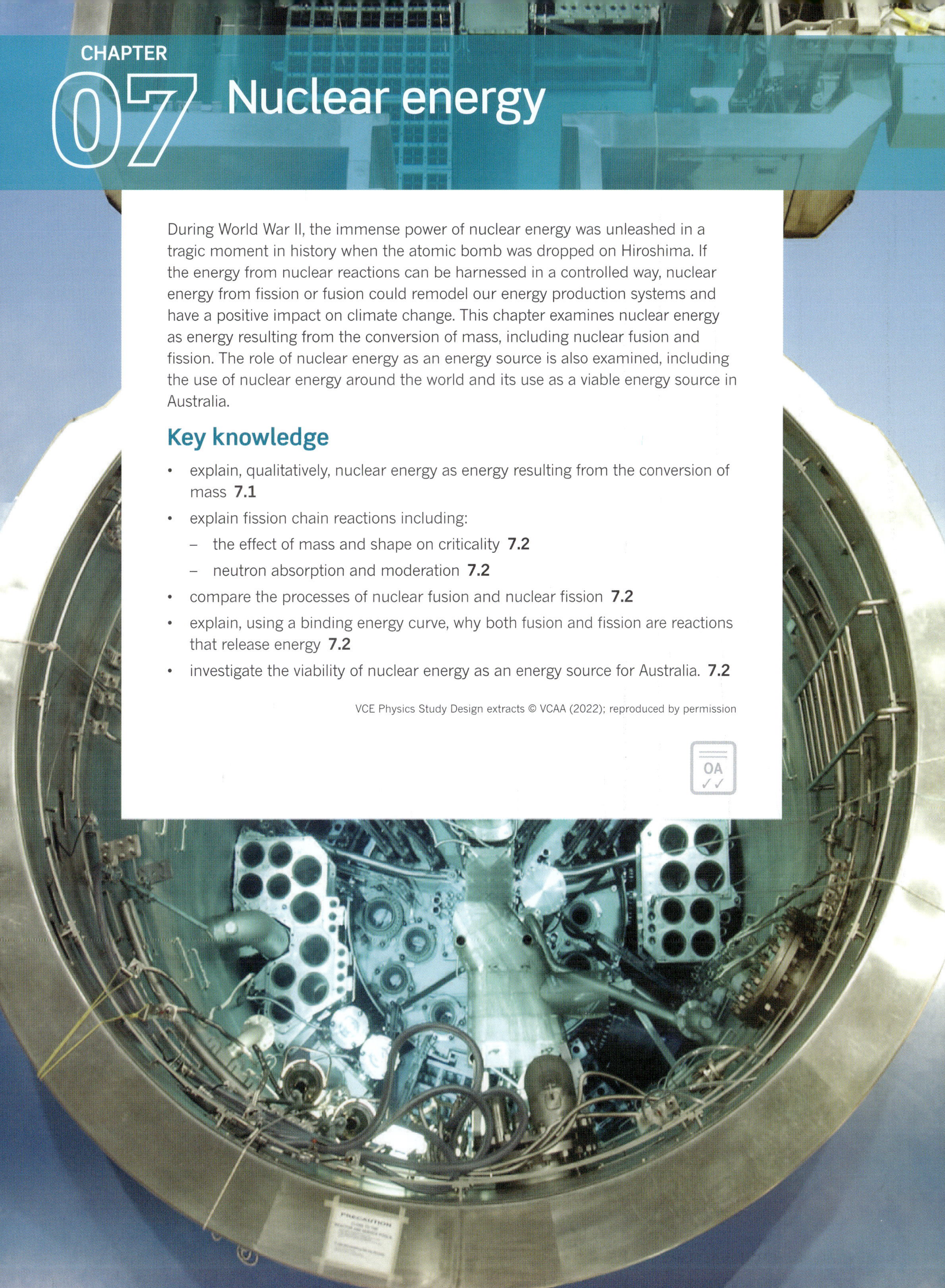

CHAPTER

07 Nuclear energy

During World War II, the immense power of nuclear energy was unleashed in a tragic moment in history when the atomic bomb was dropped on Hiroshima. If the energy from nuclear reactions can be harnessed in a controlled way, nuclear energy from fission or fusion could remodel our energy production systems and have a positive impact on climate change. This chapter examines nuclear energy as energy resulting from the conversion of mass, including nuclear fusion and fission. The role of nuclear energy as an energy source is also examined, including the use of nuclear energy around the world and its use as a viable energy source in Australia.

Key knowledge

- explain, qualitatively, nuclear energy as energy resulting from the conversion of mass **7.1**
- explain fission chain reactions including:
  - the effect of mass and shape on criticality **7.2**
  - neutron absorption and moderation **7.2**
- compare the processes of nuclear fusion and nuclear fission **7.2**
- explain, using a binding energy curve, why both fusion and fission are reactions that release energy **7.2**
- investigate the viability of nuclear energy as an energy source for Australia. **7.2**

OA

7.1 Energy from mass

In the previous chapter you learnt about nuclear stability, with reference to the forces in the nucleus, as well as nuclear transformations and radioactive decay. Recall that in every decay equation, energy is always released as a result of the parent nucleus becoming more stable and creating a daughter nucleus. The stability of the nucleus is dependent on the number of protons and neutrons. Due to the electrostatic force, the protons repel each other. But the strong nuclear force acts as a force of attraction over short distances between every nucleon. The amount of energy released during radioactive decay depends on how tightly the nucleons are held together. In this section, you will explore the amount of energy released during nuclear reactions and see that some of the mass of the nucleons is transformed into pure energy by the use of the most famous equation in science:

$$E = mc^2$$

THE ELECTRON VOLT

The energy of moving objects such as cars and tennis balls is measured in joules. However, nuclei, subatomic particles and radioactive emissions have such small amounts of energy that the joule is inappropriate.

The energy of subatomic particles and radiation is usually given in **electron volts (eV)**. One electron volt is an extremely small amount of energy.

An electron volt is the energy that an electron would gain if it were accelerated by an electrical potential difference of 1 volt and is equal to 1.6×10^{-19} J.

To convert from eV to joules, multiply by 1.6×10^{-19} J.

To convert from joules to eV, divide by 1.6×10^{-19} J.

In nuclear physics, most electron energies are described in keV (one thousand, 10^3, eV), MeV (one million, 10^6, eV), GeV (one billion, 10^9, eV) and TeV (one trillion, 10^{12}, eV).

Note that the eV is not an SI unit, which means that you will need to convert energy in eV to J before making any substitutions into formulas that use SI units.

Table 7.1.1 show the comparisons of various energetic processes in joules and in electron volts. This will give you an idea of how small 1 eV is.

TABLE 7.1.1 Energy processes and their equivalent energies

Energy process	Energy in joules	Energy in electron volts	Notes
energy consumed by a light bulb in one second	10	6.24×10^{19}	single 10-watt LED light bulb for one second
highest energy radio waves	2.0×10^{-22}	0.0012	the least energetic form of electromagnetic radiation
a falling apple	1	6.2×10^{18}	energy attained by an apple falling 1 metre on Earth
car travelling on a highway	200 000	1.2×10^{24}	the kinetic energy of a 500 kg car moving at 100 km h^{-1}

Worked example 7.1.1

THE ELECTRON VOLT

Radium-226 is a radioisotope that decays into radon-222 via alpha radiation. The reaction is:

$$^{226}_{88}Ra \rightarrow \, ^{222}_{86}Rn + \, ^{4}_{2}He$$

a If 4.9 MeV of energy is released from each radium-226 nucleus, state this in joules.

Thinking	Working
Convert 4.9 MeV into joules. First change 4.9 MeV into eV by multiplying by 10^6.	4.9 MeV = 4.9×10^6 eV
To convert eV into J, multiply by 1.6×10^{-19}.	Energy = $4.9 \times 10^6 \times 1.6 \times 10^{-19}$ = 7.8×10^{-13} J

b If 6.8×10^{24} nuclei of radium-226 nuclei decay, determine the amount of energy that would be released in eV and in joules.

Thinking	Working
Each reaction releases 4.9 MeV of energy, so, 6.8×10^{24} reactions will release $6.8 \times 10^{24} \times$ 4.9 MeV of energy.	Total amount of energy = $6.8 \times 10^{24} \times 4.9 \times 10^6$ eV = 3.3×10^{31} eV
Multiply 3.3×10^{31} by 1.6×10^{-19} to find the answer in joules.	Total amount of energy = $3.3 \times 10^{31} \times 1.6 \times 10^{-19}$ = 5.3×10^{12} J = 5.3 TJ

Worked example: Try yourself 7.1.1

THE ELECTRON VOLT

The radioisotope technetium-99m becomes stable by emitting gamma radiation.

If the energy of the gamma ray emitted by the parent nucleus is 2.3×10^{-14} J, determine how much energy this is in eV, keV and MeV.

MASS DEFECT AND BINDING ENERGY

Burning 1 kg of coal typically produces 30 MJ of energy. Burning 1 kg of hydrogen can produce up to 140 MJ of energy, but 'burning' 1 kg of uranium-238 produces 80.6 TJ of energy, or more than 2 million times more energy than burning coal. Of course, the uranium-238 isn't actually burnt as it would be in a chemical reaction, but instead it is involved in a nuclear reaction. All nuclear reactions give off millions of times more energy than chemical reactions. To explain this, we need to understand how mass and energy are connected and how a nucleus is held together.

If a given, intact nucleus could be disassembled into its individual protons and neutrons, then the sum of the masses of the individual protons and neutrons would be slightly higher than the mass of the nucleus itself. The difference between the mass of the nucleus as a whole and the sum of the individual nucleons is called the **mass defect**, Δm.

The mass defect is calculated as:

Δm = total mass of individual protons and neutrons − actual mass of nucleus.

This 'missing' mass might seem to violate conservation laws. However, when the individual protons and neutrons come together to form a nucleus, some of their mass is converted into the energy needed to bind them into a nucleus. This is called the **binding energy**, ΔE. It is defined as the energy that would need to be given to a nucleus to separate it into its individual nucleons.

Binding energy is shown diagrammatically in Figure 7.1.1.

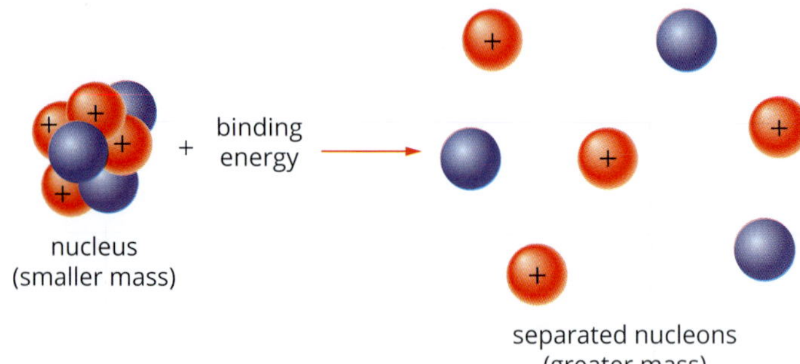

FIGURE 7.1.1 Mass is converted to energy to hold together a nucleus. This energy, called the binding energy, is defined as the energy needed to completely separate a nucleus into its individual protons and neutrons.

In many chemical reactions, for example the combustion of methane (natural) gas with oxygen, energy is released when the bonds between atoms in reactants break and new bonds form to make the products, but no mass is lost. In a nuclear reaction, however, it is parts of the protons and neutrons themselves that are converted directly into energy. For example, for the isotope fluorine-19, the sum of the masses of the 19 nucleons is greater than the mass of the fluorine-19 nucleus:

$$\begin{aligned}\Delta m &= (\text{mass of 9 protons} + \text{mass of 10 neutrons}) - \text{mass of fluorine-19 nucleus}\\ &= (9 \times m_p + 10 \times m_n) - 3.15 \times 10^{-26}\\ &= (9 \times 1.67262 \times 10^{-27} + 10 \times 1.67493 \times 10^{-27}) - 3.15 \times 10^{-26}\\ &= 3.03 \times 10^{-28}\,\text{kg}\end{aligned}$$

This is illustrated in Figure 7.1.2. The difference in mass has been converted into binding energy to hold together the fluorine-19 nucleus.

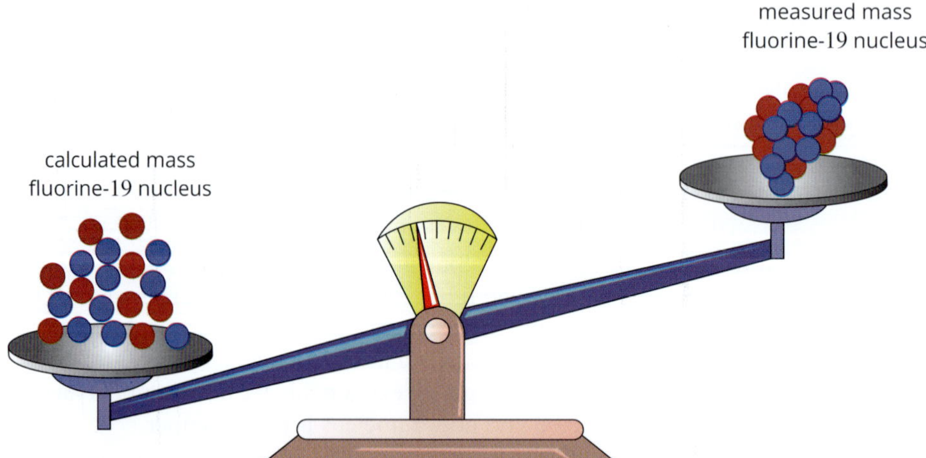

FIGURE 7.1.2 The sum of the masses of the 9 protons and 10 neutrons is larger than the mass of the fluorine-19 nucleus. The difference in mass has been converted into binding energy.

The mass–energy equivalence

Einstein developed a mathematical rule that linked energy and matter, based on the fact that matter can be converted entirely into energy, and energy can be converted entirely into matter. His equation is known as the mass–energy equivalence relationship. It has many applications, but in this case, it relates the binding energy, ΔE, to the mass defect, Δm, by:

$$\Delta E = \Delta mc^2$$

where c is the speed of light in a vacuum ($3.0 \times 10^8\,\text{m}\,\text{s}^{-1}$).

In Unit 4 you will see this written simply as $E = mc^2$.

In the example of fluorine on the previous page, the binding energy can be calculated as:

$$\begin{aligned}\Delta E &= \Delta mc^2 \\ &= 3.03 \times 10^{-28}\,\text{kg} \times (3.0 \times 10^8)^2 \\ &= 2.8 \times 10^{-13}\,\text{J}\end{aligned}$$

The following worked example takes you through calculations of mass defect and binding energy. The numbers involved are so small that, because of rounding, you may notice that the answers here are slightly different from published binding energies for these isotopes.

The mass–energy equivalence relationship:

$$\Delta E = \Delta mc^2$$

where ΔE is energy (J)

Δm is the mass defect (kg)

c is the speed of light in a vacuum ($3.0 \times 10^8\,\text{m}\,\text{s}^{-1}$).

Worked example 7.1.2

CALCULATING BINDING ENERGY

Calculate the binding energy of a nickel-62 nucleus, given its atomic mass is 1.028×10^{-25} kg. Give your answer in both J and MeV.

Use $m_p = 1.673 \times 10^{-27}$ kg and $m_n = 1.675 \times 10^{-27}$ kg and refer to a periodic table as necessary.

Thinking	Working
Recall Einstein's energy–mass equivalence rule.	$\Delta E = \Delta mc^2$
Determine the number of protons and neutrons in nickel-62.	From the periodic table, nickel has an atomic number of 28, so it has 28 protons. Therefore it has 62 – 28 = 34 neutrons.
Calculate the mass defect.	$\Delta m = (28 \times m_p + 34 \times m_n) - 1.028 \times 10^{-25}$ $= (28 \times 1.673 \times 10^{-27} + 34 \times 1.675 \times 10^{-27}) - 1.028 \times 10^{-25}$ $= 9.940 \times 10^{-28}$ kg
Calculate the binding energy.	$\Delta E = 9.940 \times 10^{-28} \times (3.0 \times 10^8)^2$ $= 8.946 \times 10^{-11}$ J
Convert to MeV	$\dfrac{8.946 \times 10^{-11}}{1.6 \times 10^{-19}} = 559.1$ MeV

Worked example: Try yourself 7.1.2

CALCULATING BINDING ENERGY

Calculate the binding energy of a uranium-235 nucleus, given its atomic mass is 3.902×10^{-25} kg. Give your answer in both J and MeV.

Use $m_p = 1.673 \times 10^{-27}$ kg and $m_n = 1.675 \times 10^{-27}$ kg and refer to a periodic table as necessary.

7.1 Review

OA ✓✓

SUMMARY

- The electron volt (eV) is a unit of energy equal to 1.6×10^{-19} J. It is a non-SI unit, but is preferred for use in nuclear physics as it is more convenient than joules.
- The mass of a nucleus is always smaller than the sum of the masses of the individual protons and neutrons (the nucleons) that make it up. The difference in these masses is called the mass defect, Δm.
- The mass defect is converted into energy that holds the nucleus together. This is called the binding energy, ΔE. It is equivalent to the energy required to break a nucleus completely into its constituent protons and neutrons.
- The binding energy can be calculated by using Einstein's famous mass–energy equivalence relationship: $\Delta E = \Delta mc^2$.

KEY QUESTIONS

Knowledge and understanding

1. Convert 5.0 MeV into joules.
2. Convert 6.0×10^{-15} J into eV and MeV.
3. State the meaning of 'mass defect'.
4. Describe the term 'binding energy'.

Analysis

5. Calculate the binding energy of tin-120, given the atomic mass is 1.991×10^{-25} kg. Give your answer in both J and MeV. Use $m_p = 1.673 \times 10^{-27}$ kg and $m_n = 1.675 \times 10^{-27}$ kg and refer to a periodic table as necessary.

7.2 Fission, fusion and the future of nuclear energy in Australia

The idea that mass and energy are equivalent through the equation $\Delta E = \Delta mc^2$ led to the realisation that vast amounts of energy lie unharnessed within the nuclei of atoms. The ramifications of Einstein's work and the discovery of nuclear fission were realised in 1945 with the explosion of the first atomic bomb in the desert near Alamogordo in New Mexico, USA (Figure 7.2.1). In this section, nuclear fission and fusion—and the energy that they can unleash—will be explored, as will the role of nuclear energy in Australia.

FIGURE 7.2.1 An atomic bomb explosion and its associated mushroom cloud

FISSION

The neutron was discovered by James Chadwick in 1932. This discovery enabled scientists to explore the behaviour of larger atomic nuclei. Up until then, physicists such as Enrico Fermi had been firing α-particles (helium nuclei) at target nuclei and analysing the results. Chadwick found that with larger target nuclei, the positive α-particles were too strongly repelled from the positively charged nuclei and collisions did not occur.

The advantage of a neutron is that it is neutral (has no charge) and so is not repelled by any target nucleus. The bombarding neutrons can be absorbed into the nucleus of the target atom, as shown in Figure 7.2.2. This makes neutrons very useful as a form of radiation. They are used in many experiments to artificially transmutate (change the form of) different isotopes.

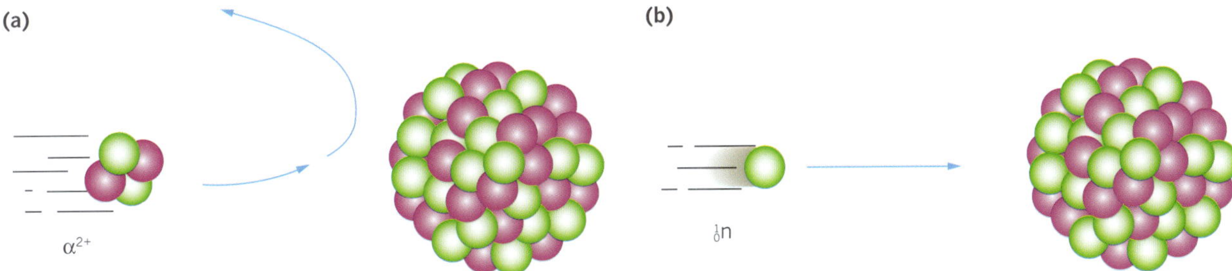

FIGURE 7.2.2 (a) Charged α-particles are repelled by a nucleus. (b) Uncharged neutrons are able to smash into a nucleus.

Nuclear fission is the process in which a large nucleus is forced to split into at least two approximately equal-sized fragments with the release of neutrons and energy.

Nuclear **fission** occurs when an atomic nucleus splits into two or more pieces of roughly equal size known as the daughter nuclei or **fission fragments**. This is usually triggered by the absorption of a neutron, as shown in Figure 7.2.3. The sum of the masses of the fission fragments is smaller than the mass of the original nucleus, so some mass appears to have been lost. However, this mass has been converted into energy. The fission fragments have kinetic energy, which comes from converting a small amount of the mass of the original atom into energy. The energy is related to the change in mass using the equation $\Delta E = \Delta mc^2$ in the same way that the binding energy of a nucleus can be found from its mass defect.

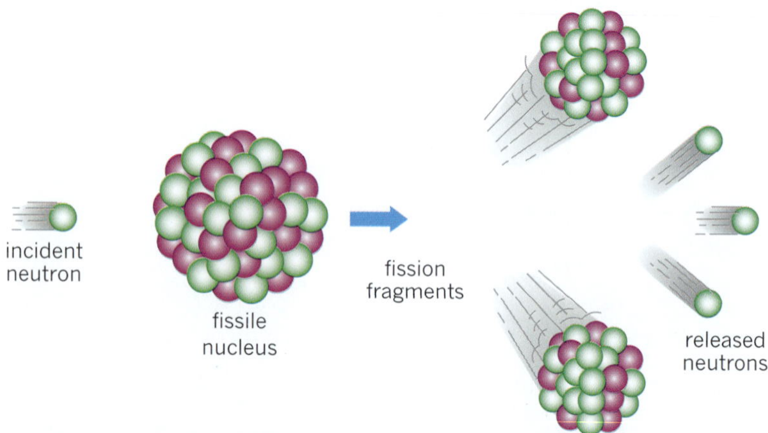

FIGURE 7.2.3 Nuclear fission is the splitting of a nucleus.

Nuclides that are capable of undergoing nuclear fission after absorbing a neutron are said to be **fissile**. Only a handful of fissile nuclides exist in nature. Fission does not happen independently—it requires the addition of energy to overcome the binding energy of the parent nucleus. For example, uranium-235 and plutonium-239 (the nuclides most commonly used in nuclear reactors and weapons) are fissile and can be made to split when bombarded by a slow-moving neutron. Uranium-238 and thorium-232 require a very high-energy neutron to induce fission, so they are regarded as fissionable, but non-fissile. Fission is more likely to be induced by a slow-moving neutron because it is more easily captured.

A fissile nucleus may split in many different ways—usually into two roughly equal-sized pieces plus two or three neutrons. When a sample of uranium-235 undergoes fission, a wide variety of fission products are produced. Figure 7.2.4 shows one outcome, but many others are possible. The fission of uranium-235 releases an average of 2.47 neutrons.

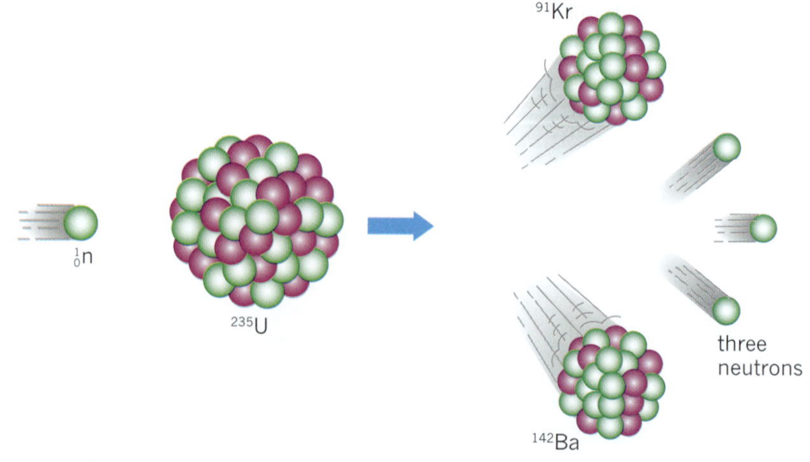

FIGURE 7.2.4 One possible outcome for the neutron-induced fission of uranium-235. This example shows three neutrons released along with krypton-91 and barium-142 daughter nuclides.

A typical fission reaction for uranium-235 is:

$$^{1}_{0}\text{n} + ^{235}_{92}\text{U} \rightarrow ^{236}_{92}\text{U} \rightarrow ^{91}_{36}\text{Kr} + ^{142}_{56}\text{Ba} + 3^{1}_{0}\text{n}$$

In this case, krypton-91 and barium-142 are the fission fragments, and three neutrons are freed from this uranium nucleus when it splits. Energy is emitted mostly in the form of the kinetic energy of the products. Note that, as for radioactive decay, both the atomic number, Z, and the mass number, A, are conserved in these nuclear reactions. For the reaction equation shown, the atomic numbers on either side of the arrows add up to 92 and the mass numbers add up to 236.

The decay products of nuclear fission are often radioactive themselves. It is these radioactive fission fragments that comprise the bulk of the high-level waste produced by nuclear reactors.

Plutonium-239 will also undergo fission in a variety of ways. It releases an average of 2.89 neutrons per fission, slightly more than uranium-235.

Worked example 7.2.1

NEUTRONS RELEASED IN FISSION

Plutonium-239 is a fissile material. When a plutonium-239 nucleus is struck by and absorbs a neutron, it can split in many different ways. Consider the example of a nucleus that splits into barium-145 and strontium-93 and releases some neutrons.

The nuclear equation for this is:

$$^{1}_{0}\text{n} + ^{239}_{94}\text{Pu} \rightarrow ^{145}_{56}\text{Ba} + ^{93}_{38}\text{Sr} + a^{1}_{0}\text{n}$$

How many neutrons are released during this fission process, i.e. what is the value of a?

Thinking	Working
Analyse the mass numbers (A).	$1 + 239 = 145 + 93 + (a \times 1)$ $a = (1 + 239) - (145 + 93)$ $= 2$ Two neutrons are released during fission.

Worked example: Try yourself 7.2.1

NEUTRONS RELEASED IN FISSION

Plutonium-239 is a fissile material. When a plutonium-239 nucleus is struck by and absorbs a neutron, it can split in many different ways. Consider the example of a nucleus that splits into lanthanum-143 and rubidium-94 and releases some neutrons.

The nuclear equation for this is:

$$^{1}_{0}\text{n} + ^{239}_{94}\text{Pu} \rightarrow ^{143}_{57}\text{La} + ^{94}_{37}\text{Rb} + a^{1}_{0}\text{n}$$

How many neutrons are released during this fission process, i.e. what is the value of a?

Chain reactions

As already noted, an important feature of nuclear fission is the production of extra neutrons, which can go on to react with other fissile nuclei to trigger more fission reactions. For example, if three neutrons are released in a single fission reaction, those neutrons can trigger three more fission reactions by bombarding more fissile material. Each of those fission reactions then produces three more neutrons, releasing 9, then 27, then 81 neutrons, and so on. This exponential cascade of more neutrons initiating more fission reactions is called a **chain reaction**, and is shown in Figure 7.2.5.

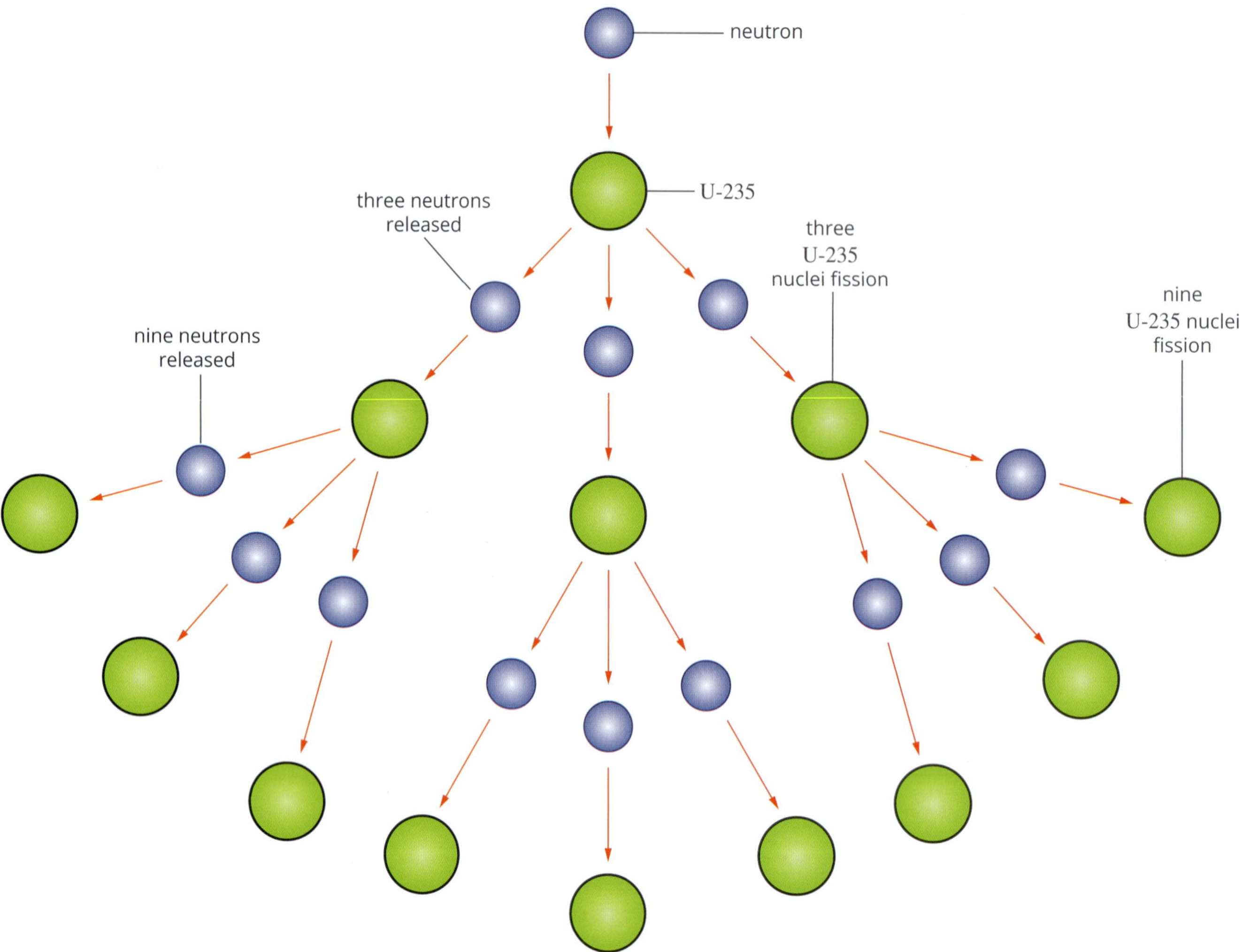

FIGURE 7.2.5 The start of a chain reaction involving uranium-235. Each fission reaction produces three neutrons, which then go on to react with three other uranium-235 nuclei to create nine new neutrons, and so on, in a runaway series of reactions producing very large amounts of energy.

A chain reaction occurs when a single nuclear reaction causes an exponential increase in the number of successive nuclear reactions.

After only 13 series of reactions, there will be over one million neutrons available to trigger that many fission reactions. The number of fission reactions occurring at one given time during a chain reaction is increasing exponentially, so the amount of energy produced also increases exponentially.

Chain reactions continue for as long as there is enough material available. The amount of material needed to sustain a chain reaction is known as the **critical mass**. The critical mass of a material can be modified by changing certain attributes, including the shape of the material. For a chain reaction to continue, neutrons need to encounter another nucleus. This is more likely in a spherical piece of radioactive material than in a flat piece. In the flat piece, a larger proportion of neutrons would escape as the surface area is quite high in comparison to its volume; therefore, a sphere of radioactive material equal to the critical mass would be not be critical if that same mass were in a flat piece.

A chain reaction can be controlled or uncontrolled. An uncontrolled chain reaction creates an explosion such as in a nuclear weapon (i.e. fission bomb), and Figure 7.2.6 shows the result. A controlled reaction (such as in a nuclear reactor) keeps the reaction rate relatively constant and produces a steady power output.

FIGURE 7.2.6 The unmistakable mushroom cloud formed by an uncontrolled chain reaction of uranium-235 or plutonium-239 in a nuclear bomb. This photograph was taken in 1946 at Bikini Atoll in the Marshall Islands. The explosion was part of the nuclear weapons testing conducted by the United States known as 'Operation Crossroads'.

In a nuclear fission reactor, the neutrons released are travelling at very high speeds, making it less likely that they will collide with other atoms to continue the chain reaction. A moderator is used to slow the neutrons and increase the likelihood that neutrons will collide with other fissionable atoms. Water is commonly used as the moderator in many nuclear reactors.

The rate of a chain reaction in a fission reactor can be controlled using control rods to absorb neutrons as necessary. These rods are made of a material that easily absorbs neutrons, such as graphite or cadmium. The rods are then either withdrawn or inserted further into the reactor core to either increase or decrease the rate of the reaction. There is still, however, a chain reaction. The use of control rods is a key safety feature of nuclear reactors. An issue with the control rods was to blame for one of the worst nuclear power plant disasters in the world at the Chernobyl Nuclear Power Plant in Ukraine, because of an uncontrolled nuclear chain reaction releasing enormous amounts of energy.

CASE STUDY

Enrico Fermi, Lise Meitner and nuclear fission

Enrico Fermi (Figure 7.2.7) was born in Italy in 1901. He completed his doctorate and postdoctorate work in physics at the University of Pisa and in Germany. Fermi had emigrated to the USA by the time the nuclear age dawned in the 1930s. The neutron had been discovered in 1932, which enabled scientists to fire neutral particles at atomic nuclei for the first time. Fermi was at the forefront of this research.

FIGURE 7.2.7 Enrico Fermi

Fermi bombarded uranium-238 atoms with neutrons and found that uranium-238 nuclei absorbed the neutrons and formed a radioactive isotope of uranium. This isotope then decayed by emitting a beta-minus particle to become neptunium, which then emitted another beta-minus particle to become plutonium, two completely undiscovered elements. Fermi had successfully produced the world's first artificial and transuranic (i.e. after uranium) elements. The nuclear reactions for this process are:

$$^{1}_{0}n + ^{238}_{92}U \rightarrow ^{239}_{92}U$$

$$^{239}_{92}U \rightarrow ^{239}_{93}Np + ^{0}_{-1}\beta$$

$$^{239}_{93}Np \rightarrow ^{239}_{94}Pu + ^{0}_{-1}\beta$$

In 1938, following on from Fermi's work, two German scientists, Otto Hahn and Fritz Strassmann, were also bombarding uranium ($Z = 92$) in an attempt to produce transuranic elements ($Z > 92$). They found that, rather than producing larger elements, they were getting isotopes of barium ($Z = 56$). Hahn wrote to his colleague physicist Lise Meitner (Figure 7.2.8), about this unexpected result. Meitner discussed the result with her nephew, Otto Frisch, also a physicist. Together, they realised that the bombarding neutrons were causing the uranium nuclei to split. If barium ($Z = 56$) was one of the products, then krypton ($Z = 36$) must be another. This was found to be the case. It was Frisch who coined the term 'fission' and Meitner who proposed that energy would be released during this process.

FIGURE 7.2.8 Lise Meitner

After the start of World War II, Enrico Fermi was commissioned by President Roosevelt to design and build a device that would sustain the fission process in the form of a chain reaction. In 1942, Fermi succeeded in this task. A squash court at the University of Chicago was used as the site for the world's first nuclear reactor. It produced less than 1 W of power—not even enough to power a small light globe! This sounds like a bit of a failure, but in fact, achieving fission for the first time was a very important breakthrough. The reactor was later modified to produce about 200 W. Fermi died of cancer in 1954. One year after his death, the element with atomic number 100 was artificially produced and named fermium, Fm, in his honour.

FUSION

Nuclear fusion is a process that has been occurring inside the Sun and other stars for billions of years. **Fusion** involves the combining of small nuclei such as hydrogen and helium to form a larger nucleus. The amount of energy per nucleon released with fusion is greater than with fission and there is no radioactive waste produced—its main product is helium.

Nuclear fusion is the process of two nuclei combining to form a single nucleus.

Energy during fusion

Nuclear fusion occurs when two light nuclei are combined to form a larger nucleus. The example of nuclei fusing to form a helium atom is shown in Figure 7.2.9(a).

As in nuclear fission, the mass of the reactants is slightly greater than the mass of the products when the nuclei combine during fusion. This mass difference is represented by the unbalanced scales shown in Figure 7.2.9(b). In this figure, two isotopes of hydrogen with one proton and one neutron fuse to form a helium nucleus. In this case, the mass of the helium nucleus is less than the sum of the masses of the two hydrogen isotopes.

The energy created by this missing mass can again be determined from:

$$\Delta E = \Delta mc^2$$

where ΔE = energy (J)

Δm = mass defect (kg)

$c = 3.0 \times 10^8\,\text{m s}^{-1}$.

Nuclear fusion is a very difficult process to re-create. The main problem is that nuclei are positively charged. They exert an electrostatic force of repulsion on each other; that is, they push each other away. As such, it is not easy to force the nuclei together. Remember that the electrostatic force is a long-range force and the strong nuclear force of attraction only acts at much shorter distances.

As two nuclei approach each other, the electrostatic force will cause them to be repelled. Slow-moving nuclei with relatively small amounts of kinetic energy will not be able to get close enough for the strong nuclear force to come into effect. Fusion will not take place, as can be seen in Figure 7.2.10(a).

If the nuclei are travelling towards each other at higher speeds, as shown in Figure 7.2.10(b), they may have enough kinetic energy to overcome the repulsive force. The nuclei can now get close enough for the strong nuclear force to start acting. If this happens, fusion will occur.

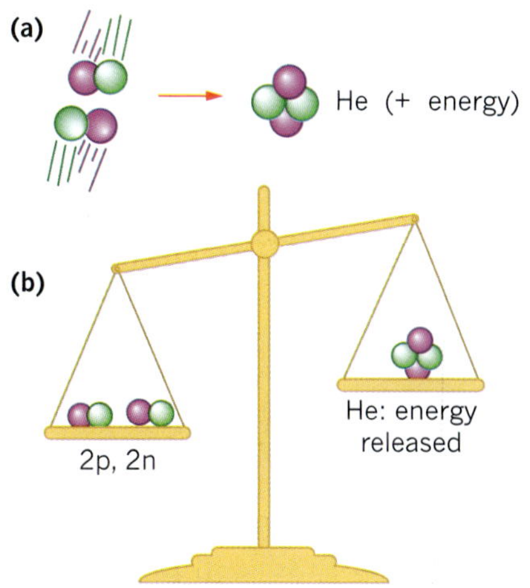

(a) He (+ energy)

(b) 2p, 2n — He: energy released

FIGURE 7.2.9 (a) When two isotopes of hydrogen fuse to form a helium nucleus, energy is released. (b) The loss in mass, m, can be calculated using $E = mc^2$.

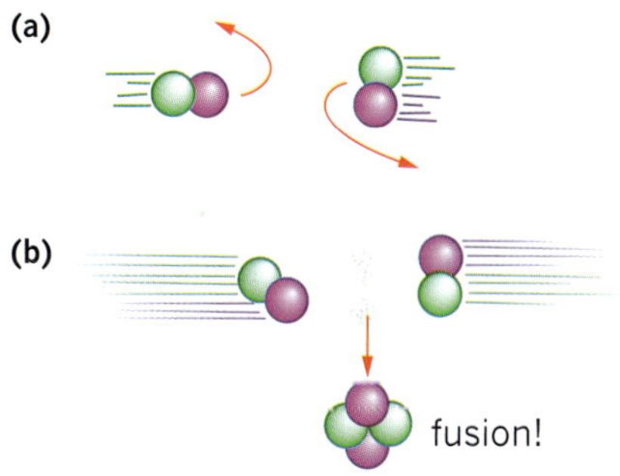

(a)

(b) fusion!

FIGURE 7.2.10 (a) Slow-moving nuclei do not have enough energy to fuse together. The electrostatic forces cause them to repel each other. (b) If the nuclei have sufficient kinetic energy, then they will overcome the repulsive forces and move close enough together for the strong nuclear force to come into effect. At this point, fusion will occur and energy will be released.

PHYSICSFILE

Hydrogen bomb

In 1952, a fission reaction was used to power a fusion reaction in the world's first hydrogen bomb. It had five times the destructive power of all the conventional bombs that were dropped during the whole of World War II. The high temperature achieved by a fissile fuel explosion was used to initiate the fusion reaction. In other words, an atomic bomb was used as the fuse for the hydrogen fusion bomb.

The first test of a hydrogen bomb, conducted by the United States in 1952, Marshall Islands

The graph in Figure 7.2.11 shows the effect of the electrostatic force and the strong nuclear force on the potential energy of a pair of deuterium ($^{2}_{1}H$) nuclei. At large separation distances, the electrostatic force dominates and the nuclei repel each other (shown to the right of the energy barrier in the graph). At small separation distances, the strong nuclear force dominates and the nuclei can fuse together. However, to get the nuclei to this point, they need an enormous amount of energy. Temperatures in the hundreds of millions of degrees are required. This enormous amount of energy enables the nuclei to overcome the energy barrier shown in the graph and fuse together.

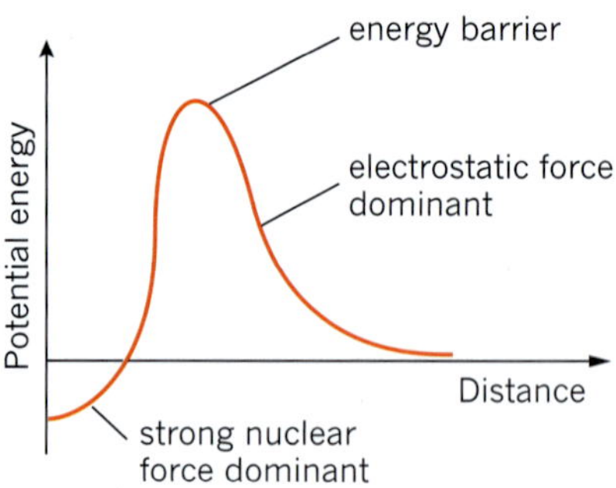

FIGURE 7.2.11 If two hydrogen-2 (deuterium) nuclei are to get close enough for the strong nuclear force to act, they must overcome the energy barrier presented by the electrostatic force.

As in fission, in any fusion reaction the atomic numbers and mass numbers on either side of the equation are conserved. The fusion of two hydrogen-2 (deuterium) nuclei is shown:

$$^{2}_{1}H + ^{2}_{1}H \rightarrow ^{3}_{1}H + ^{1}_{1}H$$

The atomic numbers add up to two and the mass numbers add up to 4 on both sides of the equation. However, the total mass of the reactants will be greater than the total mass of the products. The mass defect is converted into energy according to $\Delta E = \Delta mc^2$.

FISSION, FUSION AND BINDING ENERGY

Nuclear fission reactions produce energy when heavy nuclei split into two or more lighter nuclei. Nuclear fusion reactions produce energy when two lighter nuclei combine to produce a heavier nucleus. In both cases there is a change in mass during the reaction (the mass defect) and this mass defect is linked to energy by Einstein's equation: $\Delta E = \Delta mc^2$.

- During nuclear fusion, in which a heavier nucleus forms, mass is lost and converted into energy.
- During nuclear fission, in which an unstable nucleus splits into fragments, mass is lost and converted into energy.
- The energy released per nucleon is greater during fusion than during fission.

> The binding energy per nucleon is equal to the binding energy of a nucleus divided by the number of nucleons in the nucleus. It can be used to determine the stability of a nucleus.

Recall that the binding energy of the nucleus indicates how much energy is needed to separate the nucleus into individual protons and neutrons. As the number of nucleons in a nucleus increases, so does the mass of the nucleus and the binding energy of the nucleus. In order to compare the binding energies of nuclei, the binding energy is divided by the mass number. This is known as the **binding energy per nucleon** and it is an indication of how strongly a nucleus is held together. Nuclei with high binding energies per nucleon are held together more tightly than nuclei with lower binding energies per nucleon because it takes more energy to completely separate them. The binding energy per nucleon for fluorine-19 is:

$$\Delta E = \frac{147.8}{19} = 7.779\,\text{MeV per nucleon}$$

Most stable isotopes have a binding energy of about 8 MeV per nucleon. For example, iron is one of the most stable nuclei and has a binding energy per nucleon of 8.8 MeV.

Each isotope has its own binding energy per nucleon value. The graph of binding energy per nucleon against mass number shown in Figure 7.2.12 allows a comparison of nuclear stabilities.

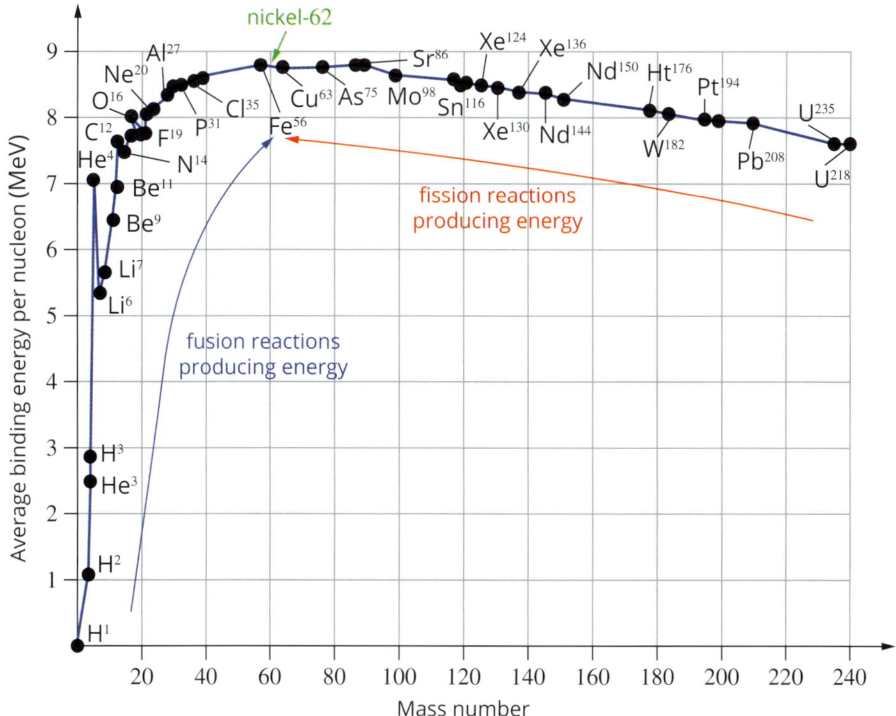

FIGURE 7.2.12 Fusion reactions produce energy when the product is lighter than nickel-62, and fission reactions produce energy when the product is heavier than nickel-62.

Figure 7.2.12 can be analysed to better understand fission and fusion.

- Small nuclei have lower binding energy per nucleon values, indicating that they are easier to break apart than the larger nuclei. Helium-4 has a relatively high value, indicating that it is stable.
- As very small nuclei fuse together, the binding energy per nucleon increases. This is the energy released during fusion.
- Elements with mass numbers between 40 and 80 have nuclei that are tightly bound. It takes more energy to break these nuclei apart than it does for the nuclei of elements with mass numbers outside this range. These are the most stable nuclei. These elements have the highest binding energy per nucleon values on the graph.
- Larger nuclei have lower binding energy per nucleon values, indicating relatively lower stabilities.
- If a large nucleus such as uranium splits into two fragments, the binding energy per nucleon of the fragments again increases. This is the energy released during fission.
- Iron (Fe), with a mass number of 56, has the most stable nucleus.
- Nuclei smaller than iron undergo fusion and release energy. Nuclei larger than iron undergo fission and release energy.

NUCLEAR ENERGY IN AUSTRALIA

The use of nuclear power and the replacement of fossil fuels have long been debated all over the world. Currently, more than 50 countries use some form of nuclear fission reaction to produce electricity or heating and, worldwide, about 10% of electricity is generated by nuclear fission. France has the world's largest proportion of power generated by nuclear power, with almost three-quarters of its electricity being produced by nuclear fission reactors.

Australia has approximately one-third of the world's uranium deposits and is one of the world's largest producers of uranium. However, despite the abundance of available uranium, Australia has never had a nuclear power station, instead relying on low-cost coal and natural gas reserves (fossil fuels). As a result, Australia's production of electricity is this country's highest contributor to enhanced global warming from the production of carbon dioxide.

Nuclear fission, however, produces almost no carbon dioxide so, in that aspect, nuclear fission will not harm the environment. Fission power is also much more efficient than the burning of fossil fuels: 1 kg of uranium-235 fuel in a nuclear reactor produces millions of times more energy than the same mass of coal, oil or natural gas.

The risks involved in harnessing nuclear power are the major reasons why many countries have not adapted nuclear fission as a source of energy. Some of these factors are discussed.

Waste products

Fission reactions produce many isotopes and some of these are very radioactive and have long half-lives. A major issue with these isotopes is the environmental impact and concern, as the materials can remain radioactive and dangerous to humans and the environment for thousands of years. See Chapter 6 for the impacts of radioactive sources to the human body. Some waste products from a fission reactor may include uranium-234, plutonium-238 and americium-241. These products have half-lives ranging from a few hundred years to over a hundred thousand years.

Australia does not currently have a nuclear reactor that is used for energy purposes; the only reactor is the Open Pool Australian Lightwater (OPAL) reactor, which is a research facility. The radioactive waste that is produced in Australia in large part comes from medicine, scientific research, industry and agriculture. Currently, radioactive waste in Australia is managed by the Australian Radioactive Waste Agency, which coordinates the storage of radioactive waste products.

Security

Countries that use fission require uranium as a fuel, which they may have to import from another country. For some countries, there is a fear that uranium sold for use in a reactor could be enriched to become weapons-grade. Importation also creates perceived security issues with the fear that a third party might intercept shipments during transportation. Although the fuel used in nuclear reactors cannot easily be modified to make nuclear weapons without specialised knowledge and equipment, it could cause harm if it were combined with conventional explosives to make a 'dirty bomb', which could spread radioactive material over a large area.

Risk of accidents

Nuclear fission is generally safe; but when accidents occur, the consequences are disastrous and can last for centuries. There have been many nuclear reactor incidents (e.g. Chernobyl in the Ukraine in 1986, Fukushima in Japan in 2011) in which the cooling systems of the reactor failed, causing the nuclear fuel to burn through the floor. At the Chernobyl incident this produced what is known as the Elephant's foot (Figure 7.2.13), which burned through reinforced concrete and will remain highly radioactive for decades or centuries.

When the cooling systems of the reactor failed, the fission reactions became uncontrolled to a point where explosions released a large amount of radioactive material into the environment.

The containment and clean-up of waste from these accidents is an extremely large project taking many years and at great financial cost.

FIGURE 7.2.13 The structure known as the 'Elephant's Foot' at Chernobyl is a solidified radioactive 'lava' of nuclear fuel, melted concrete and other materials.

Fusion as a power source

The fusion processes created so far have only lasted for a few seconds, and it is not expected that a fusion reactor will successfully operate for a few more decades. Scientists are working on experimental fusion reactors such as the ITER (International Thermonuclear Experimental Reactor) in France, and the National Ignition Facility in the USA, which is shown in Figure 7.2.14. In late 2021, a machine called JET (the Joint European Torus) in the UK set the current world record for energy released in a fusion reaction, producing 59 MJ of energy over 5 seconds, which is more than double the previous record. The challenge is to produce more energy than is needed for the fusion reaction. As described above, a large amount of energy needs to be provided to the deuterium and tritium nuclei to overcome the electrostatic repulsion of the protons. However, the results at JET suggest that the much-larger ITER will be able to produce more energy than is provided.

Research is continuing in this area as a functional nuclear fusion reactor would give Australia and other nations a cleaner way of producing energy than by the use of fossil fuels. The Australian Nuclear Science and Technology Organisation (ANSTO) is a contributor to the ITER research.

FIGURE 7.2.14 The experimental fusion reactor at the National Ignition Facility in the USA

PHYSICSFILE

Nuclear submarines

Nuclear technology is used to increase the operational period for a submarine to remain underwater. These submarines are known as nuclear submarines. They are not necessarily carrying nuclear weapons, rather they use a nuclear reactor to power the submarine (as opposed to other fuel sources, such as diesel-electric). A key advantage of nuclear submarines is that they do not need to surface frequently and can operate at high speeds for long time periods, the limit on their voyage times underwater often dictated by the need to restock food or other essential items, or to change crew.

The USS Nautilus, an American submarine, was the world's first operational nuclear-powered submarine.

7.2 Review

OA ✓✓

SUMMARY

- Nuclear fission occurs when a nucleus is made to split and release a number of neutrons. This can be induced by striking a fissile nucleus with a neutron. A relatively large amount on energy is released during this process.
- When fission occurs, the mass of the fission fragments is always less than the mass of the original particles. This decrease in mass is proportional to the energy emitted, as given by $\Delta E = \Delta mc^2$.
- In a fission reaction, usually two or three neutrons per fission are produced. A chain reaction occurs when the free neutrons collide with further fissile nuclei, resulting in an exponential increase in the number of fission reactions and energy released.
- Nuclear fusion is the combining of light nuclei to form heavier nuclei. Extremely high temperatures are required for fusion to occur.
- When fusion occurs, the mass of the combined nucleus is less than the original, separate nuclei. This decrease in mass is proportional to the energy emitted, as given by $\Delta E = \Delta mc^2$.
- The amount of energy released per nucleon is greater for fusion than for fission.
- The binding energy per nucleon is equal to the binding energy of a nucleus divided by the number of nucleons in that nucleus. It is a way of describing the stability of a nucleus.
- If a large nucleus such as uranium splits into two fragments, the binding energy per nucleon increases and so the daughter nuclei become more stable.
- The binding energy per nucleon increases dramatically when very small nuclei fuse together and so the product becomes more stable.

KEY QUESTIONS

Knowledge and understanding

1 Define the following terms.
 a fissile
 b chain reaction

2 How does fusion differ from fission?

3 Fission and fusion reactions are different processes, but energy is released in both. Explain why.

4 Give an example of a nuclide that will undergo nuclear fission.

5 Give an example of a nuclide that will undergo fusion.

6 Why is the waste produced by nuclear reactors dangerous?

Analysis

7 Create a short summary of the advantages and disadvantages of utilising nuclear energy in Australia. Would you recommend it as an energy source for Australia?

8 Create a news article updating the public on the current, and possible future, use of nuclear energy in Australia.

Chapter review

KEY TERMS

binding energy
binding energy per nucleon
chain reaction
critical mass
electron volt (eV)
fissile
fission
fission fragments
fusion
mass defect

REVIEW QUESTIONS

Knowledge and understanding

1 Which of these nuclides below are fissile and which are non-fissile?
cobalt-60, uranium-235, uranium-238, plutonium-239

2 Compare the waste and the energy per nucleon produced by fusion reactors with those of fission reactors.

3 What happens to the binding energy per nucleon and the stability of the nucleus when a uranium-238 nucleus splits apart to form two smaller nuclei?

4 The binding energy per nucleon for iron (mass number 56) is higher than for other elements. What does this mean for the stability of iron nuclei?

5 The energy released per nucleon is much higher for a fusion reaction than for a fission reaction. However, a single fission reaction releases more energy than a single fusion reaction. Explain why this is the case.

6 Are all nuclides fissile? Give examples to support your answer.

7 Two slow-moving protons are travelling directly towards each other. Explain whether the protons will collide and fuse together.

8 Two fast-moving protons are travelling directly towards each other. The protons collide and fuse together. Explain why this happens.

Application and analysis

9 Molybdenum-99 is a beta emitter with a half-life of 67 hours. It decays to form metastable technetium-99m, which has a half-life of 6 hours. This, in turn, decays to form technetium-99, which has a half-life of about 250 000 years.

a Identify the most stable of the three isotopes. Explain your choice.

b Identify the least stable of the three isotopes. Explain your choice.

10 Determine the number of neutrons (x) released in this fission reaction:

$$^{1}_{0}n + ^{235}_{92}U \rightarrow ^{127}_{50}Sn + ^{104}_{42}Mo + x^{1}_{0}n$$

11 A typical fusion reaction is $^{2}_{1}H + ^{2}_{1}H \rightarrow ^{3}_{1}H + ^{1}_{1}H$.
Why are high temperatures such as 100 million degrees needed for this reaction to occur?

12 Consider this fusion reaction:

$$^{1}_{1}H + ^{3}_{2}He \rightarrow ^{4}_{2}He + ^{0}_{+1}e + \nu$$

Hydrogen and helium-3 are being fused together and a helium-4 nucleus is being created along with a positron and a neutrino (ν) and 21 MeV of energy is released.

a How does the combined mass of the hydrogen and helium-3 nuclei compare with the combined mass of the helium-4 nucleus, positron and neutrino?

b Where has the energy that was released come from?

c Convert the energy into joules.

13 What happens to the binding energy per nucleon and the stability of two hydrogen-2 nuclei when they are fused together to form helium-4?

14 The binding energy of the samarium-147 nuclide, $^{147}_{62}Sm$, is 1217 MeV. Determine the binding energy per nucleon of $^{147}_{62}Sm$.

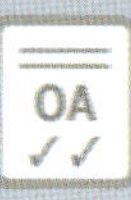

REVIEW QUESTIONS

How is energy from the nucleus utilised?

Multiple-choice questions

1 Which atomic particles are affected by the strong nuclear force?
- **A** electrons
- **B** neutrons
- **C** protons
- **D** neutrons and protons

2 What is the term for when a single nuclear reaction causes an exponential increase in the number of successive nuclear reactions?
- **A** chain reaction
- **B** nuclear fission
- **C** nuclear fusion
- **D** nuclear meltdown

3 Which of the following forms of radioactive decay has no effect on the charge of the parent particle?
- **A** α
- **B** β
- **C** γ
- **D** All forms of radioactive decay have charge.

4 If a particular atom in the sample has not decayed during the first half-life, which one of the following statements best describes what will occur during the second half-life?
- **A** It will definitely decay during the second half-life.
- **B** If it does not decay during the first half-life, it will not decay at all.
- **C** The probability that it will decay cannot be determined.
- **D** It has a 50% chance of decaying during the second half-life.

The following information relates to questions 5 and 6.

Tritium (hydrogen-3) is radioactive and its decay equation is ${}^{3}_{1}\text{H} \rightarrow {}^{0}_{-1}\text{X} + \text{Y}$.

5 Which element is the daughter nuclide Y?
- **A** hydrogen
- **B** helium
- **C** lithium
- **D** beryllium

6 Which of the following best describes the nature of X in the decay equation?
- **A** It is a neutron.
- **B** It is a proton.
- **C** It is an electron.
- **D** It is a positron.

7 Which of the following sources contribute the most background radiation to the average Australian annually?
- **A** cosmic radiation
- **B** rocks, air and water
- **C** manufactured radiation
- **D** radioactive food and drink

8 Which of the following factors is **not** a feature of radioactive tracers?
- **A** They have a long half-life to allow for ease of transport to medical facilities.
- **B** They emit only γ radiation that is suitable for detection by a γ camera.
- **C** They do not emit any α or β radiation as this would become trapped in the patient's tissue.
- **D** They have an activity level that is high enough to be detected.

The following information relates to questions 9 and 10.

A nuclear scientist has prepared equal quantities of two radioisotopes of bismuth, ${}^{211}\text{Bi}$ and ${}^{215}\text{Bi}$. These isotopes have half-lives of 2 minutes and 8 minutes respectively. Assume when answering these questions that each sample has the same number of atoms.

9 Which one of the following statements best describes the initial activities of these samples?
- **A** Bismuth-211 initially has twice the activity of bismuth-215.
- **B** Bismuth-215 initially has twice the activity of bismuth-211.
- **C** Bismuth-211 initially has four times the activity of bismuth-215.
- **D** Bismuth-215 initially has four times the activity of bismuth-211.

10 How will the activity of these samples compare after 8 minutes?
- **A** The samples will have the same activity.
- **B** Bismuth-215 will be the only sample with activity remaining.
- **C** Bismuth-211 will have twice the activity of bismuth-215.
- **D** Bismuth-215 will have twice the activity of bismuth-211.

11 The binding energy for uranium-235 is approximately 1783 MeV. What is the approximate binding energy per nucleon?
- **A** 5.5 MeV
- **B** 7.6 MeV
- **C** 12.5 MeV
- **D** 19.4 MeV

12 Calculate the energy in MeV of an alpha particle with energy 1.4×10^{-12} J.

A 1.1 MeV

B 2.2 MeV

C 4.4 MeV

D 8.8 MeV

Short-answer questions

13 As a result of the disaster at the Fukushima nuclear power plant in 2011, the radioactive isotopes caesium-137 and iodine-131 were released into the atmosphere. Use a periodic table to determine the number of protons, neutrons and nucleons contained in a nuclide of each radioisotope.

14 A radioisotope of potassium, ^{40}K, has a half-life of 1.3×10^9 years and decays to a stable isotope of argon, ^{40}Ar, which is used for dating rocks. Copy and complete the following table.

Time ($\times 10^9$ years)	Number of K nuclei	Number of Ar nuclei	Ratio K:Ar
0	1000	0	–
1.3			
2.6			
3.9			

15 One of the oldest samples of Aboriginal rock art, a painting of a kangaroo from the Kimberley region, was aged by using radiocarbon dating techniques on fossilised wasp nests that were found under and over the painting. The age of wasp nests under the painting was compared to the age of wasp nests over the painting to find a reliable date range of the painting. When the wasp nests were first built, the organic material they contained would have had the same number of carbon-14 and carbon-12 nuclei. Over time, the number of carbon-14 nuclei will have decayed but the number of carbon-12 nuclei, which are stable isotopes, will have remained constant.

a A sample of the wasp nests would have initially contained 800 carbon-14 and 800 carbon-12 nuclei. The wasp nests were found to have an approximate carbon-14 to carbon-12 ratio of 1:8. Create a similar table to the one in question 14 and use it to estimate how many half-lives have occurred.

b Carbon-14 has a half-life of 5730 years. Estimate how old the wasp nests, and by extension, the painting, are.

c The kangaroo painting was created using red ochre, which is created by grinding up rock minerals. Using your knowledge of radiocarbon dating, explain why wasp nests needed to be used to date the painting rather than the red ochre paint.

16 Given that gold is an alpha emitter, use a periodic table to help you write a decay equation for this isotope. Use the isotope $^{185}_{79}$Au.

17 Sodium-26 is a beta minus (β^-) emitter. Write the nuclear equation for its decay.

The following information relates to questions 18 and 19.

The graph below shows the data obtained in an experiment to determine the half-life of sodium-26.

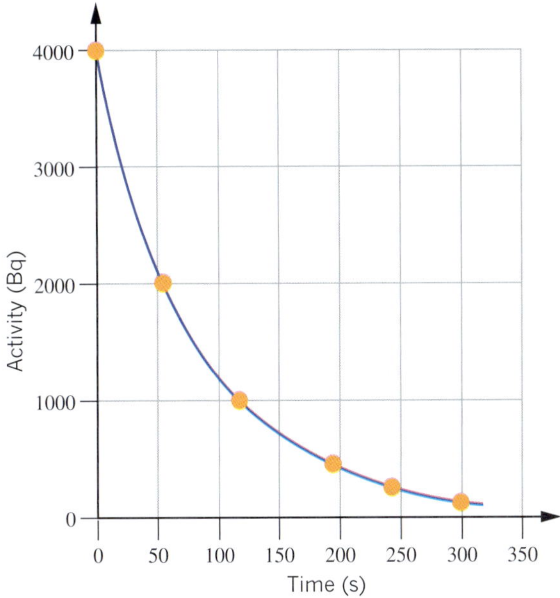

18 Use the graph to work out the approximate half-life of sodium-26.

19 If the initial sample contained 150 g of sodium-26, approximately how much of this radioisotope will remain after 5 minutes?

20 Linear accelerators (linacs) are used in Australia to treat cancer by targeting tumour sites with high-energy X-ray beams. Why might high-energy X-rays be used by these devices?

21 A 20 g tumour absorbs 4.0×10^{-3} J of energy from the X-rays (QF = 1) produced by the linac. Calculate the equivalent dose the patient receives from this treatment.

22 Calculate the annual effective dose that the oesophagus (W = 0.05) would receive from an equivalent dose of 80 mSv per year.

23 Compare the effective dose from question 22 with a similar equivalent dose in the lungs (W = 0.12). Comment on the probability of cancer developing in these organs from this exposure.

The following information relates to questions 24 and 25.

The following diagram indicates the decay series for ^{238}U through the various decay products until the stable isotope ^{206}Pb is reached.

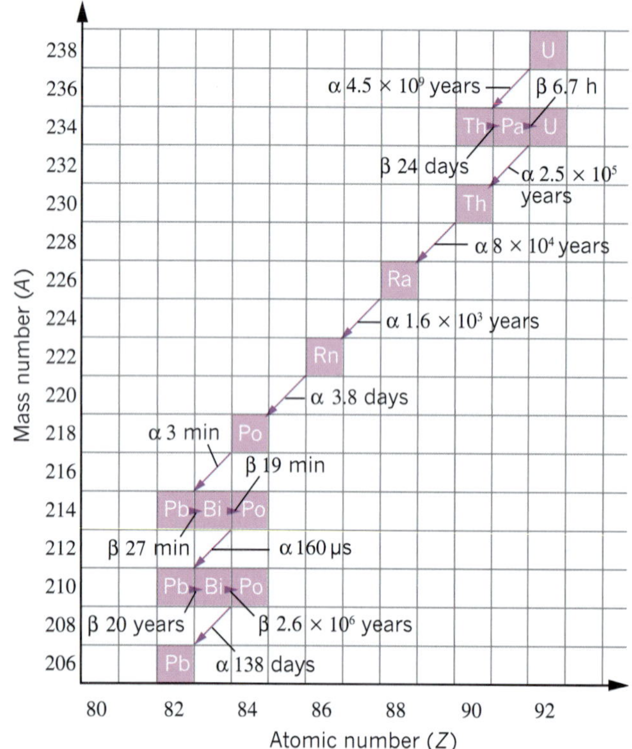

24 Explain why Th, Pa and U can have the same mass number, but still be distinct elements.

25 As can be seen in the decay series for ^{238}U, one of the natural radioisotopes produced is ^{222}Rn. Radon is a radioactive gas that has no smell, colour or taste, and decays through alpha emission.

a Explain what health problems this might pose for workers in a uranium mine.

b Propose two safety regulations that might help minimise the danger of ^{222}Rn exposure.

The following curve should be used to answer questions 26–28.

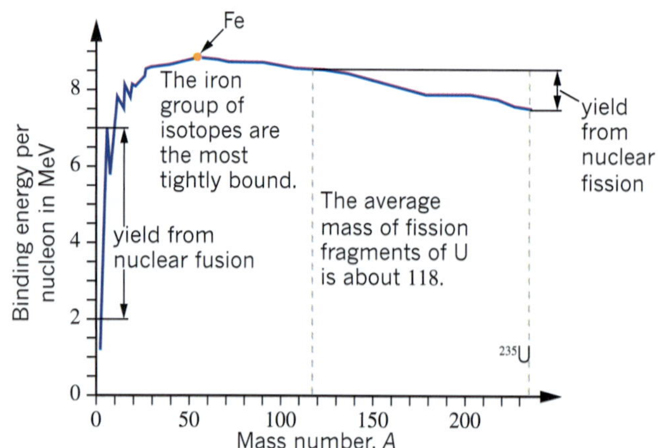

26 The element Fe is considered to be the most stable nucleus. Explain why this is the case with reference to the above binding energy curve.

27 Fission of $^{235}_{92}U$ results in fission fragments of average mass number around 118. By referring to the binding energy per nucleon for the fuel and the fragments, explain why there is a net energy release in a fission reaction.

28 Estimate the energy yield of the fission of a ^{235}U nucleus to form two fission fragments of mass approximately 118.

29 Fusion and fission are both processes by which energy is released from the nucleus of an atom. Identify some of the similarities and differences between these two nuclear processes.

30 Nuclear fission reactors use control rods which can be inserted and withdrawn from the reactor to absorb neutrons. Explain how the absorption of neutrons controls the rate of nuclear fission in the reactor.

31 Discuss one advantage of the use of nuclear power in Australia.

32 Discuss one safety concern of using nuclear power and propose a solution that would help minimise this risk.

CHAPTER

08 Electrical physics

Every object around you is made up of charged particles. When these particles move relative to one another, we experience a phenomenon known as 'electricity'. This chapter looks at the fundamental concepts, such as current and voltage, that scientists have developed to explain electrical phenomena. This will provide the foundation for studying practical electric circuits in the following chapter.

Key knowledge

- apply concepts of charge (Q), electric current (I), potential difference (V), energy (E) and power (P), in electric circuits **8.1, 8.2, 8.3**
- analyse and evaluate different analogies used to describe electric current and potential difference **8.2, 8.3**
- investigate and analyse theoretically and practically electric circuits using the relationships: $I = \frac{Q}{t}$, $V = \frac{E}{Q}$, $P = \frac{E}{t} = VI$ **8.2, 8.3**
- justify the use of selected meters (ammeter, voltmeter, multimeter) in circuits **8.3**
- model resistance in series and parallel circuits using:
  - current versus potential difference (I–V) graphs **8.4**
  - resistance as the potential difference to current ratio, including R = constant for ohmic devices. **8.4**

WS 17

OA ✓✓

8.1 Behaviour of charged particles

All matter in the universe is made of tiny particles. These particles have a property called **charge** that can be positive, negative or neutral. Usually, the numbers of positive and negative charges balance out so perfectly that we are completely unaware of them. However, when significant numbers of these charged particles become separated or move relative to each other, it results in **electricity**.

In order to understand electricity, it is important to first understand the way charged particles interact with each other.

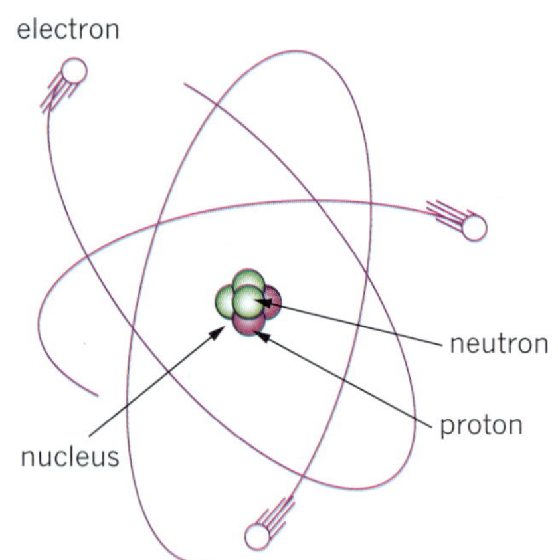

FIGURE 8.1.1 A simple model of an atom

EXISTENCE OF CHARGE CARRIERS

Recall from Section 6.1 on page 152 that the tiny particles that make up matter are called atoms. Every atom contains a nucleus at its centre. A nucleus is made up of positively charged particles called protons and neutral particles called neutrons. The nucleus, which is positively charged due to the protons, is surrounded by negatively charged electrons. A simplified model of an atom is shown in Figure 8.1.1. This is referred to as a planetary model, as it shows electrons orbiting the nucleus much like the planets orbit the Sun.

Particles with like charges repel each other, but particles with opposite charges attract each other. In an atom, the negatively charged electrons are attracted to the positively charged nucleus but stay in orbits due to their velocity.

This is an important rule to remember when thinking about the interaction of charged particles. Table 8.1.1 summarises this.

TABLE 8.1.1 Like charges repel each other, but particles with opposite charges will attract.

Charge	Positive	Negative
positive	repel	attract
negative	attract	repel

In neutral atoms, the number of electrons is the same as the number of protons. The magnitude of charge on a proton is equal to that on an electron and so their charges balance each other out, leaving the atom electrically neutral.

It is difficult to remove a proton from the nucleus of an atom. In comparison, electrons are loosely held to their respective atoms, and it is relatively easy for them to be removed.

When electrons move from one object to another, each object is said to have gained a **net charge** (a positive or negative sum of charges). The object that loses the electrons will gain a net positive charge, since it will now have more positive protons than negative electrons. The object that gains electrons will gain a net negative charge. When an atom has gained or lost electrons, we say it has been **ionised** or has become an **ion**.

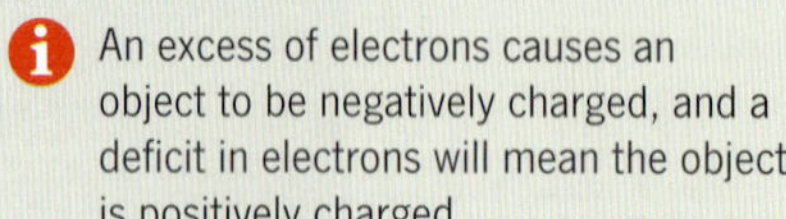

An excess of electrons causes an object to be negatively charged, and a deficit in electrons will mean the object is positively charged.

MEASURING CHARGE

In order to measure the actual amount of charge on a charged object, a 'natural' unit would be the magnitude (size) of the charge on one electron or proton. This fundamental charge is often referred to as the **elementary charge** and is given the symbol e. A proton therefore has a charge of $+e$ and an electron has a charge of $-e$.

The size of the elementary charge is very small. For most practical situations, it is more convenient to use a larger unit to measure charge. The SI (standard) unit of charge is known as the **coulomb** (symbol C). It is named after Charles-Augustin de Coulomb, who was the first scientist to measure the forces of attraction and repulsion between charges.

A coulomb is quite a large unit of charge: +1 coulomb (1 C) is equivalent to the combined charge of 6.25×10^{18} protons. Therefore, the charge on a single proton is $+1.6 \times 10^{-19}$ C. Similarly, −1 C is equivalent to the combined charge of 6.25×10^{18} electrons and the charge on a single electron is -1.6×10^{-19} C. The amount of charge, Q, is equal to the number of electrons or protons multiplied by the elementary charge for the respective particle.

The elementary charge, e, of a proton is equal to 1.6×10^{-19} C.
The elementary charge, $-e$, of an electron is equal to -1.6×10^{-19} C.

PHYSICSFILE

Electron models

The way an electron moves around the nucleus of an atom is more complex than the simple planetary model would suggest. An individual electron is so small that its exact position at any point in time is impossible to measure. Recent models of the structure of the atom describe an electron in terms of the probability of finding it in a certain location. In diagrams of atoms, this is often represented by drawing the electrons around the nucleus as a fuzzy cloud, rather than as points or solid spheres. The nucleus of an atom occupies only about 10^{-12} of the volume of the atom, yet it contains more than 99% of its mass. Atoms are mostly empty space.

Worked example 8.1.1

THE AMOUNT OF CHARGE ON A GROUP OF ELECTRONS

Calculate the charge, in coulombs, carried by 6 billion electrons.

Thinking	Working
Express 6 billion in scientific notation.	1 billion = 10^9 6 billion = 6×10^9
Calculate the charge, Q, in coulombs by multiplying the number of electrons by the charge on an electron (-1.6×10^{-19} C).	$Q = (6 \times 10^9) \times (-e)$ $= (6 \times 10^9) \times (-1.6 \times 10^{-19}\,\text{C})$ $= -9.6 \times 10^{-10}\,\text{C}$

Worked example: Try yourself 8.1.1

THE AMOUNT OF CHARGE ON A GROUP OF ELECTRONS

Calculate the charge, in coulombs, carried by 4.0 million electrons.

Worked example 8.1.2

THE NUMBER OF ELECTRONS IN A GIVEN AMOUNT OF CHARGE

The net charge on an object is $-3.0\,\mu\text{C}$ ($1\,\mu\text{C}$ = 1 microcoulomb = 10^{-6} C). Calculate the number of extra electrons on the object.

Thinking	Working
Express $-3.0\,\mu\text{C}$ in scientific notation.	$Q = -3.0\,\mu\text{C}$ $= -3.0 \times 10^{-6}\,\text{C}$
Find the number of electrons by dividing the charge on the object by the charge on an electron (-1.6×10^{-19} C).	$n_e = \frac{Q}{-e}$ $= \frac{3.0 \times 10^{-6}\,\text{C}}{-1.6 \times 10^{-19}\,\text{C}}$ $= 1.9 \times 10^{13}$ electrons

Worked example: Try yourself 8.1.2

THE NUMBER OF ELECTRONS IN A GIVEN AMOUNT OF CHARGE

The net charge on an object is $-4.8\,\mu\text{C}$ ($1\,\mu\text{C}$ = 1 microcoulomb = 10^{-6} C). Calculate the number of extra electrons on the object.

ELECTRICAL CONDUCTORS AND INSULATORS

Electrons are much easier to move than protons and this movement occurs more freely in some materials than in others. The materials in which electrons have freedom of movement are known as **metals** and they are classified as **conductors**. Conductors readily allow electric charge to flow through them.

The atoms in a metal are tightly packed in a lattice structure. Some electrons, the valence shell (outer) electrons, are far enough away from the nucleus of the atom that they experience a much smaller force of attraction than the inner-shell electrons and have varying degrees of mobility between the metallic nuclei. These are referred to as 'free electrons' because they are not bound to a single nucleus but can freely move among the metal nuclei. When a force is applied to the electrons, such as by a battery connected to each end of a conductor, these free electrons accelerate (Newton's second law) and gain kinetic energy, colliding with the nuclei and other electrons in the metal lattice and producing what we call an electric current.

PHYSICSFILE

Separating positive and negative charges

Electrons can be transferred (moved) from one object to another by simply rubbing two objects together. The objects need to be made of different materials. If you rub a balloon against your hair and then slowly move the balloon away, you will notice that your hair seems to stick to the balloon. This is because electrons have rubbed off your hair and transferred onto the balloon. This causes the balloon to gain a net negative charge and your hair to gain a net positive charge—the balloon and your hair are attracted to each other.

An **insulator** is a material in which the chemical bonding and atomic packing results in no such free electrons, hence these materials have no ability to conduct electricity.

Figure 8.1.2 shows diagrammatically how loosely held electrons in a metal can jump from one atom to another and move freely throughout the metal.

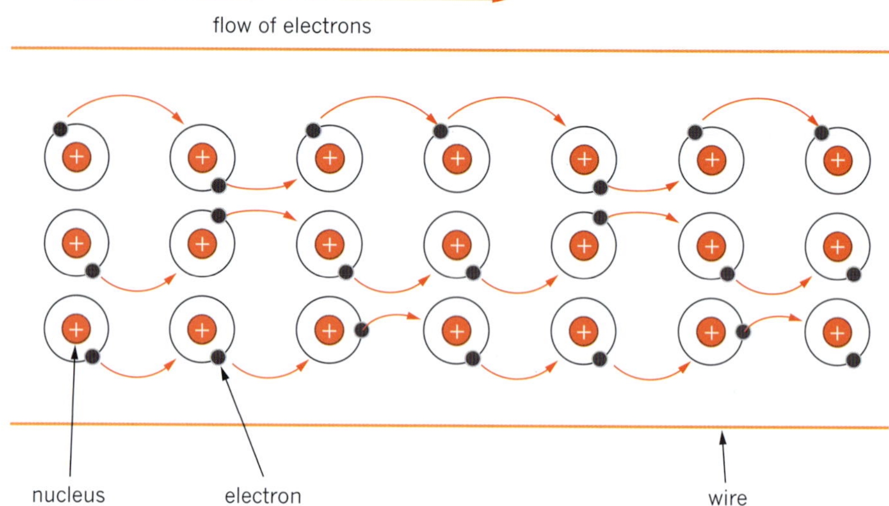

FIGURE 8.1.2 Electrons moving through a conductor. The electrons are free to move throughout the lattice of positive ions.

FIGURE 8.1.3 These copper wires conduct electricity by allowing the movement of charged particles.

Copper is an example of a very good conductor. For this reason, it is used in telecommunications and electrical and electronic products (Figure 8.1.3).

In comparison, the electrons in **non-metals** are very tightly bound to their respective nuclei and cannot readily move from one atom to another. Non-metals do not conduct electricity very well and are known as insulators. A list of common conductors and insulators is given in Table 8.1.2.

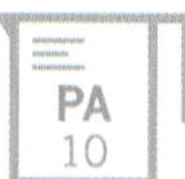

TABLE 8.1.2 Some common conductors and insulators

Conductors	Insulators
Good	
all metals, especially silver, gold, copper and aluminium any ionic solution	plastics polystyrene dry air glass porcelain cloth (dry)
Moderate	
water earth semiconductors, e.g. silicon, germanium skin	wood paper damp air ice, snow

8.1 Review

OA ✓✓

SUMMARY

- Like charges repel; unlike charges attract.
- When an object loses electrons, it develops a positive net charge; when it gains electrons, it develops a negative net charge.
- The letter Q is used to represent the amount of charge. The SI unit of charge is the coulomb (C).
- The elementary charge (e), the charge on a proton, is equal to 1.6×10^{-19} C. The elementary charge, $-e$, of an electron is -1.6×10^{-19} C.
- Electrons move easily through conductors, but not through insulators. This is because the electrons in materials that are good conductors are weakly attracted to the nucleus, and electrons in insulators are more strongly attracted to the nucleus.

KEY QUESTIONS

Knowledge and understanding

1. Plastic strip A, when rubbed, is found to attract plastic strip B. Strip C is found to repel strip B. Explain what will happen when strip A and strip C are brought close together.
2. Explain why electric circuits often consist of wires that are made from copper and coated in protective plastic.
3. Explain why metals are good conductors of electricity and plastics are poor conductors.

Analysis

4. Calculate how many electrons make up a charge of -5.0 C.
5. Calculate the charge, in coulombs, of 4.2×10^{19} protons.

8.2 Electric current and circuits

A flow of electric charge is called electric **current**. Current can be carried by moving electrons in a conductor when under the influence of a force field, or by ions in solution. This section explores current in wires in electric circuits.

Electric circuits are involved in much of the technology used every day and are responsible for many familiar sights (Figure 8.2.1). To construct electric circuits, you must know about the components of a circuit and be able to read circuit diagrams.

FIGURE 8.2.1 Electric circuits are responsible for lighting up whole cities.

ELECTRIC CIRCUITS

An **electric circuit** is a path made of conductive material through which charges can flow in a closed loop. This flow of charges is called an electric current. The most common conductors used in circuits are metals, such as copper wire. The charges that flow around the circuit within the wire are negatively charged electrons. The movement of electrons in the wire is called **electron flow**.

A simple example of an electric circuit is shown in Figure 8.2.2. The light bulb is in contact with the positive terminal (end) of the battery; a copper wire joins the negative terminal of the battery to one end of the filament in the light bulb. This arrangement forms a closed loop that allows electrons within the circuit to flow from the negative terminal towards the positive terminal of the battery. The battery is a source of energy. The light bulb converts (changes) this energy into other forms of energy, such as heat and light, when the circuit is connected.

If a switch is added to the circuit in Figure 8.2.2, the light bulb can be turned off and on. When the switch is closed, the circuit is complete. The current follows a loop along a path made by the conductors and then returns to the battery.

When the switch is open, there is a break in the circuit and the current ceases. This is what happens when you turn off the switch for a lamp or TV. A circuit in which the conducting path is broken is often called an open circuit.

A switch on a power point or an appliance allows you to break the circuit. A break in the circuit occurs when two conductors in the switch are no longer in contact. This stops the current and the appliance will not work.

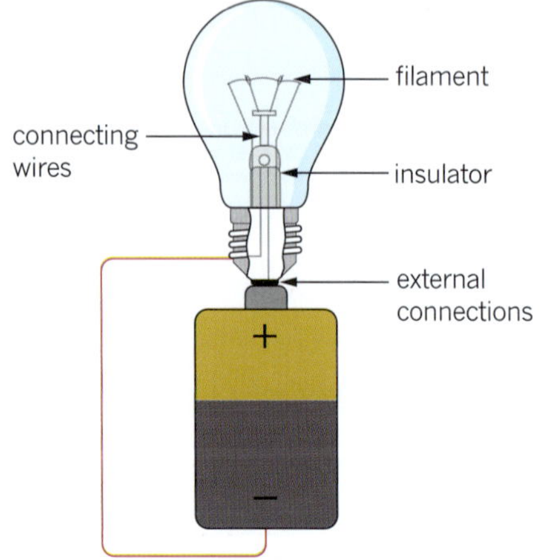

FIGURE 8.2.2 When there is a complete conduction path from the positive terminal of a battery to the negative terminal, there is a current in the circuit.

There is current in a circuit only when the circuit forms a continuous (closed) loop from one terminal of a power supply to the other terminal.

REPRESENTING ELECTRIC CIRCUITS

A number of different components can be added to a circuit. It is not necessary to be able to draw detailed pictures of these components; simple symbols are much clearer. The common symbols used to represent the electrical components in electric circuits are shown in Figure 8.2.3.

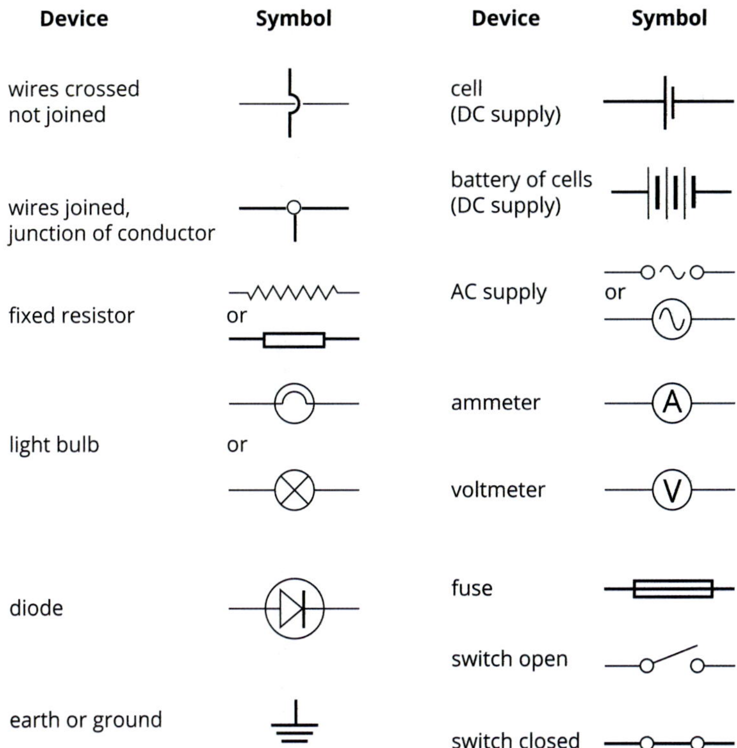

FIGURE 8.2.3 Some commonly used electrical devices and their symbols

Circuit diagrams

When building anything, it is important that the builder has a clear set of instructions from the designer. This is as much the case for electric circuits as it is for a tall building or a motor vehicle.

Circuit diagrams are used to clearly show how the components of an electric circuit are connected. They simplify the physical layout of the circuit into a diagram that is recognisable by anyone who knows how to interpret it. You can use the list of common symbols for electrical components (Figure 8.2.3) to interpret any circuit diagrams.

The circuit diagram in Figure 8.2.4(b) shows how the components of the torch shown in Figure 8.2.4(a) are connected in a circuit. The circuit can be traced by following the straight lines representing the connecting wires.

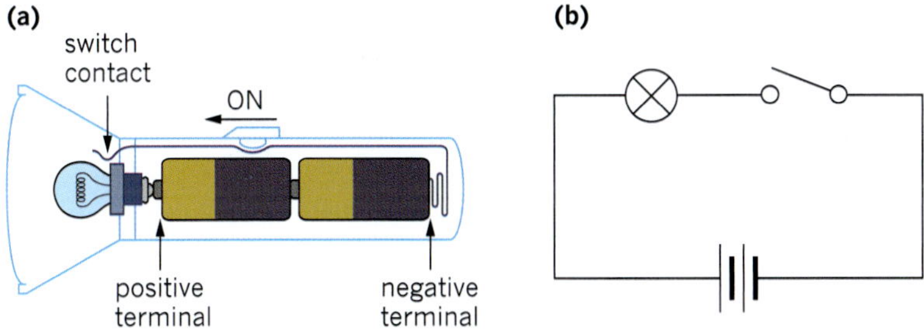

FIGURE 8.2.4 (a) A battery and light bulb connected by conductors in a torch constitute an electric circuit. (b) The torch's circuit can be represented by a simple circuit diagram.

Conventional current (or current), I, models charge as flowing from the positive terminal of a power supply to the negative terminal.
Electron flow (or electron current) refers to the flow of electrons from the negative terminal to the positive terminal of a power supply.

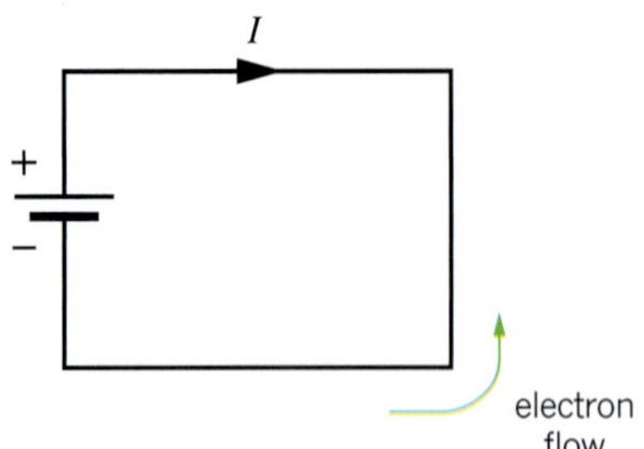

FIGURE 8.2.5 Conventional current (I) and electron flow are in opposite directions. The long terminal of the battery is positive.

PHYSICSFILE

Conventional current

Note that a flow of 1 C of positive charge to the right in one second is equivalent to a flow of 1 C of negative charge to the left in one second. That is, both situations represent a conventional current of 1 A (which is 1 C in one second) to the right.

CONVENTIONAL CURRENT VS ELECTRON FLOW

When electric currents were first studied, it was (incorrectly) thought the charges that flowed in circuits were positive. Based on this, scientists traditionally talked about electric current as if charge flowed from the positive terminal of the battery to the negative terminal. This convention is still used today, even though we know now that it is actually the negative charges (electrons) that flow around a circuit.

On a circuit diagram, current is indicated by a small arrow and the symbol I. This is called **conventional current** or just current. The direction of conventional current is opposite to the direction of electron flow (Figure 8.2.5).

QUANTIFYING CURRENT

One common misconception about current is that charges are used up or lost when current flows around a circuit. However, the charge carriers (electrons) are conserved at all points in a circuit.

In any piece of conducting material, such as copper wire, electrons are present throughout the material. If there is no current flowing, this means there is no net flow of electrons, but the electrons are still present.

When you connect a piece of conducting material to the negative terminal of a battery, the negative terminal tries to 'push' the electrons away. However, the electrons will not flow if the circuit is open. This is because the electrons at the open part of the circuit have effectively reached a dead end, like cars stopped at a road-block. This prevents all the other electrons in the material from flowing, like a long traffic jam caused by the road-block. When you close the circuit, you create a clear pathway for the electrons to flow through. This means the electron closest to the negative terminal forces the next electron to move, and so on, all the way around the circuit. Therefore, all electrons move almost simultaneously throughout the circuit so that electrical devices, such as light bulbs, seem to turn on immediately after you flick the switch.

When current flows, electrons travel into the wire at the negative terminal of the battery. As electrons flow around a circuit, they remain within the metal conductor. They flow through the circuit and return to the battery at the positive terminal; they are not lost in between.

In common electric circuits, a current consists of electrons flowing within a copper wire, as shown in Figure 8.2.6.

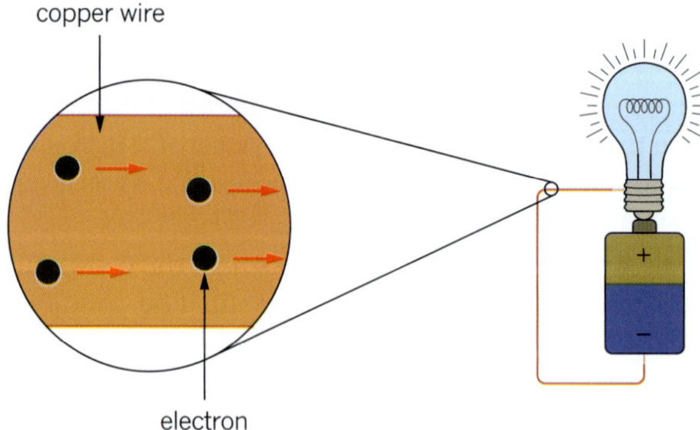

FIGURE 8.2.6 A current consists of electrons flowing within a copper wire. Because the electrons do not leave the wire, current is conserved in all parts of the circuit.

This current, I, can be defined as the amount of charge that passes through a point in the conductor per second:

$$I = \frac{Q}{t}$$

where I is the current (in A)

Q is the amount of charge (in C)

t is the number of seconds that have passed.

Current is measured in amperes, or amps (A). One ampere is equivalent to one coulomb per second ($C\,s^{-1}$).

Current is equal to the number of electrons (n_e) that flow through a particular point in the circuit multiplied by the charge on one electron ($q_e = -1.6 \times 10^{-19}$ C) divided by the time that has elapsed in seconds (t):

$$I = \frac{Q}{t} = \frac{n_e q_e}{t}$$

A typical current in a circuit powering a small DC motor would be about 50 mA. Even with this seemingly small current, approximately 3×10^{17} electrons flow past any point on the wire each second.

Worked example 8.2.1

USING $I = \frac{Q}{t}$

Calculate the number of electrons that flow past a particular point each second in a circuit that carries a current of 0.50 A.

Thinking	Working
Rearrange the equation $I = \frac{Q}{t}$ to make Q the subject.	$I = \frac{Q}{t}$ $I \times t = \left(\frac{Q}{t}\right) \times t$ $I \times t = Q$ So $Q = I \times t$.
Calculate the amount of charge that flows past the point in question by substituting the values given.	$Q = 0.50 \times 1$ $= 0.50$ C
Find the number of electrons by dividing the charge by the charge on an electron (1.6×10^{-19} C).	$n_e = \frac{Q}{q_e}$ $= \frac{0.5}{1.6 \times 10^{-19}}$ $= 3.1 \times 10^{18}$ electrons

Worked example: Try yourself 8.2.1

USING $I = \frac{Q}{t}$

Calculate the number of electrons that flow past a particular point each second in a circuit that carries a current of 0.75 A.

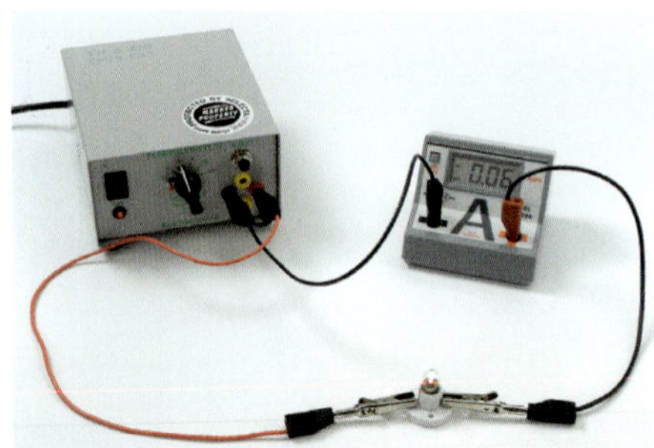

FIGURE 8.2.7 A digital ammeter (labelled with an A) measures current in a circuit.

TABLE 8.2.1 Typical values for electric current

Situation	Current
lightning	10 000 A
starter motor in car	200 A
fan heater	10 A
toaster	3 A
light bulb	400 mA
pocket calculator	5 mA
nerve fibres in body	1 μA

Measuring current: the ammeter

Current is commonly measured by a device called an **ammeter**.

Figure 8.2.7 shows the ammeter connected along the same path taken by the current flowing through the light bulb. This is referred to as connecting the ammeter 'in series'. Series circuits are covered in more detail in Chapter 9. The positive terminal of the ammeter is connected so that it is closest to the positive terminal of the power supply. The negative terminal of the ammeter is closest to the negative terminal of the power supply.

Measuring the current is possible because charge is conserved at all points in a circuit. This means that the amount of current into a light bulb is the same as the amount of current out of the light bulb. An ammeter can therefore be connected before or after the bulb in series to measure the current. Table 8.2.1 lists some typical values for electric current in common situations.

ANALOGIES FOR ELECTRIC CURRENT

Since we cannot see the movement of electrons in a wire, it is sometimes helpful to use analogies or 'models' to visualise or explain the way an electric current behaves. It is important to remember that no analogy is perfect: it is only a representation, and there will be situations in which the electric current does not act as you would expect from the analogy.

Water model

A common model is to think of electric charges as water being pumped around a pipe system, as shown in Figure 8.2.8. The battery pushes electrons through the wires just like a pump pushes water through the pipes. Since water cannot be compressed, the same amount of water flows in every part of a pipe, just as the electric current is the same in every part of a wire. Light bulbs in an electric circuit are like turbines: whereas the turbine converts the gravitational energy of the water into kinetic energy, a light bulb converts electrical energy into heat and light. The water that has flowed through the turbine flows back to the pump that provides the energy needed for it to keep flowing.

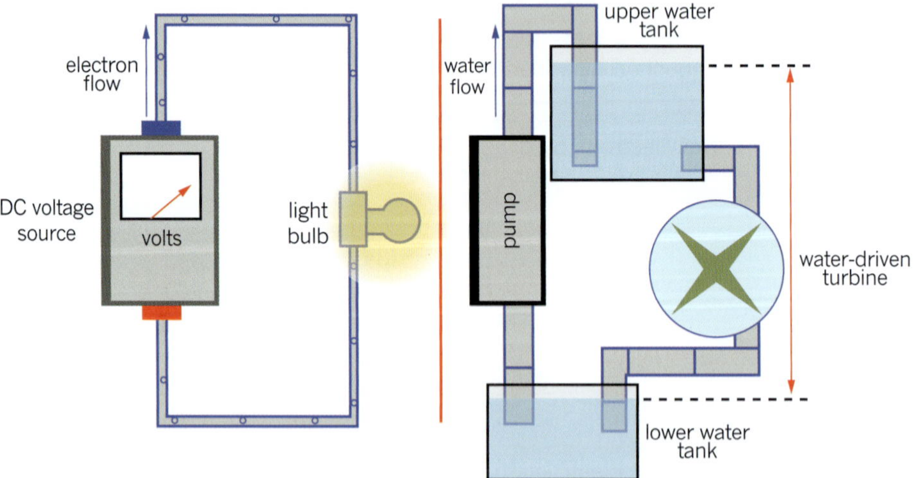

FIGURE 8.2.8 An electric current can be compared to water flowing through a pipe system.

This model explains the energy within a circuit quite well.

- The power supply transfers energy to the electrons and so the electrons gain potential energy.
- The energy of the electrons is converted into other forms when the electrons pass through the components in the circuit.

One of the limitations of the water model is that you usually cannot see water moving through a pipe and so you have to imagine what is happening in the pipe and then compare it to the motion of electrons in the wires.

Bicycle chain model

Although electrons move relatively slowly through a conductor, electric effects are almost instantaneous. For example, the delay between flicking a light switch and the light coming on is too small to be noticed. One way of understanding this is to compare an electric current to a bicycle chain, as shown in Figure 8.2.9.

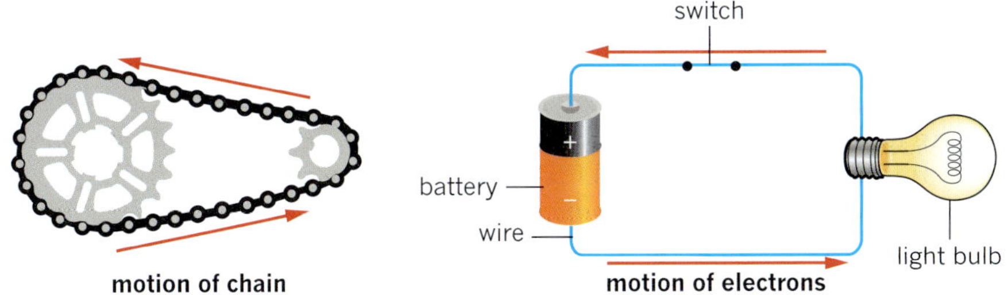

FIGURE 8.2.9 Electrons in a wire are like the links of a bicycle chain. Just like the links of a bicycle chain, electrons move together in a conductor.

A wire is full of electrons that all repel each other, so moving one electron affects all the others around it. An electric current is like a bicycle chain: even if the cyclist pedals slowly, the links in the chain mean that energy is instantly transferred from the pedals to the wheel.

In this model, the pedals of the bicycle are like the battery of the electric circuit: the pedals provide the energy that causes the chain to move.

This model reinforces a number of important characteristics of an electric current.

- Effects of an electric current are nearly instantaneous, just as there is no delay between turning the pedals and the back wheel of the bicycle turning.
- Charges in an electric current are not 'consumed' or 'used up', just as links in the chain are not used up.
- The amount of energy provided by an electric current is not entirely dependent on the current. This is like when a cyclist changes gears to give the same amount of energy to the bicycle while pedalling at different rates.

Although the bicycle chain model can be a helpful analogy, there are a number of important differences between a bicycle chain and an electric circuit.

- The number of charges flowing in an electric current is much larger than the number of links in a bicycle chain.
- Electrons in a wire do not touch one another like the links in a chain.

8.2 Review

OA ✓✓

SUMMARY

- Current will flow in a circuit only when the circuit forms a continuous (closed) loop from one terminal of a power supply to the other terminal.
- When there is an electric current, electrons all around the circuit move towards the positive terminal, at the same time. This is called electron flow.
- Conventional current in a circuit is from the positive terminal to the negative terminal.
- Current, I, is defined as the amount of charge, Q, that passes through a point in a conducting wire per second. It has the unit amperes or amps (A), which are equivalent to coulombs per second:

$$I = \frac{Q}{t} = \frac{n_e q_e}{t}$$

- Current is measured with an ammeter connected along the same path as the current (i.e. in series) within the circuit.

KEY QUESTIONS

Knowledge and understanding

1 What are the requirements for current in a circuit?

2 List the electrical components shown in the circuit diagram.

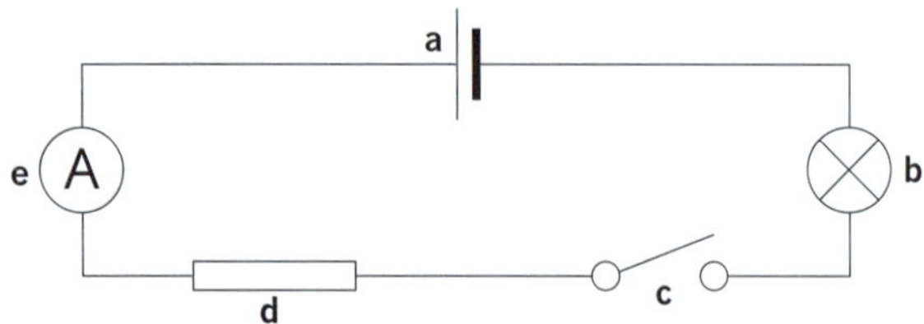

3 Why do scientists refer to conventional current as being from positive to negative?

A Protons flow from the positive terminal of a battery to the negative terminal.

B Electrons flow from the positive terminal of a battery to the negative terminal.

C Originally, scientists thought charge carriers were positive.

4 State which of the circuits shown below would enable you to measure the current passing through both light bulbs when the switch is closed. Explain why you chose this answer.

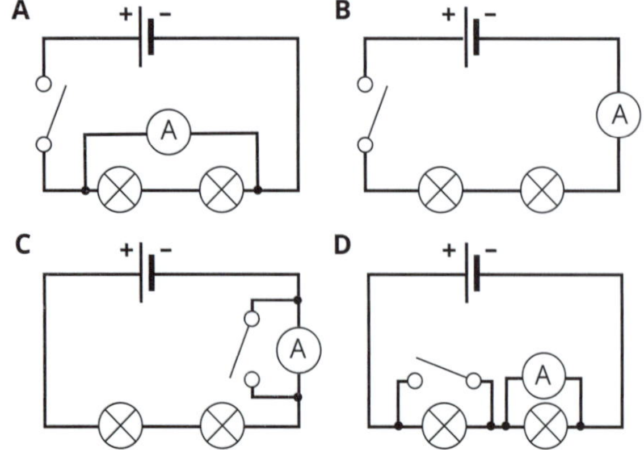

Analysis

5 Calculate the current in a light bulb through which a charge of 30 C flows in:

a 10 seconds

b 1 minute

c 1 hour.

6 A car headlight may draw a current of 5 A. Calculate how much charge will have flowed through it in:

a 1 second

b 1 minute

c 1 hour.

7 **a** In a solution of salt water, a total positive charge of +15 C moved past a point to the right in 5 s, and in the same time a total negative charge of –30 C moved to the left. What was the current through the solution during this time?

b Some time later it was found that in 5 s a total of +5 C had moved to the right while –15 C had moved to the right as well. What was the current during this time?

8 Using the values given in Table 8.2.1 on page 218, find the amount of charge that would flow through a:

a pocket calculator in 10 min

b car starter motor in 5 s

c light bulb in 1 h.

9 10^{20} electrons flow past a point in 4 seconds. Calculate:

a the amount of charge, in coulombs, that moves past a point in this time

b the current, in amps.

10 3.2 C flow past a point in 10 seconds. Calculate:

a the number of electrons that move past a point in this time

b the current, in amps.

8.3 Energy in electric circuits

Electrons won't move around a circuit unless they are given energy. Inside every battery, a chemical reaction will take place when a circuit is connected between the two ends of the battery (Figure 8.3.1). The chemical reaction provides potential energy to the electrons inside the battery. When a circuit is connected between the two ends of the battery, the potential energy of the electrons in both the battery and the circuit is converted into kinetic energy and the electrons move through the wire.

Chemical reactions in the battery release electrons towards its negative terminal. The electrons at the negative terminal repel each other. This repulsion moves them into the wire. At the positive terminal, electrons in the wire are attracted to the positive charges created by the deficiency of electrons. This attraction causes them to move into the battery. The net effect of electrons flowing into the wire at one end and out of it at the other end is that an electric current flows through the wire.

FIGURE 8.3.1 Chemical energy is stored within batteries.

ENERGY TRANSFERS AND TRANSFORMATIONS

Chemical energy stored inside a battery is **transformed** (changed) into **electrical potential energy**. This potential energy is stored as a separation of charge between the two terminals of the battery. This can be visualised as a 'concentration' of charge at either end of the battery. One terminal (the negative terminal) has a concentration of negative charges; the other terminal (the positive terminal) has a concentration of positive charges. Once the battery is connected within a device, chemical reactions will, for some time, maintain this difference in charge between the two terminals.

The difference in charge between the two terminals of a battery can be quantified (given a numerical value) as a difference in the electrical potential energy per unit charge. This is commonly called **potential difference** (V) and is measured in volts (V).

It is this potential difference at the terminals of the battery that provides the energy to a circuit. The energy is then **transferred** (passed) to different components in the circuit. At each component the energy is transformed into a different type of energy. For example, the energy could be transformed into heat and light if the component is a light bulb. If the component is a fan, the energy is transformed into motion (kinetic energy) and some heat and sound.

PHYSICSFILE

Potential difference between conductors

If we use a conductor to link two bodies between which there is a potential difference, charges will flow through the conductor until the potential difference is equal to zero. For the same reasons, when a conductor is charged, charges will move through it until the potential difference between any two points in the conductor is equal to zero.

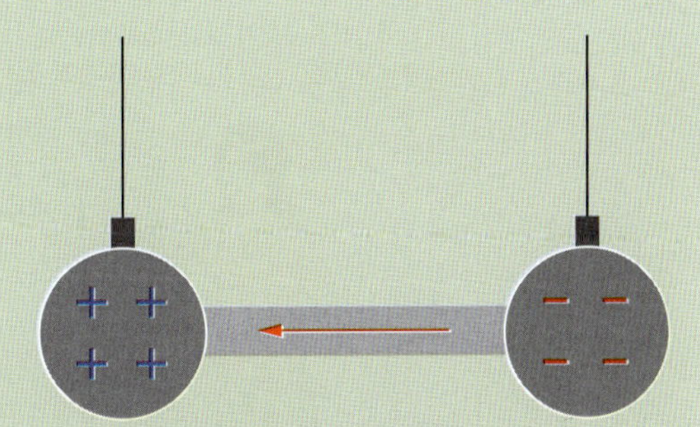

A potential difference exists between these two objects due to the difference in charge concentration. Electrons flow from the negative object to the positive object, as shown by the arrow, until the potential difference is zero.

PHYSICSFILE

Volts and voltage

Somewhat confusingly, scientists use the symbol 'V' for *both* the quantity potential difference *and* its unit of measurement, the volt. For the quantity potential difference, we use italics: *V*. For the unit volts, the symbol is not in italics: V. The context usually makes it clear which meaning is intended.

For this reason, potential difference is often referred to as 'voltage'.

Energy transfers and transformations in a torch

A torch is a simple example of how energy is transformed and transferred within a circuit. In the torch shown in Figure 8.3.2, chemical energy in the battery is transformed to electrical potential energy. There are two batteries connected, so a bigger potential difference is available. Energy can be transferred to the light bulb once the end terminals of the batteries have been connected to the torch's circuit; that is, when the torch is switched on.

The electrical potential difference between the battery's terminals causes electrons within the circuit to move. The electrons flow through the wires of the torch. These electrons collide with the atoms within the small wire (filament) in the torch's light bulb and transfer kinetic energy to them. This transfer of kinetic energy means that the particles inside the filament move faster and faster and the filament gets very hot. When it is hot, the filament emits visible light.

The energy changes can be summarised:

- chemical energy $\xrightarrow{\text{transformed}}$ electrical potential energy
- electrical potential energy $\xrightarrow{\text{transformed}}$ kinetic energy (electrons)
- kinetic energy (electrons) $\xrightarrow{\text{transferred}}$ kinetic energy (filament atoms)
- kinetic energy (filament atoms) $\xrightarrow{\text{transformed}}$ thermal energy + light

Eventually, when most of the chemicals within the battery have reacted, the battery is no longer able to provide enough potential difference to power the torch. This is because the chemical reaction has slowed and too few electrons are being produced at the negative terminal to provide sufficient repulsion to move electrons through the circuit. The torch stops working and the batteries are said to have 'gone flat'.

Similar energy transfers and transformations take place every time electrical energy is used.

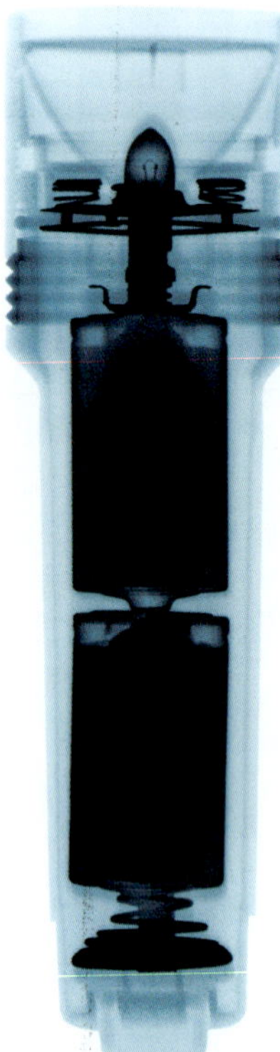

FIGURE 8.3.2 An X-ray image of the internal structure of a torch. The bulb and two batteries are clearly visible.

PHYSICSFILE

Cells and batteries

A single cell generates electricity by converting chemical energy to electrical potential energy. If a series of cells are added together, it is called a battery. Often a series of cells are packaged in a way that makes it look like a single device (see below), but inside is a battery of cells connected together. The terms 'battery' and 'cell' can be used interchangeably, as the term 'battery' is frequently used in common language to describe a cell.

A mobile-phone battery. The term 'battery' actually refers to a group of electric cells connected together.

EXPLAINING POTENTIAL DIFFERENCE

When charges are separated in a battery, each charge gains electrical potential energy. In a similar way, if a mass is lifted above the ground it gains gravitational potential energy. The change in the electrical potential energy of each charge is known as the potential difference (V).

As you can see in Figure 8.3.3, when you lift an object to some height above the ground (in this case, from point A to point B), you have done some work and have moved the object to a point in the field where it has more gravitational potential energy (E_g).

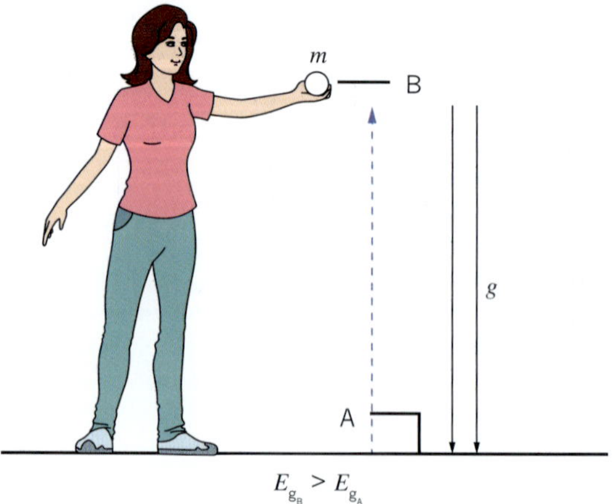

FIGURE 8.3.3 Lifting a mass above ground level increases its gravitational potential energy, just as moving an electron to the negative terminal of an electric cell gives it electrical potential energy.

The term 'potential difference' can often cause confusion. Stated simply, it means that there is a difference in electrical potential energy per charge between two points. The potential difference of a battery refers to the difference in electrical potential energy per charge at either terminal of the battery. You may see potential difference used to describe what happens on either side of a component, such as a light bulb. If you measure the potential energy of charges on one side of a bulb and compare it to the potential energy of the charges on the other side, you would find that there is a difference. The difference equates to the energy transferred from the charges to the bulb to light it up.

Quantifying potential difference: voltage

As with other forms of energy, it is useful to be able to quantify the amount of potential difference in a given situation. Potential difference is formally defined as the amount of electrical potential energy given to each coulomb of charge:

$$V = \frac{E}{Q}$$

where V is potential difference (V)

E is electrical potential energy (J)

Q is charge (C).

Since energy is measured in joules and charge in coulombs, the potential difference is measured in joules per coulomb (J C^{-1}). This quantity has been assigned a unit, the **volt** (V). So a potential difference of $1\,\text{J C}^{-1}$ is equal to 1V.

Worked example 8.3.1

DEFINITION OF POTENTIAL DIFFERENCE

Calculate the amount of electrical potential energy carried by 5.0C of charge at a potential difference of 10V.

Thinking	Working
Recall the definition of potential difference.	$V = \frac{E}{Q}$
Rearrange this to make energy the subject.	$E = VQ$
Substitute in the appropriate values and solve.	$E = 10 \times 5.0$ $= 50\,\text{J}$

Worked example: Try yourself 8.3.1

DEFINITION OF POTENTIAL DIFFERENCE

A car battery can provide 3600C charge at 12.0V. How much electrical potential energy is stored in the battery?

Measuring voltage: the voltmeter

Voltage is usually measured by a device called a **voltmeter**.

Voltmeters are wired into circuits differently from ammeters. Unlike an ammeter, which measures the current passing through a wire, a voltmeter measures the change in voltage (potential difference) as current passes through a particular component. This means that one wire of the voltmeter is connected to the circuit before the component and the other wire is connected to the circuit after the component. This is called connecting the voltmeter 'in parallel', making a **parallel circuit**.

PHYSICSFILE

Birds on a wire

Birds can sit on power lines and not get electrocuted even though the wires are not insulated.

For a current to flow through a bird on a wire, there would have to be a potential difference between its two feet. Since the bird has both feet touching the same wire, which might be at a very high potential (voltage), there is no potential *difference* between the bird's feet. If the bird could stand on the wire and touch any other object such as the ground or another wire, then it would get a big electric shock. This is because there would be a potential difference between the wire and the other object and current would flow.

There is no potential difference between the bird's feet.

PHYSICSFILE

Energy in a battery

When a battery is labelled 9V, this means that the battery provides 9J of energy to each coulomb of charge.

In Figure 8.3.4, the voltmeter is connected to the circuit on either side of the light bulb, that is, in parallel. This is so that it can measure the voltage drop (potential difference) across the light bulb. As for the ammeter, it is important to connect the voltmeter with the positive terminal closest to the positive terminal of the power supply. The voltmeter's negative terminal is connected closest to the negative terminal of the power supply.

FIGURE 8.3.4 A voltmeter measures the voltage change (in this case, 6.23 V) across a light bulb.

WORK DONE BY A CIRCUIT

In electric circuits, electrical potential energy is converted into other forms of energy. When energy is changed from one form to another, work is done. (Work is covered in more detail in Chapter 13.) The amount of energy provided by a particular circuit can be calculated using the definitions for potential difference and current:

$$V = \frac{E}{Q} \text{ and } I = \frac{Q}{t}$$

Rearranging the definition of voltage gives:

$$E = VQ$$

Using the definition of current:

$$Q = It$$

Therefore:

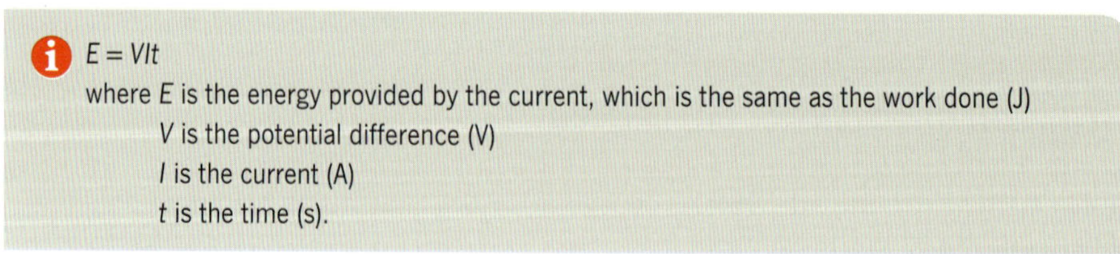

$E = VIt$

where E is the energy provided by the current, which is the same as the work done (J)

V is the potential difference (V)

I is the current (A)

t is the time (s).

This gives us a practical way to calculate the energy used in a circuit from measurements we can make.

Worked example 8.3.2

USING $E = VIt$

A potential difference of 12 V is used to generate a current of 750 mA to heat water for 5.0 minutes. Calculate the energy transferred to the water in that time.

Thinking	Working
Convert the quantities to SI units.	$\frac{750\,\text{mA}}{1000} = 0.750\,\text{A}$ $5.0\,\text{min} \times 60\,\text{s} = 300\,\text{s}$
Substitute values into the equation and calculate the amount of energy in joules.	$E = VIt$ $= 12 \times 0.750 \times 300$ $= 2.7 \times 10^3\,\text{J}$

Worked example: Try yourself 8.3.2

USING $E = VIt$

A potential difference of 12.0 V is used to generate a current of 1750 mA to heat water for 7 minutes 30 seconds. Calculate the energy transferred to the water in that time.

Rate of doing work: power

If you wanted to buy a new kettle, you might wonder how you could determine how quickly different kettles boil water.

Printed on all appliances is a rating for the **power** of that device. Power is a measure of how quickly energy is converted by the appliance. In other words, power is the rate at which energy is transformed by the components within the device. This can also be described as the rate at which work is done:

$$P = \frac{\text{energy transformed}}{\text{time}} = \frac{E}{t}$$

where P is the power in joules per second (J s^{-1}). One joule per second is 1 watt (W).

The more powerful the appliance, the faster it can do a given amount of work. In other words, an appliance that draws more power can do the same amount of work in a shorter amount of time. If you want something done quickly, then you need an appliance that has a higher power rating.

Rearranging the previous relationship:

$$E = VIt \text{ to } \frac{E}{t} = VI$$

and combining this with the power expression gives:

$$P = \frac{E}{t} = VI$$

This expression enables us to calculate the energy transformations in a circuit by measuring voltage and current across circuit components. The power dissipated by those components can be calculated in watts (W).

MULTIMETERS

The internal circuitry of a voltmeter, used to measure voltage, is significantly different from that of an ammeter, used to measure current. Electricians and scientists who work with electric circuits often find it inconvenient to keep collections of voltmeters and ammeters, each only able to measure a given range, so they use a single device known as a multimeter that has a bundle of circuitry and several ranges for both types of meter. It is also able to measure other electrical parameters such as resistance. A multimeter is shown in Figure 8.3.5.

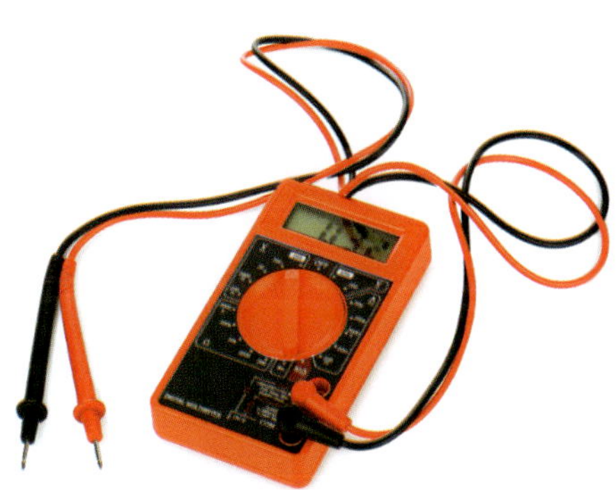

FIGURE 8.3.5 A digital multimeter can be used as either an ammeter or a voltmeter.

This is much more convenient because the multimeter can be quickly switched from being an ammeter to being a voltmeter as needed. However, when a multimeter is switched from one mode to another, it is important to make a corresponding change to the way it is connected to the circuit being measured. An ammeter is connected in series and a voltmeter in parallel. In fact, if a multimeter is working as an ammeter and it is connected in parallel like a voltmeter, it may draw so much current that its internal circuitry will be burnt out and the multimeter will be destroyed.

Worked example 8.3.3

USING $P = VI$

An appliance running on 230 V draws a current of 4.0 A. Calculate the power used by this appliance.

Thinking	Working
Identify the relationship needed to solve the problem.	$P = VI$
Identify the known values from the question, substitute and calculate.	$P = 230 \times 4.0$ $= 920\,\text{W}$

Worked example: Try yourself 8.3.3

USING $P = VI$

An appliance running on 120 V draws a current of 6.0 A. Calculate the power used by this appliance.

ANALOGIES FOR POTENTIAL DIFFERENCE

The analogies used for electric current in Section 8.2 'Electric current and circuits' can also be used to understand the concept of potential difference.

In the water model (page 218), potential difference is similar to the water pressure in the pipe. If the water is pumped into a raised water tank, as in Figure 8.2.8 on page 218, potential difference can also be compared to the gravitational potential energy given to each drop of water.

In the bicycle chain analogy (page 219), potential difference is related to how hard the bicycle is being pedalled. If the cyclist is pedalling hard, this would correspond to a high voltage in which each link in the chain is carrying a larger amount of energy than if the cyclist is pedalling slowly.

In both analogies the overall rate of energy output—the power—is related to both the current and the potential difference. In the water analogy, the pressure in the pipe could be very high but the rate of energy transfer will depend on how quickly the water is flowing. Similarly, a cyclist can work at the same rate by pedalling hard with the chain moving slowly or pedalling more easily but with the chain moving more quickly.

CASE STUDY ANALYSIS

Lightning

Lightning (Figure 8.3.6) is a sudden electrical discharge that occurs during an electrical storm due to an imbalance of static charge within storm clouds, or between the storm clouds and the ground. The exact details of the charging process are still being studied, but it is thought that charge is transferred in collisions between tiny hailstone-like material, called grauphel, in the upper atmosphere and the tiny ice crystals in the

FIGURE 8.3.6 Lightning bolts over a city skyline

upward-flowing warmer air. It is thought that the ice crystals become positively charged, and rise to the top of the cloud, while the heavier grauphel, which is negatively charged, stays at the bottom of the cloud. As this charge builds up, the potential difference between the cloud and the ground increases until a discharge, the lightning, occurs. There can be a second, smaller, positively charged region at the bottom of the cloud due to precipitation and warmer temperatures, as seen in Figure 8.3.7.

A potential difference in the order of 300 megavolts exists between these oppositely charged regions. This creates a strong electric field in which any charged particle will experience a force. If this force is sufficiently strong, electrons can be stripped from the various molecules in the air, causing them to become ionised. These free electrons and positive ions will gain a large amount of kinetic energy and collide with more molecules, starting an 'avalanche of charges'. The ionised gas molecules and free electrons move together at such a high speed (30000 km h^{-1} or 1.08×10^4 m s^{-1}) that they cannot recombine. The result is the flash of light, seen either within the cloud or between the ground and the cloud, along with the generation of a huge amount of heat with temperatures of up to 30000 K, hotter than the surface of the Sun. Most flashes of light are within the cloud; only a relatively small number strike the ground. This superheated matter, comprising the ionised gas molecules and electrons, is called a plasma, a fourth state of matter. It will take the path of least resistance through the air, resulting in the lightning being spikes rather than a straight line. The flow of charge, the lightning, stops once the negative charge on the cloud has been discharged. The ionised gas atoms and electrons in the plasma recombine to re-form their original atmospheric molecules.

During the development of the lightning bolt from the negative cloud, the charge at a point on the ground becomes more positive, ionising the molecules in air between it and the descending lightning. Plasma channels of ionised air molecules strike out from the ground to meet the lightning bolt coming down. This is referred to as a streamer.

Lightning is the most extreme example of an electric current. A typical lightning bolt transfers about 10 C or more of negative charge in approximately 70 to 100 microseconds. A moderate thundercloud with a few flashes per minute generates several hundred megawatts of electrical power, the equivalent of a small power station.

It is estimated that at any one time there are 2000 lightning storms around the bulb, which create more than 100 lightning strikes every second. The energy from just one large thunderstorm would be enough to power all the homes in Australia for a few hours.

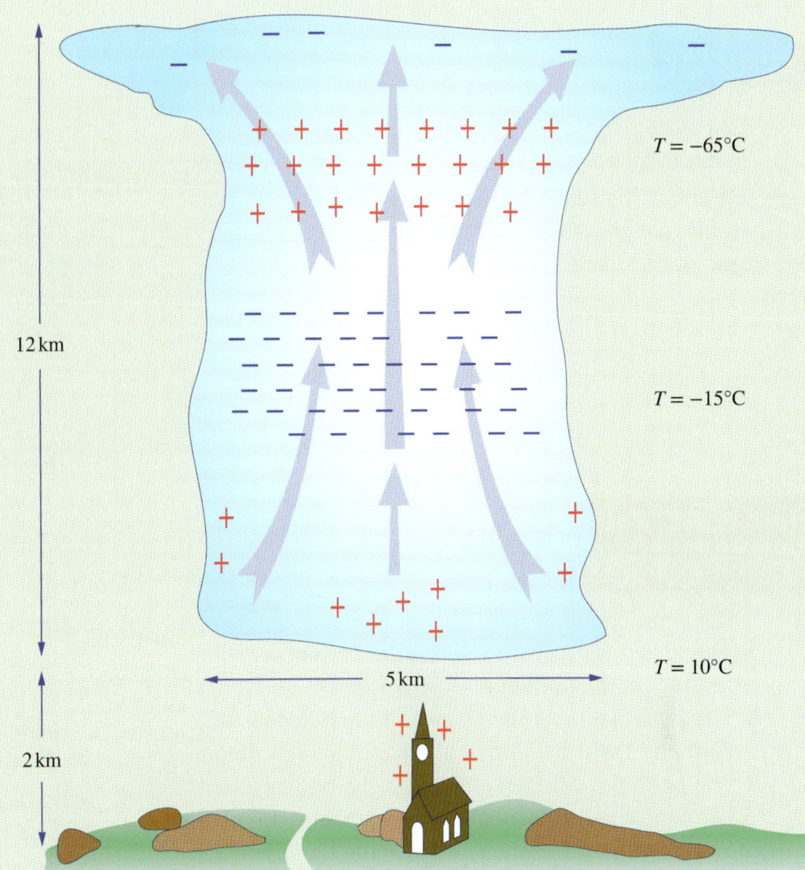

FIGURE 8.3.7 A thundercloud can be several kilometres wide and well over 10 km high. Strong updrafts drive the electrical processes that lead to the separation of charge. The strong negative charge of the lower region of the cloud will induce positive charges on tall objects on the ground. This may lead to a discharge, which can form a conductive path for lightning.

Analysis

1 Explain what is meant by the term 'static charge'.
2 Discuss what would be occurring in the collision between the ice particles and grauphel to make one of them positive and the other negative.
3 Describe how plasma in lightning is formed.
4 Using data provided in the article, calculate:
   a the kinetic energy of the electrons in the plasma
   b the current in a lightning bolt that lasted for 60 μs
   c the power available in such a bolt.

8.3 Review

OA ✓✓

SUMMARY

- Electrical potential difference measures the difference in electrical potential energy available per unit charge.
- Potential difference can be defined as the work done to move a charge against an electric field between two points, using the equation:

$$E = VQ \text{ or } V = \frac{E}{Q}$$

- In an electric circuit, the energy required for charge separation is provided by a cell or battery. The chemical energy within the cell is transformed into electrical potential energy.
- Power is the rate at which energy is transformed in a circuit component. It is defined and quantified by the relationships:

$$P = \frac{E}{t} = VI$$

KEY QUESTIONS

Knowledge and understanding

1 Under what conditions will charge flow between two bodies linked with a rod? Choose the correct response from the following options.
  - **A** The potential difference between the bodies is not zero and the rod is made of a conducting material.
  - **B** The potential difference between the bodies is not zero and the rod is made of an insulating material.
  - **C** The potential difference between the bodies is equal to zero and the rod is made of a conducting material.
  - **D** The potential difference between the bodies is equal to zero and the rod is made of an insulating material.

2 In comparing the electrical energy obtained from a battery to the energy of water stored in a hydroelectric dam in the mountains, to what could the potential difference of the battery be likened?

3 Andy wishes to measure the current and potential difference for a light bulb. He has set up a circuit as shown.

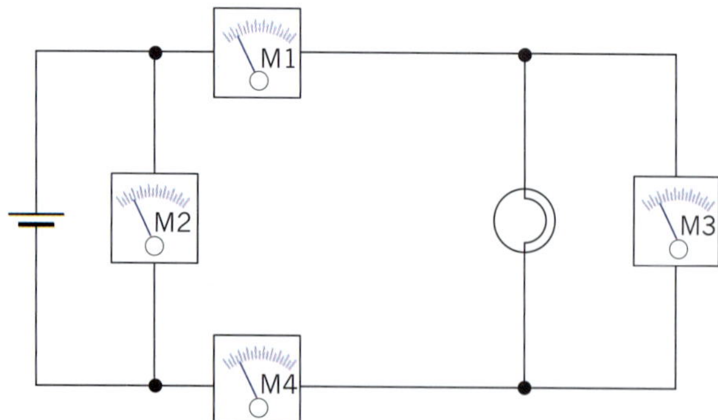

In which positions (M1, M2, M3 or M4) can he place:
  - **a** a voltmeter?
  - **b** an ammeter?

Clearly explain why you made these choices.

Analysis

4 A freezer has a power rating of 460 W and it is designed to be connected to 230 V.
  - **a** Calculate the work performed by the freezer in 5.00 minutes.
  - **b** What is the current flowing through the freezer?

5 A battery is capable of delivering 40 J of energy to a charge of 10 C in 10 s to a circuit.
  - **a** Determine the potential difference of the battery.
  - **b** Calculate the current that flowed in this circuit.

6 A charge of 5.0 C flows from a battery through an electric water heater and delivers 100 J of heat to the water. What was the potential difference of the battery?

7 How much charge must have flowed through a 12 V car battery if 2.0 kJ of energy was delivered to the starter motor?

8 A light bulb that is connected to 240 V uses 3.6 kJ of electrical potential energy in one minute.
  - **a** Into what type(s) of energy has the electrical energy been transformed?
  - **b** Calculate the power of the lamp.
  - **c** Calculate the current flowing through the lamp.

8.4 Resistance

Resistance is an important concept because it links the ideas of potential difference and current. **Resistance** is a measure of how difficult it is for charge to flow through a particular material. Conductors allow current to pass through easily, so they are said to have low resistance. Insulators have a high resistance because they 'resist' or limit the flow of charges through them.

For a particular object or material, the amount of resistance can be quantified (given a numerical value). This means that the performance of electric circuits can be studied and predicted with a high degree of confidence.

- Resistance is a measure of how hard it is for charge to flow through a particular material.
- Resistance is measured in ohms (Ω).

RESISTANCE TO THE FLOW OF CHARGE

Energy is required to create and maintain a flow of charge that we refer to as the electric current. For electrons to move from one place to another, they need to first be separated from their atoms and then given energy to move. In metals the amount of energy required for this is negligible (almost zero), but in poorer conductors a much larger amount of energy may be required.

Once the electrons are moving through the material, energy is also required to keep them moving at a constant speed. Consider an electron travelling through a piece of copper wire. It is common to imagine the wire as an empty pipe or hose through which electrons flow. However, a piece of copper wire is not empty—it is full of copper ions. These ions are packed tightly together in a lattice arrangement. As an electron moves through the wire, it will 'bump' into the ions. The electron needs constant 'energy boosts' to keep it moving in the right direction. This is why an electrical device will stop working as soon as the energy source (e.g. battery) is disconnected.

PHYSICSFILE

Electron movement

Even when there is no current, free electrons tend to move around inside a piece of metal with a kinetic energy, due to thermal effects. The free electrons are rushing around at random with great speed. The net speed of an electron through a wire, however, is quite slow. The diagram below compares the random motion of an electron when there is no current (AB) to the motion of the electron when there is a current (AB′). The difference between the two paths is only small. However, the combined effect of countless electrons moving together in this way represents a significant net movement of charge.

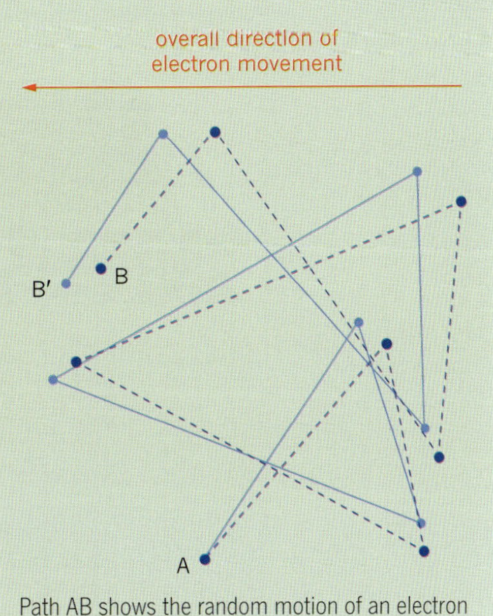

Path AB shows the random motion of an electron due to thermal effects. Path AB′ shows the path of the same electron when there is an electric current in the direction indicated.

CASE STUDY

Variables that affect resistance

Understanding the way electrons move through a wire can help us make some predictions about the resistance of a conductor.

Effect of cross-sectional area and length on resistance

As electrons move through a wire, they will 'bump' into the ions of the wire. In a longer piece of wire, the electrons bump into more ions along the way, so more energy would be needed for the electrons to travel from one end to the other. In other words, a longer piece of wire would provide greater 'resistance' to the flow of charge than a shorter piece of the same wire.

Similarly, a thicker piece of wire allows more electrons to flow through it at the same time, much like a dual-lane highway allows faster traffic flow than a single lane. In practice, the cross-sectional area of the wire (its area when viewed end on) is important. The greater the cross-sectional area of the wire, the lower its resistance per unit length will be.

Calculating the effect of length and area on resistance

The relationship between the resistance of a conductor and its length and thickness follows a mathematical relationship. There is a direct relationship between resistance and length: doubling the length of the conductor doubles its resistance. There is an inverse relationship between resistance and the cross-sectional area of the conductor. These relationships are captured in the equation:

$$R = \frac{\rho L}{A}$$

where R is resistance, L is length, A is cross-sectional area and ρ is resistivity, a property of the material of which the conductor is made.

Temperature and resistance

Another factor that affects the resistance of a material is its temperature. The temperature of an object is a measure of the average kinetic energy of its particles. The temperature of a solid is an indication of how quickly its particles are vibrating.

Increasing the temperature of a piece of copper wire means that the copper ions will vibrate back and forth more quickly. This makes it more likely that an electron will collide with the ion as it moves past it. Therefore, increasing the temperature of the wire also increases the resistance of the wire.

Similarly, current passing through a conductor can cause it to heat up. Think again of an electron moving through a copper wire: when the electron collides with a copper ion, it loses some of its kinetic energy. However, due to this collision, the copper ion gains kinetic energy, causing it to vibrate more quickly. An increase in the kinetic energy of the copper ion means that the temperature has increased, so the copper wire heats up.

The relationship between electric current and temperature is put to use in many household electrical heating devices such as toasters, kettles and electric heaters. It is also an unwanted product in light bulbs, especially incandescent and fluorescent bulbs but less so for LED bulbs as any heat produced is wasted energy.

This is one of the reasons why personal computers contain cooling fans, as shown in Figure 8.4.1. Electrical components are packed very tightly together on the computer motherboard. Cooling the components and the conductors that connect them prevents the computer from overheating. It also reduces the resistance of the components and helps them to run more efficiently.

FIGURE 8.4.1 The fan in this computer motherboard circulates air around the electrical components to cool them down.

OHM'S LAW

Georg Ohm (1789–1854) discovered that when the temperature of a metal wire was kept constant, the current through it was directly proportional to the potential difference across it: mathematically, $I \propto V$. This relationship is known as Ohm's law. This relationship means that if the potential difference across a wire is doubled, for example, then the current through the wire must also double. If the potential difference is tripled, then the current would also triple.

Ohm's law is usually written as:

$\Delta V = IR$ (or just $V = IR$) = constant

where V is the potential difference in volts (V)

I is current in amps (A)

R is the constant of proportionality called resistance, in ohms (Ω).

This equation can be transposed to give a quantitative (mathematical) definition for resistance:

$$R = \frac{V}{I}$$

The relationship $V = IR$ is often referred to as Ohm's law. However, strictly speaking, it is only Ohm's law when R is constant.

If an identical voltage produces currents of two different sizes when separately connected to two light bulbs, then the resistance of the two light bulbs must differ. A higher current would mean a lower resistance of the light bulb, according to Ohm's law. This is because, when a conductor provides less resistance, more charge can flow.

Worked example 8.4.1

USING OHM'S LAW TO CALCULATE RESISTANCE

When a potential difference of 3 V is applied across a piece of wire, it has a current of 5 A. Calculate the resistance of the wire.

Thinking	Working
Ohm's law is used to calculate resistance.	$V = IR$
Rearrange the equation to find R.	$R = \frac{V}{I}$
Substitute in the values for this situation.	$R = \frac{3}{5}$ $= 0.6\,\Omega$

Worked example: Try yourself 8.4.1

USING OHM'S LAW TO CALCULATE RESISTANCE

An electric bar heater draws 10 A of current when connected to a 240 V power supply. Calculate the resistance of the element in the heater.

OHMIC AND NON-OHMIC CONDUCTORS

Conductors that obey Ohm's law are known as **ohmic** conductors. Resistors are ohmic conductors.

An ohmic conductor maintained at constant temperature will give the same voltage divided by current value for all sets of data obtained by measuring the current through a conductor when varying voltages are applied across it. This value would be R, the resistance.

For an ohmic device, at constant temperature, R = constant.

Worked example 8.4.2

USING OHM'S LAW TO CALCULATE RESISTANCE, CURRENT AND POTENTIAL DIFFERENCE

The table below shows pairs of measurements for the potential difference and corresponding current for an ohmic conductor.

V [V]	0	2	4	V_2
I [A]	0	0.25	I_1	0.75

Determine the missing results, I_1 and V_2.

Thinking	Working
Determine the factor by which potential difference has increased from the second column to the third column.	$\frac{4}{2} = 2$ The potential difference has doubled.
Apply the same factor increase to the current in the second column, to determine the current in the third column (I_1).	$I_1 = 2 \times 0.25$ $= 0.50\,\text{A}$
Determine the factor by which current has increased from the second column to the fourth column.	$\frac{0.75}{0.25} = 3$ The current has tripled.
Apply the same factor increase to the potential difference in the second column, to determine the potential difference in the fourth column (V_2).	$V_2 = 3 \times 2$ $= 6\,\text{V}$

Worked example: Try yourself 8.4.2

USING OHM'S LAW TO CALCULATE RESISTANCE, CURRENT AND POTENTIAL DIFFERENCE

The table below shows pairs of measurements for the potential difference and corresponding current for an ohmic conductor.

V [V]	0	3.0	9.0	V_2
I [A]	0	0.20	I_1	0.80

Determine the missing results, I_1 and V_2.

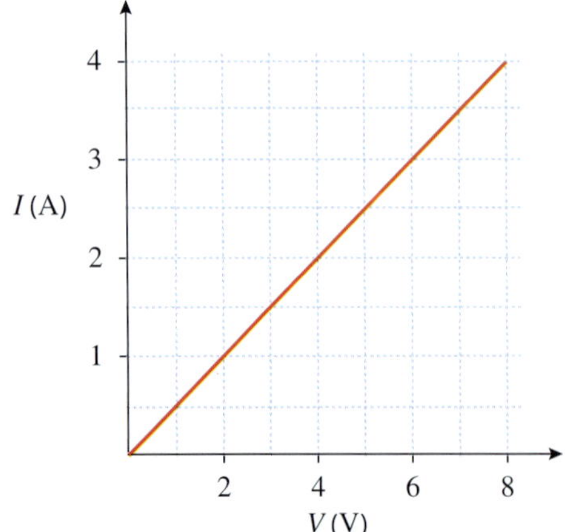

FIGURE 8.4.2 As the resistance of an ohmic conductor is constant, the I–V graph is a straight line.

I–V GRAPHS

The data from an experiment in which the current and potential difference is measured for a device is usually plotted on an I–V graph. The current that is measured (the dependent variable) is plotted against the various set values of voltage (the independent variable). If the conductor is ohmic, this graph will be a straight line, as can be seen in Figure 8.4.2.

The gradient of this graph is $\frac{I}{V}$. The resistance of the ohmic conductor (or resistor) can be found from the gradient of the I–V graph; resistance is the inverse of the gradient. For Figure 8.4.2 this is:

$$\frac{1}{R} = \frac{\text{rise}}{\text{run}} = \frac{4-1}{8-2} = \frac{3}{6}$$

So $R = \frac{6}{3} = 2\,\Omega$.

However, not all conductors are ohmic. The I–V graphs for **non-ohmic** conductors are not straight lines (Figure 8.4.3). Incandescent light bulbs and semiconductor diodes, which includes LEDs, are examples of non-ohmic conductors.

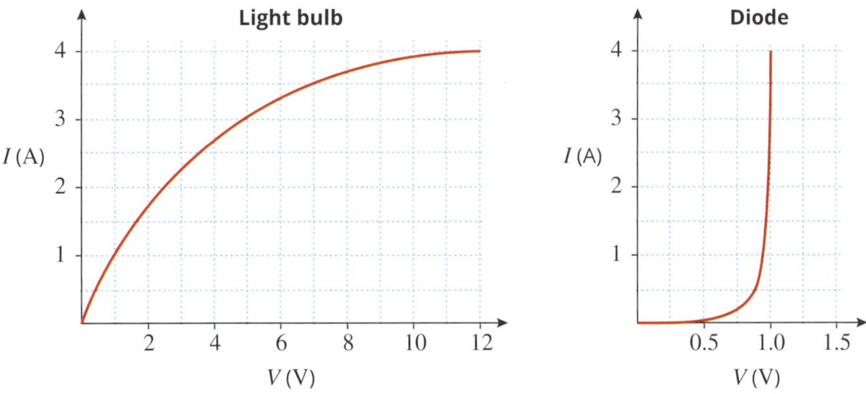

FIGURE 8.4.3 The I–V graph for a non-ohmic conductor is not a straight line.

Using I–V graphs to determine resistance

The inverse of resistance is defined as the ratio $\frac{I}{V}$. For an ohmic conductor, this value will be a constant regardless of the potential difference across the conductor. However, the resistance of a non-ohmic conductor will vary. The resistance of a non-ohmic conductor for a particular potential difference can be found by determining the current in the conductor at this value and using the relationship $R = \frac{V}{I}$. This value will not be constant.

Worked example 8.4.3

CALCULATING RESISTANCE FOR A NON-OHMIC CONDUCTOR

Calculate the resistance of the light bulb with the I–V graph shown when the potential difference is 5.0 V.

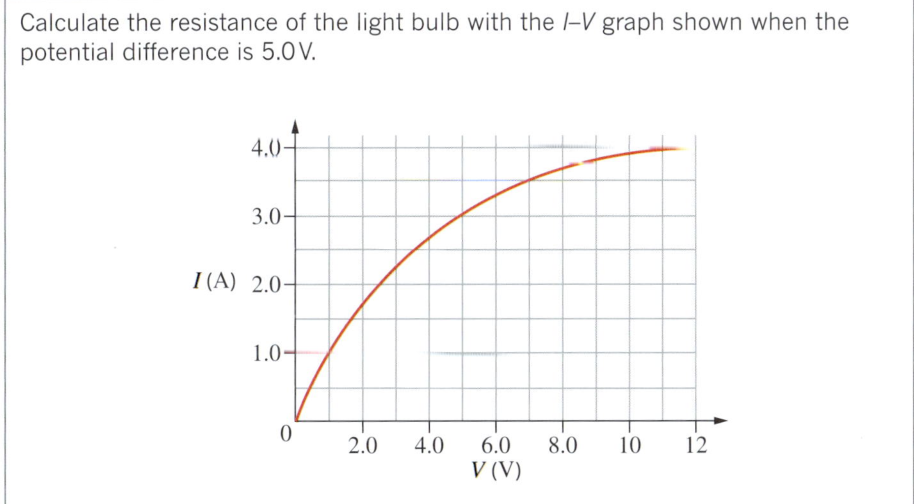

Thinking	Working
From the graph, determine the current at the required potential difference.	At $V = 5.0\,V$, $I = 3.0\,A$.
Substitute these values into Ohm's law to find the resistance.	$R = \frac{V}{I}$ $= \frac{5.0}{3.0}$ $= 1.7\,\Omega$

Worked example: Try yourself 8.4.3

CALCULATING RESISTANCE FOR A NON-OHMIC CONDUCTOR

A 240 V, 60 W incandescent light bulb has the *I–V* characteristics shown in the graph. Calculate the resistance of the light bulb at 175 V.

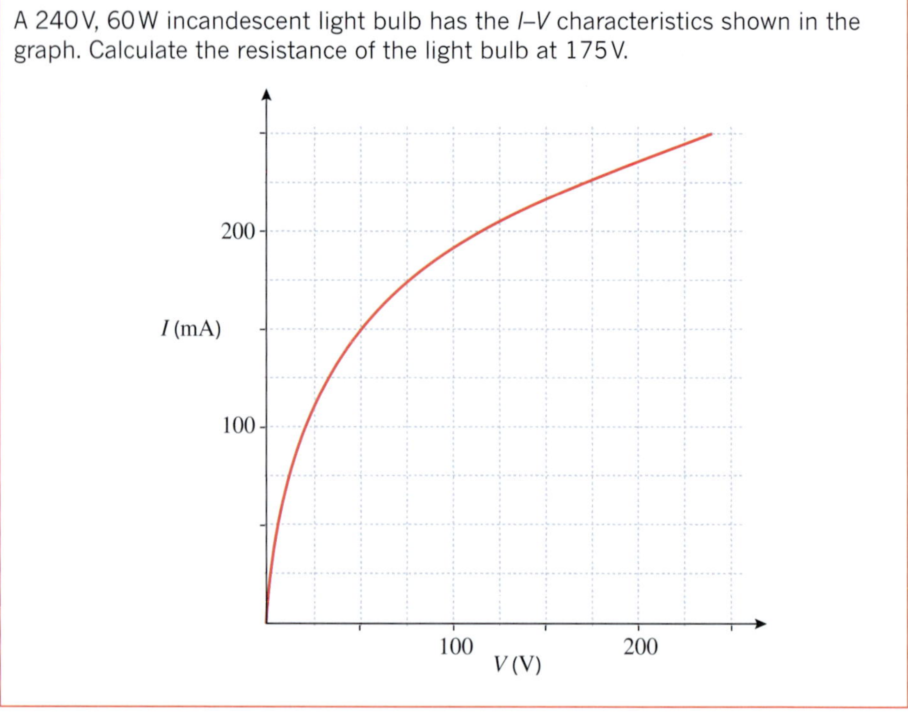

RESISTORS IN SIMPLE CIRCUITS

Ohmic **resistors** are often used to control the amount of current in a particular circuit. Resistors can be manufactured to produce a relatively constant resistance over a range of temperatures. A colour-coding system is used on resistors to explain the amount of resistance they provide, including a percentage tolerance (precision). Figure 8.4.4 shows a resistor that uses the colour-coding system.

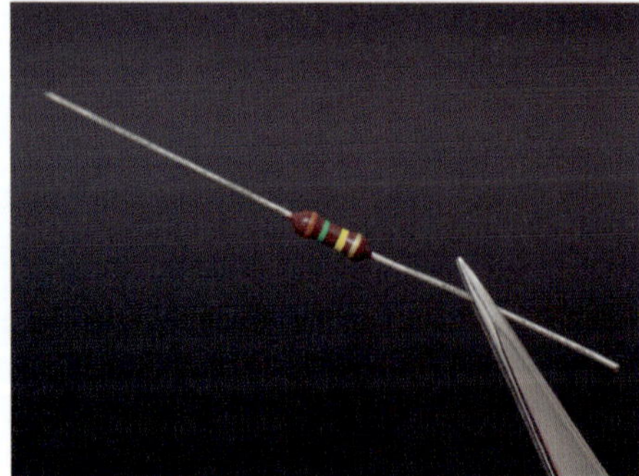

FIGURE 8.4.4 Common resistors are electrical devices with a known resistance. The coloured bands indicate the resistor's resistance and tolerance.

Ohm's law can be used to determine the current in a resistor when a particular potential difference is applied across it. Similarly, if the current and resistance are known, the potential difference across the resistor can be calculated.

PHYSICSFILE

Colour-coded resistors

A resistor is typically a small piece of equipment that does not allow enough room to clearly print information about the resistor in the form of numbers. A colour-coding system is used on many resistors to convey detailed information in a small space about the resistance and tolerance of the resistor. The diagram below explains how to interpret this colour-coding system.

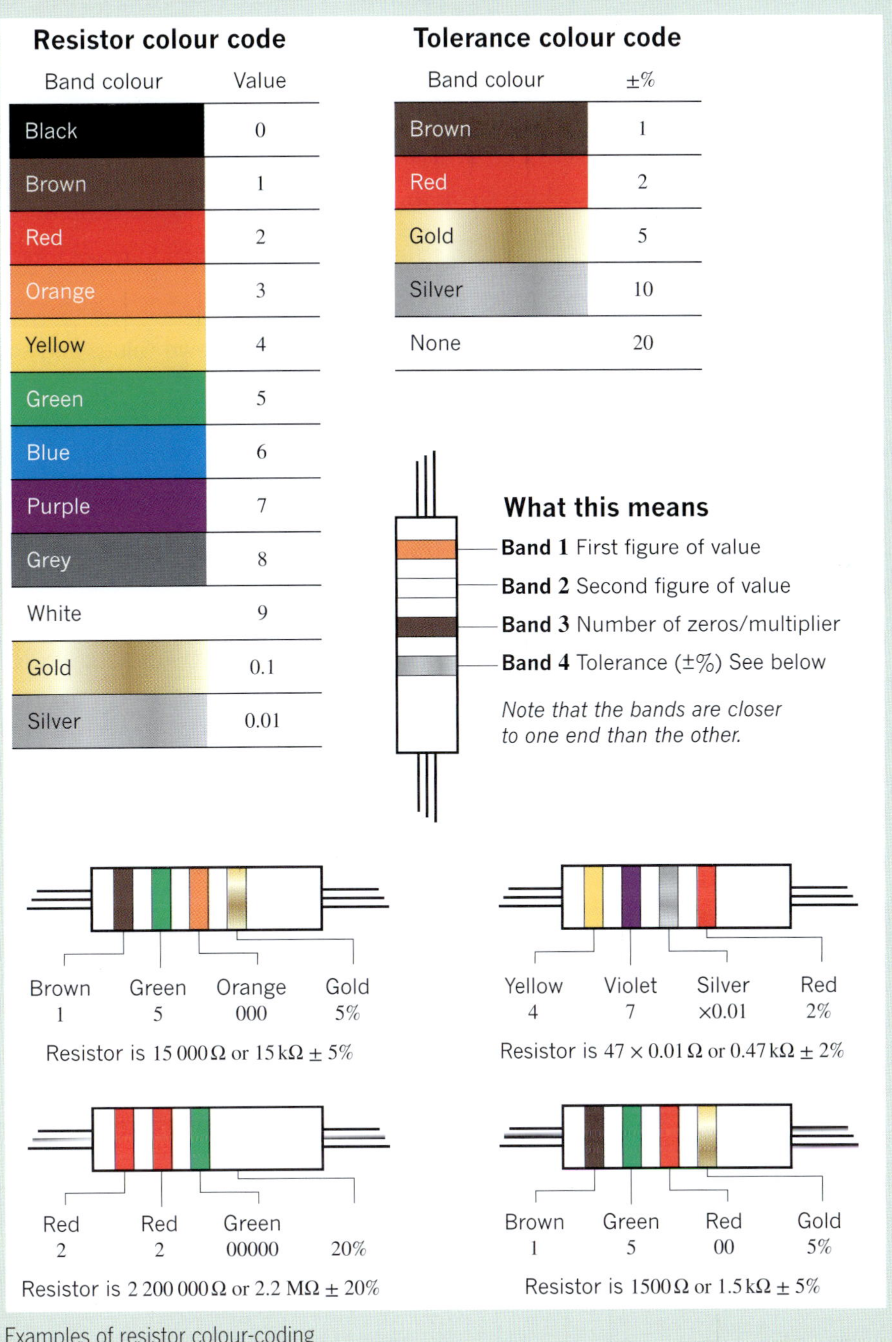

Resistor colour code

Band colour	Value
Black	0
Brown	1
Red	2
Orange	3
Yellow	4
Green	5
Blue	6
Purple	7
Grey	8
White	9
Gold	0.1
Silver	0.01

Tolerance colour code

Band colour	±%
Brown	1
Red	2
Gold	5
Silver	10
None	20

Examples of resistor colour-coding

Worked example 8.4.4

USING OHM'S LAW TO FIND CURRENT

A 100 Ω resistor is connected to a 12 V battery. Calculate the current (in mA) in the resistor.

Thinking	Working
Recall Ohm's law.	$V = IR$
Rearrange the equation to make I the subject.	$I = \frac{V}{R}$
Substitute in the known values and solve.	$I = \frac{12}{100}$ $= 0.12\,\text{A}$
Convert the answer to the required units.	$I = 0.12\,\text{A}$ $= 0.12 \times 10^3\,\text{mA}$ $= 120\,\text{mA}$

Worked example: Try yourself 8.4.4

USING OHM'S LAW TO FIND CURRENT

The element of a bar heater has a resistance of 25 Ω. Calculate the current (in mA) in this element when it is connected to a 240 V supply.

Worked example 8.4.5

USING OHM'S LAW TO FIND POTENTIAL DIFFERENCE

A 22 Ω resistor draws a current of 0.25 A. Calculate the voltage across the resistor. Give your answer correct to one decimal place.

Thinking	Working
Recall Ohm's law.	$V = IR$
Substitute in the values for this problem and solve.	$V = 0.25 \times 22$ $= 5.5\,\text{V}$

Worked example: Try yourself 8.4.5

USING OHM'S LAW TO FIND POTENTIAL DIFFERENCE

The light bulb of a torch has a resistance of 5.7 Ω when it draws 700 mA of current. Calculate the potential difference across the bulb.

8.4 Review

OA ✓✓

SUMMARY

- Resistance is a measure of how hard it is for charge to flow through a particular material. The unit of resistance is the ohm (Ω).
- The resistance of a material depends on its length, cross-sectional area and temperature.
- Ohm's law describes the relationship between current, potential difference and resistance:

$$V = IR$$

- Ohmic conductors have a constant resistance at a constant temperature. The resistance of non-ohmic conductors varies for different potential differences.

KEY QUESTIONS

Knowledge and understanding

1 An experiment is conducted to gather data about the relationship between current and potential difference for three ohmic devices, labelled A, B and C. The data is used to plot an I–V graph for each device.

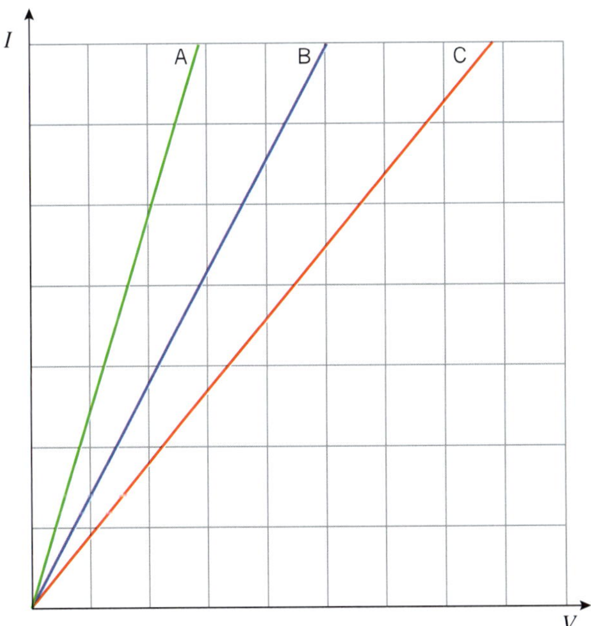

a For a given potential difference, list the devices in order of highest current to lowest current.

b List the devices in order of highest resistance to lowest resistance.

Analysis

2 The table below shows measurements for the potential difference and corresponding current for an ohmic conductor.

V [V]	0	2	3	V_2
I [A]	0	0.25	I_1	0.60

Determine the missing results, I_1 and V_2.

3 A student obtains a graph of the current–voltage characteristics of a piece of resistance wire.

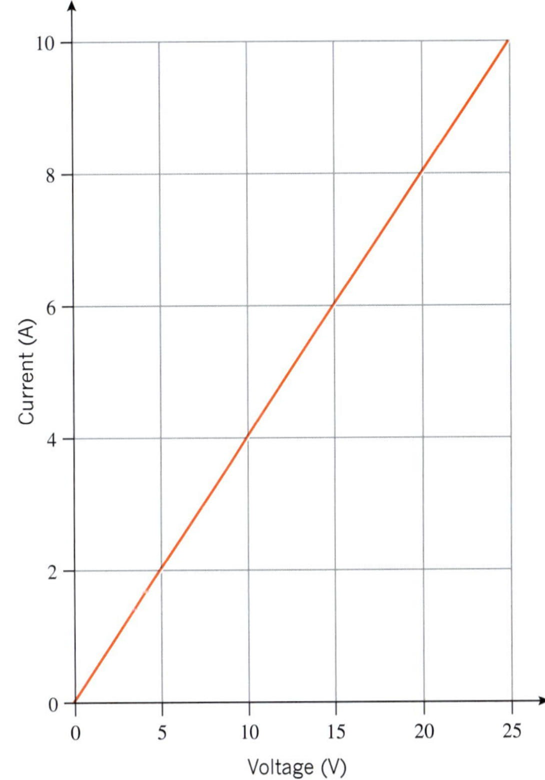

a Explain whether this piece of wire is ohmic or non-ohmic.

b What current is in this wire at a voltage of 7.5 V?

c What is the resistance of this wire?

4 A student finds that the current through a resistor is 3.5 A when a voltage of 2.5 V is applied to it.

a What is the resistance?

b The voltage is then doubled and the current is found to increase to 7.0 A. Is the resistor ohmic or non-ohmic?

continued over page

8.4 Review *continued*

5 Rose and Rachel are trying to find the resistance of an electrical device. They find that at 5 V it draws a current of 200 mA and at 10 V it draws a current of 500 mA. Rose says that the resistance is 25 Ω, but Rachel maintains that it is 20 Ω. Who is right and why?

6 Nick has an ohmic resistor to which he has applied 5 V. He measures the current as 45 mA. He then increases the voltage to 8 V. What current will he find now?

7 Lisa finds that when she increases the voltage across an ohmic resistor from 6 V to 10 V, the current increases by 2 A.

a What is the resistance of this resistor?

b What current does it draw at 10 V?

8 A strange electrical device has the *I–V* characteristics shown in this graph.

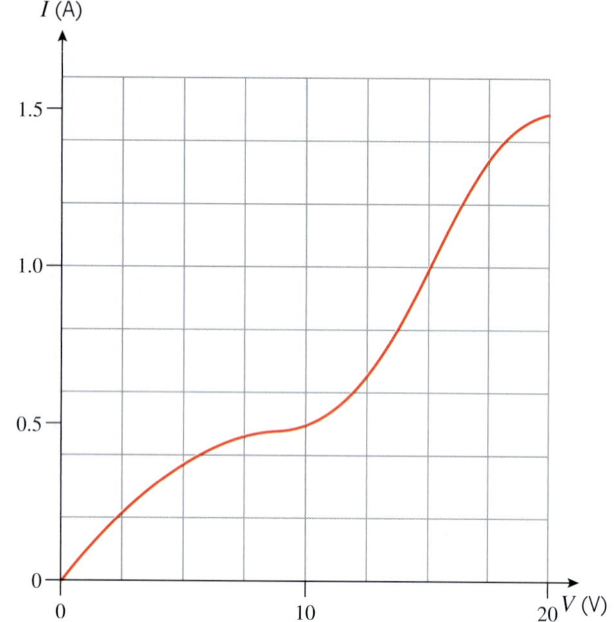

a Is it an ohmic or non-ohmic device? Explain your answer.

b What current is drawn when a voltage of 10 V is applied to it?

c What voltage would be required to double the current drawn at 10 V?

d Calculate the resistance of the device at:

i 10 V

ii 20 V.

Chapter review

KEY TERMS

ammeter
charge
conductor
conventional current
coulomb
current
electric circuit
electrical potential energy
electricity
electron flow
elementary charge
insulator
ion
ionised
metal
net charge
non-metal
non-ohmic
ohmic
parallel circuit
potential difference
power
resistance
resistor
transfer
transform
volt
voltmeter

REVIEW QUESTIONS

Knowledge and understanding

1 Approximately how many electrons make up a charge of –3 C?

2 What will be the approximate charge on 4.2×10^{19} protons?

3 Which charged particles are moving when electricity flows in a closed circuit?
   A negatively charged electrons
   B positively charged electrons
   C positively charged protons
   D both negative and positive charges

4 The analogy of electron flow to the links in a bicycle chain is used on page 219. Explain what this analogy shows.

5 Which of the choices below lists the quantities you would need to measure to calculate the amount of electrical energy required to heat water using an electric kettle?
   A potential difference, resistance and current
   B time, current and charge
   C current, time and potential difference
   D potential difference and current

6 Compare the meaning of the terms ‘conventional current’ and ‘electron flow’.

7 Explain why even a good conductor such as copper wire provides some resistance to current.

Application and analysis

8 Calculate the current that flows when 0.23 C of charge passes a point in a circuit each minute.

9 An alpha particle consists of two protons, two neutrons and no electrons. Calculate the charge on an alpha particle.

10 A current of 1.6 A flows for 100 seconds. Calculate:
   a the amount of charge, in coulombs, that moves past a point in this time
   b the number of electrons that move past a point in this time.

11 A current of 0.04 A flows for a certain length of time. In this time 5×10^{18} electrons move past a point. Calculate:
   a the amount of charge, in coulombs, that moves past this point
   b the length of time that the current is flowing.

12 A phone battery has a voltage of 3.8 V. If 2 C of charge is drawn from the battery, what amount of energy would this provide?

13 A battery does 2 J of work on a charge of 0.5 C to move it from point A to point B. Calculate the potential difference between points A and B.

14 How much power does an appliance use if it does 2500 J of work in 30 minutes?

15 A battery gives a single electron 1.4×10^{-18} J of energy. Calculate the potential difference supplied by the battery to two significant figures.

16 A 230 V appliance consumes 2000 W of power. The appliance is left on for 2 hours. What current flows through the appliance?

17 A student finds that the current through a wire is 5 A while a voltage of 2.5 V is applied to it. Calculate the resistance of the wire.

18 A 10 W LED bulb produces the same amount of light as a 60 W incandescent bulb when connected to a 240 V power supply.
   a Calculate the current drawn by the incandescent bulb.
   b Determine the number of LED bulbs that could be connected in parallel in a circuit with a 10 A circuit breaker.

continued over page

19 Calculate the resistance at 175 mA of the non-ohmic conductor with the *I–V* graph shown.

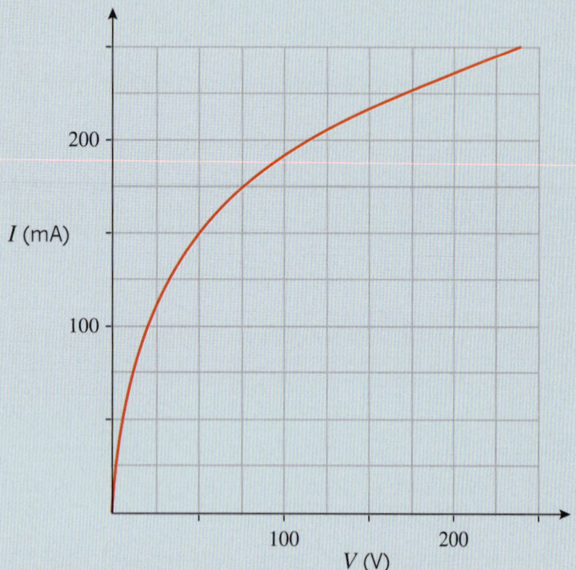

20 A current of 0.25 A flows through an 80 Ω resistor. Calculate the voltage across it.

21 When 1.5 V is applied across a particular resistor, the current through the resistor is 50 mA. What is the resistance of the resistor?

22 Calculate the resistance at the following voltages of the non-ohmic conductor with the *I–V* graph shown.

a 1.0 V
b 7.0 V
c 12.0 V

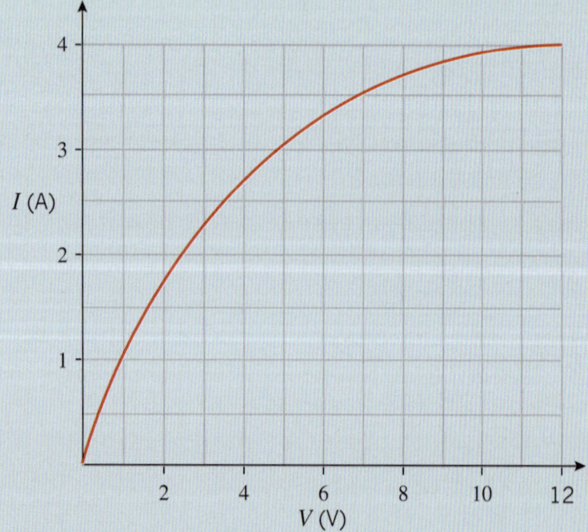

23 A potential difference of 240 V is used to generate a current of 12 A in a hot water heater for 15 minutes. Calculate the energy in MJ transferred to the water in that time.

24 Justine is measuring the current through a non-ohmic conductor as she varies the potential difference of the power source. The results are shown in the graph. Describe the general trend in the resistance of the conductor as the voltage increases. Use data from the graph to support your answer.

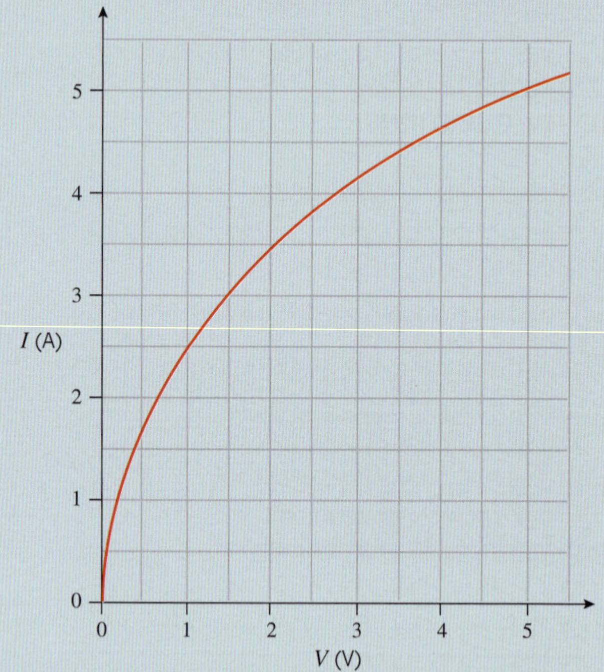

25 The power supply for a laptop computer has a rated output of 19 V and 980 mA. Calculate the maximum power this can deliver.

26 An electric heater is rated at 1600 W when connected to a 240 V power supply. Calculate the resistance of the coil of the heater.

27 A 240 V lamp draws 7.5 A when cold but only 0.6 A when hot. Calculate the resistance of the lamp at each temperature.

28 If 0.6 A of current flows through a light bulb, calculate how many electrons enter the light bulb each second.

29 A 4.5 V torch with a 0.4 A bulb is switched on for 2 minutes.

a How much charge has travelled through the filament of the bulb in this time?
b How much energy has been used?
c Where has this energy come from?

OA ✓✓

CHAPTER

09 Electric circuits

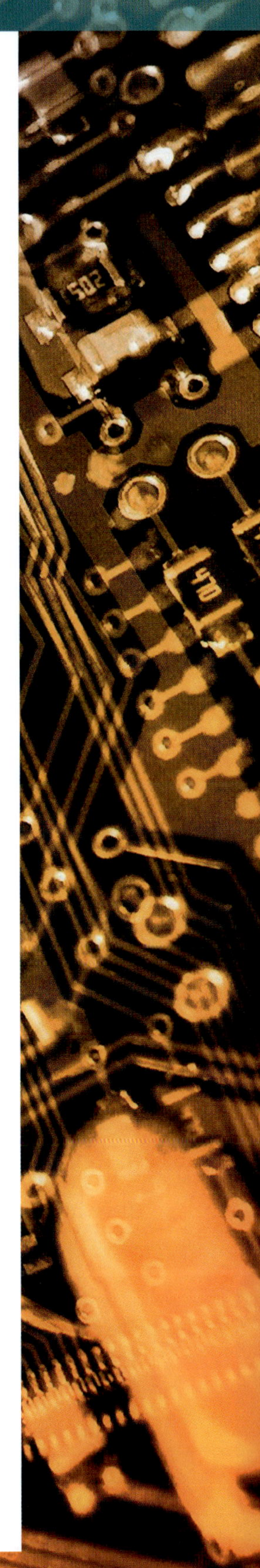

Electric circuits are the basis of much of our modern society. This chapter introduces a range of circuits, from simple series circuits to the complex parallel wiring systems that make up a modern home. Electric circuits can be used to perform energy transfers and transformations through devices such as light bulbs, thermistors, light dependent resistors and light-emitting diodes. It is essential that anyone working with electricity learns how to do so safely in the home and the laboratory. Several safety mechanisms are examined that minimise the effect of current on humans.

Key knowledge

- apply the kilowatt-hour (kWh) as a unit of energy **9.3**
- model resistance in series and parallel circuits using:
  - equivalent resistance in arrangements in
    - series: $R_{\text{equivalent}} = R_1 + R_2 + \ldots + R_n$ and **9.1**
    - parallel: $\frac{1}{R_{\text{equivalent}}} = \frac{1}{R_1} + \frac{1}{R_2} + \ldots + \frac{1}{R_n}$ **9.1**
- calculate and analyse the equivalent resistance of circuits comprising parallel and series resistance **9.1**
- analyse circuits comprising voltage dividers **9.1**
- model household (AC) electrical systems as simple direct current (DC) circuits **9.3**
- compare power transfers in series and parallel circuits **9.1**
- explain why the circuits in homes are mostly parallel circuits **9.1, 9.3**
- investigate and apply theoretically and practically concepts of current, resistance, potential difference (voltage drop) and power to the operation of electronic circuits comprising resistors, light bulbs, diodes, thermistors, light dependent resistors (LDRs), light-emitting diodes (LEDs) and potentiometers (quantitative analysis restricted to use of $I = \frac{V}{R}$ and $P = VI$) **9.1**
- investigate practically the operation of simple circuits containing resistors, variable resistors, diodes and other non-ohmic devices **9.1**
- describe energy transfers and transformations with reference to resistors, light bulbs, diodes, thermistors, light dependent resistors (LDRs), light-emitting diodes (LEDs) and potentiometers in common devices **9.2**
- model household electricity connections as a simple DC circuit comprising fuses, switches, circuit breakers, loads and earth **9.3**
- compare the operation of safety devices including fuses, circuit breakers and residual current devices (RCDs) **9.3**
- describe the causes, effects and first aid treatment of electric shock and identify the approximate danger thresholds for current and duration. **9.3**

9.1 Series and parallel circuits

When a circuit contains more than one resistor, Ohm's law alone is not sufficient to predict the current through and the potential difference across each resistor. Additional concepts such as Kirchhoff's rules and the idea of equivalent resistance can be used to analyse these complex, multi-component circuits.

No matter how complex a circuit, it can always be broken up into sections in which the circuit elements are in series or parallel. This section investigates the difference between these two types of circuits.

RESISTORS IN SERIES

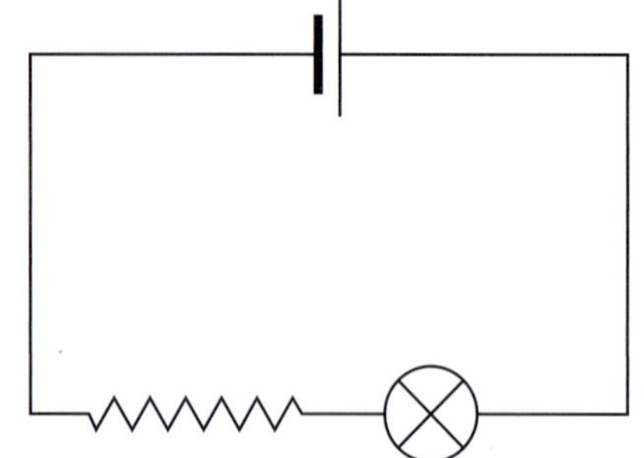

FIGURE 9.1.1 This circuit has a resistor and light bulb connected in series.

Some circuits contain more than one electrical component. When these components are connected one after another in a continuous loop, this is called a **series circuit**. Components connected in this way are said to have been connected 'in series'. The circuit shown in Figure 9.1.1 shows a resistor and a light bulb connected in series with an electric cell.

Series circuits are very easy to construct, but they have some disadvantages. As every component is connected one after the other, the components are dependent on each other. If one component is removed or breaks down, the circuit is no longer a closed loop and it won't work. Figure 9.1.2 shows that removing a light bulb from a series circuit interrupts the entire circuit. Because of this characteristic, series circuits with more than one component are not commonly used in the home.

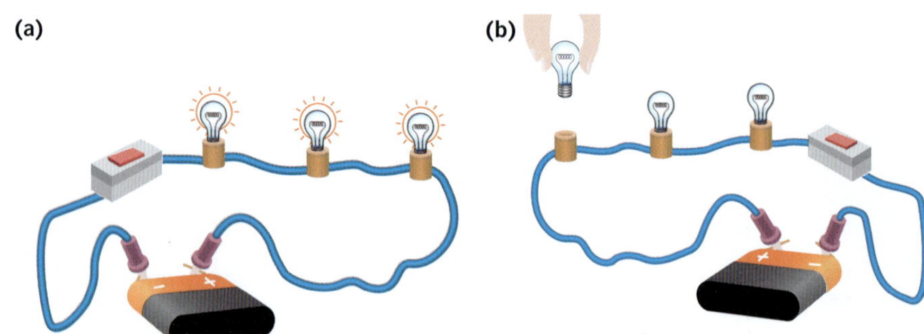

FIGURE 9.1.2 (a) All components in the circuit are intact and so the circuit is a closed loop. (b) When one light bulb is removed, the whole circuit is interrupted.

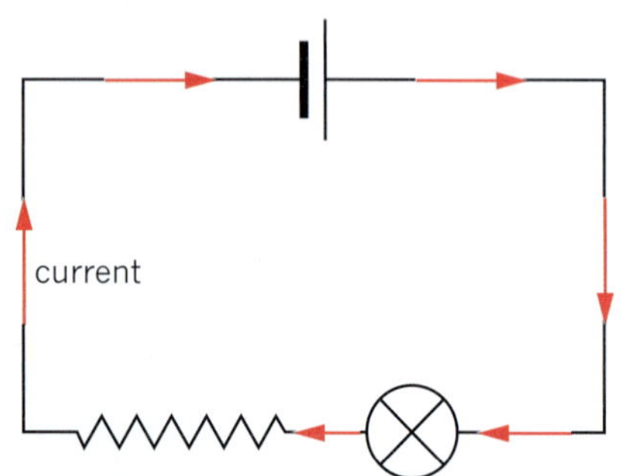

FIGURE 9.1.3 In a series circuit, the current in each component is the same.

The current in a series circuit is the same in every part of the circuit.

The energy given to the charges (potential gain) must be equal to the energy lost by the charges (potential drop). In a series circuit, the energy loss will be spread across a number of different components.

Conservation of charge

When analysing a series circuit, such as the one as shown in Figure 9.1.3, it is important to understand that every part of the circuit has the same amount of current. Since electric charges are not created or destroyed within an electric circuit, the number of charges flowing out of the cell must be the same as the number flowing through the bulb, which is also the same as the number flowing through the resistor. This means the current is unchanged throughout the circuit. Thus the charge flowing from the cell through the circuit and back into the cell is the same.

Remember that, by convention, the direction of current is from the positive terminal of the cell to the negative terminal. Electrons move in the opposite direction.

Kirchhoff's loop rule

Kirchhoff's loop rule says that the sum of the potential differences across all the elements around any closed circuit loop must be zero. This means that the total potential drop around a closed circuit must be equal to the total potential gain in the circuit. For example, if a battery provides 9 V to a circuit, then the sum of all of the potential drops across the components must add to 9 V.

This rule is essentially another version of the law of conservation of energy.

Figure 9.1.4 shows how the voltage provided by the battery is shared across a resistor and a light bulb. The power supply in the figure is labelled emf. Devices that are a source of energy for a circuit are referred to as sources of emf or electromotive force. emf, measured in volts (V), is another term for the work done on charges to provide a potential difference between the terminals of the power supply.

There are a number of ways to visualise the energy changes in this circuit. One common analogy is to think of the charges as water being pumped around an elevated water course. The water gains potential energy as it is pumped higher, and as it flows back down the potential energy is converted into other forms. The diagram in Figure 9.1.5 shows how the analogy works with the energy changes that occur in a circuit. The battery acts as a 'pump' that pushes electrons up to a higher energy level and the electrons gain potential energy. As the electrons pass 'down' through components in the circuit, their energy is transformed into other forms.

The change in electrical energy available to electrons can also be represented graphically, as shown in Figure 9.1.6.

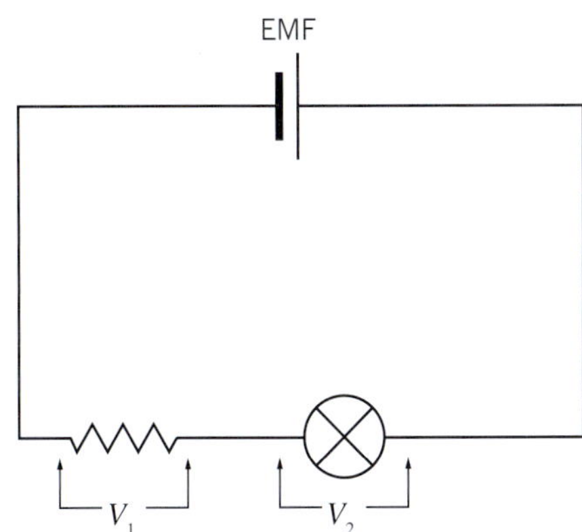

FIGURE 9.1.4 Kirchhoff's loop rule. In this series circuit, the sum of the potential drops across the resistor and the light bulb (i.e. $V_1 + V_2$) will be equal to the potential difference (emf) provided by the battery.

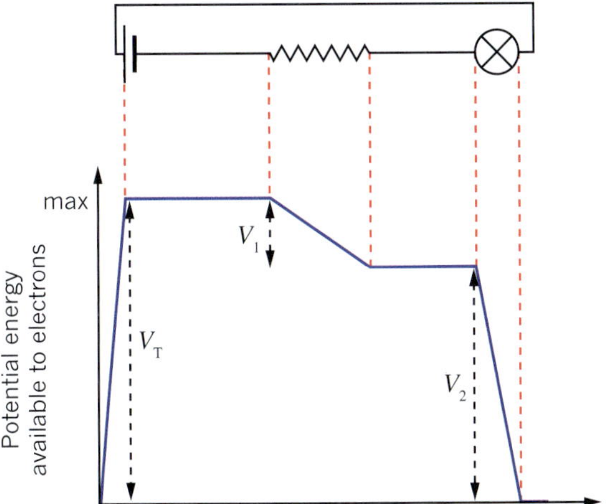

FIGURE 9.1.6 The electrical potential energy of an electron changes as it moves around the circuit. Some of this energy is lost as the electrons pass through the resistor. The remaining energy is lost as the electrons pass through the light bulb. In this circuit, the bulb has more resistance than the resistor.

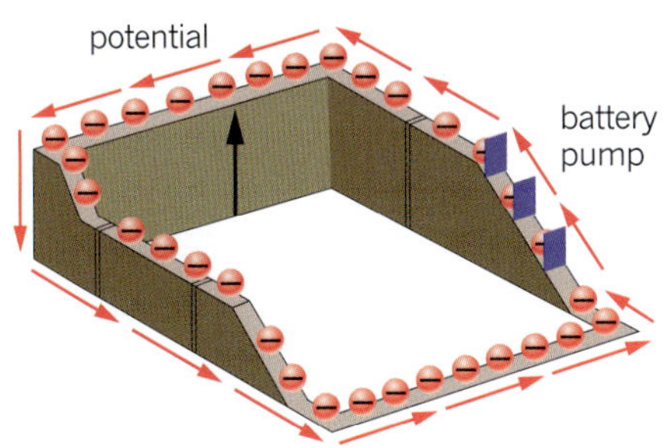

FIGURE 9.1.5 An analogy for analysing a circuit. The battery acts as a 'pump' that transfers potential energy to electrons. The electrons lose potential energy as they flow 'down' through components in the circuit.

Equivalent series resistance

Consider the circuit in Figure 9.1.7. If the resistance of the fixed resistor is R_1, the resistance of the light bulb is R_2 and the current through both of them is I, then Ohm's law gives:

$$V_1 = I \times R_1$$

and

$$V_2 = I \times R_2$$

The total voltage drop (V_T) across the two components is:

$$V_T = V_1 + V_2 = IR_1 + IR_2 = I \times (R_1 + R_2)$$

This equation shows the relationship between the potential difference supplied by the cell and the potential differences of the light bulb and resistor. The last part of the equation also shows that the bulb and resistor can be replaced with a single resistor, without changing the current in the circuit. The single resistor needs to have a resistance of $R_1 + R_2$.

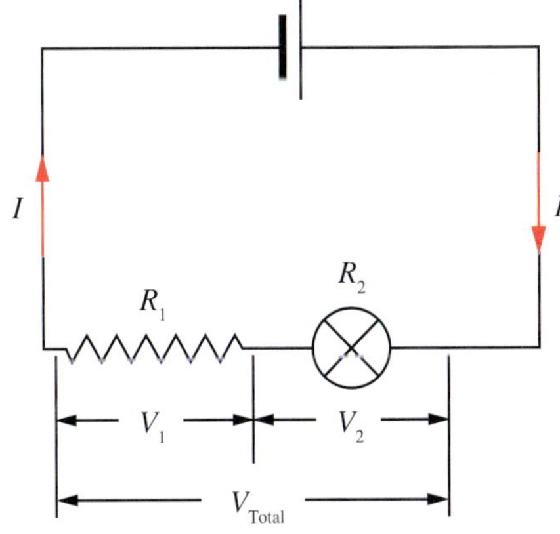

FIGURE 9.1.7 Ohm's law shows the relationship between the potential difference supplied by the cell, V_{Total}, the current in the circuit, I, and the resistances of the two components R_1 and R_2.

In general, a number of individual resistors connected in series can be replaced by an **equivalent resistance** ($R_{equivalent}$) equal to the sum of the individual resistances. Figure 9.1.8 shows that two resistors can be replaced with a single one and have the same effect in a circuit.

$R_{equivalent} = R_1 + R_2 + \ldots + R_n$

where $R_{equivalent}$ is the equivalent series resistance and $R_1, R_2, \ldots, R_n$ are the individual resistances.

Equivalent resistances can be used in circuit analysis to simplify a complicated circuit diagram so that current and potential difference can be determined.

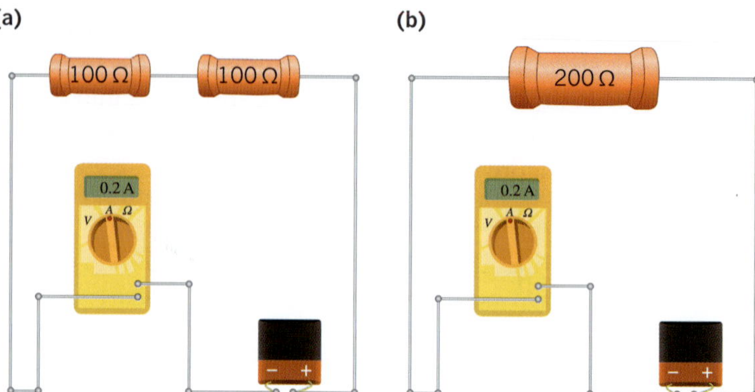

FIGURE 9.1.8 Two 100 Ω resistors in series (a) can be replaced with a single 200 Ω resistor (b) to have the same effect in a circuit.

Worked example 9.1.1

CALCULATING AN EQUIVALENT SERIES RESISTANCE

A 100 Ω resistor is connected in series with a 690 Ω resistor and a 1.2 kΩ resistor. Calculate the equivalent series resistance.

Thinking	Working
Recall the formula for equivalent series resistance.	$R_{equivalent} = R_1 + R_2 + \ldots + R_n$
Substitute in the given values for resistance. Make sure to convert kΩ to Ω. Solve to find the equivalent series resistance.	$R_{equivalent} = 100 + 690 + 1200$ $= 1990\,\Omega$

Worked example: Try yourself 9.1.1

CALCULATING AN EQUIVALENT SERIES RESISTANCE

A string of fairy lights consists of 20 light bulbs connected in series. Each bulb has a resistance of 8.0 Ω. Calculate the equivalent series resistance of the lights.

Worked example 9.1.2

USING EQUIVALENT SERIES RESISTANCE FOR CIRCUIT ANALYSIS

Use an equivalent series resistance to calculate the current in the series circuit below, and the potential difference across each resistor.

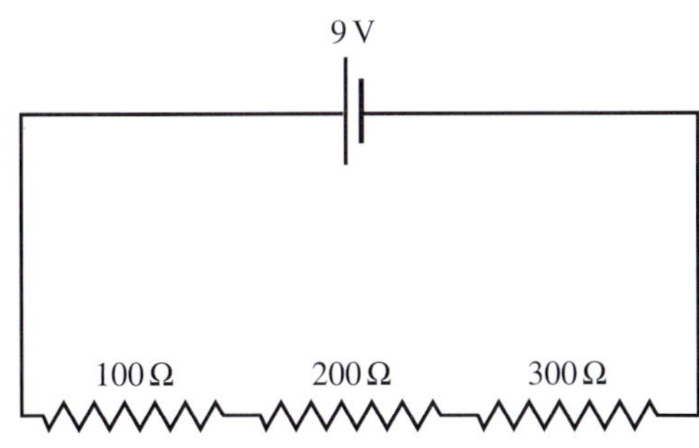

Thinking	Working
Recall the formula for equivalent series resistance.	$R_{\text{equivalent}} = R_1 + R_2 + R_3 + \ldots + R_n$
Find the equivalent (total) resistance in the circuit.	$R_{\text{equivalent}} = 100 + 200 + 300$ $= 600\,\Omega$
Use Ohm's law to calculate the current in the circuit. Whenever calculating current in a series circuit, use $R_{\text{equivalent}}$ and the voltage of the power supply.	$I = \frac{V}{R}$ $= \frac{9}{600}$ $= 0.015\,\text{A}$
Use Ohm's law to calculate the potential difference across each separate resistor.	$V = IR$ So: $V_1 = 0.015\,\text{A} \times 100\,\Omega = 1.5\,\text{V}$ $V_2 = 0.015\,\text{A} \times 200\,\Omega = 3.0\,\text{V}$ $V_3 = 0.015\,\text{A} \times 300\,\Omega = 4.5\,\text{V}$
Use the loop rule to check the answer.	$V_T = V_1 + V_2 + V_3$ $= 1.5\,\text{V} + 3.0\,\text{V} + 4.5\,\text{V}$ $= 9.0\,\text{V}$ Since this is the same as the voltage provided by the cell, the answer is reasonable.

Worked example: Try yourself 9.1.2

USING EQUIVALENT SERIES RESISTANCE FOR CIRCUIT ANALYSIS

Use an equivalent series resistance to calculate the current in the series circuit below, and the potential difference across each resistor.

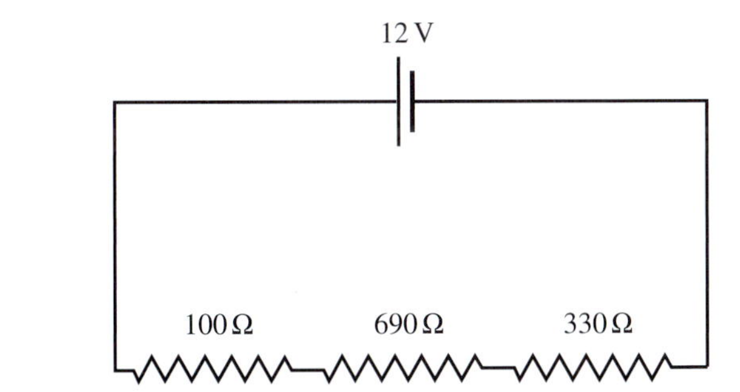

RESISTORS IN PARALLEL

One of the disadvantages of series circuits is that if a switch is opened or a device disconnected, then the circuit is broken and current stops. In everyday life, we often want to switch devices on and off independently. **Parallel circuits** allow us to do this.

The circuit diagram in Figure 9.1.9 shows a simple parallel circuit. If switch A is open (as shown), bulb B will still light up as it is part of an unbroken complete circuit that includes the battery. Similarly, if switch A is closed and switch B is opened, current will light up bulb A but not bulb B. Alternatively, both switches could be closed to light up both bulbs or both switches could be opened to switch both bulbs off.

FIGURE 9.1.9 In this parallel circuit, bulb A will be off and bulb B will be on.

In a series circuit, all the components are in the same loop and therefore each component has the same current. In comparison, each loop of a parallel circuit acts like an independent circuit with its own current.

Consider again the water analogy. When water flows through pipes, it is not lost. If the pipe branches in two, some of the water flows in one pipe, and the remaining water will flow in the other pipe. If the two sections rejoin, the water comes back together again, just as it was before the pipe was split. The same occurs with charges flowing in a parallel circuit. Figure 9.1.10 shows that the charges go through one bulb or the other. This means that, although the current in the main part of the circuit remains constant, in the parallel section the current is divided between the branches. The readings on ammeters A_1 and A_2 will be the same.

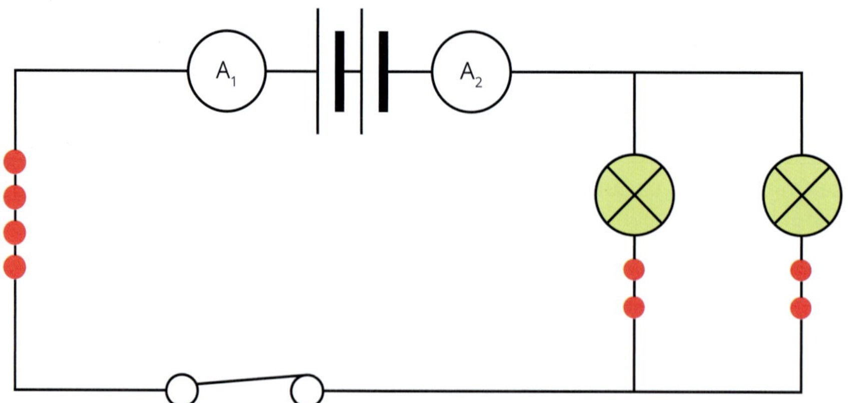

FIGURE 9.1.10 Unlike a series circuit, charges flowing in a parallel circuit have a choice of path.

The current in the main part of a parallel circuit is the sum of the currents in each branch of the circuit.

$$I_T = I_1 + I_2 + \ldots I_n$$

The voltage is the same across each branch of a parallel circuit.

Unlike a series circuit, in a parallel circuit, the voltage is not shared between resistors; the voltage is the same across each branch. This is because, although the charges take different pathways, they have the same amount of energy no matter which path they take. An advantage of parallel circuits is that bulbs connected in this way are brighter than if they were connected in series to the same voltage. The potential energy of the charges is not shared between the bulbs.

Kirchhoff's junction rule

Parallel circuits have **junctions** with more than one path for current. The behaviour of current at these points is predicted by Kirchhoff's junction rule.

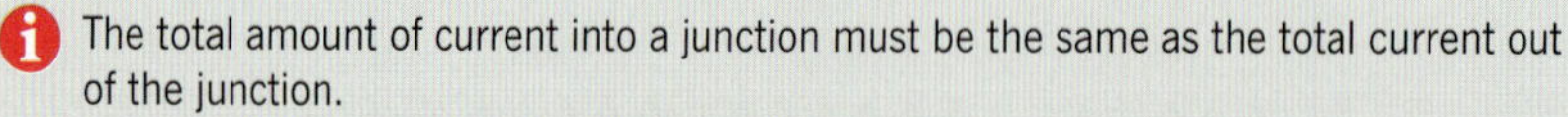

The total amount of current into a junction must be the same as the total current out of the junction.

This rule is just an extension of the idea of conservation of charge; that is, that charges cannot be created or destroyed. Although the number of electrons flowing into a junction might be very large, electrons are not created or destroyed in the junction, so the same number of electrons must flow out again. This is illustrated in Figure 9.1.11. Kirchhoff's junction rule explains how current splits in a parallel circuit. It explains why the current in the main part of the circuit is the sum of the currents in each parallel branch.

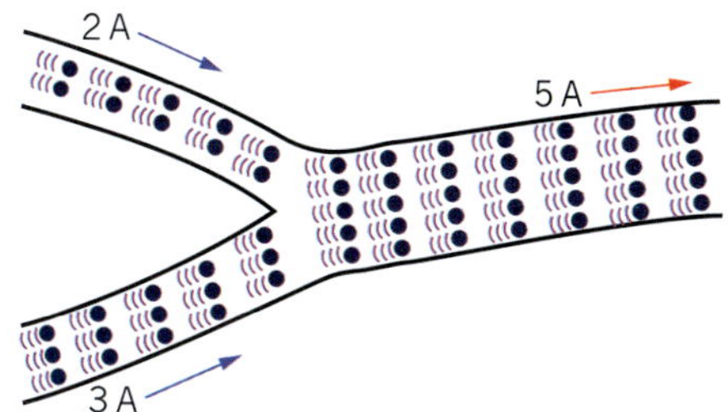

FIGURE 9.1.11 The current into any junction must be equal to the current out of it.

Equivalent parallel resistance

When additional resistors are added in a series circuit, the total resistance of the circuit increases. Higher resistance means that the circuit has less current.

In contrast, adding an additional resistor in parallel means that there is more current through the circuit because another path for charges has been added. This means that the total resistance for the circuit decreases.

Parallel circuits are more complicated than series circuits, hence the formula used to calculate the equivalent (total) resistance is more complicated.

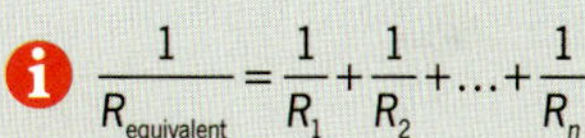

$$\frac{1}{R_{\text{equivalent}}} = \frac{1}{R_1} + \frac{1}{R_2} + \ldots + \frac{1}{R_n}$$

where $R_{\text{equivalent}}$ is the equivalent resistance, and R_1, R_2, ... R_n are the individual resistances.

Worked example 9.1.3

CALCULATING AN EQUIVALENT PARALLEL RESISTANCE

A 100 Ω resistor is connected in parallel with a 300 Ω resistor. Calculate the equivalent parallel resistance.

Thinking	Working
Recall the formula for equivalent resistance.	$\frac{1}{R_{\text{equivalent}}} = \frac{1}{R_1} + \frac{1}{R_2} + \ldots + \frac{1}{R_n}$
Substitute in the given values for resistance.	$\frac{1}{R_{\text{equivalent}}} = \frac{1}{100} + \frac{1}{300}$
Solve for $R_{\text{equivalent}}$.	$\frac{1}{R_{\text{equivalent}}} = \frac{3}{300} + \frac{1}{300}$ $= \frac{4}{300}$ $R_{\text{equivalent}} = \frac{300}{4}$ $= 75.0\,\Omega$

Worked example: Try yourself 9.1.3

CALCULATING AN EQUIVALENT PARALLEL RESISTANCE

A 20.0 Ω resistor is connected in parallel with a 50.0 Ω resistor. Calculate the equivalent parallel resistance.

Notice that in Worked example 9.1.3, the equivalent resistance was smaller than the smallest individual resistance. This is because adding a resistor provides an additional pathway for current. As there is more current, the resistance of the circuit has been effectively reduced.

PHYSICSFILE

Kirchhoff's contributions

Both the junction rule discussed here and the loop rule described earlier were first discovered by the German physicist Gustav Kirchhoff (1824–87). Kirchhoff also made important contributions in the fields of spectroscopy, thermochemistry and the study of black-body radiation. He worked with Robert Bunsen, the German chemist who developed the Bunsen burner.

Gustav Kirchhoff discovered the rules that underpin our understanding of how electric circuits work.

If you consider the smallest resistor in any parallel combination, say, the 20 Ω resistor in Worked example: Try yourself 9.1.3, the addition of the 50 Ω resistor in parallel with it allows the current an extra pathway and therefore it is easier for the charges to flow through the combination. The equivalent resistance of the pair must be less than the 20 Ω alone.

The equivalent (total) resistance of a set of resistors connected in parallel will always be smaller than the smallest resistor in the set.

In a parallel circuit:

$$R_{\text{equivalent}} < R_{\text{smallest resistor}}$$

Worked example 9.1.4

USING EQUIVALENT PARALLEL RESISTANCE FOR CIRCUIT ANALYSIS

Find an equivalent parallel resistance to calculate the current supplied by the 12 V cell in the parallel circuit shown. Also find the current in each resistor.

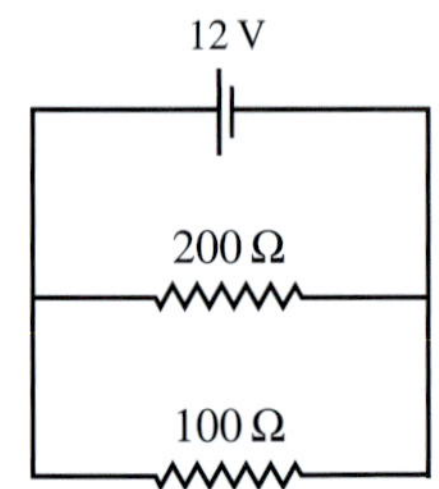

Thinking	Working
Recall the formula for equivalent parallel resistance.	$\frac{1}{R_{\text{equivalent}}} = \frac{1}{R_1} + \frac{1}{R_2} + \ldots + \frac{1}{R_n}$
Substitute in the given values for resistance.	$\frac{1}{R_{\text{equivalent}}} = \frac{1}{100} + \frac{1}{200}$
Solve for $R_{\text{equivalent}}$.	$\frac{1}{R_{\text{equivalent}}} = \frac{2}{200} + \frac{1}{200}$ $= \frac{3}{200}$ $R_{\text{equivalent}} = \frac{200}{3}$ $= 66.7\,\Omega$
Use Ohm's law to calculate the current in the circuit. To calculate I, use the voltage of the power supply and the total resistance.	$I_{\text{circuit}} = \frac{V}{R} = \frac{12}{66.7} = 0.18\,\text{A}$
Use Ohm's law to calculate the current through each separate resistor. Remember that the voltage through each resistor is the same as the voltage of the power supply, 12 V in this case.	100 Ω resistor: $I_{100} = \frac{V}{R} = \frac{12}{100} = 0.12\,\text{A}$ 200 Ω resistor: $I_{200} = \frac{V}{R} = \frac{12}{200} = 0.060\,\text{A}$
Use the junction rule to check the answers.	$I_{\text{circuit}} = I_{100} + I_{200}$ $0.18\,\text{A} = 0.12\,\text{A} + 0.060\,\text{A}$ This is correct, so the answers are reasonable.

Worked example: Try yourself 9.1.4

USING EQUIVALENT PARALLEL RESISTANCE FOR CIRCUIT ANALYSIS

Use an equivalent parallel resistance to calculate the current in the parallel circuit below and through each resistor of the circuit.

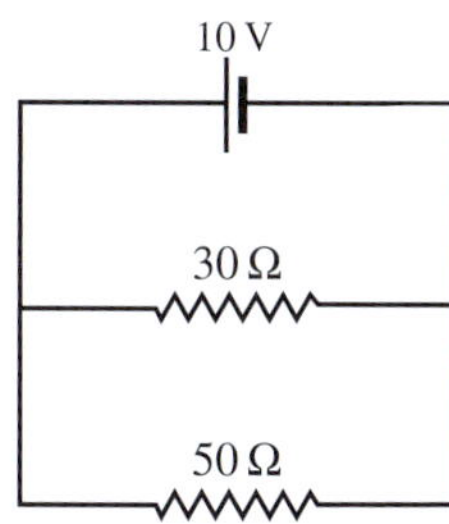

COMPLEX CIRCUIT ANALYSIS

Some circuits combine elements of series wiring and parallel wiring. A general strategy for analysing these circuits is to reduce the complex circuit to a single equivalent resistance to determine the current drawn by the circuit. It is then possible to step back through the process of simplification to analyse each section of the circuit as needed.

Worked example 9.1.5

COMPLEX CIRCUIT ANALYSIS

Calculate the potential difference across and current through each resistor in the circuit below.

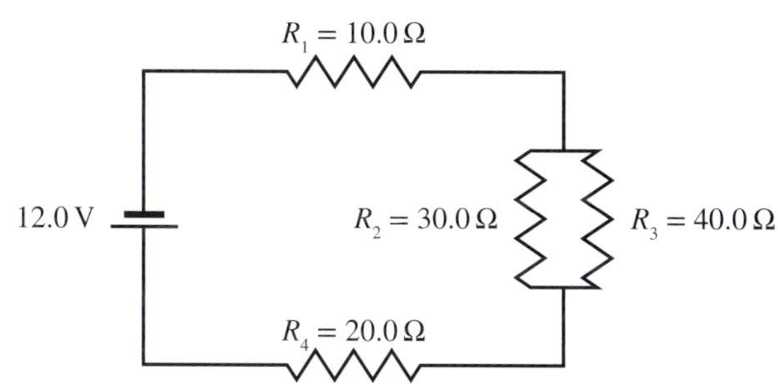

Thinking	Working
Find an equivalent parallel resistance for the parallel resistors. The equivalent resistance of these should be less than the smaller resistor; that is, smaller than 30 Ω.	$\frac{1}{R_{\text{equivalent}}} = \frac{1}{R_2} + \frac{1}{R_3}$ $= \frac{1}{30.0} + \frac{1}{40.0}$ $= \frac{4}{120} + \frac{3}{120}$ $= \frac{7}{120}$ $R_{\text{equivalent}} = \frac{120}{7}$ $= 17.1\,\Omega$
Find an equivalent series resistance for the circuit, as the circuit can now be thought of as three resistors in series: 10.0 Ω, 17.1 Ω and 20.0 Ω.	$R_{\text{equivalent}} = 10.0\,\Omega + 17.1\,\Omega + 20.0\,\Omega$ $= 47.1\,\Omega$

Thinking (*continued*)	Working (*continued*)
Use Ohm's law to calculate the current in the circuit. Use the supply voltage and total series resistance to do this calculation.	$V = IR$ $I = \frac{V}{R} = \frac{12.0}{47.1}$ $= 0.255\text{ A}$
Use Ohm's law to calculate the potential difference across each resistor (or parallel group of resistors) in series. (Note that the potential difference across R_2 is the same as that across R_3 as they are in parallel.)	$V = IR$ $V_1 = 0.255 \times 10.0 = 2.55\text{ V}$ $V_{2\text{-}3} = 0.255 \times 17.1 = 4.36\text{ V}$ $V_4 = 0.255 \times 20.0 = 5.10\text{ V}$ Check: $2.55 + 4.36 + 5.10 \approx 12.0\text{ V}$ (with some slight rounding error) This confirms that the loop rule holds for this circuit.
Use Ohm's law where necessary to calculate the current through each resistor.	$I_1 = I_4 = 0.255\text{ A}$ $I = \frac{V}{R}$ $I_2 = \frac{4.36}{30.0} = 0.145\text{ A}$ $I_3 = \frac{4.36}{40.0} = 0.109\text{ A}$ Check: $0.145 + 0.109 \approx 0.255\text{ A}$ (with some slight rounding error) This confirms that the junction rule holds for this section.

PHYSICSFILE

Identical resistors in parallel

Where identical resistors are placed in parallel, the total resistance of the combination can be found by simply dividing the value of one of the resistors by the number of resistors.

For example, three 12 Ω resistors connected in parallel would have an equivalent resistance of 4 Ω.

$$R_{\text{equivalent}} = \frac{12}{3} = 4\,\Omega$$

The three 12 Ω resistors in parallel could be replaced with a single 4 Ω resistor.

Similarly, the equivalent resistance of two 10 Ω resistors placed in parallel would be 5 Ω.

Worked example: Try yourself 9.1.5

COMPLEX CIRCUIT ANALYSIS

Calculate the potential difference across and the current through each resistor in the circuit below.

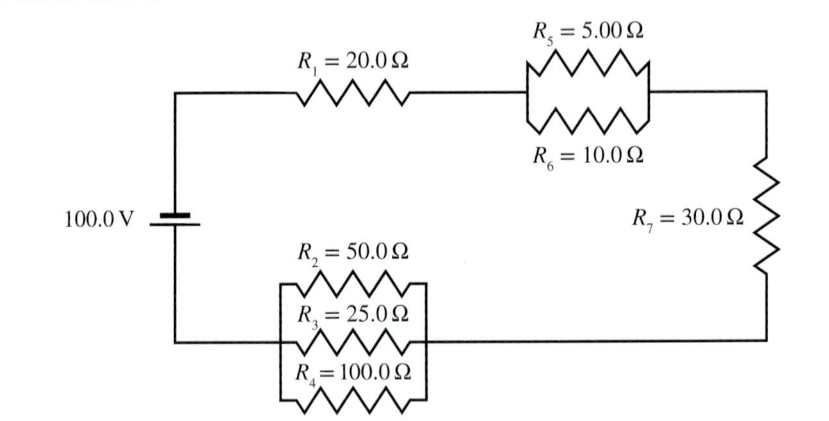

CASE STUDY

Superconductors

Materials are classified as conductors, semiconductors or insulators when their electrical properties are taken into consideration. Scientists of the late 1800s also knew that the resistivity, hence resistance, of a material was temperature dependent and that this included some materials that were insulators at room temperature.

Further research on this led to the discovery of a new material classification of superconductors. Superconductivity occurs at very low temperatures at which the resistivity of a material is found to be zero and no interior magnetic field exists within it. Present research is trying to find materials that are superconductors at higher and near-room temperatures.

In 1908, Dutch physicist Kamerlingh Onnes (1853–1926) was the first to liquefy helium at 4.2 K (–269°C). Using this, he began investigating the resistance of pure metals at very low temperatures.

He immersed a solid wire of mercury into liquid helium and found that its resistance was indeed zero at 4.2 K. He called this the superconducting state. Theoretically, once an electric current was started in a loop maintained at such temperatures it would circulate indefinitely. For this work Onnes was awarded the 1913 Nobel Prize in Physics.

Other metals were soon found to become superconductors at extremely low temperatures; for example, aluminium at 1.2 K and lead at 7.9 K. However, not all metals exhibit superconductivity.

The mechanism by which a metal's electrical resistance drops to zero was not understood until 1957 when the American physicists John Bardeen, Leon Cooper and Robert Schrieffer published a paper that used quantum mechanics, a theory not known in Onnes's era, to explain the phenomenon. This is now known as the BCS theory, and for their work Bardeen, Cooper and Schrieffer were awarded the 1972 Nobel Prize in Physics.

Materials classified as high temperature (above 77 K) superconductors have also been found. These use liquid nitrogen instead of liquid helium as it is cheaper to produce, much easier to handle and more readily available. The eventual goal, and the focus of current research, is to discover a material that will become superconducting at room temperature. Such a material would have many applications and would not require all the equipment and expense required to maintain it at low, or extremely low, temperatures.

The application of the extremely powerful and special magnetic properties of many superconducting materials has seen technological advancement in many areas. Some examples include transportation using magnetic levitation (Maglev trains, Figure 9.1.12), non-invasive medical diagnosis (MRI, Figure 9.1.13), cancer treatment using proton therapy, determining atomic structures of molecules (NMR), magnetic confinement fusion reactors, land-mine detectors, electricity generation and transmission, as beam steering and focusing magnets in particle accelerators (the Large Hadron Collider, Figure 9.1.14) and in the development of supercomputers.

FIGURE 9.1.12 Maglev train. This part of Japan's rail system uses superconductors to provide the magnetic levitation of the train and carriages.

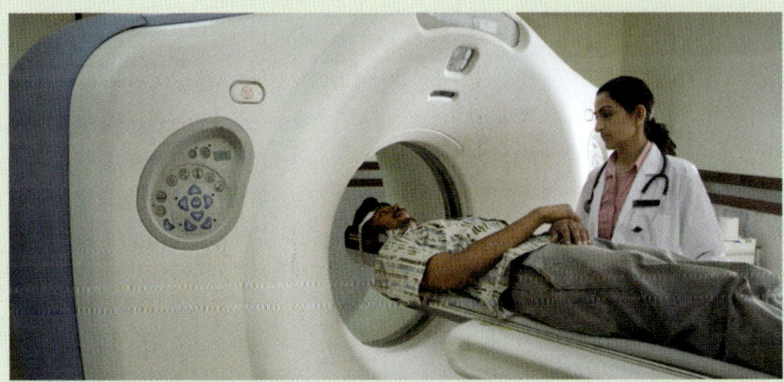

FIGURE 9.1.13 Magnetic resonance imaging (MRI) devices use superconducting wire and electromagnets cooled with liquid helium to produce detailed images.

FIGURE 9.1.14 Two LHC magnets before connection. The blue cylinders contain the magnets and a liquid helium system that cools them so they become superconducting.

RESISTORS AND POWER

A particular combination of resistors will draw different amounts of power, depending on whether the resistors are wired in series or parallel. In general, since resistors in parallel circuits will draw more current than resistors in series circuits, parallel circuits use more power than series circuits containing the same resistors.

Recall from Chapter 8 that the equation for power is:

$$P = VI$$

where P is the power (W), V is the voltage (V) and I is the current (A).

Worked example 9.1.6

COMPARING POWER IN SERIES AND PARALLEL CIRCUITS

Consider a 100 Ω and a 300 Ω resistor wired in parallel with a 12 V cell. Calculate the power drawn by these resistors. Compare this to the power drawn by the same two resistors when wired in series.

Thinking	Working
Calculate the equivalent resistance for the parallel circuit.	$\frac{1}{R_{equivalent}} = \frac{1}{R_1} + \frac{1}{R_2}$ $= \frac{1}{300} + \frac{1}{100}$ $= \frac{1}{300} + \frac{3}{300}$ $= \frac{4}{300}$ $R_{equivalent} = \frac{300}{4}$ $= 75.0\,\Omega$
Calculate the total current drawn by the parallel circuit.	$V = IR$ $I = \frac{V}{R} = \frac{12}{75} = 0.16\,A$
Use the power equation to calculate the power drawn by the parallel circuit.	$P = VI$ $= 12 \times 0.16 = 1.92\,W$
Calculate the equivalent resistance for the series circuit.	$R_{equivalent} = R_1 + R_2 + \ldots + R_n$ $= 100 + 300$ $= 400\,\Omega$
Calculate the total current drawn by the series circuit.	$V = IR$ $I = \frac{V}{R} = \frac{12}{400} = 0.03\,A$
Use the power equation to calculate the power drawn by the series circuit.	$P = VI$ $= 12 \times 0.03 = 0.36\,W$
Compare the power drawn by the two circuits.	$\frac{P_{parallel}}{P_{series}} = \frac{1.92}{0.36} = 5.33$ The parallel circuit draws over 5 times as much power as the series circuit.

Worked example: Try yourself 9.1.6

COMPARING POWER IN SERIES AND PARALLEL CIRCUITS

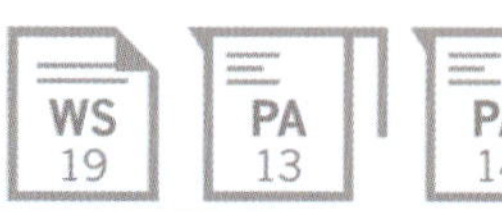

Consider a 200 Ω and an 800 Ω resistor wired in parallel with a 12 V cell. Calculate the power drawn by these resistors. Compare this to the power drawn by the same two resistors when wired in series.

CASE STUDY ANALYSIS

High power–low power

Many 240 V household electrical appliances, such as electric stoves, electric blankets and portable electric heaters, have switches or controls that allow for a variety of heat settings. Electric heaters and electric blankets often have a three-position switch with settings for low, medium and high. Rather than making different heating elements for each of the three heat settings in these appliances, the manufacturer can use two, usually identical, elements in different series and parallel combinations to obtain the three heat settings.

Given these appliances are plugged in to a 240 V supply, it is a simple matter to work out the relative power being used for each of the three settings. If it is assumed that the resistance (R) of the two elements is the same and does not change appreciably with temperature, the equivalent resistance in the three cases can be calculated.

Analysis

An electric blanket is a blanket with a pattern of high-resistance wires embedded in its structure and a hand control unit to adjust the amount of heat the blanket produces. These units usually have four settings, one at each position of a switch on the hand control. The settings for the blanket are the same as those outlined in Figure 9.1.15.

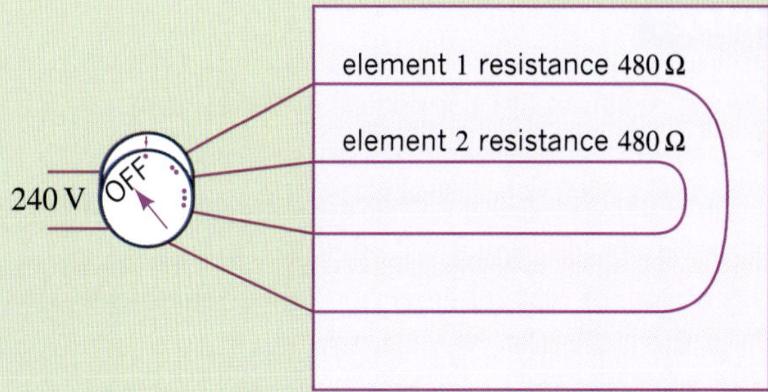

FIGURE 9.1.15 An example of the use of series and parallel combinations of resistors to achieve three heat settings for an electric blanket. These settings are as follows:
OFF: Not connected to power.
• LOW: Resistive elements connected in series.
• • MEDIUM: Only one resistive element is connected.
• • • HIGH: Two resistive elements connected in parallel.

1 Write a formula for the total resistance ($R_{equivalent}$) of the electric blanket in each of the following cases.
   a Low heat
   b Medium heat
   c High heat
2 The switch is in the position 'High' to produce maximum heat.
   a Calculate the resistance of the electric blanket.
   b Determine the power dissipated by the blanket.
   c Calculate the current drawn by the blanket.
3 Determine the ratio of the power on the highest heat setting to the power on the lowest heat setting.

PHYSICSFILE

Parallel connections

All household appliances and lights are connected in parallel. This is done for two reasons.

Figure 9.1.16 shows a TV, a washing machine, an air-conditioner and a heater connected in series. Each of these devices is designed to operate at 240 V.

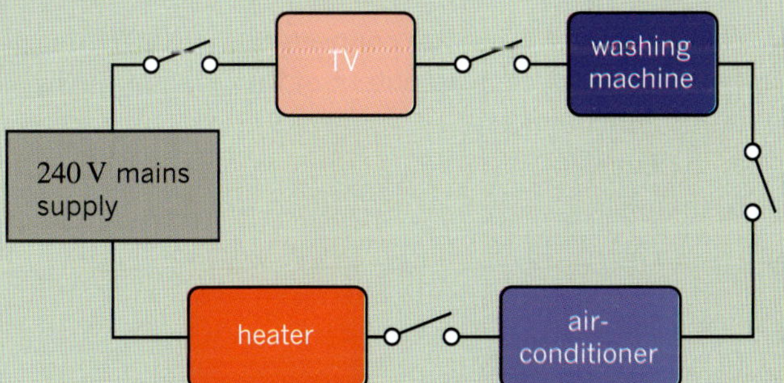

FIGURE 9.1.16 Household appliances connected in series

A circuit in which the appliances are in series poses many problems. First, all of the devices need to be switched on for the circuit to operate.

Second, the 240 V supplied to the circuit needs to be shared among all the components. Each component in the circuit would receive far less than the 240 V it requires to operate. Also, as more and more devices are added to the circuit, the share of the 240 V would become smaller. This system could never be practical.

The circuit diagram in Figure 9.1.17 shows how same devices connected in parallel.

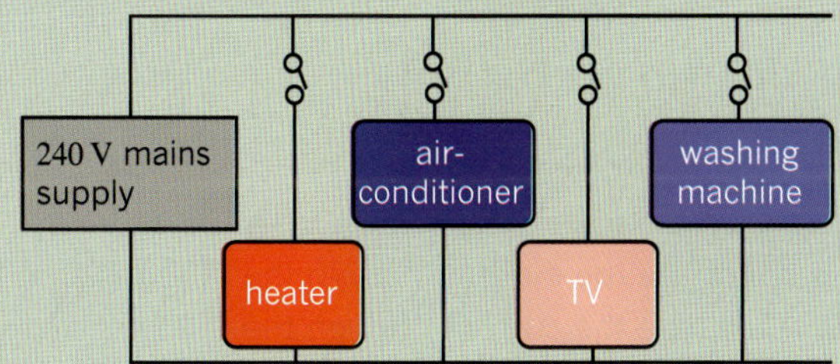

FIGURE 9.1.17 Household appliances connected in parallel

Each device in the parallel circuit receives the same voltage, 240 V. Each device can be independently switched on or off without affecting the others, and more devices can be added to this system without affecting the operation of the others.

9.1 Review

SUMMARY

- When resistors are connected in series, the:
  - current through each resistor is the same
  - sum of the potential differences is equal to the potential difference provided to the circuit
  - equivalent resistance, $R_{equivalent}$, is equal to the sum of the individual resistances.
- Parallel circuits allow individual components to be switched on and off independently.
- When resistors are connected in parallel, the:
  - voltage across each resistor is the same
  - current is shared between the resistors
  - equivalent resistance is given by the equation:

$$\frac{1}{R_{parallel}} = \frac{1}{R_1} + \frac{1}{R_2} + \ldots + \frac{1}{R_n}$$

- Complex circuit analysis may require the calculation of both equivalent series and equivalent parallel resistances.
- A parallel circuit generally draws more power than a series circuit using the same resistors.

KEY QUESTIONS

Knowledge and understanding

1 Two 20 Ω resistors are connected in series with a 6.0 V battery. What is the voltage drop across each resistor?

A 0.3 V
B 3.0 V
C 6.0 V
D 12 V

2 Consider the series circuit below.

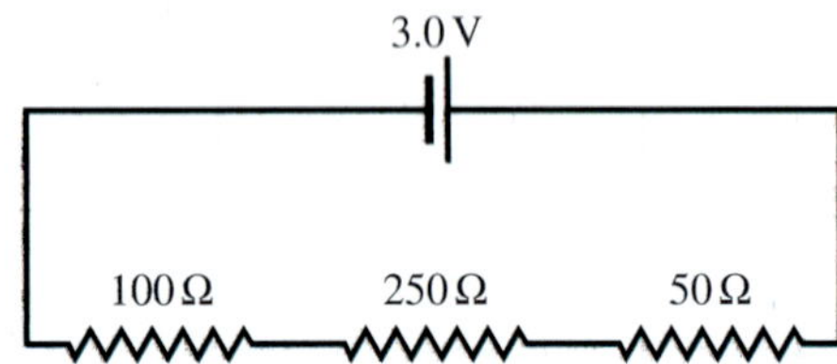

a Calculate the current in the circuit. Give your answer correct to two significant figures.
b Calculate the potential difference across the 100 Ω resistor in the circuit.

3 Two equal resistors are connected in parallel and are found to have an equivalent resistance of 68 Ω. Calculate the resistance of each resistor.

Analysis

4 A 20 Ω resistor and a 10 Ω resistor are connected in parallel to a 5.0 V battery. Give your answers correct to two decimal places.

a Calculate the current drawn from the battery.
b Calculate the current through the 20 Ω resistor.
c Calculate the current through the 10 Ω resistor.

5 A 40 Ω resistor and a 60 Ω resistor are connected in parallel to a battery, with 300 mA through the 40 Ω resistor.

a Calculate the voltage of the battery.
b Calculate the current through the 60 Ω resistor.

6 Calculate the potential difference across, and the current through, each resistor in the circuit below.

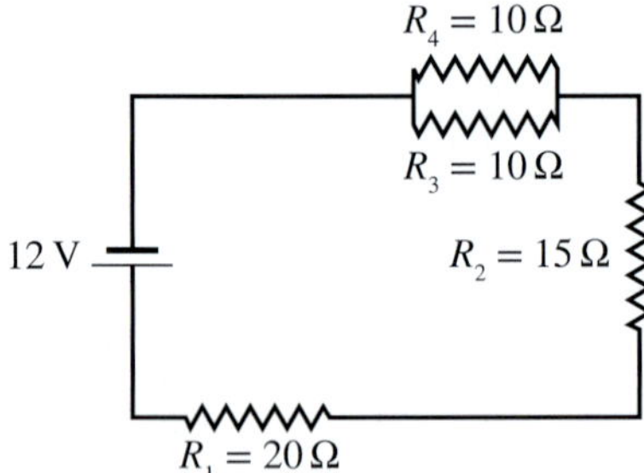

7 Calculate the equivalent resistance of the combination of resistors shown below.

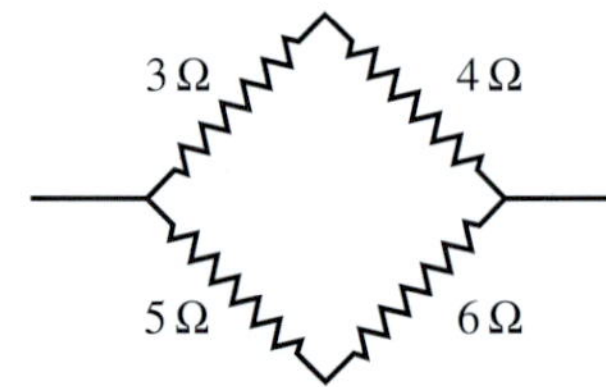

8 Four 20 Ω light bulbs are connected to a 10 V battery. What is the total power output of the circuit if the light bulbs are connected:

a in series?
b in parallel?

9.2 Using electricity

Electricity is a convenient and versatile form of energy. The electrical appliances in homes transform electrical energy into other forms. Devices such as televisions, laptops, tablet computers and mobile phones contain complex electronic circuitry. However, the components that make up those circuits are reasonably simple. Understanding how the individual components work is essential to understanding how more complex circuits operate.

TRANSDUCERS

A **transducer** is a device that receives a signal in the form of one type of energy and converts it into another form of energy. For example, a microphone is a transducer that converts sound energy into an electrical signal. A microphone is a type of input transducer because it takes one form of energy (sound) and converts it into electrical energy. An output transducer is one that converts the electrical energy back into another form of energy.

Most complex electric circuits can be understood as a combination of input and output transducers separated by some form of signal-processing circuitry. The flow chart in Figure 9.2.1 explains the process in a simple way.

For example, the input transducer could be a simple battery with a switch while the output transducer is a light bulb. The wires connecting the battery, switch and light bulb are the signal-processing circuitry; in this case, the signal is simply carried from one transducer to another without really being processed.

Other, more complex, circuits may use sophisticated input transducers that can produce a range of electrical outputs. The signal-processing circuitry may amplify the signal or change it in some way while the output transducer could produce light, sound or almost any other type of energy imaginable.

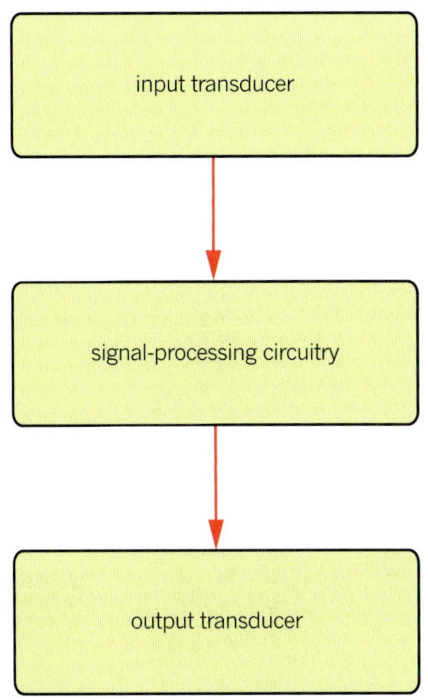

FIGURE 9.2.1 An electric circuit can be modelled as an input transducer followed by signal processing circuitry and then an output transducer.

A transducer is a device that transforms energy from one form to another.

Signal-processing components

One of the simplest forms of signal-processing is to vary the amount of current or voltage in a circuit using a variable resistor or **potentiometer**. Although potentiometers can take a variety of forms (as seen in Figure 9.2.2), the basic design is always the same, and consists of a three-terminal resistor with a sliding or rotating contact called the wiper.

If the potentiometer is connected using just one end and the wiper, it acts as a simple variable resistor: the further the wiper slides or rotates, the greater the resistance value.

A potentiometer can also be used to divide voltage. This means that if a potential difference is applied across the two ends, then the wiper can be used to access any voltage between these two extremes. For example, if the ends of a potentiometer were connected to a 12 V battery, then setting the wiper in the middle position would give a 6 V difference between it and either end; that is it would divide the total resistance of the potentiometer in half.

Just as a potentiometer can divide voltage, so too can a **voltage divider** circuit. A voltage divider circuit, like the one shown in Worked example 9.2.1 on page 256, is simply a series circuit with two or more components. The circuit is called a voltage divider because the voltage supplied to the circuit is shared (or divided) between the components in the circuit. In this worked example the components are two fixed resistors, but these could be replaced with any of the variable resistors described in this section.

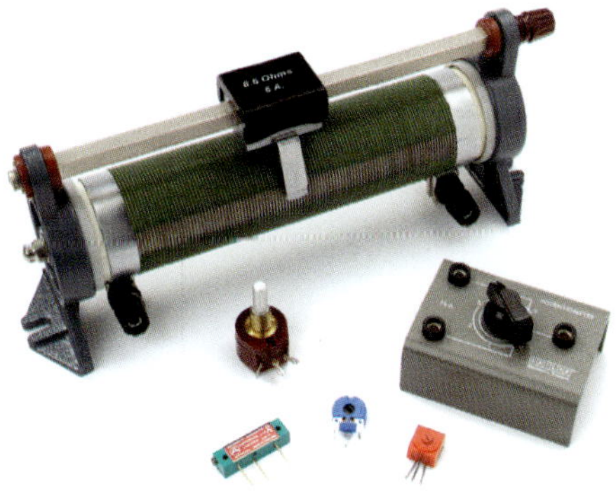

FIGURE 9.2.2 Different types of potentiometers

Worked example 9.2.1

VOLTAGE DIVIDER

A voltage divider is constructed from a 12 V battery and two 20 Ω resistors as shown. Calculate the voltage output, V_{out}, of the circuit.

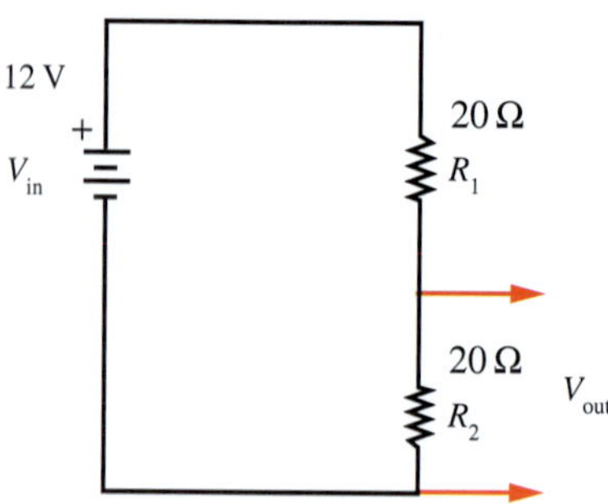

Thinking	Working
Calculate the total resistance of the circuit.	$R_{equivalent} = 20 + 20$ $= 40\,\Omega$
Calculate the current in the circuit.	$I = \frac{V}{R}$ $= \frac{12}{40}$ $= 0.3\,A$
Calculate the potential difference across the second resistor.	$V_{out} = IR$ $= 0.3 \times 20$ $= 6\,V$

Worked example: Try yourself 9.2.1

VOLTAGE DIVIDER

A voltage divider is constructed from a 12 V battery, a 40 Ω resistor and a 20 Ω resistor as shown. Calculate the voltage output, V_{out}, of the circuit.

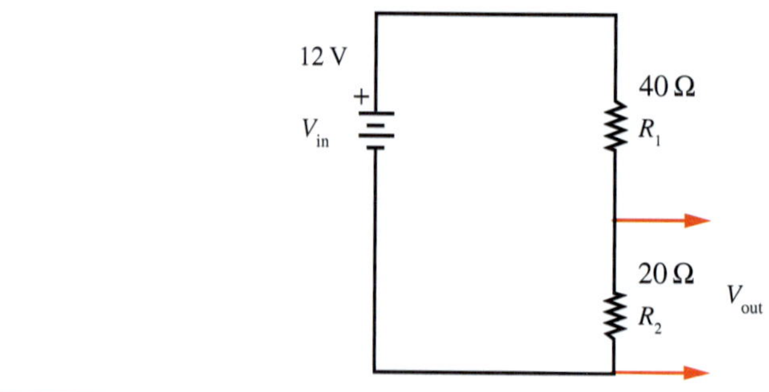

Potentiometers are widely used as light dimmer switches, and for volume control on sound systems.

Consider Worked example 9.2.1. Each of the resistors provides exactly the same amount of resistance to the circuit. They divide the supply voltage exactly in half. If, instead, one of the resistors was larger than the other, the potential difference (voltage drop) would be greater over the larger resistor ($V = IR$).

If you completed Worked example: Try yourself 9.2.1, you would have seen that the voltage was not divided equally between the resistors as they were not equal in size.

If one of the fixed resistors is replaced with a variable resistor, it is possible to constantly change the way the supply voltage is divided. This is what happens in a dimmer switch.

Figure 9.2.3 shows what happens to the brightness of a light bulb connected in series with a potentiometer set at low resistance and then at high resistance. When the potentiometer is set to its lowest resistance, the potential difference across the bulb will be large and the bulb will glow brightly (Figure 9.2.3(a)). When the resistance of the potentiometer increases, the potential difference across the light bulb decreases and the bulb is dim (Figure 9.2.3(b)).

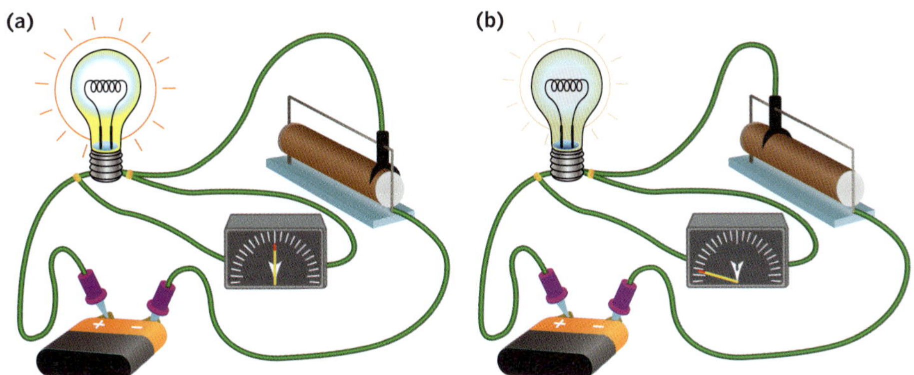

FIGURE 9.2.3 (a) When the potentiometer is at its lowest resistance, the bulb glows brightly. (b) As the resistance of the potentiometer increases, the bulb glows less brightly.

It is important to note that, as well as voltage, the current in a circuit is also affected by changing the resistance of the potentiometer. In the circuit shown in Figure 9.2.3(b), when the resistance of the potentiometer is very high, the current in the circuit decreases.

Input transducers

A **thermistor**, such as the one shown in Figure 9.2.4, is a type of variable resistor. The resistance of the thermistor, however, depends on its temperature.

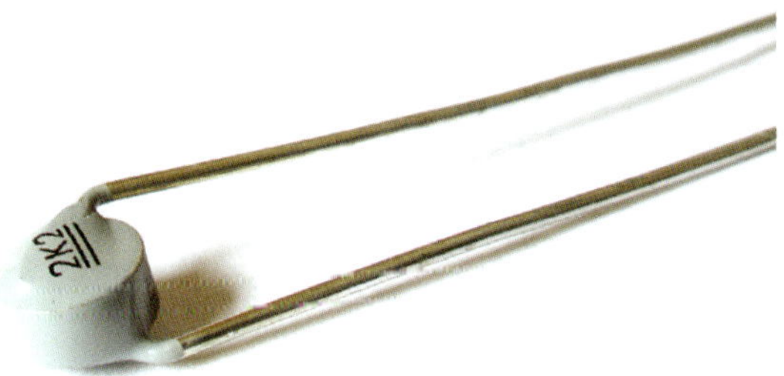

FIGURE 9.2.4 The resistance of a thermistor changes with its temperature.

Thermistors are usually categorised as either NTC (negative temperature coefficient) or PTC (positive temperature coefficient).

As the labels imply, the resistance of an NTC thermistor decreases as its temperature increases. The resistance of a PTC thermistor increases as its temperature increases. An incandescent light bulb is an example of a PTC thermistor. As its filament gets hotter with increasing current, its resistance increases.

Figures 9.2.5(a) and 9.2.5(b) show how the resistances of these devices varies with temperature. In comparison, Figure 9.2.5(c) shows how a fixed (non-variable) resistor, in ideal conditions, will maintain constant resistance even when the temperature changes.

NTC thermistors are semiconductor devices and are the most common type used in applications where electronic control is needed in response to temperature change. Thermistors are used in a wide variety of temperature-control applications such as in refrigerators, toasters, coffee-makers, electric circuits or engines.

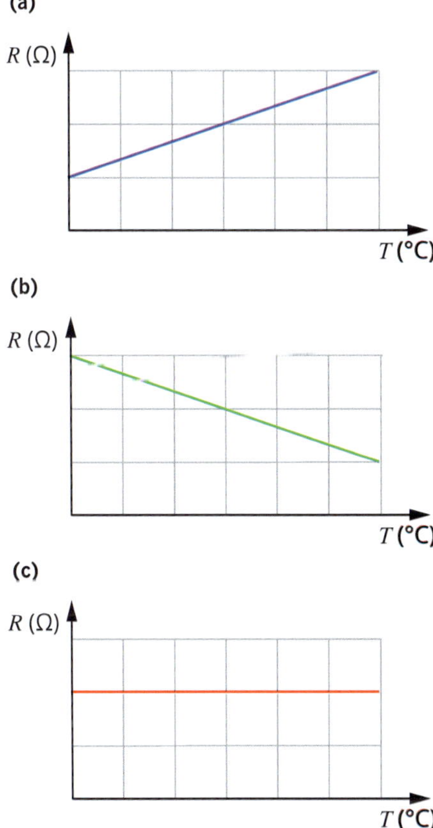

FIGURE 9.2.5 (a) The resistance of a PTC thermistor (e.g. a light bulb) increases with increasing temperature. (b) The resistance of an NTC thermistor decreases with increasing temperature. (c) The resistance of an ideal resistor is independent of the temperature.

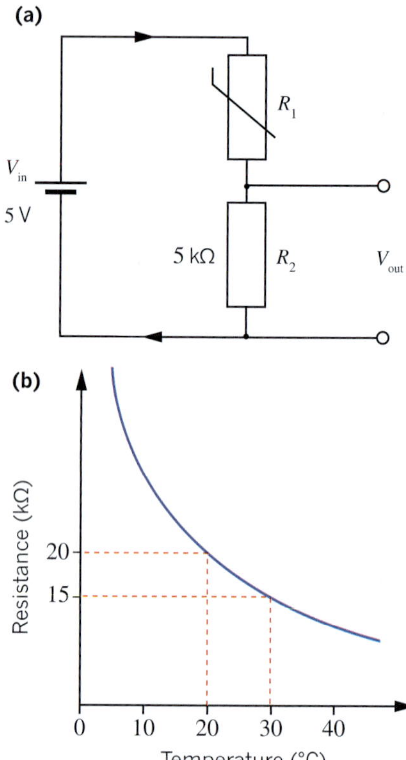

FIGURE 9.2.6 (a) A thermistor (R_1) is connected in series with a fixed resistor (R_2). This is a voltage divider circuit as the voltage supplied by the battery will be divided between the two resistors. (b) The characteristic curve of the thermistor shows how its resistance varies with temperature.

Transducers in voltage divider circuits

Voltage divider circuits, like the one in Worked example 9.2.1 on page 256, can contain fixed or variable resistors. Pages 256–257 described how a potentiometer used in this kind of circuit could act as a dimmer switch.

Thermistors can also be connected in voltage divider circuits with other resistors. To know how a thermistor changes its resistance with temperature, manufacturers provide a characteristic curve. Figure 9.2.6 shows how the thermistor can be connected in a voltage divider circuit, as well as the characteristic curve for a particular thermistor.

The circuit shown in Figure 9.2.6(a) can be used as a temperature sensor. When the voltage over the fixed resistor reaches a given value, it can signal for a heater or cooling system to switch on or off.

Worked example 9.2.2

THERMISTOR IN A VOLTAGE DIVIDER CIRCUIT

A voltage divider circuit includes a thermistor (R_1) and a fixed resistor (R_2). The characteristic curve of the thermistor and the circuit are shown in Figure 9.2.6. Using the graph and the information included on the circuit diagram, determine:

a the resistance of the thermistor at 30°C	
Thinking	**Working**
The resistance of the thermistor can be read straight from the graph at the point where the temperature is 30°C.	$R = 15\,\text{k}\Omega$

b the current in the circuit	
Thinking	**Working**
Find the equivalent resistance of the circuit. Note that the fixed resistor is 5 kΩ.	$R_{\text{equivalent}} = 15\,\text{k}\Omega + 5\,\text{k}\Omega$ $= 20\,\text{k}\Omega$
Find the current.	$I = \frac{V}{R}$ $= \frac{5.0}{20\,000}$ $= 0.25\,\text{mA}$

c the output potential difference, V_{out}.	
Thinking	**Working**
Use Ohm's law to calculate the voltage across the fixed resistor.	$V = IR$ $= 0.00025 \times 5000$ $= 1.25\,\text{V}$

Worked example: Try yourself 9.2.2

THERMISTOR IN A VOLTAGE DIVIDER CIRCUIT

A voltage divider circuit includes a thermistor (R_1) and a fixed resistor (R_2). The characteristic curve of the thermistor and the circuit are shown in Figure 9.2.6. Using the graph and the information included on the circuit diagram, determine:

a the resistance of the thermistor at 20°C

b the current in the circuit

c the output potential difference, V_{out}.

Light dependent resistor

FIGURE 9.2.7 The resistance of a light dependent resistor changes depending on how much light is falling on it.

The device shown in Figure 9.2.7 is an input transducer that provides a variable resistance in the circuit. As its name suggests, a **light dependent resistor** (also known as an LDR or photoresistor) is a transducer whose resistance depends on the amount of light falling on it. An LDR is usually designed to have a very high resistance in the dark (for example a few million ohms). In light, their resistance can drop to as low as a few hundred ohms.

As the resistance of the LDR changes, it has the effect of changing the potential difference (voltage drop) as well. As the resistance of the LDR increases, so does the potential difference across it. A decrease in resistance means the voltage drop decreases as well. This is because in any device with high resistance, more voltage (electrical potential energy) is needed to make the charges flow through the device.

Increasing the resistance of a device also affects the current in the circuit. It is much harder for charges to flow through a device with a high resistance. So, as the resistance of the LDR increases, the current through it decreases. As current is the same at every point of a series circuit, the current in the entire circuit will decrease.

LDRs have a range of applications in devices such as camera light meters, street lights and night lights. They can also be used to detect when a person blocks a beam of light shining across a doorway, triggering a buzzer, doorbell chime or stopping lift doors from closing.

The circuit diagrams in Figure 9.2.8 show how an LDR can be connected in a circuit to achieve different outputs. When the LDR and the light bulb are connected in series, the bulb switches on only if light falls on the LDR. This is because they are connected in series and charge must flow through the LDR in order for it to also flow through the bulb. The only way charge will flow through the LDR is if its resistance is low, and this happens when light falls upon it.

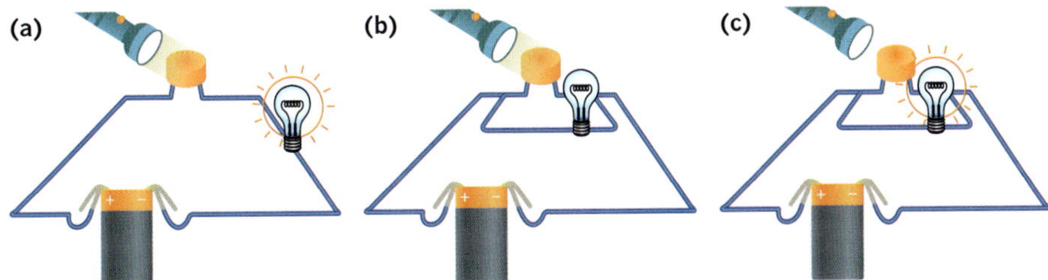

FIGURE 9.2.8 (a) When connected in series, the light bulb lights up when the LDR is illuminated. (b) and (c) When connected in parallel, a light bulb lights up when the LDR is not illuminated.

When the LDR and the light bulb are connected in parallel, the bulb switches on only if the LDR is not illuminated (lit up). The current divides between the branches of parallel circuits but if one path has very high resistance, the charges will 'choose' the other path: the path of least resistance. When the LDR is not illuminated, it has very high resistance and so the current passes through the other branch, lighting the bulb.

It is more likely that an LDR is used in a circuit in which a light bulb switches on when the ambient (surrounding) light is low. In these cases, the LDR needs to be connected in parallel with the bulb.

When provided with a characteristic curve, the operation of an LDR in a voltage divider circuit can be analysed. The characteristic curve provides information about the resistance of the LDR; then using Ohm's law, current and voltage can be calculated.

Worked example 9.2.3

LDR IN A VOLTAGE DIVIDER CIRCUIT

A voltage divider circuit includes an LDR (R_1) and a fixed resistor (R_2).

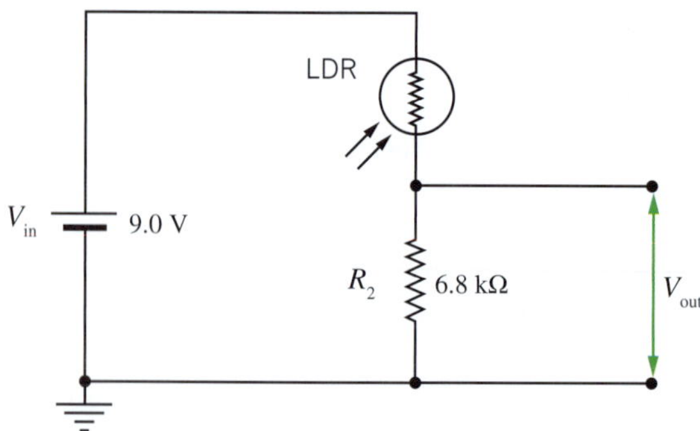

Using the information included on the circuit diagram and the fact that the LDR has a resistance of 20 kΩ at a light intensity of 100 mW m^{-2}, determine:

a the total resistance of the circuit	
Thinking	**Working**
The total resistance equals the resistance of the LDR plus the 6.8 kΩ fixed resistor.	$R_{equivalent} = 20 + 6.8 = 26.8\,\text{k}\Omega$

b the current in the circuit	
Thinking	**Working**
Find the current using Ohm's law.	$I = \frac{V}{R}$ $= \frac{9.0}{26800}$ $= 0.34\,\text{mA}$

c the output potential difference, V_{out}.	
Thinking	**Working**
Use Ohm's law to calculate the voltage across the fixed resistor.	$V = IR$ $= 0.00034 \times 6800$ $= 2.3\,\text{V}$

Worked example: Try yourself 9.2.3

LDR IN A VOLTAGE DIVIDER CIRCUIT

A voltage divider circuit includes an LDR (R_1) and a fixed resistor (R_2).

Using the information included on the circuit diagram in Worked example 9.2.3 and the fact that the LDR has a resistance of 3.0 kΩ at a light intensity of 2000 mW m^{-2}, determine:

a the total resistance of the circuit

b the current in the circuit

c the output potential difference, V_{out}.

Diodes

A **diode** is a semiconductor device that has the special property that it will only allow electric current through it in one direction. Several types of diodes are shown in Figure 9.2.9. This simple function has a wide range of applications. Diodes are often used to protect current-sensitive components or circuits from stray currents. They are used in circuits that convert alternating current (AC) into direct current (DC), such as in in phone and computer chargers. This is a particularly important application since the household mains supply (available through a power point) is alternating current while most electronic devices operate on direct current. Another use is to prevent current moving in reverse through a faulty solar panel and overheating it.

It is important to note that diodes are non-ohmic; that is, their I–V graphs are not linear (not a straight line). The resistance across a diode is not constant for all potential differences. Worked example 9.2.4 shows a typical I–V graph for a diode.

FIGURE 9.2.9 Different types of diodes

Worked example 9.2.4

DIODES

A student is investigating the current–voltage characteristics of a diode using the circuit shown in diagram (a). The I–V graph for this diode is illustrated in (b). The current in the circuit is measured as 4.5 mA.

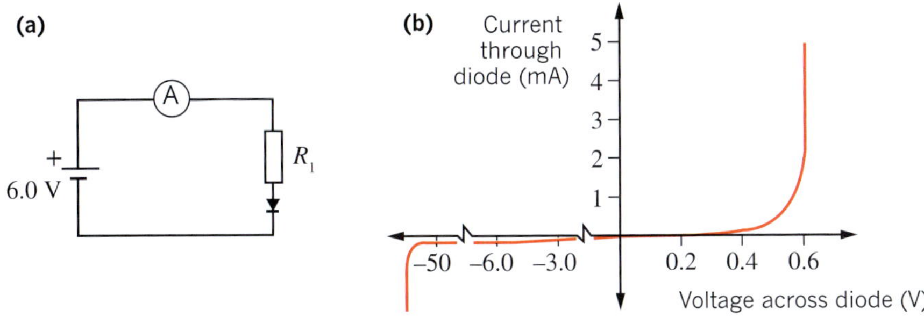

a What is the potential difference across the diode?	
Thinking	**Working**
The potential difference across the diode can be read directly from the I–V graph.	At 4.5 mA the voltage across the diode will be 0.6 V.

b What is the value of R_1?	
Thinking	**Working**
The battery voltage is 6.0 V, the voltage drop across the diode is 0.6 V and the current is 4.5 mA. Use this information to find R_1.	Voltage across $R_1 = 6.0 - 0.6$ $= 5.4\text{ V}$ $R_1 = \frac{V}{I}$ $= \frac{5.4}{4.5 \times 10^{-3}}$ $= 1200\,\Omega$

Worked example: Try yourself 9.2.4

DIODES

A student is investigating the current–voltage characteristics of a diode using the circuit shown in diagram (a). The I–V graph for this diode is illustrated in (b). The current in the circuit is measured as 5.0 mA.

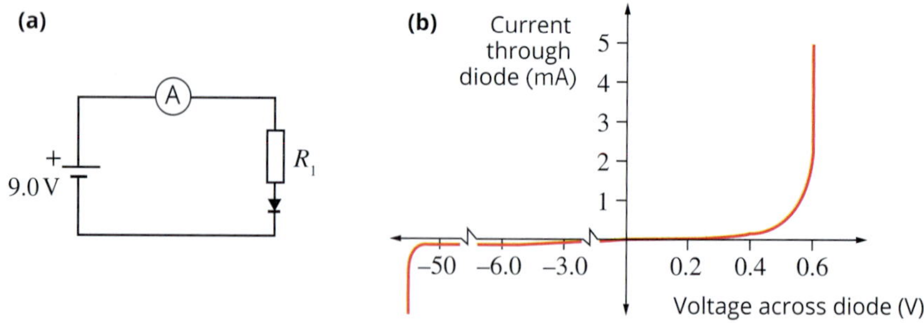

a What is the potential difference across the diode?

b What is the value of R_1?

Output transducers

A **light-emitting diode** (LED) is a specialised diode that emits light when a current passes through it (Figure 9.2.10). As with other types of diodes, LEDs conduct electricity in only one direction. The difference with LEDs is that when correctly connected in the circuit (called the forward-biased state) the recombination of charges at the semiconductor junction releases energy as photons of visible light. LEDs are available in all colours of the spectrum from infrared to ultraviolet. The colour of the light emitted depends on the type and ratio of the exotic compounds used to make the semiconductor materials.

FIGURE 9.2.10 LEDs are available in many colours.

FIGURE 9.2.11 A group of LEDs can be used to produce the same amount of light as a traditional incandescent light bulb, but use much less energy to produce it.

LEDs have many lighting applications from infrared of TV remote controls to car headlights, through household lighting to high-intensity spotlights. They are now commonly used to replace incandescent light bulbs, halogen bulbs and fluorescent tubes, as they are much more efficient and have a much longer operating life. When a comparison is made between bulb types with the same light output, LEDs use about 10–15% of the electricity of incandescent bulbs and about 75% of the energy of fluorescent tubes. LEDs are solid state devices, so they are much smaller and more durable than comparable light sources and have a lifetime of about 20 times that of an incandescent bulb and 5 times that of a fluorescent tube, with little or no environmental waste issues at their end of life (Figure 9.2.11).

Many of the common LEDs have a switch-on voltage (the voltage at which the current through the diode increases very rapidly) in the range from 1.8 to 3.6 V and a forward current rating of 10 to 25 mA. The switch-on voltage depends on the semiconductor material used (which determines the colour of the LED). Infrared and red LEDs have a lower switch-on voltage than blue and ultraviolet LEDs

Figure 9.2.12 shows a typical circuit used to activate an LED. Note there is a resistor in series with the LED that is acting as a voltage divider. This is a current-limiting resistor, the value of which will be determined to ensure the forward-biased current is within the maximum allowed rating for the diode. If the current exceeds the maximum allowed for a particular diode, permanent damage to it will occur.

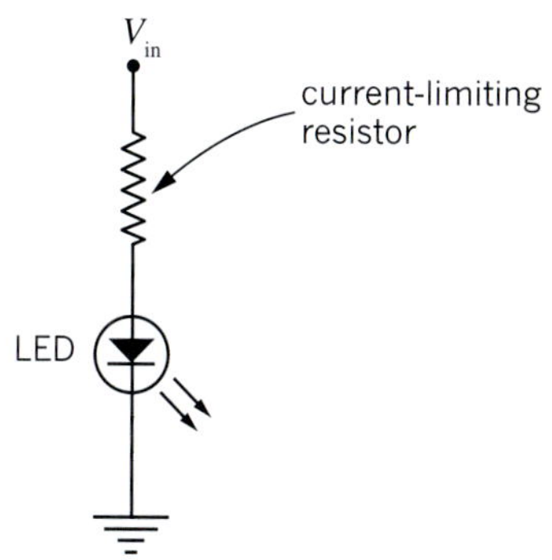

FIGURE 9.2.12 An LED driver circuit

Worked example 9.2.5

LEDS

An LED has the following optimum operating characteristics:

$I_d = 20\,\text{mA}$ when $V_d = 1.8\,\text{V}$

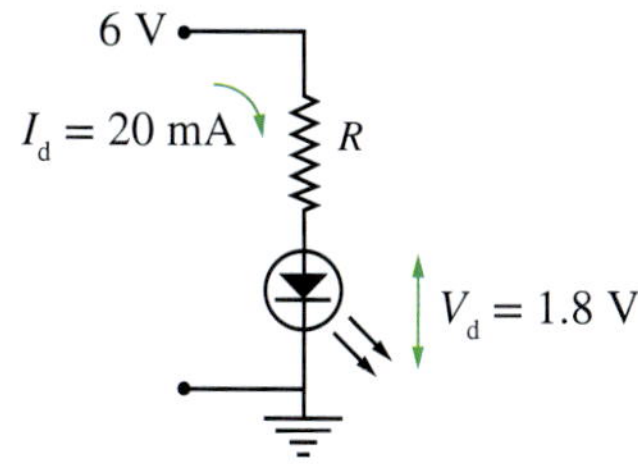

Determine the value of the current-limiting resistor (R) if the current through the diode is to be limited to 20 mA when powered by a 6 V battery.

Thinking	Working
The 6 V is divided between the resistor R and the LED. Work out the voltage drop across R.	$V_R = 6 - 1.8 = 4.2\,\text{V}$
Use Ohm's law to calculate the value of R.	$R = \frac{V}{I}$ $= \frac{4.2}{20 \times 10^{-3}}$ $= 210\,\Omega$

Worked example: Try yourself 9.2.5

LEDS

An LED has the following optimum operating characteristics:

$I_d = 30\,\text{mA}$ when $V_d = 1.9\,\text{V}$

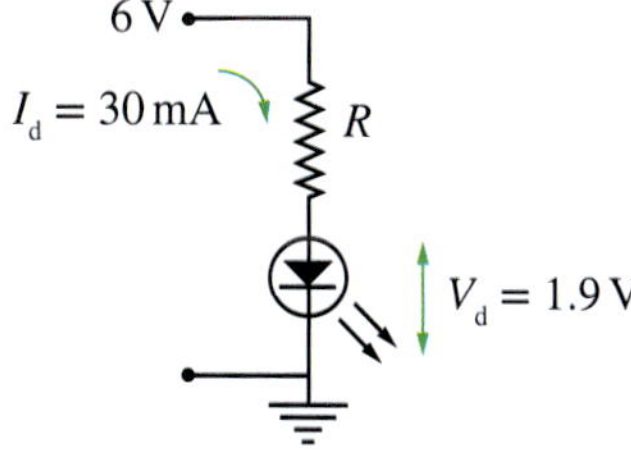

Determine the value of the current-limiting resistor (R) if the current through the diode is to be limited to 30 mA when powered by a 6 V battery.

PHYSICSFILE

Diodes and LEDs

Diodes and LEDs only allow current to pass in one direction. As such, it is essential that they are placed in circuits in the correct way. When a diode is placed in the circuit in the correct way, it is said to be forward biased. If a diode is connected the wrong way around, it will shut down part or all of a circuit; therefore, manufacturers are careful to make clear the way in which these components are to be orientated.

The circuit symbol for a diode consists of a triangle and a line. The LED circuit symbol consists of a diode symbol with two arrows. The direction of the arrows shows the only direction for conventional current through the device.

The red arrow under the diode symbol shows the direction of current.

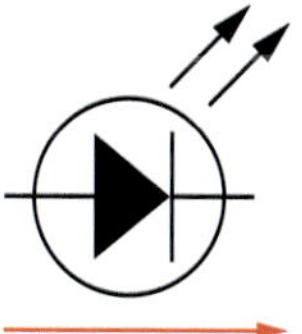

An LED circuit symbol. The direction of current in the LED is given by the red arrow.

9.2 Review

OA ✓✓

SUMMARY

- Most complex electric circuits can be understood as a combination of input and output transducers separated by signal-processing circuitry.
- A potentiometer is a three-terminal resistor with a sliding or rotating contact called the wiper.
- Potentiometers can be used to divide voltage and are used in light dimmer switches and for volume control on sound systems.
- A voltage divider circuit is a series circuit with two or more components. The voltage supplied to the circuit is shared (or divided) between the components in the circuit.
- The resistance of a thermistor depends on its temperature. Thermistors are used in a wide variety of temperature-control applications such as in refrigerators, toasters, coffee makers, electric circuits and engines.
- A light dependent resistor (LDR) is a transducer whose resistance depends on the amount of light falling on it. Usually an LDR is designed to have a very high resistance in the dark and a very low resistance in light.
- LDRs have a range of applications (e.g. camera light meters, street lights, night lights). They can be used to detect when a person's body blocks a beam of light shining across a doorway, triggering a buzzer or doorbell chime.
- A diode is a semiconductor device that will only allow electric current through it in one direction. Diodes are widely used in circuits that convert alternating current into direct current.
- A light-emitting diode (LED) is a type of diode that emits light as current passes through it.
- LEDs are increasingly used as highly energy-efficient replacements for traditional incandescent light bulbs.

KEY QUESTIONS

Knowledge and understanding

1 Copy the table below and classify the following components under the appropriate heading: diode, LED, LDR, light bulb, microphone, potentiometer, speaker, thermistor

Input transducer	Signal-processing component	Output transducer

2 What type of transducer is used in a light dimmer switch? Explain why you made this choice.

3 A student constructs a circuit with an LDR and light bulb in parallel with each other, and a battery. Explain the conditions under which the bulb will light up.

4 What type of transducers are used to produce light? Explain the advantage these have over other alternatives that produce light using electricity.

Analysis

5 A 10 V battery is connected to a voltage divider circuit that consists of a 100 Ω resistor and a 300 Ω resistor.

a Determine the voltage across the 300 Ω resistor.

b Calculate the voltage drop across the 100 Ω resistor.

6 A 24 V battery is connected to a voltage divider circuit consisting of a 1 kΩ resistor and a 2 kΩ resistor. Determine the voltage across the 2 kΩ resistor.

7 The LED in the circuit below has a switch-on voltage (V_s) of 2.0 V and an operating current of 20 mA. Determine the value of R_L for the LED to be operating correctly.

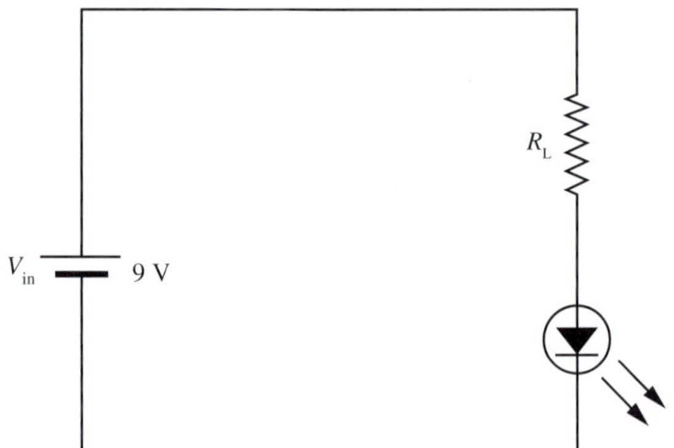

8 The following graph shows the resistance–temperature characteristics of a thermistor. A circuit uses this thermistor as part of a temperature sensor that can activate an LED whenever the temperature rises above a certain level.

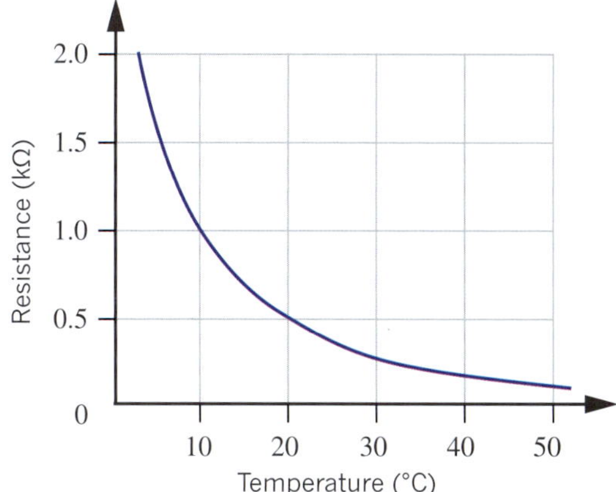

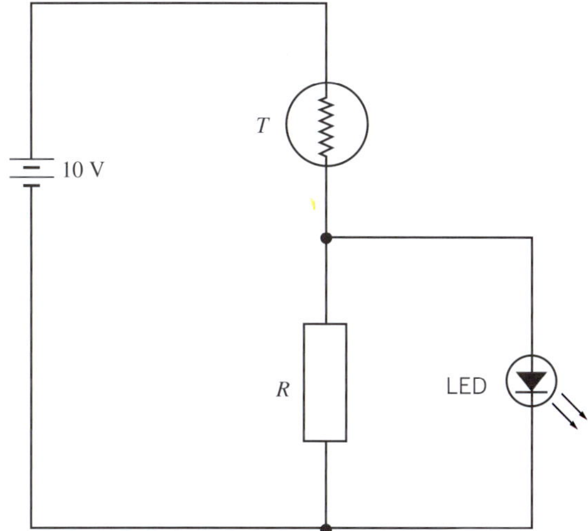

a What is the resistance of the thermistor at 20°C?

b The potential difference across the LED at 20°C is 2.5 V and the current through it is 11 mA. What is the value of *R*?

c The LED is activated by a minimum potential difference of 2.0 V across it which gives a current through it of 4.8 mA. What is the minimum temperature that will activate the LED?

9.3 Electrical safety

Homes, schools and workplaces are filled with all sorts of electrical appliances. You use these appliances every day. A scientific understanding of electricity can help you to understand how they work and how to make sure you use them safely and effectively.

ELECTRICITY IN THE HOME

The wiring for a house is much more complicated than the relatively simple series and parallel circuits considered so far. Figure 9.3.1 shows the basic structure of the electrical wiring in a house. Most appliances and power points are wired in parallel to allow them to be switched on and off independently of each other. The neutral wires from all switches and power points are connected to the same bar as the earth connection.

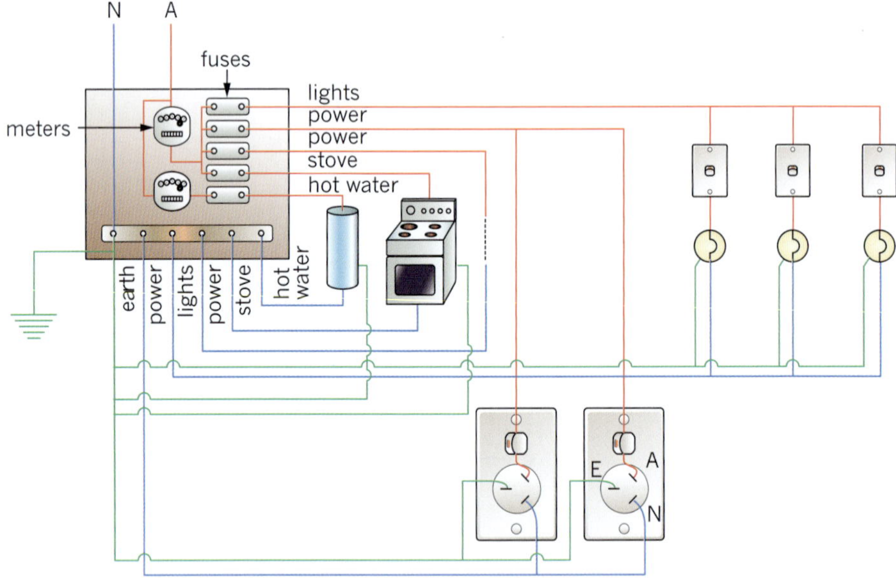

FIGURE 9.3.1 A household wiring diagram includes active (A), neutral (N) and earth (E) wires. The neutral and earth wires are connected to the same bar.

Power points in Australia have three pins. Each of these pins is connected to a different wire. Figure 9.3.1 shows the arrangement of the active, neutral and earth pins on a typical power point. The wires carrying the electric current to and from the appliance are known as the active wire (usually red or brown) and the neutral wire (usually black or blue). The third wire is an important safety feature called the earth wire (usually green or green and yellow). Figure 9.3.2 shows the corresponding active, neutral and earth pins in a power cord.

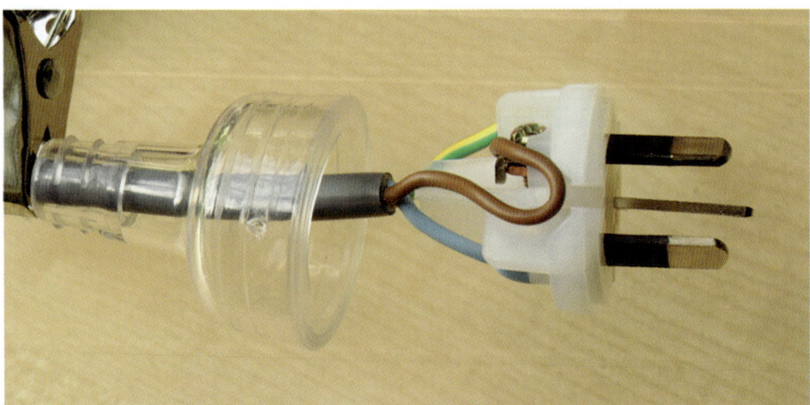

FIGURE 9.3.2 This three-pin plug shows the correct colour code used: brown for active, blue for neutral and green and yellow for earth.

The purpose of household electric circuits is to enable electrical energy to be transferred to electrical appliances, where it is transformed into a range of other useful forms of energy. For example, an electric oven converts electrical energy into heat, and fans convert electrical energy into kinetic energy. Power points give users the option of connecting their own appliances and therefore choosing the type of energy produced.

Figure 9.3.3 shows that, before being distributed to various parts of the house, the active wire passes through a meter that measures the amount of electrical energy supplied to the household.

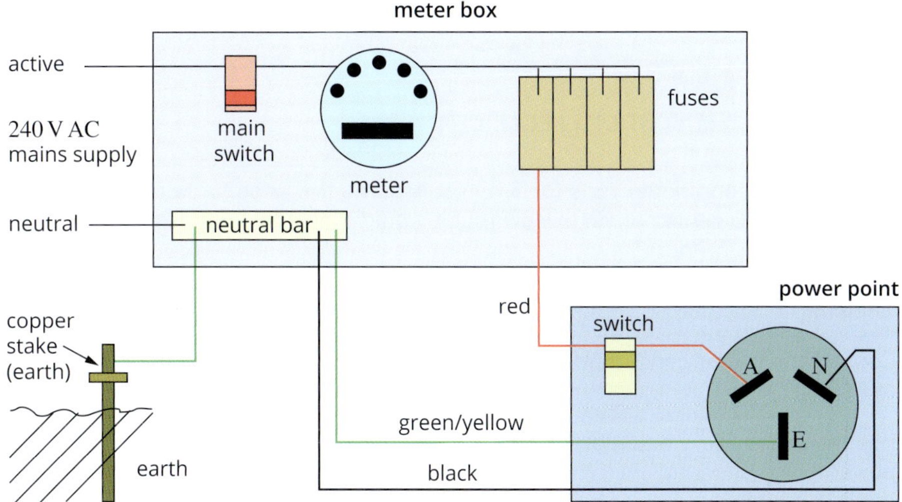

FIGURE 9.3.3 The active wire passes through the meter box before being connected to power points throughout the house. The circuit is completed by the neutral wire, which returns through the neutral bar that is also connected to earth.

AC/DC

The electric current that comes out of a power point is very different from the current that comes from a battery or electric cell (Figure 9.3.4).

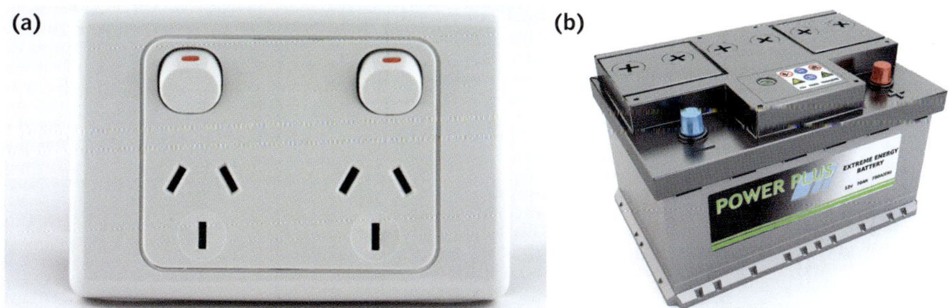

FIGURE 9.3.4 (a) A power point provides alternating current (AC) whereas (b) a car battery provides direct current (DC).

For example, the potential difference between the active and neutral pins of a household power point is 240 V. This is much higher than most electric cells or batteries can provide. In fact, 240 V is the equivalent of putting twenty 12 V car batteries together in series.

Also, because of the way household electrical energy is generated, it is delivered as an **alternating current** (AC). This means that the electrons in the wire oscillate backwards and forwards in the wire. In comparison, a battery provides **direct current** (DC), which means the electrons travel directly from one terminal of the battery to the other, in one direction only.

Fortunately, most household AC systems can be modelled using simple DC circuits.

Electricity bills

The energy consumed by a household appliance is measured in joules. It is the power, P, in watts multiplied by the time, t, in seconds.

$$E = Pt$$

If you have a look at your home electricity bill, you'll see the energy used is measured in **kilowatt hours (kW h)**. This gives a convenient number, without scientific notation, that is easy to write on your bill.

To calculate how much an appliance costs to run, multiply its power consumption in kW by the number of hours the appliance operates. Then take this number in kW h and multiply it by the cost of electricity per kW h. Complete Worked example 9.3.1 below to calculate the cost of running of some household appliances.

Worked example 9.3.1

CALCULATING THE COST OF ELECTRICITY

A 2000 W air conditioner runs for 5 hours. Assume the price for household electricity is 26 cents per kW h. How much would it cost to run this air conditioner for 5 hours?

Thinking	Working
Convert the power consumption of the appliance to kW.	$\frac{2000\text{ W}}{1000} = 2\text{ kW}$
Use the appropriate equation to multiply the power of the appliance in kW by the number of hours it operates.	$E = Pt$ $= 2 \times 5$ $= 10\text{ kW h}$
Multiply the number of kW h by the cost per kW h.	Cost $= 10 \times 0.26$ $= \$2.60$

Worked example: Try yourself 9.3.1

CALCULATING THE COST OF ELECTRICITY

A 2500 W iron is used for 2.5 hours. Assume the price for household electricity is 26 cents per kW h. How much would it cost (to the nearest cent) to use this iron for 2.5 hours?

ELECTRICAL SAFETY DEVICES

Household electrical wires can carry large amounts of energy. This means that they have the potential to do a lot of harm. The inherent danger associated with the use of electricity can be reduced using various safety devices.

Fuses and circuit breakers

Because wires heat up when current passes through them, there is a limit to how much current the wires in a house or building can safely carry. Household wiring systems are designed to prevent wires from becoming **overloaded**. Appliances that draw a lot of current, such as ovens, hot-water systems and air-conditioners, are put on separate circuits to lights and power points, and often have larger diameter wiring.

Despite these precautions, overloading can still occur, most often due to a **short circuit**. A short circuit occurs when an electric circuit contains very little resistance. This can occur in an electrical appliance when the insulation between the active and neutral wires becomes damaged and these wires are in direct contact. In household circuits, short circuits are always dangerous situations. Large amounts of current means that wires will heat up, causing insulation to melt or catch alight.

An electric current will always take the path of least resistance. The bulb in Figure 9.3.5(a) is on because the current has no alternative but to pass through the high resistance of the bulb. The bulb in Figure 9.3.5(b) does not light because the closed switch provides a zero-resistance alternative pathway for the current: a short circuit.

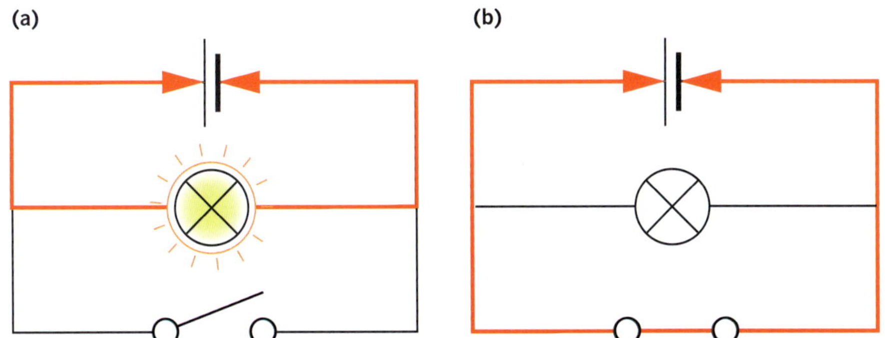

FIGURE 9.3.5 The bulb in (a) functions normally, while (b) shows how a short circuit forms through the closed switch, so the bulb does not light.

Every domestic electric circuit contains either a **fuse** or a **circuit breaker**. The function of both components is to interrupt the current if it exceeds a certain value. Unlike a fuse, a circuit breaker can be easily reset after it has been activated, whereas a fuse needs to be physically replaced once it has melted through. Both fuses and circuit breakers can be chosen for different amounts of current, since appliances such as ovens and hot-water systems might typically draw much larger amounts of current than regular power points.

Earth wires

Many household electrical appliances such as kettles, toasters and ovens have metal cases. If the active wire inside the appliance becomes loose and touches the case, then the whole case becomes electrically live. If anyone touches the case, the current will travel through their body, with possibly fatal consequences.

To prevent this, an **earth** wire is permanently connected to the metal case of the appliance, as shown in Figure 9.3.6. When the appliance is plugged in, this wire is connected via the household wiring system to the earth. This means that if the active wire touches the case, a short circuit will be created and current will immediately travel to earth. The large amount of current in this situation should trip the fuse or circuit breaker, alerting users of the appliance to the problem.

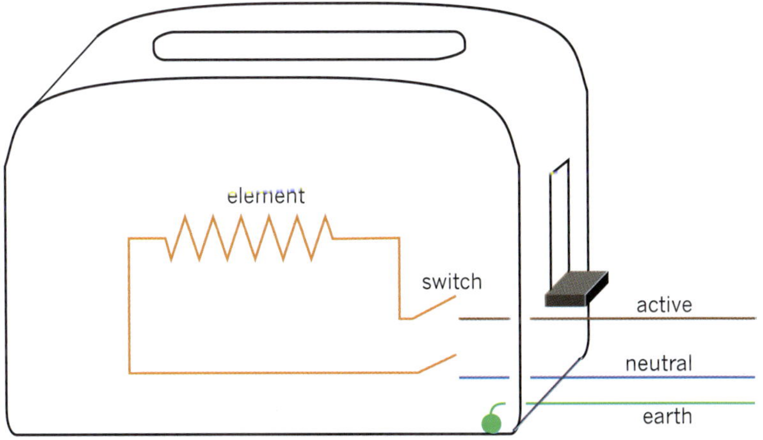

FIGURE 9.3.6 The earth wire inside a metal toaster is permanently connected to the casing.

PHYSICSFILE

Power board overload

Another common reason for circuits overloading is the overuse of power boards and double adaptors. Most power boards are designed to carry a maximum of 10 A of current. If too many high-current appliances, such as heaters, kettles and irons, are plugged into a power board, it can overheat. This may cause the insulation around the wires to melt, causing a short circuit or even a fire.

Overuse of power boards and double adaptors can cause overloading of circuits. This can cause wires and components to overheat and start a fire.

FIGURE 9.3.7 The double square symbol indicates that this appliance has double insulation and does not use an earth connection.

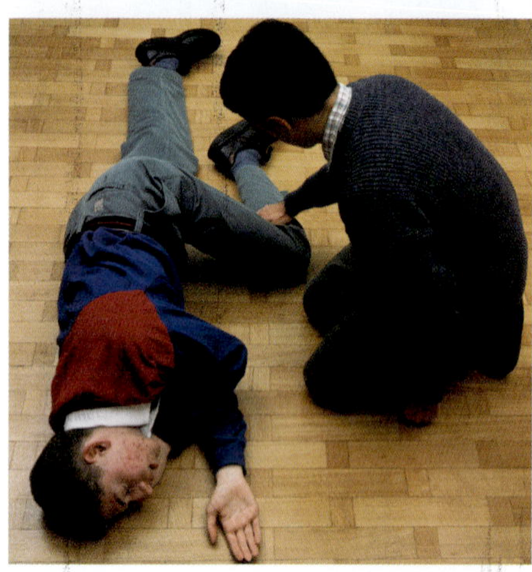

FIGURE 9.3.8 Before attempting first aid, make sure that the person is no longer in contact with the source of the electric current.

Double insulation

Another way of protecting against the possibility of a loose active wire inside an appliance is to use double insulation. This process involves using two insulating barriers to protect users. Often this is done by making the case of the appliance out of plastic. This acts as an insulating layer in addition to the plastic insulation surrounding the active wire inside the appliance. Double-insulated appliances do not need an earth wire, so their electrical plugs only have two pins. They also usually have a double square symbol on their cases, indicating that they are double insulated (Figure 9.3.7).

Residual current devices

Residual current devices, also known as RCDs or earth leakage systems, detect any difference between the current in the active wire and the current in the neutral wire. In a properly operating circuit, these two currents should be exactly the same, but in opposite directions. The most likely reason for a difference between the active and neutral currents is that some current is going to earth through a fault or, in a worst case, through a person. If this happens, the RCD is able to switch off the supply in about 20 milliseconds, hopefully preventing any serious harm.

ELECTRIC SHOCK

Despite all the safety features of modern electrical systems and appliances, each year approximately 50 Australians are killed in electrical accidents. The effects of **electric shock** depend on a number of factors including:

- the amount of current passing through the body
- its duration
- the path it takes through the body.

When attending to a victim of electrocution, it is important to first check whether they are still in contact with the electrical source. First aid should only be administered when it is safe to approach the victim (Figure 9.3.8).

The amount of current through the body if it is in good contact with a 240 V source is well above the level that will cause death. Anything that improves the contact, such as wet hands or bare feet, lowers the resistance and increases the current, potentially to life-threatening levels. Although rubber boots and gloves will increase resistance and lower the current, these must *not* be used as a form of protection in place of other sensible electrical safety precautions.

As our bodies are relatively poor conductors, electrical energy passing through our bodies is quickly converted into heat and can cause terrible internal and external burns. Table 9.3.1 shows the likely effect on the human body of a half-second electric shock at different currents.

TABLE 9.3.1 The effect of a half-second electric shock on the body

Current (mA)	Effect on the body
1	able to be felt
3	easily felt
10	painful
20	muscles paralysed—cannot let go
50	severe shock
90	breathing upset
150	breathing very difficult
200	death likely
500	serious burning, breathing stops, death inevitable

The amount of electrical energy that enters the body depends on the duration of electrocution. This is why the high voltage spark from a Van de Graaff generator is harmless: the duration of the current is about a microsecond and the total energy delivered is tiny. Table 9.3.2 shows the likely effect on the human body of a 50 mA shock for different time periods.

The path that the current takes through the body is also important in determining its effect. Our bodies are controlled by electrical impulses along the nerves, so any current into the body from an external source may interfere with our vital functions. In particular, any current passing from one arm to the other may cause the chest muscles to contract and breathing to stop. A current through the heart region can cause the muscles to become uncoordinated and the heart function stops. A brief current of about 80 mA is sufficient to cause fibrillation (irregular contraction of the heart muscle) if it is directly through the heart. This is the cause of most electrical fatalities.

TABLE 9.3.2 The effect of time on the severity of a shock

Time (s)	Effect on the body
less than 0.2	noticeable but usually not dangerous
0.2–4	significant shock, possibly dangerous
more than 4	severe shock, possible death

9.3 Review

OA ✓✓

SUMMARY

- Electrical energy use in the home is usually measured in kilowatt hours, where 1 kWh = 1000 W × 1 h.
- The effect of electrocution depends on the amount of current, its duration and the path the current takes through the body.
- The danger associated with the use of electricity in the home can be managed by using fuses, circuit breakers, double insulation and residual current devices.

KEY QUESTIONS

Knowledge and understanding

1. Explain how a fuse or circuit breaker increases household electrical safety.
2. How does double insulation of an electrical appliance increase household electrical safety?
3. What is the function of the 'earth stake' that will normally be found near a meter box?
4. A toaster cable with conductors coloured red, black and green is to be joined to another cable with brown, blue and green-yellow conductors. Peter has joined the red and blue, black and brown, and green and green-yellow. Will the toaster work normally when it is plugged in and turned on? Why is the way he has connected the cables dangerous?
5. An appliance was mistakenly wired between the active and earth instead of between the active and neutral. Explain why that is a very dangerous thing to do, even though the appliance will appear to work normally.

Analysis

6. Convert 10 kWh into J.
7. One of the values given in the information below is incorrect. Use your knowledge of kWh to determine which value it is.
   A 750 W air conditioner uses 0.75 kWh of energy in 1 hour. A typical price for household electricity is 27 cents per kWh. Therefore this air conditioner would cost approximately $10 to run for 5 hours.
8. Determine the current through a person with dry hands and a total contact resistance of 100 kΩ when they touch a 240 V live wire.

Chapter review

KEY TERMS

alternating current
circuit breaker
diode
direct current
earth
electric shock
equivalent resistance
fuse
junction
kilowatt hour (kW h)
light dependent resistor
light-emitting diode
overload
parallel circuit
potentiometer
residual current device
series circuit
short circuit
thermistor
transducer
voltage divider

REVIEW QUESTIONS

Knowledge and understanding

1 Two resistors, R_1 and R_2, are wired in series. Which of the following gives the equivalent series resistance for these two resistors?

A $R_{equivalent} = R_1 + R_2$

B $\frac{1}{R_{equivalent}} = \frac{1}{R_1} + \frac{1}{R_2}$

C $R_{equivalent} = R_1 - R_2$

D $R_{equivalent} = R_1 \times R_2$

2 Explain how a circuit breaker improves electrical safety in the home.

3 Sketch a circuit diagram showing how three 20 Ω resistors can be connected to have a total equivalent resistance of 30 Ω.

4 An LDR is classified as a transducer. State what type of transducer it is and explain how it functions as one.

5 Which of the following would be most likely to cause serious electrocution harm to a human being? Explain your answer.

A high voltage spark from a Van de Graaff generator; duration = 1 ms

B 3 mA current; duration = 0.5 s

C 50 mA current; duration = 0.1 s

D 50 mA current; duration = 4.5 s

6 Define the term 'power' in relation to a component of an electric circuit. Two identical resistors are connected first in series and then in parallel, with both circuits connected to the same power supply and switched on for the same length of time. Identify which circuit would consume more energy and explain why you made this choice.

7 Why is the shock received when a finger touches a live wire likely to be less severe than the shock received by a person who touches a live wire with a pair of uninsulated pliers?

8 It is said that a fuse protects property and a safety switch or RCD protects lives. Explain why this statement is true.

9 Explain why the components in household electric circuits are wired in parallel. You must use a diagram in your answer.

10 Explain why an LED requires a series resistor connected to it.

11 Which of the following components—LED, thermistor, LDR, diode—would be used to control the temperature in a refrigerator? Explain your choice.

12 Why is it that there are only two cables coming into the house from the street and yet power points always have three connections?

13 The function of a fuse is to burn out, and thus turn off the current, if the circuit is overloaded. Explain why is it always placed in the active wire at the meter box.

Application and analysis

14 An electric circuit is constructed as shown in the diagram. Use the information given below about the circuit to answer the following questions.

The electric cell provides 6.0 V.

The equivalent resistance $R_{equivalent}$ of the circuit is 10.8 Ω.

R_2 has a resistance of 8.0 Ω.

The equivalent resistance of resistors R_1 and R_2 is 4.8 Ω.

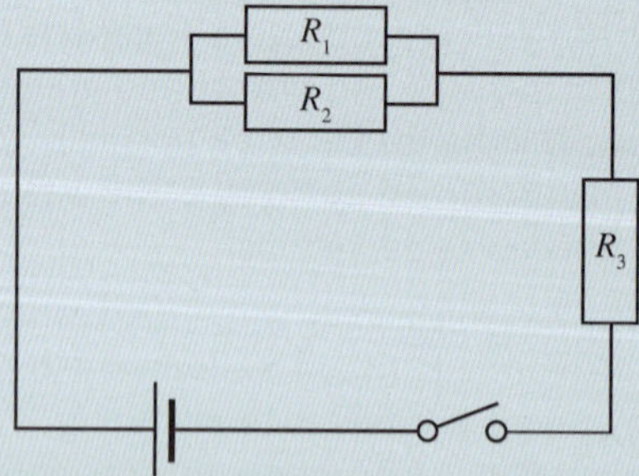

a Find the value of R_3.

b Find the current through R_3.

c Find the potential difference across the parallel pair R_1 and R_2.

d Find the current through R_2.

e Find the current through R_1.

f Find the value of R_1.

15 Eight equal-value resistors are connected between points P and Q. The value of each of these resistors is 18.0Ω.

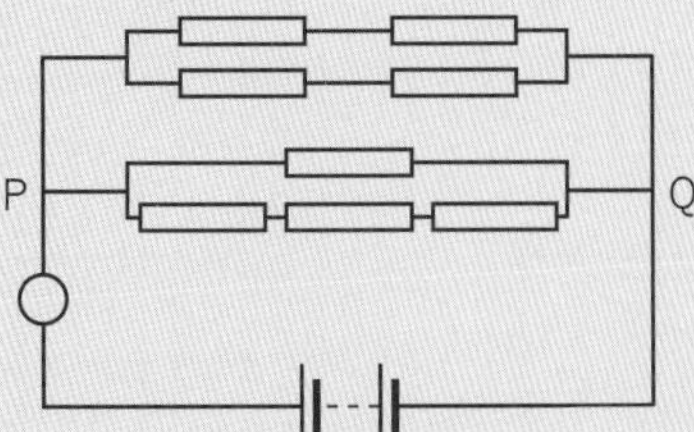

a The circle in the circuit diagram represents a meter. Identify which type of meter this should be and what property of the circuit it would measure.

b Calculate the total equivalent resistance of the circuit, ignoring any effects from the battery and the meter.

16 A 12V battery is connected to a voltage divider consisting of three resistors with values of 400Ω, 700Ω and 900Ω. What would be the voltage across the 700Ω resistor?

17 A voltage divider is constructed using a 20V battery, a 400Ω resistor and a variable resistor, *R*, as shown in the diagram. Determine the resistance of *R* when V_{out} = 5V.

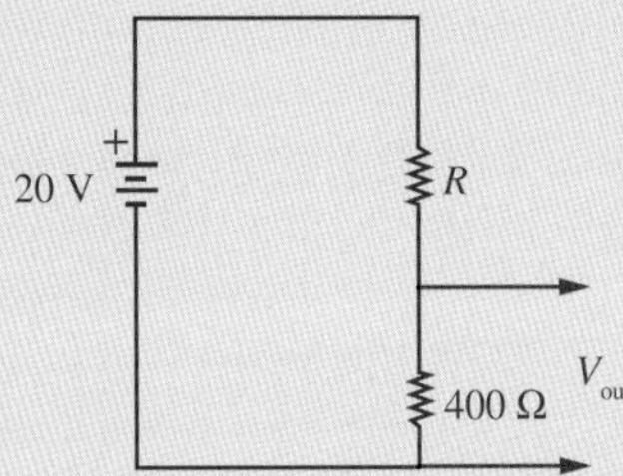

18 A set of Christmas tree lights consists of sixteen 30Ω light bulbs connected in series. A householder has two of these sets that can be used to decorate their Christmas tree and one 24V power supply.

a Calculate the power consumed by one of these sets of bulbs when connected to the power supply.

The householder wants to use both sets of bulbs connected to the one power supply but is unsure whether to connect them in series or in parallel.

b Calculate the power they would consume if the two strands are connected in series with each other.

c Calculate the power they would consume if the two strands are connected in parallel with each other.

d Explain, with reasons, which of the connections would you advise them to use.

19 A 3kW heating unit runs for 4 hours. If household electricity costs 30 cents per kWh, how much does it cost to run the heater for this time?

20 Consider the following circuit in which three resistors, R_1, R_2 and R_3, are connected in parallel. Assume that R_1 = 100Ω, R_2 = 200Ω and R_3 = 600Ω. The battery supplies 120V.

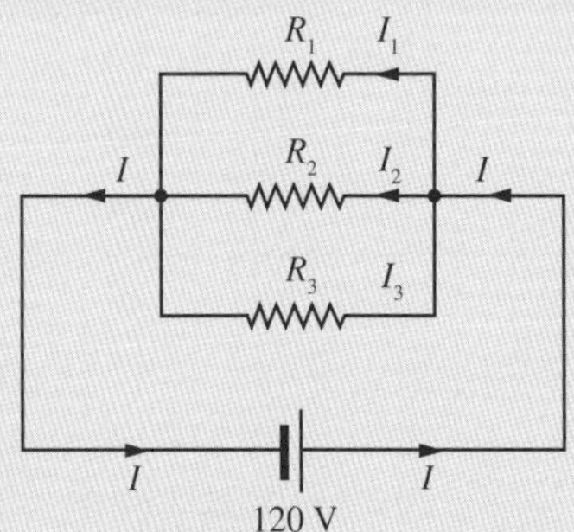

a Calculate the equivalent resistance, $R_{equivalent}$, in the circuit.

b Calculate the line current, *I*, in the circuit.

c Determine the branch currents I_1, I_2 and I_3.

d What is the power output, *P*, of the battery?

e Calculate the total power consumed by all of the resistors in the circuit.

21 In the simple LDR light-detector circuit shown below, V_{out} is to be used to activate an alarm when the ambient light reaches a certain level. At this particular light level, the resistance of the LDR is 200Ω. The alarm activates whenever V_{out} is above the trigger level.

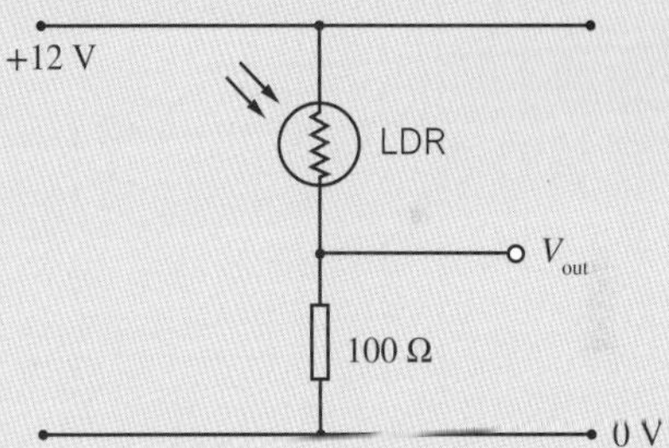

a What is the value of V_{out} at which the alarm should activate?

b Will the alarm activate when the light is above or below the particular level of concern? Explain your answer.

c When it is very dark, what would you expect V_{out} to become?

continued over page

22 A student measures the resistance, R, of a thermistor for various temperatures, T, and records the results in the table below.

T (°C)	R (kΩ)	T (°C)	R (kΩ)
2.5	20.0	30	2.5
5.0	15.0	40	2.0
10.0	10.0	45	1.8
20.0	5.0	50	1.0

a What is a thermistor?

b Plot a graph of R versus T and describe whether the relationship between resistance and temperature is linear.

23 The graph below shows the resistance versus temperature characteristics for a particular thermistor. Determine the temperature of the thermistor in the circuit if V_{out} is 1 V.

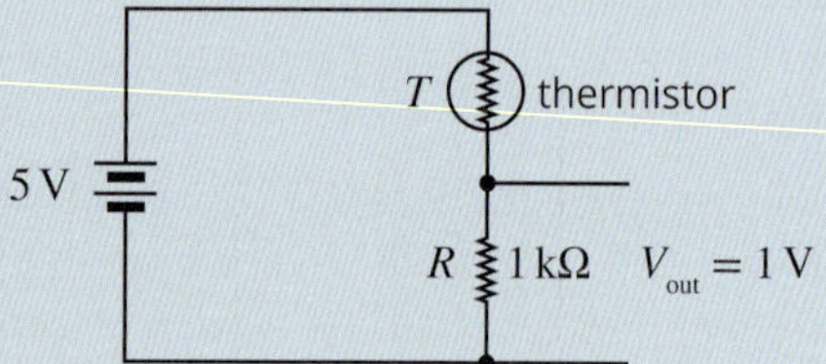

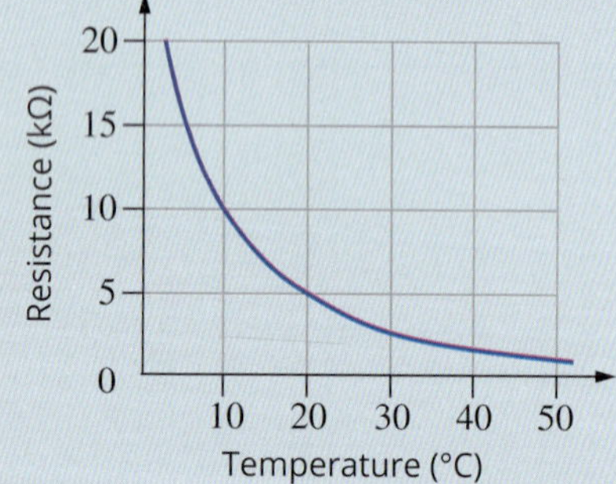

24 Three 100 Ω resistors are connected as shown. The maximum power that can safely be dissipated in any one resistor is 25 W.

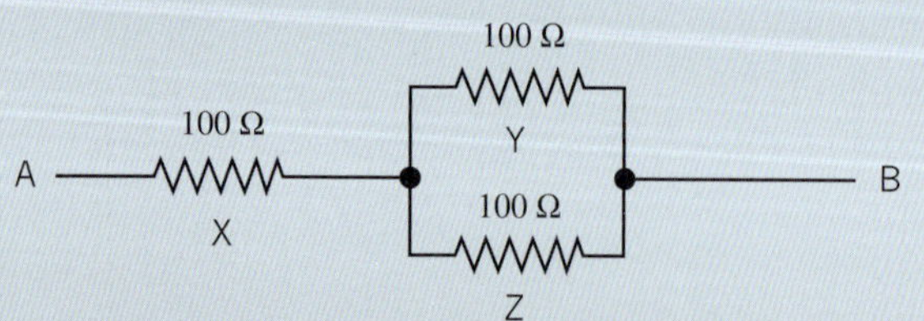

a What is the maximum potential difference that can be applied between points A and B?

b What is the maximum power that can be dissipated in this circuit?

25 All LEDs in circuits (i) and (ii) below are identical. Each LED has a switch-on voltage (V_s) of 2 V and draws a current of 20 mA for optimal light production. (If the current is much smaller, the LED light output is too dim. If the current is much larger, the LEDs overheat and burn out.)

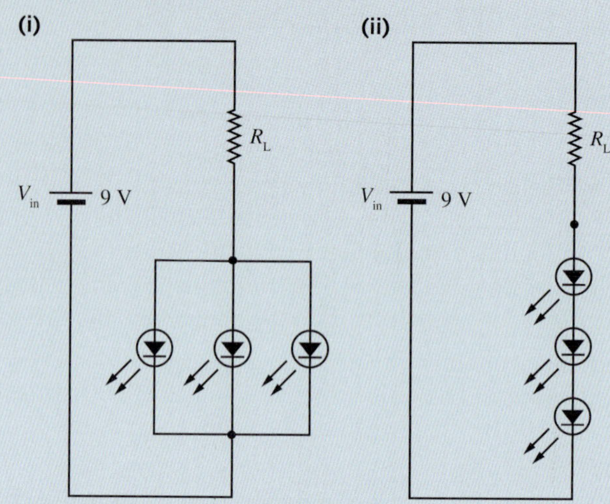

a For each circuit, determine the R_L that gives optimum operation for all of the LEDs.

b Which combination of LEDs (circuit (i) or (ii)) emits the more light with these particular values of R_L?

c Assuming circuits (i) and (ii) have exactly the same ideal battery power supply, determine which circuit will emit light (i.e. with the LEDs operating at their optimum level) for the longest time. Explain your answer.

26 This diagram shows the three sockets numbered 1, 2 and 3 when looking directly at a power point.

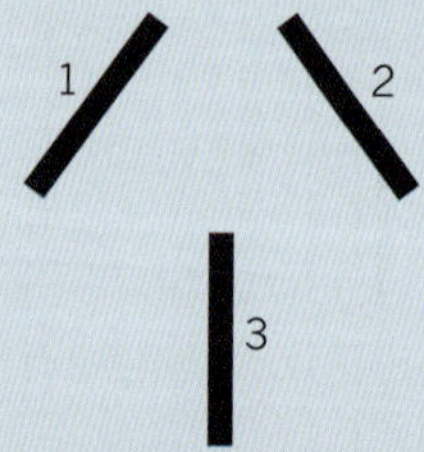

a Which number is the earth socket?

b Which number is the active socket?

c Which number is the neutral socket?

OA ✓✓

REVIEW QUESTIONS

How can electricity be used to transfer energy?

Multiple-choice questions

1 A current in a copper wire primarily relates to the movement of
- **A** protons.
- **B** neutrons.
- **C** electrons.
- **D** atoms.

2 When a voltmeter is used to measure the potential difference across a circuit element, it should be placed
- **A** in series with the circuit element.
- **B** in parallel with the circuit element.
- **C** either in series or in parallel with the circuit element.
- **D** neither in series nor in parallel with the circuit element.

3 When an ammeter is used to measure the current through a circuit element, it should be placed
- **A** in series with the circuit element.
- **B** in parallel with the circuit element.
- **C** either in series or in parallel with the circuit element.
- **D** neither in series nor in parallel with the circuit element.

4 When two circuit elements are placed in series
- **A** the power dissipated in each element must be the same.
- **B** the current through each element must be the same.
- **C** the voltage across each element must be the same.
- **D** the equivalent resistance of the combination will double.

5 When two circuit elements are placed in parallel
- **A** the power dissipated in each element must be the same.
- **B** the current through each element must be the same.
- **C** the voltage across each element must be the same.
- **D** the equivalent resistance of the combination will halve.

6 When a person appears to have been electrocuted, the rescuer should
- **A** pull them away from the electrical device.
- **B** call 000.
- **C** commence CPR immediately.
- **D** switch off the power first.

7 The *I–V* graph for a light bulb is shown. Determine the resistance of the bulb when it has a current of 4.0 A.

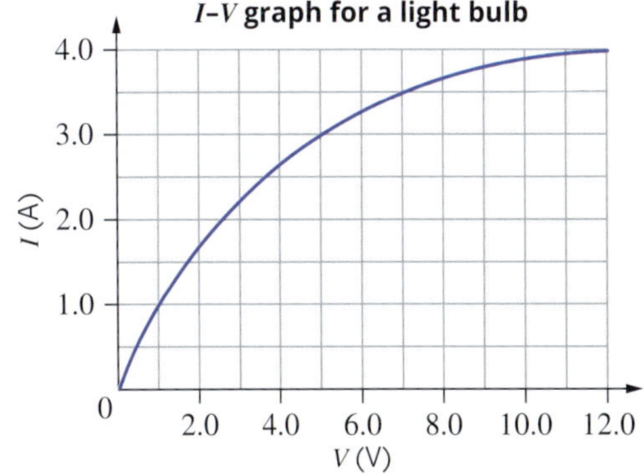

- **A** 3.0 W
- **B** 3.0 Ω
- **C** 0.33 Ω
- **D** 0.33 W

8 The potential difference between the two terminals on a battery is 12.0 V. Calculate the amount of work required to transfer 8.0 C of charge across the terminals.
- **A** 1.5 J
- **B** 96 J
- **C** 0.66 J
- **D** 12.0 J

9 A heating radiator is connected to a 240 V power supply. The heater dissipates 2.4 MJ of energy in 20 min. Calculate the power dissipated by the heating element.
- **A** 0.50 kW
- **B** 0.50 W
- **C** 2.0 kW
- **D** 2.0 W

The following information applies to questions 10–13.

Three identical light bulbs, L_1, L_2 and L_3, are connected into a circuit as shown in the diagram below.

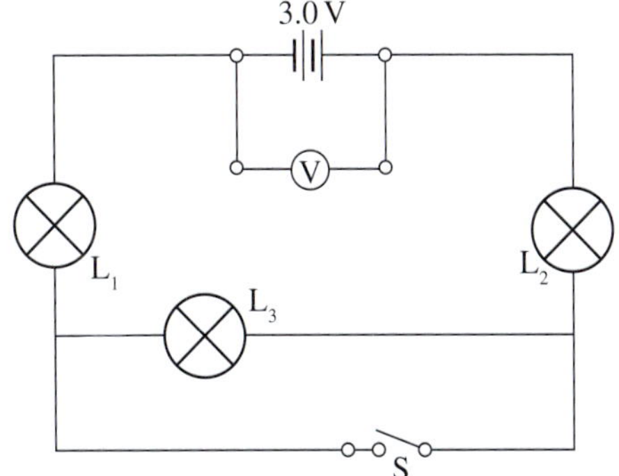

All bulbs are operating normally. The brightness of a bulb increases when the current through it increases. Assume that all connecting wires have negligible resistance and the battery has no internal resistance.

10 When switch S is closed, the brightness of L_1 will
- **A** be greater than previously.
- **B** be less than previously but greater than zero.
- **C** remain the same.
- **D** be zero.

11 When switch S is closed, the brightness of L_3 will
- **A** be greater than previously.
- **B** be less than previously but greater than zero.
- **C** remain the same.
- **D** be zero.

12 When switch S is closed, the reading on voltmeter V will
- **A** be greater than previously.
- **B** be less than previously but greater than zero.
- **C** remain the same.
- **D** be equal to zero.

13 When switch S is closed, the power output of the battery will
- **A** be greater than previously.
- **B** be less than previously but greater than zero.
- **C** remain the same.
- **D** be equal to zero.

The following information applies to questions 14–16.

In the circuit shown, the current through R_1 is I_1 and the current through ammeter A is I_A. The voltmeter V has extremely high resistance and ammeter A has negligible resistance.

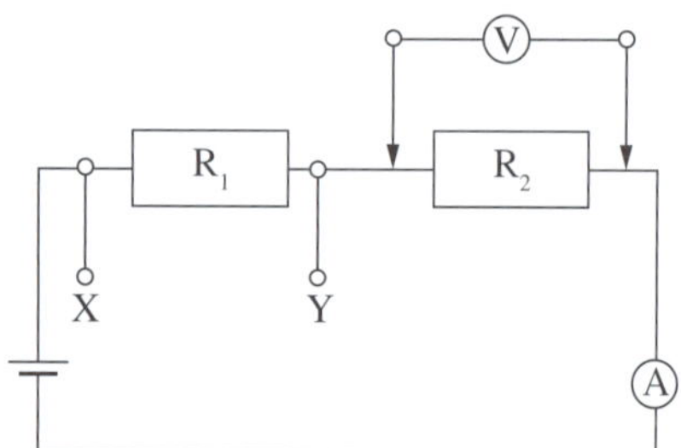

14 Which of the following statements is true?
- **A** $I_1 = I_A$
- **B** $I_1 > I_A$
- **C** $I_1R_1 = I_A(R_1 + R_2)$
- **D** $I_1R_1 = I_AR_2$

15 If another resistor, R_3, was connected between X and Y, what would happen to the meter readings?
- **A** The reading on V would increase; the reading on A would decrease.
- **B** The reading on V would decrease; the reading on A would increase.
- **C** They would both decrease.
- **D** They would both increase.

16 If another resistor, R_3, was connected between X and Y, what would happen to the power output of the battery?
- **A** It would increase.
- **B** It would decrease.
- **C** It would remain the same.

Short-answer questions

17 Birds can sit unharmed on a high-voltage power line. However, on occasion large birds such as eagles are electrocuted when they make contact with more than one line. Explain why this occurs.

18 A student needs to construct a circuit in which there is a potential difference (voltage drop) of 5.0 V across a combination of resistors, and a total current of 2.0 mA in the combination. She has a 4.0 kΩ resistor that she wants to use and proposes to add another resistor in parallel with it.
- **a** Calculate the value for the second resistor she should use.
- **b** Calculate the equivalent resistance of the combination.

19 A light bulb rated as 6.0 W and 0.50 A is to be operated from a 20 V supply using a limiting resistor, R. The circuit is represented below.

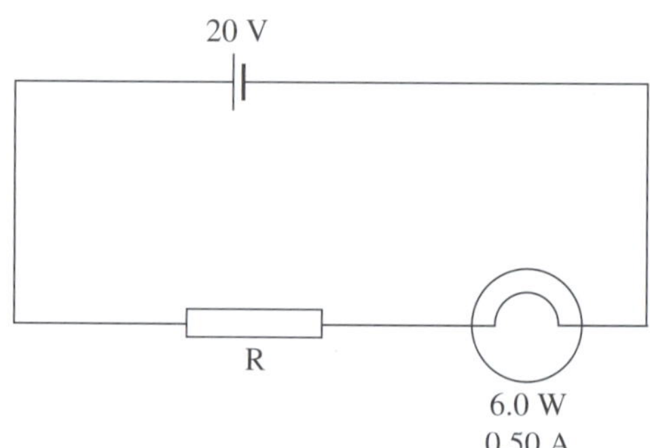

- **a** Calculate the required voltage for the light bulb to operate correctly.
- **b** Explain what voltage drop there must be across R when the circuit is operating correctly.
- **c** Calculate the required resistance of R.

20 The following figure describes a simple electric circuit in which a length of resistance wire is connected to a battery ($q_e = 1.6 \times 10^{-19}$ C).

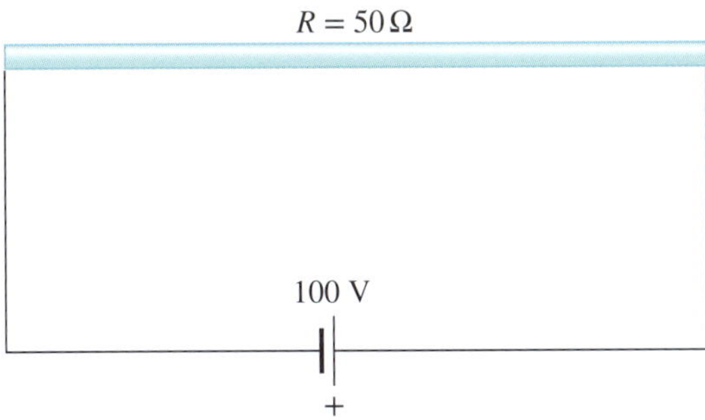

a Calculate the number of electrons passing through the wire every second.

b Calculate the electrical energy each electron loses as it moves through the wire.

c Describe what happens to the electrical energy of the electrons as they move through the wire.

d Calculate the amount of power being dissipated in the resistance wire.

e Calculate the total energy being supplied to all the electrons passing through the wire each second.

f Calculate the power output of the battery.

g Discuss the significance of your answers to parts d and f.

21 The diagram below shows the current–voltage graph for a section of platinum wire. A potential difference of 9.0 V is established across the section of wire.

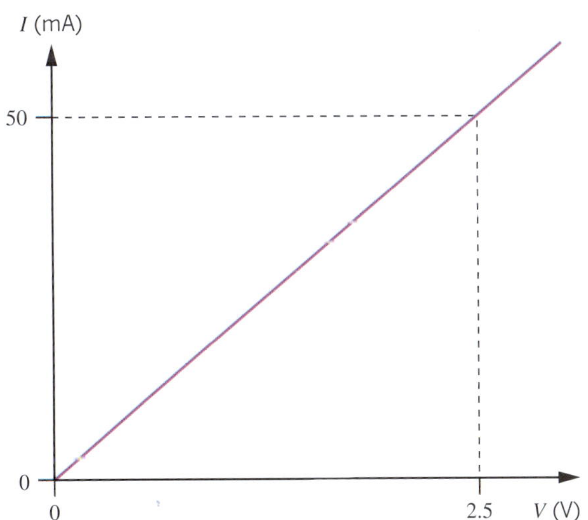

a Calculate the resistance of the section of wire.

b Calculate the amount of energy an electron loses in travelling through the wire.

c Calculate the power dissipated in the wire.

d Determine how many electrons pass through the wire in 10 s.

22 Three resistors, $R_1 = 100\,\Omega$, $R_2 = 200\,\Omega$ and $R_3 = 600\,\Omega$, in a circuit are connected in parallel with a 120 V cell.

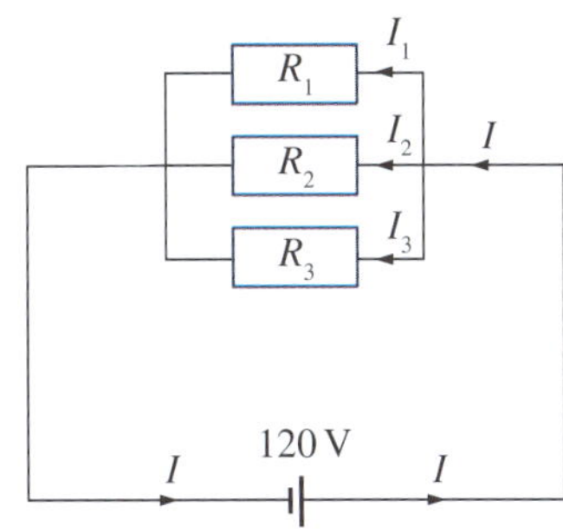

a Calculate the equivalent resistance in the circuit.

b Determine the total current in the circuit.

c Calculate the branch currents I_1, I_2 and I_3.

d Calculate the power output of the battery.

e Calculate the total power consumed by all the resistors in the circuit.

23 Four identical 200 Ω resistors and a 12 V power supply are arranged as shown in a circuit. Calculate the current through R_3, and the power dissipated by R_2. Give the power to two significant figures.

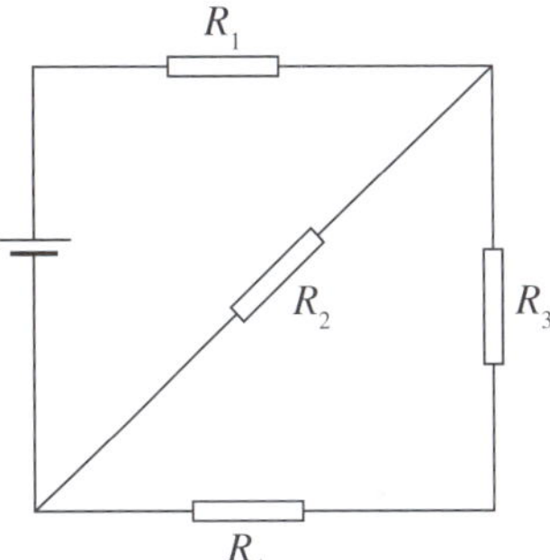

24 A special type of circuit device, X, has the I–V characteristic depicted in the following graph.

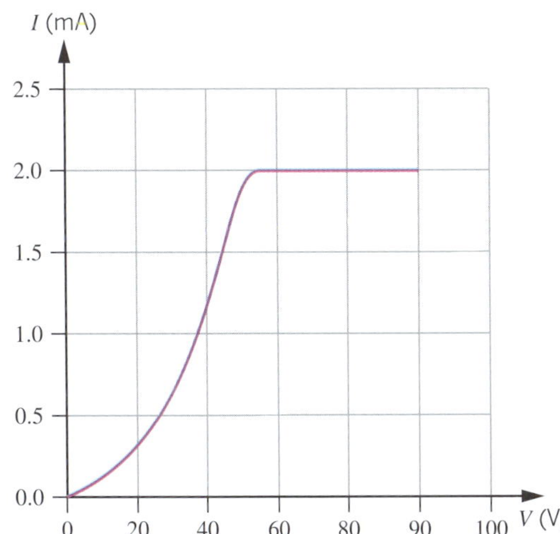

When circuit device X is connected into a circuit, as shown in the circuit diagram below, the reading on the ammeter is 2.0 mA.

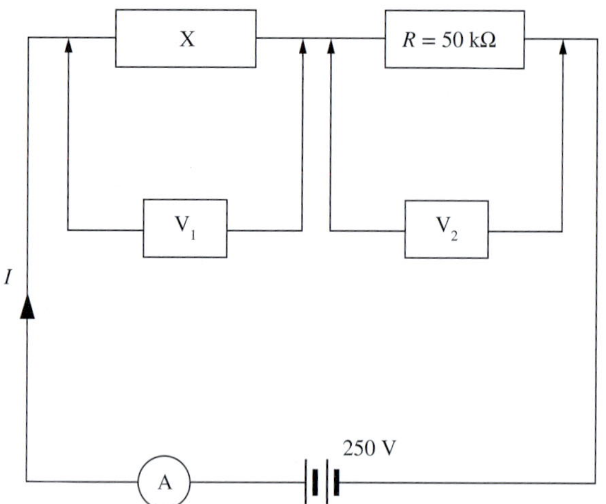

a Describe the behaviour of device X and suggest its purpose.

b Describe the resistance of such a device in the voltage range 60–90 V.

c Calculate the expected readings on voltmeters V_1 and V_2.

d Calculate the power being consumed by the circuit device X.

e Calculate the power being consumed by the resistor.

f Calculate the power output of the 250 V battery.

25 Bill and Mary are discussing the lighting for their living room. At present they have four 60 W, 240 V light bulbs in parallel. Bill suggests that they could save money on their power bills if they replace these with four bulbs wired in series.

a If this was to be done, calculate the voltage and power rating of each of the new bulbs they would need in order to produce the same amount of power.

b Calculate the total current in the circuit and compare this to the total current when the original parallel bulbs were used.

c Bill says that they would save on electricity bills because the current is going through all four bulbs and therefore being used more effectively. Mary says this is not right and that the power bill would be exactly the same. Explain who is correct.

d Mary argues that it would be a disadvantage to change the wiring of the lights to a series circuit. Justify why Mary might argue this.

26 Explain the difference between a fuse and a residual current device in terms of their operation and the kind of protection that they provide.

27 A household circuit is fitted with a 15.0 A circuit breaker. Susan connects the 800 W toaster, the 1200 W dishwasher and the 2000 W kettle in the same circuit.

a Explain whether the devices in a household circuit are connected in series or in parallel and give a reason for this wiring.

b Describe a reason for the installation of the circuit breaker.

c Calculate how much current the toaster draws.

d Calculate the resistance of the kettle.

e Perform a calculation to predict whether or not the circuit breaker will trip the circuit when all three above devices are operating simultaneously.

f The dishwasher cycle is 2.5 hours. If Susan uses the dishwasher 6 times a week, what is the annual energy used in kW h?

28 A thermistor is a semiconductor device whose resistance depends on the temperature. The graph shows the resistance versus temperature characteristic for a particular thermistor. Calculate the temperature of the thermistor in the circuit if V_{out} is 1 V.

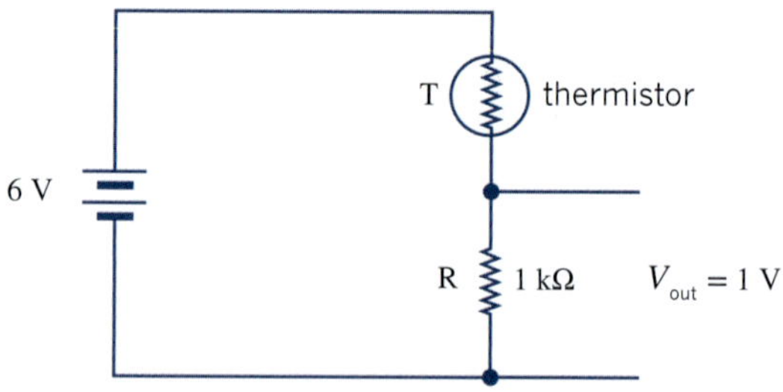

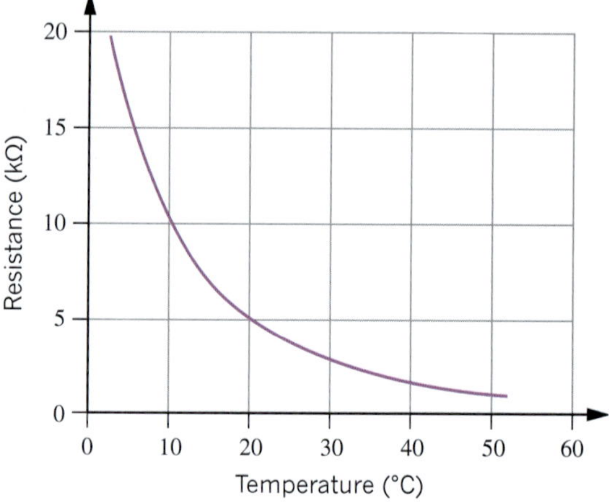

29 Consider the potentiometer circuit shown below, and explain how it may be used to control the current through the load.

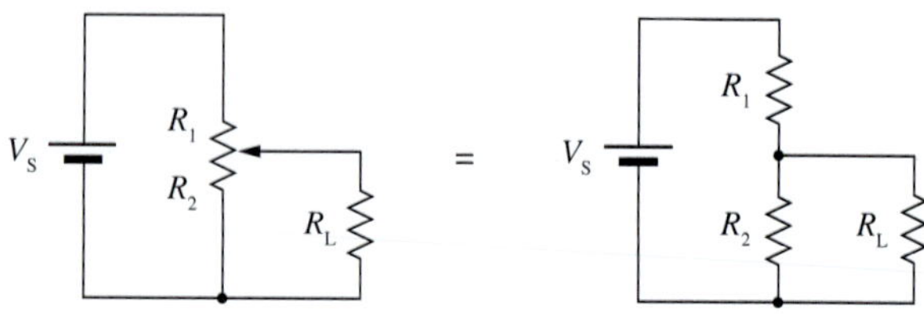

30 Explain why both a microphone and a speaker would be considered transducers, and describe the energy transformations occurring when each device operates.

31 Explain why a short circuit may trigger a circuit breaker.

32 Lixin is puzzled because he sees that some household plugs have three prongs and some only two. He is concerned that the two-pronged plugs may be unsafe. Explain how this works to put his mind at rest.

33 The current–voltage characteristics of two circuit elements were independently measured, then graphed on the same set of axes below.

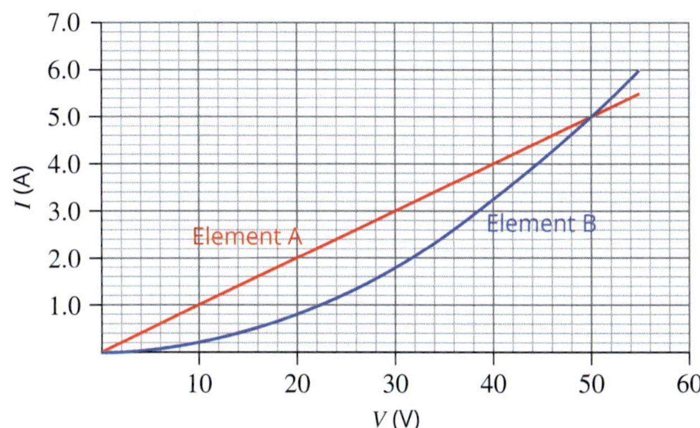

a The elements are first connected in parallel in a circuit. The voltage across element A is measured to be 20 V.

i Explain why the voltage drop across element B is also 20 V.

ii Determine the current that would be measured in each circuit element.

b The elements are disconnected, then reconnected in series. The current through element B is now measured to be 3.2 A.

i Explain what current that would be measured in element A.

ii Determine the total voltage supplied to the elements.

c The elements are disconnected, then placed in parallel again. It is noticed that both elements, under these conditions, have the same resistance.

i Calculate the total current supplied to the circuit.

ii Calculate the equivalent resistance of the circuit.

Two students are discussing the *I–V* graphs of the elements.

Evie states that both circuit elements are ohmic, as each circuit element has a constant value of resistance.

Nick states that the elements are only ohmic when the graphs intersect.

d Evaluate the statement of each student.

34 A piece of copper wire and a piece of plastic are placed across two terminals side by side, and a current is passed through them.

a It is observed that the wire gets hot if the current is left on for any length of time, but the plastic does not. Explain why this is so.

b Using the wire and plastic as an example, define 'electrical resistance'.

c Design a method that could be used to determine the resistance of the copper wire. Include safe laboratory practices.

d Another student in your class obtained the following results in order to predict the current that will be used when 24 V is applied across the copper wire.

Voltage (*V*)	Current (*I*)
0.5	8
1.0	10
1.5	25

Give the student two pieces of advice to improve their results.

35 A circuit is created using a 4.0 Ω resistor, two 3.0 Ω resistors and a 2.0 Ω resistor, as shown below. When connected to a power supply, 2.0 A of current goes through the 4.0 Ω resistor.

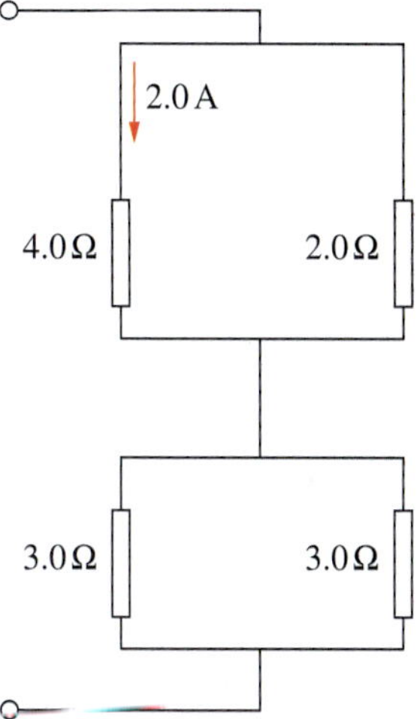

a Calculate the total current that is supplied to the circuit.

b Calculate the equivalent resistance of this circuit.

c Calculate the voltage supplied to the circuit.

d Draw a circuit diagram that would allow the current and voltage of the 2.0 Ω resistor to be found using an ammeter and a voltmeter.

e Justify why an electrician might alternatively use a multimeter to find the quantities described in part d.

f With the same resistors and power supply, design a parallel circuit that has the minimum current in the branch of the circuit going through the 2.0 Ω resistor.

g Using all of the same resistors, design a circuit that has an equivalent resistance of 3.0 Ω.

UNIT 2 How does physics help us to understand the world?

To achieve the outcomes in Unit 2, you will draw on key knowledge outlined in each area of study and the related key science skills on pages 11 and 12 of the study design. The key science skills are discussed in Chapter 1 of this book.

AREA OF STUDY 1

How is motion understood?

Outcome 1: On completion of this unit the student should be able to investigate, analyse, mathematically model and apply force, energy and motion.

AREA OF STUDY 2

Options: How does physics inform contemporary issues and applications in society?

Outcome 2: On completion of this unit the student should be able to investigate and apply physics knowledge to develop and communicate an informed response to a contemporary societal issue or application related to a selected option.

AREA OF STUDY 3

How do physicists investigate questions?

Outcome 3: On completion of this unit the student should be able to draw an evidence-based conclusion from primary data generated from a student-adapted or student-designed scientific investigation related to a selected physics question.

CHAPTER

10 Scalars and vectors

Scalars and vectors are mathematical representations of quantities that are used in physics. An understanding of scalars and vectors is essential to learning concepts involving forces and motion.

By the end of this chapter, you will be able to distinguish between scalar and vector quantities. You will be able to use arrows to represent vectors, and to add and subtract vectors in one and two dimensions.

Key knowledge

- identify parameters of motion as vectors or scalars **10.1**
- apply the vector model of forces, including vector addition and components of forces, to readily observable forces including the force due to gravity, friction and normal forces. **10.2, 10.3**

10.1 Scalars and vectors

You will come into contact with many physical quantities in the natural world every day. For example, time, mass and distance are all physical quantities. Each of these physical quantities has units with which to measure them. For example, seconds, kilograms and metres.

Some measurements only make sense if a direction is also included. For example, a GPS system tells you when to turn and in which direction. Without both of these instructions, the information is incomplete.

All physical quantities can be divided into two broad groups based on what information you need for the quantity to make sense. These groups are called scalars and vectors. Often vectors are represented by arrows. Both of these types of measures will be investigated throughout this section.

SCALARS

There are a number of properties in nature that can be measured or determined, and described using only a number and unit. For example, if the time taken for a student to travel to school is measured, you need the **magnitude** (size) and the **units** in order to understand the journey. It may take 90 minutes or one and a half hours—the number is important and so too are the units.

Quantities that require magnitude and units are called **scalar** quantities. Scalars do not need direction.

Examples of scalar quantities are:

- time
- distance
- volume
- speed
- temperature.

VECTORS

Vector quantities require magnitude, units and direction in order to make sense.

Examples of vectors include:

- position
- displacement
- velocity
- acceleration
- force
- momentum.

These measures are discussed in more detail in the coming chapters.

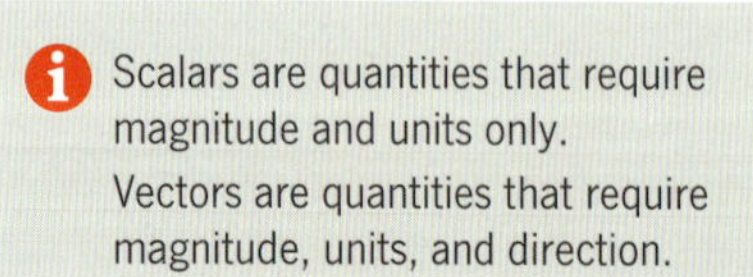

Scalars are quantities that require magnitude and units only.
Vectors are quantities that require magnitude, units, and direction.

Vectors as arrows

A vector is a measurement that has both a magnitude and a direction. A vector can be visually represented as an arrow.

Figure 10.1.1 shows two vector diagrams. In a **vector diagram**, the length of the arrow indicates the magnitude of the vector. The arrowhead shows the direction of the vector. The direction of the vector is always from its tail to its head.

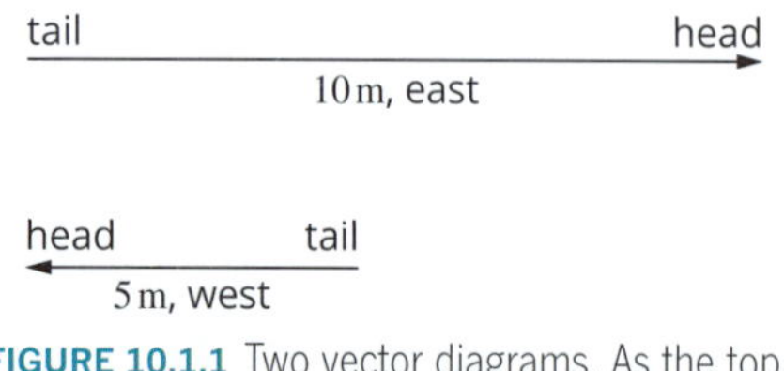

FIGURE 10.1.1 Two vector diagrams. As the top vector is twice as long as the bottom vector, it represents a measure twice the magnitude of the bottom vector. The arrowheads indicate that the vectors are in opposite directions.

A force is a push or a pull and the unit of measure for force is the newton (N). If you push a book to the right, it will respond differently from the way it would respond if you push the book to the left. Therefore, a force is only described properly if a direction is included, and so force is considered to be a vector. Forces are described in more detail on page 356. Force is an important concept to understand in physics, so many of the examples in this chapter refer to forces.

In most vector diagrams, the length of the arrow is drawn to scale so that it accurately represents the magnitude of the vector.

In the vector diagrams shown in Figure 10.1.2, force and displacement vectors are shown acting on a variety of objects. The arrows are labelled with the magnitude, unit and direction.

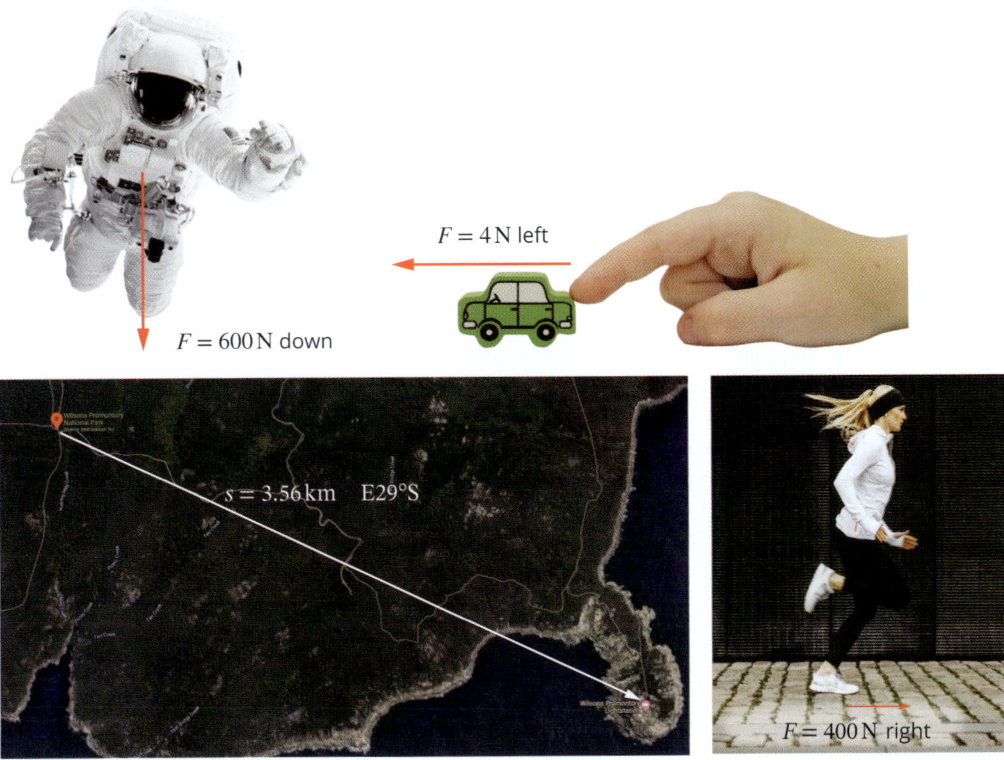

FIGURE 10.1.2 Examples of force and displacement vectors

An exact scale for the magnitude is not always used. However, it is important that vectors are drawn relative to one another. For example, a vector of 50 m north should always be about half as long as a vector of 100 m north.

Point of application of arrows

Vector diagrams may be presented slightly differently depending on what they are depicting. If the vector represents a force, the tail end of the arrow is drawn at the point where the force is applied to the object. If it is a displacement vector, attach the tail of the arrow to the position where the object starts. Friction vectors are drawn at the point where they act between an object and a surface.

Figure 10.1.3 shows a force applied by an ice hockey stick to a puck (5 N right) and an opposing friction force (0.5 N left).

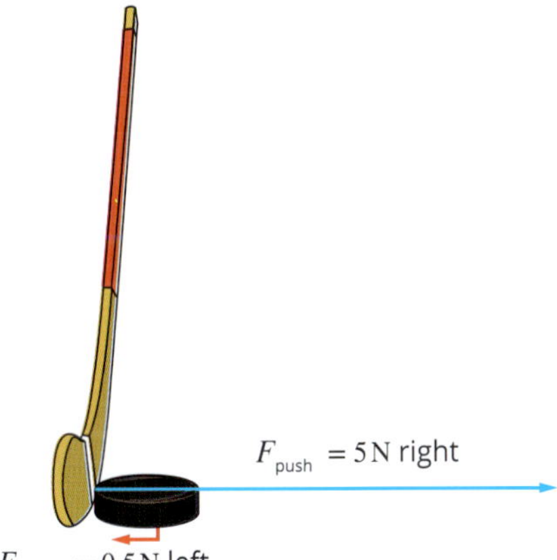

FIGURE 10.1.3 The force on the puck acts at the point of contact between the stick and the puck. The friction force acts between the puck and the ice. The push force, as indicated by the length of the arrow, is larger than the friction force.

DIRECTION CONVENTIONS

Vectors need a direction in order to make sense. However, for any description of a vector quantity to be useful, there needs to be a **direction convention** (a way of describing the direction that everyone understands and agrees upon).

Vectors in one dimension

For vector problems in one dimension, there are a number of direction conventions that can be used. For example:

- forwards or backwards
- upwards or downwards
- left or right
- north or south
- east or west.

As you can see, for vectors in one dimension there are only two directions possible. The two directions must be in the same **dimension** (along the same line). The direction convention used should be presented graphically in all vector problems. This is shown in Figure 10.1.4. Arrows like these are placed near the vector diagram so that it is clear which convention is being used.

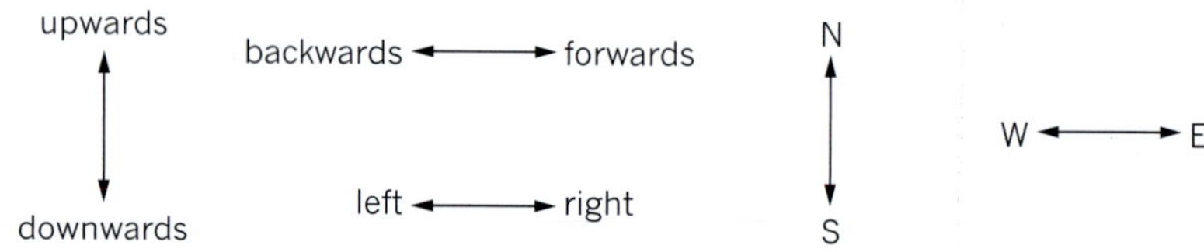

FIGURE 10.1.4 Some common one-dimensional direction conventions

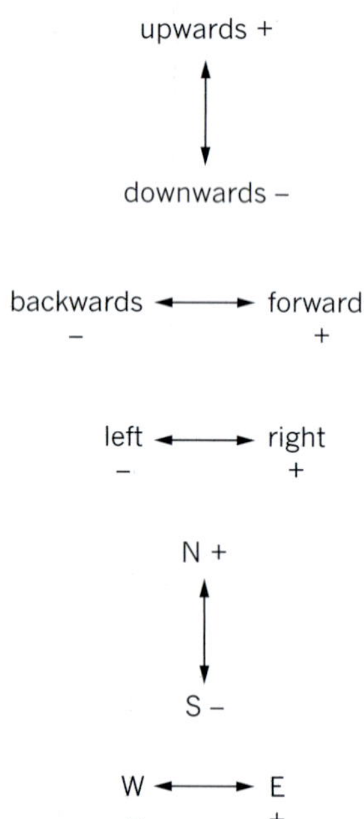

FIGURE 10.1.5 One-dimensional direction conventions can also be expressed as sign conventions.

In calculations involving one-dimensional vectors, a sign convention can also be used to convert physical directions to the mathematical signs of positive and negative. For example, forwards can be positive and backwards can be negative, or right can be positive and left can be negative. A vector of 100 m upwards can be described as +100 m, provided the relationship between sign and direction conventions are clearly indicated in a legend or key. Some examples are provided in Figure 10.1.5.

The advantage of using a sign convention is that the signs of positive and negative can be entered into a calculator, but the words 'upwards' and 'right' cannot. This is useful when adding vectors together, and will be discussed in the next section.

Worked example 10.1.1

DESCRIBING VECTORS IN ONE DIMENSION

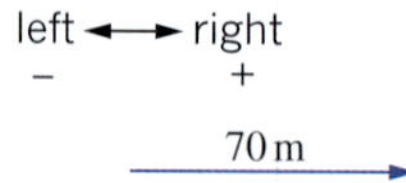

Describe the vector above using:

a the direction convention shown	
Thinking	**Working**
Identify the direction convention being used in the vector.	In this case, the vector is pointing to the right according to the direction convention.
Note the magnitude, unit and direction of the vector.	In this example, the vector is 70 m right.

b an appropriate sign convention.	
Thinking	**Working**
Convert the physical direction to the corresponding mathematical sign.	The physical direction of right is positive and left is negative. In this example, the arrow is pointing right, so the mathematical sign is +.
Represent the vector with a mathematical sign, magnitude and unit.	This vector is +70 m.

Worked example: Try yourself 10.1.1

DESCRIBING VECTORS IN ONE DIMENSION

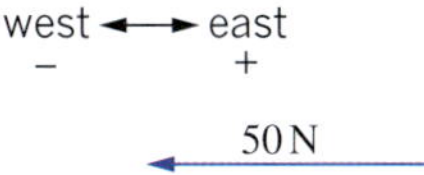

Describe the vector above using:

a the direction convention shown
b an appropriate sign convention.

Vectors in two dimensions

When vectors are in one dimension, it is relatively simple to understand direction. However, some vectors will require a description in a two-dimensional plane. These planes could be:

- horizontal, which can be defined using north, south, east and west
- vertical, which can be defined in a number of ways including forwards, backwards, upwards, downwards, left and right.

The description of the direction of these vectors is more complicated. Therefore, a more detailed convention is needed for identifying the direction of a vector. A variety of conventions are used, but they all describe a direction as an angle from a known reference point.

For a horizontal, two-dimensional plane, there are two common methods used for describing the direction of a vector:

- full circle (or true) bearing. A 'full circle bearing' describes north as zero degrees true. This is written as 0°T. In this convention, all directions are given as a clockwise angle from north. For example, 95°T is 95° clockwise from north.
- quadrant bearing. An alternative method is to provide a 'quadrant bearing', in which all angles are referenced from the nearest compass point (north, south, east or west) and are between 0° and 90° towards the next-nearest compass point. In this method, 30°T becomes N30°E, which can be read as 'from north 30° towards the east'. A bearing of 120°T becomes E30°S, which can be read as 'from east 30° towards the south'.

Using these two conventions, north-west (NW) would be 315°T using a full circle bearing, or N45°W (or W45°N) using a quadrant bearing. Figure 10.1.6 demonstrates these two methods.

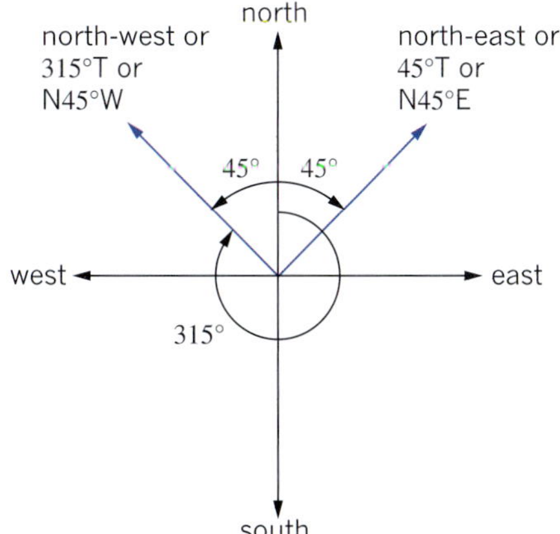

FIGURE 10.1.6 Two horizontal vector directions, viewed from above, using full circle bearings and quadrant bearings

For a vertical, two-dimensional plane the directions are referenced to vertical (upwards and downwards) or horizontal (left and right) and are between 0° and 90° clockwise or anticlockwise. For example, a vector direction can be described as '60° clockwise from the left direction'. The same vector direction could be described as '30° anticlockwise from the upwards direction'. The opposite direction to this vector would be '60° clockwise from the right direction'. This example is illustrated in Figure 10.1.7.

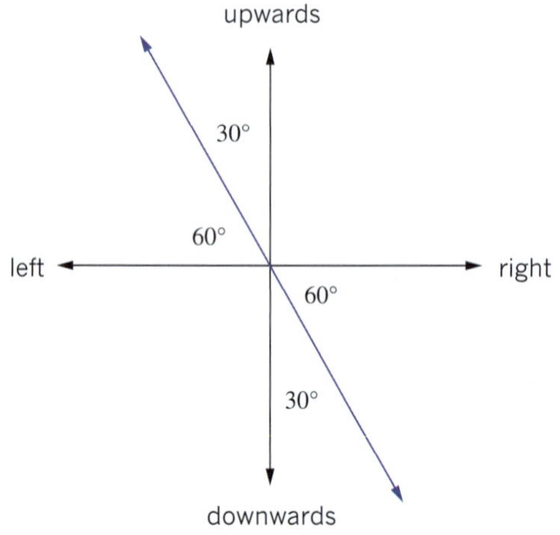

FIGURE 10.1.7 Two vectors in the vertical plane

PHYSICSFILE

Vector graphics and rasters

If you look closely at a computer screen, you will notice that it is made up of millions of tiny dots called pixels. Your computer can display all kinds of images by lighting up these pixels with different colours, and with different levels of brightness These images can be stored in your computer as a list of instructions for each pixel, starting from the top-left, continuing across the screen, then down to the next row and so on until finishing at the bottom right. This image is known as a raster image after the Latin word *rastrum*—rake. Examples are a .jpeg, .png or .gif. One problem with raster images is that they become very blurry and pixelated when you zoom in. A different kind of image is a vector image (.svg, .eps or .pdf). This uses instructions, like vectors, to locate points on the screen and then connects those points with geometric shapes such as lines, squares, polygons, arcs or circles. The advantages of these images are that there is a reduction in the size of the file as there is less information to store, and the image remains sharp as you zoom in.

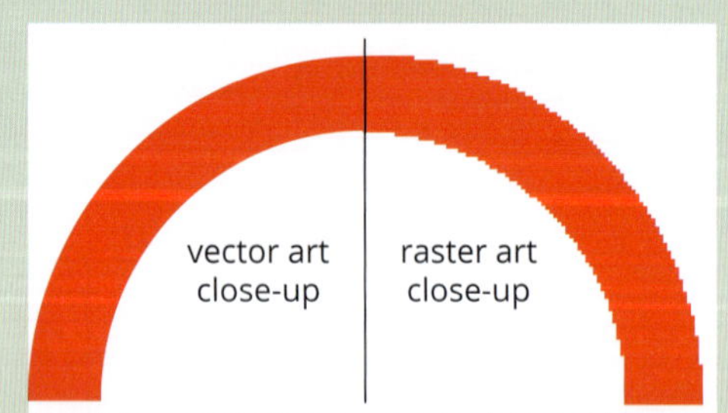

One advantage of vector graphics over raster images is that the image remains sharp even when you zoom in.

Worked example 10.1.2

DESCRIBING TWO-DIMENSIONAL VECTORS

Describe the direction of the following vector using an appropriate method.

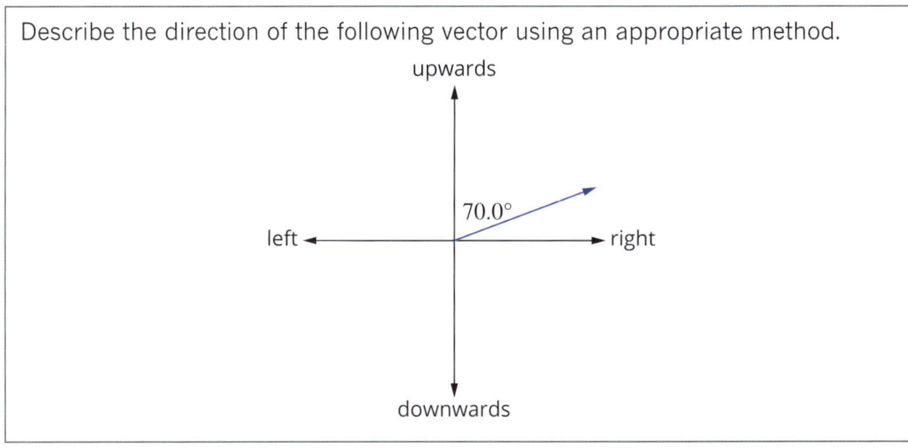

Thinking	Working
Choose the appropriate points to reference the direction of the vector. In this case using the vertical reference makes more sense, as the angle is given from the vertical.	The vector can be referenced to the vertical.
Determine the angle between the reference direction and the vector.	In this example, there is 70.0° from vertically upwards to the vector.
Determine the direction of the vector from the reference direction.	From vertically upwards, the vector is clockwise.
Describe the vector using the sequence: angle, clockwise or anticlockwise from the reference direction.	This vector is 70.0° clockwise from the upwards direction.

Worked example: Try yourself 10.1.2

DESCRIBING TWO-DIMENSIONAL VECTORS

Describe the direction of the following vector using an appropriate method.

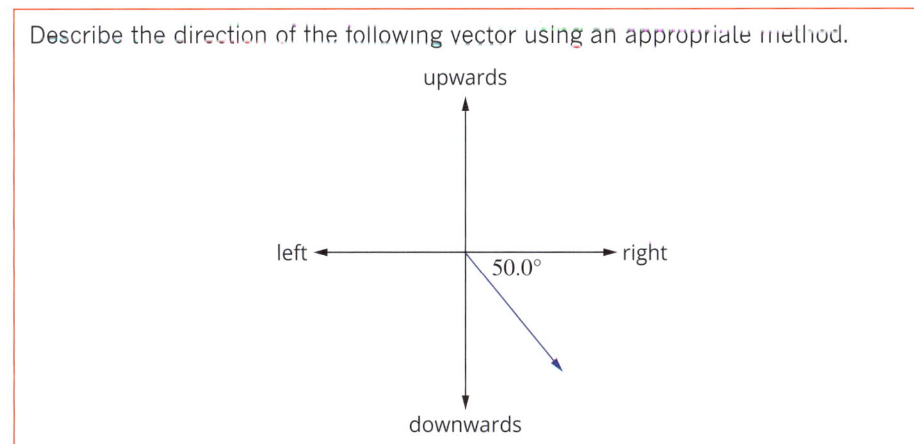

10.1 Review

SUMMARY

- Scalar quantities require a magnitude and a unit to make sense. No direction is required for scalar quantities.
- Distance, time, speed and mass are examples of scalar quantities.
- Vectors require magnitude, units and direction to make sense.
- Displacement, velocity, acceleration and force are examples of vectors.
- Arrows are used to represent vectors.
  - The length of the arrow represents the magnitude of the vector.
  - The direction the arrow is pointing indicates the direction of the vector.
  - Vector arrows can be drawn to scale or drawn relative to each other.
  - Force vectors are drawn with their tails attached to the point of application on the object.
  - Displacement vectors are drawn from the start of the journey towards the end of the journey.
- One-dimensional vectors use direction conventions and sign conventions to describe the direction of the vector. Examples include left and right, upwards and downwards, + and –.
- The direction of two-dimensional vectors in the horizontal plane can be described using a full circle bearing or a quadrant bearing. Vectors in the vertical plane can be described using angles measured clockwise and anticlockwise from the vertical or horizontal.

KEY QUESTIONS

Knowledge and understanding

1 What information is required to fully describe a scalar measure?

2 What information is required to fully describe a vector measure?

3 Copy and complete the table to classify each of the following quantities as scalar or vector.
time, force, acceleration, distance, position, displacement, volume, momentum, speed, velocity, temperature

Scalar	Vector

4 Give the opposing direction to each of the following one-dimensional descriptions.

a upwards
b north
c backwards
d downwards
e west
f negative

Analysis

5 Decide which of the following vector magnitudes provided describes each vector diagram. Note: one of the vector magnitudes is not required.
5.4 N; 2.7 N; 9.0 N; 8.1 N

(a) ⟶

(b) ⟶

(c) ⟶

6 Decide which of the following vector magnitudes provided describes each vector diagram. Note: one of the vector magnitudes is not required.
10.8 N; −2.7 N; −5.4 N; 16.2 N

(a) ⟶

(b) ⟵

(c) ⟶

7 Describe the following vector using an appropriate convention.

– ⟷ +

8 Describe the following vectors using:

i full circle bearings

ii quadrant bearings.

a

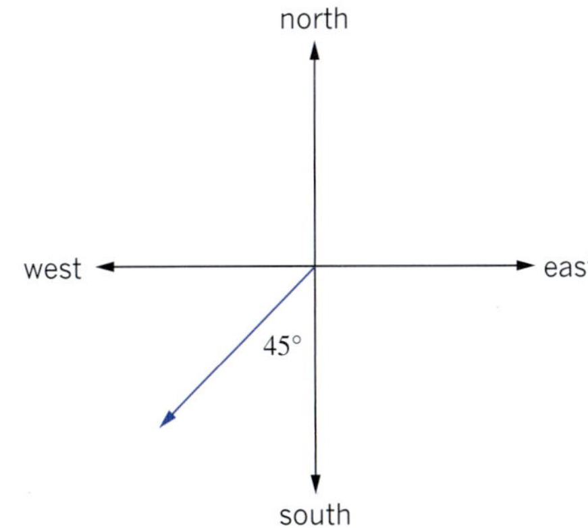

b

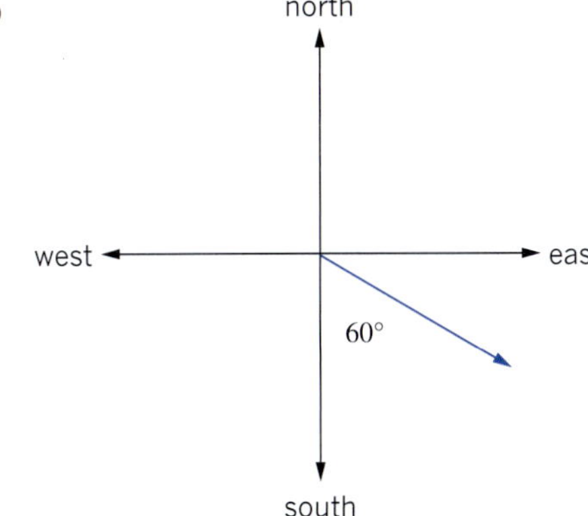

10.2 Adding and subtracting vectors in one and two dimensions

In real situations, more than one vector may act on an object. In such cases, it is helpful to analyse the associated vector diagrams to find the overall or combined effect of the vectors.

Vectors can be combined by adding or subtracting them. To do this, you can use the techniques of drawing vectors and using a sign convention that you learnt in the previous section. This can be done graphically by drawing arrows to represent vectors and finding the resultant vector. Alternatively, a sign convention can be applied and vectors can then be added or subtracted algebraically.

This can be done in one dimension (when the vectors are all in the same line) or two dimensions (when the vectors are all in the same plane).

ADDING AND SUBTRACTING VECTORS IN ONE DIMENSION

When two or more vectors are in the same dimension, it means that the vectors are either pointing in the same direction or in opposite directions. They are **collinear** (in line with each other). For example, the vector measurements 10 m west, 15 m east and 25 m west are all in one dimension. They are all in the same or opposite directions to each other.

Graphical method of adding vectors

Vectors are added head to tail. The resultant vector is drawn from the tail of the first vector to the head of the last vector.

Vector diagrams, such as those shown in Figure 10.2.1, are convenient for adding vectors. To combine vectors in one dimension, draw the first vector, then start the second vector with its tail at the head of the first vector. Continue adding arrows 'head to tail' until the last vector is drawn. The sum of the vectors, the **resultant** vector, is drawn from the tail of the first vector to the head of the last vector.

In Figure 10.2.1, the two vectors s_1 (15 m east) and s_2 (5 m east) are drawn separately. The two vectors are then drawn with s_1 and s_2 connected head to tail. The resultant vector s_R is drawn from the tail of s_1 to the head of s_2. The magnitude (size) of the resultant vector can be deduced from the magnitudes of the separate vectors: 15 m + 5 m = 20 m.

Alternatively, vectors can be drawn to scale, for example: 1 m = 1 cm. The resultant vector is then directly measured from the scale diagram. The direction of the resultant vector is the same as the direction from the tail of the first vector to the head of the last vector.

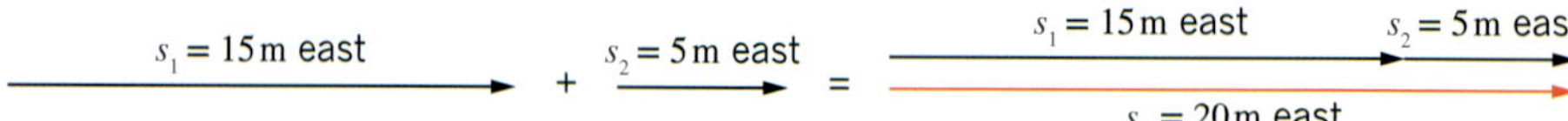

FIGURE 10.2.1 Adding vectors head to tail. This particular diagram represents the addition of 15 m east and 5 m east. The resultant vector, shown in red, is 20 m east.

Algebraic method of adding vectors

To add vectors in one dimension algebraically, use a sign convention to represent the direction of the vectors (Figure 10.2.2). When applying a sign convention, it is important to provide a key explaining the convention used.

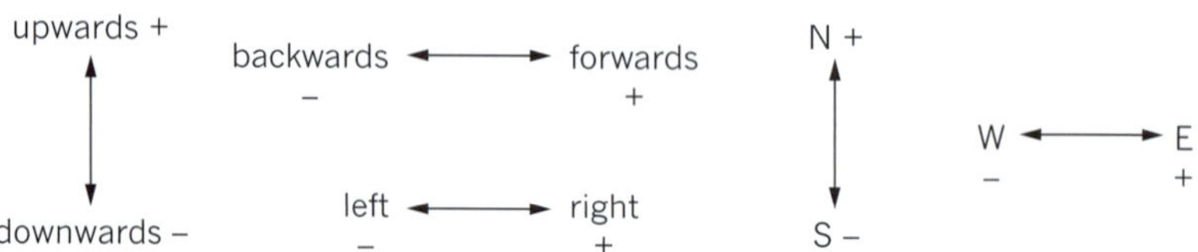

FIGURE 10.2.2 Common sign and direction conventions

The sign convention allows you to enter the signs and magnitudes of vectors into a calculator. The sign of the final magnitude gives the direction of the resultant vector.

Worked example 10.2.1

ADDING VECTORS IN ONE DIMENSION USING ALGEBRA

Use the sign and direction conventions shown in Figure 10.2.2 to determine the resultant vector of a student who walks 25.0 m west, 16.0 m east, 44.0 m west and then 12.0 m east.

Thinking	Working
Apply the sign conventions to change each of the directions to signs.	25.0 m west = −25.0 m 16.0 m east = +16.0 m 44.0 m west = −44.0 m 12.0 m east = +12.0 m
Add the magnitudes and their signs together.	Resultant vector = (−25.0) + (+16.0) + (−44.0) + (+12.0) = −41.0 m
Refer to the sign and direction conventions to determine the direction of the resultant vector.	Negative is west. ∴ Resultant vector = 41.0 m west

Worked example: Try yourself 10.2.1

ADDING VECTORS IN ONE DIMENSION USING ALGEBRA

Use the sign and direction conventions shown in Figure 10.2.2 to determine the resultant force on a box that has the following forces acting on it: 16.0 N upwards, 22.0 N downwards, 4.0 N upwards and 17.0 N downwards.

Graphical method of subtracting vectors

Velocity is a quantity that gives an indication of how fast an object is moving. It is a vector because the direction is important when stating the velocity of an object. For example, the velocity of the tennis ball moving towards the racquet in Figure 10.2.3 is different from the velocity of the tennis ball as it leaves the racquet. The concept of velocity is covered in more detail in Chapter 11, but it is useful to use the example of velocity now when discussing the subtraction of vectors. The processes applied to the subtraction of velocity vectors works for all other vectors.

To subtract velocity vectors in one dimension using a graphical method, determine which vector is the initial velocity and which is the final velocity. The final velocity is drawn first. The initial velocity is then drawn, but in the opposite direction to its original form. The sum of these vectors, or the resultant vector, is drawn from the tail of the final velocity to the head of the reversed initial velocity. This resultant vector is the difference between the two velocities, or Δv.

In Figure 10.2.4, the two separate velocity vectors, v_1 (9 m s^{-1} east) and v_2 (3 m s^{-1} east) are drawn separately. The initial velocity, v_1, is then drawn again in the opposite direction: $-v_1$ or 9 m s^{-1} west.

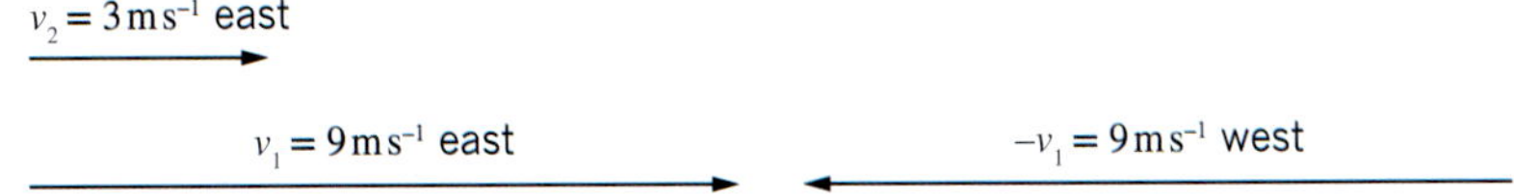

FIGURE 10.2.4 Subtracting vectors using the graphical method

Figure 10.2.5 illustrates how the difference between the vectors is found. First, the final velocity, v_2, is drawn. Then the opposite of the initial velocity, $-v_1$, is drawn head to tail. The resultant vector, Δv, is drawn from the tail of v_2 to the head of $-v_1$.

FIGURE 10.2.3 As velocity is a vector, direction is important. The tennis ball has a different velocity when it leaves the racquet from when it travelled towards the racquet.

To find the difference between, or change in, vectors, subtract the initial vector from the final vector. Graphically, vectors are subtracted by adding the opposite of the initial vector to the final vector.

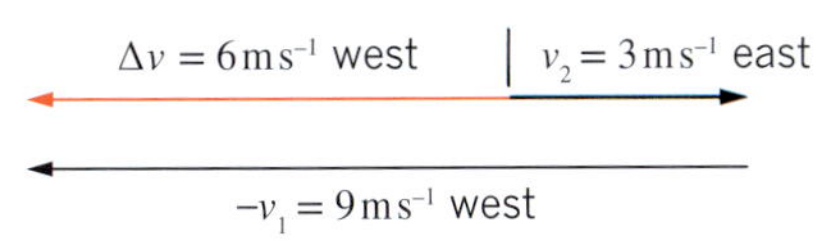

FIGURE 10.2.5 Subtracting vectors using the graphical method

The magnitude of the resultant vector, Δv, can be calculated from the magnitudes of the two vectors. Alternatively, you could draw the vectors to scale and then measure the resultant vector against that scale, for example $1\,m\,s^{-1} = 1\,cm$.

The direction of the resultant vector, Δv, is the same as the direction from the tail of the final velocity, v_2, to the head of the opposite of the initial velocity, $-v_1$.

Algebraic method of subtracting vectors

To subtract velocity vectors in one dimension algebraically, a sign convention is used to represent the direction of the velocities. Some examples of one-dimensional directions include east and west, north and south and upwards and downwards. These options are replaced by positive (+) or negative (–) signs when calculations are performed.

The equation for finding the change in velocity is:

$$\text{change in velocity} = \text{final velocity} - \text{initial velocity}$$

$$\Delta v = v_2 - v_1$$

Since the change in velocity is a vector, it will consist of a sign and a magnitude. The sign of the answer can be compared with the sign and direction convention (see Figure 10.2.2 on page 292) to determine the direction of the change in velocity.

PHYSICSFILE

Minus a negative

It is important to differentiate between the terms 'subtract', 'minus', 'take away' or 'difference between' and the term 'negative'. The terms 'minus', 'take away' or 'difference between' are different terms for 'subtract'. This is a mathematical process like add, multiply and divide. These are grouped together on your calculator. The term 'negative' is a property of a number that means that it is opposite to positive. There is a separate button on your calculator for this property.

When a negative number is subtracted from a positive number, the two numbers are added together. For example, $(5) - (-2) = 7$.

Worked example 10.2.2

SUBTRACTING VECTORS IN ONE DIMENSION USING ALGEBRA

Use the sign and direction conventions shown in Figure 10.2.2 on page 292 to determine the change in velocity of a plane as it changes from $255\,m\,s^{-1}$ west to $160\,m\,s^{-1}$ east.

Thinking	Working
Apply the sign and direction convention to change the directions to signs.	$v_1 = 255\,m\,s^{-1}$ west $= -255\,m\,s^{-1}$ $v_2 = 160\,m\,s^{-1}$ east $= +160\,m\,s^{-1}$
Use the formula for change in velocity to calculate the magnitude and the sign of Δv.	$\Delta v = v_2 - v_1$ $= (+160) - (-255)$ $= +415\,m\,s^{-1}$
Refer to the sign and direction convention to determine the direction of the change in velocity.	Positive is east. $\therefore \Delta v = 415\,m\,s^{-1}$ east

Worked example: Try yourself 10.2.2

SUBTRACTING VECTORS IN ONE DIMENSION USING ALGEBRA

Use the sign and direction conventions shown in Figure 10.2.2 on page 292 to determine the change in the velocity of a rocket as it changes from $212\,m\,s^{-1}$ upwards to $834\,m\,s^{-1}$ upwards.

ADDING AND SUBTRACTING VECTORS IN TWO DIMENSIONS

To add or subtract vectors in two dimensions, all of the vectors must be in the same plane. The vectors can go in any direction within the plane, and can be separated by any angle. The examples in this section illustrate vectors in the horizontal plane, but the same strategies apply to adding and subtracting vectors in the vertical plane.

The horizontal plane is one that is looked down on from above. Examples include looking at a house plan or map placed on a desk. The direction conventions that suit this plane best are the north, south, east and west convention, or the forwards, backwards, left and right convention, as shown in Figure 10.2.6. When two vectors are in the horizontal plane, the angles between them can be right-angled, acute or obtuse.

An example of a vector changing in two dimensions is the change in velocity that can occur when turning a corner. For example, walking at $3\,m\,s^{-1}$ west, then turning to travel at $3\,m\,s^{-1}$ north. Although the magnitude of the velocity is the same, the direction is different.

> When a change in a vector occurs, the magnitude and/or the direction of the vector can change.

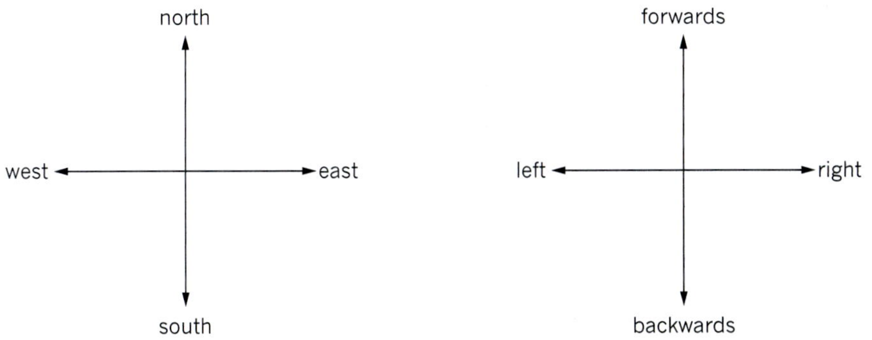

FIGURE 10.2.6 The direction conventions for the horizontal plane

Graphical method of adding vectors in two dimensions

The magnitude and direction of a resultant vector can be determined by measuring an accurately drawn, scaled vector diagram. There are two main ways to do this:

- head to tail method
- parallelogram method.

Head to tail method

To add vectors at right angles to each other using a graphical method, use an appropriate scale and then draw each vector head to tail. The resultant vector is the vector that starts at the tail of the first vector and ends at the head of the last vector. To determine the magnitude and direction of the resultant vector, measure the length of the resultant vector and compare it to the scale, then measure and describe the direction appropriately.

In Figure 10.2.7, two vectors, 30 m east and 20 m south, are added head to tail. The resultant vector, shown in red, is measured to be about 36 m according to the scale provided. Using a protractor, the resultant vector is measured to be in the direction 34° south of east. This represents a direction of E34°S when using quadrant bearings.

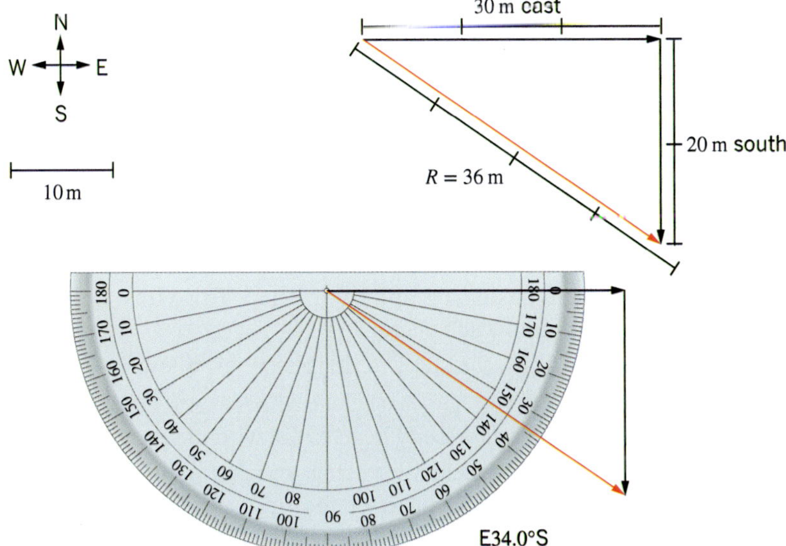

FIGURE 10.2.7 Adding two vectors at right angles, using the graphical method

If the two vectors are at angles other than 90° to each other, the graphical method is ideal for finding the resultant vector. In Figure 10.2.8, the vectors 15 N east and 10 N S45°E are added head to tail. The magnitude of the resultant vector is measured to be about 23 N. The direction of the resultant vector is measured by a protractor from east to be 18° towards the south, which should be written as E18°S.

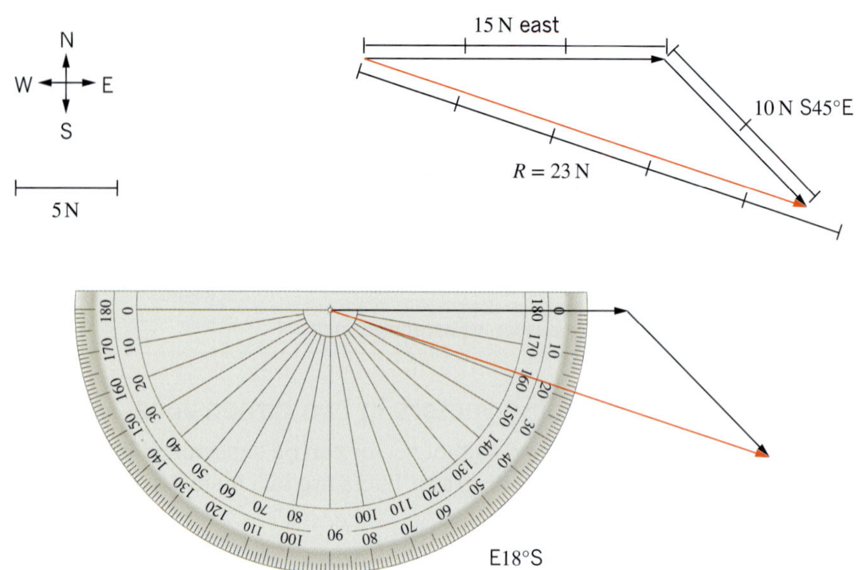

FIGURE 10.2.8 Adding two vectors not at right angles, using the graphical method

Parallelogram method

An alternative method for determining a resultant vector is to construct a parallelogram of vectors. In this method, the two vectors to be added are drawn tail to tail. Next, a parallel line is drawn for each vector, as shown in the two different examples in Figure 10.2.9. In this figure, the parallel lines have been drawn as dotted lines. The resultant vector is drawn from the tails of the two vectors to the intersection of the dotted parallel lines.

FIGURE 10.2.9 Parallelogram of vectors method for adding two vectors

Graphical method of subtracting vectors in two dimensions

Vectors can also be subtracted using a graphical method. To do this, use a direction convention and a scale and draw each vector.

Using velocity as an example, the steps to do this are as follows:

- Draw in the final velocity first.
- Draw the opposite of the initial velocity head to tail with the final velocity vector.
- Draw the resultant change in velocity vector, starting at the tail of the final velocity vector and ending at the head of the opposite of the initial velocity vector.
- Measure the length of the resultant vector and compare it to the scale to determine the magnitude of the change in velocity.
- Measure an appropriate angle to determine the direction of the resultant vector.

Figure 10.2.10 shows the velocity vectors for travelling $3\,m\,s^{-1}$ west and then turning and travelling $3\,m\,s^{-1}$ north. The opposite of the initial velocity is drawn as $3\,m\,s^{-1}$ east.

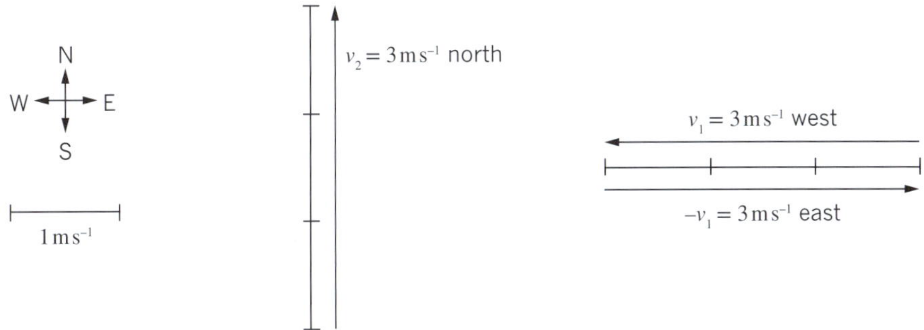

FIGURE 10.2.10 Subtracting two vectors at right angles, using the graphical method

To determine the change in velocity, the final velocity vector ($3\,m\,s^{-1}$ north) is drawn first, then from its head, the opposite of the initial velocity ($3\,m\,s^{-1}$ east) is drawn. This is shown in Figure 10.2.11. The magnitude of the change in velocity (resultant vector) is shown in red. It is measured to be about $4.3\,m\,s^{-1}$ according to the scale provided. Using a protractor, the resultant vector is measured to be in the direction N45.0° E.

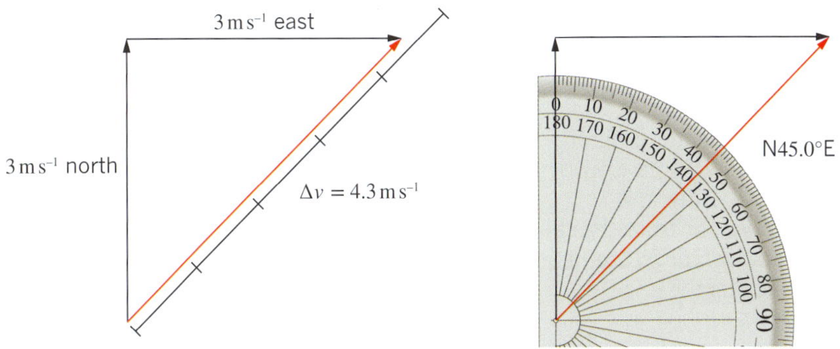

FIGURE 10.2.11 Subtracting two vectors at right angles, using the graphical method

Geometric method of adding vectors in two dimensions

Graphical methods of adding vectors in two dimensions only give approximate results, as they rely on comparing the magnitude of the resultant vector to a scale and measuring the direction with a protractor. A more accurate method to resolve vectors is to use Pythagoras' theorem and trigonometry. These techniques are referred to as geometric methods. Geometric methods can be used to calculate the magnitude of the vector and its direction. However, Pythagoras' theorem and trigonometry can only be used for finding the resultant vector of two vectors that are at right angles to each other.

Pythagoras' theorem is $a^2 + b^2 = c^2$, where c is the hypotenuse (the longest side) and a and b are the two shorter sides of a right triangle. The hypotenuse is opposite the right angle of the triangle.

In Figure 10.2.12, two vectors, 30.0 m east and 20.0 m south, are added head to tail. The resultant vector, shown in red, is calculated using Pythagoras' theorem to be 36.1 m. The resultant vector is calculated to be in the direction E33.7°S. This result is more accurate than the answer determined earlier in this section.

> Most students learn the mnemonic SOHCAHTOA in their maths classes. It is often pronounced soh-cah-toa and provides a way to remember the trigonometric ratios:
>
> $\sin\theta = \frac{\text{opposite}}{\text{hypotenuse}}$
>
> $\cos\theta = \frac{\text{adjacent}}{\text{hypotenuse}}$
>
> $\tan\theta = \frac{\text{opposite}}{\text{adjacent}}$

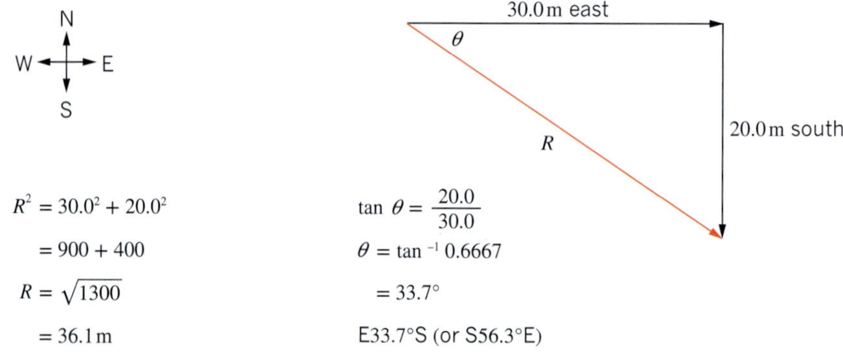

FIGURE 10.2.12 Adding two vectors at right angles, using the geometric method

Worked example 10.2.3

ADDING VECTORS IN TWO DIMENSIONS USING GEOMETRY

Determine the resultant vector that represents a child running 25.0 m west and 16.0 m north.

Thinking	Working
Construct a vector diagram showing the vectors drawn head to tail. Draw the resultant vector from the tail of the first vector to the head of the last vector.	
As the two vectors to be added are at 90° to each other, apply Pythagoras' theorem to calculate the magnitude of the resultant vector.	$R^2 = 25.0^2 + 16.0^2$ $= 625 + 256$ $R = \sqrt{881}$ $= 29.7\text{ m}$
Using trigonometry, calculate the angle from the west vector to the resultant vector.	$\tan\theta = \frac{16.0}{25.0}$ $\theta = \tan^{-1} 0.640$ $= 32.6°$
Determine the direction of the vector relative to north or south.	90° − 32.6° = 57.4° The direction is N57.4°W.
State the magnitude and direction of the resultant vector.	$R = 29.7\text{ m}$, N57.4°W

Worked example: Try yourself 10.2.3

ADDING VECTORS IN TWO DIMENSIONS USING GEOMETRY

Determine the resultant force on a tree when forces of 5.00 N east and 3.00 N north act on the tree.

Geometric method of subtracting vectors in two dimensions

Vectors can also be subtracted more accurately using Pythagoras' theorem and trigonometry.

Figure 10.2.13 shows how to calculate the resultant velocity when changing from 25.0 m s^{-1} east to 20.0 m s^{-1} south. The initial velocity of 25.0 m s^{-1} east and the final velocity of 20.0 m s^{-1} south are drawn. Then the opposite of the initial velocity is drawn as 25.0 m s^{-1} west. The final velocity vector is drawn first, then from its head the opposite of the initial velocity is drawn. The resultant velocity vector, shown in red, is calculated to be 32.0 m s^{-1}. The resultant vector is calculated to be in the direction S51.3°W.

The resultant vector is 32.0 m s^{-1} S51.3°W.

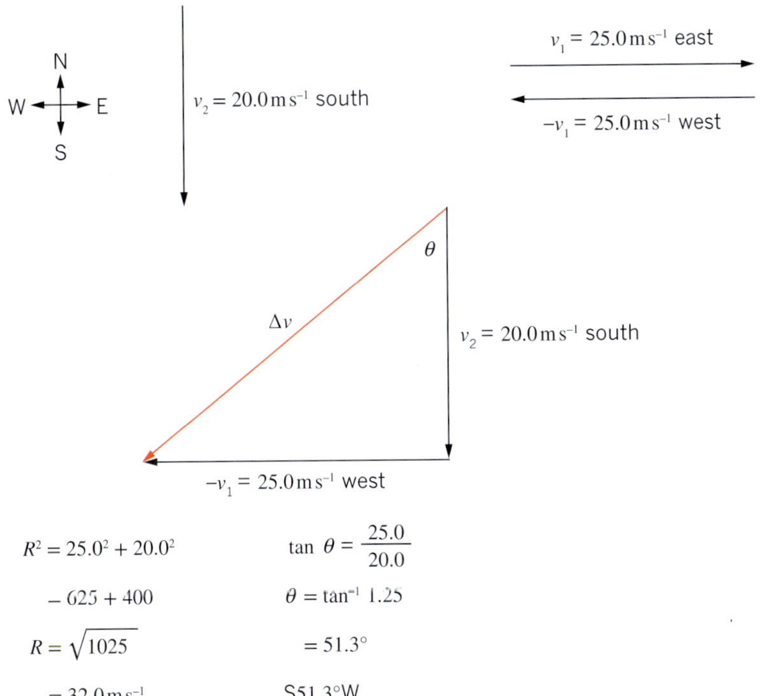

FIGURE 10.2.13 Subtracting two vectors at right angles, using the geometric method

Worked example 10.2.4

SUBTRACTING VECTORS IN TWO DIMENSIONS USING GEOMETRY

Determine the change in velocity of Clare's scooter as she turns a corner if she approaches it at 18.7 m s^{-1} west and exits at 16.6 m s^{-1} north.

Thinking	Working
Draw the final velocity vector, v_2, and the initial velocity vector, v_1, separately. Then draw the initial velocity in the opposite direction.	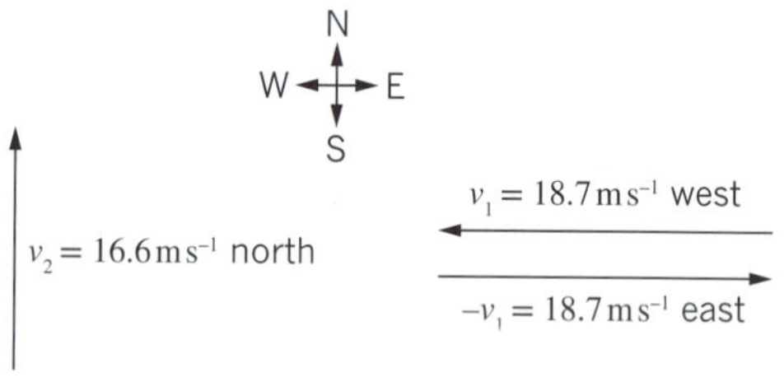 N W E S v_2 = 16.6 m s^{-1} north v_1 = 18.7 m s^{-1} west $-v_1$ = 18.7 m s^{-1} east
Construct a vector diagram drawing v_2 first and then from its head draw the opposite of v_1. The change of velocity vector is drawn from the tail of the final velocity to the head of the opposite of the initial velocity.	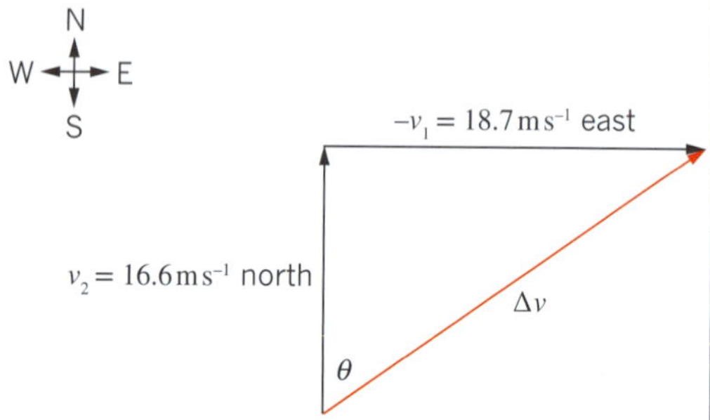 N W E S $-v_1$ = 18.7 m s^{-1} east v_2 = 16.6 m s^{-1} north Δv θ
As the two vectors to be added are at 90° to each other, apply Pythagoras' theorem to calculate the magnitude of the change in velocity.	$R^2 = 16.6^2 + 18.7^2$ $= 275.26 + 349.69$ $R = \sqrt{625.25}$ $= 25.0\text{ m s}^{-1}$
Calculate the angle from the north vector to the change in velocity vector.	$\tan\theta = \dfrac{18.7}{16.6}$ $\theta = \tan^{-1} 1.16$ $= 48.4°$
State the magnitude and direction of the change in velocity.	$\Delta v = 25.0\text{ m s}^{-1}$ N48.4°E

Worked example: Try yourself 10.2.4

SUBTRACTING VECTORS IN TWO DIMENSIONS USING GEOMETRY

Determine the change in velocity of a ball as it bounces off a wall. The ball approaches at 7.00 m s^{-1} south and rebounds at 6.00 m s^{-1} east.

CASE STUDY **ANALYSIS**

Surveying

Surveyors use technology to measure, analyse and manage data about the shape of the land and the exact location of landmarks and buildings. They take many measurements, including angles and distances, and use them to calculate more advanced data such as vectors, bearings, coordinates, elevations, maps etc. Surveyors typically use theodolites (see Figure 10.2.14 and Figure 10.2.15), GPS survey equipment, laser range finders and satellite images to map the land in three dimensions.

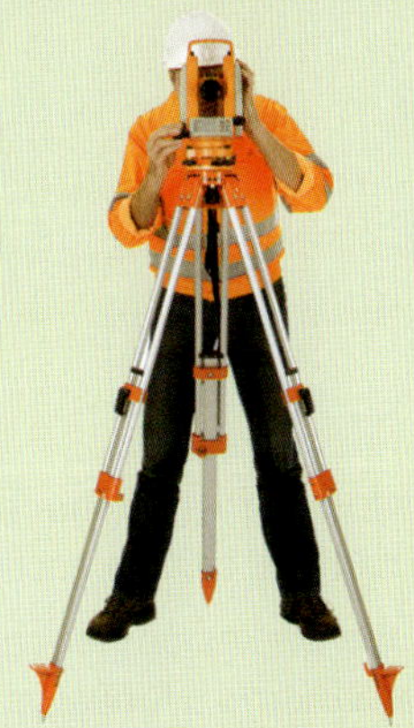

FIGURE 10.2.14 Surveying the land with a theodolite

Surveyors are often the first professionals on a building site to ensure that the boundaries of the property are correct. They also ensure that the building is built in the correct location. Surveyors must liaise closely with architects both before and during a building project, as they provide position and height data for walls and floors.

FIGURE 10.2.15 Surveying equipment being used on a building site

Analysis

1 A surveyor is hired to find the average slope of a plot of land for a client. The figure below shows that the distance between the two measuring rods is 7.320 m. The instrument is horizontally level with the 2.920 m mark on the downhill rod and with the 0.400 m mark on the uphill rod. Determine the slope of the land as an angle up from horizontal.

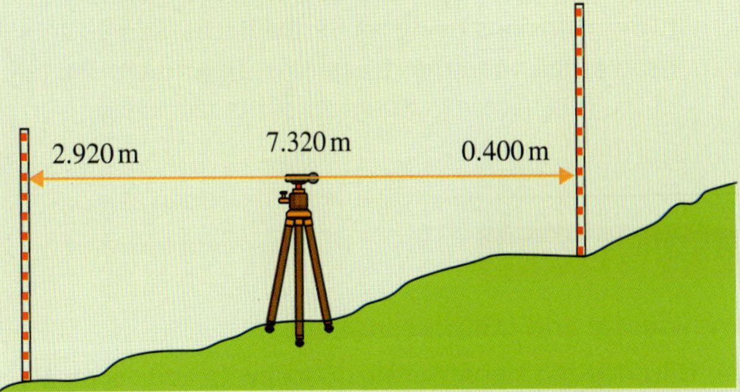

2 The diagram below shows the shape of a building site. A theodolite is located at a reference point at the centre of the protractor provided. Using the protractor and the scale, determine the angle and distance from the reference point to each corner of the building site labelled A, B, C and D.

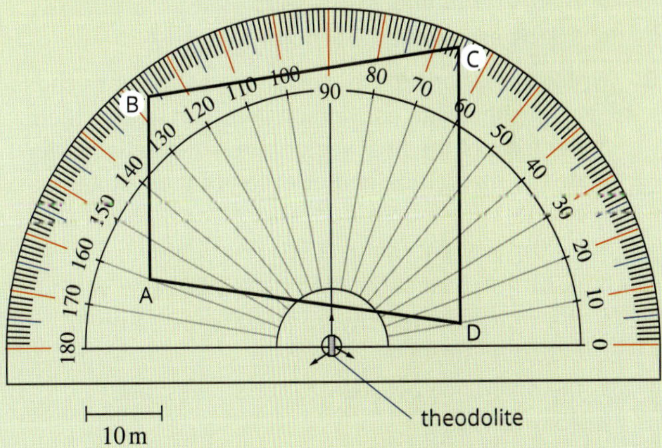

3 Using a protractor and a suitable scale, construct the shape of the concrete pad of a new home, if there are five corners that have the following distances and angles from a reference point.

Corner	Angle from reference point	Distance from reference point (m)
A	88°	7.0
B	165°	32
C	133°	56
D	69°	58
E	38°	36

10.2 Review

OA ✓✓

SUMMARY

- The process of combining vectors is known as adding vectors.
- One-dimensional vector addition refers to vectors in a line, while two-dimensional vector addition refers to vectors on a plane.
- Adding or subtracting vectors in one dimension can be done algebraically by applying positive or negative signs to the directions.
- Adding vectors can be done graphically in one or two-dimensions using vector diagrams. Add vectors head to tail, then the resultant is drawn from the tail of the first vector to the head of the last vector.
- Adding or subtracting vectors in two dimensions can be:
  - estimated graphically with a scale and a protractor
  - calculated using Pythagoras' theorem for the magnitude and trigonometry for the direction.
- Alternatively, vectors can be added or subtracted in two dimensions by constructing a parallelogram of vectors.
- To find the difference between, or change in, vectors, subtract the initial vector from the final vector.

KEY QUESTIONS

Knowledge and understanding

1 Describe the steps you would follow to add vector B to vector A using the head to tail method.

2 Draw vector diagrams using the scale of 1 cm = 2 N for the following additions. (Draw and label the resultant force using a different colour.) State the magnitude and direction of the resultant vector.
 a 10.0 N right and 6.0 N right
 b 8.00 N north and 2.00 N south
 c 12.0 N east and 8.00 N north

3 Write expressions for the following vector subtractions by applying a sign convention and then writing them in the form of the final vector minus the initial vector. State the magnitude and direction of the resultant vector.
 a 7.00 m s^{-1} right minus 2.00 m s^{-1} right
 b 9.00 m s^{-1} east subtracted from 4.00 m s^{-1} west
 c the change in velocity from 6.00 m s^{-1} south to 6.00 m s^{-1} north
 d the difference between 17.0 m s^{-1} west and 45.0 m s^{-1} east

4 Draw vector diagrams using the scale of 1 cm = 2 m s^{-1} for the following vector subtractions. (Draw and label the resultant force using a different colour.) State the magnitude and direction of the resultant vector.
 a 9.0 m s^{-1} left minus 2.0 m s^{-1} right
 b 5.0 m s^{-1} west subtract 9.0 m s^{-1} west
 c the change in velocity from 12.0 m s^{-1} east to 12.0 m s^{-1} south
 d the difference between 10.0 m s^{-1} north and 8.00 m s^{-1} west

Analysis

5 Find the resultant vector when the following are combined: 2.00 m west, 5.00 m east and 7.00 m west.

6 Jamelia applies the brakes on her car and changes her velocity from 22.2 m s^{-1} forwards to 8.20 m s^{-1} forwards. Calculate the change in velocity of the car.

7 Calculate the magnitude of the resultant vector when 30.0 m south and 40.0 m west are added.

8 A yacht tacks during a race, changing its velocity from 7.05 m s^{-1} south to 5.25 m s^{-1} west. Calculate the change in the velocity of the yacht.

10.3 Vector components

Section 10.2 explored how vectors can be combined to find a resultant vector. In this section, the process is reversed. In physics there are times when it is useful to break one vector up into two vectors that are at right angles to each other. For example, if a force vector is acting at an angle up from the horizontal, as shown in Figure 10.3.1, this vector can be considered to consist of two independent vertical and horizontal components.

The components of a vector can be found using trigonometry.

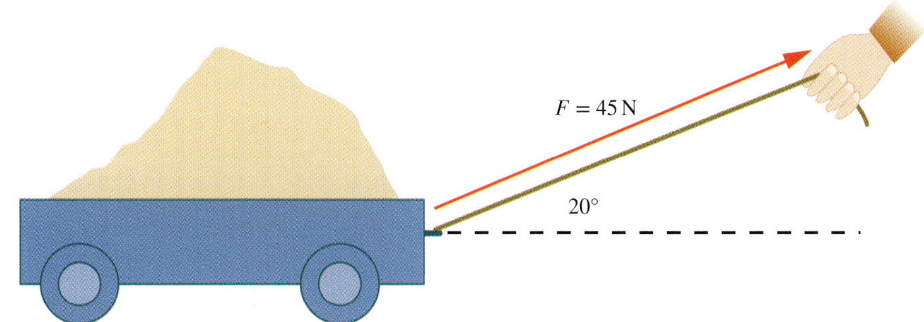

FIGURE 10.3.1 The pulling force acting on the cart has a component in the horizontal direction and a component in the vertical direction.

FINDING PERPENDICULAR COMPONENTS OF A VECTOR

Vectors at an angle are more easily dealt with if they are broken up into perpendicular **components**; that is, two components that are at right angles to each other. These components, when added together, give the original vector. To find the components of a vector, a right-angled triangle is constructed with the original vector as the hypotenuse. This is shown in Figure 10.3.2. The hypotenuse is always the longest side of a right-angled triangle and is opposite the 90° angle. The other two sides of the triangle are each shorter than the hypotenuse and form the 90° angle with each other. These two sides are the perpendicular components of the original vector.

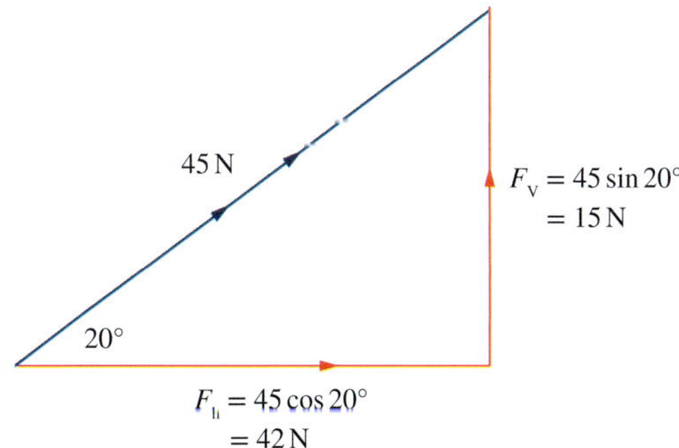

FIGURE 10.3.2 The perpendicular components (shown in red) of the original vector (shown in blue). The original vector is the hypotenuse of the triangle.

Geometric method of finding vector components

The geometric method of finding the perpendicular components of a vector is to construct a right-angled triangle using the original vector as the hypotenuse. This was illustrated in Figure 10.3.2. The magnitude and direction of the components are then determined using trigonometry. A good rule to remember is that no component of a vector can be larger than the vector itself. In a right-angled triangle, no side is longer than the hypotenuse. The original vector must be the hypotenuse and its components must be the other two sides of the triangle.

Figure 10.3.3 shows a force vector of 50.0 N (drawn in black) acting on a box in a direction 30.0° upwards from horizontal to the right. The horizontal and vertical components of this force must be found in order to complete further calculations.

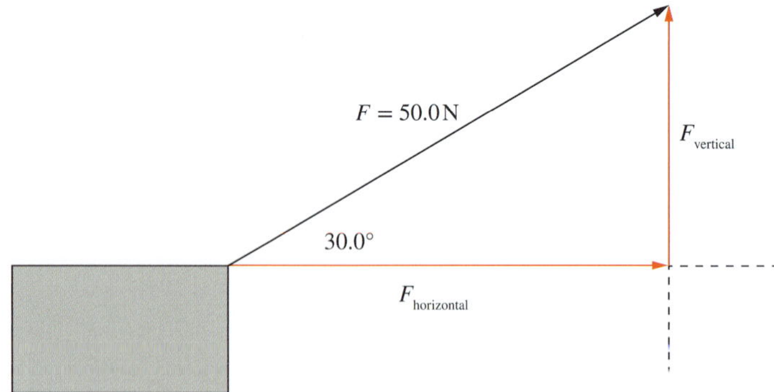

FIGURE 10.3.3 Finding the horizontal and vertical components of a force vector

The horizontal component vector is drawn from the tail of the 50.0 N vector towards the right, with its head directly below the head of the original 50.0 N vector. The vertical component vector is drawn from the head of the horizontal component to the head of the original 50.0 N vector.

Using trigonometry, the horizontal component of the force is calculated to be 43.3 N horizontally to the right. The vertical component is calculated to be 25.0 N vertically upwards. The calculations are shown below:

$$\cos\theta = \frac{\text{adjacent}}{\text{hypotenuse}} = \frac{F_{\text{horizontal}}}{F}$$

$$F_{\text{horizontal}} = 50.0 \times \cos 30.0°$$
$$= 43.3\text{ N horizontal to the right}$$

$$\sin\theta = \frac{\text{opposite}}{\text{hypotenuse}} = \frac{F_{\text{vertical}}}{F}$$

$$F_{\text{vertical}} = 50.0 \times \sin 30.0°$$
$$= 25.0\text{ N vertically upwards}$$

Worked example 10.3.1

CALCULATING THE PERPENDICULAR COMPONENTS OF A FORCE

Use the direction conventions to determine the perpendicular components of a 235 N force acting on a bike in a direction of 17.0° north of west.

Thinking	Working
Draw F_W from the tail of the 235 N force along the horizontal direction, then draw F_N from the horizontal vector to the head of the 235 N force.	N, W, E, S; F_N; $F = 235\text{ N}$; 17.0°; F_W
Calculate the west component of the force, F_W, using $\cos\theta = \frac{\text{adjacent}}{\text{hypotenuse}}$	$\cos\theta = \frac{\text{adjacent}}{\text{hypotenuse}} = \frac{F_W}{235}$ $F_W = 235 \times \cos 17.0°$ $= 224.7\text{ N west}$
Calculate the north component of the force, F_N, using $\sin\theta = \frac{\text{opposite}}{\text{hypotenuse}}$	$\sin\theta = \frac{\text{opposite}}{\text{hypotenuse}} = \frac{F_N}{235}$ $F_N = 235 \times \sin 17.0°$ $= 68.7\text{ N north}$

Worked example: Try yourself 10.3.1

CALCULATING THE PERPENDICULAR COMPONENTS OF A FORCE

Use the direction conventions to determine the perpendicular components of a 3540 N force acting on a trolley in a direction of 26.5° anticlockwise from the left direction.

10.3 Review

OA ✓✓

SUMMARY

- A vector can be resolved into two component vectors that are perpendicular (at right angles to each other).
- Any component vectors must be smaller in magnitude than the original vector.
- A right-angled triangle vector diagram can be drawn with the original vector as the hypotenuse and the perpendicular components drawn from the tail of the original to the head of the original.
- The perpendicular components can be determined using trigonometry.

KEY QUESTIONS

Knowledge and understanding

1 Copy each of the vectors shown below and draw the two component vectors that each could be resolved into (use a different colour for the components).

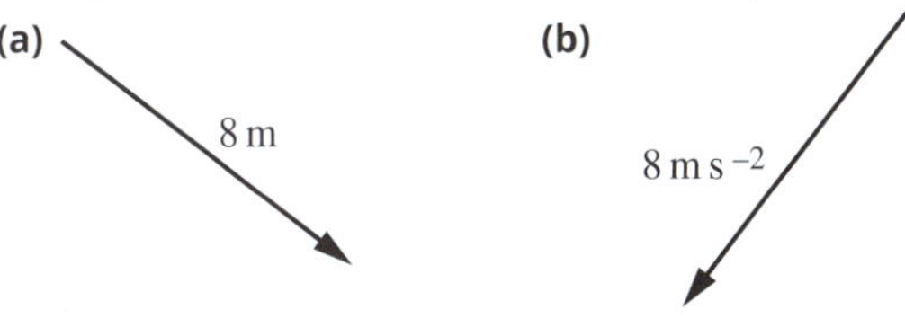

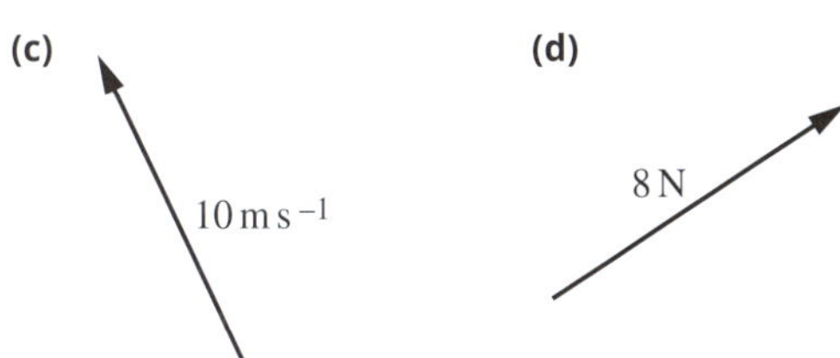

Analysis

2 Rayko applies a force of 462 N on the handle of a mower in a direction of 35.0° clockwise downwards from the right direction.

a What is the downwards force applied?

b What is the rightwards force applied?

3 A force of 25.9 N acts in the direction of S40.0°E. Find the perpendicular components of the force.

4 A ferry is transporting students to Phillip Island. At one point in the journey the ferry is travelling at 18.3 $m s^{-1}$ S75.6°E. Calculate its velocity in the southerly direction and in the easterly direction at that time.

5 Zehn walks 47.0 m in the direction of S66.3°E across a hockey field. Calculate the change in Zehn's position down the field and across the field during that time.

6 A cargo ship has two tugs attached to it by ropes. One of the tugs is pulling directly north, while the other tug is pulling directly west. The pulling forces of the tugboats combine to produce a total force of 235 000 N in a direction of N62.5°W. Calculate the force that each tug boat applies to the cargo ship.

7 Resolve the following forces into their perpendicular components around the north–south line. In part **d**, use the horizontal and vertical directions.

a 100 N S60°E

b 60.0 N north

c 300 N 160°T

d 3.00×10^5 N 30.0° upwards from the horizontal

8 What are the horizontal and vertical components of a 455 N force that is applied at 63.0° upwards from horizontal by a rope used to drag an object across a yard?

Chapter review

10

KEY TERMS

collinear
components
dimension
direction convention
magnitude
resultant
scalar
units
vector
vector diagram

REVIEW QUESTIONS

Knowledge and understanding

1 Select the scalar quantities in the list below. (There may be more than one answer.)
  A force
  B time
  C acceleration
  D mass

2 Select the vector quantities in the list below. (There may be more than one answer.)
  A displacement
  B distance
  C volume
  D velocity

3 Why is it sometimes appropriate to rename direction conventions with a positive or negative sign—for example, + instead of north or − instead of left?

4 A basketballer applies a force with his hand to bounce the ball. Describe how a vector can be drawn to represent this situation.

5 Vector arrow A is drawn twice the length of vector arrow B. What does this mean?

6 A car travels at 15.0 m s^{-1} north and then later travels at 20.0 m s^{-1} south. Why is a sign convention often used to describe vectors like these?

7 When finding the change in velocity graphically between an initial velocity of 34.0 m s^{-1} south and a final velocity of 12.5 m s^{-1} east, which two vectors need to be added together?

8 If the vector 20.0 N forwards is written as 20.0 N, how would you write a vector representing 80.0 N backwards?

9 Describe the following vector direction.

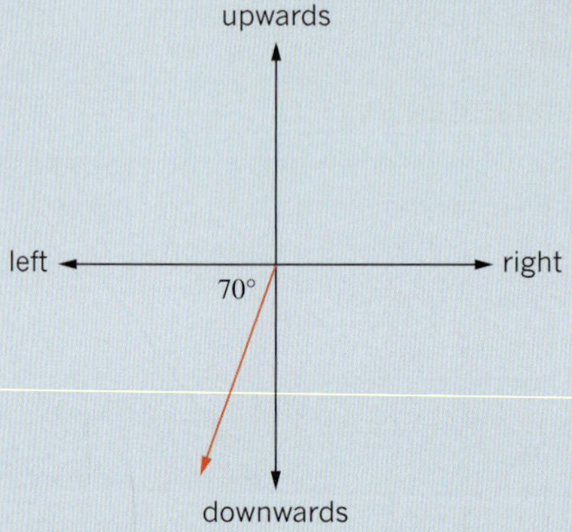

10 Describe the following vector using appropriate conventions.

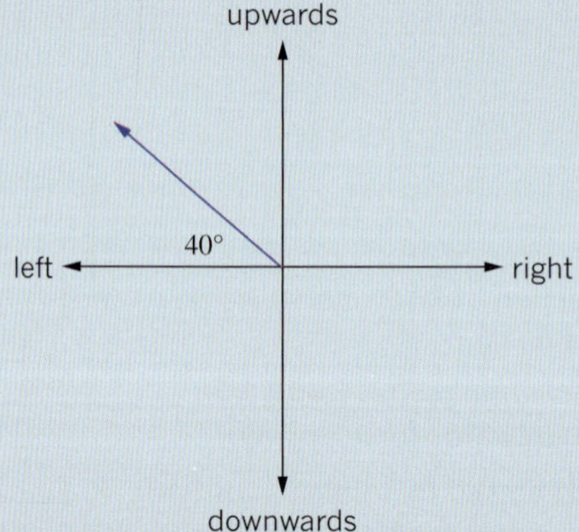

Application and analysis

11 Add the following force vectors using a number line: 3.00 N left, 2.00 N right, 6.00 N right. Then also draw and describe the resultant force vector.

12 Add the following vectors to find the resultant vector: 3.00 m upwards, 2.00 m downwards and 3.00 m downwards.

13 Determine the resultant vector of the following combination: 45.0 m forwards, 70.5 m backwards, 34.5 m forwards, 30.0 m backwards.

14 Determine the change in velocity of a runner who changes from running at 4.00 m s^{-1} to the right on grass to running 2.00 m s^{-1} to the right in sand.

15 A student throws a ball up into the air at 4.00 m s^{-1}. A short time later the ball is travelling back downwards to hit the ground at 3.00 m s^{-1}. Determine the change in velocity of the ball during this time.

16 Determine the change in velocity of a bird that changes from flying 3.00 m s^{-1} to the right to flying 3.00 m s^{-1} to the left.

17 Find the vector that results from the addition of 36.0 m south and 55.0 m west.

18 Add the following vectors: 481 N north and 655 N east. Give answers to three significant figures.

19 Describe the magnitude and direction of the resultant vector, drawn in red, in the following diagram.

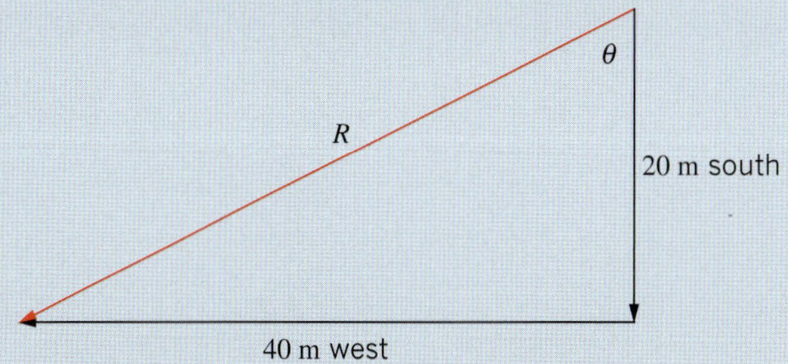

20 Forces of 2000 N north and 6000 N east act on an object. What is the resultant force acting?

21 Determine the resultant vector of a toy train that is made to move in these directions: 23.0 m forwards, 16.0 m backwards, 7.00 m forwards and 3.00 m backwards.

22 A car that was initially travelling at a velocity of 3.00 m s^{-1} west is later travelling at 5.00 m s^{-1} east. What is the difference between the two vectors?

23 A car makes a turn, changing its velocity from 13.0 m s^{-1} south to 18.7 m s^{-1} west. Calculate the change in the velocity vector, Δv, of the car, to three significant figures.

24 Bill hits a cricket ball so that the velocity of the ball changes from 38.8 m s^{-1} east to 55.5 m s^{-1} north. Calculate the change in the velocity vector, to three significant figures.

25 Tom hits a tennis ball against a wall. If the ball travels towards the wall at 35.0 m s^{-1} north and rebounds at 32.5 m s^{-1} south, calculate the change in velocity of the ball.

26 A jet plane makes a turn after taking off, changing its velocity from 345 m s^{-1} south to 406 m s^{-1} west. Calculate the change in the velocity of the jet.

27 Yvette hits a golf ball that strikes a tree and changes its velocity from 42.0 m s^{-1} east to 42.0 m s^{-1} north. Calculate the change in the velocity of the golf ball.

28 A force of 45.5 N acts in the direction of S60.0°E. Find the eastern and southern components of this force. Give your answers to three significant figures.

OA
✓✓

CHAPTER

11 Linear motion

Motion, from the simple to the complex, is a fundamental part of everyday life. The motion of a gymnast performing a floor routine is a complex form of motion. An Olympic snowboarder competing in a half-pipe event also exhibits a complex form of motion. Simpler examples include a skier travelling in a straight line down a ski run, a train pulling into a station and a swimmer completing a lap of a pool.

Key knowledge

- analyse graphically, numerically and algebraically, straight-line motion under constant acceleration:
  $v = u + at$, $v^2 = u^2 + 2as$, $s = \frac{1}{2}(u+v)t$, $s = ut + \frac{1}{2}at^2$, $s = vt - \frac{1}{2}at^2$ **11.1, 11.2, 11.3, 11.4, 11.5**
- analyse, graphically, non-uniform motion in a straight line **11.3**
- model the force due to gravity, F_g, as the force of gravity acting at the centre of mass of a body, $F_{\text{on body by Earth}} = mg$, where g is the gravitational field strength ($9.8\,\text{N kg}^{-1}$ near the surface of Earth). **11.5**

OA ✓✓

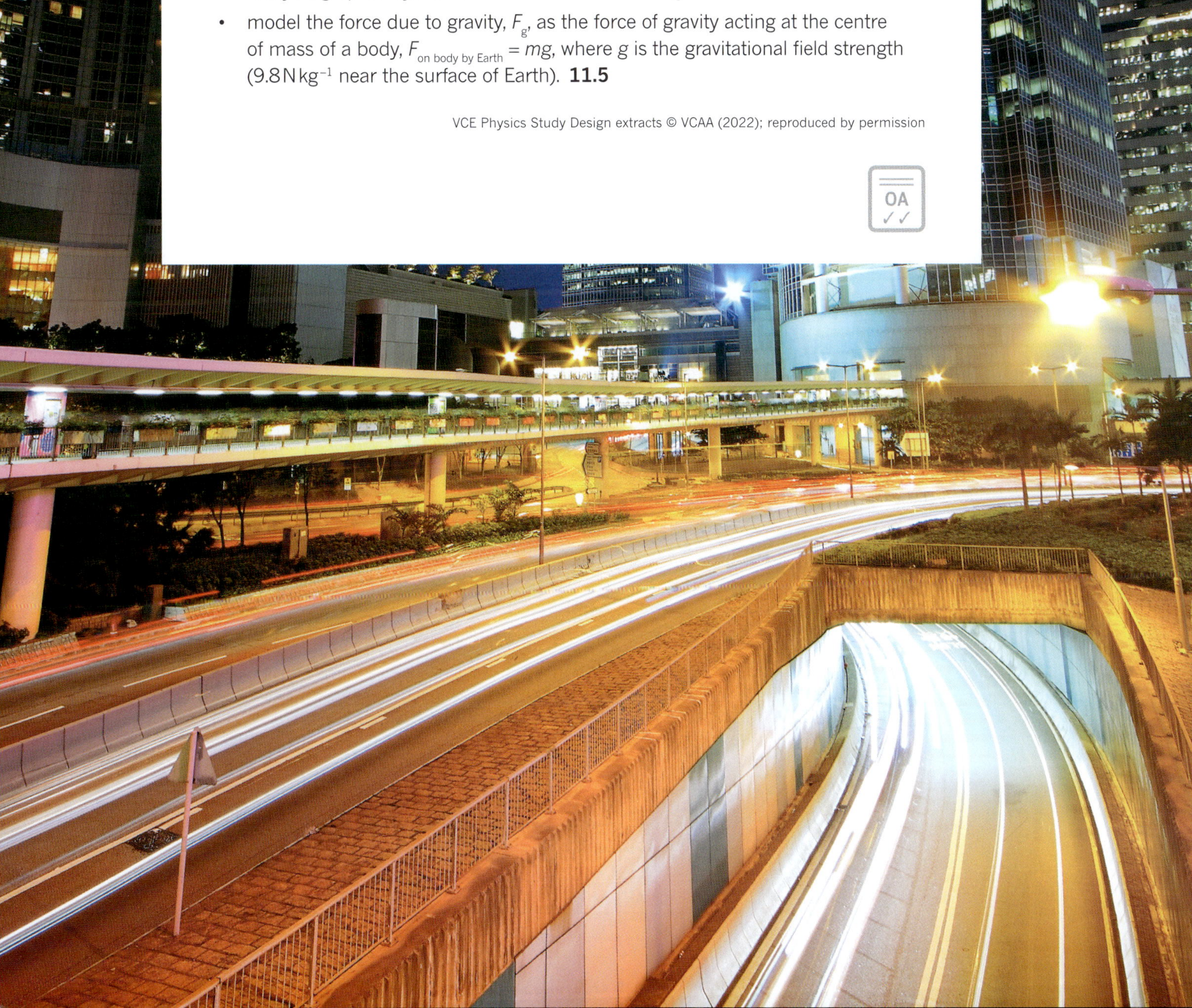

11.1 Displacement, speed and velocity

(a)

(b)

(c)

(d)

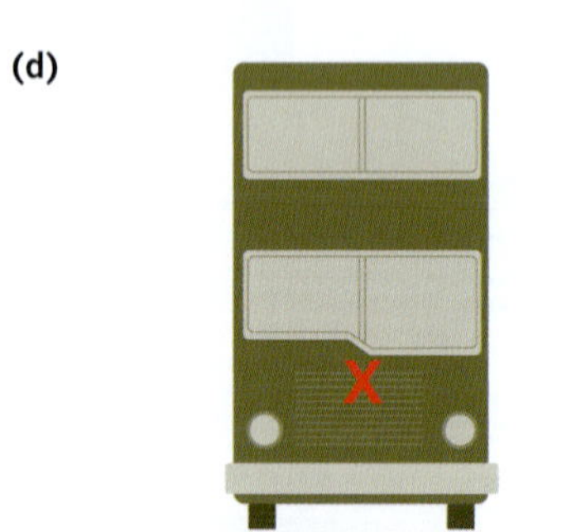

FIGURE 11.1.1 The centre of mass of each object is indicated by a cross.

In order to analyse and communicate ideas about motion, it is important to understand the terms used to describe motion—even in its simplest form. In this section you will learn some of the terms used to describe straight-line motion, such as position, distance, displacement, speed and velocity.

CENTRE OF MASS

When analysing motion, things are often more complicated than they first appear. As a freestyle swimmer travels at a constant speed of $2.00\,m\,s^{-1}$, their head and torso move forwards at this speed. The motion of their arms, however, is more complex. At times their arms move forwards through the air faster than $2.00\,m\,s^{-1}$, and at other times they are actually moving backwards through the water.

It is beyond the scope of this course to analyse such a complex motion. However, the motion of the swimmer can be simplified by treating the swimmer as a simple object located at a single point called the **centre of mass** or centre of gravity. The centre of mass is the balance point of an object. For a person, the centre of mass is located near the waist. The centres of mass of some everyday objects are shown in Figure 11.1.1.

POSITION

One important term to understand when analysing straight line motion is **position**.

- Position describes the location of an object at a certain point in time with respect to the origin.
- Position is a vector quantity and therefore requires a direction.

Consider a swimmer, Sophie, doing laps in a 50.0 m pool, as shown in Figure 11.1.2. To simplify her motion, Sophie is treated as a simple point object. The pool can be treated as a one-dimensional number line, with the starting block as the origin. The direction to the right of the starting block is taken to be positive.

Sophie's position as she is warming up behind the starting block in Figure 11.1.2(a) is −10.0 m. The negative sign indicates the direction from the origin; that is, to the left. Her position could also be given as 10.0 m to the left of the starting block.

At the starting block (Figure 11.1.2(b)), Sophie's position is 0.0 m, then after swimming half a length of the pool she is +25.0 m or 25.0 m to the right of the origin (Figure 11.1.2(c)).

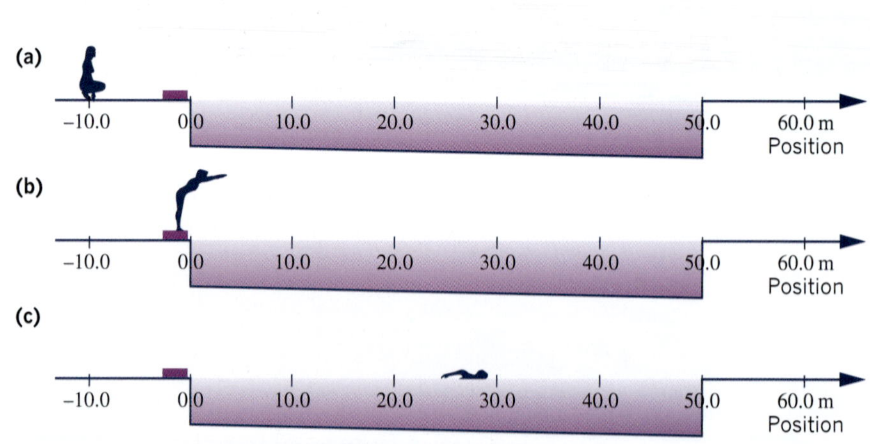

FIGURE 11.1.2 The position of the swimmer is given with reference to the starting block. (a) While warming up, Sophie is at −10.0 m. (b) When she is on the starting block, her position is zero. (c) After swimming for a short time, she is at a position of +25.0 m.

DISTANCE TRAVELLED

Position describes where an object is at a certain point in time. However, **distance travelled** is how far a body travels during a journey.

- Distance travelled, *d*, describes the length of the path covered by an object over the entire journey.
- Distance travelled is a scalar quantity and is measured in metres (m).

For example, if Sophie completes three lengths of the pool, the distance travelled during her swim will be 50.0 + 50.0 + 50.0 = 150.0 m.

The distance travelled is not affected by the direction of the motion. That is, the distance travelled by an object always increases as it moves, regardless of its direction. The tripmeter or odometer of a car or bike measures distance travelled.

DISPLACEMENT

Displacement, *s*, is defined as the change in position of an object. Displacement takes into account only where the motion starts and finishes. The route taken between these points has no effect on displacement. The sign of the displacement indicates the direction in which the position has changed from the start to the end.

- Displacement is the change in position of an object in a given direction.
- Displacement s = final position – initial position.
- Displacement is a vector quantity and is measured in metres (m).

Consider the example of Sophie completing one length of the pool. During her swim, the distance travelled is 50 m. Her final position is +50.0 m and her initial position is 0 m. Her displacement is:

$$\begin{aligned} s &= \text{final position} - \text{initial position} \\ &= 50.0 - 0.0 \\ &= +50.0\,\text{m or } 50.0\,\text{m in a positive direction.} \end{aligned}$$

Notice that magnitude, units and direction are required for a vector quantity. The distance will be equal to the magnitude of the displacement only if the body is moving in a straight line and does not change direction. If Sophie swims two lengths, her distance travelled will be 100.0 m: 50.0 m out and 50.0 m back. However, her displacement during this swim will be:

$$\begin{aligned} s &= \text{final position} - \text{initial position} \\ &= 0.0 - 0.0 \\ &= 0.0\,\text{m} \end{aligned}$$

Even though Sophie has swum 100.0 m, her displacement is zero because the initial and final positions are the same.

The above formula for displacement is useful if you already know the initial and final positions of a body's motion. An alternative method to determine total displacement, if you know the displacement of each section of the motion, is to add up the individual displacements for each section of motion.

total displacement = sum of individual displacements

It is important to remember that displacement is a vector and so, when adding displacements, you must obey the rules of vector addition (discussed in Chapter 10).

In the example above, in which Sophie completed two laps, overall displacement could have been calculated by adding the displacement of each lap:

$$\begin{aligned} s &= \text{sum of displacements for each lap} \\ &= 50.0\,\text{m in the positive direction} + 50.0\,\text{m in the negative direction} \\ &= 50.0 + (-50.0) \\ &= 0.0\,\text{m} \end{aligned}$$

SPEED AND VELOCITY

For thousands of years, humans have tried to travel at ever greater speeds. This desire has contributed to the development of all sorts of competitive activities, as well as major advances in engineering and design. World records for some of these pursuits are given in Table 11.1.1 (page 312).

TABLE 11.1.1 World record speeds for a variety of sports or modes of transport

Activity	World record speed ($m\,s^{-1}$)	World record speed ($km\,h^{-1}$)
luge	42	150
train (Japanese maglev train)	167.5	603
tennis serve	73.1	263
waterskiing	63.4	230
cricket delivery	44.7	161
racehorse	19.7	71

Speed and velocity are both quantities that give an indication of how quickly the position of an object is changing. Both terms are in common use and are often assumed to have the same meaning. In physics, however, these two terms have different definitions.

- **Speed** is defined in terms of the rate at which the distance is travelled. Like distance, speed is a scalar. A direction is not required when describing the speed of an object.
- **Velocity** is defined in terms of the rate at which displacement changes, and so is a vector quantity. A direction should always be given with a velocity.
- The SI unit for speed and velocity is metres per second ($m\,s^{-1}$), but kilometres per hour ($km\,h^{-1}$) is also commonly used.

Instantaneous speed and velocity

Instantaneous speed and instantaneous velocity give a measure of how fast something is moving at a particular point in time. The speedometer on a car or bike indicates instantaneous speed.

If a speeding car is travelling north and is detected on a police radar gun at 150.0 $km\,h^{-1}$, it indicates that this car's instantaneous speed is 150.0 $km\,h^{-1}$, while its instantaneous velocity is 150.0 $km\,h^{-1}$ north. Notice that the instantaneous speed is equal to the magnitude of the instantaneous velocity. This is always the case for instantaneous speed and velocity.

Average speed and velocity

Average speed and average velocity both give an indication of how fast an object is moving over a time interval.

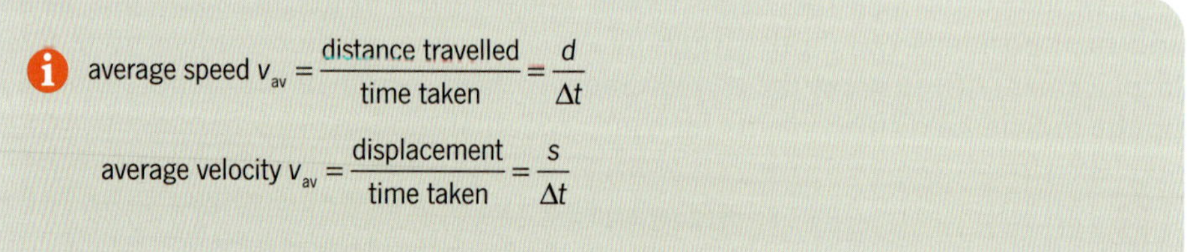

$$\text{average speed } v_{av} = \frac{\text{distance travelled}}{\text{time taken}} = \frac{d}{\Delta t}$$

$$\text{average velocity } v_{av} = \frac{\text{displacement}}{\text{time taken}} = \frac{s}{\Delta t}$$

Average speed is equal to instantaneous speed only when a body is moving in uniform motion (i.e. if it moves at a constant speed).

The average speed of a car that takes 30.0 minutes to travel 20.0 km from St Kilda to Dandenong is 40.0 $km\,h^{-1}$. However, this does not mean that the car travelled the whole distance at this speed. In fact it is more likely that the car was moving at 60.0 $km\,h^{-1}$ for a significant amount of time, while for some of the time the car would not be moving at all.

A direction (such as north, south, upwards, downwards, left, right, positive, negative) must be given when describing a velocity. The direction will always be the same as that of the displacement. Similar to the relationship between distance and displacement, average speed will be equal to the magnitude of average velocity only if the body is moving in a straight line and does not change direction.

For example, in a race around a circular track, such as the velodrome shown in Figure 11.1.3, regardless of the average speed for a complete lap, the magnitude of the average velocity will be zero because the displacement is zero.

FIGURE 11.1.3 Anna Meares won the UCI world championship in 2013. She rode 500 m in a world record time of 32.836 s. Her average speed was 55.6 $km\,h^{-1}$ but her average velocity was zero.

CASE STUDY **ANALYSIS**

Alternative units for speed and distance

Metres per second is the standard unit for measuring speed because it is derived from the standard unit for distance (metres) and the standard unit for time (seconds). However, alternative units are often used to better suit a certain application.

Road speed limits are better described in kilometres per hour, as cars tend to travel for greater distances, and for longer times.

Converting km h^{-1} to m s^{-1}

The speed limit for most freeways and country roads in Victoria is $100\,\text{km h}^{-1}$. Cars that maintain this speed would travel 100 km in 1 hour. Since there are 1000 m in 1 km and 3600 seconds in 1 hour, then $100\,\text{km h}^{-1}$ is the same as travelling 100 000 m in 3600 s.

$$\begin{aligned} 100\,\text{km h}^{-1} &= 100 \times 1000\,\text{m h}^{-1} \\ &= 100\,000\,\text{m h}^{-1} \\ &= \frac{100\,000}{3600}\,\text{m s}^{-1} \\ &= 27.8\,\text{m s}^{-1} \end{aligned}$$

So km h^{-1} can be converted to m s^{-1} by multiplying by $\frac{1000}{3600}$ (or dividing by 3.6).

Converting m s^{-1} to km h^{-1}

A champion Olympic sprinter can run at an average speed of close to $10.0\,\text{m s}^{-1}$. Each second, the athlete will travel approximately 10.0 m. At this rate, in 1 hour the athlete would travel $10.0 \times 3600 = 36\,000$ m or 36 km.

$$\begin{aligned} 10.0\,\text{m s}^{-1} &= 10 \times 3600\,\text{m h}^{-1} \\ &= 36\,000\,\text{m h}^{-1} \\ &= \frac{36\,000}{1000}\,\text{km h}^{-1} \\ &= 36.0\,\text{km h}^{-1} \end{aligned}$$

So m s^{-1} can be converted to km h^{-1} by multiplying by $\frac{3600}{1000}$ or 3.6.

When converting a speed from one unit to another, it is important to think about the speed to ensure that your answer makes sense. The diagram in Figure 11.1.4 summarises the conversion between units for speed.

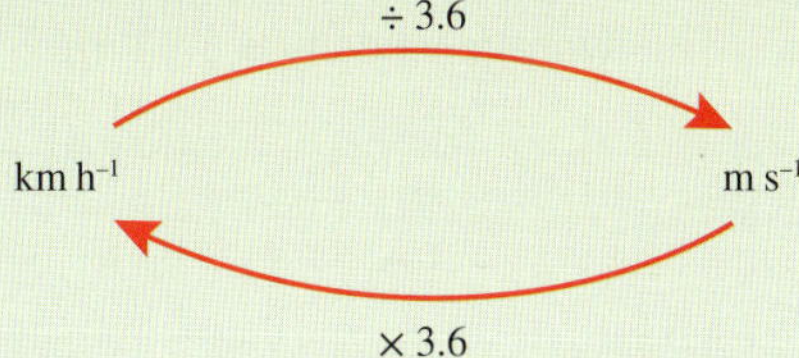

FIGURE 11.1.4 Rules for converting between m s^{-1} and km h^{-1}

Alternative units

The speed of a ship is usually measured in knots, where 1 knot = $0.514\,\text{m s}^{-1}$. This unit originated in the nineteenth century, when the speed of sailboats would be measured by allowing a rope to trail in the water and be dragged by the water through a sailor's hands. The rope would have knots tied at regular intervals of 14.4 m, and by counting the number of knots that passed over a period of 28.0 seconds, you would know the speed of the ship in knots.

The speed of very fast aeroplanes, such as the one in Figure 11.1.5, can be measured in Mach numbers. One Mach (referred to as Mach 1) is equal to the speed of sound, which is $340\,\text{m s}^{-1}$. Mach 2 is equal to $680\,\text{m s}^{-1}$, or twice the speed of sound.

FIGURE 11.1.5 Modern fighter aeroplanes are able to fly at speeds above Mach 1.

The light-year is an alternative unit for measuring distance. The speed of light in a vacuum is close to $300\,000\,\text{km s}^{-1}$. One light-year is the distance that light travels in one year. Because distances between objects in the universe are so large, astronomers use the light-year to measure distances. It takes over 4 years for light to travel from our nearest star (Alpha Centauri) to Earth. That means the distance from Earth to our nearest star is over 4 light-years. Light takes approximately 8.5 minutes to travel from the Sun to Earth, so it could be said that the Sun is 8.5 light-minutes away.

Analysis

1 Calculate the following speeds in the desired unit.
   - **a** $40.0\,\text{km h}^{-1}$ in m s^{-1}
   - **b** $60.0\,\text{km h}^{-1}$ in m s^{-1}
   - **c** $25.0\,\text{m s}^{-1}$ in km h^{-1}
   - **d** $215\,\text{m s}^{-1}$ in km h^{-1}

2 Modern yachts like the AC75, which is a foiling monohull, are capable of incredible speeds. The fastest recorded speed of 53.3 knots was measured in an America's Cup semifinal in 2021. Determine the yacht's equivalent speed in m s^{-1} and km h^{-1}.

continued over page

CASE STUDY ANALYSIS *continued*

3 The minimum speed required for a rocket to be able to launch a satellite into orbit is 28000 km h^{-1}. Calculate what this speed equates to in Mach.

4 In early 2021, a radio signal coming from the Voyager 1 space probe, launched in 1977, took approximately 21 hours, 3 minutes and 42 seconds to reach Earth. Calculate the distance at that time to Voyager 1 from Earth in kilometres.

PHYSICSFILE

Reaction time

Drivers are often distracted by loud music or phone calls. These distractions result in many accidents and deaths on the road. If cars are travelling at high speeds, they will travel a large distance in the time that the driver takes just to apply the brakes. A short reaction time is very important for all road users. This is easy to understand given the relationship between speed, distance and time:

$$\text{distance travelled} = v \times t$$

For example, consider a car travelling at 100 km h^{-1} = 28 m s^{-1}. An extra 1 s added to the reaction time to brake would result in the car travelling an extra 28 × 1 = 28 m.

Worked example 11.1.1

AVERAGE VELOCITY AND CONVERTING UNITS

Sam is an athlete performing a training routine by running back and forth along a straight stretch of running track. He jogs 100.0 m north in a time of 20.0 s, then turns and walks 50.0 m south in a further 25.0 s before stopping.

a What is Sam's average velocity in m s^{-1}?	
Thinking	**Working**
Calculate the displacement. Remember total displacement is the sum of individual displacements. Sam's total journey consists of two displacements: 100.0 m north and then 50.0 m south. Take north to be the positive direction.	s = sum of displacements = 100.0 m north + 50.0 m south = 100.0 + (−50.0) = +50.0 m or 50.0 m north S ← → N start, finish, 0, 20.0, 100.0 m start, finish, 0, 45.0, 50.0 m
Work out the total time taken for the journey.	20.0 + 25.0 = 45.0 s
Substitute the values into the velocity equation.	Displacement is 50.0 m north. Time taken is 45.0 s. Average velocity $v_{av} = \frac{s}{\Delta t}$ $= \frac{50.0}{45.0}$ = 1.11 m s^{-1}
Velocity is a vector, so a direction must be given.	Average velocity = 1.11 m s^{-1} north

b What is the magnitude of Sam's average velocity in $km\,h^{-1}$?	
Thinking	**Working**
Convert from $m\,s^{-1}$ to $km\,h^{-1}$ by multiplying by 3.6.	$v_{av} = 1.11\,m\,s^{-1}$ $= 1.11 \times 3.6$ $= 4.00\,km\,h^{-1}$ north
As the magnitude of the velocity is needed, direction is not required in this answer.	Magnitude of $v_{av} = 4.0.0\,km\,h^{-1}$.

c What is Sam's average speed in $m\,s^{-1}$?	
Thinking	**Working**
Calculate the distance. Remember distance is the length of the path covered in the entire journey. The direction does not matter. Sam travels 100 m in one direction and then 50 m in the other direction	$d = 100.0 + 50.0$ $= 150.0\,m$
Work out the total time taken for the journey.	$20.0 + 25.0 = 45.0\,s$
Substitute the values into the speed equation.	Distance is 150.0 m. Time taken is 45.0 s. Average speed $v_{av} = \frac{d}{\Delta t}$ $= \frac{150.0}{45.0}$ $= 3.33\,m\,s^{-1}$

d What is Sam's average speed in $km\,h^{-1}$?	
Thinking	**Working**
Convert from $m\,s^{-1}$ to $km\,h^{-1}$ by multiplying by 3.6.	Average speed $v_{av} = 3.33\,m\,s^{-1}$ $= 3.33 \times 3.6$ $= 12.0\,km\,h^{-1}$

Worked example: Try yourself 11.1.1

AVERAGE VELOCITY AND CONVERTING UNITS

Sally is an athlete performing a training routine by running back and forth along a straight stretch of running track. She jogs 100.0 m west in a time of 20.0 s, then turns and walks 160.0 m east in a further 45.0 s before stopping.

a What is Sally's average velocity in $m\,s^{-1}$?

b What is the magnitude of Sally's average velocity in $km\,h^{-1}$?

c What is Sally's average speed in $m\,s^{-1}$?

d What is Sally's average speed in $km\,h^{-1}$?

CASE STUDY

How police measure the speeds of cars

Road accidents cause the deaths of about 1200 people in Australia each year and many times this number are seriously injured. Numerous steps have been taken to reduce the number of road fatalities, including random alcohol and drug testing, the compulsory wearing of bicycle helmets and the zero blood alcohol level for probationary drivers.

One of the main causes of road trauma is speeding. In their efforts to combat speeding motorists, police employ a variety of speed-measuring devices such as speed cameras. One such device is shown in Figure 11.1.6.

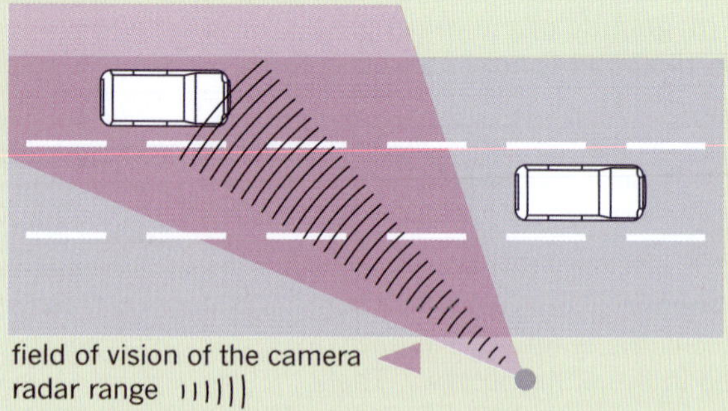

FIGURE 11.1.7 Diagram showing the visual range of a speed camera

FIGURE 11.1.6 Speed cameras on poles

Speed camera radar

Speed cameras emit a microwave radar signal with a precisely known frequency of 24.15 GHz (2.415×10^{10} Hz). The radar antenna has a parabolic reflector that enables the unit to produce a directional radar beam that is 5° wide, allowing individual vehicles to be targeted. The radar range and field of vision for a camera is shown in Figure 11.1.7. The radar signal allows speeds to be determined by the Doppler principle, as the reflected radar signal from a vehicle heading towards a radar unit has a higher frequency than the original signal. Conversely, the reflected signal from a vehicle heading away from the radar unit has a lower frequency. This change in frequency or 'Doppler shift' is processed by the unit and gives a measurement of the instantaneous speed of the target vehicle. Camera radar units are capable of targeting a single vehicle up to 1.2 km away.

Laser speed guns

Speed guns are used by police to obtain an instant measure of the speed of an approaching or receding vehicle. The unit is usually handheld and is aimed directly at a vehicle using a target sight. It emits a pulse of infrared radiation frequency of 331 THz (3.31×10^{14} Hz). In a similar way to the speed camera units, the speed is determined by the Doppler shift produced by the target vehicle. The infrared pulse is very narrow and directional, just 0.17° wide. This allows vehicles to be targeted with great precision. Handheld units can be used at distances up to 800 m.

Fixed speed cameras

Fixed speed cameras take their readings using three strips with piezoelectric sensors in them placed under the bitumen across the road (Figure 11.1.8). The strips are triggered by pressure as the car drives over them to create an electrical pulse that is detected. By knowing the distance between the strips, and measuring the time that the car takes to travel across them, the speed of the car can be determined.

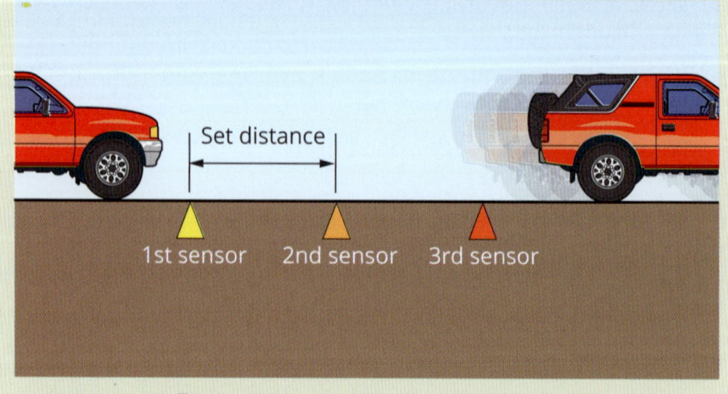

FIGURE 11.1.8 Fixed speed cameras record the speed of a car twice by measuring the time the car takes to travel over a series of three sensor strips embedded in the roadway.

CASE STUDY ANALYSIS

Breaking the speed limit

Over the past 100 years, advances in engineering and technology have led to the development of faster machines. Cars, planes and trains can now move people at speeds that were thought to be both unattainable and life-threatening a century ago.

The fastest production car in the world is the 2020 Tuatara Hypercar, which beat the speed record set by the Koenigsegg Agera RS. The Tuatara reached an average speed of $508.73\,\text{km}\,\text{h}^{-1}$ ($141.31\,\text{m}\,\text{s}^{-1}$) over two runs in opposite directions on a public road.

The fastest jet plane ratified by the International Federation of Aeronautics is the Lockheed SR-71 Blackbird. In 1976 it reached a speed of $3529.6\,\text{km}\,\text{h}^{-1}$ ($980.44\,\text{m}\,\text{s}^{-1}$), which is almost 2.9 times the speed of sound.

The fastest speed recorded by a conventional train is $574.8\,\text{km}\,\text{h}^{-1}$ ($159.7\,\text{m}\,\text{s}^{-1}$) in 2007 by the French TGV.

In 2007, Markus Stoeckl of Austria set a new speed record for production mountain bikes. He reached a speed of $210\,\text{km}\,\text{h}^{-1}$ ($58.3\,\text{m}\,\text{s}^{-1}$) racing down a snow-covered ski slope in Chile. He is pictured in Figure 11.1.9.

World speed records are achievable by anyone with an imagination. For example, the fastest average speed over a 100-metre run by a person wearing swim fins is $24.2915\,\text{km}\,\text{h}^{-1}$ ($6.74764\,\text{m}\,\text{s}^{-1}$).

FIGURE 11.1.9 Markus Stoeckl setting a new speed record for mountain biking in 2007

Analysis

1 Suggest a reason why the production car speed record should be an average determined from two runs in opposite directions.

2 The distance from Melbourne to Sydney is 713 km by plane. If you were flying a Lockheed SR-71 Blackbird at its top speed, how many seconds would it take to make that trip?

3 Japan is currently building a maglev train system from Tokyo to Nagoya, which is a journey of 285.6 km. With an operating speed of $505\,\text{km}\,\text{h}^{-1}$ this is slower than the French TGV. Research the maglev system and suggest what advantages a maglev train would have over conventional trains.

11.1 Review

OA ✓✓

SUMMARY

- Position defines the location of an object with respect to a defined origin.
- Distance travelled, *d*, tells us how far an object has actually travelled. Distance travelled is a scalar.
- Displacement, *s*, is a vector and is defined as the change in position of an object in a given direction: *s* = final position – initial position.
- The average speed of a body, v_{av}, is defined as the rate of change of distance and is a scalar quantity:

$$\text{average speed } v_{av} = \frac{\text{distance travelled}}{\text{time taken}} = \frac{d}{\Delta t}$$

- The average velocity of a body, v_{av}, is defined as the rate of change of displacement and is a vector quantity:

$$\text{average velocity } v_{av} = \frac{\text{displacement}}{\text{time taken}} = \frac{s}{\Delta t}$$

- To convert from $m\,s^{-1}$ to $km\,h^{-1}$, multiply by 3.6.
- To convert from $km\,h^{-1}$ to $m\,s^{-1}$, divide by 3.6.
- The SI unit for both speed and velocity is metres per second ($m\,s^{-1}$).

KEY QUESTIONS

Knowledge and understanding

1 A girl swims ten lengths of a 25.0 m pool. Describe the distance travelled and her displacement from the starting point.

2 An ant is walking back and forth along a metre ruler, as show in the figure below. Taking the right as positive, determine both the size of the displacement and the distance travelled by the ant as it travels on the following paths.

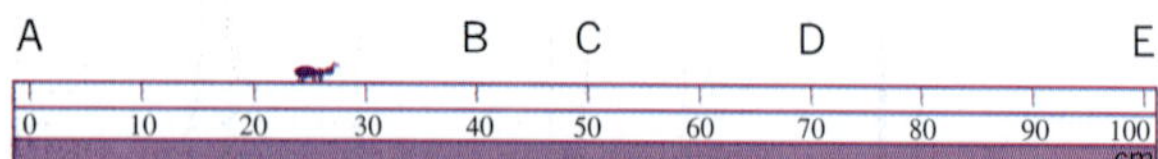

a A to B
b C to B
c C to D
d C to E and then to D

3 During a training ride, a cyclist rides 50.0 km north then 30.0 km south.

a What is the distance travelled by the cyclist during the ride?
b What is the displacement of the cyclist for this ride?

4 A lift in a city building, shown in the figure on the right, carries a passenger from the ground floor down to the basement, then up to the top floor.

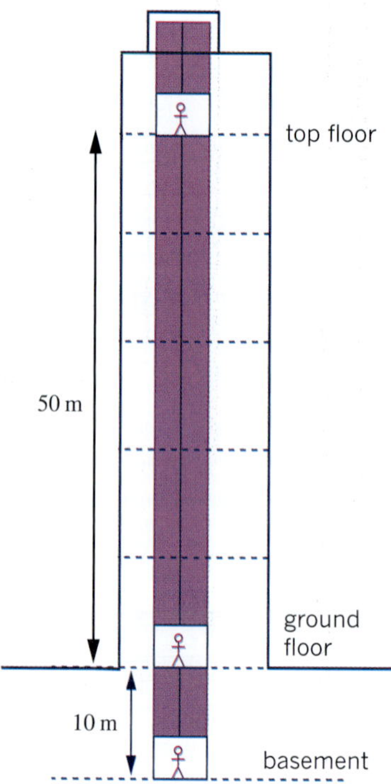

a What is the displacement of the lift as it travels from the ground floor to the basement?
b What is the displacement of the lift as it travels from the basement to the top floor?
c What is the distance travelled by the lift during this entire trip?
d What is the displacement of the lift during this entire trip?

Analysis

5 A car travelling at a constant speed was timed over 400.0 m and found to cover the distance in 12.0 s.

a What was the average speed of the car?

b The driver was distracted and his reaction time was 0.750 s before applying the brakes. How far did the car travel in this time?

6 A cyclist travels 25.0 km in 90.0 minutes.

a What is her average speed in $km\,h^{-1}$?

b What is her average speed in $m\,s^{-1}$?

7 Liam pushes his toy truck 5.00 m east, then stops it and pushes it 4.00 m west. The entire motion takes 10.0 seconds.

a What is the truck's average speed?

b What is the truck's average velocity?

8 An athlete in training for a marathon runs 10.0 km north along a straight road before realising that she has dropped her drink bottle. She turns around and runs back 3.00 km to find her bottle, then resumes running in the original direction. After running for 1.50 h, the athlete reaches 15.0 km from her starting position and stops.

a What is the distance travelled by the athlete during the run?

b What is the athlete's displacement during the run?

c What is the average speed of the athlete in $km\,h^{-1}$?

d What is the athlete's average velocity in $km\,h^{-1}$?

11.2 Acceleration

If you have been on a train as it pulled out of the station, you have experienced acceleration. If you have been in an aeroplane as it took off along a runway, you will have experienced a much greater acceleration. Astronauts and fighter pilots experience enormous accelerations that would make an untrained person lose consciousness. **Acceleration** is a measure of how quickly velocity changes, and will be discussed in this section.

FINDING THE CHANGE IN VELOCITY AND SPEED

The velocity and speed of everyday objects are changing all the time. Examples of these are a car as it moves away as the traffic lights turn green, a tennis ball as it bounces or you travelling on a rollercoaster. If the initial and final velocity of an object are known, its change in velocity can be calculated.

To find the change, Δ (Greek symbol 'delta'), in any physical quantity, including speed and velocity, the initial value is taken away from the final value.

Change in speed or velocity is the final value minus the initial value:

$$\Delta v = v - u$$

where u is the initial speed or velocity (in m s^{-1})
v is the final speed or velocity (in m s^{-1})
Δv is the change in speed or velocity (in m s^{-1}).

Change in speed is a simple algebraic subtraction, but since velocity is a vector, this should be done by performing a vector subtraction. As for all vectors, direction is required.

Vector subtraction was covered in detail in Section 10.2 on page 292.

Worked example 11.2.1

CHANGE IN SPEED AND VELOCITY PART 1

A golf ball is dropped onto a concrete floor and strikes the floor at 5.00 m s^{-1}. It then rebounds at 5.00 m s^{-1}.

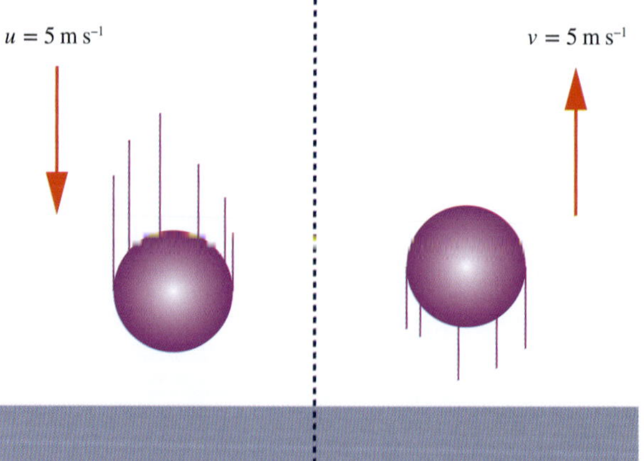

a What is the change in the speed of the ball?	
Thinking	**Working**
Note the values for the initial speed and the final speed of the ball.	$u = 5.00$ m s^{-1} $v = 5.00$ m s^{-1}
Substitute the values into the change in speed equation: $\Delta v = v - u$	$\Delta v = v - u$ $= 5.00 - 5.00$ $= 0$ m s^{-1}

b What is the change in the velocity of the ball?	
Thinking	**Working**
Velocity is a vector. Apply the sign convention to replace the directions.	$u = 5.00\,\mathrm{m\,s^{-1}}$ down $= -5.00\,\mathrm{m\,s^{-1}}$ $v = 5.00\,\mathrm{m\,s^{-1}}$ up $= +5.00\,\mathrm{m\,s^{-1}}$
Substitute the values into the change in velocity equation: $\Delta v = v - u$	$\Delta v = v - u$ $= (+5.00) - (-5.00)$ $= +10.0\,\mathrm{m\,s^{-1}}$
Apply the sign convention to describe the direction.	$\Delta v = 10.0\,\mathrm{m\,s^{-1}}$ up

Worked example: Try yourself 11.2.1

CHANGE IN SPEED AND VELOCITY PART 1

A golf ball is dropped onto a concrete floor and strikes the floor at $9.00\,\mathrm{m\,s^{-1}}$. It then rebounds at $7.00\,\mathrm{m\,s^{-1}}$.

a What is the change in the speed of the ball?

b What is the change in the velocity of the ball?

CALCULATING ACCELERATION

Consider the following information about the velocity of a car that starts from rest, as shown in Figure 11.2.1.

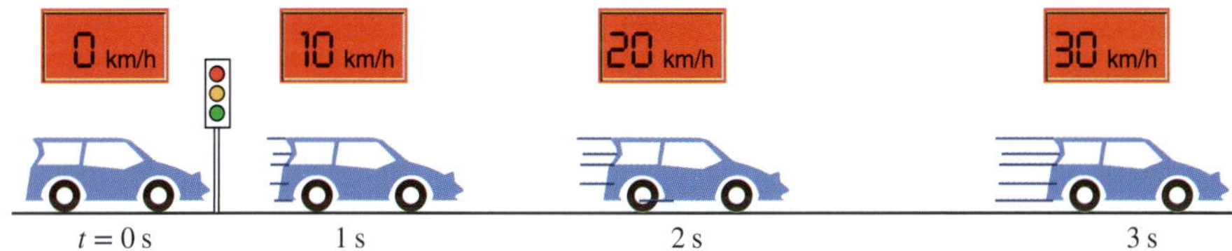

FIGURE 11.2.1 A car's acceleration as it increases in velocity from $0\,\mathrm{km\,h^{-1}}$ to $30\,\mathrm{km\,h^{-1}}$

The velocity of the car in Figure 11.2.1 increases by $10\,\mathrm{km\,h^{-1}}$ each second. In other words, its velocity changes by $+10\,\mathrm{km\,h^{-1}}$ per second. This is stated as an acceleration of +10 kilometres per hour per second or $+10\,\mathrm{km\,h^{-1}\,s^{-1}}$. More commonly in physics, velocity information is given in metres per second.

The athlete in Figure 11.2.2 takes 3 seconds to come to a stop at the end of a race.

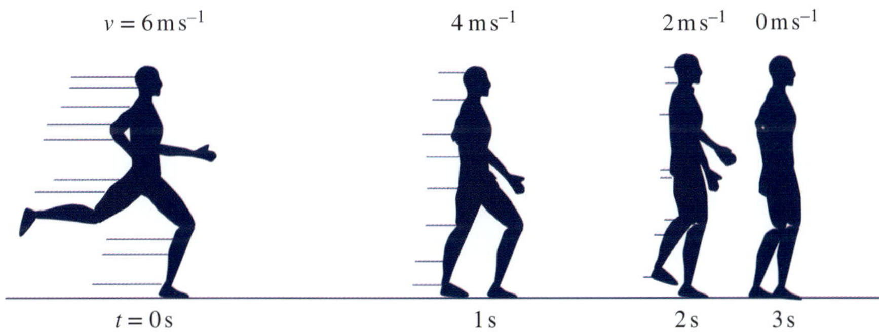

FIGURE 11.2.2 The velocity of the athlete changes by $-2\,\mathrm{m\,s^{-1}}$ each second. The acceleration is $-2\,\mathrm{m\,s^{-2}}$.

The velocity of the athlete changes by $-2\,\text{m}\,\text{s}^{-1}$ each second, so the acceleration is −2 metres per second per second. This is usually expressed as −2 metres per second squared or $-2\,\text{m}\,\text{s}^{-2}$.

A negative acceleration can mean that the object is slowing down in the direction of travel, as is the case with the athlete in Figure 11.2.2. A negative acceleration can also mean speeding up but in the opposite direction.

As acceleration is a vector quantity, vector diagrams can be used to calculate resultant accelerations of an object. Vector diagrams were covered in Chapter 10.

Average acceleration

As with speed and velocity, the average acceleration of an object can also be calculated.

Average acceleration, a_{av}, is the rate of change of velocity:

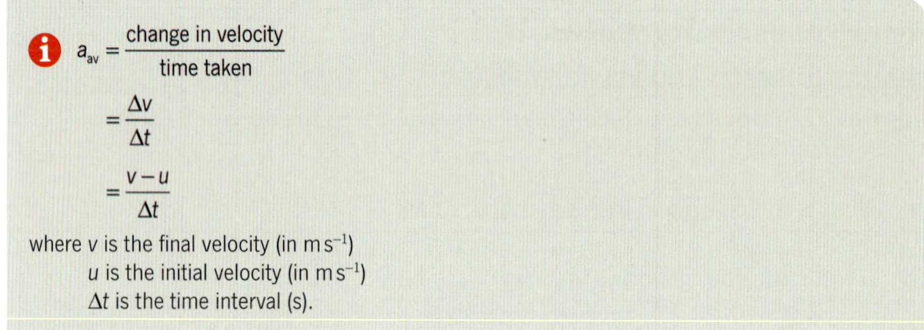

$$a_{av} = \frac{\text{change in velocity}}{\text{time taken}} = \frac{\Delta v}{\Delta t} = \frac{v - u}{\Delta t}$$

where v is the final velocity (in $\text{m}\,\text{s}^{-1}$)
u is the initial velocity (in $\text{m}\,\text{s}^{-1}$)
Δt is the time interval (s).

Worked example 11.2.2

CHANGE IN SPEED AND VELOCITY PART 2

A golf ball is dropped onto a concrete floor and strikes the floor at $5.00\,\text{m}\,\text{s}^{-1}$. It then rebounds at $5.00\,\text{m}\,\text{s}^{-1}$. The contact with the floor lasts for 25.0 ms.

What is the average acceleration of the ball during its contact with the floor?

Thinking	Working
Note the values you will need in order to find the average acceleration: initial velocity, final velocity and time. Convert ms into s by dividing by 1000. (Note that Δv was calculated for this situation in the previous worked example.)	$u = -5.00\,\text{m}\,\text{s}^{-1}$ $v = 5.00\,\text{m}\,\text{s}^{-1}$ $\Delta v = 10.00\,\text{m}\,\text{s}^{-1}$ upwards $\Delta t = 25.0\,\text{ms}$ $= 0.0250\,\text{s}$
Substitute the values into the average acceleration equation.	$a_{av} = \frac{\text{change in velocity}}{\text{time taken}}$ $= \frac{\Delta v}{\Delta t}$ $= \frac{10.0}{0.0250}$ $= 400\,\text{m}\,\text{s}^{-2}$
Acceleration is a vector, so you must include a direction in your answer.	$a_{av} = 4.00 \times 10^2\,\text{m}\,\text{s}^{-2}$ upwards

Worked example: Try yourself 11.2.2

CHANGE IN SPEED AND VELOCITY PART 2

A golf ball is dropped onto a concrete floor and strikes the floor at $9.00\,\text{m}\,\text{s}^{-1}$. It then rebounds at $7.00\,\text{m}\,\text{s}^{-1}$. The contact time with the floor is 35.0 ms.

What is the average acceleration of the ball during its contact with the floor?

CASE STUDY ANALYSIS

Human acceleration

In the 1950s, the United States Air Force used the Sonic Wind Number 1 rocket sled to determine the effect of extremely large accelerations on humans. One of these sleds is shown in Figure 11.2.3. The aim was to find out the greatest accelerations that humans could safely withstand in order to develop ejector seats for pilots.

FIGURE 11.2.3 The rocket-powered sled used to test the effects of acceleration on humans

The testing site consisted of a 915.0 m long railway track and a sled with nine rocket motors. A volunteer, Colonel John Stapp, was strapped into the sled and accelerated to speeds of more than 1017.11 km h^{-1} in a very short time. This made Stapp the fastest person to travel on land at the time. To put this speed into perspective, Stapp was travelling so fast that he would have overtaken a 45-calibre bullet fired from a gun. Water scoops were used to stop the sled abruptly in just 1.10 s, causing an average deceleration of 256.85 m s^{-2}. However, the deceleration was not constant over the 1.10 s time. The maximum deceleration Stapp experienced was 452.76 m s^{-2}.

The effects of these massive accelerations are evident on his face (Figure 11.2.4).

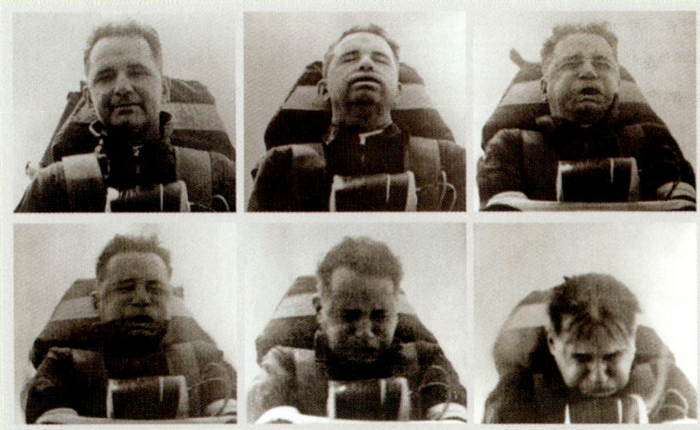

FIGURE 11.2.4 Photos showing the distorted face of Colonel John Stapp

Colonel John Stapp was a human guinea pig who suffered a great deal of discomfort so that other pilots would benefit. In all, Stapp rode the Sonic Wind rocket sled 29 times. Safer ejector seats and non-human crash test dummies were developed because of these experiments.

Analysis

1. Calculate how many times the acceleration due to gravity divides into the average deceleration Colonel Stapp experienced. This value is known as the deceleration in gs.
2. Compare the maximum deceleration, in gs, of Colonel Stapp to the average deceleration from question 1.
3. Stapp was keen to boost the speed of the sled to 1609 km h^{-1} by adding more rockets; however, the Air Force authorities thought he would not survive the stop. If the rocket were to stop over 1.10 seconds, calculate the average deceleration in m s^{-2} and in gs.

11.2 Review

OA ✓✓

SUMMARY

- Change in speed is a scalar calculation:
  Δv = final speed − initial speed = $v - u$
- Change in velocity is a vector calculation:
  Δv = final velocity − initial velocity = $v - u$
- Acceleration is a vector. The average acceleration of a body, a_{av}, is defined as the rate of change of velocity:

$$a_{av} = \frac{\text{change in velocity}}{\text{time taken}} = \frac{\Delta v}{\Delta t} = \frac{v - u}{\Delta t}$$

- Acceleration is measured in metres per second per second ($m\,s^{-2}$).

KEY QUESTIONS

Knowledge and understanding

1 A radio-controlled car is travelling east at $10.0\,km\,h^{-1}$. It hits some sand and slows down to $3.00\,km\,h^{-1}$ east. What is its change in speed?

2 A lump of Blu Tack is falling vertically at $5.00\,m\,s^{-1}$ and as it hits the floor it stops dead. What is its change in velocity during the collision?

3 A ping pong ball is falling vertically at $5.00\,m\,s^{-1}$. It hits the floor and rebounds at $3.00\,m\,s^{-1}$ upwards. What is its change in velocity during the bounce?

Analysis

4 While playing soccer, Ashley runs north at $7.50\,m\,s^{-1}$, then slides along the ground and stops in 1.50 s. What is his average acceleration as he slides to a stop?

5 Olivia launches a model rocket vertically and it reaches a speed of $150\,km\,h^{-1}$ after 3.50 s. What is the magnitude of its average acceleration in $km\,h^{-1}\,s^{-1}$?

6 A squash ball travelling east at $25.0\,m\,s^{-1}$ strikes the front wall of the court and rebounds at $15.0\,m\,s^{-1}$ west. The contact time between the wall and the ball is 0.0500 s. Use vector diagrams, where appropriate, to help you with your calculations.

a What is the change in speed of the ball?

b What is the change in velocity of the ball?

c What is the magnitude of the average acceleration of the ball during its contact with the wall?

7 A greyhound starts from rest and accelerates uniformly. Its velocity after 1.20 s is $8.00\,m\,s^{-1}$ south.

a What is the change in speed of the greyhound?

b What is the change in velocity of the greyhound?

c What is the magnitude of the acceleration of the greyhound?

11.3 Graphing position, velocity and acceleration over time

At times, even the motion of an object travelling in a straight line can be complicated. The object may travel forwards or backwards, speed up or slow down, or even stop. Where the motion remains in one dimension, the information can be presented in graphical form.

The main advantage of a graph compared with a table is that it allows the nature of the motion to be seen clearly. Information that is contained in a table is not as readily accessible or as easy to interpret as information presented graphically. This section examines position–time, velocity–time and acceleration–time graphs.

POSITION-TIME (x–t) GRAPHS

A position–time graph indicates the position, x, of an object at any time, t, for motion that occurs over an extended time interval. However, the graph can also provide additional information.

Consider Sophie, shown in Figure 11.3.1, swimming laps of a 50 m pool. Her position–time data is shown in Table 11.3.1. The starting point is treated as the origin for this motion.

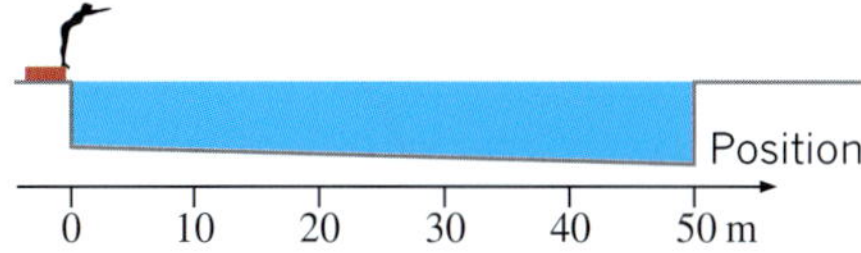

FIGURE 11.3.1 Sophie standing at the end of a 50 m swimming pool

TABLE 11.3.1 Positions and times of a swimmer completing 1.5 lengths of a pool

Time (s)	0.0	5.0	10	15	20	25	30	35	40	45	50	55	60
Position (m)	0.0	10	20	30	40	50	50	50	45	40	35	30	25

Analysis of Table 11.3.1 reveals several features of Sophie's swim. For the first 25 s, she swims at a constant rate. Every 5.0 s she travels 10 m in a positive direction; that is, her velocity is $+2.0\,\text{m s}^{-1}$. Then, from 25 s to 35 s, her position does not change. She seems to be resting, as she is stationary for this 10 s interval. Finally, from 35 s to 60 s, she swims back towards the starting point, in a negative direction. On this return lap, she maintains a more leisurely rate of 5.0 m every 5.0 s, so her velocity is $-1.0\,\text{m s}^{-1}$. However, Sophie does not complete this lap but ends 25 m from the start. This data is shown more conveniently on the position–time graph in Figure 11.3.2.

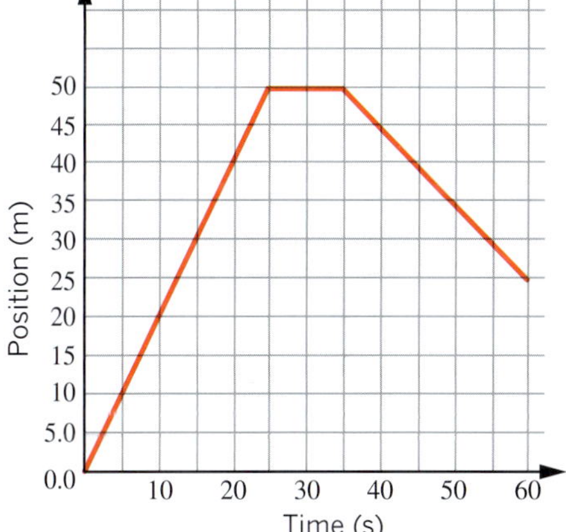

FIGURE 11.3.2 This position–time graph represents the motion of Sophie travelling 50 m along a pool, then resting and swimming back towards the starting position. Sophie finishes halfway along the pool.

Sophie's displacement, s, can be determined by comparing the initial and final positions. Her displacement between 20 s and 60 s is, for example:

$$\begin{aligned} s &= \text{final position} - \text{initial position} \\ &= 25 - 40 \\ &= -15\,\text{m} \end{aligned}$$

By further examining the graph above, it can be seen that during the first 25 s, Sophie has a displacement of +50 m. Thus her average velocity during this time is $+2.0\,\text{m s}^{-1}$; that is, $2.0\,\text{m s}^{-1}$ to the right. This value can also be obtained by finding the gradient of this section of the graph (Figure 11.3.3).

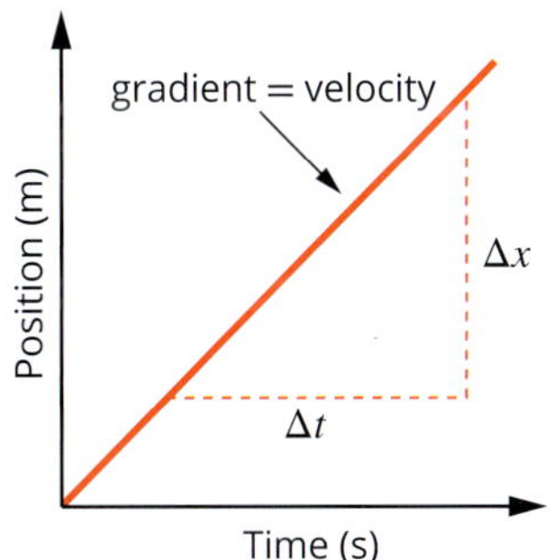

FIGURE 11.3.3 Position–time graph with gradient

Gradient of x–t graph = velocity

A positive velocity indicates that the object is moving in a positive direction and negative velocity indicates motion in a negative direction.

To confirm that the gradient of a position–time graph is a measure of velocity you can use **dimensional analysis**, for which you use the units in a graph or formula to check that the derived term is correct:

$$\text{Gradient of } x\text{–}t \text{ graph} = \frac{\text{rise}}{\text{run}} = \frac{\Delta x}{\Delta t}$$

The units of this gradient will be metres per second (m s^{-1}), so gradient is a measure of velocity. Note that the rise in the graph is the change in position, which is the definition of displacement; that is, $\Delta x = s$.

Non-uniform velocity

For motion with uniform (constant) velocity, the position–time graph will be a straight line, but if the velocity is non-uniform the graph will be curved. If the position–time graph is curved, the instantaneous velocity will be the gradient of the tangent to the line at the point of interest; the average velocity will be the gradient of the chord between two points. This is illustrated in Figure 11.3.4.

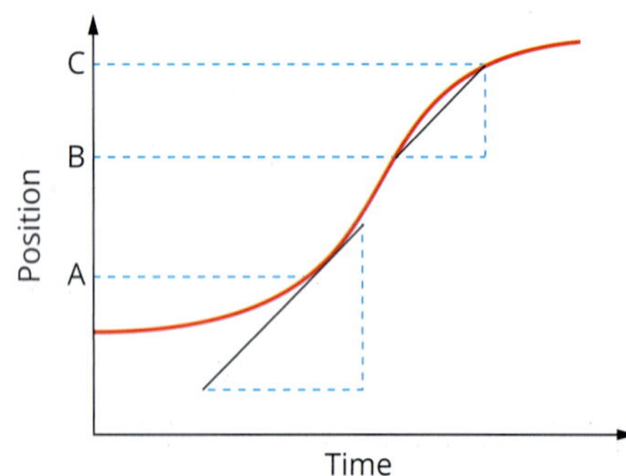

FIGURE 11.3.4 The instantaneous velocity at point A is the gradient of the tangent at that point. The average velocity between points B and C is the gradient of the chord between these points on the graph.

Worked example 11.3.1

ANALYSING A POSITION–TIME GRAPH

The motion of a cyclist is represented by the position–time graph shown, with important features of the motion labelled A, B, C, D, E and F.

Motion of a cyclist (position–time)

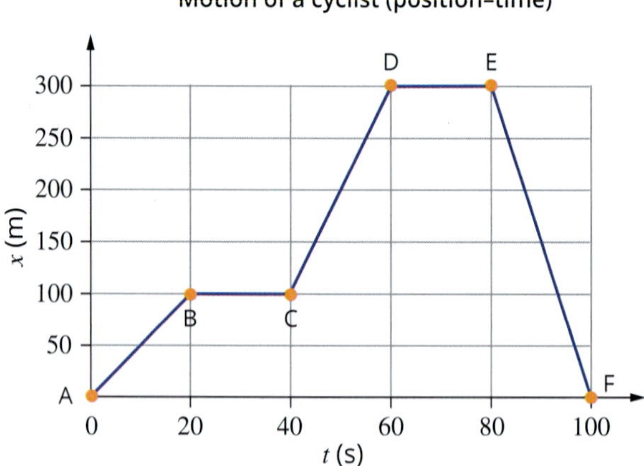

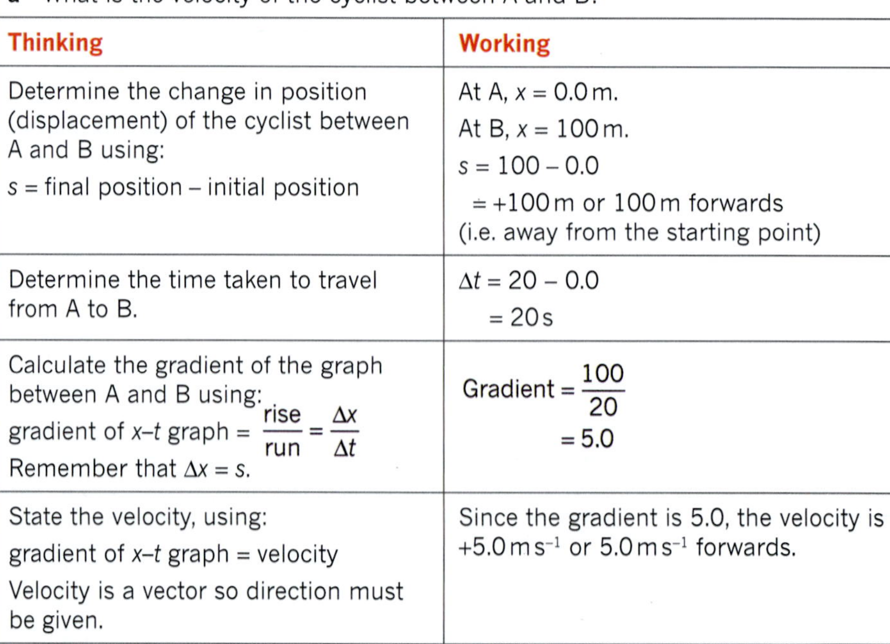

a What is the velocity of the cyclist between A and B?	
Thinking	**Working**
Determine the change in position (displacement) of the cyclist between A and B using: s = final position – initial position	At A, $x = 0.0$ m. At B, $x = 100$ m. $s = 100 - 0.0$ $= +100$ m or 100 m forwards (i.e. away from the starting point)
Determine the time taken to travel from A to B.	$\Delta t = 20 - 0.0$ $= 20$ s
Calculate the gradient of the graph between A and B using: gradient of x–t graph $= \frac{\text{rise}}{\text{run}} = \frac{\Delta x}{\Delta t}$ Remember that $\Delta x = s$.	Gradient $= \frac{100}{20}$ $= 5.0$
State the velocity, using: gradient of x–t graph = velocity Velocity is a vector so direction must be given.	Since the gradient is 5.0, the velocity is $+5.0\text{ m s}^{-1}$ or 5.0 m s^{-1} forwards.

b Describe the motion of the cyclist between B and C.	
Thinking	**Working**
Interpret the shape of the graph between B and C.	The graph is flat between B and C, indicating that the cyclist's position is not changing for this time, so the cyclist is not moving. If the cyclist is not moving, the velocity is $0\,\text{m s}^{-1}$.
You can confirm the result by calculating the gradient of the graph between B and C using: gradient of x–t graph = $\frac{\text{rise}}{\text{run}} = \frac{\Delta x}{\Delta t}$ Remember that $\Delta x = s$.	$\text{Gradient} = \frac{0.0}{20}$ $= 0.0$

Worked example: Try yourself 11.3.1

ANALYSING A POSITION–TIME GRAPH

Use the graph shown in Worked example 11.3.1 to answer the following questions.

a What is the velocity of the cyclist between E and F?

b Describe the motion of the cyclist between D and E.

VELOCITY-TIME (v–t) GRAPHS

A graph of velocity, v, against time, t, shows how the velocity of an object changes with time. This type of graph is useful for analysing the motion of an object moving in a complex manner.

Consider the example of Aliyah in Figure 11.3.5. Aliyah is running back and forth along an aisle in a supermarket. A study of the velocity–time graph in Figure 11.3.5 reveals that Aliyah is moving with a positive velocity (i.e. in a positive direction) for the first 6.0 s. Between the 6.0 s mark and the 7.0 s mark she is stationary, then she runs in the reverse direction (i.e. she has negative velocity) for the final 3.0 s.

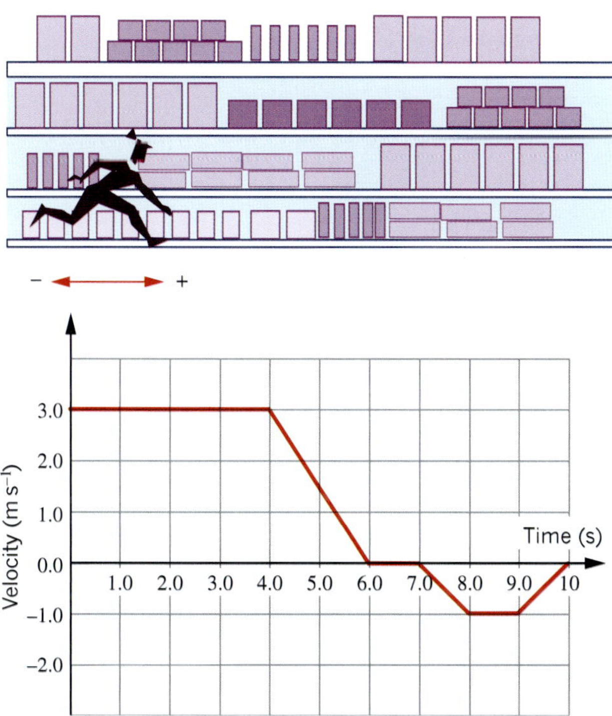

FIGURE 11.3.5 Diagram and v–t graph for Aliyah running along an aisle

The graph shows Aliyah's velocity at each instant in time. She moves in a positive direction with a constant speed of 3.0 m s^{-1} for the first 4.0 s. From 4.0 s to 6.0 s, she continues moving in a positive direction but slows down. At 6.0 s, she comes to a stop for 1.0 second. During the final 3.0 s, she accelerates in the negative direction for 1.0 s then travels at a constant velocity of −1.0 m s^{-1} for 1.0 s. She then slows down and comes to a stop at 10 s. Remember that whenever the graph is below the time axis, velocity is negative, which indicates travel in the reverse direction. So Aliyah is travelling in the reverse direction for the last 3.0 s of her journey.

Finding displacement

A velocity–time graph can also be used to find the displacement of the object under consideration.

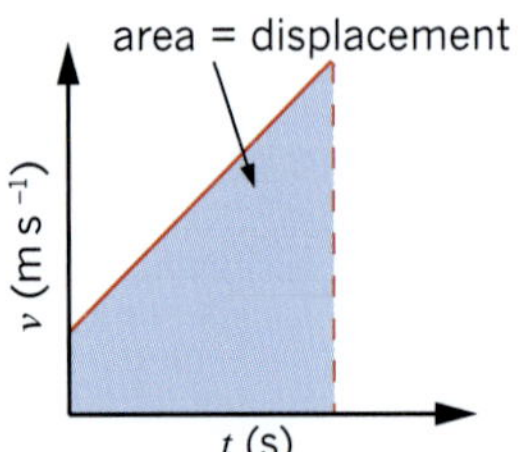

FIGURE 11.3.6 The area under a *v–t* graph gives displacement.

Displacement, *s*, is given by the area under a velocity–time graph (Figure 11.3.6); it is the area between the graph and the time axis. It is important to note that an area below the time axis indicates a negative displacement (i.e. motion in a negative direction).

It is easier to see why the displacement is given by the area under the *v–t* graph when velocity is constant. For example, the graph in Figure 11.3.7 shows that in the first 6.0 s of motion, Aliyah moves with a constant velocity of +3.0 m s^{-1} for 4.0 s. Note that the area under the graph for this period of time is a rectangle. Her displacement, *s*, during this time can be determined by rearranging the formula for velocity:

$$
\begin{aligned}
v &= \frac{s}{\Delta t} \\
\therefore \quad s &= v \times \Delta t \\
&= \text{height} \times \text{base} \\
&= \text{area under } v\text{–}t \text{ graph}
\end{aligned}
$$

Aliyah then slows from 3.0 m s^{-1} to zero in the next 2.0 s. In order to understand why the displacement for this period of time is given by the triangular area under the graph requires more complicated mathematics known as calculus, which is beyond the scope of this book.

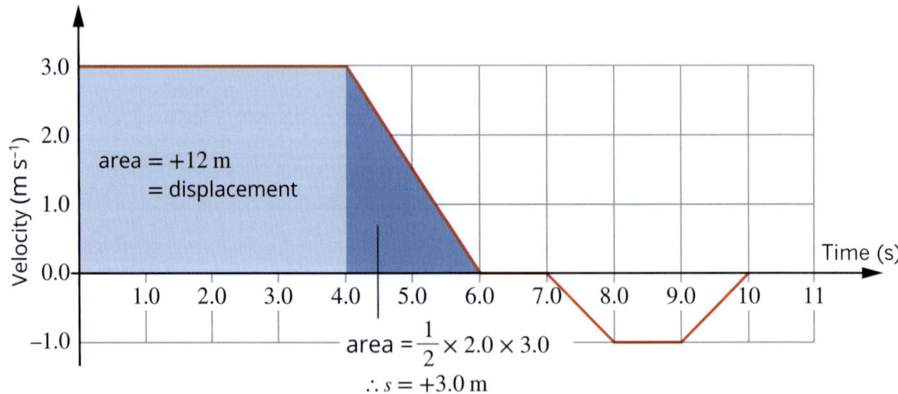

FIGURE 11.3.7 Area values as shown in a *v–t* graph

From Figure 11.3.7, the area under the graph for the first 4.0 s gives Aliyah's displacement during this time, +12 m. The displacement from 4.0 s to 6.0 s is represented by the area of the darker blue triangle and is equal to +3.0 m. The total displacement during the first 6.0 s is +12 m + 3.0 m = +15 m.

Worked example 11.3.2

ANALYSING A VELOCITY–TIME GRAPH

The motion of a radio-controlled car initially travelling east in a straight line across a driveway is represented by the graph below.

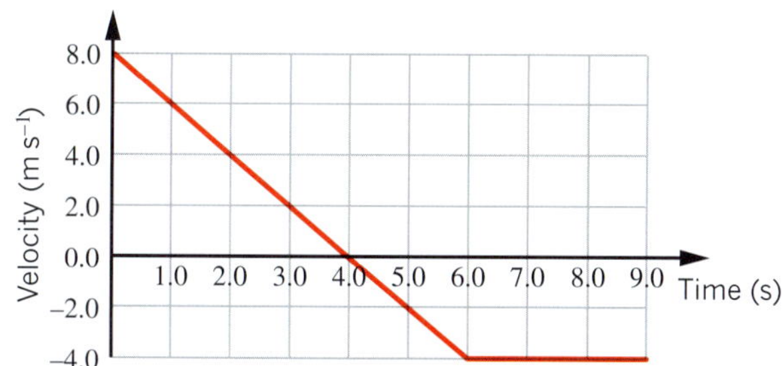

a What is the displacement of the car during the first 4.0 seconds?

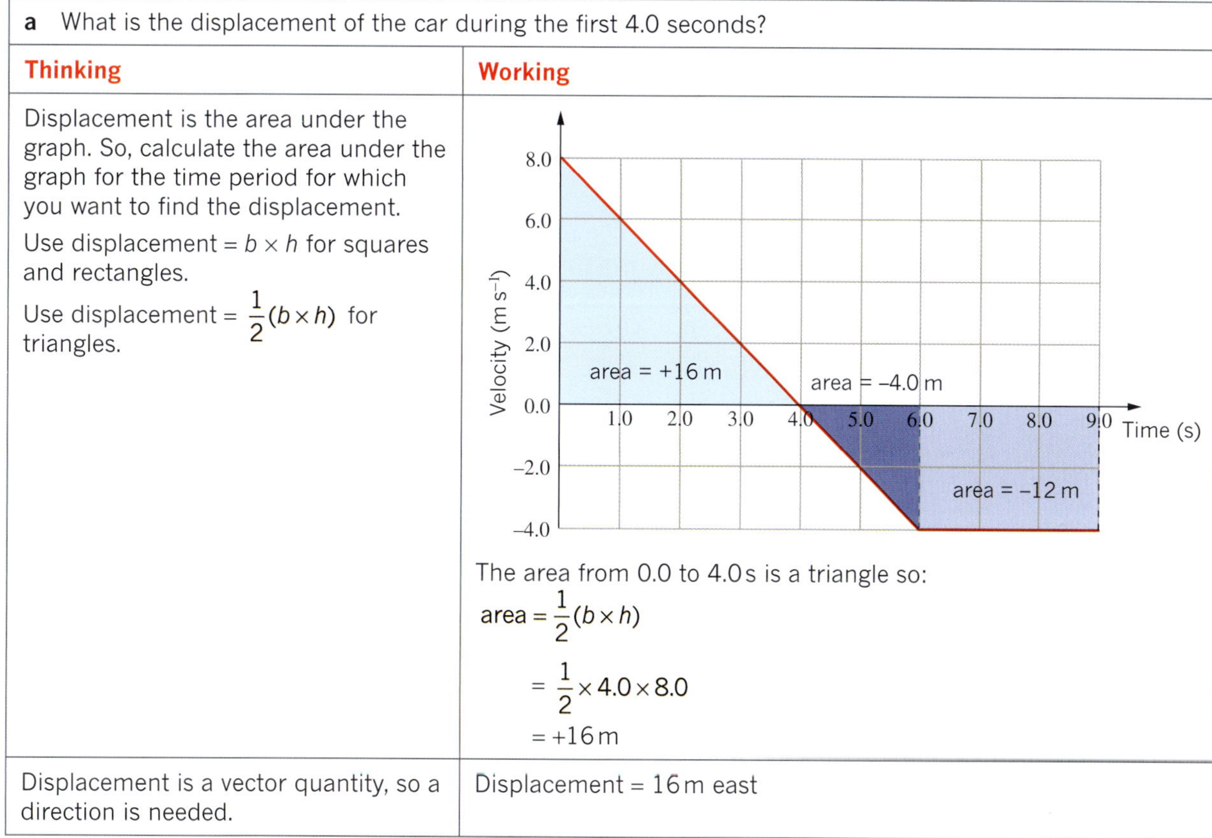

Thinking	Working
Displacement is the area under the graph. So, calculate the area under the graph for the time period for which you want to find the displacement. Use displacement = $b \times h$ for squares and rectangles. Use displacement = $\frac{1}{2}(b \times h)$ for triangles.	The area from 0.0 to 4.0 s is a triangle so: $\text{area} = \frac{1}{2}(b \times h)$ $= \frac{1}{2} \times 4.0 \times 8.0$ $= +16\,\text{m}$
Displacement is a vector quantity, so a direction is needed.	Displacement = 16 m east

b What is the average velocity of the car for the first 4.0 seconds?

Thinking	Working
Identify the equation and variables, and apply the sign convention.	$v = \frac{s}{\Delta t}$ $s = +16\,\text{m}$ $\Delta t = 4.0\,\text{s}$
Substitute values into the equation: $v = \frac{s}{\Delta t}$	$v = \frac{s}{\Delta t}$ $= \frac{+16}{4.0}$ $= +4.0\,\text{m s}^{-1}$
Velocity is a vector quantity, so a direction is needed.	$v_{av} = 4.0\,\text{m s}^{-1}$ east

Worked example: Try yourself 11.3.2

ANALYSING A VELOCITY–TIME GRAPH

Use the graph shown in Worked example 11.3.2 to answer the following questions.

a What is the displacement of the car from 4.0 to 6.0 seconds?

b What is the average velocity of the car from 4.0 to 6.0 seconds?

> The gradient of a velocity–time graph gives the average acceleration of the object over the time interval (Figure 11.3.8).

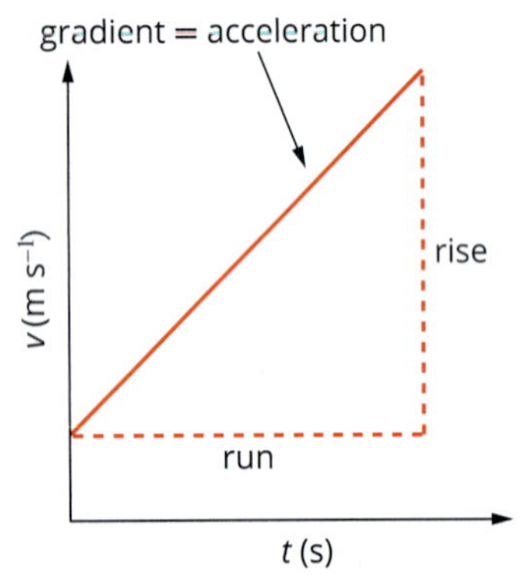

FIGURE 11.3.8 Gradient as displayed in a *v–t* graph

Acceleration from a (*v*–*t*) graph

The acceleration of an object can also be found from a velocity–time graph.

Consider the motion of Aliyah in the 2.0 s interval between 4.0 s and 6.0 s on the graph in Figure 11.3.9. She is moving in a positive direction but slowing down from 3.0 m s^{-1} to rest.

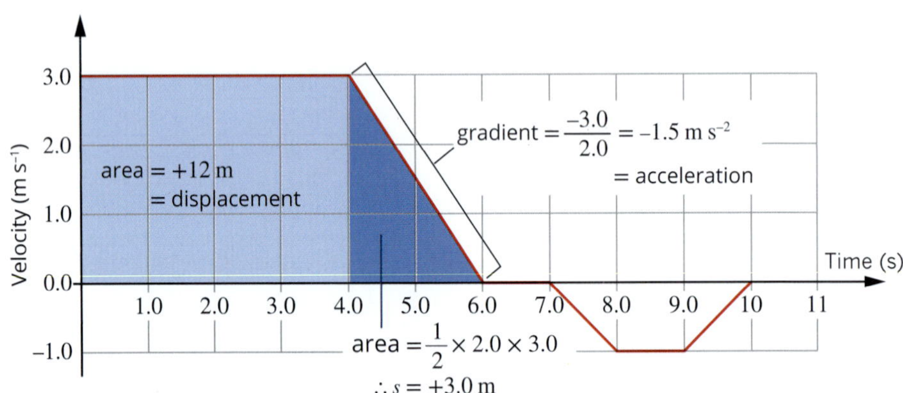

FIGURE 11.3.9 Acceleration as displayed in a *v–t* graph

Aliyah's acceleration is:

$$a = \frac{\Delta v}{\Delta t} = \frac{v-u}{\Delta t} = \frac{0.0-3.0}{2.0} = -1.5\,\text{m}\,\text{s}^{-2}$$

Since acceleration is the velocity change divided by time taken, it is given by the gradient of the v–t graph. As can be seen from Figure 11.3.9, the gradient of the line between 4.0 s and 6.0 s is −1.5 m s^{-2}.

Worked example 11.3.3

FINDING ACCELERATION USING A VELOCITY–TIME GRAPH

Consider the motion of the same radio-controlled car initially travelling east in a straight line across a driveway, as shown by the graph below.

Velocity (m s−1): 8.0, 6.0, 4.0, 2.0, 0.0, −2.0, −4.0
Time (s): 1.0, 2.0, 3.0, 4.0, 5.0, 6.0, 7.0, 8.0, 9.0

What is the acceleration of the car during the first 4.0 s?

Thinking	Working
Acceleration is the gradient of a v–t graph. Calculate the gradient using: $\text{gradient} = \frac{\text{rise}}{\text{run}}$	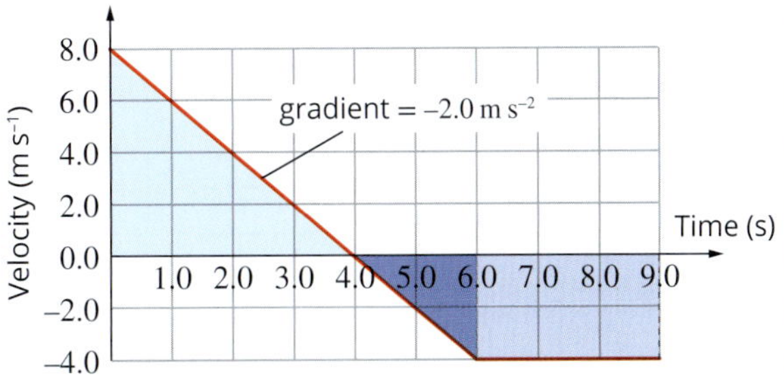 Velocity (m s^{-1}) 8.0 6.0 4.0 2.0 0.0 −2.0 −4.0 Time (s) 1.0 2.0 3.0 4.0 5.0 6.0 7.0 8.0 9.0 gradient = $-2.0\ \text{m s}^{-2}$ Gradient from 0.0 to 4.0 $= \frac{\text{rise}}{\text{run}}$ $= \frac{-8.0}{4.0}$ $= -2.0\ \text{m s}^{-2}$
Acceleration is a vector quantity, so a direction is needed. Interpret the minus sign in the answer.	From the graph for 0.0 s to 4.0 s, you can see that the magnitude of the velocity has changed from 8.0 to 0.0 m s^{-1}, so the car must be slowing down. Therefore the minus sign indicates that the car is slowing down while it travelling east. Acceleration = $-2.0\ \text{m s}^{-2}$ east.

Worked example: Try yourself 11.3.3

FINDING ACCELERATION USING A VELOCITY–TIME GRAPH

Use the graph shown in Worked example 11.3.3 to answer the following question.
What is the acceleration of the car during the period from 4.0 to 6.0 seconds?

Distance travelled

A velocity–time graph can also be used to calculate the distance travelled by a moving object. The process of determining distance requires you to calculate the area under the v–t graph, similar to when calculating displacement. However, since distance travelled by an object always increases as the object moves, regardless of direction, you must add up all the areas between the graph and the time axis, regardless of whether the area is above or below the axis.

For example, Figure 11.3.10 (page 332) shows the velocity–time graph of the radio-controlled car from Worked example 11.3.3. The area above the time axis, which corresponds to motion in the positive direction, is +16 m, while the area below the axis, which corresponds to negative motion, consists of −4.0 m and −12 m. To calculate the total displacement, add up each displacement:

$$\begin{aligned}\text{total displacement} &= 16 + (-4.0) + (-12)\\ &= 16 - 16\\ &= 0.0\ \text{m}\end{aligned}$$

To calculate the total distance, add up the magnitude of the areas, ignoring whether they are positive or negative:

$$\begin{aligned}\text{total distance} &= 16 + 4.0 + 12\\ &= 32\ \text{m}\end{aligned}$$

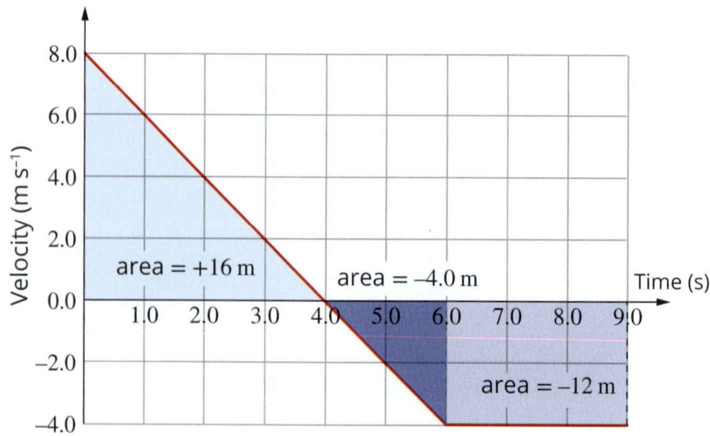

FIGURE 11.3.10 Both distance and displacement can be calculated by using the area under the velocity–time graph.

Non-uniform acceleration

For motion with uniform (constant) acceleration, the velocity–time graph will be a straight line. For non-uniform acceleration the velocity–time graph will be curved. If the velocity–time graph is curved, the instantaneous acceleration will be the gradient of the tangent to the line at the point of interest; the average acceleration will be the gradient of the chord between two points. The displacement can still be calculated by finding the area under the graph; however, you will need to make some estimations.

ACCELERATION-TIME (a–t) GRAPHS

An acceleration–time graph simply indicates the acceleration of the object as a function of time. The area under an acceleration–time graph is found by multiplying an acceleration, a, and a value for a period of time, Δt. The area gives a value for the change in velocity, Δv:

$$\text{area} = a \times \Delta t = \Delta v$$

In order to establish the actual velocity of the object, its initial velocity must be known. Figure 11.3.11 shows both Aliyah's velocity versus time (v–t) graph and her acceleration versus time (a–t) graph.

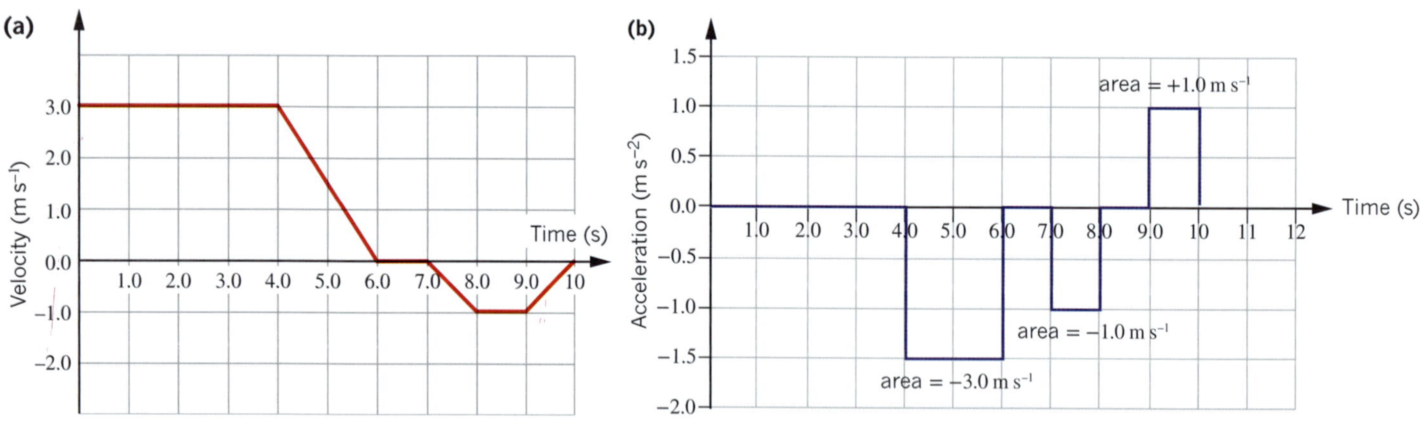

FIGURE 11.3.11 (a) Aliyah's velocity versus time (v–t) graph and (b) Aliyah's acceleration versus time (a–t) graph

From 4.0 s to 6.0 s, the area shows a Δv of $-3.0\,\text{m}\,\text{s}^{-1}$. This indicates that Aliyah has slowed down by $3.0\,\text{m}\,\text{s}^{-1}$ during this time. Her v–t graph confirms this fact. Her initial speed is $3.0\,\text{m}\,\text{s}^{-1}$, so she must be stationary ($v = 0.0$) after 6.0 s. This calculation could not be made without knowing Aliyah's initial velocity.

CASE STUDY **ANALYSIS**

Analysing performance in sport

The workload of elite athletes needs to be constantly monitored to ensure they perform at the highest level throughout the season. To help the coaches manage their players, sport scientists use Global Positioning System (GPS) devices to track and record position, velocity and acceleration data for each player on the team. The tracking devices are inserted into a small pocket in the nape of the player's jersey (Figure 11.3.12), or in a halter top prior to the game, and are removed after the game so that the data can be downloaded and analysed.

Figure 11.3.13 is a graphical representation of GPS data for an AFLW player. This data is a simplified version of the actual data provided to the sports scientists. Use this graph to answer the questions.

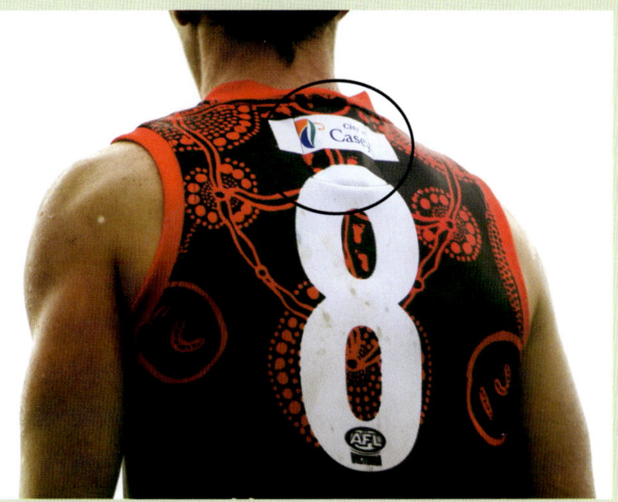

FIGURE 11.3.12 A GPS tracker is inserted into the nape of a player's jersey.

Analysis

1 Estimate the distance travelled by the player between the 2.0 and 6.0 second marks.

2 Estimate the acceleration of the player during the following time periods.
   a 0.0 to 1.0 second
   b 6.0 to 7.0 seconds
   c 9.0 to 11.0 seconds

3 Estimate the total distance travelled by the player during the 16-second period of time.

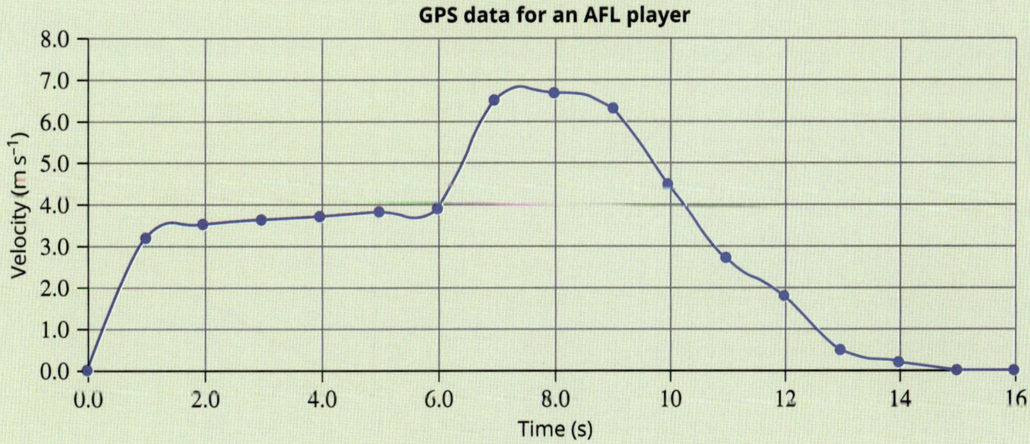

FIGURE 11.3.13 Data such as this is downloaded and analysed for every player after the game.

11.3 Review

OA ✓✓

SUMMARY

- A position–time graph can be used to determine the location of an object at any given time. Additional information can also be derived from the graph:
  - Displacement is given by the change in position of an object.
  - The velocity of an object is given by the gradient of the position–time graph.
  - If the position–time graph is curved, the gradient of the tangent at a point gives the instantaneous velocity. The gradient of the chord between two points on the graph gives the average velocity between those points.
- The gradient of a velocity–time graph is the acceleration of the object.
- The area under a velocity–time graph is the displacement of the object.
- The area under an acceleration–time graph is the change in velocity of the object.

KEY QUESTIONS

Knowledge and understanding

1 Which property does the gradient of a position–time graph represent?

2 The graph represents the straight-line motion of a radio-controlled toy car.

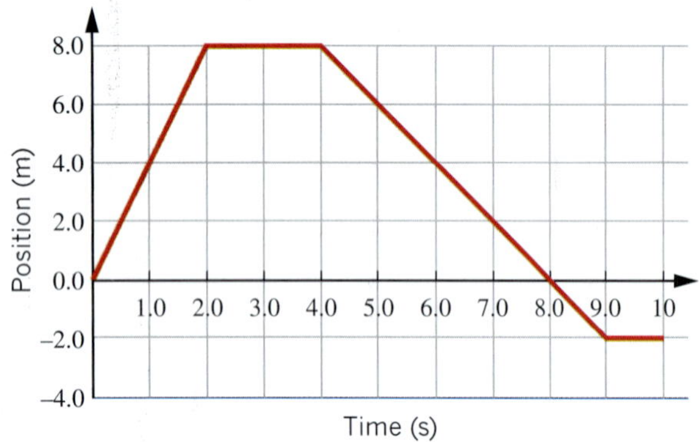

a Describe the motion of the car in terms of its position.

b What was the position of the toy car after:
  i 2.0 s?
  ii 4.0 s?
  iii 6.0 s?
  iv 10 s?

c When did the car return to its starting point?

d What was the velocity of the toy car:
  i during the first 2.0 s?
  ii at 3.0 s?
  iii from 4.0 s to 8.0 s?
  iv at 8.0 s?
  v from 8.0 s to 9.0 s?

e During its 10 s motion, what was:
  i the distance travelled by the car?
  ii the displacement of the car?

Analysis

3 This position–time graph for a cyclist travelling north along a straight road is shown. Calculate the following information about the cyclist's motion.

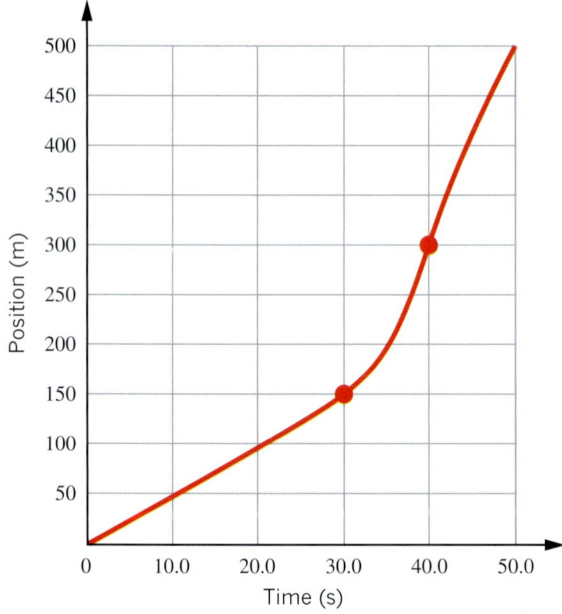

a What was the average speed of the cyclist during the first 30.0 s?

b What was the average velocity of the cyclist during the final 10.0 s?

c What was the average velocity of the cyclist for the whole trip?

4 The graph below shows the motion of a dog running along a footpath.

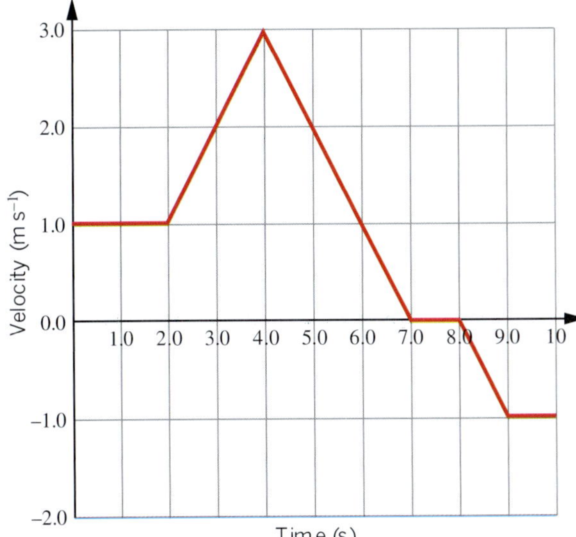

a What is the magnitude of the acceleration of the dog at t = 1.0 s?

b What is the magnitude of the acceleration of the dog at t = 5.0 s?

c What is the magnitude of the displacement of the dog for the first 7.0 s?

d What is the magnitude of the average velocity of the dog over the first 7.0 s?

5 The graph shows the position of a motorbike along a straight stretch of road as a function of time. The motorcyclist starts 200 m north of an intersection.

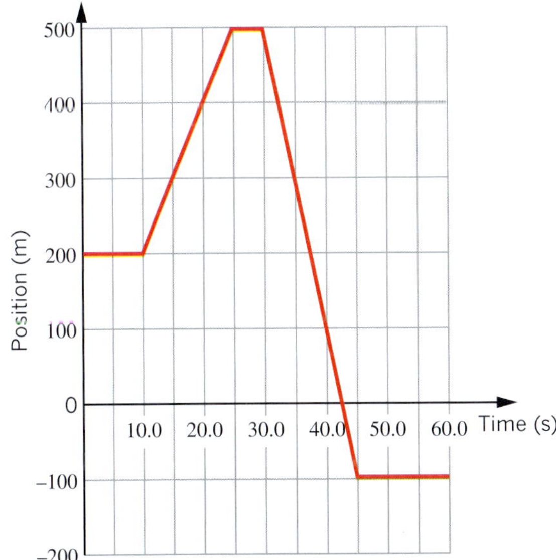

Calculate the instantaneous velocity of the motorcyclist at each of the following times.

a 15.0 s

b 35.0 s

6 The straight-line motion of a high-speed intercity train is shown in the graph below.

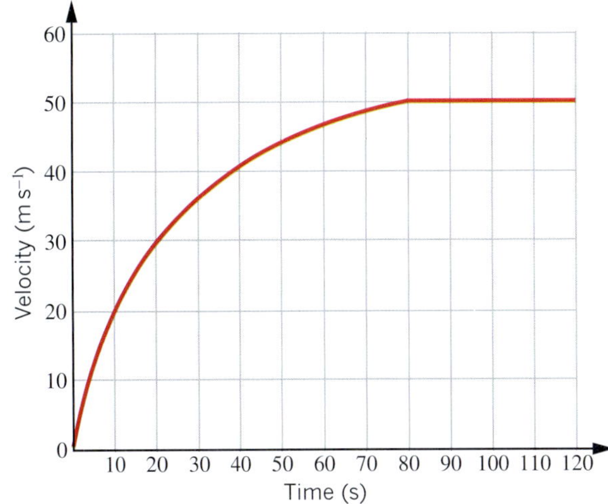

a How long does it take the train to reach its cruising speed?

b What is the acceleration of the train 10 s after starting?

c What is the acceleration of the train 40 s after starting?

d By counting squares, or by another suitable method, approximate the displacement (in km) of the train after 120 s.

7 The velocity–time graphs for a bus and a bicycle travelling along the same straight stretch of road are shown below. The bus is initially at rest and starts moving as the bicycle passes it.

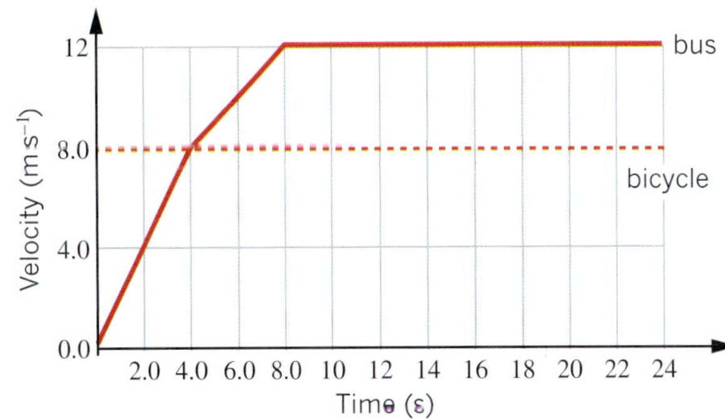

a What is the magnitude of the initial acceleration of the bus?

b At what time does the bus overtake the bicycle?

c How far has the bicycle travelled before the bus catches it?

d What is the magnitude of the average velocity of the bus during the first 8.0 s?

8 a Draw an acceleration–time graph for the bus discussed in question 7.

b Use your acceleration–time graph to determine the change in velocity of the bus over the first 8.0 s.

11.4 Equations for uniform acceleration

A graph is an excellent way of representing motion because it provides a great deal of information that is easy to interpret. However, a graph is time-consuming to draw and sometimes values have to be estimated rather than calculated precisely.

In the previous section, graphs of motion were used to determine quantities such as displacement, velocity and acceleration. This section examines a more powerful and precise method of solving problems involving *constant* or *uniform acceleration*. This method involves the use of a series of equations that can be derived from the basic definitions developed earlier.

DERIVING THE EQUATIONS

Consider an object moving in a straight line with an initial velocity, u, and a uniform acceleration, a, for a time interval, Δt. As u, v and a are vectors, and the motion is limited to one dimension, the sign and direction convention of right as positive and left as negative can be used. After a period of time, Δt, the object has changed its velocity from an initial velocity of u and is now travelling with a final velocity of v. Its acceleration will be given by:

$$a = \frac{\Delta v}{\Delta t} = \frac{v-u}{\Delta t}$$

If the initial time is 0 s, and the final time is t s, then $\Delta t = t$. The above equation can then be rearranged as:

$$v = u + at \qquad \text{(i)}$$

The average velocity of the object is:

$$v_{\text{av}} = \frac{\text{displacement}}{\text{time taken}} = \frac{s}{\Delta t}$$

When acceleration is uniform, average velocity, v_{av}, can also be found as the average of the initial and final velocities:

$$v_{\text{av}} = \frac{1}{2}(u+v)$$

This relationship is shown graphically in Figure 11.4.1.

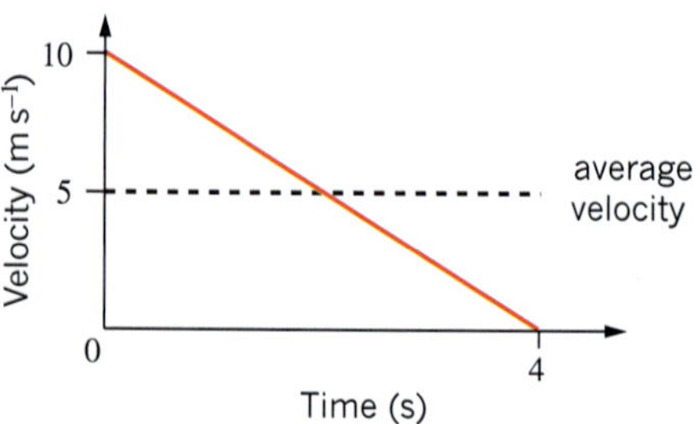

FIGURE 11.4.1 Uniform acceleration as displayed by a v–t graph

So:

$$\frac{s}{t} = \frac{1}{2}(u+v)$$

This gives:

$$s = \frac{1}{2}(u+v)t \qquad \text{(ii)}$$

A graph describing constant acceleration motion is shown in Figure 11.4.2. For constant acceleration, the velocity is increasing by the same amount in each time interval, so the gradient of the v–t graph is constant.

The displacement, s, of the body is given by the area under the velocity–time graph. The area under the velocity–time graph, as shown in Figure 11.4.2, is given by the combined area of the rectangle and the triangle:

$$\text{Area} = s = ut + \frac{1}{2} \times (v - u) \times t$$

As $a = \dfrac{v-u}{t}$

then $v - u = at$, and this can be substituted for $v - u$:

$$s = ut + \frac{1}{2} \times at \times t$$

$$s = ut + \frac{1}{2}at^2 \qquad \text{(iii)}$$

Making u the subject of equation (i) gives:

$$u = v - at$$

You might like to derive another equation yourself by substituting this into equation (ii). You will get:

$$s = vt - \frac{1}{2}at^2 \qquad \text{(iv)}$$

Rewriting equation (i) with t as the subject gives:

$$t = \frac{v-u}{a}$$

Now, if this is substituted into equation (ii):

$$s = \frac{1}{2}(u + v)t$$

$$= \frac{u+v}{2} \times \frac{v-u}{a}$$

$$= \frac{v^2 - u^2}{2a}$$

Finally, transposing this gives:

$$v^2 = u^2 + 2as \qquad \text{(v)}$$

Equations (i)–(v) are commonly used to solve problems in which acceleration is constant.

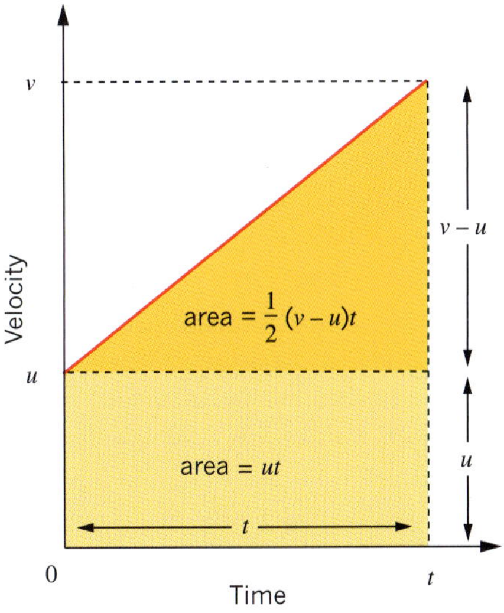

FIGURE 11.4.2 The area under a v–t graph can be broken up into a rectangle and a triangle.

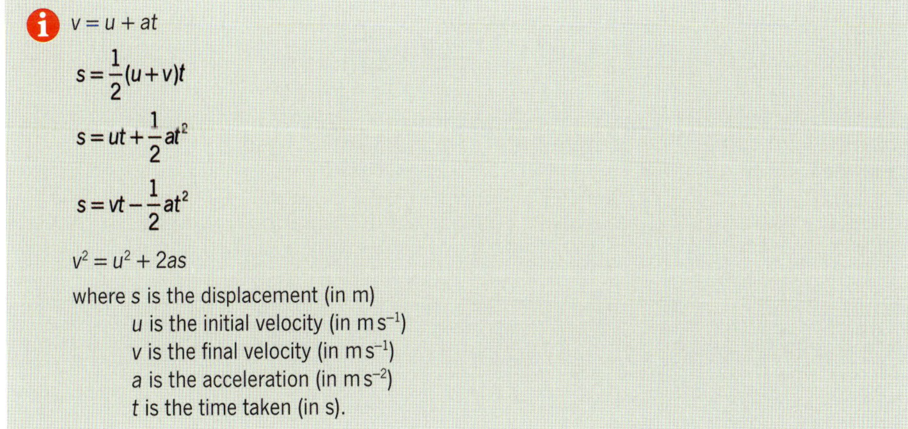

$v = u + at$

$s = \frac{1}{2}(u + v)t$

$s = ut + \frac{1}{2}at^2$

$s = vt - \frac{1}{2}at^2$

$v^2 = u^2 + 2as$

where s is the displacement (in m)
u is the initial velocity (in m s^{-1})
v is the final velocity (in m s^{-1})
a is the acceleration (in m s^{-2})
t is the time taken (in s).

PHYSICSFILE

Some general rules for SI units

The SI (*Système international*) has some general rules that should be followed.

1 The value of a variable should be written as a number, followed by a space, then the symbol for the unit, e.g. 13.1 m.

2 There is no space between a prefix and the unit, and any unit raised to a power includes the prefix, e.g. km (not k m), and cm^3 is the same as $(cm)^3$.

3 Derived units with two or more symbols are written with a space between the symbols, e.g. N m, or kg m s^{-2}. Note that the unit for velocity is m s^{-1}, while ms^{-1} is 'per millisecond', a unit of frequency.

SOLVING PROBLEMS USING THE EQUATIONS OF MOTION

When solving problems using these equations, it is important to think about the problem and try to visualise what is happening. Follow these steps.

Step 1 Draw a simple diagram of the situation.

Step 2 Write down the information that has been given in the question. You might like to use the word 'suvat' as a memory trick to help you remember to list the variables in the order *s*, *u*, *v*, *a*, and *t*. Use a sign convention to assign positive and negative values to indicate directions. Convert all units to SI form.

Step 3 Select the equation that matches your data. It should include three values that you know, and the one value that you want to solve.

Step 4 Use the appropriate number of significant figures in your answer.

Step 5 Include units with the answer and specify a direction if the quantity is a vector.

Worked example 11.4.1

USING THE EQUATIONS OF MOTION

A snowboarder in a race is travelling 10.0 m s^{-1} north as he crosses the finishing line. He then decelerates uniformly, coming to a stop over a distance of 20.0 m.

a What is his acceleration as he comes to a stop?	
Thinking	**Working**
Write down the known quantities as well as the quantity you are finding. Apply the sign convention that north is positive and south is negative.	Take all the information that you can from the question: • Acceleration is constant, so use equations for uniform acceleration. • 'Coming to a stop' means that the final velocity is zero. $s = +20.0$ m $u = +10.0$ m s^{-1} $v = 0$ m s^{-1} $a = ?$
Identify the correct equation to use.	$v^2 = u^2 + 2as$
Substitute known values into the equation and solve for *a*. Include units with the answer.	$v^2 = u^2 + 2as$ $0^2 = 10.0^2 + 2 \times a \times 20.0$ $0 = 100 + 40.0a$ $-100 = 40.0a$ $a = \frac{-100}{40.0}$ $= -2.50$ m s^{-2}
Use the sign convention to state the answer with its direction.	$a = 2.50$ m s^{-2} south

b How long does he take to come to a stop?	
Thinking	**Working**
Write down the known quantities as well as the quantity you are finding. Apply the sign convention that north is positive and south is negative. Use the value of acceleration you calculated in part (a).	Take all the information that you can from the question: • Acceleration is constant, so use equations for uniform acceleration. • 'Coming to a stop' means that the final velocity is zero. $s = +20.0\,m$ $u = +10.0\,m\,s^{-1}$ $v = 0\,m\,s^{-1}$ $a = -2.50\,m\,s^{-2}$ $t = ?$
Identify the correct equation to use. Since you now know four values, any equation involving t will work.	$v = u + at$
Substitute known values into the equation and solve for t. Include units with the answer.	$v = u + at$ $0 = 10.0 + (-2.50) \times t$ $-10.0 = -2.50t$ $t = \frac{-10.0}{-2.50}$ $= 4.00\,s$

c What is the average velocity of the snowboarder as he comes to a stop?	
Thinking	**Working**
Write down the known quantities as well as the quantity that you are finding. Apply the sign convention that north is positive and south is negative.	Take all the information that you can from the question: • Acceleration is constant, so we only need to find the average of the final and initial speeds. $u = +10.0\,m\,s^{-1}$ $v = 0\,m\,s^{-1}$ $v_{av} = ?$
Identify the correct equation to use.	$v_{av} = \frac{1}{2}(u+v)$
Substitute known values into the equation and solve for v_{av}. Include units with the answer.	$v_{av} = \frac{1}{2}(u+v)$ $= \frac{1}{2}(0+10.0)$ $= 5.00\,m\,s^{-1}$
Use the sign convention to state the answer with its direction.	$v_{av} = 5.00\,m\,s^{-1}$ north

Worked example: Try yourself 11.4.1

USING THE EQUATIONS OF MOTION

A snowboarder in a race is travelling $15.0\,m\,s^{-1}$ east as she crosses the finishing line. She then decelerates uniformly until coming to a stop over a distance of 30.0 m.

a What is her acceleration as she comes to a stop?

b How long does she take to come to a stop?

c What is the average velocity of the snowboarder as she comes to a stop?

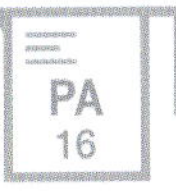

11.4 Review

SUMMARY

- The equations in the right column can be used for situations in which there is a constant acceleration, where:
  - s = displacement (m)
  - u = initial velocity (m s^{-1})
  - v = final velocity (m s^{-1})
  - a = acceleration (m s^{-2})
  - t = time (s).
- A sign and direction convention for motion in one dimension needs to be used with these equations.

- $v = u + at$
- $s = \frac{1}{2}(u+v)t$
- $s = ut + \frac{1}{2}at^2$
- $s = vt - \frac{1}{2}at^2$
- $v^2 = u^2 + 2as$
- $v_{av} = \frac{s}{t} = \frac{u+v}{2}$

KEY QUESTIONS

Knowledge and understanding

1 A cyclist has a uniform acceleration as he rolls down a hill. His initial speed is 5 m s^{-1}, he travels a distance of 30 m and his final speed is 18 m s^{-1}. Which one equation could be used to determine his acceleration?

2 A stone is dropped vertically into a lake. Describe the direction of its motion and the direction of its acceleration at the instant it enters the water.

Analysis

3 A new-model Subaru travels with a uniform acceleration on a racetrack. It starts from rest and covers 400.0 m in 16.0 s.

a What is the magnitude of its average acceleration during this time?

b What is the final speed of the car in m s^{-1}?

c What is the car's final speed in km h^{-1}?

4 A Prius hybrid car starts from rest and accelerates uniformly in a positive direction for 8.00 s. It reaches a final speed of 16.0 m s^{-1}.

a What is the magnitude of the acceleration of the Prius?

b What is the magnitude of the average velocity of the Prius?

c What is the distance travelled by the Prius?

5 During its launch phase, a space rocket accelerates uniformly from rest to 160.0 m s^{-1} upwards in 4.00 s, then travels with a constant speed of 160.0 m s^{-1} for the next 5.00 s.

a What is the initial acceleration of the rocket?

b How far (in km) does the rocket travel in this 9.00 s period?

c What is the final speed of the rocket in km h^{-1}?

d What is the average speed of the rocket during the first 4.00 s?

e What is the average speed of the rocket during the 9.00 s motion?

6 A car is travelling along a straight road at 75.0 km h^{-1} east. In an attempt to avoid an accident, the motorist has to brake suddenly and stop the car.

a What is the car's initial speed in m s^{-1}?

b If the reaction time of the motorist is 0.250 s, what distance does the car travel while the driver is reacting to apply the brakes? While the driver is reacting, assume there is no change in velocity.

c Once the brakes are applied, the car has an acceleration of -6.00 m s^{-2}. How far does the car travel while coming to a stop?

d What total distance does the car travel from the time the driver first notices the danger to when the car comes to a stop?

7 A billiard ball rolls from rest down a smooth ramp that is 8.00 m long. The acceleration of the ball is constant at 2.00 m s^{-2}.

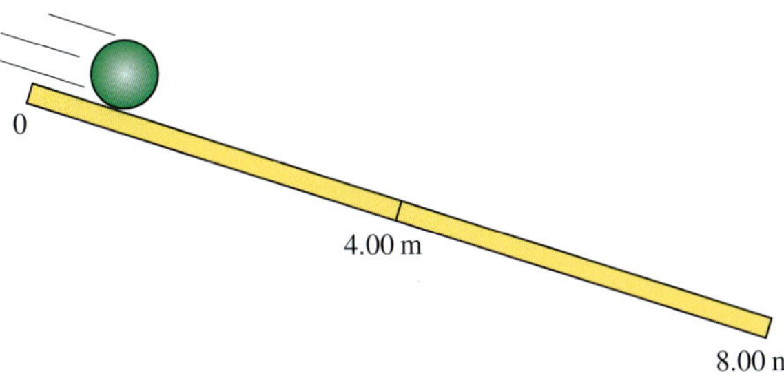

a What is the speed of the ball when it is halfway down the ramp?

b What is the final speed of the ball?

c How long does the ball take to roll the first 4.00 m?

d How long does the ball take to travel the final 4.00 m?

8 A cyclist, Anna, is travelling at a constant speed of 12.0 m s^{-1} when she passes a stationary bus. The bus starts moving just as Anna passes, and it accelerates uniformly at 1.50 m s^{-2}.

a When does the bus reach the same speed as Anna?

b How long does the bus take to catch Anna?

c What distance has Anna travelled before the bus catches up?

11.5 Vertical motion

When studying the motion of objects moving vertically, it is important to be able to understand the effects of the force due to gravity.

Until 500 years ago, it was widely believed that the heavier the object was, the faster it would fall. This was the theory of Aristotle, and it lasted for 2000 years until the end of the Middle Ages. In the seventeenth century, the Italian scientist Galileo conducted experiments that showed that the mass of the object did not affect the rate at which it fell, as long as air resistance was not a factor.

Many people still mistakenly think that heavier objects fall faster than light objects. This confusion arises because of the effects of air resistance. This section examines the motion of falling objects.

PHYSICSFILE

Defining a kilogram

The kilogram is currently defined using a constant called 'Planck's constant'. It is a constant that you will become more familiar with in your studies of electromagnetic energy. Planck's constant has a precisely known value of $6.626\,070\,15 \times 10^{-34}$ J s. These units can be converted to the SI units of $\text{kg}\,\text{m}^2\,\text{s}^{-1}$, which more clearly shows the link to the unit of mass. Importantly this value can be measured in any laboratory using a Kibble balance.

A Kibble balance can be used to accurately recreate the standard kilogram anywhere in the world.

MASS AND THE FORCE DUE TO GRAVITY

Mass (the amount of matter) is a scalar quantity. In scientific contexts, mass is measured in kilograms (kg). Since the late 1700s, the kilogram has been defined in terms of an amount of a standard material. At first, 1 litre of water at 4°C was used to define the kilogram. Later an international mass standard was introduced. This standard was a 1 kg cylinder of platinum–iridium alloy that is kept in Paris (Figure 11.5.1). For 143 years copies were made from this standard and sent around the world to calibrate balances. More recently the standard of mass has been defined using Planck's constant ($6.626\,070\,15 \times 10^{-34}\,\text{kg}\,\text{m}^2\,\text{s}^{-1}$).

FIGURE 11.5.1 Up until 20 May 2019 all mass in the world was compared to this small piece of platinum–iridium alloy held in a sealed vault in Paris.

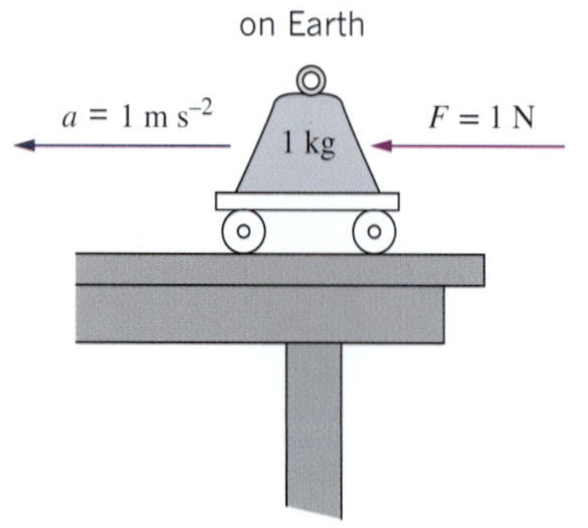

FIGURE 11.5.2 A force applied to an object can cause it to accelerate.

To gain a better understanding of mass through the effect of a force on a massive body, think about a mass resting on a frictionless surface. If a force, F, is applied to the mass, m, in the horizontal direction, an acceleration, a, is produced, as shown in Figure 11.5.2. That is, when an object experiences a force (push or pull) it will begin to move and while the force is acting, its velocity will change. When the velocity of an object changes, the object is said to be accelerating.

The more mass an object has, the greater the force required to make it accelerate. If the same force is applied to two different masses, the smaller mass will accelerate more than the greater mass. For this reason, mass can be seen as the property of a body that resists the change in motion caused by a force.

If the above experiment was repeated on the Moon with the same horizontal force acting on the body on a frictionless surface, the same acceleration would result, as shown in Figure 11.5.3. This is because the mass of the body is the same on Earth and on the Moon. Mass is a property of the body and it is not affected by its environment.

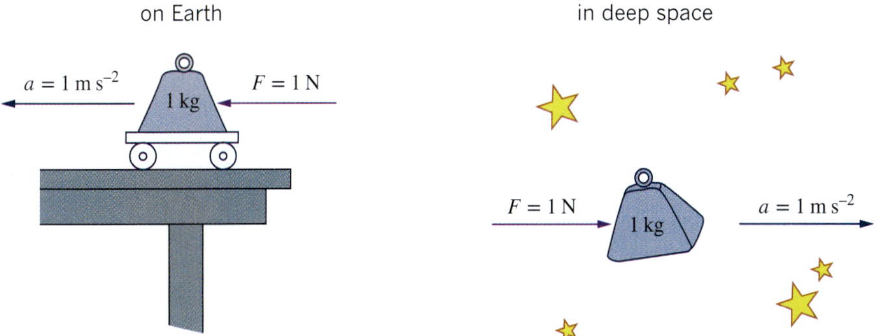

FIGURE 11.5.3 Mass is a property of the object and not of its surroundings.

Gravitational force

In the late 1500s, Galileo was able to show that all objects that are dropped near the surface of Earth accelerate at the same rate, *g*, towards the centre of Earth. The force that produces this acceleration is the force due to gravity. The force due to gravity is an attractive force that exists between all masses. In other words, it is a 'pulling' force that exists between everything that has a mass. It is one of the fundamental forces that acts over a distance, which means that the two masses do not need to be in contact in order for the force to exist.

Gravitational forces result from a mass creating a gravitational field that spreads throughout the space around the mass. Any other mass that is within this field will experience a force towards the mass creating the field. The object that is in the gravitational field also has mass, so it too has a gravitational field around it that attracts the original mass with an equal and opposite force.

The gravitational field extends through space in an inverse-squared relationship. This means that if you double the distance from the mass creating the field, then the force will be one-quarter the size.

Here on Earth, you are strongly affected by Earth's gravitational field (as well as by fields from the Sun, the Moon and other objects in the solar system). Even if you were not on Earth, you could still measure the effect of Earth's gravitational field. There is no place in the universe where Earth's gravitational field will not reach. At the 'edge' of the universe the effect will be very small, but it can be calculated. The closer you or any mass is located to Earth, the larger the gravitational force of attraction towards Earth. At a height above Earth's surface that is equal to the radius of Earth, the force due to gravity on a mass will be one-quarter of that at Earth's surface. At two Earth radii above Earth's surface, the gravitational force will be one-ninth of the force experienced at Earth's surface.

The force on a body due to gravity, F_g, is a vector quantity. Like other forces, it is measured in newtons (N).

Figure 11.5.4 shows a bowling ball falling through the air. As it falls, it accelerates downwards due to Earth's gravitational field strength *g*, which near the surface of Earth is $9.8\,N\,kg^{-1}$ down.

As the force due to gravity acting on the ball is a vector, it can be represented with an arrow drawn downwards (towards the centre of Earth) with its tail beginning at the ball's centre of mass. The centre of mass is the point where the mass can be considered to be 'concentrated'. In an object of uniform density, there is as much mass above the centre of mass as there is below it, as much mass to the left as there is to the right, and as much mass in front as there is behind it.

PHYSICSFILE

Strength of gravity

The acceleration due to gravity, *g*, on Earth varies slightly from $9.8\,m\,s^{-2}$ according to the location. The reasons for this will be studied in Physics Unit 3. On the Moon, the strength of gravity, *g*, is much weaker than on Earth and falling objects accelerate at $1.6\,m\,s^{-2}$. Other planets and bodies in the solar system have different values of g depending on their mass and size. The value of g at various locations is provided in the table below.

In Physics Unit 2, you can assume that acceleration due to gravity is always $9.8\,m\,s^{-2}$.

Acceleration due to gravity at different locations on Earth, and on other bodies in the solar system

Location	Acceleration due to gravity ($m\,s^{-2}$)
Melbourne	9.800
South Pole	9.832
Equator	9.780
Moon	1.600
Mars	3.600
Jupiter	24.600
Pluto	0.670

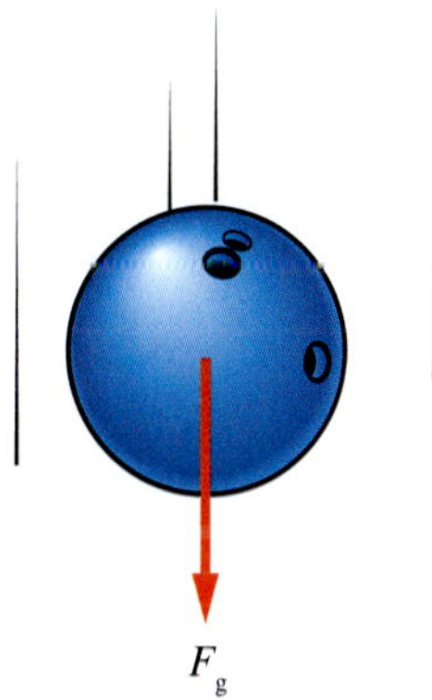

FIGURE 11.5.4 When the ball is in mid-air, there is an unbalanced force due to gravity acting on it, so it accelerates towards the ground. The vector representing force due to gravity is drawn from the centre of mass of the object and points downwards.

The term 'acceleration due to gravity' is equivalent to the term 'gravitational field strength'. The units for gravitational field strength ($N\,kg^{-1}$) are also equivalent to the units for acceleration ($m\,s^{-2}$).

The *force due to gravity* of a body, F_g, (in N) is defined as the force of attraction on a body due to gravity and is calculated using the equation:

$F_g = mg$

where F_g is the force of gravity acting at the centre of mass of a body (in N)
m is the mass of the body (in kg)
g is the gravitational field strength, which is $9.8\,N\,kg^{-1}$ near the surface of Earth.

ANALYSING VERTICAL MOTION

Air resistance is the force on an object caused by collisions with air molecules. Some falling objects are affected by air resistance to a large extent; for example, feathers and balloons. This is why these objects do not speed up much as they fall. However, if air resistance is negligible or zero, all free-falling bodies near Earth's surface will move with an equal downwards acceleration. An object is said to be in **free fall** if the only force acting on it is the force due to gravity. The stroboscopic image in Figure 11.5.5 clearly shows an apple accelerating as it falls, since the distance travelled by the apple between each photograph increases. In a vacuum, this rate of acceleration would be the same for a feather, a bowling ball, or any other object. The mass of the object does not matter.

FIGURE 11.5.5 A stroboscopic image of an apple in free fall. The time elapsed between each image of the apple is the same but the distance it travels increases, which shows the apple is accelerating. Without air resistance, this rate of acceleration is the same for all objects.

At Earth's surface, the acceleration due to gravity, g, is $9.8\,m\,s^{-2}$ down and does not depend on whether the body has been thrown upwards or is falling down.

As an example, a coin that is dropped from rest will be moving at $9.8\,m\,s^{-1}$ after 1 s, $19.6\,m\,s^{-1}$ after 2 s, and so on. Each second its speed increases by $9.8\,m\,s^{-1}$. The motion of a falling coin is illustrated in Figure 11.5.6.

However, if the coin was launched straight up at $19.6\,m\,s^{-1}$, then after 1 s its speed would be $9.8\,m\,s^{-1}$, and after 2 s it would be stationary. In other words, each second the coin would slow down by $9.8\,m\,s^{-1}$. The motion of a coin thrown vertically upwards is shown in Figure 11.5.7.

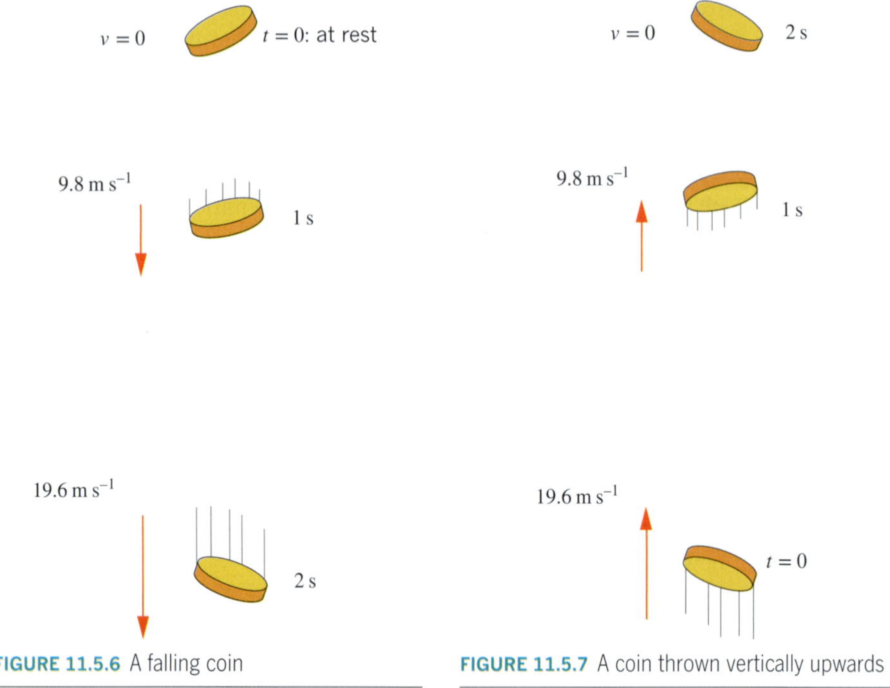

FIGURE 11.5.6 A falling coin

FIGURE 11.5.7 A coin thrown vertically upwards

So, regardless of whether the coin is falling or is tossed, its speed changes at the same rate. The speed of the falling coin *increases* by $9.8\,m\,s^{-1}$ each second and the speed of the rising coin *decreases* by $9.8\,m\,s^{-1}$ each second. That means that the acceleration of the coin due to gravity is $9.8\,m\,s^{-2}$ downwards in both cases.

Since the acceleration of a free-falling body is constant, it is appropriate to use the equations for uniform acceleration that were studied in Section 11.4. It is necessary to specify whether upwards or downwards is positive when doing these problems. You can simply follow the mathematical convention of regarding upwards as positive, which would mean the acceleration due to gravity would always be $-9.8\,\text{m}\,\text{s}^{-2}$.

PHYSICSFILE

Galileo's experiment on the Moon

In 1971, David Scott went to great lengths to show that Galileo's prediction was correct. As an astronaut on the Apollo 15 Moon mission, he took a hammer and a feather on the voyage. He stepped onto the lunar surface, held the feather and hammer at the same height and dropped them together. As Galileo had predicted 400 years earlier, in the absence of any air resistance the two objects fell side by side as they accelerated towards the Moon's surface. This experiment was later recreated by Professor Brian Cox using feathers and a bowling ball in the world's largest vacuum chamber. An internet search for the video of this demonstration would be worth your while.

The feathers fall at the same rate as a bowling ball in a vacuum.

CASE STUDY ANALYSIS

Theories of motion—Aristotle and Galileo

Aristotle was a Greek philosopher who lived in the fourth century BCE. He was such an influential person that his ideas on motion were generally accepted for nearly 2000 years. Aristotle did not do experiments as we know them today, but simply thought about different bodies in order to arrive at a plausible explanation for their motion.

Aristotle spent a lot of time classifying animals, and adopted a similar approach in his study of motion. His theory gave inanimate objects, such as rocks and rain, similar characteristics to living things. Aristotle organised objects into four terrestrial groups or elements: earth, water, air and fire (Figure 11.5.8). He said that any object was a mixture of these elements in a certain proportion.

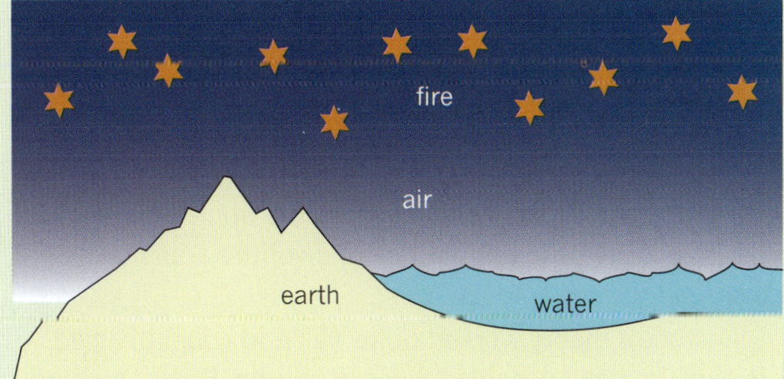

FIGURE 11.5.8 Aristotle's four elements of the universe; earth, water, air and fire

continued over page

CASE STUDY ANALYSIS *continued*

FIGURE 11.5.9 Galileo Galilei

According to Aristotle, a body would move because of a tendency that could come from inside or outside the body. An internal tendency would cause 'natural' motion and result in a body returning to its proper place. For example, if a rock, which is an earth substance, is held in the air and released, its natural tendency would be to return to Earth. This explains why it falls down. Similarly, fire was thought to head upwards in an attempt to return to its proper place in the universe.

In the Aristotelian model, an external push that acts when something is thrown or hit was the cause of 'violent' motion. An external push acted to take a body away from its proper place. For example, when an apple is thrown into the air, a violent motion carries the apple away from Earth, but then the natural tendency of the apple takes over and it returns to its home.

Aristotle's theory worked quite well and could be used to explain many observed types of motion. However, there were also many examples that it could not successfully explain, such as why some solids floated instead of sinking.

Aristotle explained the behaviour of a falling body by saying that its speed depended on how much earth element it contained. This suggested that a 2 kg rock would fall twice as fast and in half the time as a 1 kg rock dropped from the same height.

Many centuries later, Galileo Galilei (Figure 11.5.9) noticed that, at the start of a hailstorm, small hailstones arrived at the same time as large hailstones. This caused Galileo to doubt Aristotle's theory and so he set about finding an explanation for the motion of freely falling bodies.

A famous story in science is that of Galileo dropping different masses from the Leaning Tower of Pisa in Italy. This story may or may not be true, but Galileo did perform a very detailed analysis of falling bodies. Galileo used inclined planes because freely falling bodies moved too fast to analyse. He completed extensive and thorough experiments that showed conclusively that Aristotle was incorrect.

By using a water clock to time balls as they rolled down different inclines, he was able to show that the balls were accelerating and that the distance they travelled was proportional to the square of the time: $d \propto t^2$.

Galileo found that this also held true when he inclined the plane at larger and larger angles, allowing him to conclude that freely falling bodies actually fall with a uniform acceleration.

Analysis

1 In terms of air and earth substances, how would Aristotle explain that a feather falls more slowly that a rock through the air?

2 Scientific understanding progresses when observations cause us to question the existing theory. New hypotheses are created and tested, and a more scientific theory takes precedence. What observation made Galileo question Aristotle's theory?

3 Using one of the equations of accelerated motion, show how it relates to Galileo's discovery that the distance that the accelerating balls rolled down the ramp was proportional to the time squared.

Worked example 11.5.1

VERTICAL MOTION

A construction worker accidentally knocks a brick from a building so that it falls vertically a distance of 50 m to the ground. Use $g = -9.8\ \text{m s}^{-2}$ and ignore air resistance when answering these questions.

a How long does the brick take to fall halfway to 25 m?	
Thinking	**Working**
Write down the values of the quantities that are known and what you are finding. Apply the sign convention that upwards is positive and downwards is negative.	The brick starts at rest, so $u = 0$. $s = -25\ \text{m}$ $u = 0\ \text{m s}^{-1}$ $a = -9.8\ \text{m s}^{-2}$ $t = ?$
Select the equation for uniform acceleration that best fits the data you have.	$s = ut + \frac{1}{2}at^2$
Substitute known values into the equation and solve for t. Think about whether the value seems reasonable.	$-25 = 0 \times t + \frac{1}{2} \times (-9.8) \times t^2$ $-25 = -4.9t^2$ $t = \sqrt{\frac{-25}{-4.9}}$ $= 2.26\ \text{s}$ $= 2.3\ \text{s}$

b How long does the brick take to fall all the way to the ground?	
Thinking	**Working**
Write down the values of the quantities that are known and what you are finding. Apply the sign convention that upwards is positive and downwards is negative.	$s = -50\ \text{m}$ $u = 0\ \text{m s}^{-1}$ $a = -9.8\ \text{m s}^{-2}$ $t = ?$
Identify the correct equation of uniform acceleration to use.	$s = ut + \frac{1}{2}at^2$
Substitute known values into the equation and solve for t. Think about whether the value seems reasonable. Notice that the brick takes 2.3 s to travel the first 25 m and only 0.9 s to travel the final 25 m. This is because it is accelerating.	$-50 = 0 \times t + \frac{1}{2} \times (-9.8) \times t^2$ $-50 = -4.9t^2$ $t = \sqrt{\frac{-50}{-4.9}}$ $= 3.19\ \text{s}$ $= 3.2\ \text{s}$

c What is the velocity of the brick as it hits the ground?	
Thinking	**Working**
Write down the values of the quantities that are known and what you are finding. Apply the sign convention that upwards is positive and downwards is negative. Use the value for time (t) that you calculated in part (b).	$s = -50\,\text{m}$ $u = 0\,\text{m s}^{-1}$ $v = ?$ $a = -9.8\,\text{m s}^{-2}$ $t = 3.19\,\text{s}$
Identify the correct equation to use. Since you now know four values, any equation involving v will work.	$v = u + at$
Substitute known values into the equation and solve for v. Think about whether the value seems reasonable.	$v = 0 + (-9.8) \times 3.19$ $= -31.3$ $= -31\,\text{m s}^{-1}$
Use the sign and direction convention to describe the direction of the final velocity.	$v = -31\,\text{m s}^{-1}$ or $31\,\text{m s}^{-1}$ downwards

Worked example: Try yourself 11.5.1

VERTICAL MOTION

A construction worker accidentally knocks a hammer from a building so that it falls vertically a distance of 60 m to the ground. Use $g = -9.8\,\text{m s}^{-2}$ and ignore air resistance when answering these questions.

a How long does the hammer take to fall halfway to 30 m?

b How long does it take the hammer to fall all the way to the ground?

c What is the speed of the hammer as it hits the ground?

When an object is thrown vertically up into the air, it will eventually reach a point where it stops momentarily before returning back down. The velocity of the object decreases as the object rises, becomes zero at the maximum height, and then increases again as the object falls. Throughout this motion, however, the object is still in the same gravitational field, so g remains at $-9.8\,\text{m s}^{-2}$ throughout the journey. Knowing that the velocity of an object thrown in the air is zero at the top of its flight allows you to calculate the maximum height reached.

Worked example 11.5.2

MAXIMUM HEIGHT PROBLEMS

On winning a tennis match, Michael smashes the ball vertically into the air at $27.5\,m\,s^{-1}$. In this example, air resistance can be ignored and the acceleration due to gravity will be taken as $-9.8\,m\,s^{-2}$.

a Determine the maximum height reached by the ball.	
Thinking	**Working**
Write down the values of the quantities that are known and what you are finding. At the maximum height the velocity is zero. Apply the sign convention that upwards is positive and downwards is negative.	$u = 27.5\,m\,s^{-1}$ $v = 0$ $a = -9.8\,m\,s^{-2}$ $s = ?$
Select an appropriate formula.	$v^2 = u^2 + 2as$
Substitute known values into the equation and solve for *s*.	$0 = (27.5)^2 + 2 \times (-9.8) \times s$ $s = \frac{-756.25}{-19.6}$ $= +38.6$ $\therefore s = +39\,m$ The ball reaches a height of 39 m.

b Calculate the time that the ball takes to return to its starting position.	
Thinking	**Working**
To work out the time for which the ball is in the air, it is often necessary to first calculate the time that it takes to reach its maximum height. At the maximum height the velocity will be $0\,m\,s^{-1}$. Write down the values of the quantities that are known and what you are finding.	$u = 27.5\,m\,s^{-1}$ $v = 0$ $a = -9.8\,m\,s^{-2}$ $s = 38.6$ $t = ?$
Select an appropriate formula.	$v = u + at$
Substitute known values into the equation and solve for *t*.	$0 = 27.5 + (-9.8) \times t$ $9.8t = 27.5$ $t = 2.80\,s$ $\therefore t = 2.8\,s$ The ball takes 2.8 s to reach its maximum height. It will therefore take 2.8 s to fall from this height back to its starting point, and so the whole trip will last for 5.6 s.

Worked example: Try yourself 11.5.2

MAXIMUM HEIGHT PROBLEMS

On winning a cricket match, a fielder throws a cricket ball vertically into the air at $15.0\,m\,s^{-1}$. In this example, air resistance can be ignored and the acceleration due to gravity will be taken as $-9.8\,m\,s^{-2}$.

a Determine the maximum height reached by the ball.

b Calculate the time that the ball takes to return to its starting position.

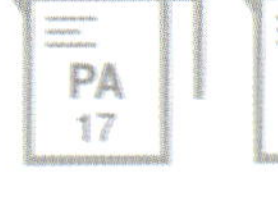

11.5 Review

OA ✓✓

SUMMARY

- The standard mass is measured in kilograms and is defined using Planck's constant.
- The mass of an object relates to its ability to resist changes in motion.
- Mass is a scalar quantity and is a property of a body that is not influenced by external environmental factors.
- Gravity is a force of attraction between masses that extends throughout space.
- The force due to gravity is a vector and requires a magnitude and direction. It is measured in newtons.
- The centre of mass is the point at which the entire mass of a body is considered to be concentrated. All external forces can be applied at this point.
- If air resistance can be ignored, all bodies falling freely near Earth will move with the same constant acceleration.
- The gravitational field strength is represented by g and is equal to $9.8\,\text{N kg}^{-1}$ in the direction towards the centre of Earth. This is equivalent to the acceleration due to gravity of $9.8\,\text{m s}^{-2}$ down.
- The equations for uniform acceleration can be used to solve problems involving vertical motion. It is necessary to specify whether upwards or downwards is positive.

KEY QUESTIONS

For these questions, ignore the effects of air resistance and assume that the gravitational field strength is $9.8\,\text{N kg}^{-1}$ unless instructed otherwise.

Knowledge and understanding

1 Would your force due to gravity be greater on Earth or on the Moon? Explain your answer.

2 Angus inadvertently drops an egg while baking a cake and the egg falls vertically towards the ground. Describe how the acceleration and the velocity of the egg changes as it falls.

3 Max is an Olympic trampolinist and is practising some routines. Describe Max's acceleration and velocity when he is at the highest point of the bounce. Assume that his motion is vertical.

Analysis

4 Mary's mass is 75 kg. What is the force of gravity acting on her if g is $9.8\,\text{N kg}^{-1}$?

5 A super ball is bounced so that it travels straight up into the air, reaching its highest point after 1.50 s.

- **a** What is the initial speed of the ball just as it leaves the ground?
- **b** What is the maximum height reached by the ball?

6 A book is knocked off a bench and falls vertically to the floor. The book takes 0.40 s to fall to the floor.

- **a** What is the book's speed as it lands?
- **b** What is the height from which the book fell?
- **c** How far did the book fall during the first 0.20 s?
- **d** How far did the book fall during the final 0.20 s?

7 While celebrating her birthday, Bindi pops a party popper. The lid travels vertically into the air. Bindi is a keen physics student and notices that the lid takes 4.00 s to return to its starting position.

- **a** How long does the lid take to reach its maximum height?
- **b** How fast was the lid travelling initially?
- **c** What was the maximum height reached by the lid?
- **d** What was the velocity of the lid as it returned to its starting point?

8 Two physics students conduct the following experiment from a very high bridge. Thao drops a 1.50 kg shot-put from a vertical height of 60.0 m, while at exactly the same time Benjamin throws a 100.0 g mass with an initial downwards velocity of $10.0\,\text{m s}^{-1}$ from a point 10.0 m above Thao.

- **a** How long does it take the shot-put to reach the ground?
- **b** How long does it take the 100.0 g mass to reach the ground?

Chapter review

KEY TERMS

acceleration
air resistance
centre of mass
dimensional analysis
displacement
distance travelled
free fall
mass
position
speed
velocity

REVIEW QUESTIONS

For the following questions, the acceleration due to gravity is $9.8\,m\,s^{-2}$ downwards and air resistance is considered to be negligible, unless indicated otherwise.

Knowledge and understanding

1 A car travels at $95.0\,km\,h^{-1}$ along a freeway. What is its speed in $m\,s^{-1}$?

2 A cyclist travels at $15.0\,m\,s^{-1}$ during a sprint finish. What is this speed in $km\,h^{-1}$?

3 A ping pong ball is falling vertically at $6.00\,m\,s^{-1}$ as it hits the floor. It rebounds at $4.00\,m\,s^{-1}$ up. What is its change in speed during the bounce?

4 A car is moving in a positive direction. The car approaches a red light and slows down. Describe its acceleration and velocity as it slows down.

5 A girl tosses a marble straight up into the air at $5.00\,m\,s^{-1}$ and then catches it at the same height from which it was thrown. Ignore air resistance.

- **a** Is the acceleration of the marble on the way up the same as, less than or greater than its acceleration on the way down? Justify your answer.
- **b** Is the launch speed of the marble the same as, less than or greater than its landing speed? Justify your answer.

Application and analysis

6 An athlete in training for a marathon runs 15.0 km north along a straight road before realising that she has dropped her drink bottle. She turns around and runs back 5.00 km to find her bottle, then resumes running in the original direction. After running for 2.00 hours, the athlete reaches a point 20.0 km from her starting position and stops.

- **a** Calculate her average speed in $km\,h^{-1}$.
- **b** Calculate her average velocity in:
  - **i** $km\,h^{-1}$
  - **ii** $m\,s^{-1}$

7 Mihi rides her bicycle to school and travels 2.50 km south in 15.0 min.

- **a** Calculate her average speed in kilometres per hour ($km\,h^{-1}$).
- **b** What was her average velocity in metres per second ($m\,s^{-1}$)?

8 A skier is travelling along a horizontal ski run at a speed of $15.0\,m\,s^{-1}$. The skier falls over and takes 2.50 s to come to rest. Calculate the skier's average acceleration.

9 The graph below shows the position of a motorbike along a straight stretch of road as a function of time. The motorcyclist starts 200 m north of an intersection.

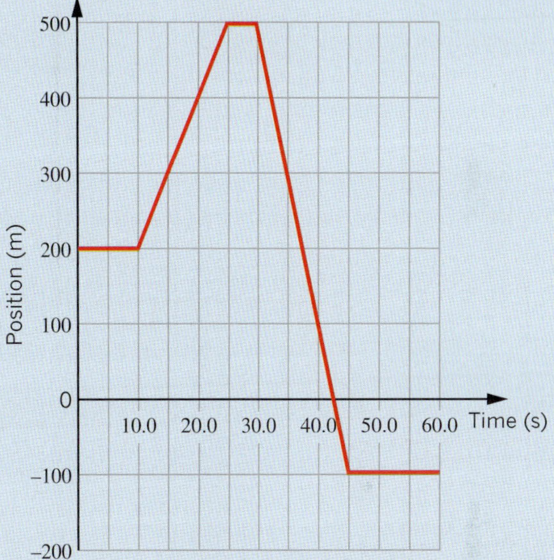

- **a** During what time interval is the motorcyclist travelling in a northerly direction?
- **b** During what time interval is the motorcyclist travelling in a southerly direction?
- **c** During what time intervals is the motorcyclist stationary?
- **d** At what time is the motorcyclist passing back through the intersection?

10 A car goes through a set of traffic lights that has a fixed speed camera installed and a speed limit of $70\,km\,h^{-1}$. In this instance the distance between the first two strips is 1.5000 m and the distance between the last two strips is 1.5500 m. If the time measured for the first interval is 0.080 597 s and the time for the second interval is 0.096 207 s, calculate the average speed between the first two set of strips and the last two set of strips in $km\,h^{-1}$. Determine if the car was speeding and whether it was slowing down or speeding up.

continued over page

11 For each of the activities below, indicate which of the following velocity–time graphs best represents the motion involved.

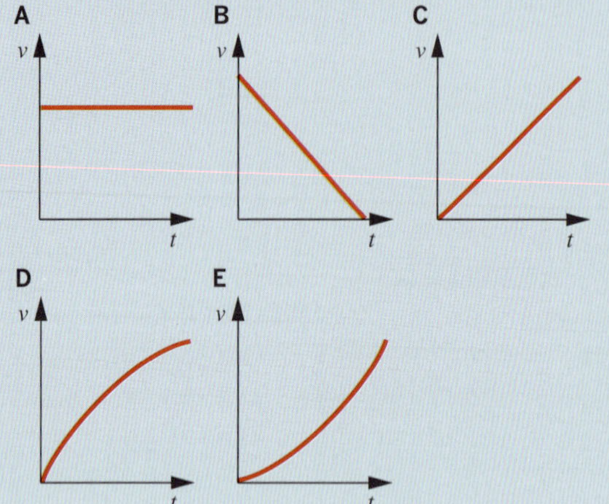

a A car comes to a stop at a red light.

b A swimmer is travelling at a constant speed.

c A motorbike starts from rest with uniform acceleration.

12 This velocity–time graph is for an Olympic road cyclist as he travels, initially north, along a straight section of track.

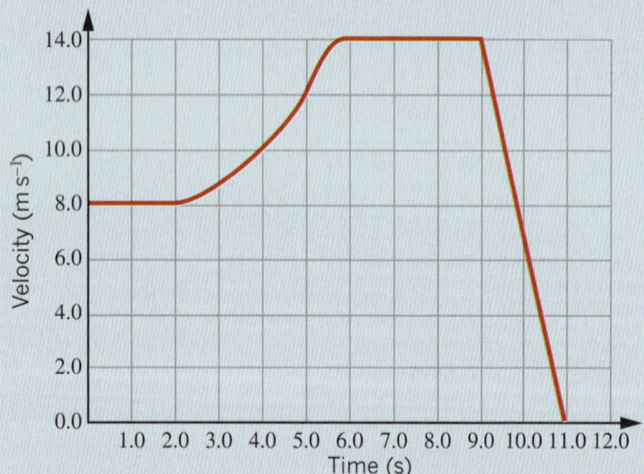

a Calculate the displacement of the cyclist during his journey.

b Calculate the magnitude, to three significant figures, of the average velocity of the cyclist during this 11.0 s interval.

c Calculate the acceleration of the cyclist at $t = 1.0$ s.

d Calculate the acceleration of the cyclist at $t = 10.0$ s.

e Which one or more of the following statements correctly describes the motion of the cyclist?

A He is always travelling north.

B He travels south during the final 2.0 s.

C He is stationary at $t = 8.0$ s

D He returns to the starting point after 11.0 s.

13 A car starts from rest and has a constant acceleration of 3.50 m s^{-2} for 4.50 s. What is its final speed?

14 A pumpkin has a mass of 10 kg on Earth. What is the force due to gravity on the pumpkin on Earth?

15 On the surface of Earth, a geological hammer has a mass of 1.5 kg. Determine its mass and the force due to gravity on the hammer on Mars, where $g = 3.6$ m s^{-2}.

16 A jet-ski starts from rest and accelerates uniformly away from the beach. It travels 2.00 m in its first second of motion. Calculate:

a its acceleration

b its speed at the end of the first second

c the distance the jet-ski travels in its second second of motion.

17 A skater is travelling along a horizontal skate rink at a speed of 10.0 m s^{-1}. The skater falls over and comes to rest in 10.0 m. Calculate, to three significant figures:

a the average acceleration of the skater

b how long it takes the skater to come to a stop.

18 The graph shows the position of Candice dancing across a stage.

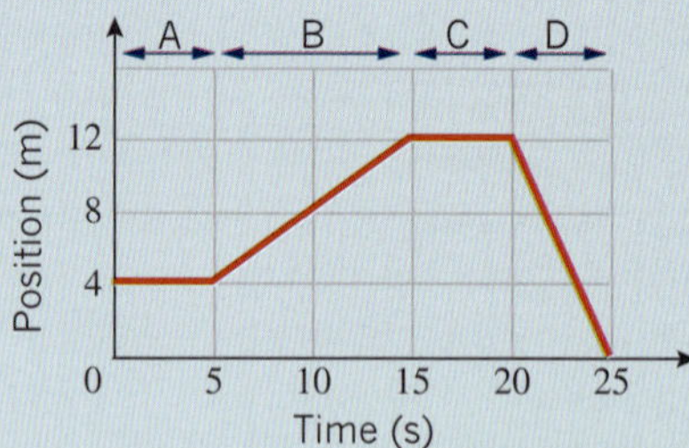

a What is Candice's starting position?

b In which of the sections (A–D) is Candice at rest?

c In which of the sections (A–D) is Candice moving in a positive direction, and what is her velocity?

d In which of the sections (A–D) is Candice moving with a negative velocity and what is the magnitude of this velocity?

e Calculate Candice's average speed during the 25 s motion.

19 A slingshot is used to launch a marble vertically into the air at 39.2 m s^{-1}. Discuss the velocity and acceleration of the marble as it travels to its maximum height. Indicate the time that it takes to reach the top. Consider upwards as positive.

20 Ben is overtaking another cyclist and increases his speed uniformly from 4.20 m s^{-1} to 6.70 m s^{-1} east over a time interval of 0.500 s.

a What is the magnitude of Ben's average acceleration during this time?

b How far does Ben travel while overtaking?

c What is Ben's average speed during this time?

21 A diver plunges headfirst into a diving pool while travelling at 28.0 m s^{-1} downwards. The diver enters the water and stops after a distance of 4.00 m. Consider the diver to be a single point located at her centre of mass and assume her acceleration through the water to be uniform.

a What is the magnitude and direction of the average acceleration of the diver as she travels through the water?

b How long does the diver take to come to a stop?

c What is the velocity of the diver after she has dived through 2.00 m of water?

22 A golfer mis-hits a golf ball straight up into the air. Which statement describes the acceleration of the ball while it is on its way up, when it is at its maximum height, and when it on its way down?'

A The acceleration of the ball decreases as it travels upwards, becoming zero as it reaches its highest point.

B The acceleration is constant as the ball travels upwards, then the acceleration reverses direction as the ball falls down again.

C The acceleration of the ball is greatest when the ball is at the highest point.

D The acceleration of the ball is constant throughout.

23 Steph tosses a rock vertically into the air. Copy the following paragraph and fill in the blanks with 'upwards', 'zero' or 'downwards' to complete the statement about the rock's motion.

On its way upwards, the rock has ________________ velocity and ________________ acceleration. At the highest point, the rock has ________________ velocity and ________________ acceleration. On its way downwards, the rock has ________________ velocity and ________________ acceleration.

24 Claire hits a tennis ball vertically into the air at 30 m s^{-1}. The v–t and a–t graphs for the tennis ball are shown. Use the graphs or the equations for uniform acceleration to answer the following questions. Use $g = 10$ m s^{-2} for these questions. Assume the motion in question is symmetrical, starting and ending at the same point.

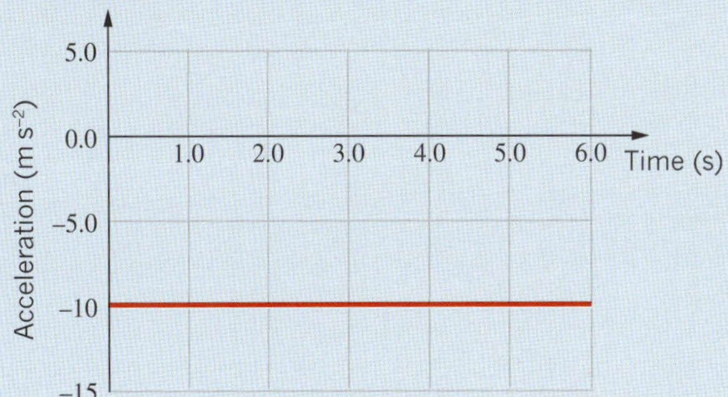

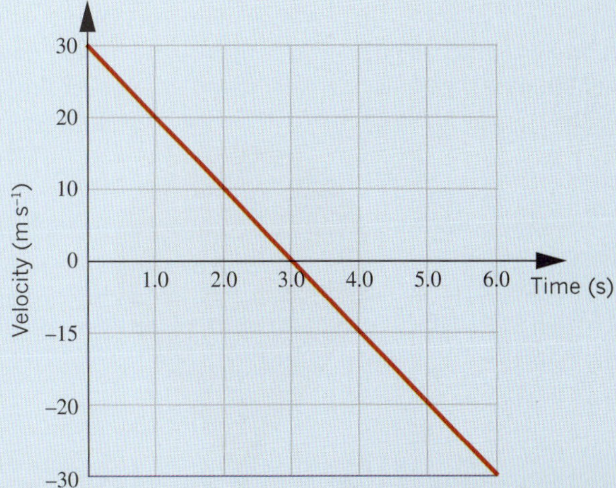

a What is the maximum height reached by the ball?

b How long does the ball take to return to its starting position?

c What is the velocity of the ball 5.0 s after Claire hits it?

d What is the acceleration of the ball at its maximum height?

25 A window cleaner working on a tower accidently drops her mobile phone. The phone falls vertically towards the ground with an acceleration of 9.8 m s^{-2}.

a Determine the speed of the phone after 3.00 s.

b How fast is the phone moving after it has fallen 30.0 m?

c What is the average velocity of the phone during a fall of 30.0 m?

26 A skateboard has a force due to gravity of 20.6 N acting on it on Earth. What is its mass?

27 a What is the mass of an 85 kg astronaut on the surface of Earth where g is −9.8 m s^{-2}?

b What is the mass of an 85 kg astronaut on the surface of the Moon where g is −1.6 m s^{-2}?

c What is the force due to gravity on an 85 kg astronaut on the surface of Mars where g is −3.6 m s^{-2}?

28 Given the values in question 27, order the force due to gravity of a 1 kg object from greatest force due to gravity to least force due to gravity when it is on the Moon, on Mars and on Earth.

continued over page

29 A hot-air balloon is 80.0 m above the ground and travelling vertically downwards at 8.00 m s^{-1} when one of the passengers accidentally drops a coin over the side.

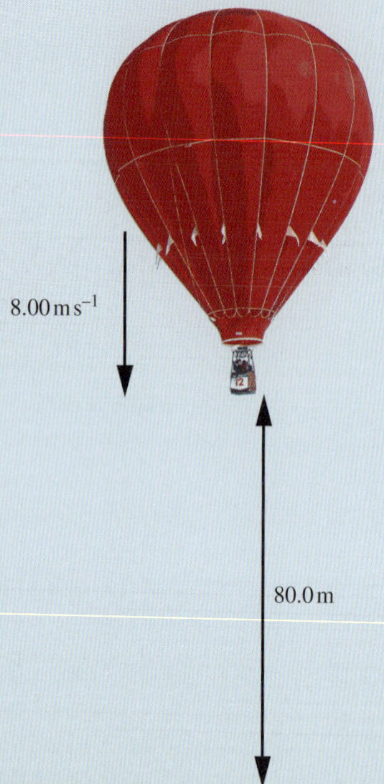

a How long does the balloon take to reach the ground?

b What is the speed of the coin as it reaches the ground?

c How long after the coin reaches the ground does the balloon touch down?

30 During a game of minigolf, Renee putts a ball so that it hits an obstacle and travels straight up into the air, reaching its highest point after 1.50 s.

a What was the initial velocity of the ball as it launched into the air?

b Calculate the maximum height reached by the ball.

31 At the start of a football match, the umpire bounces the ball so that it travels vertically upwards and reaches a height of 15.0 m.

a How long does the ball take to reach this maximum height?

b One of the ruckmen is able to leap and reach to a height of 4.00 m with his hand. How long after the bounce should this ruckman try to make contact with the ball?

OA ✓✓

CHAPTER

12 Momentum and force

In the seventeenth century, Sir Isaac Newton published three laws that explain why objects in our universe move as they do. These laws became the foundation of a branch of physics called mechanics: the science of how and why objects move. They have become commonly known as Newton's three laws of motion.

Using Newton's laws, this chapter will describe the relationship between the forces acting on an object and its motion. It will also discuss the relationship between force, period of time and change in momentum (impulse).

Key knowledge

- apply concepts of momentum to linear motion: $p = mv$ **12.4**
- explain changes in momentum as being caused by a net force: $\Delta p = F_{net}\Delta t$ **12.6**
- model the force due to gravity, F_g, as the force of gravity acting at the centre of mass of a body, $F_{\text{on body by Earth}} = mg$, where g is the gravitational field strength ($9.8\,\text{N kg}^{-1}$ near the surface of Earth) **12.3**
- model forces as vectors acting at the point of application (with magnitude and direction), labelling these forces using the convention 'force on A by B' or $F_{\text{on A by B}} = -F_{\text{on B by A}}$ **12.3**
- apply Newton's three laws of motion to a body on which forces act: $a = \frac{F_{net}}{m}$, $F_{\text{on A by B}} = -F_{\text{on B by A}}$ **12.1, 12.2, 12.3**
- apply the vector model of forces, including vector addition and components of forces, to readily observable forces including the force due to gravity, friction and normal forces **12.1, 12.3, 12.4, 12.5**
- analyse impulse in an isolated system (for collisions between objects moving in a straight line): $F\Delta t = m\Delta v$ **12.5**
- investigate and analyse theoretically and practically momentum conservation in one dimension. **12.4**

12.1 Newton's first law

Chapter 11 developed the concepts and ideas needed to describe the motion of a moving body. In this chapter, rather than simply describe the motion, you will investigate the forces that cause the motion to occur.

FORCE

In simple terms, a **force** can be thought of as a push or a pull, but forces exist in a wide variety of situations in your life and are fundamental to the nature of matter and the structure of the universe. Consider each of the photographs in Figure 12.1.1. For each situation a force—a push or pull—is acting.

FIGURE 12.1.1 (a) At the moment of impact, both the tennis ball and the racquet strings are distorted by the forces acting at this instant. (b) The rock climber is relying on the frictional force between his hands and feet and the rock face. (c) A continual force causes the clay to deform into the required shape. (d) The gravitational force between Earth and the Moon is responsible for two high tides each day. (e) The globe is suspended in mid-air because of the magnetic forces of repulsion and attraction.

In each of the situations depicted in Figure 12.1.1, forces are acting. Some are applied directly to an object and some act on a body without touching it. Forces that act directly on a body are called **contact forces**, because the body will only experience the force while contact is maintained. Forces that act on a body at a distance are **non-contact forces**.

Contact forces are easier to understand and include the simple pushes and pulls that are experienced daily in people's lives. Examples of these include the forces between colliding billiard balls, and the forces that act between you and your chair as you sit reading this book. Friction and drag forces are also contact forces.

Non-contact forces occur when the object causing the push or pull is physically separated from the object that experiences the force. These forces are said to 'act at a distance'. Gravitation, and magnetic and electric forces are examples of non-contact forces.

> A force is a push or a pull that is measured in newtons (N). It is a vector and so it requires a magnitude and a direction to describe it fully.

The amount of force acting can be measured using the SI unit called the newton, which is given the symbol N. The unit, which will be defined later in the chapter, honours Sir Isaac Newton (1642–1727), who is still considered to be one of the most significant physicists to have lived, and whose first law is the subject of this section. A force of one newton, 1 N, is approximately the force you have to exert when holding a 100 g mass against the downwards pull of gravity. In everyday life this is about the same as holding a small apple.

If more than one force acts on a body at the same time, the body behaves as if only one force—the vector sum of all the forces—is acting. The vector sum of the forces is called the resultant or **net force**, F_{net}. (Note: vectors were covered in detail in Chapter 10.)

> The net force acting on a body experiencing a number of forces acting simultaneously is given by the vector sum of all the individual forces:
> $F_{net} = F_1 + F_2 + \ldots + F_n$

NEWTON'S FIRST LAW OF MOTION

Inertia and Newton's first law are closely related; in fact, some people call Newton's first law the law of inertia. Inertia is the tendency of an object to maintain its velocity. This tendency is related to the mass of an object, so that the greater the mass, the harder it is to get it moving or to stop it from moving.

Newton's first law can be stated as:

> An object will maintain a constant velocity unless an unbalanced, external force acts on it.

This statement needs to be analysed in more detail by first examining some of the key terms used. The term 'maintain a constant velocity' implies that, if the object is moving, then it will continue to move with a velocity that has the same magnitude and direction. For example, if a car is moving at 12.0 m s^{-1} south, then some time later it will still be moving at 12.0 m s^{-1} south (Figure 12.1.2). It should also be noted that zero velocity can also be constant, so if the car is moving at 0 m s^{-1}, then some time later it will still be moving at 0 m s^{-1}.

FIGURE 12.1.2 A car maintaining a constant velocity

The use of the term 'unbalanced' in relation to the acting force implies that there must be a net force acting on the object. If the forces are balanced, then the object's velocity will remain constant. If the forces are unbalanced, then the velocity will change, or will not remain constant. Balanced and unbalanced forces are illustrated in Figure 12.1.3.

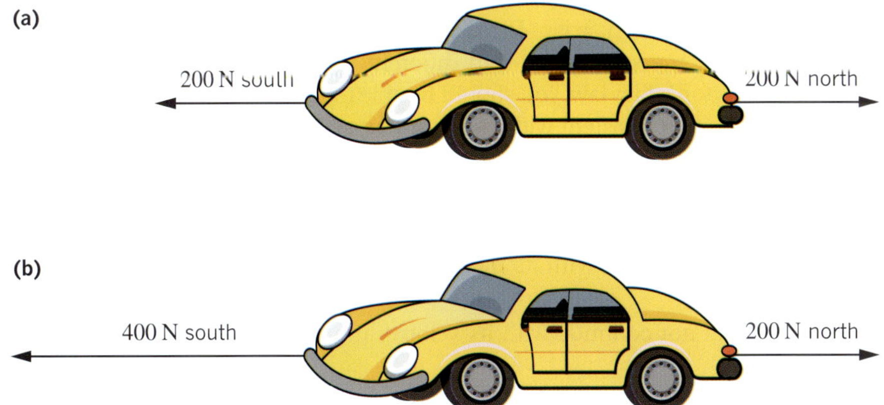

FIGURE 12.1.3 (a) The forces on the car are balanced, and so it will maintain a constant velocity. (b) The forces on the car are unbalanced: it has a net force in the forwards direction, so its velocity will change.

The term 'external', in relation to forces, implies that the forces are not internal. When forces are internal, they will have no effect on the motion of the object. For example, if you are sitting in a car and push forwards on the steering wheel then the car will not move forwards due to this force. In order for you to push forwards on the steering wheel, you must push backwards on the seat. Both the steering wheel and the seat are attached to the car, therefore there are two forces acting on the car that are equal and in the opposite direction to each other, as shown in Figure 12.1.4. All internal forces must result in balanced forces on the object and therefore they will not change the velocity of the object.

FIGURE 12.1.4 This driver is applying internal forces on a car. These internal forces will balance and cancel each other.

Applying a force to an object can cause it to speed up, to slow down, to start moving, to stop moving, or to change its direction. The effect depends on the direction of the force in relation to the direction of the velocity vector of the object experiencing the force. The effects of external forces are summarised in Table 12.1.1.

TABLE 12.1.1 The effect of the application of a force, depending on the relationship between the direction of the force and the velocity

Relationship between velocity and force	Effect of force
force applied to object at rest	object starts moving
force in same direction as velocity	magnitude of velocity increases (object speeds up)
force in opposite direction to velocity	magnitude of velocity decreases (object slows down)
force perpendicular to velocity	direction of velocity changes (object turns)

In all cases, the effect of a force is to change the velocity of an object, whether it is the magnitude of the velocity, the direction of the velocity, or both that change.

Stating Newton's first law in different ways

Ludwig Wittgenstein, an Austrian–British philosopher, suggested that 'understanding means seeing that the same thing said different ways is the same thing'. To truly understand Newton's first law, you should be able to state it in different ways yet still recognise it as being consistent with Newton's first law.

All of these statements are consistent with Newton's first law:

- An object will maintain a constant velocity unless an unbalanced, external force acts on it.
- An object will continue with its motion unless an unbalanced, external force is applied.
- An object will not continue with its velocity if an unbalanced, external force is applied.
- A body will either remain at rest or continue with constant speed in a straight line (i.e. constant velocity) unless it is acted on by a net force.
- If an unbalanced, external force is applied, then an object's velocity will change.
- If a net force is applied, then the object's velocity will change.

- If no net force is applied, the object will not accelerate.
- If a net force is applied, an acceleration will result.
- Net forces cause acceleration.
- No force, no acceleration.
- Constant velocity means no net force is applied.

CASE STUDY ANALYSIS

Terminal velocity of raindrops

In Chapter 11, it was stated that in the absence of air resistance, all objects accelerate towards the surface of Earth at a constant rate of $9.8\,m\,s^{-2}$. However, in practice, air resistance acts against the force due to gravity.

Newton's first law can be used to explain how air resistance causes a falling object such as a raindrop to experience terminal velocity. As the raindrop begins falling towards Earth the only external force is the force due to gravity. This gravitational force causes the raindrop to accelerate at $9.8\,m\,s^{-2}$. As the raindrop gets faster, air resistance pushes upwards. This reduces the magnitude of the net force, which decreases the acceleration of the raindrop. Eventually the air resistance becomes so great that it exactly balances the force due to gravity. According to Newton's first law, the raindrop maintains a constant velocity when all the external forces are balanced: this velocity is terminal velocity.

The amount of air resistance on a falling object depends on factors such as its speed, shape and size. Therefore, the terminal velocity will be different for different objects. Table 12.1.2 lists the terminal velocities for raindrops and hailstones of different sizes.

TABLE 12.1.2 The size of a raindrop or hailstone affects its terminal velocity.

Type of drop	Drop size (mm)	Terminal velocity ($m\,s^{-1}$)
small raindrop	1.2	4.64
large raindrop	4.0	8.83
largest possible raindrop	5.0	9.09
small hailstone	10	10.0
large hailstone	40	20.0

Analysis

1 Construct a graph showing the relationship between drop size and terminal velocity for raindrops and hailstones. Is this relationship best described as a direct relationship or an inverse relationship?

2 Use the data to estimate the terminal velocity of a 3.0 mm raindrop.

3 Raindrops are typically formed at an altitude of around 800 m. If a raindrop did not have a terminal velocity and, instead, accelerated at $9.8\,m\,s^{-2}$ for the entire time that it was falling, how fast would it be going when it hit Earth's surface?

INERTIA

Inertia is considered to be the resistance to a change in motion of an object. It is related to the mass of the object. As the mass of the object increases, the inertia increases and therefore:

- it becomes harder to start the object moving if it is stationary, or
- it becomes harder to stop it moving, or
- it becomes harder to change the direction of motion of the object.

You can experience the effect of inertia when you push a trolley in a supermarket. If the trolley is empty, it is relatively easy to start pushing it, or to pull it to a stop once it is already moving. It is also easy to turn a corner. If you fill the trolley with heavy groceries, you notice that it becomes more difficult—that is, it requires more force—to make the trolley start moving when it is at rest, and it becomes more difficult to pull it to a stop if it is already moving. It also requires more force to change the direction of the trolley.

It is important to note that the effect of inertia is independent of gravity. Since inertia depends on mass, and the force due to gravity also depends on mass, it is a common misunderstanding to think that the effects of inertia only apply in the presence of gravity. However, even in space it would be just as difficult to change the state of motion of the trolley, as described above, as it is in the supermarket aisle.

Newton's first law and inertia

The connection between Newton's first law and inertia is very close. Due to inertia, an object will continue with its motion unless a net force acts on the object.

You experience the connection between Newton's first law and inertia if you are standing on public transport. Imagine standing on a tram that is initially at rest and then starts moving forwards. If you are not holding on to anything, you may stumble backwards as though you have been pushed backwards. However, you have not been pushed backwards; the tram has started moving forwards and, since you have inertia, your mass resists the change in motion. According to Newton's first law, your body is simply maintaining its original state of being motionless until an unbalanced force acts to accelerate it. When the tram later comes to a sudden stop, your body again resists the change by continuing to move forwards until an unbalanced force acts to bring it to a stop.

OBSERVING NEWTON'S FIRST LAW

When an object is in motion—for example, a pen sliding across a table—it will eventually stop. It may not seem obvious, but this is a very good example of Newton's first law. The motion does not continue; therefore a net force must be acting. In this case, however, the force is not an obvious one. Confusion sometimes arises if the force due to friction is overlooked. Friction is a force that always acts in the opposite direction to the motion of objects. Air resistance is also a force that is often overlooked, as is the force due to gravity. By ignoring the effect of these important forces, it can be easy to come to the incorrect conclusion that the natural state of any object is to be at rest. By considering all the external forces acting on an object, it becomes clear that the natural state of any object is to maintain whatever velocity it currently has.

Misconceptions about Newton's first law

The way in which the first law is traditionally phrased can be a little misleading because it might lead us to expect that, in everyday life, most objects should move at a constant velocity. In reality, we observe that objects tend to slow down if they don't have an external force acting on them to maintain their velocity. This is because of friction. Friction is an unbalanced force that acts on almost every object. Therefore, to get an object to travel at a constant velocity, we usually need to add a force to counterbalance the force of friction.

Another way of expressing Newton's first law that addresses this misconception is:

- An object will not maintain its velocity if an unbalanced, external force is applied.

Frictional forces

Friction is a force that opposes movement. Suppose you want to push your textbook along the table. As you start to push the book, you find that the book does not move at first. You then increase the force that you apply. Suddenly, at a certain critical value, the book starts to move.

There is a maximum frictional force that resists the start of the slide. This force is called the static friction force, F_s. Once the book begins to slide, a much lower force than F_s is needed to keep the book moving. This force is called the kinetic friction force and is represented by F_k. The graph in Figure 12.1.5 shows how the force required to move an object changes as static friction is overcome.

This phenomenon can be understood when you consider that even the smoothest surfaces are quite jagged at the microscopic level. When the book is resting on the table, the jagged points of its bottom surface have settled into the valleys of the surface of the table and this helps to resist attempts to slide the book. Once the book is moving, the surfaces do not have any time to settle into each other, so less force is required to keep the book moving.

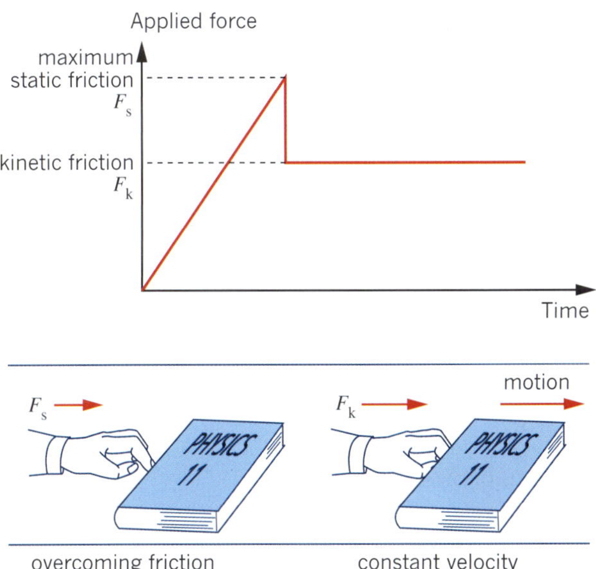

FIGURE 12.1.5 To get things moving, the static friction between an object and the surface must be overcome. This requires a force larger than that needed to maintain constant velocity.

CASE STUDY

Galileo's law of inertia

Galileo Galilei was born into an academic family in Pisa, Italy, in 1564. He made significant contributions to physics, mathematics and scientific method through intellectual rigour and the quality of his experimental design. But, more than this, Galileo helped to change the way the universe is viewed.

Galileo's most significant contributions were in astronomy. Through his development of the refracting telescope he discovered sunspots, lunar mountains and valleys, the four largest moons of Jupiter (now called the Galilean Moons, shown in Figure 12.1.6) and the phases of Venus. In mechanics, he demonstrated that projectiles move with a parabolic path and that different masses fall at the same rate (the law of falling bodies).

FIGURE 12.1.6 The major moons of Jupiter are known as the Galilean moons.

These developments were important because they changed the framework within which mechanics was understood. This framework had been in place since Aristotle had constructed it in the fourth century BCE. Aristotle's thesis was based on the observation that a moving body's natural state is at rest, and the object will come to rest unless a force is applied. However, Galileo's experiments led him to believe that the natural state of an object is not at rest. He suggested that objects maintained their state of motion. He called this tendency inertia. One might think that Galileo would have won praise from his peers for making such progress, but the Aristotelian view was so entrenched that Galileo actually lost his job as a professor of mathematics in Pisa in 1592.

Galileo held a number of opinions that were considered controversial in his lifetime. In 1630, he published a book in which he challenged the Ptolemaic (i.e. Earth centred) view of the universe and supported the new Sun-centred model proposed by Copernicus. Because of this, Galileo was summoned to Rome to face the Inquisition for heresy (opposition to the Church). The finding went against Galileo and all copies of his book had to be burned. He was sentenced to permanent house arrest for the rest of his life.

Despite this, by the time Galileo died in 1642, he had become an influential thinker across Europe. In 1992 a papal commission reversed the Catholic Church's condemnation of him.

12.1 Review

OA ✓✓

SUMMARY

- Newton's first law can be written in many ways:
  - An object will continue with its velocity unless an unbalanced force causes the velocity to change.
  - Net forces cause acceleration.
- Inertia is the tendency of an object to resist changes in motion.
- Inertia is related to mass; an object with a large mass will have a large inertia.

KEY QUESTIONS

Knowledge and understanding

1 A student observes a box sliding across a surface and slowing down to a stop. From this observation what can the student conclude about the forces acting on the box?

2 A car changes its direction as it turns a bend in the road while maintaining its speed of $16\,\text{m}\,\text{s}^{-1}$. From this, what can you conclude?

3 Passengers on commercial flights are required to be seated and have their seatbelts fastened when the plane is coming in to land. What would happen to a person who was standing in the aisle as the plane travelled along the runway during landing?

4 Consider the following situations, and name the force that causes each object to travel along a path that is not a straight line.
- **a** Earth moves in a circle around the Sun with constant speed.
- **b** An electron orbits the nucleus with constant speed.
- **c** A cyclist turns a corner at constant speed.
- **d** An athlete swings a hammer in a circle with constant speed.

Analysis

5 A young boy is using a horizontal rope to pull his billycart at a constant velocity. A frictional force of 25 N also acts on the billycart.
- **a** What force must the boy apply to the rope?
- **b** The boy's father then attaches a longer rope to the cart because the short rope is uncomfortable to use. The rope now makes an angle of 30° to the horizontal. What is the horizontal component of the force that the boy needs to apply in order to move the cart with constant velocity?
- **c** What is the tension force acting along the rope that the boy must supply?

6 A magician performs a trick in which a cloth is pulled quickly from under a glass filled with water without causing the glass to fall over or the water to spill out. Does using a full glass make the trick easier or more difficult? Explain your answer.

7 Which of these objects would find it most difficult to come to a stop: a cyclist travelling at $50\,\text{km}\,\text{h}^{-1}$, a car travelling at $50\,\text{km}\,\text{h}^{-1}$ or a fully laden semitrailer travelling at $50\,\text{km}\,\text{h}^{-1}$? Explain your answer.

8 When flying at constant speed at a constant altitude, a light aircraft has a force due to gravity of 50 kN downwards, and the thrust produced by its engines is 12 kN to the east. What is the lift force required by the wings of the plane, and how large is the drag force that is acting?

12.2 Newton's second law

Newton's second law makes the quantitative connection between force, mass and acceleration.

Newton's second law of motion states that:

The acceleration of an object is directly proportional to the net force on the object and inversely proportional to the mass of the object:

$$a = \frac{F_{net}}{m}$$

where a is the acceleration of an object (in m s^{-2})
F_{net} is the force applied to the object (in N)
m is the mass of the object (in kg).

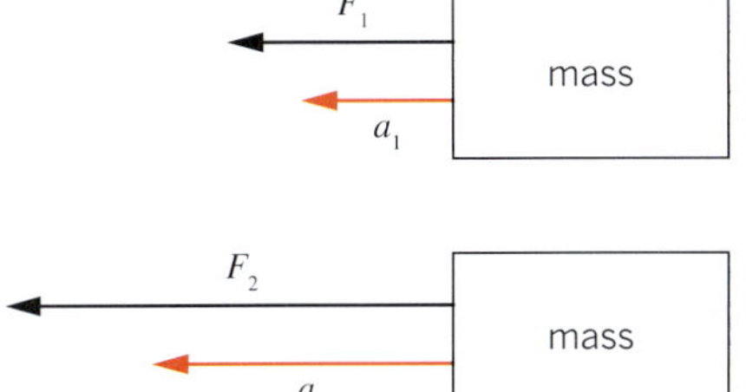

FIGURE 12.2.1 Given the same mass, a larger force will result in a larger acceleration. If the force is doubled, then the acceleration is also doubled.

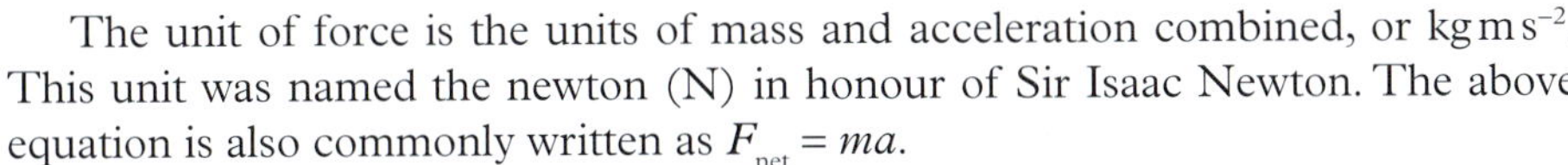

The unit of force is the units of mass and acceleration combined, or kg m s^{-2}. This unit was named the newton (N) in honour of Sir Isaac Newton. The above equation is also commonly written as $F_{net} = ma$.

By definition, 1 **newton** is the force needed to accelerate a mass of 1 kg at 1 m s^{-2}.

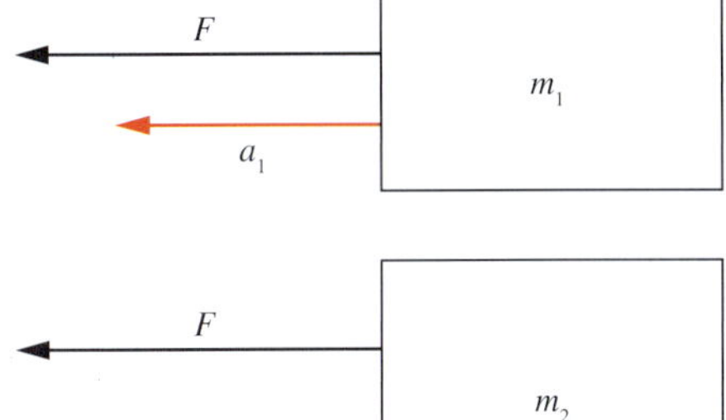

FIGURE 12.2.2 Given the same force, a larger mass will result in a lower acceleration. If the mass is doubled, then the acceleration is halved.

One of the implications of Newton's second law is that, for a given mass, a greater acceleration is achieved by applying a greater force. This is shown in Figure 12.2.1. Doubling the applied force will double the acceleration of the object. In other words, acceleration is proportional to the net force applied.

Notice also in Figure 12.2.1 that the acceleration of the object is in the same direction as the net force applied to it.

Newton's second law also explains how acceleration is affected by the mass of an object. For a given force, the acceleration of an object will decrease with increased mass. In other words, acceleration is inversely proportional to the mass of an object. This is shown in Figure 12.2.2.

PHYSICSFILE

Maximising acceleration

Dragster race cars are designed to achieve the maximum possible acceleration in order to win a race in a straight line over a relatively short distance. According to Newton's second law, acceleration is increased by increasing the applied force and by reducing the mass of the object. For this reason, dragster race cars are designed with very powerful engines that produce an enormous forwards force and an aerodynamic shape to minimise air resistance. There is not much else to the car, so this helps to minimise the mass.

Newton's second law also helps you understand why a motorcycle can accelerate away from the traffic lights at a greater rate than a car or a truck. Although the engines in cars and trucks are usually more powerful than a motorcycle engine, the motorcycle has much less mass, which allows for greater acceleration.

The aerodynamic design of motorcycles and their lower mass enables them to accelerate faster than cars and trucks.

CALCULATIONS WITH NEWTON'S SECOND LAW

Writing Newton's second law as $F_{net} = ma$ enables you to calculate the force that causes a mass to accelerate.

Worked example 12.2.1

CALCULATING THE FORCE THAT CAUSES AN ACCELERATION

Calculate the net force causing a 5.50 kg mass to accelerate at 3.75 m s^{-2} west.

Thinking	Working
Ensure that the variables are in their standard units.	$m = 5.50$ kg $a = 3.75$ m s^{-2} west
Apply the equation for force from Newton's second law.	$F_{net} = ma$ $= 5.50 \times 3.75$ $= 20.6$ N
Give the direction of the net force, which is always the same as the direction of the acceleration.	$F_{net} = 20.6$ N west

Worked example: Try yourself 12.2.1

CALCULATING THE FORCE THAT CAUSES AN ACCELERATION

Calculate the net force causing a 75.8 kg runner to accelerate at 4.05 m s^{-2} south.

The first equation for uniform acceleration, which is discussed in Section 11.4 on page 336, can be combined with Newton's second law to calculate changes in time or velocity.

The first equation for uniform acceleration is:

$$v = u + at$$

This can be rearranged to give:

$$a = \frac{v - u}{t}$$

Combining this with $F_{net} = ma$ gives:

$$F_{net} = m\left(\frac{v - u}{t}\right)$$

Worked example 12.2.2

CALCULATING THE FINAL VELOCITY OF AN ACCELERATING MASS

Calculate the final velocity of a 225 kg scooter that accelerates for 2.00 s from rest due to a force of 2430 N north.

Thinking	Working
Ensure that the variables are in their standard units.	$m = 225$ kg $t = 2.00$ s $u = 0$ m s^{-1} $F_{net} = 2430$ N north
Apply a variation of the equation for force from Newton's second law.	$F_{net} = m\left(\frac{v - u}{t}\right)$ $v - u = \frac{F_{net}t}{m}$ $v = \frac{F_{net}t}{m} + u$ $= \frac{2430 \times 2.00}{225} + 0$ $= 21.6$ m s^{-1}
Give the direction of the final velocity as being the same as the direction of the force.	$v = 21.6$ m s^{-1} north

Worked example: Try yourself 12.2.2

CALCULATING THE FINAL VELOCITY OF AN ACCELERATING MASS

Calculate the final velocity of a 307 g fish that accelerates for 5.20 s from rest due to a force of 0.250 N left.

Forces do not always act alone. Often more than one force will act on an object at any time. The overall effect of the forces depends on the direction of each of the forces. For example, some forces act together and some may oppose each other. When using Newton's second law, it is important to use the net, or resultant, force in the calculation. As forces are vectors, they can be added or combined using the techniques discussed in Chapter 10. Consider the following worked examples.

Worked example 12.2.3

CALCULATING THE ACCELERATION OF AN OBJECT WITH MORE THAN ONE FORCE ACTING ON IT

A swimmer whose mass is 75 kg applies a force of 50 N as they start a lap. The water opposes their efforts to accelerate with a drag force of 20 N. What is their initial acceleration?

Thinking	Working
Determine the individual forces acting on the swimmer, and apply the vector sign convention.	F_1 = 50 N forwards = 50 N F_2 = 20 N backwards = −20 N
Determine the net force acting on the swimmer.	$F_{net} = F_1 + F_2$ = 50 + (−20) = +30 N or 30 N forwards $F_{applied}$ 50 N + F_{drag} 20 N ΣF 30 N
Use Newton's second law to determine acceleration.	$a = \frac{\Sigma F}{m}$ $= \frac{F_{net}}{m}$ $= \frac{30}{75}$ = 0.40 m s^{-2} forwards

Worked example: Try yourself 12.2.3

CALCULATING THE ACCELERATION OF AN OBJECT WITH MORE THAN ONE FORCE ACTING ON IT

A car with a mass of 900 kg applies a driving force of 3000 N as it starts moving. Friction and air resistance oppose the motion of the car with a force of 750 N. What is the car's initial acceleration?

A common type of problem in Physics involves a situation in which the force due to gravity acting on one body is causing the acceleration of another body that is connected to it. This is known as a 'connected body' problem. When solving these problems, it is important to recognise that, although only one body is providing the force, both bodies are accelerating.

Worked example 12.2.4

CALCULATING THE ACCELERATION OF A CONNECTED BODY

A 1.5 kg trolley cart is connected by a cord to a 2.5 kg mass as shown. The cord is placed over a pulley and the mass is allowed to fall under the influence of gravity.

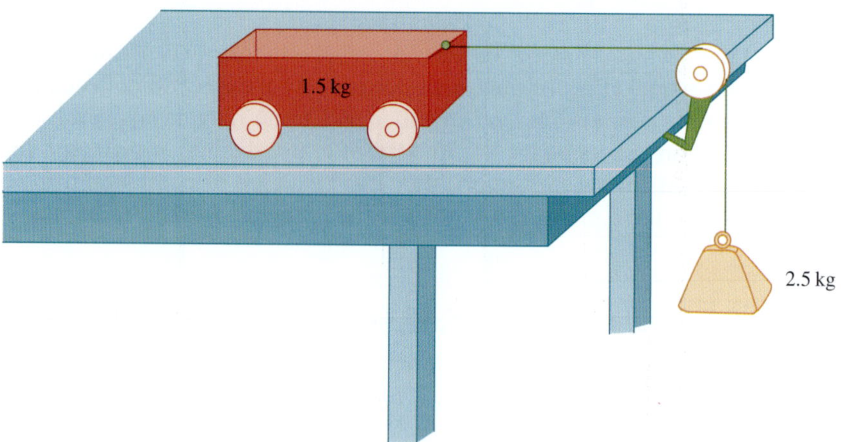

a Assuming that the cart can move over the table unhindered by friction, determine the acceleration of the cart.

Thinking	Working
Recognise that the cart and the falling mass are connected, and determine a sign convention for the motion.	As the mass falls, the cart will move to the right. Therefore, both the downwards movement of the mass and the rightwards movement of the cart will be considered positive motion.
Write down the data that is given. Apply the sign convention to vectors.	$m_1 = 2.5\,\text{kg}$ $m_2 = 1.5\,\text{kg}$ $g = 9.8\,\text{m}\,\text{s}^{-2}$ downwards $= +9.8\,\text{m}\,\text{s}^{-2}$
Determine the forces acting on the system.	The only net force acting on the combined system of the cart and mass is the force due to gravity on the falling mass, m_1. $F_{\text{net}} = F_g$ $= m_1 g$ $= 2.5 \times 9.8$ $= 24.5\,\text{N}$ in the positive direction
Calculate the mass being accelerated.	This net force has to accelerate not only the falling mass but also the cart. $m_1 + m_2 = 2.5 + 1.5$ $= 4.0\,\text{kg}$
Use Newton's second law to determine acceleration.	$a = \frac{F_{\text{net}}}{m}$ $= \frac{24.5}{4.0}$ $= 6.1\,\text{m}\,\text{s}^{-2}$ to the right

b If a frictional force of 8.5 N acts against the cart, what is the acceleration now?	
Thinking	**Working**
Write down the data that is given. Apply the sign convention to vectors.	$m_1 = 2.5\,kg$ $m_2 = 1.5\,kg$ $g = 9.8\,m\,s^{-2}$ downwards $= +9.8\,m\,s^{-2}$ $F_{fr} = 8.5\,N$ left $= -8.5\,N$
Determine the forces acting on the system.	There are now two forces acting on the combined system of the cart and mass: the force due to gravity on the falling mass and friction. $F_{net} = F_g + F_{fr}$ $= 24.5 + (-8.5)$ $= 16.0\,N$ $= 16.0\,N$ in the positive direction
Use Newton's second law to determine acceleration.	$a = \frac{\Sigma F}{m}$ $= \frac{F_{net}}{m}$ $= \frac{16.0}{4.0}$ $= 4.0\,m\,s^{-2}$ to the right

Worked example: Try yourself 12.2.4

CALCULATING THE ACCELERATION OF A CONNECTED BODY

A 0.6 kg trolley cart is connected by a cord to a 1.5 kg mass. The cord is placed over a pulley and allowed to fall under the influence of gravity.

a Assuming that the cart can move over the table unhindered by friction, determine the acceleration of the cart.

b If a frictional force of 4.2 N acts against the cart, what is the acceleration now?

THE FEATHER AND HAMMER EXPERIMENT

Newton's second law helps to resolve the misconception that many people have about the time taken for objects of different mass to fall to the ground. Many mistakenly believe that heavy objects will fall faster than lighter objects. Once again, air resistance acts to complicate the matter and results in the observation of different times for different masses to fall the same distance. However, even when air is removed, the misconception that larger masses fall faster than lighter masses still persists.

Newton's second law was dramatically demonstrated in a famous experiment by lunar astronaut David Scott, depicted in Figure 12.2.3. This showed that on the Moon, where there is no air resistance, a feather and hammer fall at the same rate. A web search for 'hammer and feather on the Moon' will enable you to view a video of this experiment. Although the image quality is quite poor, you should be able to see both objects falling with the same acceleration.

FIGURE 12.2.3 An artist's image of the famous hammer and feather experiment conducted on the Moon

Why the experiment works on the Moon

When two objects with different mass fall under the influence of the force due to gravity in the absence of air resistance, they will both fall at the same rate. That is, their accelerations will be the same. They will cover the same displacement in the same time and will hit the ground at the same time if dropped from the same height. This experiment works on the Moon because there is no atmosphere and, therefore, no air resistance.

If you understand that all objects accelerate due to gravity at the same rate in a vacuum, the common misconception is that the force due to gravity is the same on all objects. This is not true. In fact, the force due to gravity is larger on larger objects and smaller on smaller objects. You mustn't forget that the objects have different masses. A larger mass experiences a greater force due to gravity than a smaller mass, but it also has more inertia so it requires that greater force in order to achieve the same acceleration. Refer to Figure 12.2.4 to see how this works.

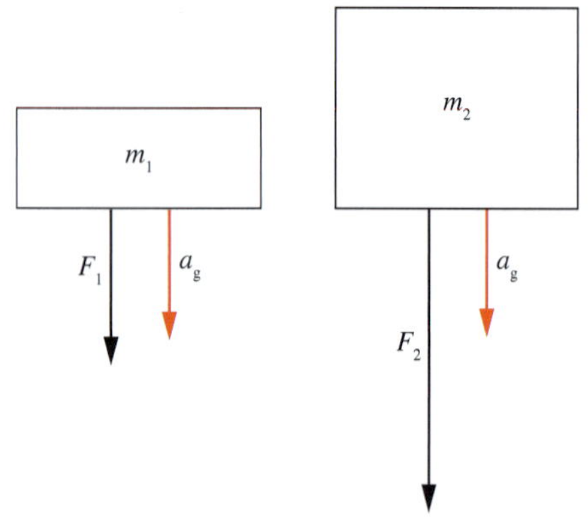

FIGURE 12.2.4 Given the same acceleration, a larger mass must have a larger force due to gravity acting on it. If the mass is doubled, then the force is doubled.

If you consider the relationship between acceleration and mass, and the relationship between acceleration and force, then:

$$a = \frac{F}{m}$$

If m_2 is ten times the mass of m_1, then the force due to gravity on m_2 is ten times the force due to gravity on m_1. Consider the acceleration of both masses:

$$a_1 = \frac{F_1}{m_1}$$

If $F_2 = 10F_1$ and $m_2 = 10m_1$, then:

$$\begin{aligned} a_2 &= \frac{F_2}{m_2} \\ &= \frac{10F_1}{10m_1} \\ &= \frac{F_1}{m_1} \\ &= a_1 \end{aligned}$$

This proof shows that the ratio of force to mass is equal for all combinations of force and mass under the same effects due to gravity. This shows that all masses will experience the same acceleration if air resistance is removed.

Why the experiment does not work on Earth

When you see a feather floating down through the air, you know that it is accelerating at a rate far less than a hammer falling from the same height. From the previous section, you will know that the hammer and the feather have forces due to gravity acting on them that are proportional to their mass. They do not fall at the same rate because of the force of air resistance. Remember, Newton's second law says the acceleration is proportional to the *net* force acting on an object, which means you must consider all the forces acting on an object to determine the acceleration.

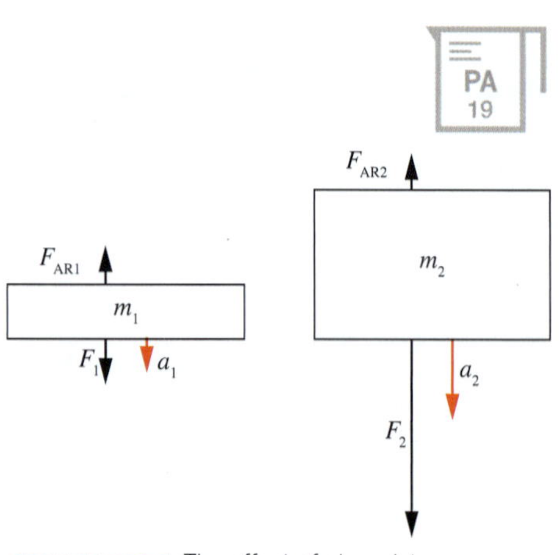

FIGURE 12.2.5 The effect of air resistance on an object depends on the surface area perpendicular to the motion and the speed of the object.

Air resistance is a force that results from air molecules colliding with the object. The faster the object moves, the greater the air resistance. In addition, the greater the surface area perpendicular to the direction of motion, the greater the air resistance. This force, which acts in the opposite direction to the motion of an object, is significant when compared with the force due to gravity on the feather, but insignificant when compared with the force due to gravity on the hammer. As a result, this force has a noticeable effect on the acceleration of the feather but makes no noticeable difference to the acceleration of the hammer.

In Figure 12.2.5, the force of air resistance is denoted as F_{AR} and is the same size on both objects. The difference between the two objects is the downwards force (due to gravity). Figure 12.2.5 also shows that the acceleration of the thinner object is much less than the acceleration of the thicker object. This is the observation that often causes misconceptions.

12.2 Review

OA ✓✓

SUMMARY

- Newton's second law states: The acceleration of an object is directly proportional to the net force on the object and inversely proportional to the mass of the object. This is commonly expressed as the formula
$$a = \frac{F_{net}}{m}$$
- Force can be calculated using the following formula:
$$F_{net} = ma$$
- This can be rewritten as:
$$F_{net} = m\left(\frac{v-u}{t}\right) \quad \Sigma F = m\left(\frac{v-u}{t}\right)$$
- Force is difficult to perceive when it acts on objects, but we can perceive mass and acceleration.
- Different forces due to gravity act on different masses to cause the same acceleration.
- Air resistance is a force that acts to decrease the acceleration of objects moving through air.

KEY QUESTIONS

Knowledge and understanding

Use $g = 9.8\,\text{m}\,\text{s}^{-2}$ when answering the following questions.

1 Describe the feather and hammer experiment and explain why its results on the Moon are different from those on Earth.

2 Calculate the final velocity of a stationary 55.9 kg mass when a net force of 56.8 N north acts on it for 3.50 seconds.

3 Calculate the acceleration of a 45.0 kg mass that has a net force of 441 N acting on it due to gravity.

4 Calculate the final velocity of a 60.0 kg mass moving at $2.67\,\text{m}\,\text{s}^{-1}$ east, when a net force of 45.5 N west acts on it for 2.80 seconds.

Analysis

5 Mary is paddling a canoe. The paddles are providing a constant driving force of 45 N south and the drag forces total 25 N north. The mass of the canoe is 15 kg and Mary has a mass of 50.0 kg.

- a Calculate the force due to gravity acting on Mary.
- b Find the net horizontal force acting on the canoe.
- c Calculate the magnitude of the acceleration of the canoe.

6 A 0.50 kg metal block is attached by a piece of string to a dynamics cart, as shown below. The block is allowed to fall from rest, dragging the cart along. The mass of the cart is 2.5 kg.

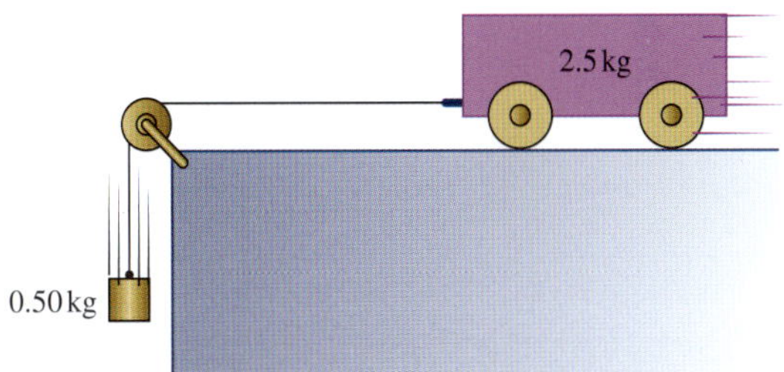

- a If friction is ignored, what is the acceleration of the block as it falls?
- b How fast will the block be travelling after 0.50 s?
- c If a frictional force of 4.3 N acts on the cart, what is the acceleration of the cart?

7 An empty truck of mass 2000 kg has a top acceleration of $2.0\,\text{m}\,\text{s}^{-2}$. The mass of one box is 300 kg. How many boxes would be loaded if the truck's top acceleration decreased to $1.25\,\text{m}\,\text{s}^{-2}$.

8 The thrust force of a rocket with a mass of 50 000 kg is 1 000 000 N. Calculate the acceleration of the rocket as it is launched from Earth's surface.

12.3 Newton's third law

Newton's first two laws of motion describe the motion of an object resulting from the forces that act on that object. Newton's third law of action and reaction is easily stated and seems to be widely known by students, but it is often misunderstood and misused. It is a very important law in physics, as it assists with the understanding of the origin and nature of forces. Newton's third law is explored in detail in this section.

ACTION-REACTION PAIRS

Newton realised that all forces exist in pairs and that each force in the pair acts on a different object. Look at Figure 12.3.1, which shows a hammer hitting a nail on the head. Both the hammer and the nail experience a force during this interaction. The nail experiences a downwards force as the hammer hits it. When the nail is hit it moves a distance into the wood. As it hits the nail, the hammer experiences an upwards force that causes the hammer to stop. These forces are known as an action–reaction pair and are shown in Figure 12.3.2.

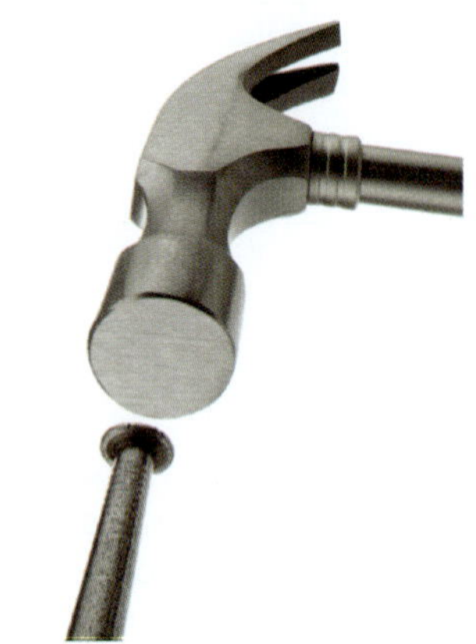

FIGURE 12.3.1 A hammer hitting a nail is a good example of an action–reaction pair and Newton's third law.

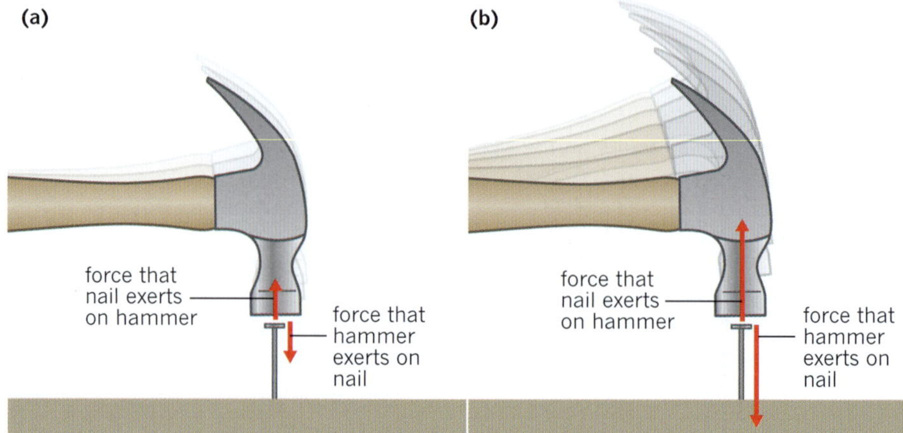

FIGURE 12.3.2 (a) As the hammer gently taps the nail, both the hammer and the nail experience small forces. (b) When the hammer smashes into the nail, both the hammer and the nail experience large forces. These forces are designated by $F_{\text{on nail by hammer}}$ and $F_{\text{on hammer by nail}}$ in both cases.

It is important to note that, regardless of whether the hammer exerts a small force or a large force on the nail, the nail will exert exactly the same-size force back on the hammer.

Newton's third law states:

 For every action (force), there is an equal and opposite reaction (force).

This means that when object A exerts a force, F, on object B, object B will exert an equal and opposite force on object A. It is important to recognise that the action force and the reaction force in Newton's third law act on different objects and so should never be added together; their effect will only be on the object on which they act. Newton's third law applies not only to forces between objects that are in direct contact, but also to non-contact forces, such as the force due to gravity between objects.

The main misconception that arises when considering Newton's third law is the belief that, if a large mass collides with a smaller mass, then the larger object exerts a larger force and the smaller object exerts a smaller force. This is not true. If you witness the collision between the car and the bus in Figure 12.3.3(c), you would see the car undergoing a large deceleration while the bus undergoes only a small acceleration.

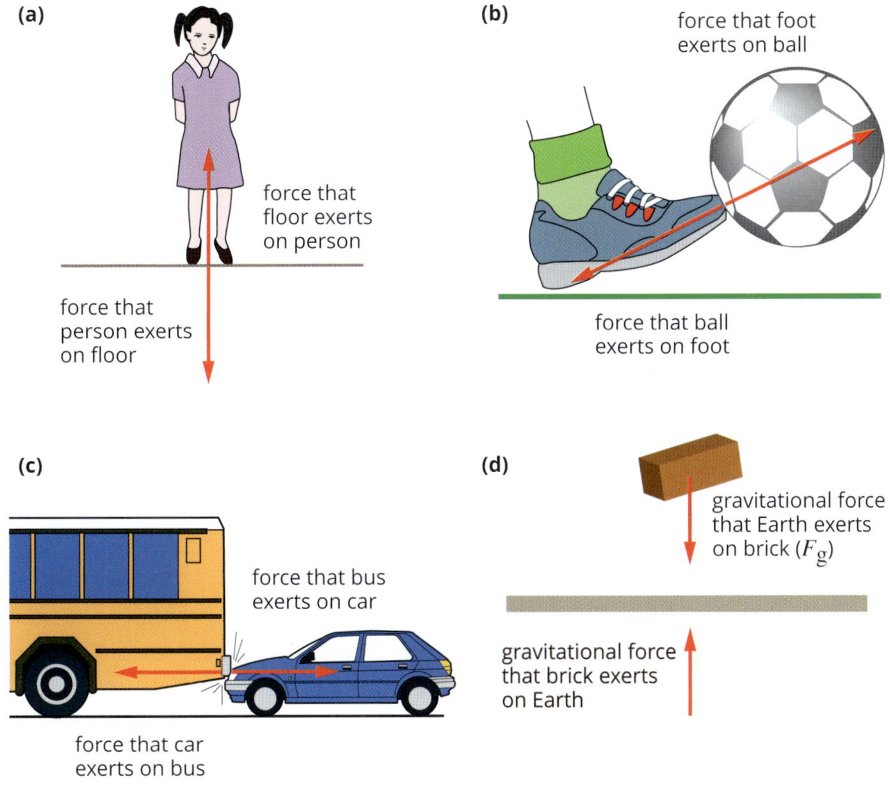

FIGURE 12.3.3 Some action–reaction pairs

From Newton's second law, you know that the same force acting on a larger mass will result in a smaller acceleration. This is the effect seen in the situation of the car colliding with the bus. Because of the car's small mass, the force acting on the car will cause the car to undergo a large deceleration. The occupants may be seriously injured as a result of this. The force acting on the bus is equal in size, but it is acting on a much larger mass. As a result, the bus will have a relatively small acceleration and the occupants will not be as seriously affected.

Identifying the action and reaction forces

When analysing a situation to determine the action and reaction forces according to Newton's third law, it is helpful to be able to label the force vectors systematically. A good strategy for labelling force vectors is to use the capital letter F to represent the force and then to include a subscript consisting of the word 'on' and the thing on which the force is acting, and then the word 'by' and the thing that is applying the force.

The equal and opposite force is then labelled with a capital F and a subscript with the objects in reverse. For example, the action and reaction force vector arrows shown in Figure 12.3.3 can be labelled as shown in Table 12.3.1.

TABLE 12.3.1 Labels of action and reaction force vectors in Figure 12.3.3

	Action vector	Reaction vector
(a)	$F_{\text{on floor by person}}$	$F_{\text{on person by floor}}$
(b)	$F_{\text{on ball by foot}}$	$F_{\text{on foot by ball}}$
(c)	$F_{\text{on bus by car}}$	$F_{\text{on car by bus}}$
(d)	$F_{\text{on Earth by brick}}$	$F_{\text{on brick by Earth}}$

PHYSICSFILE

Combining Newton's second and third laws in the classroom

You can easily observe the effect of Newton's second and third laws in the classroom if you have two dynamics carts with wheels that are free to roll on a smooth surface (such as a bench or desk). If the two carts are placed in contact with each other, and the plunger is activated on one of the carts, you will observe that both carts roll backwards. This is because of the action and reaction force pair described by Newton's third law. If the two carts have similar masses, you will observe that they accelerate apart at a similar rate. If one cart is heavier than the other, you will observe the lighter cart accelerates at a greater rate. This is because the forces acting on both carts are equal in magnitude and so, according to Newton's second law, the smaller mass will experience a greater acceleration.

It should be noted that it does not matter which force is considered the action force and which is considered the reaction force. They are always equal in magnitude and opposite in direction.

Worked example 12.3.1

APPLYING NEWTON'S THIRD LAW

In the diagram below a table-tennis bat is in contact with a table-tennis ball, and one of the forces is given.

a Label the given force using the system 'F_{on} ________________ by ________________'.

b Label the reaction force to the given force using the system 'F_{on} ________________ by ________________'.

c Draw the reaction force on the diagram, showing its size and location.

Thinking	Working
Identify the two objects involved in the action–reaction pair.	the bat and the ball
Identify which object is applying the force and which object is experiencing the force, for the force vector shown.	The force vector shown is a force from the bat on the ball.
a Use the system of labelling action and reaction forces 'F_{on} ________________ by ________________' to label the action force.	**a** $F_{\text{on ball by bat}}$
b Use the system of labelling action and reaction forces 'F_{on} ________________ by ________________' to label the reaction force.	**b** $F_{\text{on bat by ball}}$
c Use a ruler to measure the length of the action force and construct a vector arrow in the opposite direction with its tail on the point of application of the reaction force.	**c**

Worked example: Try yourself 12.3.1

APPLYING NEWTON'S THIRD LAW

In the diagram below, a bowling ball is resting on the floor, and one of the forces is given. Copy the diagram into your book and complete the following:

a Label the given force using the system 'F_{on} ________________ by ________________'.

b Label the reaction force to the given force using the system 'F_{on} ________________ by ________________'.

c Draw the reaction force on the diagram, showing its size and location.

NEWTON'S THIRD LAW AND MOTION

Newton's third law also explains how you are able to move around. In fact, Newton's third law is needed to explain all motion. Consider walking. Your leg pushes backwards on the ground with each step. This is an action force on the ground by the foot. As shown in Figure 12.3.4, a component of the force acts downwards and another component pushes backwards horizontally along the surface of the ground.

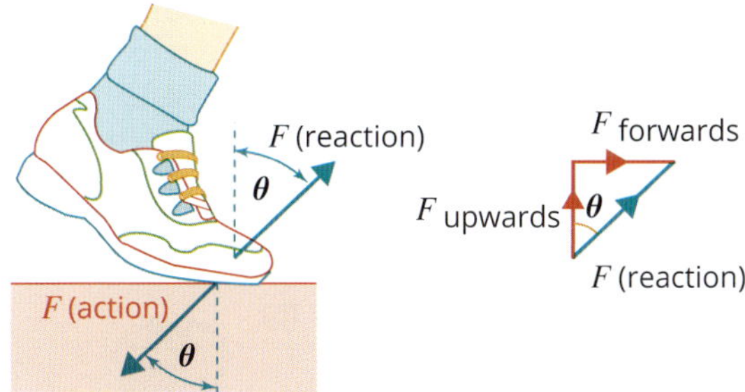

FIGURE 12.3.4 Walking relies on an action and reaction force pair in which the foot pushes down and backwards with an action force. In response, the ground pushes upwards and forwards on you.

The force is transmitted because there is friction between your shoe and the ground. In response, the ground then pushes forwards on you via your foot. This forwards component of the reaction force enables you to move forwards. In other words, it is the ground pushing forwards on you that moves you forwards. It is important to remember that in Newton's second law, $F_{net} = ma$, the net force, F_{net}, is the sum of the forces acting on the body. This does not include forces that are exerted by the body on other objects. When you push backwards on the ground, this force is acting on the ground and may affect the ground's motion. If the ground is firm, this effect is usually not noticed, but if you run along a sandy beach, the sand is clearly pushed back by your feet.

The act of walking relies on there being some friction between your shoe and the ground. Without it, there is no grip and it is impossible to supply the action force to the ground. Consequently, the ground cannot supply the reaction force needed to enable forwards motion. Walking on smooth ice is a good example of this. Mountaineers use crampons (basically, a rack of nails) attached to the soles of their boots in order to gain a better grip in icy conditions.

All motion can be explained in terms of action and reaction force pairs. Table 12.3.2 gives some examples of the action and reaction pairs in familiar motions.

TABLE 12.3.2 Action and reaction force pairs are responsible for all types of motion.

Motion	Action force	Reaction force
swimming	hand pushes backwards on water	water pushes forwards on hand
jumping	legs push downwards on Earth	Earth pushes upwards on legs
bicycle or car	tyre pushes backwards on ground	ground pushes forwards on tyre
jet aircraft and rockets	hot gas is forced backwards out of engine	gases push craft forwards
skydiving	force of gravitation on the skydiver from Earth	force of gravitation on Earth from skydiver

THE NORMAL FORCE

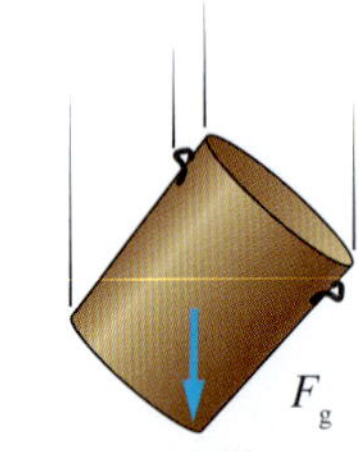

FIGURE 12.3.5 When the bin is in mid-air, there is an unbalanced force due to gravity acting on it so it accelerates towards the ground.

When an object, for example the rubbish bin shown in Figure 12.3.5, is allowed to fall under the influence of gravity, it is easy to see the effect of the force due to gravity. The action force is the force due to gravity of Earth on the bin, so the net force on the bin is equal to the force due to gravity, and the bin therefore accelerates downwards at $9.8\,m\,s^{-2}$. The reaction to this is the force due to gravity of the bin acting on the Earth. This force has the same magnitude as the force acting on the bin; however, since Earth is so massive, the acceleration that it produces on the Earth is negligible.

When the bin is placed on a table, the force due to gravity ($F_g = mg$) is still acting between Earth and the bin, as shown in Figure 12.3.6(a). However, placing the bin on the table brings a new pair of action–reaction forces into existence, as shown in Figure 12.3.6(b). The bin exerts a force, F_T, on the table (i.e. the action) and so, according to Newton's third law, the table exerts an equal and opposite force, F_N, on the bin (i.e. the reaction).

(In this situation a similar pair of action–reaction forces would exist between the table and Earth.)

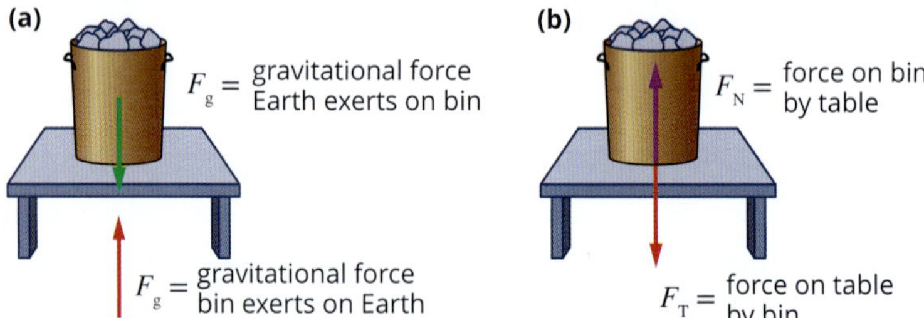

FIGURE 12.3.6 There are (a) gravitational action–reaction forces between the bin and Earth, and (b) action and reaction forces between the bin and the table.

The bin now has two forces acting on it—F_g and F_N—as shown in Figure 12.3.7. These forces are equal in magnitude but they act in opposite directions (i.e. F_g is downwards and F_N is upwards). Therefore the net force acting on the bin is zero and so the bin is at rest.

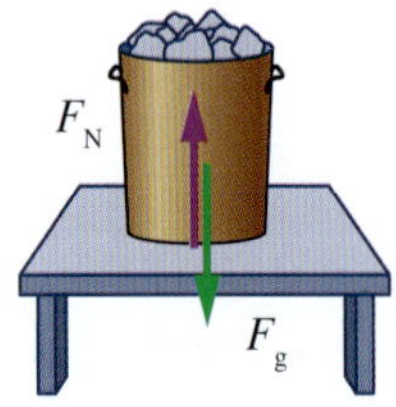

FIGURE 12.3.7 The effect of the two forces on the bin. Since these forces are equal in magnitude and opposite in direction, the bin does not accelerate.

When one surface presses on another surface, such as the bin and the table in the example above, the reaction force is called the normal force. The term 'normal' comes from the fact that this force is always perpendicular (normal) to the two surfaces. It is represented by F_N.

Note that action and reaction forces always act on different objects. In the example above, the force on the table by the bin (F_T) and force on the bin by the table (F_N) are an action-reaction pair. In comparison, F_N and F_g act on the same object (i.e. the bin), therefore they cannot be an action–reaction pair.

Worked example 12.3.2

IDENTIFYING ACTION–REACTION FORCE PAIRS

In each scenario identify the action and the reaction forces.

a A car is stationary at a set of red lights.	
Thinking	**Working**
Identify the action in the scenario.	The force due to gravity from the car on the road.
Identify the reaction force.	The force from the road on the car.

b A basketball rebounds off the backboard of the basketball net.	
Thinking	**Working**
Identify the action in the scenario.	The force from the basketball on the backboard.
Identify the reaction force.	The force from the backboard on the basketball.

Worked example: Try yourself 12.3.2

IDENTIFYING ACTION–REACTION FORCE PAIRS

In each scenario identify the action and the reaction forces.

a A beach volleyball player jumps into the air before serving.

b The top of a ladder leans against a wall.

PHYSICSFILE

The inclined plane

In the example of the bin and the table, the table's surface is horizontal. It is possible that an object could be placed on a surface that is tilted so that it makes an angle, θ, to the horizontal. In this case, the force due to gravity remains the same: $F_g = mg$ downwards. However, the normal force continues to act at right angles to the surface and will change in magnitude, getting smaller as the angle θ increases. The magnitude of the normal force is equal in size but opposite in direction to the component of the force due to gravity that is perpendicular (i.e. at right angles) to the surface. So, the normal force is $F_N = mg\cos\theta$.

The component of the force due to gravity that acts parallel to the surface will cause the mass to slide down the incline. The motion of the object along the plane will be affected by friction, if it is present. The component of the gravitational force that acts along the surface is given by $F = mg\sin\theta$.

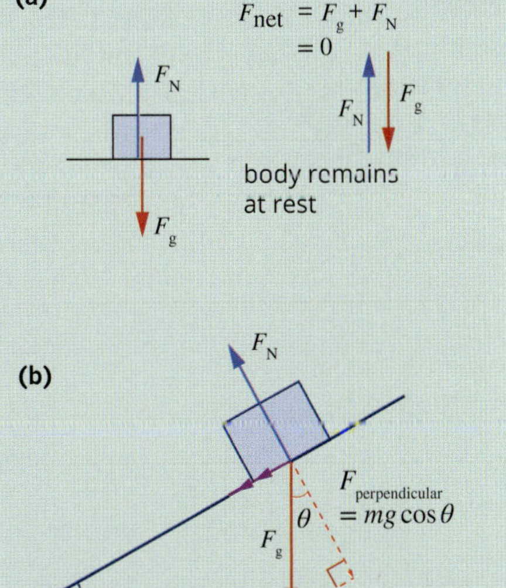

(a) Where the surface is perpendicular to the force due to gravity, the normal force acts directly upwards. (b) On an inclined plane, F_N is at an angle to F_g and is equal to $F_N = mg\cos\theta$. If no friction acts, the force that causes the object to accelerate down the plane is $F_{parallel} = mg\sin\theta$.

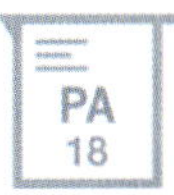

12.3 Review

OA
✓✓

SUMMARY

- For every action (force), there is an equal and opposite reaction (force). This is known as Newton's third law.
- If the action force is labelled systematically, the reaction force can be described by reversing the label of the action force.
- The action and reaction forces are equal and opposite even when the masses of the colliding objects are very different.
- The individual forces making up a Newton's third law pair act on different masses to cause different accelerations according to Newton's second law.
- When an object exerts a downwards force on a surface there is an equal and opposite Newton's third law reaction that exerts a force upwards. This is called the normal force.

KEY QUESTIONS

Knowledge and understanding

1 What forces act on a hammer and a nail when a heavy hammer hits a small nail?

2 In the figure below, one of the forces acting on an astronaut orbiting Earth is shown by the red arrow.

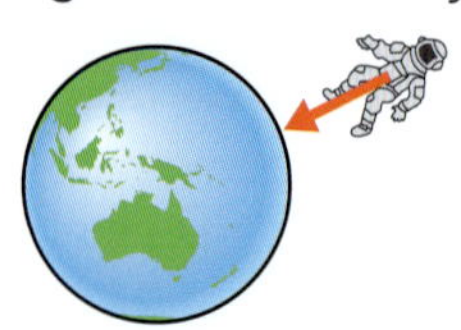

  a Name the given force using the system 'F_{on} ______________ by ______________'.

  b Name the reaction force using the system 'F_{on} ______________ by ______________'.

3 A swimmer completes a training drill in which he doesn't use his legs to kick, but only uses his stroke to move down the pool. What force causes a swimmer to move forwards down the pool?

4 When an untied inflated balloon is released it will fly around the room. What is the force that causes the balloon to move?

5 Determine the reaction force involved when a football player's head hits a ball with a force of 5 N upwards.

Analysis

6 An astronaut with a mass of 100 kg becomes untethered during a space-walk and drifts away from the spacecraft. To get back to the spacecraft, she throws her 5 kg toolkit directly away from the spacecraft with a force of 80 N.

  a What force does astronaut experience?

  b What acceleration will be given to the toolkit?

  c If the force on the toolkit lasted for 0.40 s, determine the resulting speed of both the toolkit and the astronaut.

7 Jessie and Reuben are discussing the gravitational forces that act between Earth and the Moon. Jessie states that since these forces are equal in magnitude but opposite in direction, they comprise a Newton's third law action–reaction pair. Reuben disagrees, saying that these are separate forces, each independently produced by Newton's law of universal gravitation, and that action–reaction forces can only occur between objects that are in contact with each other. Who is correct and why?

12.4 Momentum and conservation of momentum

It is possible to understand some physics concepts intuitively without knowing the physics terms or words that describe them. For example, you may know that once a heavy object gets moving it is difficult to stop it, whereas a lighter object moving at the same speed is easier to stop. In the previous sections of this chapter you have seen how Newton's laws of motion can be used to explain these observations. In this section you will explore how these observations can be related to the concept called momentum.

MOMENTUM

The **momentum** of an object relates to both its mass and its velocity. The footballers who are about to collide in Figure 12.4.1 have momentum due to their mass, and the faster they run, the more momentum they will have. For two players of similar mass, the slower moving player has less momentum than one who is moving faster. Similarly, a person with greater mass will have more momentum than a smaller, lighter person travelling at the same speed. The more momentum an object has, due to its mass or its velocity, the more momentum it has to lose before it stops.

FIGURE 12.4.1 Momentum is related to mass and velocity. The greater the mass or velocity, the harder it is to stop or start moving.

The equation for momentum, p, is the product of the object's mass, m, and its velocity, v.

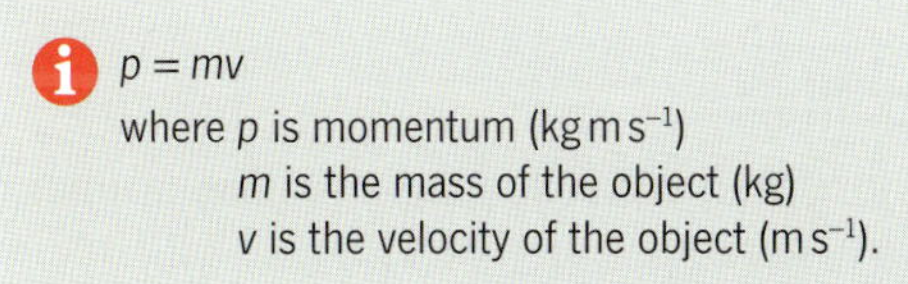

$p = mv$

where p is momentum (kg m s^{-1})

m is the mass of the object (kg)

v is the velocity of the object (m s^{-1}).

The greater an object's mass or velocity, the larger that object's momentum will be. As velocity is a vector quantity, momentum is also a vector and so it must have magnitude, units and direction. The direction of a momentum vector will always be the same as the direction of the velocity vector. For calculations of change in momentum in a single dimension, we can use the sign conventions of positive and negative.

Force is equal to the rate of change of momentum. This can be mathematically explained using Newton's second law, which was presented in Section 12.2 on page 363.

PHYSICSFILE

Momentum and inertia

The concept of momentum is very similar to the concept of inertia that was introduced in Section 12.1 on Newton's first law. There are two key differences:

- Momentum is quantitative (we can assign a value to it), whereas inertia is qualitative (we can only describe it in words).
- Although a stationary object has zero momentum (i.e. $p = mv = m \times 0 = 0$), a stationary object still has inertia (it is difficult to move).

The following derivation will show how Newton's second law relates to momentum. It results with net force, F_{net}, equal to the change in momentum, Δp, divided by the period of time, Δt, which is the rate of change of momentum:

$$a = \frac{F_{net}}{m}$$

$$F_{net} = ma$$

$$F_{net} = m\left(\frac{v-u}{\Delta t}\right)$$

$$F_{net} = \frac{mv - mu}{\Delta t}$$

$$F_{net} = \frac{\Delta p}{\Delta t}$$

This means that changes in momentum are caused by the action of a net force.

Worked example 12.4.1

MOMENTUM

Calculate the momentum of a 60.0 kg student walking at 3.50 m s^{-1} east.

Thinking	Working
Ensure that the variables are in their standard units.	$m = 60.0$ kg $v = 3.50$ m s^{-1} east
Apply the equation for momentum.	$p = mv$ $= 60.0 \times 3.50$ $= 210$ kg m s^{-1} east

Worked example: Try yourself 12.4.1

MOMENTUM

Calculate the momentum of a 1230 kg car driving at 16.7 m s^{-1} north.

CONSERVATION OF MOMENTUM

The most significant feature of momentum is that it is **conserved** in any interaction or collision between objects. This means that the total (sum of) momentum before a collision will be equal to the total (sum of) momentum after the collision for any isolated system, i.e. if no net external force is acting in the direction of motion. This is known as the **law of conservation of momentum** and can be represented by the following relationship:

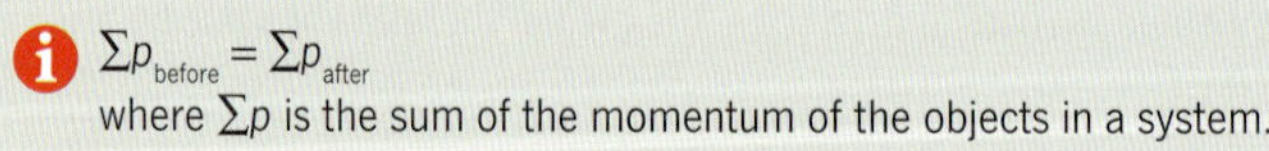

$\Sigma p_{before} = \Sigma p_{after}$

where Σp is the sum of the momentum of the objects in a system.

To find the total momentum of the objects in a system (either before or after a collision) simply find the momentum of each object, considering their masses and velocities, and then add them together.

Momentum in one-dimensional collisions

If two objects are colliding in one dimension, then the following equation applies:

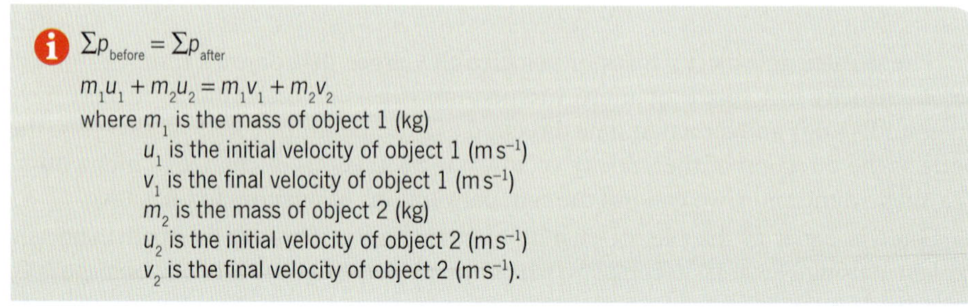

$\Sigma p_{before} = \Sigma p_{after}$

$m_1u_1 + m_2u_2 = m_1v_1 + m_2v_2$

where m_1 is the mass of object 1 (kg)

u_1 is the initial velocity of object 1 (m s^{-1})

v_1 is the final velocity of object 1 (m s^{-1})

m_2 is the mass of object 2 (kg)

u_2 is the initial velocity of object 2 (m s^{-1})

v_2 is the final velocity of object 2 (m s^{-1}).

PHYSICSFILE

The discovery of the neutron

The law of conservation of momentum was used to interpret the data from investigations that led to the discovery of the neutron. Because the neutron has no charge, it could not be investigated through the interactions of charged particles that had led to the discovery of the proton and electron. In 1932, James Chadwick investigated collisions between alpha particles and the element beryllium. However, the conservation of momentum calculations didn't add up. Chadwick knew that the law of conservation of momentum was true, so he reasoned that there was an unknown particle involved that had a mass close to the proton's mass, but without electric charge. Subsequent investigations confirmed his experiments and led to the naming of this particle as the neutron.

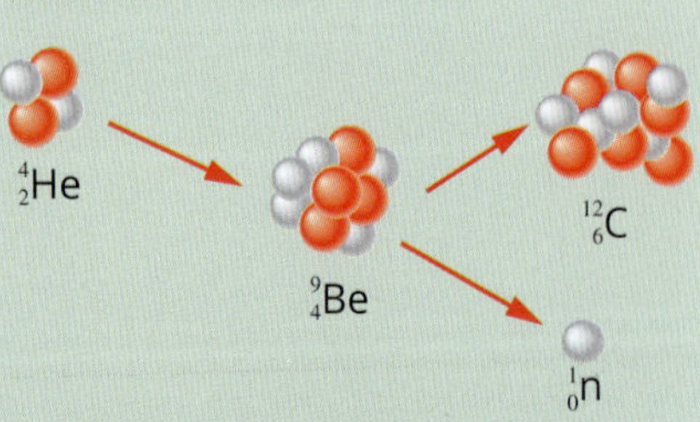

Investigating collisions between alpha particles and a beryllium nucleus led to the discovery of the neutron.

For collisions in one dimension, apply the sign convention of positive and negative directions to the velocities, and then use algebra to determine the answer to the problem.

Worked example 12.4.2

CONSERVATION OF MOMENTUM

A 2.50 kg mass is moving at 4.50 m s^{-1} west towards a 1.50 kg mass moving at 3.00 m s^{-1} east. Calculate the final velocity of the 2.50 kg mass if the 1.50 kg mass rebounds at 5.00 m s^{-1} west.

Thinking	Working
Identify the variables using subscripts. Ensure that the variables are in their standard units.	$m_1 = 2.50$ kg $u_1 = 4.50$ m s^{-1} west $v_1 = ?$ $m_2 = 1.50$ kg $u_2 = 3.00$ m s^{-1} east $v_2 = 5.00$ m s^{-1} west
Apply the sign convention to the variables.	$m_1 = 2.50$ kg $u_1 = -4.50$ m s^{-1} $v_1 = ?$ $m_2 = 1.50$ kg $u_2 = +3.00$ m s^{-1} $v_2 = -5.00$ m s^{-1}
Apply the equation for conservation of momentum.	$\Sigma p_{\text{before}} = \Sigma p_{\text{after}}$ $m_1u_1 + m_2u_2 = m_1v_1 + m_2v_2$ $(2.50 \times -4.50) + (1.50 \times 3.00)$ $= 2.50v_1 + (1.50 \times -5.00)$ $2.50v_1 = -11.25 + 4.50 - (-7.50)$ $v_1 = \frac{0.75}{2.50}$ $= 0.30$ m s^{-1}
Apply the sign convention to describe the direction of the final velocity.	$v_1 = 0.30$ m s^{-1} east

Worked example: Try yourself 12.4.2

CONSERVATION OF MOMENTUM

A 1200 kg wrecking ball is moving at 2.50 m s^{-1} north towards a 1500 kg wrecking ball moving at 4.00 m s^{-1} south. Calculate the final velocity of the 1500 kg ball if the 1200 kg ball rebounds at 3.50 m s^{-1} south.

PHYSICSFILE

Conservation of momentum and Newton's laws

The law of conservation of momentum is actually a result of combining Newton's second and third laws.

Consider two masses, m_1 with initial velocity u_1 and m_2 with initial velocity u_2, that collide with each other. When they collide, the two objects are in contact for a time, t, and m_1 experiences a force of F_1 and m_2 experiences a force of F_2. Then m_1 moves off with velocity v_1 and m_2 moves off with velocity v_2.

According to Newton's third law, $F_1 = -F_2$.

According to Newton's second law, $F = ma$; therefore, $m_1a_1 = -m_2a_2$.

Alternatively,

$$m_1\frac{(v_1 - u_1)}{t} = -m_2\frac{(v_2 - u_2)}{t}$$

Multiplying both sides by t gives

$$m_1(v_1 - u_1) = -m_2(v_2 - u_2)$$

Expanding the brackets gives

$$m_1v_1 - m_1u_1 = -m_2v_2 + m_2u_2$$

Moving the negative terms to the other side of the equal sign gives

$$m_1u_1 + m_2u_2 = m_1v_1 + m_2v_2$$

which is the law of conservation of momentum.

Momentum when masses combine

It is important to note that in the situations described in Worked example 12.4.2 (page 379), the two objects remain separate from each other. However, it is possible for two objects to combine (stick together) when they collide. If two objects combine when they collide, then the equation is modified to:

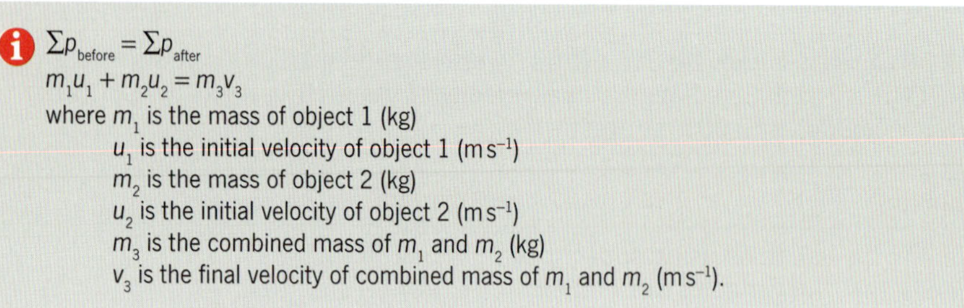

$\Sigma p_{before} = \Sigma p_{after}$

$m_1u_1 + m_2u_2 = m_3v_3$

where m_1 is the mass of object 1 (kg)

u_1 is the initial velocity of object 1 ($m\,s^{-1}$)

m_2 is the mass of object 2 (kg)

u_2 is the initial velocity of object 2 ($m\,s^{-1}$)

m_3 is the combined mass of m_1 and m_2 (kg)

v_3 is the final velocity of combined mass of m_1 and m_2 ($m\,s^{-1}$).

Worked example 12.4.3

CONSERVATION OF MOMENTUM WHEN MASSES COMBINE

A 5.00 kg lump of clay is moving at 2.00 $m\,s^{-1}$ west towards a 7.50 kg mass of clay that is moving at 3.00 $m\,s^{-1}$ east. They collide to form a single, combined mass of clay. Calculate the final velocity of the combined mass of clay.

Thinking	Working
Identify the variables using subscripts and ensure that the variables are in their standard units. Add m_1 and m_2 to get m_3.	$m_1 = 5.00$ kg $u_1 = 2.00\ m\,s^{-1}$ west $m_2 = 7.50$ kg $u_2 = 3.00\ m\,s^{-1}$ east $m_3 = 12.50$ kg $v_3 = ?$
Apply the sign convention to the variables.	$m_1 = 5.00$ kg $u_1 = -2.00\ m\,s^{-1}$ $m_2 = 7.50$ kg $u_2 = +3.00\ m\,s^{-1}$ $m_3 = 12.50$ kg $v_3 = ?$
Apply the equation for conservation of momentum.	$\Sigma p_{before} = \Sigma p_{after}$ $m_1u_1 + m_2u_2 = m_3v_3$ $(5.00 \times -2.00) + (7.50 \times 3.00) = 12.50v_3$ $v_3 = \frac{-10.0 + 22.50}{12.50}$ $= 1.00\ m\,s^{-1}$
Apply the sign convention to describe the direction of the final velocity.	$v_3 = 1.00\ m\,s^{-1}$ east

Worked example: Try yourself 12.4.3

CONSERVATION OF MOMENTUM WHEN MASSES COMBINE

An 80.0 kg rugby player is moving at 1.50 $m\,s^{-1}$ north when he tackles an opponent who has a mass of 50.0 kg and is moving at 5.00 $m\,s^{-1}$ south. Calculate the final velocity of the two players.

PHYSICSFILE

Elastic and inelastic collisions

Consider two balls, with the same mass, m, and velocity, v, colliding head-on. The momentums of the balls are mv and $-mv$, therefore the total initial momentum is zero. After the collision, there are many possible outcomes that fulfil conservation of momentum. Both balls could:

- rebound with velocity v
- rebound with velocity ½v
- stop (i.e. $v = 0$).

There is another factor that determines the collision outcome—conservation of kinetic energy. Unlike momentum, kinetic energy is not necessarily conserved in a collision, because kinetic energy can be transformed into other forms of energy, such as sound or heat.

A collision in which kinetic energy is conserved is called an elastic collision. In the example above, an elastic collision would result in both balls rebounding with velocity v. In everyday life, most collisions do not conserve kinetic energy, so they are known as inelastic collisions.

Momentum in explosive collisions

It is also possible for an object to break apart into two objects in what is known as an 'explosive collision'. If an object breaks apart when an explosive collision occurs, then the equation is modified to:

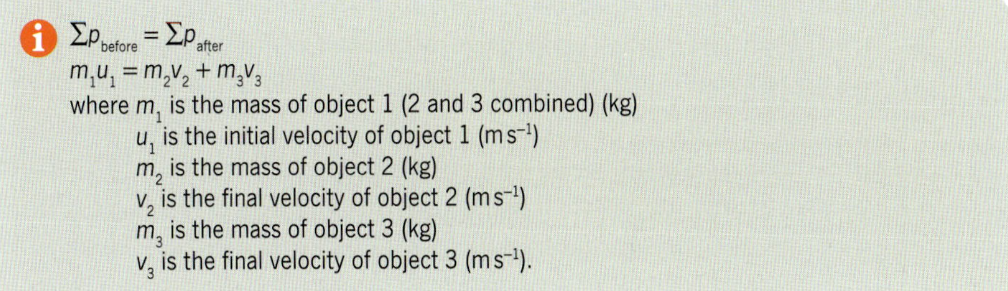

$\Sigma p_{before} = \Sigma p_{after}$

$m_1u_1 = m_2v_2 + m_3v_3$

where m_1 is the mass of object 1 (2 and 3 combined) (kg)

u_1 is the initial velocity of object 1 ($m\,s^{-1}$)

m_2 is the mass of object 2 (kg)

v_2 is the final velocity of object 2 ($m\,s^{-1}$)

m_3 is the mass of object 3 (kg)

v_3 is the final velocity of object 3 ($m\,s^{-1}$).

Worked example 12.4.4

CONSERVATION OF MOMENTUM FOR EXPLOSIVE COLLISIONS

A 90.0 kg athlete holds a 1000 g javelin. She approaches the line at 7.75 $m\,s^{-1}$ west and releases the javelin down the field. After throwing it, she continues with a velocity of 7.25 $m\,s^{-1}$ west. Calculate the velocity of the javelin just after she releases it.

Thinking	Working
Identify the variables using subscripts and ensure that the variables are in their standard units. Note that m_1 is the sum of the bodies, i.e. the athlete and the javelin.	$m_1 = 91$ kg $u_1 = 7.75\,m\,s^{-1}$ west $m_2 = 90$ kg $v_2 = 7.25\,m\,s^{-1}$ west $m_3 = 1.00$ kg $v_3 = ?$
Apply the sign convention to the variables.	$m_1 = 91$ kg $u_1 = -7.75\,m\,s^{-1}$ $m_2 = 90$ kg $v_2 = -7.25\,m\,s^{-1}$ $m_3 = 1.00$ kg $v_3 = ?$
Apply the equation for conservation of momentum for explosive collisions.	$\Sigma p_{before} = \Sigma p_{after}$ $m_1u_1 = m_2v_2 + m_3v_3$ $91.0 \times (-7.75) = 90.0 \times (-7.25) + 1.00v_3$ $v_3 = \frac{-705.25 - (-652.5)}{1.00}$ $= \frac{-52.75}{1.00}$ $= -52.8\,m\,s^{-1}$
Apply the sign convention to describe the direction of the final velocity.	$v_3 = 52.8\,m\,s^{-1}$ west

For collisions in two dimensions, problems can be solved by resolving the velocity vectors into perpendicular components and then considering the conservation of momentum in each single dimension.

Worked example: Try yourself 12.4.4

CONSERVATION OF MOMENTUM FOR EXPLOSIVE COLLISIONS

A 2000 kg cannon fires a 10 kg cannonball. The cannon and the cannonball are initially stationary. After firing, the cannon recoils with a velocity of 8.15 $m\,s^{-1}$ north. Calculate the velocity of the cannonball just after it is fired.

PHYSICSFILE

Conservation of momentum in engines

If you release an inflated rubber balloon with its neck open, it will fly off around the room. In the diagram below, the momentum of the air to the left results in the movement of the balloon to the right. Momentum is conserved.

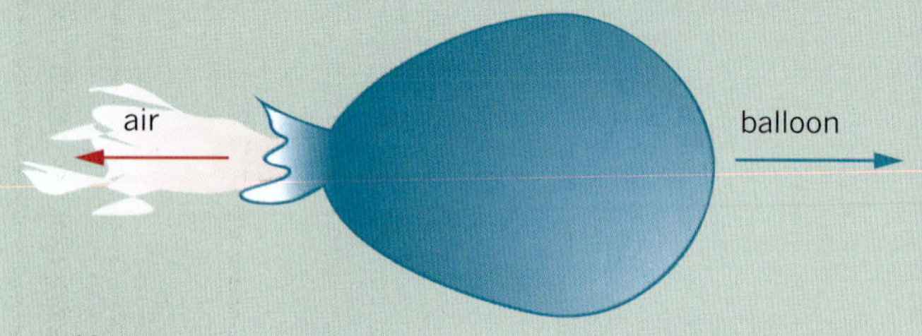

Momentum is conserved when the air is let out of a balloon.

This is the principle upon which rockets and jet engines are based. Both rockets and jet engines employ a high-velocity stream of hot gases that are vented after the combustion of a fuel–air mixture. The hot exhaust gases have a very large momentum as a result of the high velocities involved, and can accelerate rockets and jets to high velocities as they acquire an equal momentum in the opposite direction. Rockets destined for space carry their own oxygen supply, while jet engines use the surrounding air supply.

12.4 Review

OA ✓✓

SUMMARY

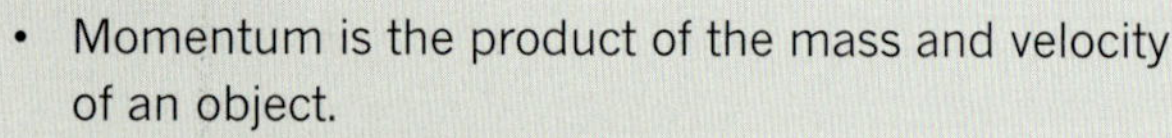

- Momentum is the product of the mass and velocity of an object.

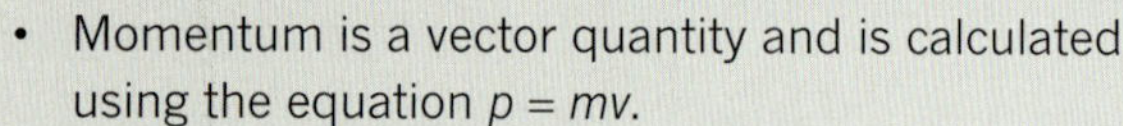

- Momentum is a vector quantity and is calculated using the equation $p = mv$.
- Force is equal to the rate of change of momentum and it can be expressed as $F_{net} = \frac{\Delta p}{\Delta t}$.
- The law of conservation of momentum can be applied to situations in which:
  - two objects collide and remain separate
    $$\Sigma p_{before} = \Sigma p_{after}$$
    $$m_1u_1 + m_2u_2 = m_1v_1 + m_2v_2$$
  - two objects collide and combine together
    $$\Sigma p_{before} = \Sigma p_{after}$$
    $$m_1u_1 + m_2u_2 = m_3v_3$$
  - one object breaks apart into two objects in an explosive collision
    $$\Sigma p_{before} = \Sigma p_{after}$$
    $$m_1u_1 = m_2u_2 + m_3v_3$$
- Momentum is conserved in an isolated system, i.e. if no net external force is acting.

KEY QUESTIONS

Knowledge and understanding

1 Object A collides with object B. After the collision, the two objects stick together to form object C. Write an equation that describes conservation of momentum for this collision in terms of the masses, *m*, and velocities, *v*, of the objects.

2 A large object A explodes into three smaller objects B, C and D. Write an equation that describes conservation of momentum for this explosion in terms of the masses, *m*, and velocities, *v*, of the objects.

3 Calculate the momentum of a 3.50 kg fish swimming at 2.50 m s^{-1} south.

4 Calculate the momentum of a 433 kg boat travelling at 22.2 m s^{-1} west.

Analysis

5 A space shuttle of mass 10 000 kg, initially at rest, burns 5.0 kg of fuel and oxygen in its rockets to produce exhaust gases ejected at a velocity of 6000 m s^{-1}. Calculate the velocity that this exchange will give to the space shuttle.

6 A small research rocket of mass 250 kg is launched vertically as part of a weather study. It sends out 50 kg of burnt fuel and exhaust gases with a velocity of 180 m s^{-1} in a 2 s initial acceleration period.

In a test, a student is asked to calculate the velocity of the rocket after the initial acceleration. They use conservation of momentum for the rocket and fuel, and get an incorrect answer. Explain why.

12.5 Momentum transfer

The previous section described the momentum of an object in terms of its velocity and its mass. For each of the different collisions described in that section, the momentum of the system was conserved. That is, when all of the objects involved in the collision were considered, the total momentum before and after the collision was the same. But for each separate object, considered in isolation, momentum may not have been conserved. In the examples explored, an object experienced a change in its velocity due to the collision.

When an object changes its velocity, its momentum will also change. An increase in velocity means an increase in momentum, while a decrease in velocity corresponds to a decrease in momentum. Change in momentum, Δp, is also called **impulse**.

You can also consider this change in momentum as a transfer of momentum. For this reason, momentum transfer is often referred to as impulse.

CHANGE IN MOMENTUM IN ONE DIMENSION

It is easy to change the velocity of an object. You can either run faster or run slower; you can press a little harder on the pedals of a bike or press a little softer. You can also bounce an object off a surface. For example, the basketball in Figure 12.5.1 experiences a change in velocity when it changes direction during the bounce. The cause of these changes in motion will be discussed in the Section 12.6. First, consider impulse or change in momentum in one dimension.

FIGURE 12.5.1 A bouncing basketball undergoes a change in momentum when it changes direction as it bounces.

The term 'impulse' means change in momentum. So the impulse or change in momentum of an object moving in one dimension is calculated using the equation:

$$\begin{aligned}\Delta p &= p_{final} - p_{initial}\\ &= mv - mu\\ &= m(v - u)\\ &= m\Delta v\end{aligned}$$

where Δp is the impulse or change in momentum ($kg\,m\,s^{-1}$)
m is the mass (kg)
v is the final velocity ($m\,s^{-1}$)
u is the initial velocity ($m\,s^{-1}$)
Δv is the change in velocity ($m\,s^{-1}$).

Worked example 12.5.1

IMPULSE OR CHANGE IN MOMENTUM

A student rides a bike to school and approaches the bike rack at $8.20\,\text{m s}^{-1}$ east. Calculate the impulse of the student during the time it takes to stop if the student and the bike have a combined mass of 80 kg and the student stops at the rack.

Thinking	Working
Ensure that the variables are in their standard units.	$m = 80\,\text{kg}$ $u = 8.20\,\text{m s}^{-1}$ east $v = 0\,\text{m s}^{-1}$
Apply the sign convention to the velocity vector.	$m = 80\,\text{kg}$ $u = +8.20\,\text{m s}^{-1}$ $v = 0\,\text{m s}^{-1}$
Apply the equation for impulse or change in momentum.	$\Delta p = m\Delta v$ $= 80 \times (0 - 8.20)$ $= -656\,\text{kg m s}^{-1}$
Apply the sign convention to describe the direction of the impulse.	$\Delta p = 656\,\text{kg m s}^{-1}$ west

As momentum is a vector quantity, the impulse or change in momentum is also a vector, so it is expressed in magnitude, units and direction.

Worked example: Try yourself 12.5.1

IMPULSE OR CHANGE IN MOMENTUM

A student of mass 75 kg is hurrying to class after lunch at $4.55\,\text{m s}^{-1}$ north. She suddenly remembers that she has forgotten her laptop and goes back to her locker at $6.15\,\text{m s}^{-1}$ south. Calculate the impulse of the student during the time it takes to turn around.

CHANGE IN MOMENTUM IN TWO DIMENSIONS

The velocity of an object can be changed not only by changing the magnitude of its velocity, but also by changing the direction of its motion. The velocity of the boat in Figure 12.5.2, for example, changes because the boat changes direction. As you saw in Chapter 10 'Scalars and vectors', a change in velocity in two dimensions can be calculated using geometry. This can then be used in the equation for impulse to find the change in momentum in two dimensions.

FIGURE 12.5.2 Changing momentum in two dimensions by changing direction

Worked example 12.5.2

IMPULSE OR CHANGE IN MOMENTUM IN TWO DIMENSIONS

A 65.0 kg mass is moving at $3.50\,m\,s^{-1}$ west and then changes to $2.00\,m\,s^{-1}$ north. Calculate the change in momentum of the mass over the period of the change.

Thinking	Working
Identify the formula for calculating a change in velocity, Δv.	Δv = final velocity – initial velocity
Draw the final velocity vector, *v*, and the initial velocity vector, *u*, separately. Then draw the initial velocity in the opposite direction, which represents the negative of the initial velocity, –*u*.	N, W, E, S; $v = 2.00\ m\ s^{-1}$ north; $u = 3.50\ m\ s^{-1}$ west; $-u = 3.50\ m\ s^{-1}$ east
Construct a vector diagram drawing *v* first and then from its head draw the opposite of *u*. The change of velocity vector is drawn from the tail of the final velocity to the head of the opposite of the initial velocity.	N, W, E, S; $v = 2.00\ m\ s^{-1}$ north; $-u = 3.50\ m\ s^{-1}$ east; θ; Δv
As the two vectors to be added are at 90° to each other, apply Pythagoras' theorem to calculate the magnitude of the change in velocity.	$\Delta v^2 = 2.00^2 + 3.50^2$ $= 4.00 + 12.25$ $\Delta v = \sqrt{16.25}$ $= 4.03\,m\,s^{-1}$
Calculate the angle from the north vector to the change in velocity vector.	$\tan\theta = \frac{3.50}{2.00}$ $\theta = \tan^{-1} 1.75$ $= 60.3°$
State the magnitude and direction of the change in velocity.	$\Delta v = 4.03\,m\,s^{-1}$ N60.3°E
Identify the variables using subscripts and ensure that the variables are in their standard units.	$m = 65.0\,kg$ $\Delta v = 4.03\,m\,s^{-1}$ N60.3°E
Apply the equation for impulse or change in momentum.	$\Delta p = mv - mu$ $= m(v - u)$ $= m\Delta v$ $= 65.0 \times 4.03$ $= 262\,kg\,m\,s^{-1}$
Apply the direction convention to describe the direction of the change in momentum.	$\Delta p = 262\,kg\,m\,s^{-1}$ N60.3°E

Worked example: Try yourself 12.5.2

IMPULSE OR CHANGE IN MOMENTUM IN TWO DIMENSIONS

A 65.0 g pool ball is moving at $0.250\,m\,s^{-1}$ south towards a cushion and bounces off at $0.200\,m\,s^{-1}$ east. Calculate the change in momentum of the ball over the period of the change.

12.5 Review

OA ✓✓

SUMMARY

- Change or transfer of momentum, Δp, is also known as impulse. It is a vector quantity.
- A change or transfer of momentum occurs when an object changes its velocity.
- The equation for impulse is $\Delta p = mv - mu = m\Delta v$.
- Change in momentum in two dimensions can be calculated using geometry.

KEY QUESTIONS

Knowledge and understanding

1 What units are used to measure impulse?

2 When an electron collides with an atom, it rebounds with the same speed as it had before the collision. Has the electron experienced a change in momentum? Explain your answer.

3 Calculate the impulse of a 9.50 kg dog that changes its velocity from 2.50 $m\,s^{-1}$ north to 6.25 $m\,s^{-1}$ south.

4 Calculate the impulse of a 6050 kg truck as it changes from moving at 22.2 $m\,s^{-1}$ west to 16.7 $m\,s^{-1}$ east.

Analysis

5 The momentum of a ball of mass 0.125 kg changes by 0.075 $kg\,m\,s^{-1}$ south. If its original velocity was 3.00 $m\,s^{-1}$ north, what is the final velocity?

6 A marathon runner with a mass of 70.0 kg is running with a velocity of 4.00 $m\,s^{-1}$ north, and then turns a corner to start running 3.60 $m\,s^{-1}$ west. Calculate the runner's change in momentum.

12.6 Momentum and net force

Section 12.2 on Newton's second law of motion discussed the quantitative connection between force, mass, time and change in velocity. This relationship is explored further in this section. The relationship between change in momentum, Δp, the period of time, Δt, and net force, F_{net}, helps to explain the effects of collisions and how to minimise those effects. It is the key to providing safer environments, including in sporting contexts such as that shown in Figure 12.6.1.

Think about what it would feel like to fall onto a concrete floor. Even from a small height it would hurt. A fall from the same height onto a foam mattress would barely be felt. In both situations speed is the same, mass has not changed and gravity provides the same acceleration, no matter the mass. Yet the two experiences would feel different.

FIGURE 12.6.1 When footballers collide, they exert an equal and opposite force on each other.

CHANGE IN MOMENTUM (IMPULSE)

According to Newton's second law, a net force will cause a mass to accelerate. A larger net force will create a faster change in the velocity of the mass. The faster the change occurs—that is, the smaller the period of time, Δt—the greater the net force that produced that change. Landing on a concrete floor changes the velocity of an object very quickly. The falling object is brought to an abrupt stop within a very short time. When landing on a foam mattress, the change occurs over a much greater timeframe. Therefore, the force needed to produce the change is smaller.

Using Newton's second law and the definition of impulse introduced in Section 12.4, the relationship between change in momentum, Δp, and the variables of force, F_{net} (often written just as F), and period of time, Δt, can be found:

Newton's second law is:

$$F_{net} = ma = m \times \frac{\Delta v}{\Delta t} = \frac{m\Delta v}{\Delta t}$$

Since $\Delta p = m\Delta v$, therefore:

$$F_{net} = \frac{\Delta p}{\Delta t}$$

Alternatively:

$$F_{net}\Delta t = \Delta p$$

or

$$\Delta p = F_{net}\Delta t$$

where Δp is the impulse or change in momentum ($kg\,m\,s^{-1}$)

F_{net} is the net force causing the change in momentum (N)

Δt is the time taken for the change in momentum (s).

These equations illustrate that for a given change in momentum or impulse, the product of force and period of time is constant.

In most situations, the mass of the object stays constant, therefore $\Delta p = m\Delta v = m(v - u)$. This means that equation above can also be written as $F_{net}\Delta t = m\Delta v$.

These relationships are key to understanding collisions. Worked examples 12.6.1 and 12.6.2 illustrate how they work.

Worked example 12.6.1

CALCULATING THE FORCE AND IMPULSE

A student drops a 105 g pool ball onto a concrete floor from a height of 2.00 m. Just before it hits the floor, the velocity of the ball is 6.26 m s^{-1} downwards. Before it bounces back up, there is an instant in time at which the ball's velocity is zero. The time it takes for the ball to change its velocity to zero is 5.02 milliseconds.

a Calculate the impulse of the pool ball.	
Thinking	**Working**
Ensure that the variables are in their standard units.	$m = 0.105$ kg $u = 6.26$ m s^{-1} downwards $v = 0$ m s^{-1}
Apply the sign and direction convention for motion in one dimension. Upwards is positive and downwards is negative.	$m = 0.105$ kg $u = -6.26$ m s^{-1} $v = 0$ m s^{-1}
Apply the equation for change in momentum.	$\Delta p = m(v - u)$ $= 0.105 \times (0 - (-6.26))$ $= 0.657$ kg m s^{-1}
Impulse is change in momentum. Refer to the sign and direction convention to determine the direction of the impulse.	$\Delta p = 0.657$ kg m s^{-1} upwards

b Calculate the average net force that acts to cause the impulse.	
Thinking	**Working**
Use the answer to part (a). Ensure that the variables are in their standard units.	$\Delta p = 0.657$ kg m s^{-1} $\Delta t = 5.02 \times 10^{-3}$ s
Apply the equation for force.	$\Delta p = F_{net}\Delta t$ $F_{net} = \frac{\Delta p}{\Delta t}$ $= \frac{0.657}{5.02 \times 10^{-3}}$ $= 131$ N
Refer to the sign and direction convention to determine the direction of the force.	$F_{net} = 131$ N upwards

Worked example: Try yourself 12.6.1

CALCULATING THE FORCE AND IMPULSE

A student drops a 56.0 g egg onto a table from a height of 60.0 cm. Just before it hits the table, the velocity of the egg is 3.43 m s^{-1} downwards. The egg's final velocity is zero as it smashes on the table. The time it takes for the egg to change its velocity to zero is 3.55 milliseconds.

a Calculate the impulse of the egg.

b Calculate the average force that acts to cause the impulse.

Worked example 12.6.2

CALCULATING THE FORCE AND IMPULSE (SOFT LANDING)

The same pool ball from Worked example 12.6.1 (i.e. mass = 105 g) is dropped onto a foam mattress (instead of a concrete floor) from a height of 2.00 m. Once again, just before the ball hits the foam mattress, its velocity is 6.26 m s^{-1} downwards and, before it bounces back up, there is an instant in time at which the ball's velocity is zero. In this situation, the time it takes for the ball to change its velocity to zero is 0.360 s.

a Calculate the impulse of the pool ball.	
Thinking	**Working**
Ensure that the variables are in their standard units.	$m = 0.105$ kg $u = 6.26$ m s^{-1} downwards $v = 0$ m s^{-1}
Apply the sign and direction convention for motion in one dimension. Upwards is positive and downwards is negative.	$m = 0.105$ kg $u = -6.26$ m s^{-1} $v = 0$ m s^{-1}
Apply the equation for change in momentum.	$\Delta p = m(v - u)$ $= 0.105 \times (0 - (-6.26))$ $= 0.657$ kg m s^{-1}
Impulse is change in momentum. Refer to the sign and direction convention to determine the direction of the impulse.	$\Delta p = 0.657$ kg m s^{-1} upwards

b Calculate the average force that acts to cause the impulse.	
Thinking	**Working**
Use the answer to part (a) and ensure that the variables are in their standard units.	$\Delta p = 0.657$ kg m s^{-1} $\Delta t = 0.360$ s
Apply the equation for force.	$\Delta p = F_{net}\Delta t$ $F_{net} = \frac{\Delta p}{\Delta t}$ $= \frac{0.657}{0.360}$ $= 1.83$ N
Refer to the sign and direction convention to determine the direction of the force.	$F_{net} = 1.83$ N upwards

Worked example: Try yourself 12.6.2

CALCULATING THE FORCE AND IMPULSE (SOFT LANDING)

A student drops a 56.0 g egg into a mound of flour from a height of 60 cm. Just before it hits the mound of flour, the velocity of the egg is 3.43 m s^{-1} downwards. The egg's final velocity is zero, as it sinks into the mound of flour. The time it takes for the egg to change its velocity to zero is 0.325 seconds.

a Calculate the impulse of the egg.

b Calculate the average force that acts to cause the impulse.

From these worked examples you should notice a number of important things.

- Regardless of the surface on which the object landed, the impulse or change in momentum remained the same.
- The period of time was the main difference between the two different surfaces. Hard surfaces resulted in a short time to stop and soft surfaces resulted in a longer time to stop.
- The effect of the period of time on the force was dramatic. A shorter time meant a greater force, while a longer time meant a much smaller force.

DETERMINING IMPULSE FROM A CHANGING FORCE

In the previous examples it was assumed that the force that acted to change the impulse over a period of time was constant during that time. This is not always the case in real situations. Often the force varies over the period of the impact, so there needs to be a way to determine the impulse as the force varies.

An illustration of this is when a tennis player strikes a ball with a racquet. At the instant the ball comes in contact with the racquet, the applied force will be small. As the strings distort and the ball compresses, the force will increase until the ball has been stopped. The force will then decrease as the ball accelerates away from the racquet. A graph of force against time is shown in Figure 12.6.2.

The impulse, Δp, affecting the ball during any time interval will be the product of applied force, F_{net}, and the period of time, Δt. The total impulse during the period of time the ball is in contact with the racquet will be:

$$\Delta p = F_{av}\Delta t$$

where F_{av} is the average net force applied during the collision and Δt is the total period of time the ball is in contact with the racquet.

In a graph showing force against time (Figure 12.6.2), the area under the line is a product of the height (force) and the width (time). Thus, the total area under the line in a force against time graph is the total impulse for any collision, even those in which the force is not constant.

The concept of impulse is appropriate when dealing with forces during any collision, since it links force and contact time as, for example, when a runner's foot hits the ground or when a ball is hit by a bat or racquet. If applied to situations where contact is over an extended period of time, the average net force involved is used since the forces are generally changing (as the ball deforms, for example). The average net applied force can be found directly from the formula for impulse. The instantaneous applied force at any particular time during the collision must be read from a graph of force against time.

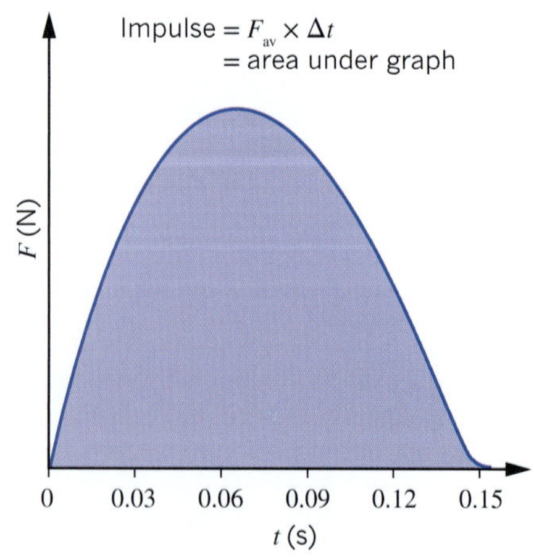

FIGURE 12.6.2 The forces acting on the tennis ball during its collision with the racquet are not constant.

Worked example 12.6.3

CALCULATING THE TOTAL IMPULSE FROM A CHANGING FORCE

A student records the force acting on a rubber ball as it bounces off a hard concrete floor over a period of time. The graph shows the forces acting on the ball during its collision with the concrete floor.

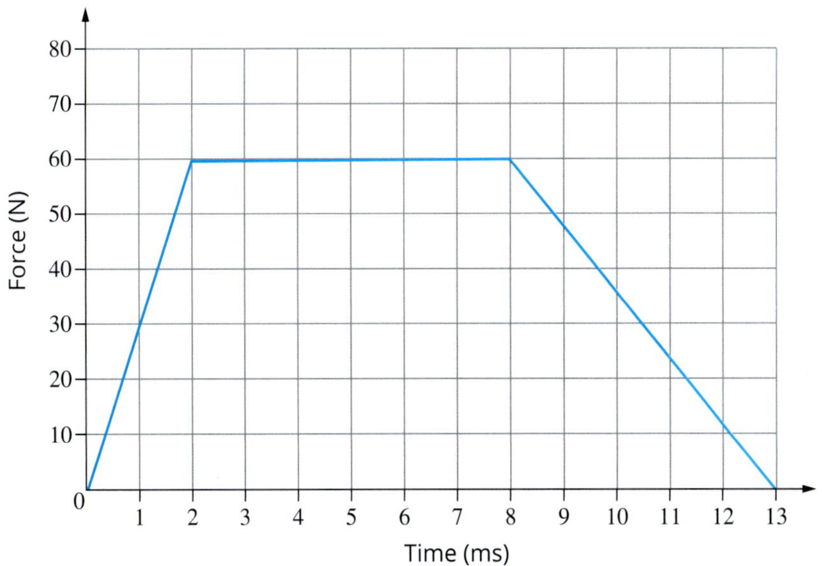

a Determine the force acting on the ball at a time of 9 ms.	
Thinking	**Working**
From the 9 ms point on the *x*-axis go up to the line of the graph, then across to the *y*-axis.	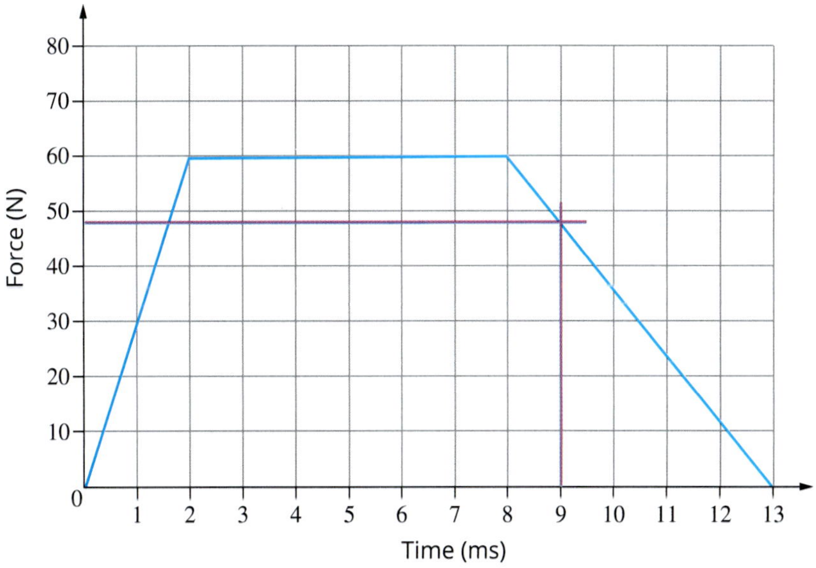
The force is estimated by reading the intercept of the *y*-axis.	$F = 48\,\text{N}$
Apply the sign and direction convention for motion in one dimension vertically.	$F = 48\,\text{N}$ upwards

b Calculate the total impulse of the ball over the 13 ms period of time.	
Thinking	**Working**
Break the area under the graph into sections for which you can calculate the area.	In this case, the graph can be broken into three sections: A, B and C.
Calculate the area of the three sections A, B and C using the equations for area of a triangle and the area of a rectangle.	Area = area of A + area of B + area of C $=\left(\frac{1}{2}b\times h\right)+(b\times h)+\left(\frac{1}{2}b\times h\right)$ $=\left[\frac{1}{2}\times(2\times10^{-3})\times60\right]+\left[(6\times10^{-3})\times60\right]+\left[\frac{1}{2}\times(5\times10^{-3})\times60\right]$ $= 0.06 + 0.36 + 0.15$ $= 0.57$
The total impulse is equal to the area.	Δp = area = 0.6 kg m s^{-1}
Apply the sign and direction convention for motion in one dimension vertically.	Δp = 0.6 kg m s^{-1} upwards

Worked example: Try yourself 12.6.3

CALCULATING THE TOTAL IMPULSE FROM A CHANGING FORCE

A student records the force acting on a tennis ball as it bounces off a hard concrete floor over a period of time. The graph shows the forces acting on the ball during its collision with the concrete floor.

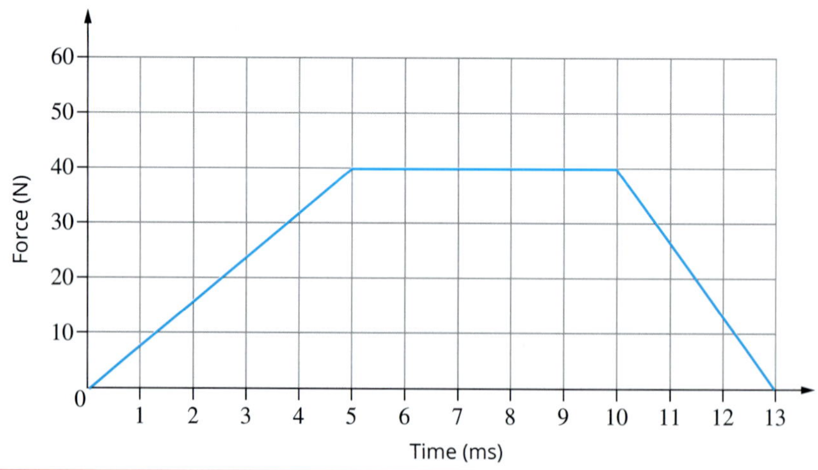

a Determine the force acting on the ball at a time of 4 ms.

b Calculate the total impulse of the ball over the 13 ms period of time.

CASE STUDY

Car safety

Designing a successful car is a complex task. A vehicle must be reliable, economical, powerful, visually appealing, secure and safe. Public perception of the relative importance of these issues varies. Magazines and newspapers concentrate on appearance, price and performance. The introduction of air-bag technology into most cars has altered the focus towards safety.

Vehicle safety is primarily about crash avoidance. Research shows potential accidents are avoided 99% of the time. The avoidance of accidents is mainly due to accident avoidance systems such as antilock brakes. When a collision does happen, passive safety features, such as the airbag, come into operation. Understanding the theory behind accidents involves primarily an understanding of impulse and force.

The airbag

The introduction of seatbelts allowed many more people to survive car accidents. However, many of these survivors sustained serious injuries. So, although seatbelts saved lives, there was also an increase in serious injuries. A further safety device was required to minimise these injuries.

The airbag in a car is designed to inflate within a few milliseconds of the occurrence of a collision to reduce secondary injuries during the collision. The airbag is designed to inflate only when the vehicle experiences an impact with a solid object at 18–20 $km\,h^{-1}$ or more. The required deceleration must be high, or accidental nudges with another car would cause the airbag to inflate. The car's computer control makes a decision within a few milliseconds to detonate the gas cylinders that inflate the airbag. The propellant detonates and inflates the airbag while, according to Newton's first law, the driver continues to move towards the dashboard. As the driver continues forwards into the airbag, the bag deflates, allowing the body to slow down over a longer time than would otherwise be possible as it moves towards the dashboard (Figure 12.6.3). The force is minimised so injury is reduced.

FIGURE 12.6.3 Airbags can prevent injuries by extending the period of time it takes the car's occupants to stop.

Calculating exactly when the airbag should inflate, and for how long, is a difficult task. Many cars have been crash tested and the results painstakingly analysed. High-speed film demonstrates precisely why the airbag is so effective. During a collision, the arms, legs and head of the occupants are restrained only by the joints and muscles. Enormous forces are involved because of the large deceleration. The shoulders and hips can, in most cases, sustain the large forces for the short duration. However, the neck is the weak link. Victims of road accidents regularly receive neck and spinal injuries. An airbag reduces the enormous forces the neck must withstand by extending the duration of the collision. This involves the direct application of the concept of impulse. A comparison of the forces applied to the occupant of a car with and without airbags is shown in Figure 12.6.4.

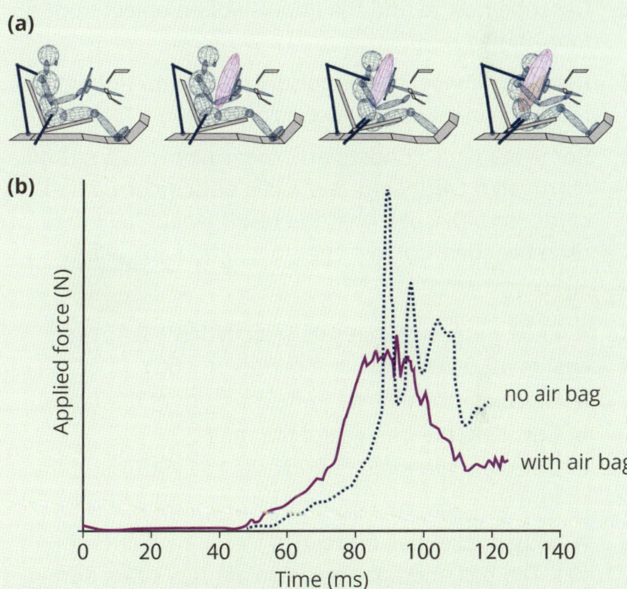

FIGURE 12.6.4 (a) The airbag extends the stopping time and distributes the force required to decelerate the mass of the driver or passenger over a larger area than a seatbelt. (b) The force withstood by the occupant of the car without an airbag is about double that felt with an airbag.

Airbags prevent the high forces caused by contact of the head with the steering wheel. The airbag ensures that the main thrust of the expansion is directed outwards instead of towards the driver. The airbag's deflation rate, governed by the size of the holes in the rear of the air bag, provides the optimum deceleration of the head for a large range of impact speeds.

The airbag is not the answer to all safety concerns associated with a collision, but it is one of many safety features that form a chain of defence in a collision.

12.6 Review

OA ✓✓

SUMMARY

- Newton's second law describes the relationship between impulse, force and the period of time:
$$\Delta p = F_{net}\Delta t$$
- A mass that has a constant change in velocity has a constant change in momentum.
- The faster a mass changes its velocity, the greater the force required to change the velocity in that period of time.
- The slower a mass changes its velocity, the smaller the force required to change the velocity in that period of time.
- Forces can change during a collision.
- The impulse over a period of time can be found by calculating the area under the line on a force versus time graph.
- During a collision, the force provided by a surface depends on how hard the surface is.
  - Hard surfaces stop objects more quickly and therefore exert a greater force.
  - Soft surfaces stop objects more slowly and therefore exert less force.

KEY QUESTIONS

Knowledge and understanding

1 Describe how to find the impulse of an object from a force–time graph.

2 Using the concept of impulse, explain how airbags can reduce injuries during a collision.

3 Some students conduct an experiment with a pile of corrugated cardboard and a pile of non-corrugated cardboard. Discuss what the result would be if they drop two identical raw eggs, one onto each pile.

Analysis

4 A 45.0 kg mass changes its velocity from 2.45 m s^{-1} east to 12.5 m s^{-1} east in a period of 3.50 s.

- **a** Calculate the change in momentum of the mass.
- **b** Calculate the impulse of the mass.
- **c** Calculate the force that causes the impulse of the mass.

5 A 200 g cricket ball (at rest) is struck by a cricket bat. The ball and bat are in contact for 0.05 s, during which time the ball is accelerated to a speed of 45 m s^{-1}.

- **a** What is the magnitude of the impulse the ball experiences?
- **b** What is the net average force acting on the ball during the contact time?
- **c** What is the net average force acting on the bat during the contact time?

6 The following graph shows the net vertical force generated as an athlete's foot strikes an asphalt running track.

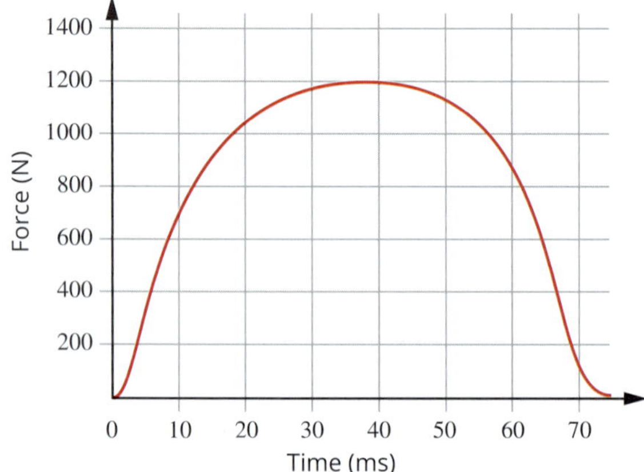

- **a** Estimate the maximum force acting on the athlete's foot during the contact time.
- **b** Estimate the total impulse during the contact time.

7 A 25 g arrow buries its head 2 cm into a target on striking it. The arrow was travelling at 50 m s^{-1} just before impact.

- **a** What change in momentum does the arrow experience as it comes to rest?
- **b** What is the impulse experienced by the arrow?
- **c** What is the average force that acts on the arrow during the period of deceleration after it hits the target?

8 Crash helmets are designed to reduce the force of impact on the head during a collision.

- **a** Explain how their design reduces the net force on the head.
- **b** Would a rigid 'shell' be as successful? Explain.

Chapter review

KEY TERMS

conserved
contact forces
force
impulse
inertia
law of conservation of momentum
momentum
net force
newton
Newton's first law
Newton's second law
Newton's third law
non-contact forces

REVIEW QUESTIONS

Knowledge and understanding

1 A student is travelling to school on a bus. When the bus turns a corner to the left, the student notices that standing passengers tend to lurch towards the right before grabbing a handrail to stop themselves from falling. Use Newton's first law to explain what is happening.

2 A bowling ball rolls along a smooth wooden floor at constant velocity. Ignoring the effects of friction and air resistance, which of the following options, relating to the force acting on the ball, is correct?

A There must be a net force acting forwards to maintain the velocity of the ball.

B There must not be an unbalanced force acting on the ball.

C The forwards force acting on the ball must be balanced by the friction that opposes the motion.

D More information is needed.

3 A box is sitting stationary on an inclined plane. Which of the following diagrams correctly indicates the forces acting on the box?

4 Calculate the acceleration of a 90.0 kg mass that has a net force of 882 N acting on it due to gravity.

5 What horizontal force must be applied to a pram for it to move at a constant speed of 1.3 $m\,s^{-1}$ along a horizontal footpath if there is a frictional force of 50 N?

6 Calculate the mass of an object if it accelerates at 9.20 $m\,s^{-2}$ east when a force of 352 N east acts on it.

7 Calculate the acceleration of a 26.8 kg mass when a net force of 185 N north acts on it.

8 Calculate the magnitude of the force required to give an object with a mass of 7.5 kg at an acceleration of 3.6 $m\,s^{-2}$ north.

9 Calculate the final velocity of a 1600 kg car moving at 11.1 $m\,s^{-1}$ south, when a net force of 7800 N north acts on it for 1.6 seconds.

10 A 85.0 kg javelin thrower applies a force of 150 N to an 800 g javelin at an angle of inclination of 45°. What force does the javelin exert on the thrower?

11 Calculate the momentum of a 5.5 kg bowling ball that has been thrown forwards at 9.3 $m\,s^{-1}$.

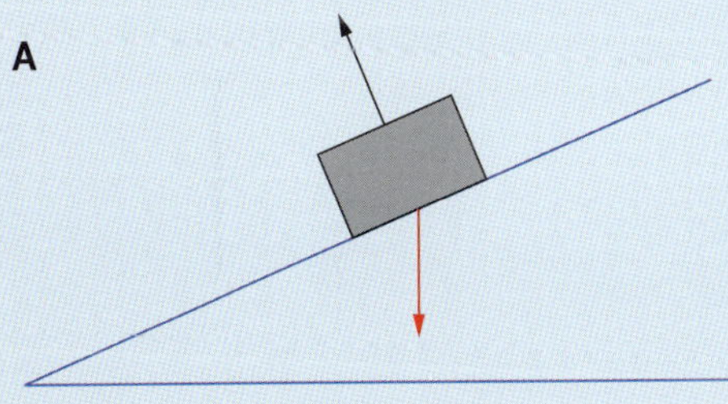

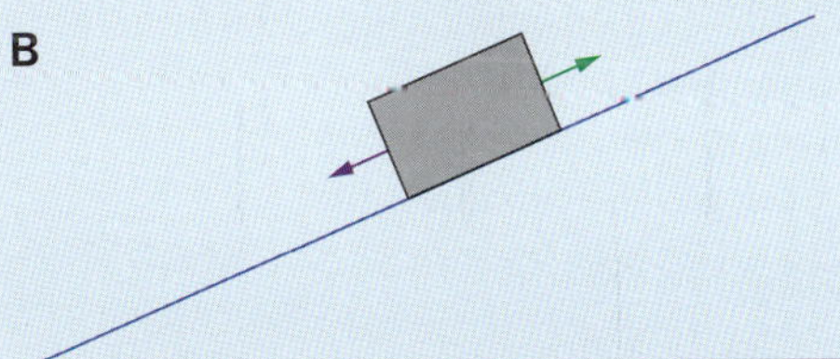

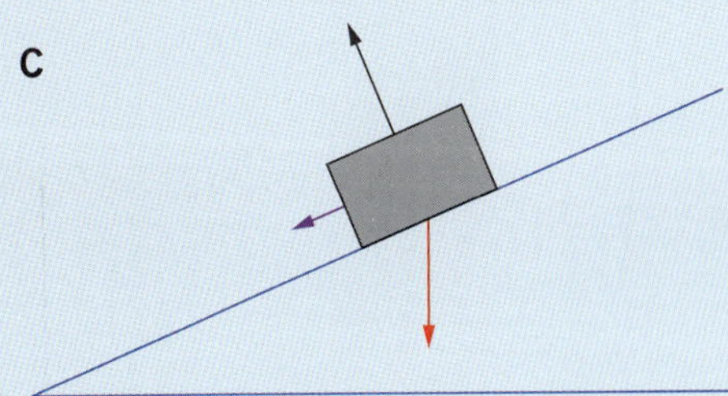

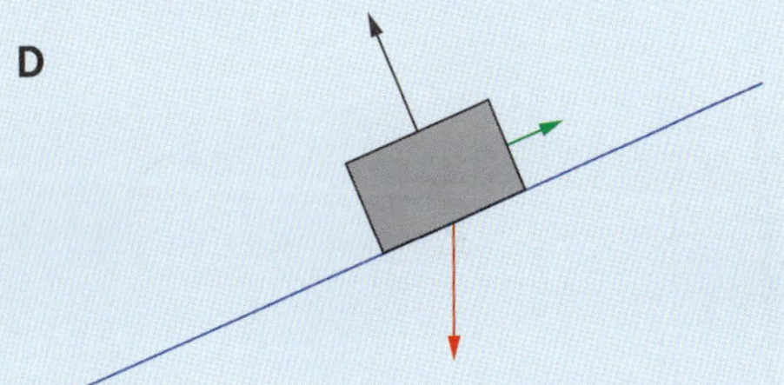

continued over page

12 Calculate the change in momentum of a 657 kg motorcycle whose velocity has changed from 27.8 m s^{-1} west to 16.7 m s^{-1} east in a period of 3.50 s.

13 When a stationary billiard ball with a mass of 260 g is hit by a cue, it moves forwards with a velocity of 3.6 m s^{-1}. Calculate the force between the cue and the billiard ball if they are in contact for 1.2 milliseconds.

14 When a 0.625 kg basketball bounces off the backboard, its velocity changes from 5.84 m s^{-1} forwards to 3.36 m s^{-1} backwards. Calculate the change in momentum of the basketball.

15 A kayaker of mass 65.0 kg steps out of a 27.0 kg kayak with a velocity of 1.40 m s^{-1} forwards onto the nearby riverbank. The kayak was initially at rest. With what velocity does the boat begin to move as the rower steps out?

16 A golf ball of mass 45.7 g is stationary on the ground when it is hit by a 330 g golf club travelling at 41.5 m s^{-1}. If the ball leaves the club at a speed of 65.3 m s^{-1}, with what speed does the club move just after hitting the ball? Give the answer to three significant figures.

Application and analysis

17 Lachy is riding his bike and producing a forwards force of 150 N. The combined mass of Lachy and the bike is 100 kg.

- **a** If there is no friction or air resistance, what is the magnitude of the acceleration of Lachy and the bike?
- **b** If friction opposes the bike's motion with a force of 45.0 N, what is the magnitude of the acceleration of the bike?
- **c** What must be the magnitude of the force of friction if Lachy's acceleration is 0.600 m s^{-2}?
- **d** Lachy now carries an additional mass of 25.0 kg due to his school bag. What must be the new forwards force he produces in order to accelerate at 0.800 m s^{-2} if friction opposes the motion with a force of 30.0 N?

18 A 70.0 kg footballer moving at 2.40 m s^{-1} south changes direction during a game to 3.20 m s^{-1} east. Calculate the change in momentum of the player.

19 A person is driving their car at 15 m s^{-1} west when they hit a telephone pole. The air bag in their car is activated and causes their head to decelerate to 0 m s^{-1} in a time of 85 milliseconds.

- **a** If the person's head has a mass of 4.8 kg, calculate the magnitude of the average force exerted on their head by the airbag.
- **b** Predict what would happen to the force on the person's head if the air bag did not activate and their head decelerated much more quickly by hitting the steering wheel.

20 Modern cars are designed with weak points that cause them to crumple in the event of a collision. Explain why these 'crumple zones' make cars safer than older cars that have a more rigid body design. Use the concept of 'impulse' in your answer.

21 Jordy is playing softball and hits a ball with her softball bat. The force versus time graph for this interaction is shown below. The ball has a mass of 170 g.

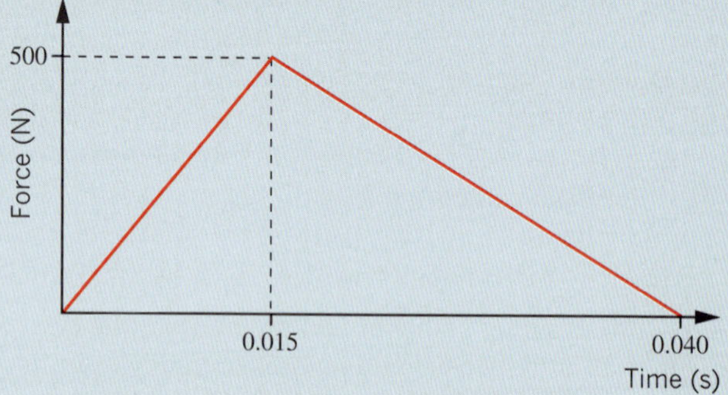

- **a** Determine the magnitude of the change in momentum of the ball.
- **b** Determine the magnitude of the change in momentum of the bat.
- **c** Determine the magnitude of the change in velocity of the ball.

OA ✓✓

CHAPTER

13 Energy and motion

Throughout this chapter you will be learn about the common thread of energy conversion that is present in so many daily activities, as well as some more extreme activities. Your own personal energy stores in the form of chemical potential energy are converted to kinetic energy and heat as you climb steps or run to catch a bus. In more thrill-seeking adventures, such as bungee jumping, gravitational potential energy is converted into kinetic and elastic potential energy. Even when jumping from a plane, the laws of physics cannot be switched off!

By the end of this chapter, you will be able to define and use the terms 'work', 'energy' and 'power'. You will use force–displacement graphs to determine the amount of work done.

Key knowledge

- apply the concept of work done by a force using:
  - work done = force × displacement: $W = Fs\cos\theta$, where force is constant **13.1**
  - work done = area under force vs distance graph **13.1**
- investigate and analyse theoretically and practically Hooke's Law for an ideal spring: $F = -kx$, where x is extension **13.2**
- analyse and model mechanical energy transfers and transformations using energy conservation:
  - changes in gravitational potential energy near Earth's surface: $E_g = mg\Delta h$ **13.2**
  - elastic potential energy in ideal springs: $E_s = \frac{1}{2}kx^2$ **13.2**
  - kinetic energy: $E_k = \frac{1}{2}mv^2$ **13.2**
- analyse rate of energy transfer using power: $P = \frac{E}{T}$ **13.3**
- calculate the efficiency of an energy transfer system: $\eta = \frac{\text{useful energy out}}{\text{total energy in}}$. **13.3**

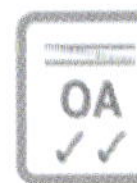

13.1 Work

The words 'energy' and 'work' are commonly used to describe a variety of everyday situations. However, these words take on quite specific definitions when used in a scientific context. They are two of the most important concepts in physics, allowing physicists to explain phenomena on a range of scales from collisions of subatomic particles to the interactions of galaxies.

ENERGY

Energy is the capacity to cause change. A moving car has the capacity to cause a change if it collides with something. Similarly, a heavy weight lifted by a crane has the capacity to cause a change if it is dropped. Energy is a scalar quantity; it has magnitude but not direction.

There are many different forms of energy. **Mechanical energy** is defined as the energy that a body possesses due to its position or motion. This category of energy can be broadly classified into two groups: kinetic energy and potential energy.

Kinetic energy is energy associated with motion. Any moving object, such as the moving car in Figure 13.1.1, has kinetic energy. In some forms of kinetic energy, the moving objects are not easily visible. An example of this is thermal energy, which is a type of kinetic energy related to the movement of particles. Table 13.1.1 lists some different types of kinetic energy and their associated moving objects.

FIGURE 13.1.1 A moving car has kinetic energy.

TABLE 13.1.1 Types of kinetic energy and their associated moving objects

Type of kinetic energy	Moving objects
translational kinetic	objects moving in a straight line
rotational kinetic	rotating objects
thermal	atoms, ions or molecules
sound	air molecules

Potential energy is energy associated with the position of objects relative to one another or within fields. For example, an object suspended by a crane has **gravitational potential energy** because of its position in Earth's gravitational field. Some examples of potential energy are listed in Table 13.1.2.

TABLE 13.1.2 Types of potential energy and their causes

Type of potential energy	Cause
gravitational	gravitational fields
chemical	relative positions of atoms
magnetic	magnetic fields
nuclear	forces within the nucleus of an atom
elastic	attractive forces between atoms

Unit of energy

The SI unit for energy, the joule (J), is named after the English scientist James Prescott Joule. He was the first person to show that kinetic energy could be converted into heat energy. The energy represented by 1 J is the equivalent to the energy needed to lift a 1 kg mass (e.g. 1 L of milk) through a height of 0.1 m or 10 cm. More commonly, scientists work in units of kilojoules (1 kJ = 1000 J) or even megajoules (1 MJ = 1 000 000 J).

WORK

Although in everyday life the word 'work' can take on a variety of meanings, in a scientific context work has a very specific meaning. In physics, when a force acts on an object and causes energy to be transferred or transformed, then work is being done on the object. For example, if a weightlifter applies a force to a barbell to lift it, then work has been done on the barbell; chemical energy within the weightlifter's body has been transformed into the gravitational potential energy of the barbell (Figure 13.1.2).

FIGURE 13.1.2 As a weightlifter lifts a barbell, chemical energy is transformed into gravitational potential energy.

Quantifying work

Work causes a change in energy, i.e. $W = \Delta E$.

More specifically, **work** is defined as the product of the force causing the energy change and the displacement of the object in the direction of the force during the energy change:

$W = Fs$

where W is work (in J)

F is force (in N)

s is the displacement in the direction of the force (in m).

Since work corresponds to a change in energy, the SI unit of work is also the joule. The definition of work allows us to find a value for a joule in terms of other SI units.

$$\text{Since } W = Fs,\ 1\,\text{J} = 1\,\text{N} \times 1\,\text{m} = 1\,\text{N\,m}.$$

A joule is equal to a newton-metre; that is, a force of 1 N acting over a distance of 1 m does 1 J of work.

Using the definition of a newton:

$$1\,\text{J} = 1\,\text{N} \times 1\,\text{m} = 1\,\text{kg\,m\,s}^{-2} \times 1\,\text{m} = 1\,\text{kg\,m}^2\,\text{s}^{-2}$$

This defines a joule in terms of fundamental units.

Although both force and displacement are vectors, work is a scalar unit. So like energy, work has no direction.

PHYSICSFILE

Units of energy

A number of non-SI units for energy are still in use. When talking about the energy content of food, it is common to use a unit called a calorie (cal). One calorie is defined as the amount of heat required to increase the temperature of 1 g of water by 1°C. This equates to 4.2 J. A calorie is a small amount, so food labels often list energies in thousands of calories or kcal. You may also see energy listed in Calories (top of image below). The uppercase C is important—a Calorie is defined as the heat energy required to raise 1 kg of water by 1°C. Therefore 1 Calorie = 1000 calorie = 1 kcal.

Electrical energy in the home is often measured in kilowatt-hours (kW h). A kilowatt-hour is a very large unit of energy:

$$1\,\text{kW\,h} = 3\,600\,000\,\text{J or } 3.6\,\text{MJ}.$$

Another, not-so-common unit of energy is the erg (from the Greek word *ergon* for energy). An erg is a very small unit of energy: $1\,\text{erg} = 10^{-7}\,\text{J}$.

Nutrition Facts

Serving Size 5 oz. (144g)
Servings Per Container 4

Amount Per Serving

Calories 310 **Calories** from Fat 100

	% Daily Value*
Total Fat 15g	**21%**
Saturated Fat 2.6g	**17%**
Trans Fat 1g	
Cholesterol 118mg	**39%**
Sodium 560mg	**28%**
Total Carbohydrate 12g	**4%**
Dietary Fiber 1g	**4%**
Sugars 1g	
Protein 24g	

Vitamin A 1% • **Vitamin C** 2%

Calcium 2% • **Iron** 5%

*Percent Daily Values are based on a 2,000 calorie diet. Your daily values may be higher or lower depending on your calorie needs:

	Calories	2,000	2,500
Total Fat	Less Than	65g	80g
Saturated Fat	Less Than	20g	25g
Cholesterol	Less Than	300mg	300mg
Sodium	Less Than	2,400mg	2,400mg
Total Carbohydrate		300g	375g
Dietary Fiber		25g	30g

Calories per gram:
Fat 9 • Carbohydrate 4 • Protein 4

The amount of energy in a serving of food is often measured in calories.

Worked example 13.1.1

CALCULATING WORK

A person pushes a heavy box along the ground for 10.0 m with a force of 30.0 N. Calculate the amount of work done.

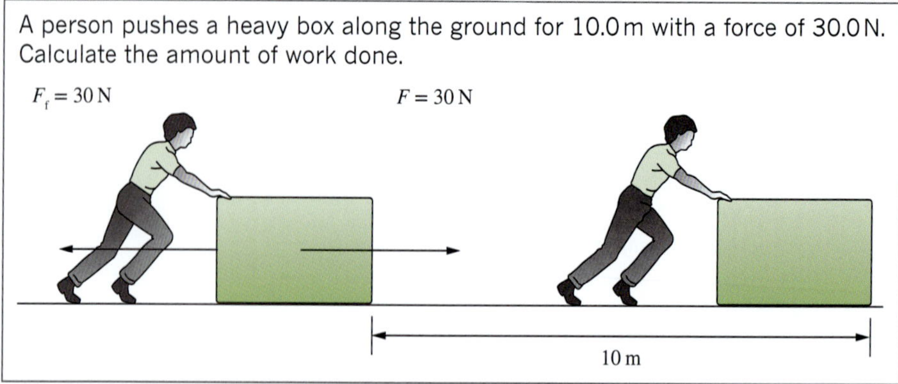

Thinking	Working
Recall the definition of work.	$W = Fs$
Substitute in the values for this situation.	$W = 30 \times 10$
Solve the problem, giving an answer with appropriate units.	$W = 300$ J

Worked example: Try yourself 13.1.1

CALCULATING WORK

A person pushes a heavy wardrobe from one room to another by applying a force of 50.0 N for a distance of 5.00 m. Calculate the amount of work done.

Work and friction

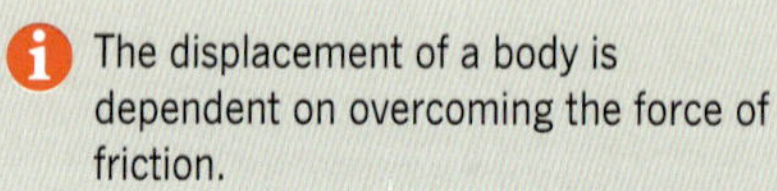
The displacement of a body is dependent on overcoming the force of friction.

The energy change produced by work is not always obvious. Consider Worked example 13.1.1, in which 300 J of work was done on a box when it was pushed 10 m. A number of energy outcomes are possible for this scenario.

- In an ideal situation, where there is no friction, all of this work would be transformed into kinetic energy and the box would end up with a higher velocity than before it was pushed.
- In most real situations, there is friction between the box and the ground, so some of the work done would become heat and sound due to friction and the rest would become kinetic energy.
- In the limiting situation in which the force applied is exactly equal to the friction, the box would slide at a constant speed. This means that its kinetic energy would not change, so all of the work done would be converted into heat and sound due to friction.

A force with no work

The mathematical definition of work has some unusual implications. One is that if a force is applied to an object but the object does not move, then no work is done.

This appears counterintuitive; that is, it goes against what you would probably expect. An example of this is shown in Figure 13.1.3. Although work is done while picking up a heavy box, holding the box at a constant height does no work on the box.

Assuming the box has a force due to gravity of 100 N and that it is lifted from the ground to a height of 1.2 m, the work done lifting the box would be: $W = Fs = 100 \times 1.2 = 120$ J. In this case, energy is being transformed from chemical energy inside the person's body into the gravitational potential energy of the box.

FIGURE 13.1.3 According to the definition of work, no work is done when the box is held at a constant height.

However, when the box is held at a constant height, the definition of work gives $W = Fs = 100 \times 0 = 0\,\text{J}$. So, no work is being done *on* the box. Although there would be energy transformations going on inside the person's body to keep their muscles working, the energy of the box does not change, therefore no work has been done on the box.

Work is done only if the net force causes a movement of one body in relation to other bodies.

Work and displacement at an angle

Sometimes, when a force is applied, the object does not move in the same direction as the force. For example, in Figure 13.1.4, when a person pulls a suitcase, the direction of the force is at an angle upwards, although the suitcase moves horizontally forwards.

FIGURE 13.1.4 When a person pulls a suitcase, the force is applied at an angle to the displacement of the suitcase.

In this case, only the horizontal component of the pull contributes to the work being done on the suitcase. The vertical component of this force pulls the suitcase upwards and is balanced by the downwards force due to gravity.

Resolving forces

In the situation of a person pulling a suitcase, the general equation $W = Fs$ applies. Recall from Section 10.3 that vectors can be resolved into perpendicular components. The person's pull can be resolved into a vertical component, $F\sin\theta$, and a horizontal component, $F\cos\theta$, (Figure 13.1.5). Substituting the horizontal component into the general definition for work gives:

$$\begin{aligned} W &= F\cos\theta \times s \\ &= Fs\cos\theta \end{aligned}$$

In situations where the force applied is not parallel to the object's displacement, work can be calculated using the general equation:

$$W = Fs\cos\theta$$

where θ is the angle between the force, F, and the displacement, s.

FIGURE 13.1.5 The force applied by the person pulling the suitcase can be resolved into a horizontal force and a vertical force.

Worked example 13.1.2

WORK WITH FORCE AND DISPLACEMENT AT AN ANGLE

A person carries a box weighing 19.6 N up a 10 m flight of stairs. Calculate the work done against gravity on the box.

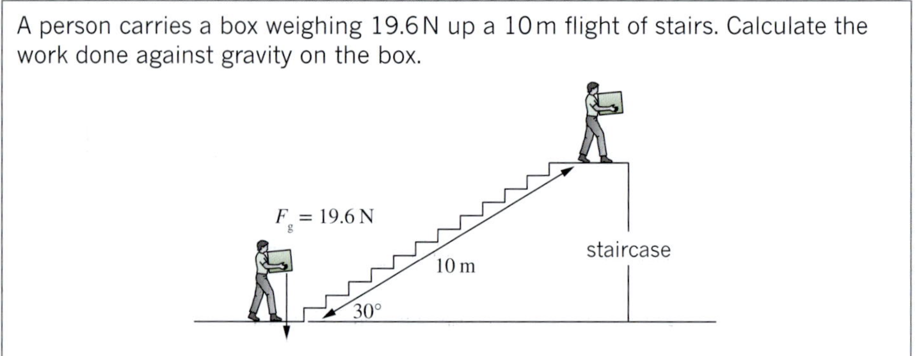

Thinking	Working
Determine values for F, s and θ. Note that the required component of the force is upwards, so the angle is not 30°. It is 90° − 30° = 60°.	Force applied to the box by the person: $F = 19.6\text{ N}$ upwards Displacement: $s = 10\text{ m}$ Angle between the force and displacement: $\theta = 60°$
Recall the work equation.	$W = Fs\cos\theta$
Substitute values into the work equation.	$W = 19.6 \times 10 \times \cos 60°$
State the answer with the correct units.	$W = 98\text{ J}$

Worked example: Try yourself 13.1.2

WORK WITH FORCE AND DISPLACEMENT AT AN ANGLE

A girl pulls her brother along in a trolley for a distance of 30 m, as shown. Calculate the work done on the box. Give your answer correct to two significant figures.

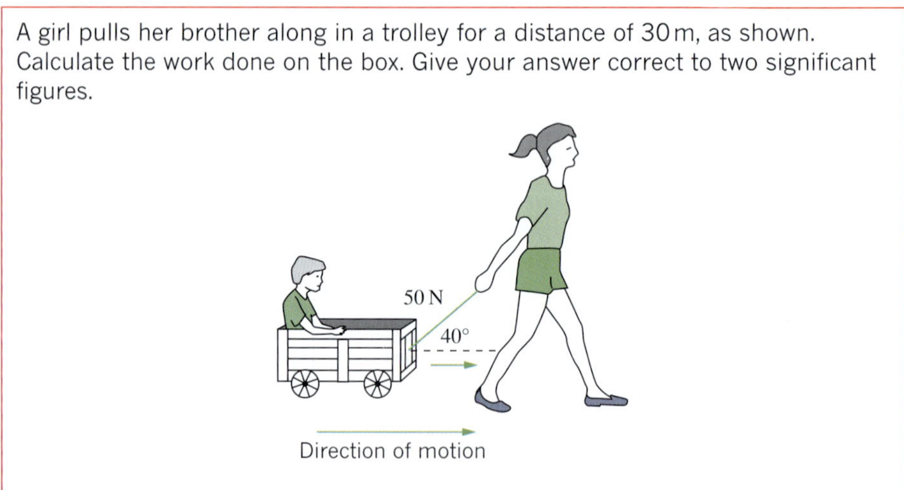

FORCE–DISPLACEMENT GRAPHS

As its name suggests, a force–displacement graph illustrates the way a force changes with displacement. For a situation in which the force is constant, this graph is simple. For example, in Figure 13.1.6, the force–displacement graph for a person lifting a weight is a flat horizontal line showing that the force applied to the barbell is constant throughout the lift.

FIGURE 13.1.6 The force–displacement graph for a person lifting a weight is a straight, horizontal line, indicating that the force applied to the barbell by the woman is constant throughout the process.

In contrast, an elastic object such as a spring obeys a relationship known as Hooke's law. Hooke's law is discussed in more detail in Section 13.2. Briefly, the law describes how, the more you stretch a spring, the greater the force required to keep stretching it. The force–displacement graph for a spring is also a straight line, but this line shows the direct relationship described by Hooke's law (Figure 13.1.7). (Note: sometimes, you will see force–displacement graphs for elastic objects labelled as force–extension graphs. In this context, the term 'extension' is the same as 'displacement'.)

Many everyday materials are only partially elastic. Their force–displacement graphs are relatively complex. For example, the force–displacement graph in Figure 13.1.8 for a sports shoe shows that the shoe is close to elastic for low displacements, but at high displacements the restoring force is relatively constant.

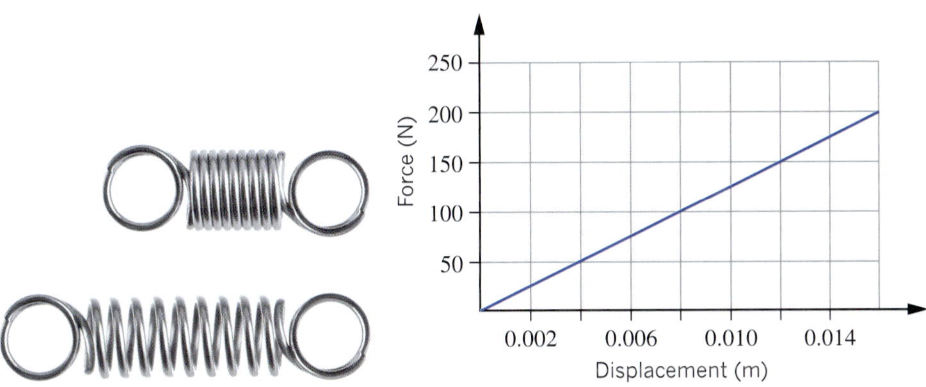

FIGURE 13.1.7 As a spring stretches, more force is required to keep stretching it. The force is proportional to the extension.

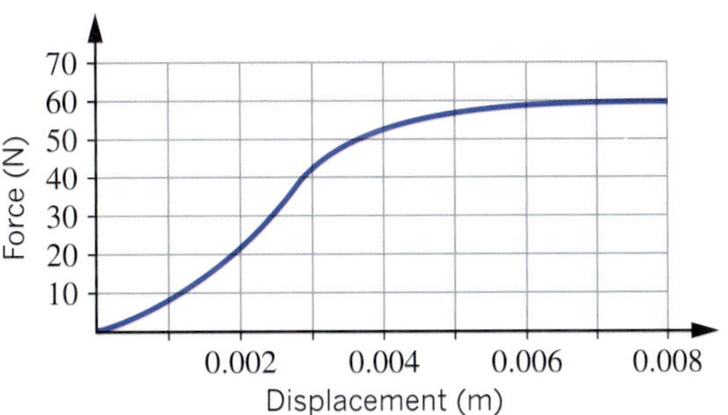

FIGURE 13.1.8 The force–displacement graph for a sports shoe is not a straight line; the change in force varies with how much the shoe has been stretched.

CALCULATING WORK FROM A FORCE-DISPLACEMENT GRAPH

> The work done by a force can be found from the area under the corresponding force–displacement graph.

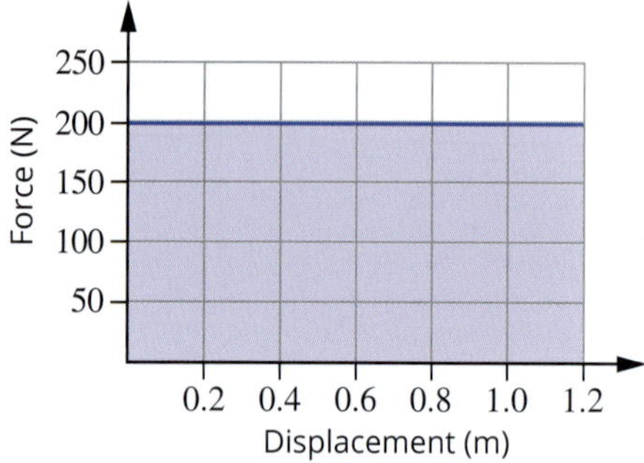

FIGURE 13.1.9 The area under a force–displacement graph gives the work done by the force.

When a force causes a displacement, the amount of work done by the force can be calculated from the area under its force–displacement graph.

For a constant force, this is very simple. Consider the earlier example of a person lifting the barbell (Figure 13.1.6 on page 403). The area can be found by counting the number of 'force times distance' squares under the line. In the example in Figure 13.1.9, there are $6 \times 4 = 24$ of these squares. Since each square has an area of $50\,\text{N} \times 0.2\,\text{m} = 10\,\text{J}$, the total work done is 240 J. This area could also be found by recognising that it is a rectangle and multiplying length by width to find the area. For Figure 13.1.9, this is $200\,\text{N} \times 1.2\,\text{m} = 240\,\text{J}$. Note that this second method is exactly the same as using the formula for work: $W = Fs = 200 \times 1.2 = 240\,\text{J}$. This relationship works in this case because the force is constant.

Similar strategies, either counting grid squares or calculating the area of the shape under the graph, can also be used when the force varies with the displacement.

Worked example 13.1.3

WORK FROM THE AREA UNDER A FORCE–DISPLACEMENT GRAPH

Use the force–extension graph for an elastic band to estimate how much work is done in stretching the elastic band 20 cm. Give your answer to the nearest 0.5 J.

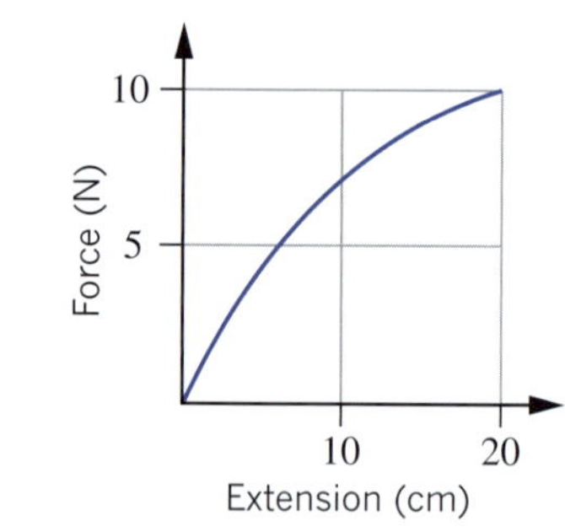

Thinking	Working
Calculate the work value of each grid square.	The dimensions of a grid square are: Force: 5 N, Extension: 10 cm = 0.1 m Area of 1 square = $5 \times 0.1 = 0.5\,\text{J}$
Count the number of grid squares under the curve.	Only count grid squares that are more than half under the curve. If the curve cuts a square in half, count every second one. Number of squares = 3
Multiply the number of grid squares under the curve by the work value of each grid square.	$W = 3 \times 0.5\,\text{J} = 1.5\,\text{J}$

Worked example: Try yourself 13.1.3

WORK FROM THE AREA UNDER A FORCE–DISPLACEMENT GRAPH

The shoes of a jogger stretch by an average of 3 mm with each step. Use the force–displacement graph for a sports shoe to estimate how much work is done on the shoe with each step. Give your answer to the nearest 0.01 J.

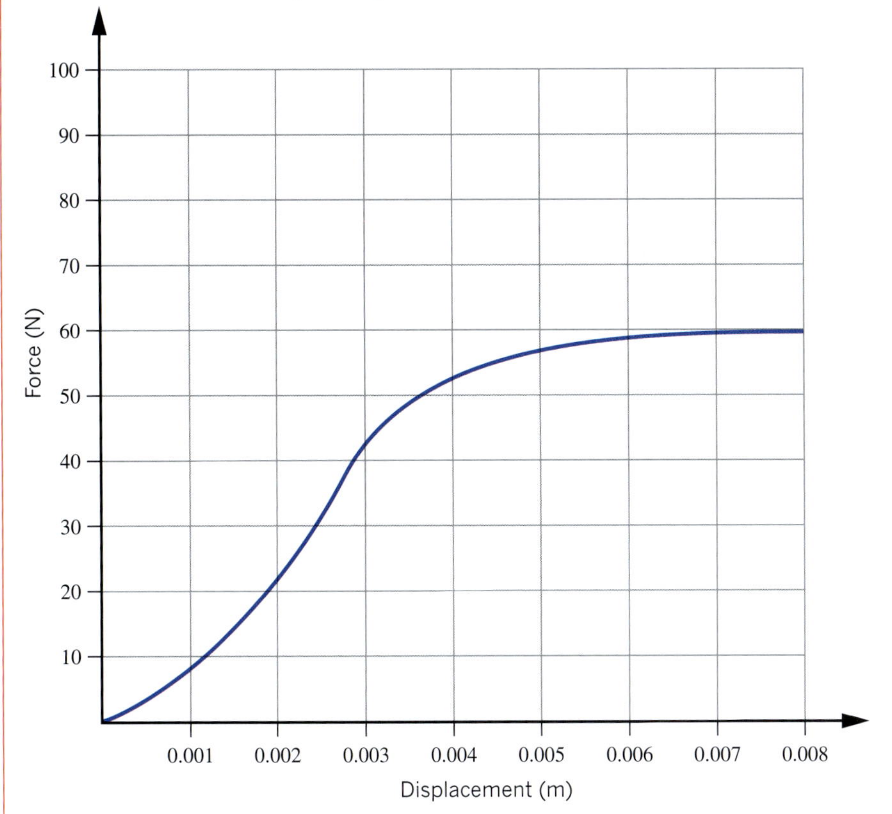

13.1 Review

OA ✓✓

SUMMARY

- Energy is the capacity to cause a change.
- Energy is conserved. It can be transferred or transformed, but not created or destroyed.
- There are many different forms of energy. These can be broadly classified as either kinetic (associated with movement) or potential (associated with the relative positions of objects).
- Work is done when energy is transferred or transformed.
- Work is done when a force causes an object to be displaced.
- Work done = force × displacement: $W = Fs\cos\theta$, where force is constant, and θ is the angle between the force and the displacement.
- When a force does not produce a displacement, or when the force and displacement are at right angles to each other, no work is done.
- Work is equal to the area under a force–displacement graph.
- A straight horizontal line in a force–displacement graph represents a constant force.
- The relationship between force and displacement for an elastic object is represented as a straight diagonal line in a force–displacement graph.

continued over page

13.1 Review *continued*

KEY QUESTIONS

Knowledge and understanding

1 A cyclist is accelerating at the beginning of a ride and applies a force of 275 N for a distance of 35 m. What is the work done by the cyclist on the bike?

2 In the case of a person leaning on a solid brick wall, explain why no work is being done.

3 A spring with this force–displacement graph is stretched as shown. Using the formula for the area of a triangle, calculate the work done to stretch the spring by 0.012 m.

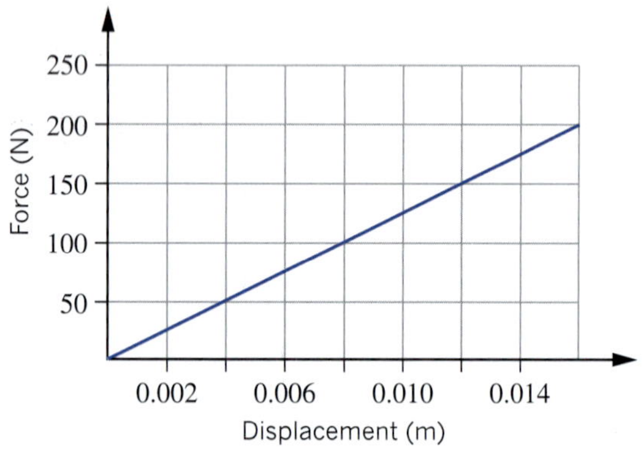

4 A cyclist does 3250 J of work when she rides her bike at a constant speed for 190 m. Calculate the average force the cyclist applies over this distance.

5 A rope at 28° to the horizontal is used to drag a heavy box along the ground for a distance of 4.5 m. Calculate the work done if the tension in the rope is 75 N. Give your answer correct to the nearest 10 J.

Analysis

6 Two people push in opposite directions on a heavy box. One person applies 35 N of force, the other applies 50 N of force. There is 15 N of friction between the box and the floor which means that the box does not move. What is the work done by the person applying 35 N of force?

7 The strings of a graphite-head tennis racquet have the force–displacement graph shown. Calculate the work done when the strings displace by 8 mm.

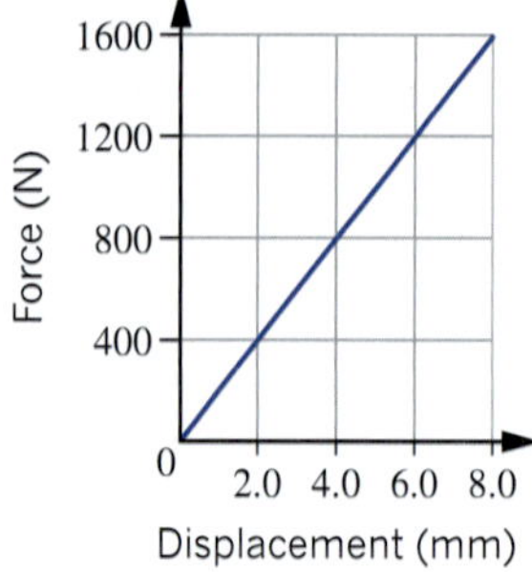

8 Three different springs have the force–displacement graphs shown. Estimate the work done by stretching each of the springs by 30 mm. Give your answers correct to two significant figures.

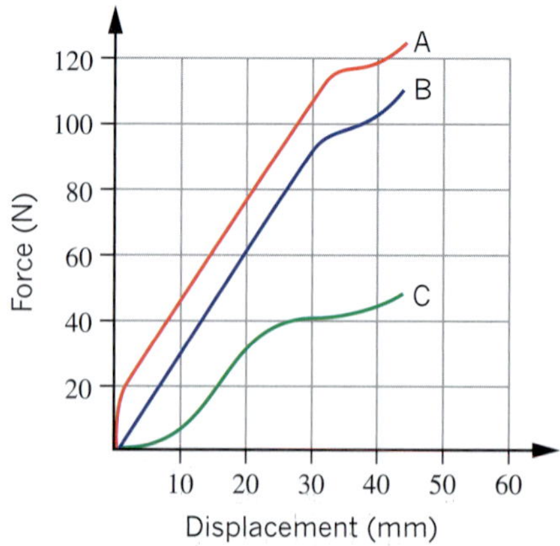

9 The following diagram is a simplified representation of the forces acting on a basketball when it bounces. The upwards arrows show the force as the basketball compresses and the downwards arrows show the force as it rebounds.

Forces acting on a bouncing basketball

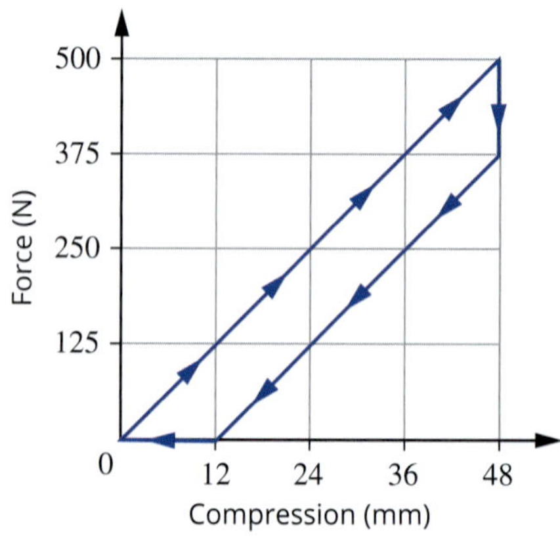

a Calculate the work done on the basketball when it compresses by 48 mm.

b Calculate the work done by the ball as it decompresses from a compression of 48 mm.

c Explain why your answers to parts a and b are different.

13.2 Mechanical energy

Mechanical energy is the energy that a body possesses due to its position or motion. Kinetic energy, gravitational potential energy and elastic potential energy are all forms of mechanical energy.

Any object that moves, such as those shown in Figure 13.2.1, has kinetic energy. Many real-life energy interactions, such as throwing a ball, involve objects with kinetic energy. Some of these interactions, for example car collisions, have life-threatening implications. Hence, it is important to be able to quantify (i.e. find numerical values for) the kinetic energy of an object.

One of the easiest forms of potential energy to study is gravitational potential energy. Any object that is lifted above Earth's surface has the capacity to cause change due to its position in Earth's gravitational field. An understanding of gravitational potential energy is essential to understanding common energy transformations.

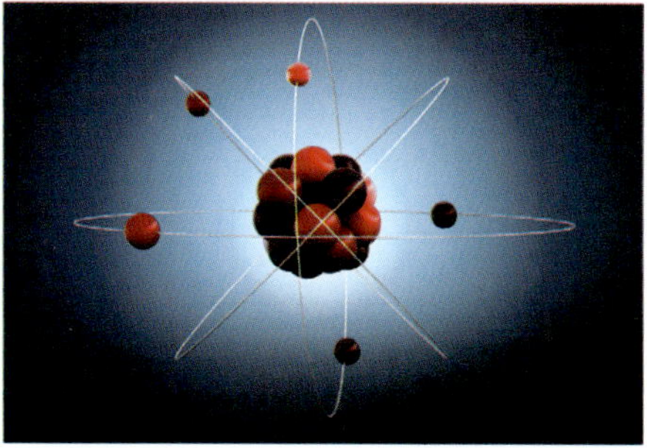

FIGURE 13.2.1 Any moving object, regardless of its size, has kinetic energy.

WORK AND KINETIC ENERGY

Kinetic energy is the energy of motion. It can be quantified by calculating the amount of work needed to give an object its velocity. We link these two concepts by introducing the expression for work from Section 13.1 in the context of velocity. Consider the dynamics cart in Figure 13.2.2. The cart has mass, m, and is starting at rest ($u = 0$). It is pushed with force, F, which acts while the cart undergoes a displacement, s, and gains a final velocity, v.

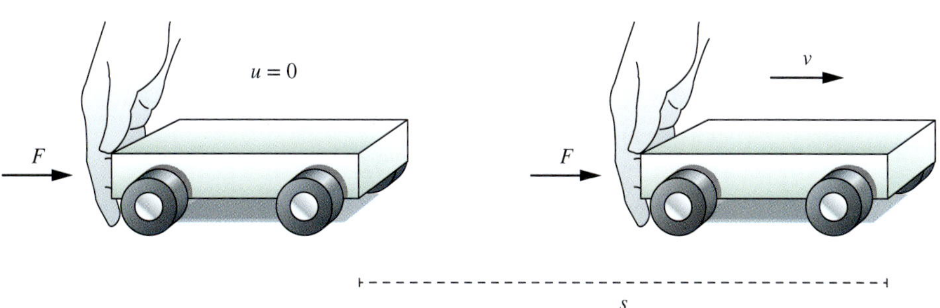

FIGURE 13.2.2 The kinetic energy of a dynamics cart can be calculated by considering the force (F) acting on it over a given displacement (s).

The work done by the force is given by the equation $W = Fs$. The force causes the cart to accelerate according to Newton's second law, $F = ma$.

Rearranging the equation of motion $v^2 = u^2 - 2as$ gives:

$$a = \frac{v^2 - u^2}{2s}$$

Combining this with $F = ma$ means that the force acting on the cart can be given by the equation:

$$F = m\left(\frac{v^2 - u^2}{2s}\right)$$

This equation can be transposed to find an expression for the amount of work (Fs) done on the cart:

$$F = \frac{m}{2s}(v^2 - u^2)$$

$$Fs = \frac{m}{2}(v^2 - u^2)$$

$$= \frac{1}{2}m(v^2 - u^2)$$

Since $W = Fs$:

$$W = \frac{1}{2}mv^2 - \frac{1}{2}mu^2$$

The work done by the force, W, causes a change in kinetic energy from its initial value $\frac{1}{2}mu^2$ to a new value of $\frac{1}{2}mv^2$.

The relationship between the work done and the change in kinetic energy can be written mathematically as:

$$W = \frac{1}{2}mv^2 - \frac{1}{2}mu^2$$

where W is work (in J)
m is mass (in kg)
u is initial velocity (in m s^{-1})
v is final velocity (in m s^{-1}).
This equation is known as the 'work–energy theorem'.

In this situation, the cart was originally at rest ($u = 0$) so:

$$W = \frac{1}{2}mv^2$$

Assuming that no energy was lost as heat or noise and that all of the work is converted into kinetic energy, this equation gives us a mathematical definition for the kinetic energy of the cart in terms of its mass and velocity:

$$E_k = \frac{1}{2}mv^2$$

where E_k is kinetic energy (in J).

Worked example 13.2.1

CALCULATING KINETIC ENERGY

A car with a mass of 1200 kg is travelling at 90 km h^{-1}. Calculate its kinetic energy at this speed.

Thinking	Working
Convert the car's speed to m s^{-1}. Recall from Section 11.1 that you can convert from km h^{-1} to m s^{-1} by dividing by 3.6.	$90\text{ km h}^{-1} = \frac{90\text{ km}}{1\text{ h}} = \frac{90\,000\text{ m}}{3600\text{ s}}$ $= 25\text{ m s}^{-1}$
Recall the equation for kinetic energy.	$E_k = \frac{1}{2}mv^2$
Substitute the values for this situation into the equation.	$E_k = \frac{1}{2} \times 1200 \times 25^2$
State the answer with appropriate units.	$E_k = 375\,000\text{ J} = 375\text{ kJ}$

Worked example: Try yourself 13.2.1

CALCULATING KINETIC ENERGY

A person crossing the street is walking at 5.0 km h^{-1}. If the person has a mass of 80 kg, calculate their kinetic energy. Give all answers correct to two significant figures.

APPLYING THE WORK-ENERGY THEOREM

The **work–energy theorem** can be seen as a definition for the change in kinetic energy produced by a force:

$$W = \frac{1}{2}mv^2 - \frac{1}{2}mu^2 = (E_k)_{final} - (E_k)_{initial} = \Delta E_k$$

Worked example 13.2.2

CALCULATING KINETIC ENERGY CHANGES

A 2 tonne truck travelling at 100 km h^{-1} slows to 80 km h^{-1} before turning a corner.

a Calculate the work done by the brakes to make this change. Give answers to two significant figures.	
Thinking	**Working**
Convert the values into SI units. Recall from Section 11.1 that you can convert from km h^{-1} to m s^{-1} by dividing by 3.6.	$u = 100\,\text{kmh}^{-1} = \frac{100\,\text{km}}{1\,\text{h}} = \frac{100\,000\,\text{m}}{3600\,\text{s}}$ $= 28\,\text{m s}^{-1}$ $v = 80\,\text{kmh}^{-1} = \frac{80\,\text{km}}{1\,\text{h}} = \frac{90\,000\,\text{m}}{3600\,\text{s}}$ $= 22\,\text{m s}^{-1}$ $m = 2$ tonne $= 2000$ kg
Recall the work–energy theorem.	$W = \frac{1}{2}mv^2 - \frac{1}{2}mu^2$
Substitute the values for this situation into the equation.	$W = \frac{1}{2}(2000 \times 22^2) - \frac{1}{2}(2000 \times 28^2)$
State the answer with appropriate units.	$W = -300\,000\,\text{J} = -3.0 \times 10^2\,\text{kJ}$ Note: the negative value indicates that the work done by the brakes has caused kinetic energy to decrease.

b If it takes 50 m for this deceleration to take place, calculate the average force applied by the truck's brakes.	
Thinking	**Working**
Recall the definition of work.	$W = Fs$
Substitute the values for this situation into the equation. Note: the negative has been ignored since work is a scalar.	$300\,000\,\text{J} = F \times 50\,\text{m}$
Iranspose the equation to find the answer.	$F = \frac{W}{s} = \frac{300\,000\,\text{J}}{50\,\text{m}} = 6.0 \times 10^3\,\text{N}$

Worked example: Try yourself 13.2.2

CALCULATING KINETIC ENERGY CHANGES

As a bus with a mass of 10 tonnes approaches a school, it slows from 60 km h^{-1} to 40 km h^{-1}.

a Calculate the work done by the brakes of the bus. Give answers to two significant figures.

b The bus travels 40 m as it decelerates. Calculate the average force applied by the truck's brakes.

Notice that the definitions for kinetic energy and change in kinetic energy have been derived entirely from known concepts: the definition of work, Newton's second law and the equations of motion. This makes kinetic energy appear a redundant concept. However, using kinetic energy calculations can often make analysis of physical interactions quicker and easier, particularly in situations where acceleration is not constant.

Worked example 13.2.3

CALCULATING SPEED FROM KINETIC ENERGY

The engine of a 1400 kg car can do 900 kJ of work in 10 s. Assuming all of this work is converted into kinetic energy, calculate the speed of the car after this time in km h^{-1}. Give your answer correct to two significant figures.

Thinking	Working
State the values in SI units.	$W = 900\,\text{kJ} = 900 \times 10^3\,\text{J}$ $m = 1400\,\text{kg}$
Recall the equation for kinetic energy.	$E_k = \frac{1}{2}mv^2$
Transpose the equation to make v the subject.	$v = \sqrt{\frac{2E_k}{m}}$
Substitute the values for this situation into the equation.	$v = \sqrt{\frac{2 \times 900 \times 10^3}{1400}} = 36\,\text{m s}^{-1}$
State the answer with appropriate units.	$v = 36 \times 3.6 = 1.3 \times 10^2\,\text{km h}^{-1}$

Worked example: Try yourself 13.2.3

CALCULATING SPEED FROM KINETIC ENERGY

A 300 kg motorbike has 150 kJ of kinetic energy. Calculate the speed of the motorbike in km h^{-1}. Give your answer correct to three significant figures.

DEFINING GRAVITATIONAL POTENTIAL ENERGY

Gravitational potential energy is a measure of the amount of energy available to an object due to its position in a gravitational field. The gravitational potential energy of an object can be calculated from the amount of work that must be done against gravity to get the object into its position.

Consider the weightlifter lifting a barbell in Figure 13.2.3. Assuming that the bar is lifted at a constant speed, then the weightlifter must apply a lifting force equal to the force due to gravity on the barbell, F_g. The lifting force, F_l, is applied over a displacement, Δh, corresponding to the change in height of the barbell.

The work done against gravity by the weightlifter is:

$$W = Fs = F_g \Delta h$$

Since the force due to gravity $F_g = mg$, the work done can be written as:

$$W = mg\Delta h$$

The work carried out in this example has resulted in the transformation of chemical energy within the weightlifter into gravitational potential energy. The change in gravitational potential energy of the barbell is:

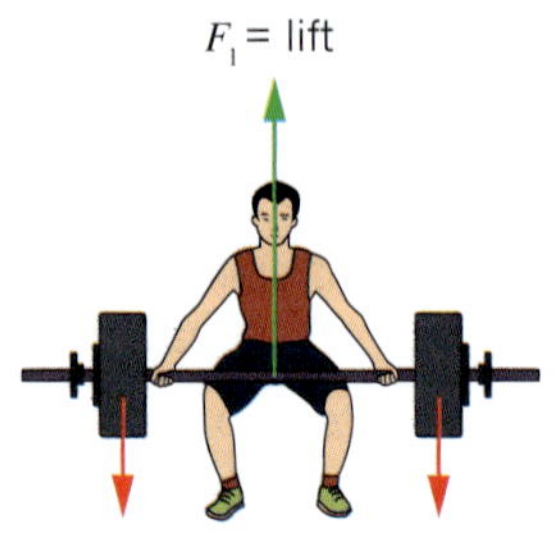

FIGURE 13.2.3 A weightlifter applies a constant force over a fixed distance to give the barbell gravitational potential energy.

$\Delta E_g = mg\Delta h$

Taking the ground as the point where the gravitational potential energy is zero (i.e. $E_g = 0$), the gravitational potential energy of an object, due to the work done against a gravitational field, is given by:

$$E_g = mg\Delta h$$

where E_g is the gravitational potential energy (in J)
m is the mass of the object (in kg)
g is the gravitational field strength (9.8 N kg^{-1} on Earth)
Δh is the change in height of the object (in m).

Worked example 13.2.4

CALCULATING GRAVITATIONAL POTENTIAL ENERGY

A weightlifter lifts a barbell that has a total mass of 80 kg from the floor to a height of 1.8 m above the ground. Calculate the gravitational potential energy of the barbell at this height. Give your answer correct to two significant figures.

Thinking	Working
Recall the formula for gravitational potential energy.	$E_g = mg\Delta h$
Substitute the values for this situation into the equation.	$E_g = 80 \times 9.8 \times 1.8$
State the answer with appropriate units and significant figures.	$E_g = 1411.2\text{ J} = 1.4\text{ kJ}$

Worked example: Try yourself 13.2.4

CALCULATING GRAVITATIONAL POTENTIAL ENERGY

A person doing their grocery shopping lifts a 5 kg grocery bag to a height of 30 cm. Calculate the gravitational potential energy of the grocery bag at this height. Give your answer correct to two significant figures.

GRAVITATIONAL POTENTIAL ENERGY AND REFERENCE LEVEL

When calculating gravitational potential energy, it is important to carefully define the level that corresponds to $E_g = 0$. Often this can be taken to be the ground or sea level; but the zero potential energy reference level is not always obvious.

It does not really matter which point is taken as the zero potential energy reference level, as long as the chosen point is used consistently throughout a particular problem (Figure 13.2.4). If objects move below the reference level, then their energies will become negative and should be interpreted accordingly.

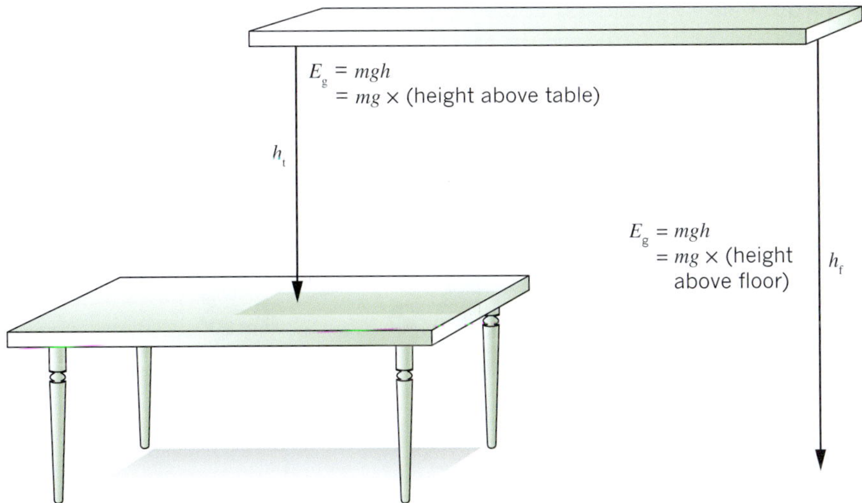

FIGURE 13.2.4 In this situation, the zero potential energy reference point could be taken as either the level of the table or the floor.

Worked example 13.2.5

CALCULATING GRAVITATIONAL POTENTIAL ENERGY RELATIVE TO A REFERENCE LEVEL

A weightlifter (m = 60 kg) lifts a 50 kg bar through a distance of 40.0 cm. Calculate the increase in gravitational potential energy of the bar with each lift. Use $g = 9.8\,\text{N kg}^{-1}$ and state your answer correct to two significant figures.

Thinking	Working
Recall the formula for gravitational potential energy.	$E_g = mg\Delta h$
Identify the relevant values for this situation. Only the mass of the bar is being lifted (the weightlifter's mass is a distractor). Take the weightlifter's body as the zero potential energy level.	$m = 50\,\text{kg}$ $g = 9.8\,\text{N kg}^{-1}$ $\Delta h = 40.0\,\text{cm} = 0.400\,\text{m}$
Substitute the values for this situation into the equation.	$E_g = 50.0 \times 9.8 \times 0.400$
State the answer with appropriate units and significant figures.	$E_g = 196\,\text{J}$ $= 2.0 \times 10^2\,\text{J}$

Worked example: Try yourself 13.2.5

CALCULATING GRAVITATIONAL POTENTIAL ENERGY RELATIVE TO A REFERENCE LEVEL

A father picks up his baby from its bed. The baby has a mass of 6.0 kg and the mattress of the bed is 70 cm above the ground. When the father holds the baby in his arms, it is 125 cm off the ground. Calculate the increase in gravitational potential energy of the baby. Use $g = 9.8\,\text{N kg}^{-1}$ and state your answer correct to two significant figures.

CASE STUDY

The high jump

Science has long been used in sport to help athletes gain a competitive edge. The concept of gravitational potential energy is of obvious importance to a high jumper. Clearly, the high jumper must do enough work in their jump to create sufficient gravitational potential energy to clear the bar.

The modern high jump technique known as the Fosbury flop gets the high jumper to bend their body as they go over the bar. This is illustrated in Figure 13.2.5.

FIGURE 13.2.5 In the Fosbury flop technique, a high jumper must bend their body over the bar.

When the technique is correctly performed, most of the mass of the jumper (e.g. their head, arms and legs) is actually lower than the bar throughout the jump. In other words, the centre of mass of the jumper passes below the bar while their body bends over it, as shown in Figure 13.2.6.

FIGURE 13.2.6 The path of the centre of mass of the high jumper (shown by the dashed curve) passes below the high jump bar.

If the jumper's technique is correct, the high jumper does not have to produce enough gravitational potential energy to lift their body above the bar. Without this technique, the world records for this event would probably be much lower than their current marks.

PHYSICSFILE

Newton's universal law of gravitation

The formula $E_g = mg\Delta h$ is based on the assumption that Earth's gravitational field is constant. Newton's universal law of gravitation predicts that Earth's gravitational field will decrease with altitude. However, this decrease only becomes significant many kilometres above Earth's surface. For everyday purposes, the assumption of a constant gravitational field is valid.

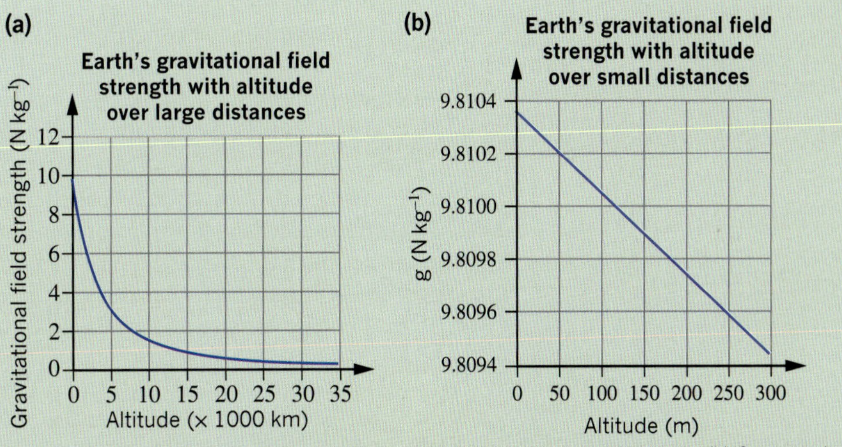

(a) Earth's gravitational field strength decreases with altitude over large distances. (b) On everyday scales, Earth's gravitational field only changes a tiny amount. If you climbed up the Eureka tower (about 300 m), the force due to gravity on you would decrease by less than 0.01%!

ELASTIC MATERIALS AND ELASTIC POTENTIAL ENERGY

The third aspect of mechanical energy under consideration here is elastic potential energy. Like gravitational potential energy, elastic potential energy occurs in situations where energy can be considered to be stored temporarily. When this energy is released, work may be done on an object.

Elastic potential energy is stored when a spring is stretched, a rubber ball is squeezed, air is compressed in a tyre, or a bungee-jumper's rope is extended during a jump. Since each object possesses energy due to its position or motion, all of these situations suit the earlier definition of mechanical energy.

Materials that have the ability to store elastic potential energy when work is done on them, and then release this energy, are called **elastic** materials. Metal springs and bouncing balls are common examples; however, many other materials are at least partially elastic. If their shape is manipulated, items such as our skin, metal hair clips and wooden rulers all have the ability to restore themselves to their original shape once released.

Materials that do not return to their original shape and release their stored potential energy as mechanical energy are referred to as plastic materials. Plasticine is an example of a very plastic material.

Ideal springs obey Hooke's law

Springs are very useful items in everyday life due to the consistent way in which many of them respond to forces and store energy. When a spring is stretched or compressed by an applied force, elastic potential energy is being stored. In order to store this energy, work must be done on the spring.

Recall that if a constant force is applied to an object, and a displacement occurs in the direction of that force, then the amount of work done can be calculated using $W = Fs$.

This formula can therefore be used when a constant force, F, has been applied to a spring and a given compression or extension, x, occurs. However, it is more interesting to examine how a spring will behave under a range of conditions.

Consider the situation in which a spring is stretched by the application of a steadily increasing force. As the force increases, the extension of the spring, x, can be graphed against the applied force, F. Well-designed springs will extend in proportion to the applied force, when the load is not too large. For example, if a 10 N force produced an extension of 6 cm, then a 20 N force would produce an extension of 12 cm. For ideal springs, the resulting graph of applied force versus extension would be linear, as shown in Figure 13.2.7.

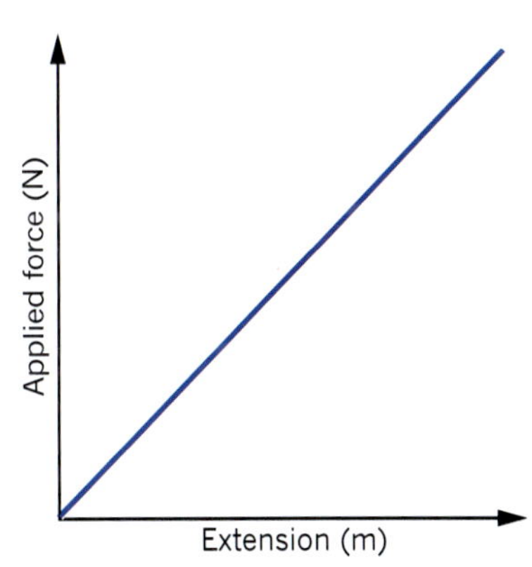

FIGURE 13.2.7 Ideal springs obey Hooke's law and so they produce a linear graph.

The gradient of a force–extension graph tells us the force, in newtons, required to produce each unit of extension in metres.

The gradient of the graph is called the **spring constant**, k, and is measured in $N m^{-1}$. The gradient indicates the stiffness of the spring. For an ideal spring this gradient has a set value (as the force–extension graph has a constant slope). A very stiff spring that is difficult to stretch would have a steep gradient; that is, a large value of k.

Although k is usually called the spring constant, it is sometimes called the stiffness constant or force constant of a spring. A spring constant of $k = 1500\,N m^{-1}$ indicates that for every metre that the spring is stretched or compressed, a force of 1500 N is required. This does not necessarily mean that the spring can be stretched by 1 m, but it indicates that the force and the change in length are in this proportion.

The relationship between the applied force and the subsequent extension or compression of an ideal spring is known as **Hooke's law**. For ideal springs, $F \propto x$ or $F = kx$.

However, when using the energy stored by stretched or compressed springs, it is appropriate to refer to the force that the distorted spring is able to exert (rather than the force that was applied to it). Newton's third law says that an extended or compressed spring in equilibrium is able to exert a restorative force equal in size but opposite in direction to the force that is being applied to it. Therefore Hooke's law is often written in the form shown below.

Hooke's law states that the force applied by a spring is directly proportional, but opposite in direction, to the spring's extension or compression. That is:

$$F = -kx$$

where F is the force applied by the ideal spring (in N)

k is the spring constant (force constant or stiffness constant) (in $N m^{-1}$)

x is the amount of extension or compression of the ideal spring (in m).

Note: the negative sign in Hooke's law indicates that the restorative force inside the spring and the extension are in opposite directions.

CALCULATING ELASTIC POTENTIAL ENERGY

Work must be done in order to store elastic potential energy, E_s, in any elastic material. Essentially, the energy is stored within the atomic bonds of the material as it is compressed or stretched. The amount of elastic potential energy stored is given by the area under the force–extension graph for the item.

For materials that obey Hooke's law (as seen in Figure 13.2.7 on page 413), an expression can be derived for the area under the force–extension graph.

$$\text{Work done} = \text{area under the } F\text{–}x \text{ graph}$$

$$W = \text{area of triangle}$$

$$= \frac{1}{2} b \times h$$

As Δx is the base, b, and F is the height, h:

$$W = \frac{1}{2} F \times x$$

But $F = kx$ so:

$$W = \frac{1}{2} kx \times x$$

$$= \frac{1}{2} kx^2$$

$$E_s = \frac{1}{2} kx^2$$

The elastic (or spring) potential energy, E_s, stored in an object is given by the area under the force–extension graph for the object. For objects that obey Hooke's law, the spring potential energy is given by:

$$E_s = \frac{1}{2}kx^2$$

where E_s is the elastic potential energy stored during the extension/compression (in J)
k is the spring constant (force constant or stiffness constant) (in N m^{-1})
x is the amount of extension or compression of the ideal spring (in m).

PHYSICSFILE

Climbing ropes

The ropes used by rock climbers have elastic properties that can save lives during climbing accidents. Ropes that were used in the nineteenth century were made of hemp, which is strong but does not stretch a lot. When climbers using these ropes fell, they stopped very abruptly. The resulting large forces acting on the climbers caused many serious injuries.

Modern ropes are made of a continuous-drawn nylon fibre core and a protective textile covering. They have a slightly lower spring constant than the older style ropes and stretch significantly (up to several metres) when stopping a falling climber. This reduces the stopping force acting on the climber. Ropes with even lower spring constants are suitable for bungee jumping. Rock climbers tend to avoid these ropes—bouncing up and down the rock face is not advisable!

Modern climbing ropes reduce the stopping force on falling climbers, thereby reducing serious injuries.

Worked example 13.2.6

CALCULATING ELASTIC POTENTIAL ENERGY

A spring with a spring constant of 75.0 N m^{-1} is stretched from its original length of 25.0 cm to 32.0 cm. Calculate the elastic potential energy stored in this spring.

Thinking	Working
Identify the variables involved and state them with their directions in their standard form.	$k = 75.0\ \text{N m}^{-1}$ $x_f = 0.320\ \text{m}$ $x_i = 0.250\ \text{m}$
Determine the extension of the spring.	$x = x_f - x_i$ $= 0.320 - 0.250$ $= 0.070\ \text{m}$
Use the equation for elastic potential energy.	$E_s = \frac{1}{2}kx^2$ $= \frac{1}{2}(75.0) \times (0.070)^2$ $= 0.184\ \text{J}$

Worked example: Try yourself 13.2.6

CALCULATING ELASTIC POTENTIAL ENERGY

A spring with a spring constant of 2050 N m^{-1} is stretched from its original length of 45.0 cm to 54.0 cm. Calculate the elastic potential energy stored in this spring.

Although many materials (at least for a small load) extend in proportion to the applied force, many materials have force–extension graphs more like that shown in Figure 13.2.8. The elastic potential energy is given by the area under the graph. To simplify this process for materials where calculating the area may be difficult, the counting squares method can be used to estimate the area.

PHYSICSFILE

Approximating the area under a graph

Sometimes a graph may have an irregular shape, and so determining the area under the curve may not be possible with a straightforward formula. Instead, the area can be approximated by using the counting squares procedure. This method involves dividing the shape up with equally spaced horizontal and vertical lines. The number of squares can be counted (including half and quarter squares) and then the total area approximated as the number of squares multiplied by the area of a single square.

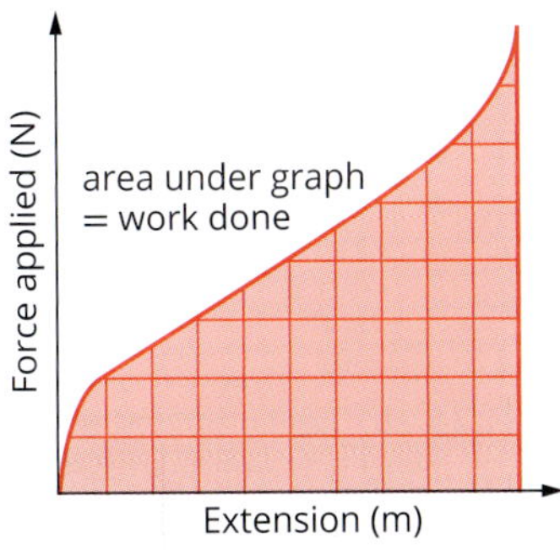

FIGURE 13.2.8 Elastic potential energy is a form of mechanical energy. The area under a graph is the elastic potential energy stored in the object.

CASE STUDY

Energy transformations

The world record for the men's pole vault is over 6 m —about as high as a single-storey house! The women's record is just over 5 metres. During the jump, a number of energy transformations take place. The athlete has kinetic energy as she runs in. This kinetic energy is used to bend the pole and carry the athlete forwards over the bar. As the pole bends, energy is stored as elastic potential energy. The athlete uses this stored energy to increase her gravitational potential energy and, hopefully, raise her centre of mass over the bar. Once the pole has been released and the bar has been cleared, the gravitational potential energy of the athlete is transformed into kinetic energy as she falls towards the mat. The energy changes are analysed by making some assumptions about the athlete and the jump.

Assume that the athlete has a mass of 60 kg and runs in at 7.0 m s^{-1}. Treat the athlete as a point mass located at her centre of mass, 1.2 m above the ground. The athlete raises her centre of mass to a height of 5.0 m as she clears the bar, and her speed at this point is just 1.0 m s^{-1}. When she plants the pole in the stop, the pole has not yet been bent and so it has no elastic potential energy. Using:

$$\Sigma E = E_k + E_g = \frac{1}{2}mv^2 + mg\Delta h$$

the vaulter's total energy at this point is 2180 J (Figure 13.2.9).

FIGURE 13.2.9 These diagrams, drawn at equal time intervals, indicate that this vaulter slows down as she nears the bar. Her initial kinetic energy is stored as elastic potential energy in the bent pole, and finally transformed into gravitational potential energy and kinetic energy, enabling her to clear the bar.

When the vaulter passes over the bar, the pole is straight again and so has no elastic potential energy. Taking the ground as zero height, and using the same relationship as above, the vaulter's total energy is now 2970 J. This does not seem consistent with the conservation of energy as there is an extra 790 J. The extra energy is from the muscles in her body. Just before the athlete plants the pole, she raises it over her head. Then, after the pole is planted but before she leaves the ground, the athlete uses her arms to bend the pole (Figure 13.2.10). She pulls downwards on the pole with one arm while the other arm pushes upwards. The effect of these forces is to do work on the pole and store some extra elastic potential energy in it. This work will be converted into gravitational potential energy later in the jump. Energy has also been put into the system by the muscles of the athlete as they do work after she has left the ground. Throughout the jump, she uses her arm muscles to raise her body higher. At the end of the jump, she is actually ahead of the pole and pushing herself up off it. In effect, she has been using the pole to push up off the ground.

FIGURE 13.2.10 As the pole is planted, the vaulter uses her arms to bend the bar. The forces are shown by the vectors. By bending the bar, the athlete has stored energy which will later be transformed into gravitational potential energy.

PA 23

13.2 Review

OA ✓✓

SUMMARY

- All moving objects have kinetic energy.
- The kinetic energy of an object is equal to the work required to accelerate the object from rest to its final velocity.
- The kinetic energy of an object is given by the equation:

$$E_k = \frac{1}{2}mv^2$$

- The work–energy theorem defines work as the change in kinetic energy:

$$W = \frac{1}{2}mv^2 - \frac{1}{2}mu^2 = \Delta E_k$$

- Gravitational potential energy is the energy an object has due to its position in a gravitational field.
- The gravitational potential energy of an object, E_g, is given by the equation:

$$E_g = mg\Delta h$$

- Gravitational potential energy is calculated relative to a zero potential energy reference level, usually the ground or sea level.
- Ideal materials extend or compress in proportion to the applied force; that is they obey Hooke's law:

$$F = -kx$$

- The elastic potential energy, E_s, stored in an object is given by the area below a force–extension graph for that object.
- The elastic potential energy, E_s, can be calculated for a spring that obeys Hooke's law using the equation:

$$E_s = \frac{1}{2}kx^2$$

KEY QUESTIONS

Knowledge and understanding

1 The total mass of a motorbike and its rider is 175 kg. If the motorbike is travelling at 68 km h^{-1}, calculate its kinetic energy.

2 A 1300 kg car is travelling at 13 m s^{-1}. How much work would its engine need to do to accelerate it to 22 m s^{-1}?

3 A cyclist has a mass of 85 kg and is riding a bicycle which has a mass of 11 kg. When riding at top speed on the bicycle, their kinetic energy is 7.0 kJ. Calculate the top speed of the cyclist to two significant figures.

4 By how much is kinetic energy increased when the mass of an object is tripled?

5 A 59 g tennis ball is thrown 13.2 m into the air. Use g = 9.8 m s^{-2}.

a Calculate the gravitational potential energy of the ball at the top of its flight.

b Calculate the gravitational potential energy of the ball when it has fallen halfway back down to Earth.

6 A 25.0 cm ideal spring has a spring constant of 550 N m^{-1}. Calculate the total length of the spring after an object is hung on it with force due to gravity of 40.0 N.

7 An ideal spring has a spring constant of 360 N m^{-1}. Calculate the elastic potential energy stored in the spring after an object that is hung on it causes the spring to be 13.00 cm longer.

Analysis

8 When climbing Mount Everest (h = 8848 m), a mountain climber stops to rest at South Base Camp (h = 5364 m). If the mountain climber has a mass of 72 kg, how much gravitational potential energy will she gain in the final section of her climb (i.e. from base camp to the summit)? For simplicity, assume that g remains at 9.8 N kg^{-1}.

9 Three springs, A, B, and C have the following spring constants: k_A = 3500 N m^{-1}, k_B = 900 N m^{-1} and k_C = 650 N m^{-1}.
Calculate the elastic potential energy stored in each spring when a force of 220 N is applied.

10 A student collects two ideal springs each with a different stiffness. She determines that spring A is stiffer than spring B. Which spring will have the steeper slope on a force–extension graph?

13.3 Using energy: power and efficiency

In many situations, energy is transformed between kinetic and gravitational potential energy. For example, when a cart travels around a rollercoaster, as shown in Figure 13.3.1, much of its kinetic energy is converted into gravitational potential energy and then back into kinetic energy again at different points on the track.

In analysing this type of situation, the concept of mechanical energy is useful. Mechanical energy is the sum of the potential energies available to an object and the kinetic energy of the object.

In situations where mechanical energy is conserved, it is possible to use this to predict the outcome of the situation. Where mechanical energy is not conserved, this can be used to help identify other important energy transformations.

When considering energy changes, the rate at which work is done is often important. For example, if two cars have the same mass, then the amount of work required to accelerate each car from a standing start to $100\,\text{km}\,\text{h}^{-1}$ will be the same. However, the fact that one car can do this more quickly than another may be an important consideration for some drivers when choosing which car to buy.

Physicists describe the rate at which work is done using the concept of power. Like work and energy, this is a word that takes on a specific meaning in a scientific context.

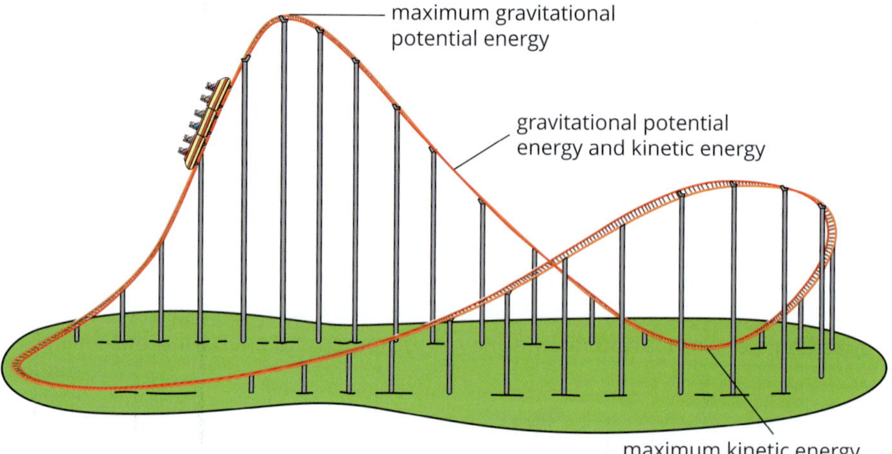

FIGURE 13.3.1 When a rollercoaster cart travels around a track, its kinetic energy and gravitational potential energy are constantly converting between one another.

MECHANICAL ENERGY

For falling objects, the mechanical energy is calculated from the sum of its kinetic and gravitational potential energies:

$$E_m = E_k + E_g = \frac{1}{2}mv^2 + mg\Delta h$$

This is a useful concept in situations where gravitational potential energy is converted into kinetic energy or vice versa. For example, consider a tennis ball with a mass of 60 g that is dropped from a height of 1.0 m (Figure 13.3.2). Initially, its total mechanical energy would comprise the kinetic energy, which would be 0 J, and the gravitational potential energy that is stored at this height (taking $g = 9.8\,\text{m}\,\text{s}^{-2}$):

$E_g = mg\Delta h = 0.060 \times 9.8 \times 1.0 = 0.59\,\text{J}$ to two significant figures.

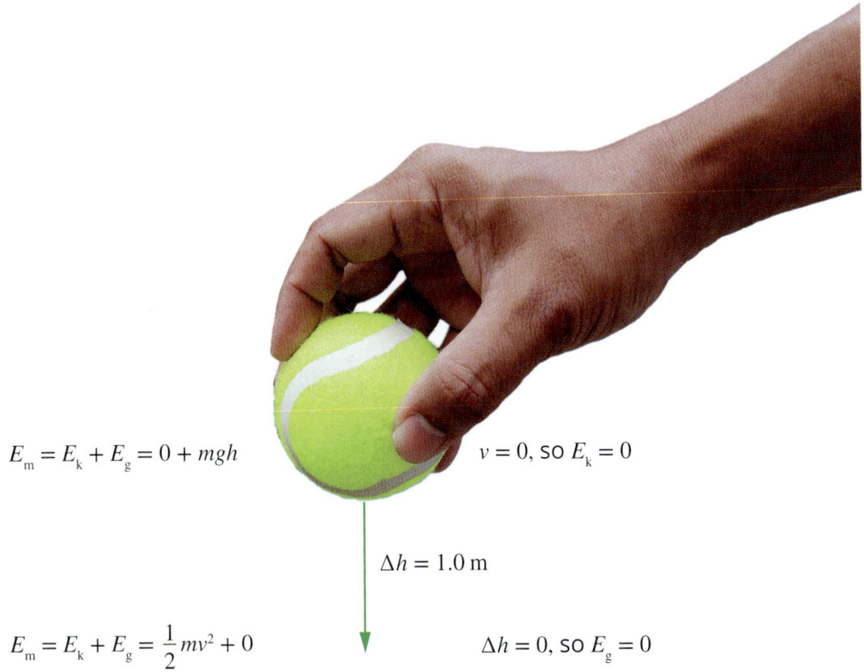

FIGURE 13.3.2 A falling tennis ball provides an example of conservation of mechanical energy.

At the instant the ball hits the ground, the total mechanical energy is the sum of the gravitational potential energy available to it and the kinetic energy just prior to hitting the ground. The gravitational potential energy is 0J because the ball is at ground level. To calculate kinetic energy, find the ball's velocity just before it hits the ground, using one of the equations of motion:

$$v^2 = u^2 + 2as$$

Since $s = -1.0\,m$, $a = -9.8\,m\,s^{-2}$ and $u = 0\,m\,s^{-1}$:

$$v^2 = u^2 + 2as$$
$$v^2 = 0^2 + 2(-9.8 \times -1.0)$$
$$v = \sqrt{19.6}$$
$$= 4.43\,m\,s^{-1}$$

Therefore the kinetic energy of the tennis ball just before it hits the ground is:

$$E_k = \frac{1}{2}mv^2 = \frac{1}{2} \times 0.060 \times 4.43^2 = 0.59\,J \text{ to two significant figures.}$$

Notice that at both the top and the bottom of the 1.0 m fall, the mechanical energy is the same. At the top:

$$E_m = E_k + E_g = 0 + 0.59 = 0.59\,J \text{ to two significant figures.}$$

At the bottom:

$$E_m = E_k + E_g = 0.59 + 0 = 0.59\,J \text{ to two significant figures.}$$

In fact, mechanical energy is constant throughout the drop. Consider the tennis ball when it has fallen halfway to the ground. At this point, $h = 0.50\,m$ and $v = 3.13\,m\,s^{-1}$:

$$E_m = E_k + E_g$$
$$= (\frac{1}{2} \times 0.060 \times 3.13^2) + (0.060 \times 9.8 \times 0.50)$$
$$= 0.294 + 0.294$$
$$= 0.59\,J \text{ to two significant figures.}$$

Notice that, at the halfway point, the mechanical energy is evenly split between kinetic energy (0.294J) and gravitational potential energy (0.294J).

Throughout the drop, the mechanical energy has been conserved.

Principle of conservation of mechanical energy

In a system of bodies, it is assumed that no other forms of energy are present except kinetic energy and potential energy. Since the total energy must be conserved, this means that the total mechanical energy of the system is constant.

Worked example 13.3.1

MECHANICAL ENERGY OF A FALLING OBJECT

A basketball with a mass of 600 g is dropped from a height of 1.2 m. Calculate its kinetic energy at the instant before it hits the ground.

Thinking	Working
Since the ball is dropped, its initial kinetic energy is zero.	$(E_k)_{initial} = 0\,J$
Calculate the initial gravitational potential energy of the ball.	$(E_g)_{initial} = mg\Delta h$ $= 0.600 \times 9.8 \times 1.2$ $= 7.1\,J$
Calculate the initial mechanical energy.	$(E_m)_{initial} = (E_k)_{initial} + (E_g)_{initial}$ $= 0 + 7.1$ $= 7.1\,J$
At the instant the ball hits the ground, its gravitational potential energy is zero.	$(E_g)_{final} = 0\,J$
Mechanical energy is conserved in this situation.	$(E_m)_{initial} = (E_m)_{final} = (E_k)_{final} + (E_g)_{final}$ $7.1 = (E_k)_{final} + 0$ $\therefore (E_k)_{final} = 7.1\,J$

Worked example: Try yourself 13.3.1

MECHANICAL ENERGY OF A FALLING OBJECT

A 6.8 kg bowling ball is dropped from a height of 0.75 m. Calculate its kinetic energy as it hits the ground.

USING MECHANICAL ENERGY TO CALCULATE VELOCITY

The speed of a falling object does not depend on its mass. This can be demonstrated using mechanical energy.

Consider an object with a mass m dropped from a height h. At the moment it is dropped, its initial kinetic energy is zero. At the moment before it hits the ground, its final gravitational potential energy is zero. Therefore, using the **conservation of mechanical energy**:

$$(E_m)_{initial} = (E_m)_{final}$$

$$(E_k)_{initial} + (E_g)_{initial} = (E_k)_{final} + (E_g)_{final}$$

$$0 + mg\Delta h = \frac{1}{2}mv^2 + 0$$

$$mg\Delta h = \frac{1}{2}mv^2$$

$$g\Delta h = \frac{1}{2}v^2$$

$$v^2 = 2g\Delta h$$

$$\therefore v = \sqrt{2g\Delta h}$$

This formula can be used to find the velocity of a falling object as it hits the ground. Note that the formula does not contain the mass of the falling object, so if air resistance is negligible, all objects with any mass will have the same final velocity when they are dropped from the same height.

PHYSICSFILE

Mechanical energy of an object falling through the air

In reality, as an object moves through the air, a very small amount of its energy is transformed into heat and sound. This means that a falling object won't quite reach the speeds that you calculate before it hits the ground. As a result, mechanical energy is not entirely conserved. However, this small effect can be considered negligible for many falling objects.

Worked example 13.3.2

FINAL VELOCITY OF A FALLING OBJECT

A basketball with a mass of 600 g is dropped from a height of 1.2 m. Calculate the speed of the basketball at the instant before it hits the ground.

Thinking	Working
Recall the formula for the velocity of the falling object.	$v = \sqrt{2g\Delta h}$
Substitute the relevant values into the formula and solve.	$v = \sqrt{2 \times 9.8 \times 1.2}$ $= 4.8\,\text{m s}^{-1}$
Interpret the answer.	The basketball will be falling at $4.8\,\text{m s}^{-1}$ just before it hits the ground.

Worked example: Try yourself 13.3.2

FINAL VELOCITY OF A FALLING OBJECT

A 6.8 kg bowling ball is dropped from a height of 0.75 m. Calculate the speed of the bowling ball just before it hits the ground.

USING CONSERVATION OF MECHANICAL ENERGY IN COMPLEX SITUATIONS

The concept of mechanical energy allows physicists to determine outcomes in non-linear situations for which the equations of linear motion cannot be used. For example, consider a pendulum with a bob of mass 400 g displaced from its mean position such that its height has increased by 20 cm, as shown in Figure 13.3.3.

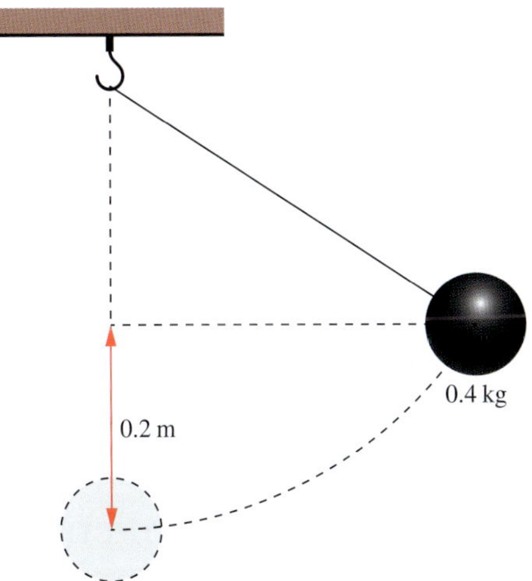

FIGURE 13.3.3 A falling pendulum provides an example of conservation of mechanical energy.

Since a falling pendulum involves gravitational potential energy being converted into kinetic energy, the conservation of mechanical energy applies to this situation. Therefore, the formula developed earlier for the velocity of a falling object can be used to find the velocity of the pendulum bob at its lowest point.

$$v = \sqrt{2g\Delta h} = \sqrt{2 \times 9.8 \times 0.2} = 2\,\text{m s}^{-1}$$

The speed of the pendulum bob will be $2\,\text{m s}^{-1}$ at its lowest point. However, unlike the falling tennis ball, the direction of the bob's motion will be horizontal instead of vertical at its lowest point. The equations of motion relate to linear motion and cannot be applied to this situation because the bob swings in a curved path.

PHYSICSFILE

Deriving a formula for velocity from the equations of linear motion

The formula for the velocity of a falling object can also be derived from the equations of linear motion (see Chapter 11). Consider an object dropped from a height, h, with an initial speed of $u = 0\,\text{m s}^{-1}$. Using the formula $v^2 = u^2 + 2as$:

$$\begin{aligned} v^2 &= u^2 + 2as \\ &= 0^2 + 2g\Delta h \\ &= 2g\Delta h \\ \therefore v &= \sqrt{2g\Delta h} \end{aligned}$$

This formula is equivalent to the result achieved using the conservation of mechanical energy.

Conservation of energy can also be used to analyse projectile motion—that is, the motion of an object thrown or fired into the air with some initial velocity. Energy is not a vector, so no vector analysis is required, even if the initial velocity is at an angle to the ground.

Worked example 13.3.3

USING MECHANICAL ENERGY TO ANALYSE PROJECTILE MOTION

A cricket ball (m = 140 g) is thrown upwards into the air at a speed of 15 m s^{-1}. Calculate the speed of the ball when it has reached a height of 8.0 m. Assume that the ball is thrown from a height of 1.5 m.

Thinking	Working
Recall the formula for mechanical energy	$E_m = E_k + E_g = \frac{1}{2}mv^2 + mg\Delta h$
Substitute in the values for the ball as it is thrown.	$(E_m)_{initial} = (E_k)_{initial} + (E_g)_{initial}$ $= \frac{1}{2}mv^2 + mg\Delta h$ $= \frac{1}{2}(0.14 \times 15^2) + (0.14 \times 9.8 \times 1.5)$ $= 18\text{ J}$
Use conservation of mechanical energy to find an equation for the final speed.	$(E_m)_{final} = (E_k)_{final} + (E_g)_{final}$ $= \frac{1}{2}mv^2 + mg\Delta h$ $18 = \frac{1}{2}(0.14)v^2 + (0.14 \times 9.8 \times 8.0)$
Solve the equation algebraically to find the final speed.	$18 = 0.07v^2 + 11$ $7 = 0.07v^2$ $v^2 = \frac{7}{0.07}$ $= 100$ $v = \sqrt{100}$ $= 10\text{ m s}^{-1}$
Interpret the answer.	The cricket ball will be moving at 10 m s^{-1} when it reaches a height of 8.0 m.

Worked example: Try yourself 13.3.3

USING MECHANICAL ENERGY TO ANALYSE PROJECTILE MOTION

An arrow with a mass of 35 g is fired into the air at 80 m s^{-1} from a height of 1.4 m. Calculate the speed of the arrow when it has reached a height of 30 m.

CASE STUDY

Ballistics pendulum

The ballistics pendulum is an example of how the law of conservation of mechanical energy can be combined with an understanding of collisions to solve a practical problem. A ballistics pendulum is a device that can be used to measure the speed of a bullet fired from a gun or rifle. It consists of a block of wood hanging at a convenient height above the ground, as shown in Figure 13.3.4.

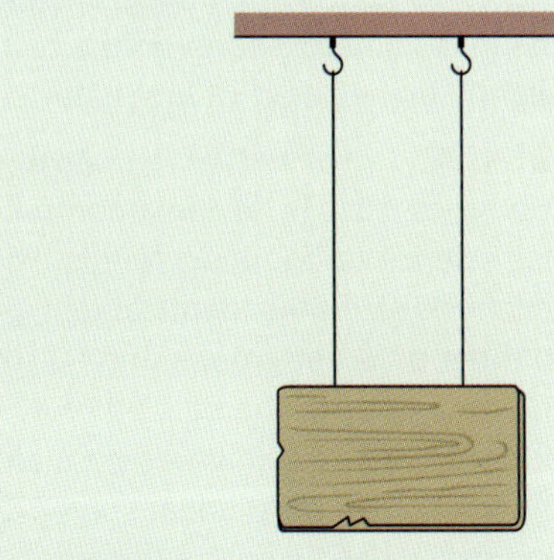

FIGURE 13.3.4 A ballistics pendulum combines an understanding of collisions and mechanical energy.

When a bullet is fired into the wooden block, an inelastic collision occurs. This means that much of the bullet's kinetic energy is converted into heat and sound and into changes made to the shape of the block. The conservation of mechanical energy does not apply for the impact of the bullet with the block.

However, the law of conservation of momentum still applies to the impact. This means that the block gains velocity from the bullet and it swings backwards and upwards as shown in Figure 13.3.5. By measuring the change in height of the block and the masses of the bullet and block, the initial speed of the bullet can be calculated. Note that conservation of mechanical energy does occur when the block swings backwards and upwards as no energy is converted into sound or heat during this part of its motion.

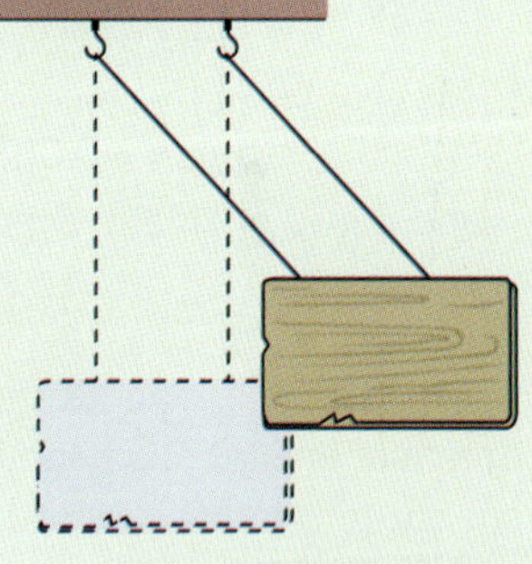

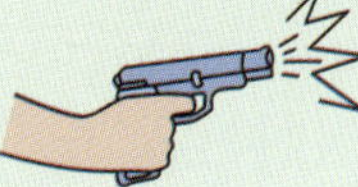

FIGURE 13.3.5 The change in height of a ballistics pendulum can be used to calculate the speed of the bullet fired into it.

LOSS OF MECHANICAL ENERGY

Mechanical energy is not conserved in every situation. For example, when a tennis ball bounces a number of times, each bounce is lower than the one before it, as shown in Figure 13.3.6.

Although mechanical energy is largely conserved as the ball moves through the air, a significant amount of kinetic energy is transformed into heat and sound when the ball compresses and decompresses as it bounces. This means that the ball does not have as much kinetic energy when it leaves the ground as it did when it landed. Therefore, the gravitational potential energy it can achieve on the second bounce will be less than the gravitational potential energy it had initially, and so the second bounce is lower.

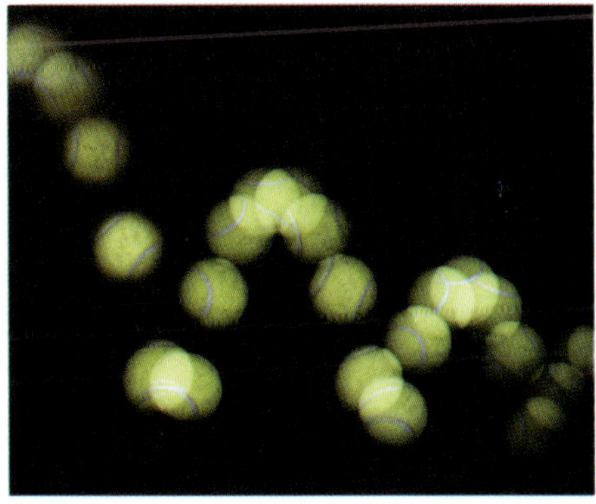

FIGURE 13.3.6 Mechanical energy is lost with each bounce of a tennis ball.

CASE STUDY ANALYSIS

Elastic potential energy

A bouncing ball involves forms of energy other than kinetic and gravitational potential energy. When the ball hits the ground, its gravitational potential energy is converted into elastic potential energy as it compresses. As the ball expands back to its original shape, some of the elastic potential energy is converted back into kinetic energy and some of it is converted into heat and sound. The amount of energy that is converted into heat and sound depends on the type of ball.

If you want the ball to reach a greater height than its original height, then instead of dropping the ball you could add to its energy by throwing it downwards with some velocity. Consider Figure 13.3.7(a) in which a tennis ball (m = 58 g) is thrown downwards at 4.0 m s^{-1} from a height of 1.0 m.

Initially, the ball has 0.57 J of gravitational potential energy:

$$mg\Delta h = 0.058 \times 9.8 \times 1.0 = 0.57\,\text{J}$$

and 0.46 J of kinetic energy:

$$\frac{1}{2}mv^2 = \frac{1}{2}(0.058 \times 4.0^2) = 0.46\,\text{J}$$

By the time it reaches the ground, the gravitational potential energy has been transformed into kinetic energy, giving it a total of 1.03 J of kinetic energy (Figure 13.3.7(b)). This is converted into elastic potential energy of 1.03 J (Figure 13.3.7(c)).

If 0.28 J of energy is lost as heat and sound as the ball expands, then the ball will have just 0.75 J of kinetic energy when it leaves the ground (Figure 13.3.7(d)). This means that it will rebound to a height of 1.3 m (Figure 13.3.7(e)):

$$\Delta h = \frac{E_g}{mg} = \frac{0.75}{0.058 \times 9.8} = 1.3\,\text{m}$$

Even though some energy has been 'lost' in the bounce, the initial kinetic energy of the ball means that it ends up slightly higher than where it started.

Analysis

A physics class is having a competition to see who can use science to bounce a tennis ball the highest. The teacher distributes identical tennis balls (with m = 58.0 g) to three different groups and tells them they can study the properties of the ball, but they only get one attempt to bounce the ball as high as possible. First, the class treats the tennis ball as an ideal spring when it compresses, and calculates its spring constant, k. Then, the teacher provides the students with a graph she prepared earlier showing the relationship between compression of the ball and energy lost (E_L) during the bounce to heat, sound and deformation. The graph is shown in Figure 13.3.8.

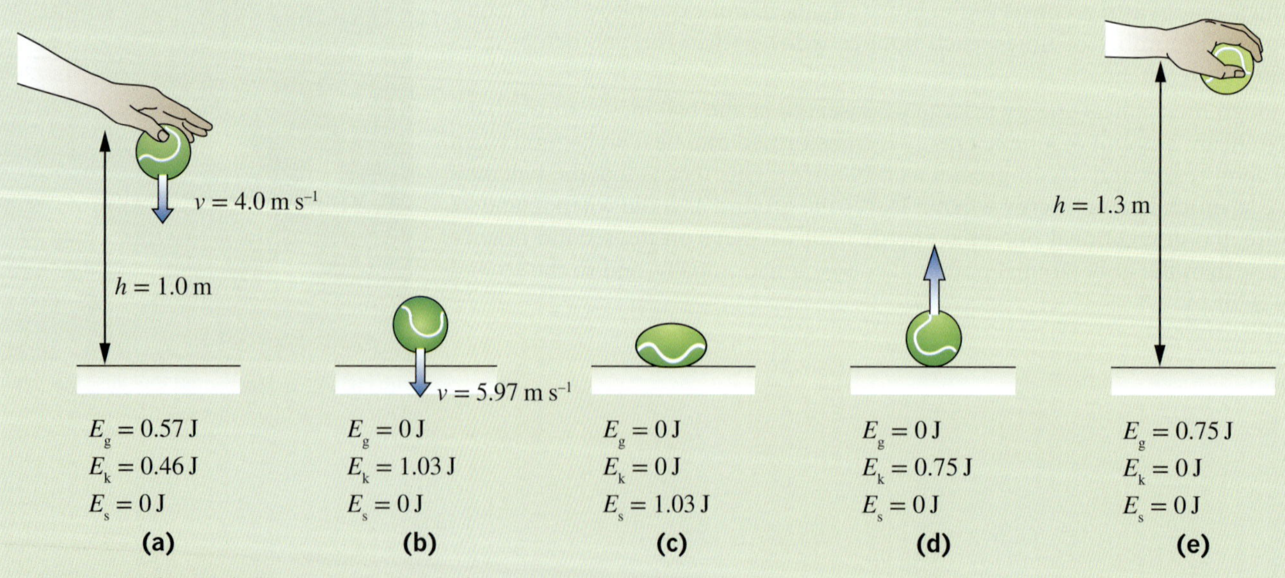

FIGURE 13.3.7 A tennis ball is thrown downwards from a height.

CASE STUDY ANALYSIS

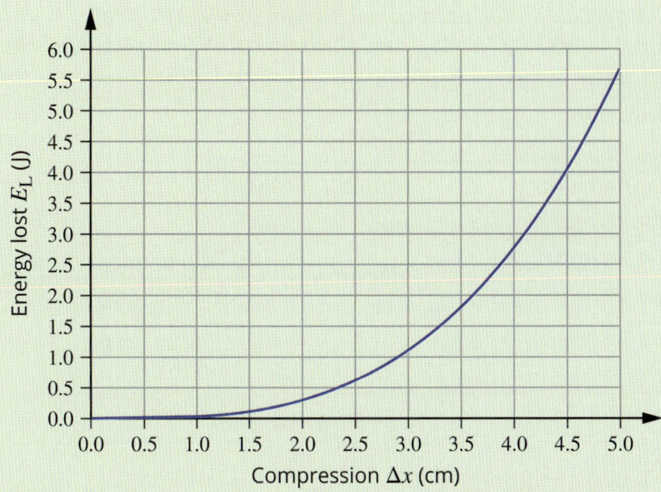

FIGURE 13.3.8 Compression of the ball and energy lost (E_L) due to heat, sound and deformation

Finally, the groups prepare their strategies. Group 1 students decide to drop their tennis ball from the fourth storey, at a height of 11.4 m. Group 2 students throw their ball down from a height of 6.00 m at a speed of 3.00 m s^{-1}. Meanwhile, the third group of students stay at ground level and throw their ball down at 5.85 m s^{-1} from a release point of 1.10 m above the ground.

1 Using the data from Figure 13.3.7, calculate the spring constant for the tennis ball, if the ball compresses by 2.00 cm in this action.
2 Discuss how the teacher may have experimentally determined the graph relating energy lost to the amount by which the ball is compressed.
3 Determine the theoretical return height of each group's tennis ball, neglecting the energy lost in the real process.
4 Estimate the real return height of each group's tennis ball, neglecting the effects of air resistance. Which group do you think will win?

EFFICIENCY OF ENERGY TRANSFORMATIONS

In the real world, energy transformations are never perfect—there is always some energy 'lost'. Because of this, for a system to continue operating (doing work), it must be constantly provided with energy. The proportion of energy that is effectively transformed by a device is called the **efficiency** of that device. This can be expressed as a decimal or a percentage. A device operating at 45% efficiency is converting 45% of its supplied energy into the useful new form. The other 55% is 'lost' or transferred to the surroundings, usually as heat and/or sound. It is not truly lost, since energy cannot be created or destroyed; rather, the form it becomes (heat and sound) is not useful.

The efficiency of a transformation from one energy form to another, as a decimal, is:

$$\text{Efficiency } (\eta) = \frac{\text{useful energy out}}{\text{useful energy in}}$$

Table 13.3.1 lists the approximate efficiencies of some common devices.

TABLE 13.3.1 Efficiencies of some common devices

Device	Energy transformation	Efficiency (%)
electric motor	electric to kinetic	90
gas heater	chemical to thermal	75
incandescent light globe	electric to light	10
compact fluorescent light	electric to light	85
LED household light	electric to light	95
steam turbine	thermal to kinetic	45
coal-fired generator	chemical to electrical	30
high-efficiency solar cell	radiation to electrical	35
car engine	chemical to kinetic	25
open fireplace	chemical to thermal	15
human body	chemical to kinetic	25

To express the efficiency as a percentage, multiply this by 100:

$$\text{Efficiency } (\eta) = \frac{\text{useful energy out}}{\text{useful energy in}} \times 100$$

Worked example 13.3.4

ENERGY EFFICIENCY

The energy input of a particular gas-fired power station is 1100 MJ. The electrical energy output is 300 MJ. What is the efficiency of the power station in achieving this energy transformation expressed as a percentage?

Thinking	Working
Recall the equation for efficiency. Substitute the given values into the equation.	output = 300 MJ input = 1100 MJ $\text{efficiency } (\eta) = \frac{\text{useful energy out}}{\text{useful energy in}} \times 100$ $= \frac{300\text{ MJ}}{1100\text{ MJ}} \times 100$
Solve the equation.	efficiency = 27%

Worked example: Try yourself 13.3.4

ENERGY EFFICIENCY

An electric kettle uses 23.3 kJ of electrical energy as it boils a quantity of water. The efficiency of the kettle is 18%. How much electrical energy is expended in actually boiling the water?

PHYSICSFILE

Coefficient of restitution

The 'bounce of the ball' is an important factor in many sports. Physicists describe the 'bounciness' of balls using a concept known as the coefficient of restitution (COR). The COR depends on both the ball and the surface it is bouncing on. A tennis ball bouncing on grass has a different COR than one bouncing on clay. This is one reason why some tennis players prefer to play on some surfaces rather than others.

DEFINING POWER

Power is a measure of the rate at which work is done. Mathematically:

$$P = \frac{W}{\Delta t}$$

Recall that when work is done, energy is transferred or transformed. So the equation can also be written as:

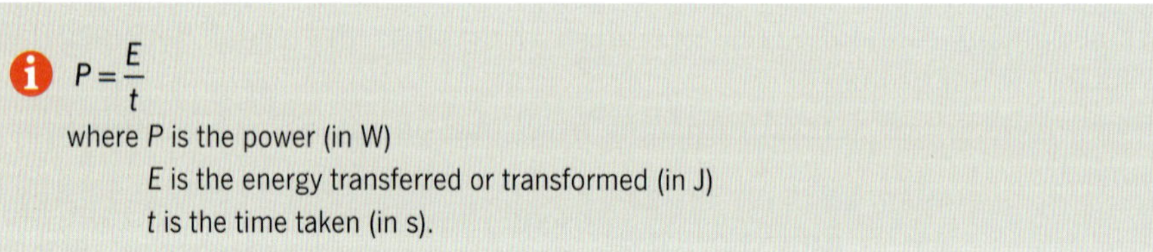

$$P = \frac{E}{t}$$

where P is the power (in W)

E is the energy transferred or transformed (in J)

t is the time taken (in s).

For example, a person running up a set of stairs does exactly the same amount of work as if they had walked up the stairs (i.e. $W = mgh$); however, the rate of energy change is faster for running up the stairs. Therefore, the runner is applying more power than the walker (Figure 13.3.9).

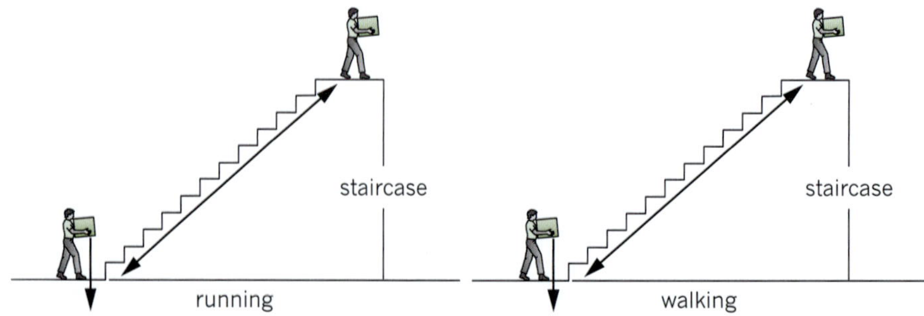

FIGURE 13.3.9 The runner and the walker both do the same amount of work, but the power output of the runner is higher than that of the walker.

Unit of power

The unit of power is named after the Scottish engineer James Watt, who is most famous for inventing the steam engine. A watt (W) is defined as a rate of work of one joule per second:

$$1\,\text{W} = \frac{1\,\text{J}}{1\,\text{s}} = 1\,\text{J s}^{-1}$$

Worked example 13.3.5

CALCULATING POWER

Calculate the power required to carry a box with a mass of 2 kg up a 5 m staircase in 20 s. (Use $g = 9.8\,\text{m s}^{-2}$.)

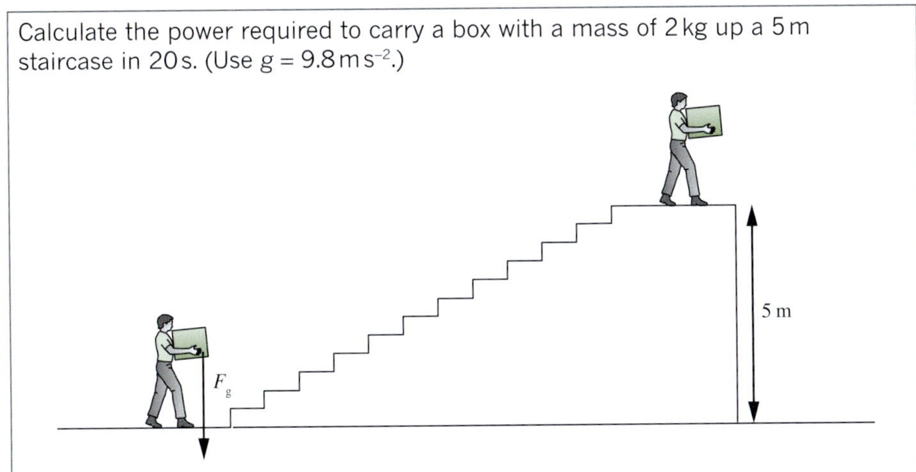

Thinking	Working
Calculate the force applied.	$F_g = mg$ $= 2 \times 9.8$ $= 19.6\,\text{N}$
Calculate the work done.	$W = Fs$ $= 19.6 \times 5$ $= 98\,\text{J}$
Recall the formula for power.	$P = \frac{W}{\Delta t}$
Substitute the appropriate values into the formula.	$P = \frac{98}{20}$
Solve.	$P = 4.9\,\text{W}$

Worked example: Try yourself 13.3.5

CALCULATING POWER

Calculate the power used by a weightlifter to lift a barbell of mass 50 kg from the floor to a height of 2.0 m above the ground in 1.4 s. (Use $g = 9.8\,\text{m s}^{-2}$.)

PHYSICSFILE

Horsepower

James Watt was a Scottish inventor and engineer. He developed the concept of horsepower as a way to compare the output of steam engines with that of horses, which were the other major source of mechanical energy available at the time. Although the unit of one horsepower (1 hp) has had various definitions over time, the most commonly accepted value today is around 750 W. This is actually significantly higher than an average horse can sustain over an extended period of time.

POWER, FORCE AND AVERAGE SPEED

In many everyday situations, a force is applied to an object to keep it moving at a constant speed; for example, pushing a wardrobe across a carpeted floor or driving a car at a constant speed. In these situations, the power being applied can be calculated directly from the force applied and the speed of the object.

Since $P = \frac{W}{\Delta t}$ and $W = Fs$, then:

$$P = \frac{Fs}{\Delta t} = F \times \frac{s}{\Delta t}$$

Since $\frac{s}{\Delta t}$ is the definition of average speed, the power equation can be written as:

$$P = Fv_{av}$$

Worked example 13.3.6

FORCE–VELOCITY FORMULATION OF POWER

A person pushes a heavy box along the ground at an average speed of $1.5\,m\,s^{-1}$ by applying a force of 40 N. What amount of power does the person exert on the box?

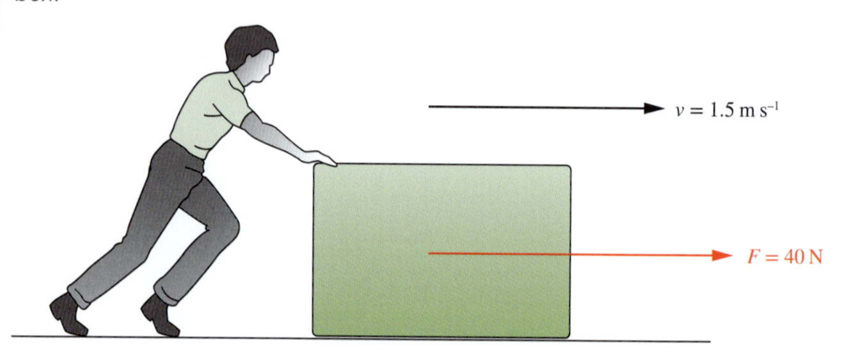

Thinking	Working
Recall the force–velocity formulation of the power equation.	$P = Fv_{av}$
Substitute the appropriate values into the formula.	$P = 40 \times 1.5$
Solve.	$P = 60\,W$

Worked example: Try yourself 13.3.6

FORCE–VELOCITY FORMULATION OF POWER

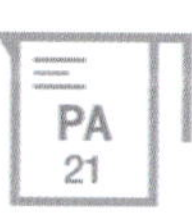

Calculate the power required to keep a car moving at an average speed of $22.0\,m\,s^{-1}$ if the force of friction (including air resistance) is 1.200 kN. Give your answer correct to three significant figures.

13.3 Review

OA ✓✓

SUMMARY

- Mechanical energy is the sum of the potential and kinetic energies of an object.
- Mechanical energy is conserved in a falling object.
- Conservation of mechanical energy can be used to predict outcomes in a range of situations involving gravity and motion.
- The final velocity of an object falling from height h can be found using the equation $v = \sqrt{2g\Delta h}$.
- When a ball bounces, some mechanical energy is transformed into heat and sound. Some energy is also lost in deformation.
- Power is a measure of the rate at which work is done: $P = \frac{W}{\Delta t} = \frac{\Delta E}{\Delta t}$.
- The power required to keep an object moving at a constant speed can be calculated from the product of the force applied and its average speed: $P = Fv_{av}$.

The efficiency of an energy transformation from one form to another (as a percentage) is:

$$\text{Efficiency } (\eta) = \frac{\text{useful energy out}}{\text{useful energy in}} \times 100$$

KEY QUESTIONS

Knowledge and understanding

1 A piano with a mass of 270 kg is pushed off the roof of a six-storey apartment block. The piano falls 3.0 m for each storey (i.e. a total of 18 m).

   a Calculate the piano's kinetic energy as it hits the ground.

   b Calculate the piano's kinetic energy as it passes the windows on the second floor, having fallen 13 m.

2 A 46 g golf ball is dropped from the roof of a six-storey apartment block. The golf ball falls 3.0 m for each storey (i.e. a total of 18 m).

   a Calculate the speed of the golf ball as it hits the ground.

   b Calculate the speed of the golf ball as it passes the windows on the second floor, having fallen 13 m.

3 A branch falls from a tree and hits the ground with a speed of 6.2 m s^{-1}. From what height did the branch fall?

4 A javelin with a mass of 620 g is thrown at an angle of inclination of 55°. It is released at a height of 1.75 m with a speed of 31.5 m s^{-1}.

   a Calculate the javelin's initial mechanical energy.

   b Calculate the speed of the javelin as it hits the ground.

5 A coal-fired generator has an efficiency of approximately 30%. If 11 kJ of energy is supplied to the generator, how much is converted into electrical energy?

Analysis

6 A rubber ball is dropped from a height of 2.5 m and loses 30% of its mechanical energy as it hits the ground. To what height will it rebound?

7 A 1210 kg car accelerates from zero to 100 km h^{-1} in 6.20 s. Calculate its average power output over this time.

8 A locomotive engine applies a force of 6000 N to keep a train moving at 33.0 m s^{-1}. Calculate the power output of the engine.

9 A 2100 kg car's engine uses 50 kW of power to maintain a constant speed of 90 km h^{-1}. Calculate the force being applied by its engine.

10 The motor of a crane has a maximum power output of 40 kW. At what average speed could it lift a concrete slab with a mass of 300 kg?

Chapter review

KEY TERMS

conservation of mechanical energy
efficiency
elastic
elastic potential energy
gravitational potential energy
Hooke's law
kinetic energy
mechanical energy
potential energy
power
spring constant
work
work–energy theorem

REVIEW QUESTIONS

Knowledge and understanding

1 State the formula used to calculate the gravitational potential energy of an object with mass *m* that is *h* metres above the chosen zero point.

2 State the name given to the combined kinetic and gravitational potential energy of an object.

3 The speed of an object is tripled. State the magnitude of the increase in its kinetic energy.

4 Describe the energy transformations that occur when a pendulum swings back and forth, and explain how this relates to the conservation of mechanical energy.

5 A car drives at a constant speed for 120 m. To overcome friction, its engine applies a force of 1850 N. Calculate the work done by the engine.

6 Estimate the total work shown in the following force–displacement graph, given that each grid square corresponds to 2.5 J.

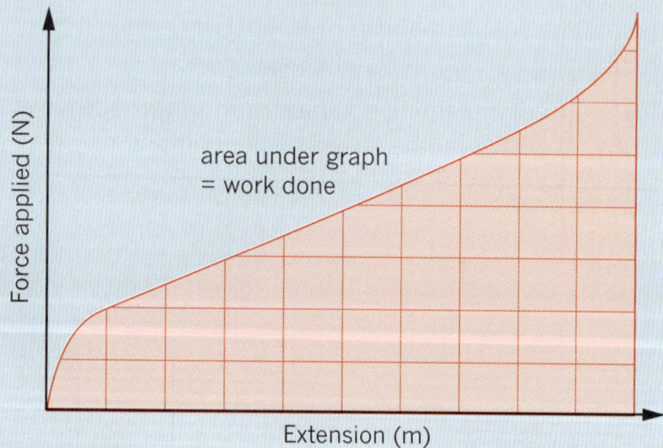

7 A crane lifts a 180 kg load from the ground to a height of 23 m. What is the work done by the crane?

8 A person walks up a flight of 17 stairs. Each step is 270 mm long and 225 mm high. If the person has a mass of 72 kg and $g = 9.8\,\text{m s}^{-2}$, what is the total amount of work done against gravity?

9 If 4000 J is used to lift a 50.0 kg object with a constant velocity, what is the theoretical maximum height to which the object can be raised?

10 A pram is pushed by the handle, which is at an angle of 35.0° to the horizontal (the direction of motion). If 1200 J of work is done pushing the pram 20.0 m, with what force was the pram pushed?

11 A cricket ball with a mass of 160 g is bowled with a speed of $161\,\text{km h}^{-1}$. What is the kinetic energy of the cricket ball?

12 If a 1700 kg car has kinetic energy of 90 kJ, what is its speed?

13 A plumber (mass 95 kg) digs a ditch 24 cm deep. By how much does the plumber's gravitational potential energy change when he steps from ground level down into the ditch?

Application and analysis

14 The force–extension graph for three different springs is shown below.

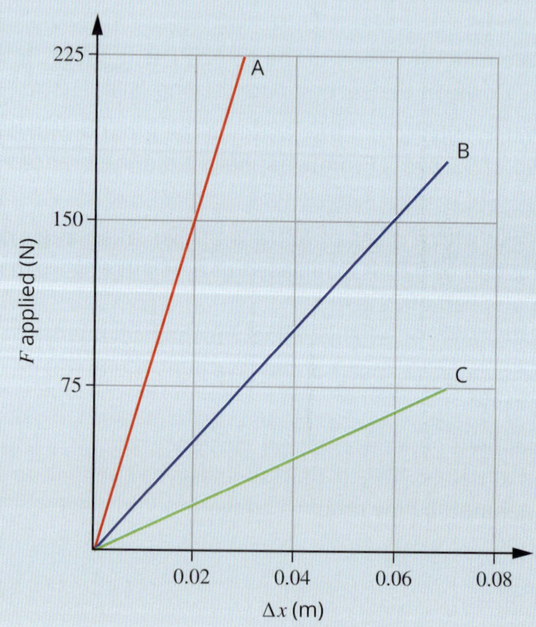

Calculate the approximate spring constant for each spring.

15 A football with a mass of 0.56 kg is kicked off the ground with a speed of 22 m s^{-1}. How fast will it be going when it hits the crossbar, which is 2.44 m above the ground?

16 A bullet of mass 12 g strikes a ballistics pendulum of mass 3.5 kg with speed v and becomes embedded in the pendulum. The block gains velocity from the bullet and swings backwards and upwards so that its height increases by 25 cm. For the questions that follow, assume that the initial gravitational potential energy of the pendulum was zero.

a What was the gravitational potential energy of the pendulum at the top of its swing?

b What was the kinetic energy of the pendulum when the bullet first embedded in it?

c What was the speed of the pendulum when it started to swing?

17 A crane can lift a load of 6.5 tonnes vertically through a distance of 30 m in 8.0 s. What is the power of the crane approximately?

18 A red Mini Cooper with a mass of 725 kg can accelerate from 0 to 100 km h^{-1} in 6.7 s. What is the average power output of the car over this time?

19 If the engine of a 1250 kg car uses 32 kW to maintain an average speed of 14 m s^{-1}, how much friction is acting on the car?

20 At the start of a 100 m race, a runner with a mass of 72 kg accelerates from a standing start to 9.0 m s^{-1} in a distance of 18 m.

a Calculate the work done by the runner's legs.

b Calculate the average force that the runner's legs apply over this distance.

21 When moving around on the Moon, astronauts find it easier to use a series of small jumps rather than to walk. If an astronaut (with a mass of 135 kg including equipment) jumps to a height of 50 cm on the Moon, where the gravitational field strength is 1.6 m s^{-2}, by roughly how much does his potential energy increase?

22 The efficiency of an appliance is known to be 65%. What energy was supplied if the output was 1700 J?

OA ✓✓

CHAPTER 14 Forces and equilibrium

In the design of buildings and other structures, engineers and architects must use their knowledge of physics to determine the forces that act within the structures they create. They must make sure that those forces are balanced so that the structure is stable.

This chapter will cover the concept of equilibrium, which describes the situation in which forces and torques are balanced. If there is equilibrium of translational (linear) forces, then there will be no net translational forces, and an object will not begin to move. If there is equilibrium of torque, then the object will not rotate.

Key knowledge

- calculate torque, $\tau = r_{\perp}F$ **14.1**
- analyse translational and rotational forces (torques) in simple structures in translational and rotational equilibrium. **14.2, 14.3**

OA ✓✓

14.1 Torque

FIGURE 14.1.1 Applying a torque to a steering wheel will cause it to turn.

Many situations involve objects that rotate about a pivot point; for example, closing a door, using a spanner, or turning a steering wheel. In these situations, a force acts to provide a turning effect or a **torque** (τ). Newton's laws use the concept of force to help understand changes in the linear (straight line) motion of an object. The concept of torque is used in exactly the same way to explain a change in the rotational (turning) motion of an object.

Consider the steering wheel in Figure 14.1.1. When a turning effect is applied to the steering wheel, a number of factors must work together to cause it to turn. For all turning objects, there must be a **pivot point** around which the object will rotate. There must be a force (F) applied to the object in such a way as to cause the object to rotate. This means that the force applied must not be aligned with the pivot point; for example, you can't turn a bolt just by pulling outwards. There must be some distance between the **line of action of the force** (an imaginary line through the force vector) and the pivot point.

PHYSICSFILE

Torque and moments

The terms 'torque' (τ) and 'moment of a force', which is usually shortened to 'moment', are terms that are used interchangeably in Physics at the high-school level.

The difference between the terms is in how they are used. Typically, torque is used for dynamic problems in which there is an angular acceleration; this means the speed at which an object is rotating is changing. This angular acceleration is caused by a net torque with a non-zero value. Moment is typically used in static problems in which there is no rotation. In these cases there is no rotation because the moments, which are often created by reaction forces or internal forces in an object, are balanced so that there is no angular acceleration.

For example, the term 'torque' might be used to describe the effect of the force applied by an axle to a car's wheel to increase its speed of rotation. The term 'moment' might be used to describe the effect of the reaction forces acting on a diving board. When you stand on the tip of a diving board, the reaction forces at the other end must not only push upwards (to counteract your force due to gravity) but they must create a moment to prevent the diving board from rotating.

Both torque and moment are calculated using the same equation. For the purposes of this chapter we can consider the two terms as interchangeable.

FORCE AND THE PIVOT POINT

FIGURE 14.1.2 The axis of rotation

When analysing a rotating system, the position of the pivot point or **axis of rotation** is an important consideration. A wheel, for example, moves in a circular path around its axle. An imaginary line along the length of the axle is called the axis of rotation and is shown in Figure 14.1.2.

The pivot point is the point on a two-dimensional representation of the object through which the axis of rotation passes. As an example, the pivot point of a wheel is shown in Figure 14.1.3.

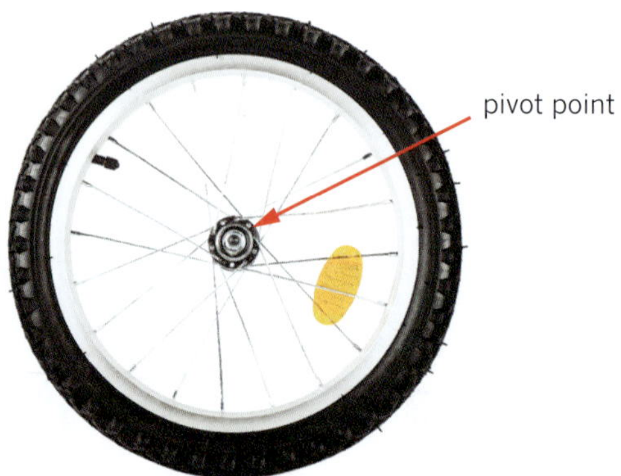

FIGURE 14.1.3 The pivot point

A force applied directly towards or directly away from the pivot point of the wheel will not create a turning effect on the wheel. So, for the example in Figure 14.1.4, if the force acted along the line labelled 'line of action', the wheel would not turn.

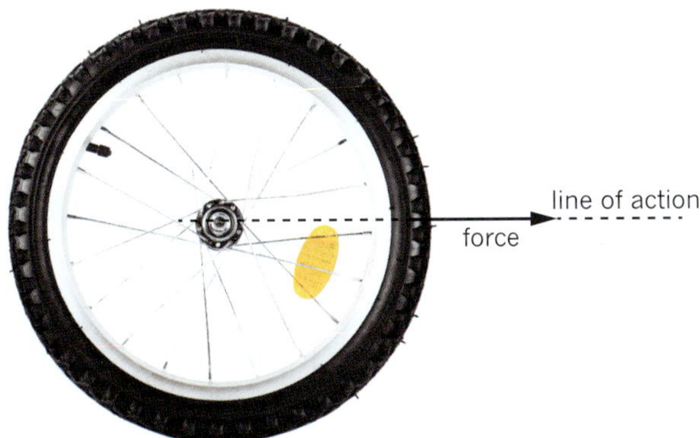

FIGURE 14.1.4 When the line of action of the force passes through the pivot point, the wheel will not turn.

Torque can be achieved by applying a force on the wheel such that the line of action of the force does not pass through the axis of rotation or the pivot point. The maximum torque is achieved when the force applied is at 90° to a line drawn from the pivot point to the point of application (the point at which the force is applied). This is shown in Figure 14.1.5.

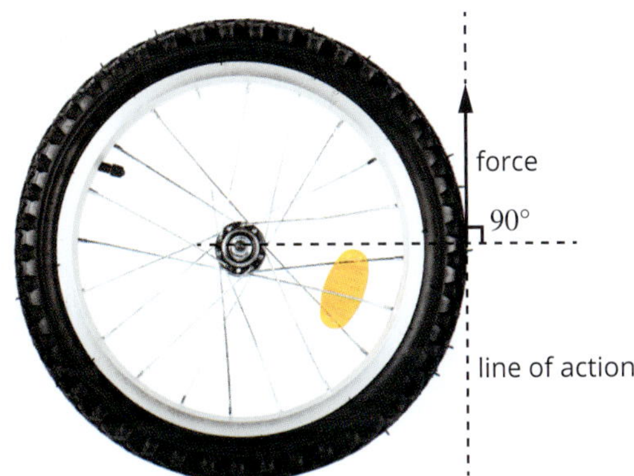

FIGURE 14.1.5 Maximum torque occurs when the force applied is perpendicular (at 90°) to a line drawn from the pivot point to the point of application.

MAGNITUDE OF THE FORCE AND TORQUE

The torque (τ) on an object is directly proportional to the magnitude of the force (F). If all other things are equal, a larger force will result in a larger torque. This is illustrated in Figure 14.1.6.

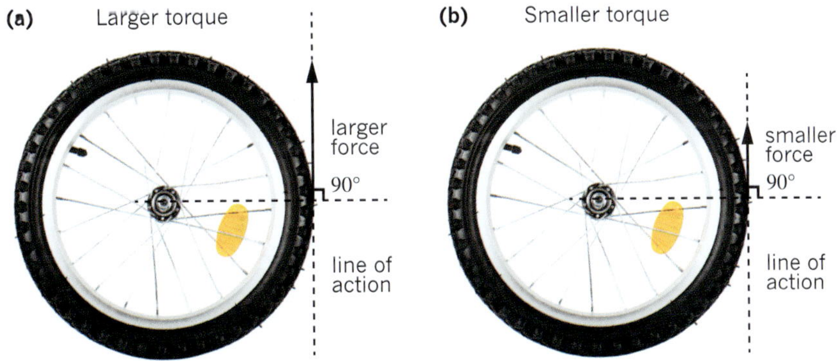

FIGURE 14.1.6 The magnitude of the force affects the torque on an object. The wheel in (a) will experience a larger torque than the wheel in (b).

DISTANCE FROM THE PIVOT POINT AND TORQUE

The amount of torque created is directly proportional to the perpendicular distance between the pivot point and the line of action of the force. This perpendicular distance is called the **force arm**. The force arm is given the symbol $r_\perp$ and is shown in Figure 14.1.7. Given that everything else is constant, then the larger the force arm or perpendicular distance ($r_\perp$), the larger the torque (τ).

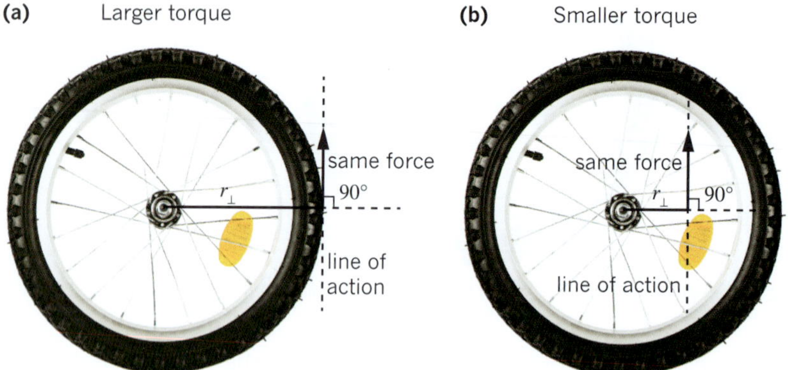

FIGURE 14.1.7 The perpendicular distance from the pivot point to the line of action of the force affects the torque on an object. The wheel in (a) will experience a larger torque than the wheel in (b).

As stated earlier, maximum torque occurs when the line of action of the force is perpendicular to a line drawn from the pivot point to the point of application. You can now use the concept of a force arm to understand this. The force arm is maximised when the line of action is perpendicular to the line between the pivot point and the point of application, and therefore the torque is maximised (Figure 14.1.8).

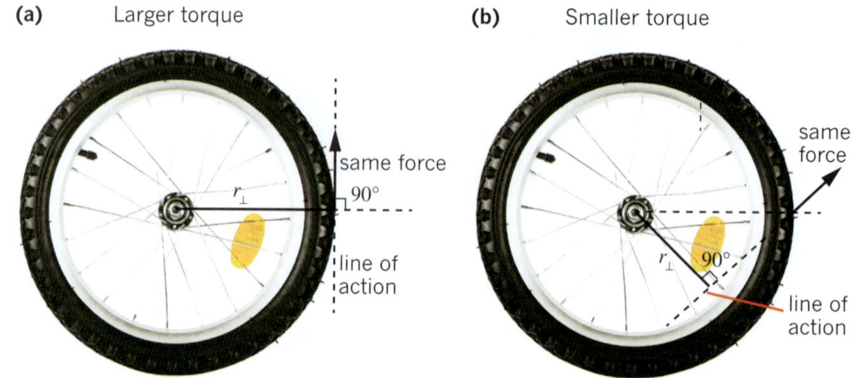

FIGURE 14.1.8 The direction of the applied force affects the size of the force arm, and therefore the torque on the object. The wheel in (a) will experience a larger torque than the wheel in (b).

THE TORQUE EQUATION

The magnitude of the torque (τ) increases or decreases as the force (F) increases or decreases. The magnitude of the torque also increases or decreases as the force arm or the perpendicular distance from the pivot to the line of action of the force ($r_\perp$) increases or decreases.

The formula for calculating torque is:

$\tau = r_\perp F$

where τ is the torque (N m)

$r_\perp$ is the force arm (m)

F is the force (N).

Torque is a vector quantity in that it must have a direction, and different torques add together as vectors (they may cancel each other out). A rotating body rotates either clockwise or anticlockwise. A clockwise rotation is considered to be positive and an anticlockwise rotation to be negative. This convention is useful when more than one torque is acting on a body and the net torque has to be found.

Worked example 14.1.1

CALCULATING TORQUE

A bus driver applies a force of 45.0 N on the steering wheel of a bus as it turns a right-hand corner. The radius of the steering wheel is 30.0 cm. If the force is applied at 90° to the radius, calculate the torque on the steering wheel.

Thinking	Working
Identify the variables involved and state them in their standard form.	$r_{\perp} = 0.300\,\text{m}$ $F = 45.0\,\text{N}$ $\tau = ?$
Apply the equation for torque.	$\tau = r_{\perp}F$ $= 0.300 \times 45.0$ $= 13.5\,\text{N m}$
State the answer with the appropriate direction.	$\tau = 13.5\,\text{N m}$ clockwise

Worked example: Try yourself 14.1.1

CALCULATING TORQUE

A force of 255 N is required to apply a torque on the steering wheel of a sports car as it turns left. The force is applied at 90° to the 15.5 cm radius of the steering wheel. Calculate the torque on the steering wheel.

TORQUE ON DIFFERENT OBJECTS

Torque doesn't need to be acting on circular objects only. Any object can rotate about a point if a force is applied such that the line of action of the force is not acting through the pivot point.

Spanners, like the one in Figure 14.1.9, apply a torque to a nut or bolt: the pivot point is the bolt and a force is applied at right angles to the spanner.

The reason a spanner is an effective hand tool is because it increases the force arm when turning a nut. If you try unscrewing a nut with your hands, you will probably find that you are unable to provide enough force to create the torque required to turn the nut. Longer spanners can apply a greater torque on a nut than shorter spanners. Some wheel nut spanners, like the one in Figure 14.1.10, have handles that can be extended so the force arm can be increased. This provides extra torque for loosening very tight nuts, or for tightening the nuts to the correct torque.

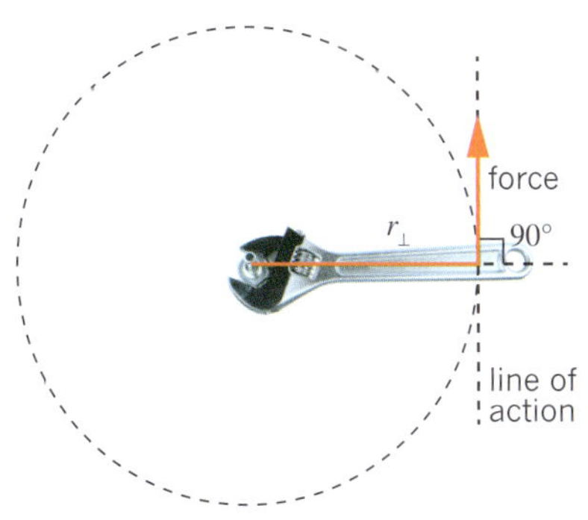

FIGURE 14.1.9 Although the adjustable spanner is not a wheel or circle, torque can still be applied to the nut.

FIGURE 14.1.10 Removing a tyre with an extended handle spanner will increase the torque on the nut.

Worked example 14.1.2

CALCULATING PERPENDICULAR DISTANCE

A car driver can apply a maximum force of 845 N on a wheel nut spanner that is adjustable up to 30.0 cm in length. The force is applied at 90° to the spanner. If the wheel nuts need a torque of 224 N m to remove them, what is the minimum length of the adjustable spanner so that the nuts can be loosened? State whether the spanner is long enough.

Thinking	Working
Identify the variables involved and state them in their standard form.	$\tau = 224$ N m $F = 845$ N $r_{\perp} = ?$
Apply the equation for torque. Rearrange if necessary.	$\tau = r_{\perp}F$ $r_{\perp} = \frac{\tau}{F}$ $= \frac{224}{845}$ $= 0.265$ m
State the answer with the appropriate units.	$r_{\perp} = 26.5$ cm
Compare the answer with the length of the spanner and state whether the spanner is or isn't appropriate for this task.	The spanner can be extended to 30.0 cm so it is long enough to provide the minimum perpendicular distance of 26.5 cm. The spanner is long enough.

Worked example: Try yourself 14.1.2

CALCULATING PERPENDICULAR DISTANCE

A truck driver can apply a maximum force of 1022 N on a large truck wheel nut spanner that has a length of 80.0 cm. The force is applied at 90° to the spanner. If the truck's wheel nuts need a torque of 635 N m to make them secure, determine whether the spanner is long enough for the job.

Doors are also good examples of torque in action, with the hinges forming the axis of rotation. If force is applied to the handle and the line of action of the force is perpendicular to the door, then the distance between the hinge and the handle represents the force arm. This is shown in Figure 14.1.11.

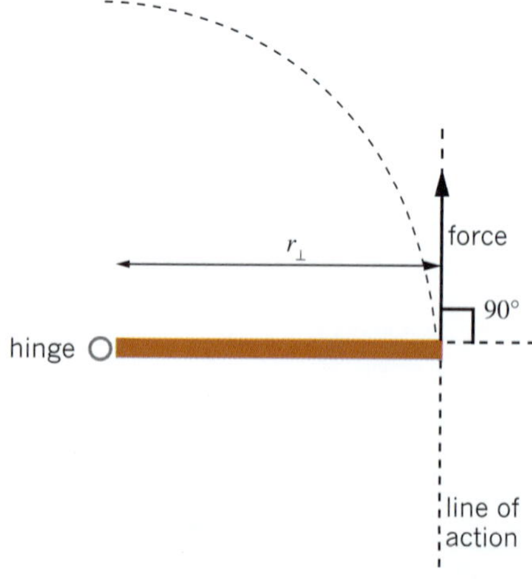

FIGURE 14.1.11 A door can have a torque applied to it, as long as the line of action of the force is not through the axis of rotation.

NON-PERPENDICULAR CALCULATIONS OF TORQUE

When the force causing a torque acts along a line that is at an angle other than 90° to an object, such as the door in Figure 14.1.12, then the torque is reduced. In these circumstances, we can calculate the torque by two approaches: either by finding the perpendicular distance from the pivot point to the line of action of the force, or by finding the component of the force acting perpendicular to the door.

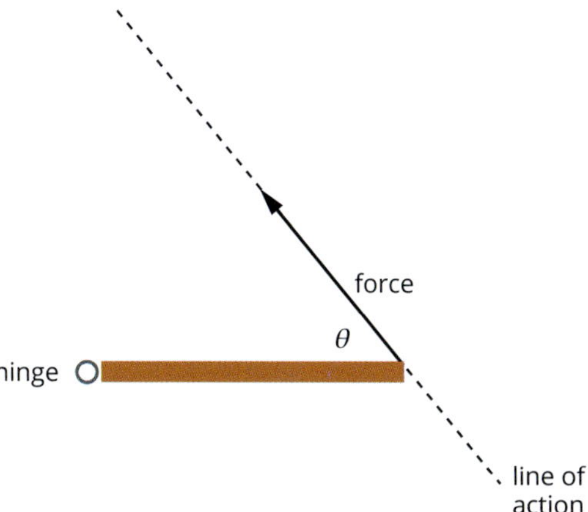

FIGURE 14.1.12 When the force causing a torque is not perpendicular to a door, the torque is reduced.

Recall that the formula for torque (τ) on an object is:

$$\tau = r_{\perp}F$$

This equation calculates the torque (τ) when the force (F) and the distance from the pivot to the line of action of the force (r) are perpendicular to each other. It really doesn't matter whether the radius is perpendicular to the line of action of the force, or if the force is perpendicular to the radius. The result is the same either way; that is, $\tau = r_{\perp}F$ and also $\tau = rF_{\perp}$.

Calculating torque using perpendicular radius

The component of any distance can be calculated using either Pythagoras' theorem or trigonometry. To find the component of a length that is perpendicular to the line of action of the force acting on a door, construct a line from the pivot point to the line of action of the force so that it intersects the line of action at right angles. An example is shown in Figure 14.1.13.

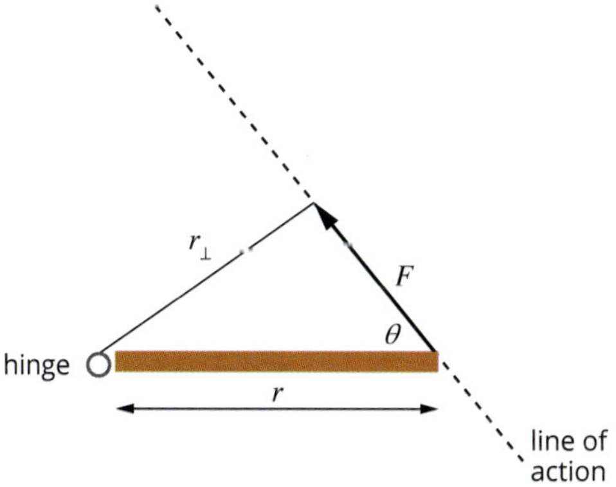

FIGURE 14.1.13 Determining the components of a distance

In this case:

$$\tau = r_{\perp}F \text{ and } r_{\perp} = r\sin\theta$$

combine to give

$$\tau = r\sin\theta\, F$$

Worked example 14.1.3

CALCULATING TORQUE FROM THE PERPENDICULAR COMPONENT OF DISTANCE

A student uses a 42.0 cm long adjustable spanner to loosen a nut on her bike. She applies a force of 65.0 N at an angle of 68.0° to the spanner.

Using the perpendicular distance, calculate the magnitude of the anticlockwise torque that the student applies to the nut.

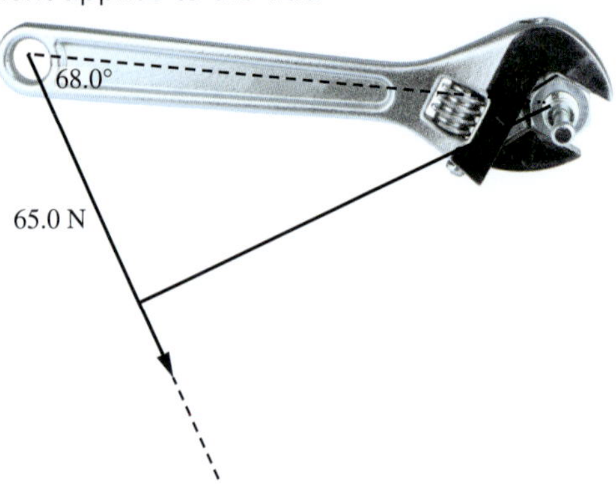

Thinking	Working
Convert variables to their standard units.	$r = 42.0\,\text{cm} = 0.420\,\text{m}$
Use the trigonometric relationship $r_{\perp} = r\sin\theta$ to determine the perpendicular distance from the pivot point to the line of action of the force.	$r_{\perp} = r\sin\theta$ $= 0.420\sin 68.0°$ $= 0.389\,\text{m}$
Apply the equation for torque: $\tau = r_{\perp}F$	$\tau = r_{\perp}F$ $= 0.389 \times 65.0$ $= 25.3\,\text{N m}$
State the answer with the appropriate units.	$\tau = 25.3\,\text{N m}$ anticlockwise

Worked example: Try yourself 14.1.3

CALCULATING TORQUE FROM THE PERPENDICULAR COMPONENT OF DISTANCE

A mechanic uses a 17.0 cm long spanner to loosen a nut on a winch. He applies a force of 104 N at an angle of 75.0° to the spanner.

Using the perpendicular distance, calculate the magnitude of the torque that the mechanic applies to the nut.

Calculating torque using perpendicular force

The components of any force can be calculated using trigonometry. To find the component of the force that is perpendicular to a door, for example, use the magnitude of the force and the angle between the door and the line of action of the force. This is shown in Figure 14.1.14.

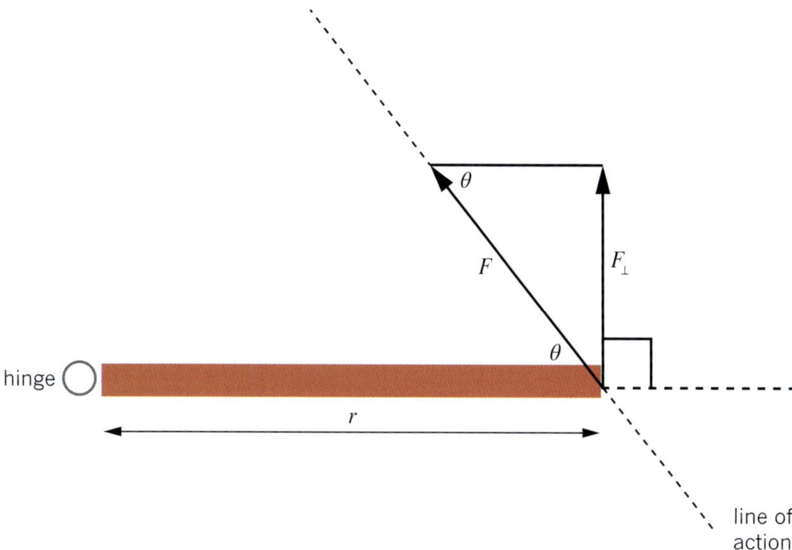

FIGURE 14.1.14 Finding the component of the force that is perpendicular to a door using the magnitude of the force and the angle between the door and the line of action of the force

In this case, $\tau = rF_{\perp}$ and $F_{\perp} = F \sin \theta$.

This combines to give $\tau = rF \sin \theta$.

The equation for torque using the perpendicular force, $\tau = rF\sin \theta$, is identical to the equation for torque using the perpendicular radius, $\tau = r\sin \theta F$. This is because for a given force and radius, the torque only depends on the angle between the two vectors. Either method would be appropriate for calculating the torque on an object when the force is not at right angles to the object. In both methods, the component of the distance or the component of the force is always going to be less than the total distance or the total force itself. This will result in a smaller torque being applied to the object. The maximum torque will always be when the line of action of the force is perpendicular to the distance from the pivot point to the point of application.

Your strategy for solving questions of this type may be to calculate the perpendicular component of force or the perpendicular radius and then apply the torque equation, or to use the combined equation. To begin with, it is recommended that you calculate the perpendicular component and then use the torque equation. When you have gained confidence with that strategy, try using the combined equation.

Worked example 14.1.4

CALCULATING TORQUE FROM THE PERPENDICULAR COMPONENT OF FORCE

A student uses a 42.0 cm long adjustable spanner to loosen a nut on her bike. She applies a force of 65.0 N at an angle of 68.0° to the spanner.

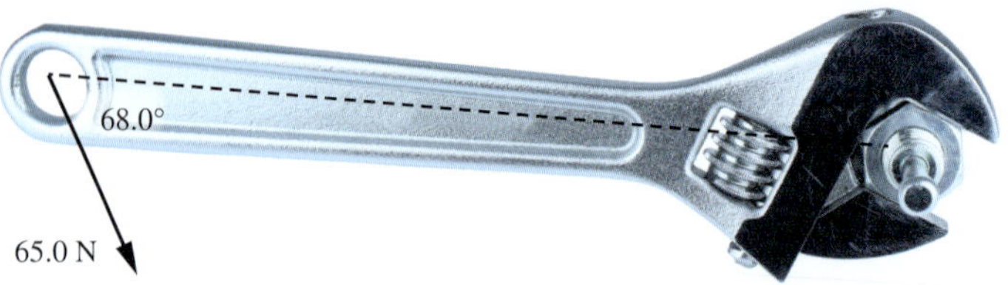

Using the perpendicular component of force, calculate the anticlockwise torque that the student applies to the nut.

Thinking	Working
Use the trigonometric relationship $F_{\perp} = F \sin \theta$ to determine the force perpendicular to the spanner.	$F_{\perp} = F \sin \theta$ $= 65.0 \sin 68.0°$ $= 60.3$ N
Convert the variables to their standard units.	$r = 42.0$ cm $= 0.420$ m
Apply the equation for torque: $\tau = rF_{\perp}$	$\tau = rF_{\perp}$ $= 0.420 \times 60.3$ $= 25.3$ N m
State the answer with the appropriate units.	$\tau = 25.3$ N m anticlockwise Note that this is the same answer as in the previous worked example.

Worked example: Try yourself 14.1.4

CALCULATING TORQUE FROM THE PERPENDICULAR COMPONENT OF FORCE

A mechanic uses a 17.0 cm long spanner to loosen a nut on a winch. He applies a force of 104 N at an angle of 75.0° to the spanner.

Using the perpendicular component of force, calculate the magnitude of the torque that the mechanic applies to the nut.

CASE STUDY

The torque wrench

The extent to which a nut or bolt is tightened can be critical to the safe operation of machinery or motors. If a nut or bolt is too loose, then it could fall out. If it is too tight, then it could either distort the part or the bolt could break off. Both of these situations could require expensive repairs. To avoid nuts and bolts being too loose or too tight, manufacturers use different tools and methods to estimate the amount of torque required to tighten a nut or bolt to the correct tightness. Some examples of these tools are shown in Figure 14.1.15.

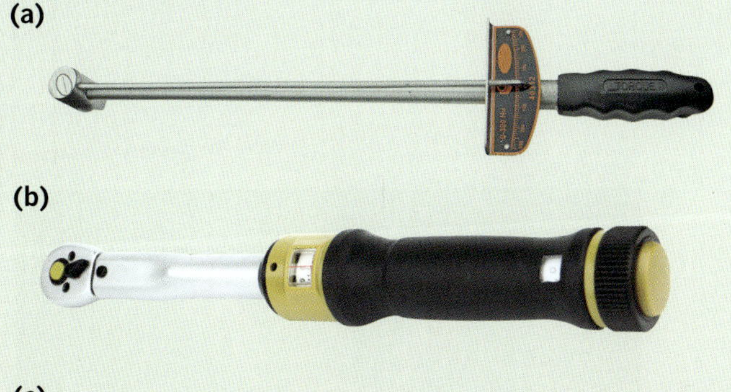

FIGURE 14.1.15 Three types of wrenches commonly used to measure the torque applied to a nut or bolt: (a) beam torque wrench, (b) a click-type torque wrench, and (c) an electronic torque wrench

The beam wrench is the simplest type of torque wrench. It has a flexible lever arm with a bar and scale separating the wrench head and handle. When torque is applied, a pointer on the scale moves to indicate the amount of torque being applied in newton metres (N m).

The click-type torque wrench can be set to apply a fixed amount of torque. When the required amount of torque has been achieved, the wrench 'clicks' and releases itself, preventing any further tightening from being applied.

More recently, electronic torque wrenches have been developed. The signal generated is converted to a torque reading (N m) and is shown on the digital readout screen. Measurements can also be stored within the instrument's memory and transferred to a computer.

14.1 Review

SUMMARY

- Torque is a measurement of the tendency of a force to cause an object to rotate around an axis.
- The formula for calculating torque is $\tau = r_{\perp}F$.
- Torque occurs when the acting force is not applied directly through the pivot point of the object.
- Maximum torque occurs when the acting force applied is perpendicular to a line drawn from the pivot point to the point of application.
- The larger the force acting on the object, the larger the torque will be.
- The longer the force arm, the greater the torque will be.
- If torque is generated by an acting force that is not perpendicular to the lever arm of the object, then either:
  - the distance from the pivot point perpendicular to the line of action of the force is used to calculate torque: $\tau = r_{\perp}F$

    or
  - the component force perpendicular to the length of the object is used to calculate torque: $\tau = rF_{\perp}$.
- Both strategies for determining torque from non-perpendicular situations equate to $\tau = rF\sin\theta$.

KEY QUESTIONS

Assume $g = 9.8\,\text{N kg}^{-1}$ when answering these questions.

Knowledge and understanding

1 Use the concept of torque to explain the following statements.

a It is easier to open a heavy door by pushing it at the handle rather than in the middle of the door.

b It is possible to move very heavy rocks in the garden by using a long crowbar.

2 Explain, in terms of torque, why pushing a door handle to the left will not cause the door to open.

3 Calculate the torque exerted on the roundabouts shown. Include the direction where appropriate.

a

b

4 The magnitude of the torque required to tighten a bicycle wheel is 15 N m. Calculate the force arm required if a 30 N force is used.

5 A student pushes a heavy door at a point that is 50 cm from the hinges such that it creates a torque of 9.0 N m. With what magnitude of force does the student push? Assume the student pushes perpendicular to the surface of the door.

6 A spanner with a length of 40 cm is used to tighten a nut on a car wheel. If the magnitude of force applied is 225 N, calculate the maximum torque that can be applied on this wheel nut.

7 A mechanic uses a spanner of length 30 cm to tighten a bolt head. The mechanic applies a force of 300 N at an angle of 30° to the length of the spanner. Calculate the torque applied.

Analysis

8 Nikki is investigating torque using a metre ruler and a 1.0 kg mass. She uses a rubber band to attach the mass to the ruler. Nikki first holds the ruler at one end so that it is horizontal, with the mass at the 50 cm mark.

a What is the size of the torque that is acting?

b She now moves the mass so that it is right at the far end of the ruler. How much torque is acting now?

c Finally, with mass remaining at the far end, she lifts the ruler so that it makes an angle of 60° to the horizontal. What is the size of the torque now?

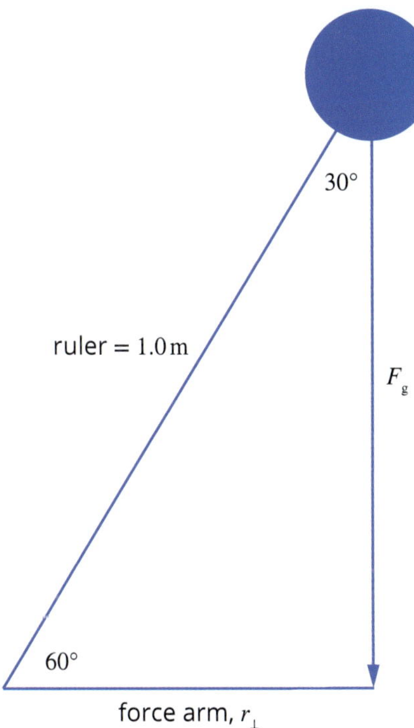

9 A crane is being used to lift a skip of concrete with a total mass of 3.5 tonnes (3500 kg). The lever arm of the crane is 25 m long and makes an angle of 37° with the vertical as shown in the diagram. Ignore the mass of the cable when answering these questions.

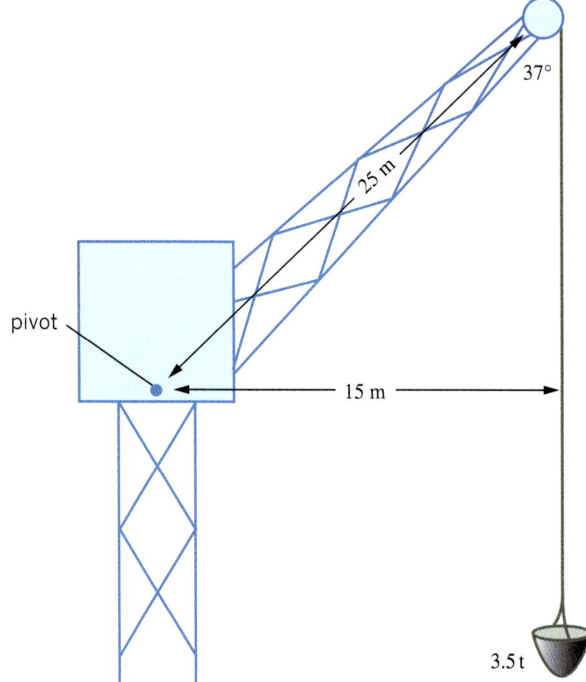

a What is the total force due to gravity on the skip?

b The skip is lifted so that it is near the top of the crane. Does the torque around the pivot created by this load increase, decrease or remain the same as the load is lifted?

c Calculate the magnitude and direction of the torque about the pivot that the skip exerts on the crane when the skip is at the highest point.

14.2 Translational equilibrium

Newton's first law states that an object will continue with its motion unless acted upon by an external unbalanced force. The velocity of an object won't change when the forces acting on it are balanced. When the forces are balanced, the forces are said to be in translational equilibrium.

FIGURE 14.2.1 When a tug-o-war starts, there is an equilibrium of forces as both teams take the strain.

An example of translation equilibrium occurs at the beginning of a game of tug-o-war, like that in Figure 14.2.1. Both teams take the strain and neither team moves. Winning a tug-o-war game involves one team applying a greater force so that there is a net force on the rope, causing the rope and teams to accelerate in the winning team's direction. When the rope and the teams are moving at a constant velocity, then an equilibrium of forces exists once again.

A **translational equilibrium** of forces occurs when the sum of the forces acting on an object add to give a zero resultant force or zero net translational force. As a net translational force causes acceleration in one direction, a zero net translational force causes no acceleration of the object (assuming there is also no rotation of the object). This condition is the defining aspect of a translational equilibrium of forces. Note that although objects in translational equilibrium may still have some rotational motion, in this section we will only analyse objects that are not rotating.

When analysing a situation involving more than one force acting on an object, translational equilibrium will exist if the sum of the forces is equal to zero: $\sum F = 0$. The sum of forces is commonly referred to as the net force, and so the condition for translational equilibrium can be expressed as: $F_{net} = 0$.

VECTOR DIAGRAMS OF AN EQUILIBRIUM OF FORCES IN ONE DIMENSION

FIGURE 14.2.2 The tug-o-war vector addition diagram shows a net force of zero, indicating an equilibrium of forces exists.

Vector diagrams can be drawn to represent the forces acting on an object when the forces are acting in one dimension. For example, if three people are pulling to the right and three people are pulling to the left in a game of tug-o-war as shown in Figure 14.2.2, then the forces are all in one dimension—left and right. These forces are added using a vector diagram by drawing all the forces from each person head to tail, as described in Chapter 10. If the tug-o-war is in translational equilibrium, then all the forces should add to give a zero net force.

CALCULATING AN EQUILIBRIUM OF FORCES IN ONE DIMENSION

To calculate whether a situation is in translational equilibrium or not, a sign convention is used to represent the direction of the force vectors. Typically, in the x-dimension left is negative and right is positive; similarly in the y-dimension upwards is positive and downwards is negative. In the z-dimension forwards is positive and backwards is negative. By applying a sign convention to the forces acting on an object, the addition of those forces, with their signs, will give a zero answer if the situation is in translational equilibrium.

$$F_{net,\ left-right} = 0 \text{ or } F_{net,\ x} = 0$$
$$F_{net,\ upwards-downwards} = 0 \text{ or } F_{net,\ y} = 0$$
$$F_{net,\ forwards-backwards} = 0 \text{ or } F_{net,\ z} = 0$$

Worked example 14.2.1

CALCULATING TRANSLATIONAL EQUILIBRIUM IN ONE DIMENSION

Three children are standing on a plank that is bridging a small stream. The plank is supported at each end by the ground. The plank has a mass of 25 kg and the children have masses of 40 kg, 30 kg and 35 kg. There is an upwards force of the left bank on the plank of 700 N. If the plank is in translational equilibrium, then calculate the force of the right bank on the plank.

Use $g = 9.8\,\text{N kg}^{-1}$ when answering this question.

Thinking	Working
Identify the variables involved and state them with their directions in their standard form.	$m_1 = 40\,\text{kg}$ $m_2 = 30\,\text{kg}$ $m_3 = 35\,\text{kg}$ $m_P = 25\,\text{kg}$ $F_{LB} = 700\,\text{N}$ upwards $g = 9.8\,\text{N kg}^{-1}$ downwards
Apply a sign convention to the vector data.	$F_{LB} = +700\,\text{N}$ $g = -9.8\,\text{N kg}^{-1}$
Identify the object that is in translational equilibrium. This is the object on which all the forces are acting.	The object experiencing translational equilibrium is the plank.
Apply the equation for translational equilibrium in one dimension.	$F_{\text{net}, y} = 0$
Expand the equation to include each of the forces acting on the plank.	$F_1 + F_2 + F_3 + F_P + F_{LB} + F_{RB} = 0$ $m_1g + m_2g + m_3g + m_Pg + F_{LB} + F_{RB} = 0$
Substitute the data into the equation and solve for the unknown.	$(40 \times -9.8) + (30 \times -9.8) + (35 \times -9.8) + (25 \times -9.8) + 700 + F_{RB} = 0$ $-392 - 294 - 343 - 245 + 700 + F_{RB} = 0$ $-574 + F_{RB} = 0$ $F_{RB} = 574\,\text{N}$
State the answer with the appropriate direction to two significant figures.	$F_{RB} = 5.7 \times 10^2\,\text{N}$ upwards

Worked example: Try yourself 14.2.1

CALCULATING TRANSLATIONAL EQUILIBRIUM IN ONE DIMENSION

Three cars are parked on a beam bridge that has a mass of 500 kg. The left pillar (labelled X) applies a force of 2.00×10^4 N upwards. If the situation is in translational equilibrium, calculate the force provided by the right-hand pillar (labelled Y).

Use $g = 9.8\,\text{N kg}^{-1}$ when answering this question.

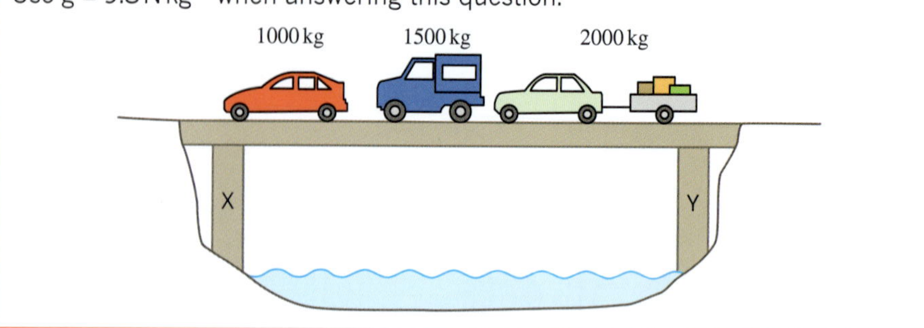

If the forces involved in the equilibrium situation are in two dimensions, there are two strategies for determining if the sum of the forces is zero.

VECTOR DIAGRAMS OF AN EQUILIBRIUM OF FORCES

Recall from Chapter 10 that in any vector addition diagram, the individual vectors are added head to tail. The net (resultant) vector is from the tail of the first vector to the head of the last vector. In a situation where the forces are in equilibrium, the vector addition diagram should end up with the head of the last vector finishing at the tail of the first vector. This means that the vector addition diagram ends up in a closed loop and therefore there is no net force. Figure 14.2.3 shows vector diagrams in which (a) the net force (F_{net}) is not zero and (b) where F_{net} is zero.

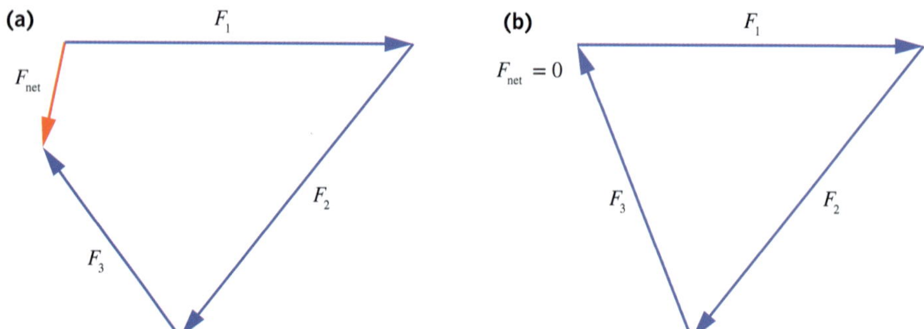

FIGURE 14.2.3 (a) A vector diagram showing three vectors added head to tail, and the resultant net force vector F_{net} in red (b) A closed loop vector addition diagram, showing an equilibrium of forces, where $F_{net} = 0$

Vector components of a translational equilibrium of forces

A vector diagram that results in a closed loop fulfils the conditions of a translational equilibrium of forces. If two forces are perpendicular, Pythagoras' theorem and trigonometry can be used to determine the magnitude and direction of a third force that will result in translational equilibrium with the two other forces.

If two forces are not perpendicular, each force is replaced with two vectors that are in perpendicular directions. These perpendicular vectors are called the **components** of a force. These force components are then added in each dimension, which results in two perpendicular resultant vectors that can be added using Pythagoras' theorem and trigonometry. The diagrams in Figure 14.2.4 show how two non-perpendicular vectors are added using vector diagrams.

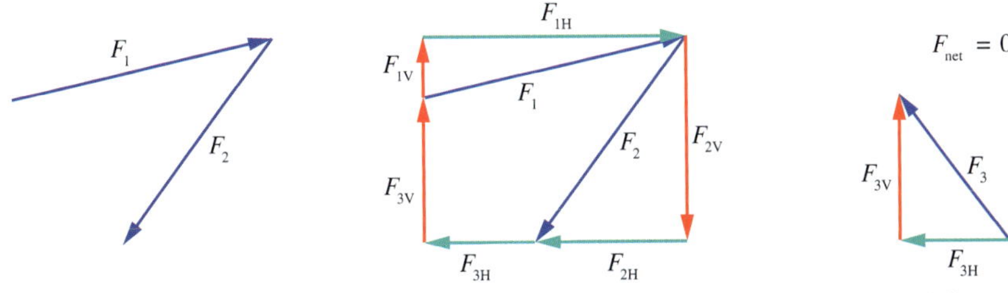

FIGURE 14.2.4 A method for finding F_3 that acts to make an equilibrium of forces with F_1 and F_2

The guiding principle behind this method is that, for the forces to be in equilibrium, the sum of the x-axis (horizontal or left–right) forces must equal zero, and the sum of the y-axis (vertical or upwards–downwards) forces must also equal zero.

$$F_{\text{net}, x} = 0 \text{ or } F_{\text{net, H}} = 0 \text{ or } F_{\text{net, left–right}} = 0$$
$$F_{\text{net}, y} = 0 \text{ or } F_{\text{net, V}} = 0 \text{ or } F_{\text{net, upwards–downwards}} = 0$$

Worked example 14.2.2

CALCULATING TRANSLATIONAL EQUILIBRIUM IN TWO DIMENSIONS

An advertising banner is hung by two cables from the ceiling of a shop. The banner has a mass of 45 kg and the cables are at an angle of 30° to the vertical, as shown in the diagram below. If the mass of the cables is ignored, calculate the tension in each cable when the sign is suspended. Give your answer to two significant figures.

Use $g = 9.8\,\text{N kg}^{-1}$ when answering this question.

Thinking	Working
Construct a vector diagram adding all of the forces together. F_{gS} is the force due to gravity on the sign. F_{TL} and F_{TR} are the tension forces in the wires on the left-hand side and right-hand side of the sign respectively.	$F_{net} = 0$ 30° F_{TL} F_{gS} 30° F_{TR}

Thinking (*continued*)	Working (*continued*)
Apply horizontal components and vertical components. F_{LH} and F_{LV} are the horizontal and vertical components of the tension force in the left-hand wire, and F_{RH} and F_{RV} are the horizontal and vertical components of the tension force in the right-hand wire.	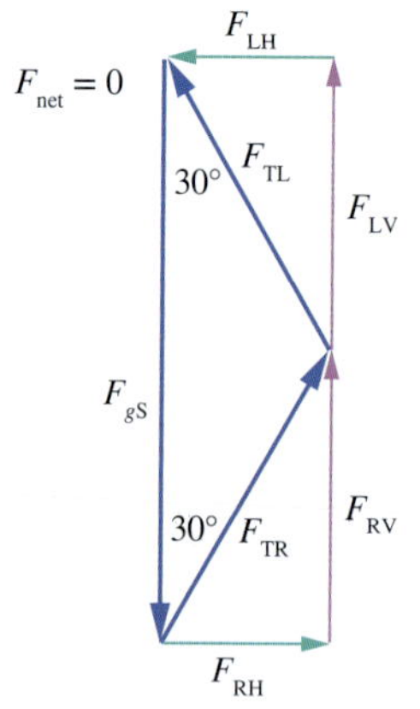 $F_{net} = 0$ F_{LH}, F_{TL}, F_{LV}, 30° F_{gS} 30°, F_{TR}, F_{RV} F_{RH}
Recognise that, in the horizontal dimension, F_{LH} is in equilibrium with F_{RH}.	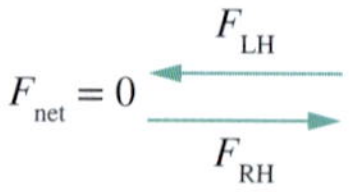 F_{LH} $F_{net} = 0$ F_{RH}
Recognise that, in the vertical dimension, F_{gS} is in equilibrium with F_{RV} and F_{LV}.	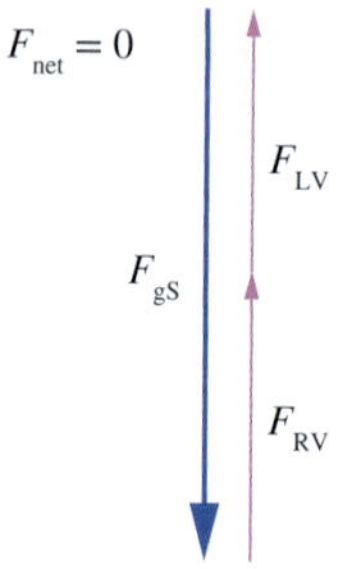 $F_{net} = 0$ F_{LV} F_{gS} F_{RV}
Apply the equation for translational equilibrium in the vertical dimension. Recognise that F_{RV} and F_{LV} are equal in magnitude and therefore each is half of F_{gS}.	$F_{net,\ V} = 0$ $F_{RV} = F_{LV}$
Expand the equation to include each of the vertical forces acting on the sign.	$F_{gS} + F_{RV} + F_{LV} = 0$ $m_S g + F_{RV} + F_{LV} = 0$
Substitute the data into the equation and solve for the unknown, ensuring to give a direction.	$45 \times (-9.8) + F_{RV} + F_{LV} = 0$ $-441 + F_{RV} + F_{LV} = 0$ $F_{RV} + F_{LV} = 441$ $F_{RV} = F_{LV} = 220.5\text{ N}$ $= 2.2 \times 10^2\text{ N}$ upwards, to two significant figures
Draw the right-angled triangle with one of the vertical components of the tension and the angle.	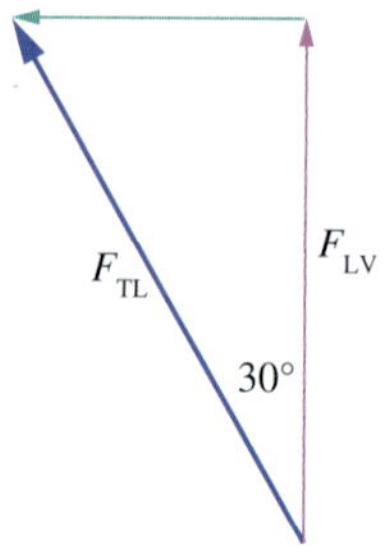 F_{TL}, F_{LV}, 30°

Use trigonometry to solve for the tension in one of the cables, which will equal the tension in the other cable as well.	$\cos\theta = \frac{F_{LV}}{F_{TL}}$ $F_{TL} = \frac{F_{LV}}{\cos\theta}$ $= \frac{220.5}{\cos 30°}$ $= 255\,N$ $F_{TL} = F_{TR} = 255\,N$ $= 2.6 \times 10^2\,N$ to two significant figures

Worked example: Try yourself 14.2.2

CALCULATING TRANSLATIONAL EQUILIBRIUM IN TWO DIMENSIONS

A concrete beam of mass 1500 kg has been lifted by cables labelled 1 and 2, and is now stationary, as shown in the diagram. The beam then moves upwards with a constant velocity of 2.0 m s^{-1}. Calculate the tension in cable 1 and cable 2. Give your answer to two significant figures.

Ignore the mass of the cable and use $g = 9.8\,N\,kg^{-1}$ when answering this question.

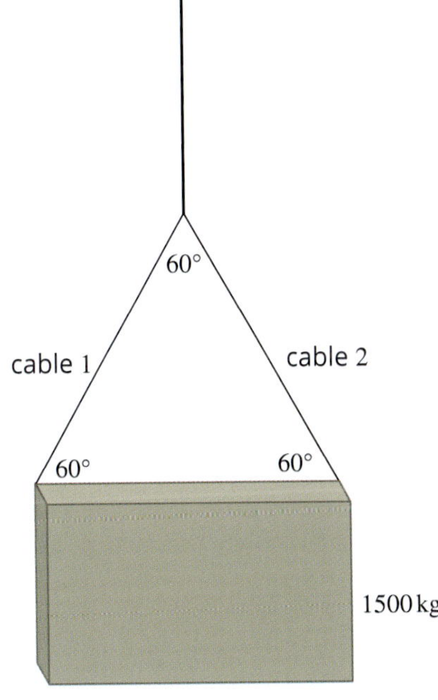

14.2 Review

SUMMARY

- A translational equilibrium of forces occurs when the sum of the forces acting on an object add to give a zero resultant force or zero net translational force.
- Translational equilibrium in one dimension can be represented mathematically as $\Sigma F = 0$ (or $F_{net} = 0$).
- Translational equilibrium in two dimensions can be represented mathematically as $F_{net,\ x} = 0$ and $F_{net,\ y} = 0$.
- In two dimensions, an equilibrium of forces can be represented in a closed vector diagram.
- By calculating the *x* and *y* components of known forces, you can use the equilibrium equation in each dimension to find the unknown force required to keep an object in equilibrium.

OA ✓✓

KEY QUESTIONS

Assume $g = 9.8\,\text{N kg}^{-1}$ when answering these questions.

Knowledge and understanding

1 Which one of the following is correct for an object to be in translational equilibrium?

A The object must be stationary.

B The object must not be experiencing rotational acceleration.

C The object must be experiencing translational acceleration.

D The object must not be experiencing translational acceleration.

2 Which one or more of the following are in translational equilibrium?

A a stationary elevator

B an elevator going up with constant velocity

C an aeroplane during take-off

D a container ship sailing with constant velocity

E a car plummeting off a cliff

3 Calculate the resultant force in each of the following vector additions.

a 200 N upwards and 50 N downwards

b 65 N west and 25 N east

c 10 N north and 10 N south

d 10 N north and 10 N west

4 A single string supports a bird-feeder that has a mass of 240 g. If the bird-feeder is in translational equilibrium, calculate the tension (upwards) force provided by the string to hold the bird-feeder.

Analysis

5 Two window cleaners working on the windows of a high-rise building work on a platform that is suspended by four cables. Alex has a mass of 88 kg, John has a mass of 75 kg and the mass of the platform is 340 kg. The platform is moving down the side of the building at a constant speed and each cable provides the same tension as the other cables. Calculate the tension in one of the cables.

6 A small car is pulled by two people using ropes. Each person supplies a force of 400 N at an angle of 40.0° to the direction in which the car travels.

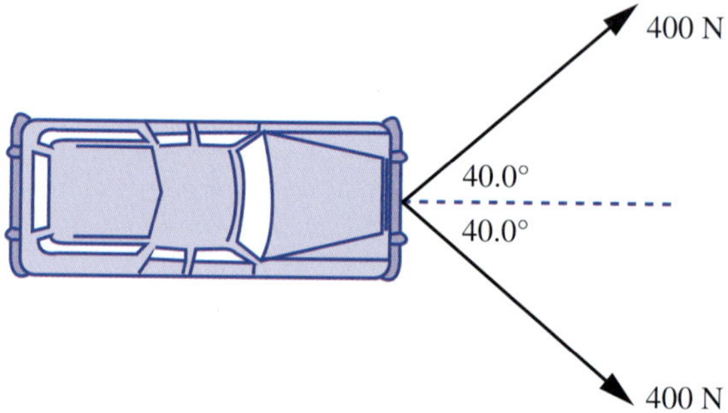

a What is the net force applied by the people to the car?

b If friction is also acting on the car so that the car is in translational equilibrium, determine the force due to friction.

7 A rectangular advertising sign is supported from the car upper corners by two cables, each making an angle of 40° to the vertical. The sign has a mass of 5.0 kg. Calculate the tension in each cable.

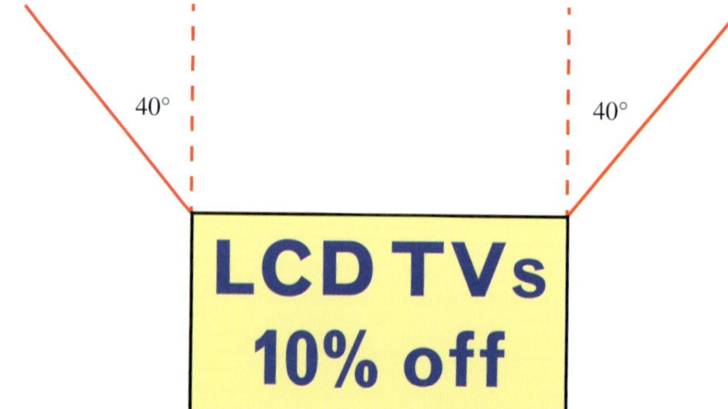

14.3 Static equilibrium

Objects are in translational equilibrium when the sum of the forces acting on the object equals zero. However, some forces may act in ways that generate torque or moment on an object. This will depend on the point of application of the force(s). Recall from Section 14.1 that a 'moment' is a term we typically use in place of torque for static situations. For an object to be in rotational equilibrium, the sum of the moments in a clockwise direction must balance the sum of the moments in an anticlockwise direction. This relationship is called the **principle of moments**. For example, when a cyclist doesn't need to pedal, they can stand on both pedals with equal force. This causes an equal torque in the clockwise and anticlockwise directions. The pedals are in rotational equilibrium.

By combining the conditions for translational equilibrium and rotational equilibrium, an object can achieve a state of **static equilibrium** (Figure 14.3.1). Static equilibrium will be explored in more detail in this section.

FIGURE 14.3.1 An object in translational and rotational equilibrium, like the prominent awning on the Melbourne Exhibition Centre, is in static equilibrium.

CENTRE OF MASS

When studying forces and motion, rigid bodies are considered as having a mass concentrated at the centre. The human body does not move in the way a more symmetrical object would. It is often helpful to treat the body as a collection of separate masses. Each part then has its own centre, at which the mass appears to be concentrated.

Think about an athlete running in a 100 m sprint. In simple terms, the athlete runs in a straight line along the track. The displacement and velocity of the athlete at any time can be calculated using the principles discussed in Chapter 11. In reality, however, the motion of the various parts of the athlete's body will differ significantly during the run. The athlete's arms and legs move in a complex manner that is not easy to analyse.

The analysis of the motion of complicated systems such as a sprinter or high-jumper can be simplified to the motion of a single point. The mass of the sprinter can be considered to be 'concentrated' into a single point that has travelled in a straight line. Recall that this single point is called the centre of mass. There is as much mass above the centre of mass as there is below it, as much mass to the left as there is to the right, and as much mass in front as there is behind it. The centre of mass is an important concept when considering an object's stability.

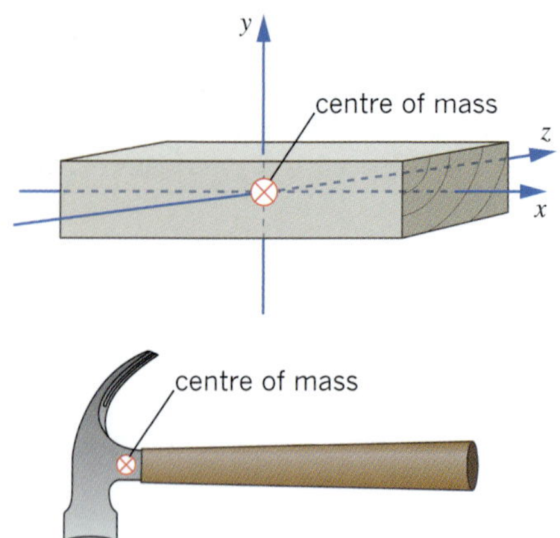

FIGURE 14.3.2 The centre of mass of a three-dimensional block of wood is shown with arrows drawn in the x, y and z dimensions illustrating the lines where there is equal mass on either side of the lines. The centre of mass of a non-uniform hammer shows that the centre of mass will be closer to the end with the higher density than the end with the lesser density.

If an object is uniform in one dimension only (e.g. a straight piece of wire), its centre of mass will be exactly half way along its length. For an object that is uniform in two dimensions (e.g. a flat piece of cardboard), the centre of mass will be the point that is half way along and half way across the object. For an object that is uniform in three dimensions (e.g. a wooden block such as that shown in Figure 14.3.2), the centre of mass will be the point that is halfway along all dimensions of the object. It is possible for the centre of mass to lie outside the body. For example, the centre of mass of a doughnut is located at the centre of the hole. The centre of mass of a person is typically just above their navel, but it will vary with the positions of the arms and legs.

A concept that is closely related to centre of mass is **centre of gravity**. Instead of being a point particle whose motion equates to the whole extended body or system, the centre of gravity is the position from which the collective force due to gravity on the body is considered to act. Consequently, the centre of gravity is the position at which the body will balance. For everyday purposes, the centre of gravity is exactly at the centre of mass.

Finding the centre of mass

The balancing point of an object is called the centre of mass. It is the point at which the mass appears to be concentrated, and so it is the point that moves as though all the net external forces acted on it.

The human body is a complex shape, and so it is often useful to analyse its centre of mass, rather than the whole body, when considering its motion. Remember that, in the case of an object that is not rigid, the mass can redistribute. Figure 14.3.3 shows how the centre of mass moves as the person raises their arms.

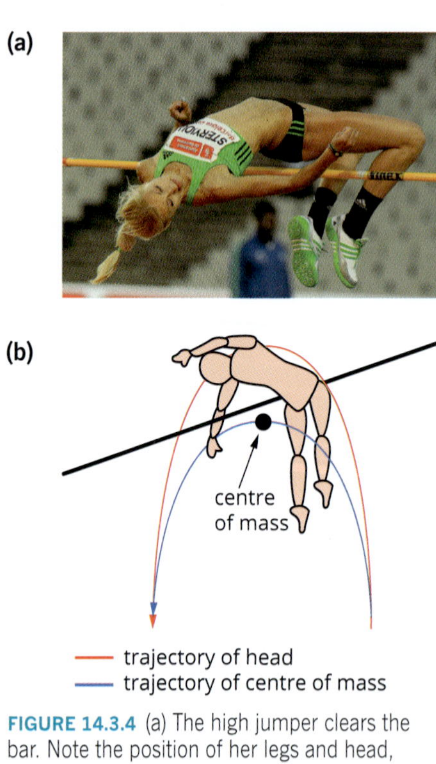

FIGURE 14.3.4 (a) The high jumper clears the bar. Note the position of her legs and head, which lowers her centre of mass and reduces the amount of energy required to complete the jump. (b) This is a diagram of the trajectory of the athlete's head and centre of mass. Note that, although her whole body clears the bar, her centre of mass passes below the bar.

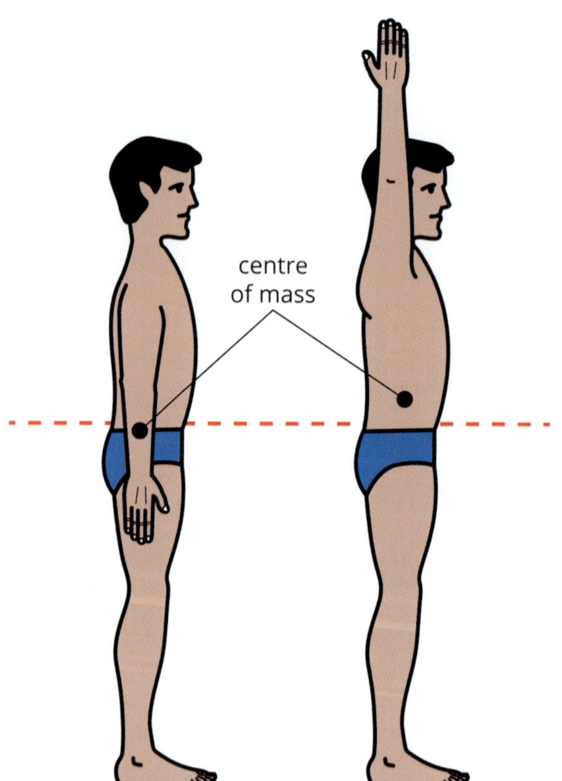

FIGURE 14.3.3 The centre of mass moves as body position changes.

The position of the centre of mass is often of interest in sports science. Consider the high jumper pictured in Figure 14.3.4. While her body needs to go over the bar, her limbs move and her back arches in such a way that her centre of mass actually moves under the bar. This was explored in more detail in Chapter 13.

Calculation of the centre of mass

To calculate the centre of mass of multiple objects, begin by considering two objects that are joined by a rigid but massless wire.

If the positions of the two objects are marked relative to an origin, as indicated in Figure 14.3.5, the position of the centre of mass is defined to be:

$$x_{cm} = \frac{m_1 x_1 + m_2 x_2}{m_1 + m_2}$$

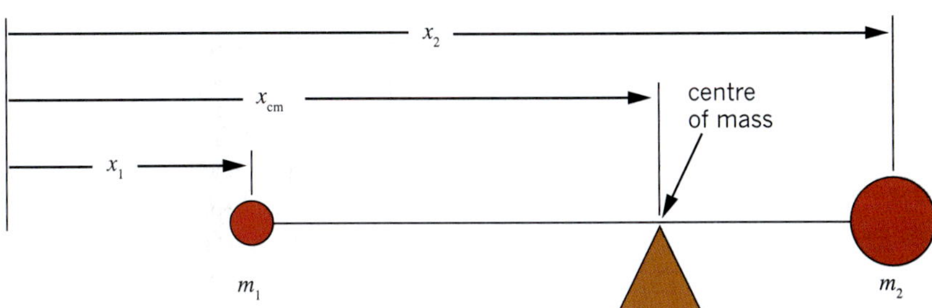

FIGURE 14.3.5 Calculation of the centre of mass: the product of mass and distance is equal for each side of the fulcrum.

In order to confirm that this definition works, suppose, for example, that $m_1 = 0$. In this case, there is only one object, of mass m_2, and the centre of mass must lie at the position of that object, which is x_2. Substituting $m_1 = 0$ into the above equation gives:

$$\begin{aligned} x_{cm} &= \frac{0 \times x_1 + m_2 x_2}{0 + m_2} \\ &= \frac{m_2 x_2}{m_2} \\ &= x_2\text{, as expected.} \end{aligned}$$

As a further confirmation, suppose that the two objects have the same mass ($m_1 = m_2$). In this case, the centre of mass would be the average of the positions of the two objects (i.e. halfway between the two objects). Replacing both m_1 and m_2 with the same mass, m, gives:

$$\begin{aligned} x_{cm} &= \frac{mx_1 + mx_2}{m + m} \\ &= \frac{m(x_1 + x_2)}{2m} \\ &= \frac{1}{2}(x_1 + x_2)\text{, as expected.} \end{aligned}$$

It is a simple extension of this principle to calculate the centre of mass for a more complicated arrangement of masses, as shown in Figure 14.3.6.

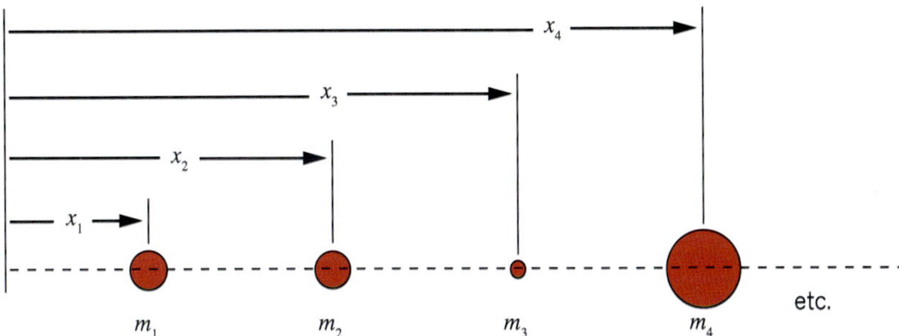

FIGURE 14.3.6 Positions and masses used to find the centre of mass of a collection of masses

Hence, the general formula for the centre of mass for a system of masses placed at positions x_1 to x_n and having masses m_1 to m_n is given by:

$$x_{cm} = \frac{m_1x_1 + m_2x_2 \ldots + m_nx_n}{m_1 + m_2 + \ldots m_n}$$

where m refers to masses 1 to n

x refers to the positions of masses 1 to n relative to the origin

x_{cm} is the position of the centre of mass relative to the origin.

A complex object may be divided into many small pieces, and a computer may be used to calculate the centre of mass to the degree of accuracy required.

Worked example 14.3.1

CALCULATING CENTRE OF MASS

Find the best position to support the object below given the values in the figure. Take the origin on the left as shown and give all responses correct to two significant figures.

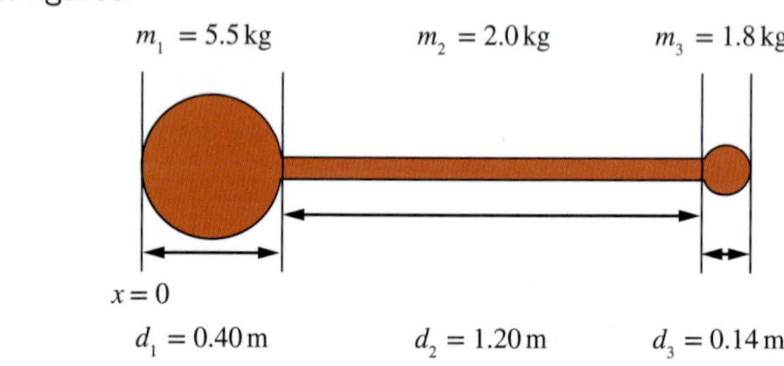

Thinking	Working
Find the centre of mass of each component part, using the symmetry of the body.	The centre of mass of m_1 is at $x_1 = 0.20$ m. The centre of mass of m_2 is halfway down its length: $x_2 = d_1 + \frac{1}{2}d_2$ $= 0.40\text{ m} + 0.60\text{ m}$ $= 1.0\text{ m}$ The centre of mass of m_3 is: $x_3 = d_1 + d_2 + \frac{1}{2}d_3$ $= 0.40\text{ m} + 1.20\text{ m} + 0.07\text{ m}$ $= 1.67$ m from the origin
Now substitute into the centre of mass equation: $x_{cm} = \frac{m_1x_1 + m_2x_2 + \ldots + m_nx_n}{m_1 + m_2 + \ldots + m_n}$	$x_{cm} = \frac{m_1x_1 + m_2x_2 + m_3x_3}{m_1 + m_2 + m_3}$ $= \frac{5.5(0.20) + 2.0(1.0) + 1.8(1.67)}{5.5 + 2.0 + 1.8}$ $= 0.66\text{ m}$
State where the object is best supported. This will be at the centre of mass.	The object is best supported at the centre of mass, 0.66 m from the origin.
Always check if your answer is reasonable.	Because there is more mass concentrated on the left-hand side, one would expect $x_{cm} = 0.66$ m to be to the left of the middle ($x = 0.87$ m), so the answer is reasonable.

Worked example: Try yourself 14.3.1

CALCULATING CENTRE OF MASS

Find the best position to support the object below given the values in the figure. Take the origin on the left as shown and give all responses correct to two significant figures.

$m_1 = 5.5\,\text{kg}$ $m_2 = 2.4\,\text{kg}$ $m_3 = 4.8\,\text{kg}$

$x = 0$

$d_1 = 0.40\,\text{m}$ $d_2 = 1.20\,\text{m}$ $d_3 = 0.30\,\text{m}$

The centre of mass approach can be useful in studying the movement of the human body. Centres of mass of body parts can be found separately, and then the relative movement of the parts can be considered. For example, the torso and the head could be considered as two separate centres of mass when studying whiplash in collision simulations.

The centre of mass and stability

For a body to be stable, the total torque on it (as defined in Section 14.1) must be zero. This condition implies that the centre of mass must lie above the base of support. When a person attempts to touch their toes, their centre of mass moves forwards. If they cannot lean backwards to compensate, their centre of mass falls in front of their feet. This produces a torque that tends to make them fall forwards. A similar example is shown in Figure 14.3.7. In general, the body is stable as long as the centre of mass lies above the base of support. As the mass distribution of our bodies changes when we are in motion, we instinctively move in an attempt to maintain our balance.

FIGURE 14.3.7 This wake-boarder is no longer able to keep their centre of mass above their feet, and so they overbalance.

(a)

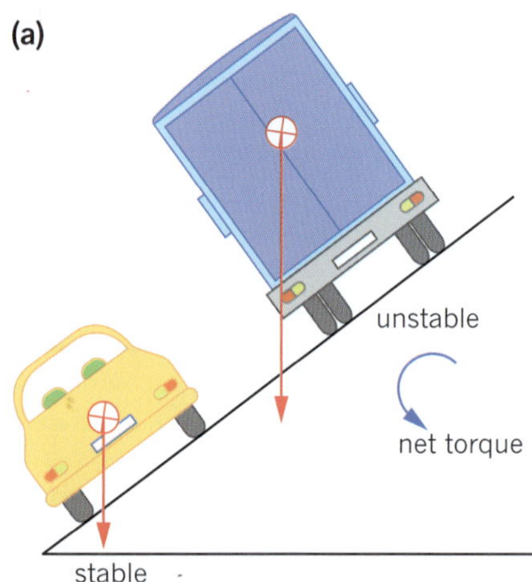

(b)

FIGURE 14.3.8 (a) The car on the incline is in stable equilibrium, while the heavily laden truck on the same incline could topple. The force due to gravity vector is outside the lower point of support for the truck, so there is no reaction force from the road to the higher wheel.
(b) Modern four-wheel drive cars and tractors have inclinometers to warn the driver if the vehicle is in danger of tipping.

STABILITY AND EQUILIBRIUM

Equilibrium refers to a state in which an object is balanced so that, as long as no external factors are changed, the object will not accelerate. Stability refers to the tendency of the system to maintain, or return to, the equilibrium state if external factors are changed. There are three types of equilibrium related to the stability of an object:

- **neutral equilibrium**—the object will remain stationary no matter where it is placed (e.g. placing a ball anywhere on a large flat surface)
- **stable equilibrium**—if the object is moved away from its equilibrium position, forces will act to return it to its equilibrium position (e.g. placing a ball in a large bowl; the equilibrium position is at the bottom of the bowl, and if the ball is moved to any other position, gravity will act to return it to the bottom of the bowl)
- **unstable equilibrium**—if the object is moved away from its equilibrium position, forces will act to push it further from the equilibrium position (e.g. placing a ball on top of a large dome; the equilibrium position is at the top of the dome, but if the ball is moved in any direction, gravity will pull it further from the top of the dome).

When designing structures, engineers and architects want to ensure that balance and stability are maintained. Whether this occurs depends on the relative positions of the centre of mass and the **base** or point of support. When a vertical line downwards from the centre of mass passes through the base of support, the object is stable. The vertical line from the centre of mass represents the direction of the force of gravity on the object.

In Figure 14.3.8, the force due to gravity on the car passes through the car's support base, between the wheels. Taking the centre of gravity to be the pivot point, the reaction forces on the left wheels provide a clockwise torque while the reaction forces on the right wheels provide an anticlockwise torque. Therefore, the torques will balance and the car will not tip over. In the case of the truck, however, the force due to gravity is directed outside the truck's base of support, so an anticlockwise torque acts to tip the truck over.

The stability of an object or structure can be increased in two ways:

- by lowering the centre of gravity
- by increasing the width of the support base.

As a result of either of these, the object has to be tipped further to make the force of gravity act outside the support base. Racing cars have a very low centre of gravity to increase their stability when cornering at high speed. Training wheels on a child's bicycle widen the support base, making it harder to tip the bicycle sideways.

ROTATIONAL EQUILIBRIUM

A **rotational equilibrium** of torque or moments occurs about a **reference point** when the sum of all the torques acting on an object in the clockwise direction is equal to the sum of all the torques acting in the anticlockwise direction.

When analysing a situation in which more than one torque acts on an object, the principle of moments applies where rotational equilibrium will exist if the net torque about a reference point is equal to zero.

$$\tau_{net} = 0$$

Recall that $\tau = r_{\perp}F$

where τ is the torque (N m)

$r_{\perp}$ is the force arm (m)

F is the force (N).

$\tau_{net} = 0$ implies that the sum of torques in the clockwise direction is balanced by the sum of torques in the anticlockwise direction. Therefore, the principle of moments is also represented as:

$$\sum\tau_{clockwise} = \sum\tau_{anticlockwise}$$

When the sum of the torques is not zero, the object will rotate about the centre of mass.

Rotational equilibrium is demonstrated in Figure 14.3.9.

FIGURE 14.3.9 To keep this mast in rotational equilibrium in relation to its base, stainless steel cables (stays) attached to it must provide opposing torques on the mast.

CONDITIONS FOR STATIC EQUILIBRIUM

When a body or a system is not accelerating or rotating, it is in both translational and rotational equilibrium.

- Translational equilibrium occurs when the sum of forces acting on an object gives zero net translational force. The object will then be travelling at constant velocity. An object in translational equilibrium may still have some net torque. For example, when turning a bolt, the bolt will not accelerate in any translational direction, but will still have some net torque applied that keeps it rotating.
- Rotational equilibrium occurs for an object when the sum of all torques acting in the clockwise direction is equal to the sum of all torques acting in the anticlockwise direction. The object will then be rotating at a constant rate. An object in rotational equilibrium may still have some net force applied to it. For example, if you push a table along the floor, it will have a net translational force to keep it moving, but it will still be in rotational equilibrium.

This situation then satisfies the conditions for static equilibrium. Static equilibrium occurs when an object is in both rotational and translational equilibrium. That is, the sum of the net forces and the net torques is zero. It is both travelling and rotating at a constant velocity.

This can be represented mathematically by:

$$F_{net} = 0 \text{ and } \tau_{net} = 0$$

which can also be represented as:

$$F_{net} = 0 \text{ and } \Sigma\tau_{clockwise} = \Sigma\tau_{anticlockwise}$$

Static equilibrium occurs when an object is in rotational and translational equilibrium, i.e.:

$$F_{net} = 0 \text{ and } \tau_{net} = 0$$

Worked example 14.3.2

CALCULATING STATIC EQUILIBRIUM

While playing in their backyard, two young children make a see-saw with a long plank. The boy sits on the see-saw 1.5 m from the pivot. The masses of the boy and girl are 20 kg and 30 kg respectively. Assume that the plank's mass is negligible. Use $g = 9.8\ \text{N kg}^{-1}$ when answering these questions.

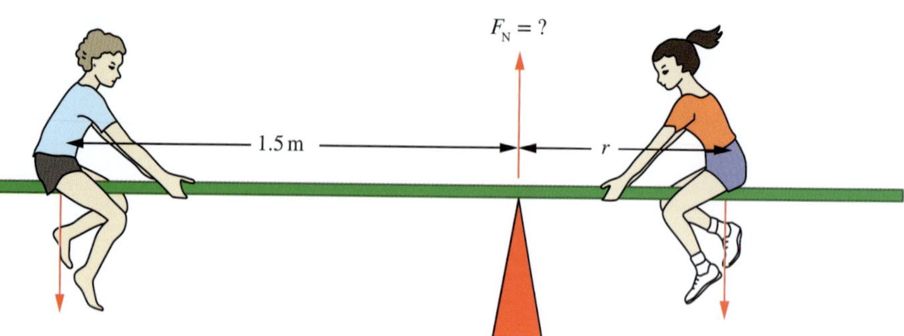

a Calculate the force applied to the plank due to the pivot point when the children are balancing on the see-saw.	
Thinking	**Working**
Identify the variables involved and state them in their standard form.	$m_g = 30\ \text{kg}$ $m_b = 20\ \text{kg}$ $g = 9.8\ \text{N kg}^{-1}$ down
Apply a sign convention to the vector data.	$g = -9.8\ \text{N kg}^{-1}$
Identify the object that is in translational equilibrium. This is the object on which all the forces are acting.	The object experiencing translational equilibrium is the see-saw.
Apply the equation for translational equilibrium in the vertical dimension.	$F_{\text{net}, y} = 0$
Expand the equation to include each of the forces acting on the plank.	$F_g + F_b + F_P = 0$ $m_g g + m_b g + F_P = 0$
Substitute the data into the equation and solve for the unknown.	$(30 \times -9.8) + (20 \times -9.8) + F_P = 0$ $(-294) + (-196) + F_P = 0$ $-490 + F_P = 0$ $F_P = 490\ \text{N}$
State the answer with the appropriate direction to two significant figures.	$F_P = 4.9 \times 10^2\ \text{N}$ upwards

b Calculate where the girl has to sit in order to balance the boy.	
Thinking	**Working**
Identify the variables involved and state them in their standard form.	$m_g = 30\ \text{kg}$ $m_b = 20\ \text{kg}$ $r_{\perp b} = 1.5\ \text{m}$ $g = -9.8\ \text{N kg}^{-1}$
Identify the object that is in rotational equilibrium. This is the object on which all the torques are acting.	The object experiencing rotational equilibrium is the see-saw.

Decide the reference point about which the torques will be calculated.	The reference point is the pivot of the see-saw.
Decide which force causes the clockwise torque and which force causes the anticlockwise torque around the chosen reference point.	The force of the girl on the see-saw provides the clockwise torque. The force of the boy on the see-saw provides the anticlockwise torque.
Apply the equation for rotational equilibrium.	$\Sigma\tau_{\text{clockwise}} = \Sigma\tau_{\text{anticlockwise}}$
Expand the equation to include each of the torques acting on the see-saw.	$r_{\perp g}F_g = r_{\perp b}F_b$
Substitute the data into the equation and solve for the unknown.	$r_{\perp g}F_g = r_{\perp b}F_b$ $r_{\perp g}m_g g = r_{\perp b}m_b g$ $r_{\perp g} \times 30 \times 9.8 = 1.5 \times 20 \times 9.8$ $= \frac{1.5 \times 20 \times 9.8}{30 \times 9.8}$ $= 1.0\,\text{m}$ The girl sits 1.0 m from the pivot.

Worked example: Try yourself 14.3.2

CALCULATING STATIC EQUILIBRIUM

A set of scales (with one longer arm) is used to measure the mass of gold. A lump of gold with a mass of 150 g is placed on the short arm, which is 10 cm long, and a standard set of masses are placed on the long arm.

Use $g = 9.8\,\text{N kg}^{-1}$ when answering these questions.

a Calculate the force applied to the scale's arm due to the pivot point if a standard mass of 50 g exactly balances the gold.

b Calculate how long the long arm should be in order to balance the gold.

In Worked example 14.3.2, the see-saw is in equilibrium because all the forces and torques are balanced. In solving the problem, it seemed obvious to choose the pivot as the reference point around which the torques are determined. But because the see-saw plank is in equilibrium, any point could have been chosen as the reference point. For example, the reference point could be where the girl is sitting (X), as shown in Figure 14.3.10 on page 462. This will mean that the torques acting on the plank would be due to the boy, and due to the see-saw pivot point. The torque due to the girl becomes zero as the lever arm distance for her will be zero.

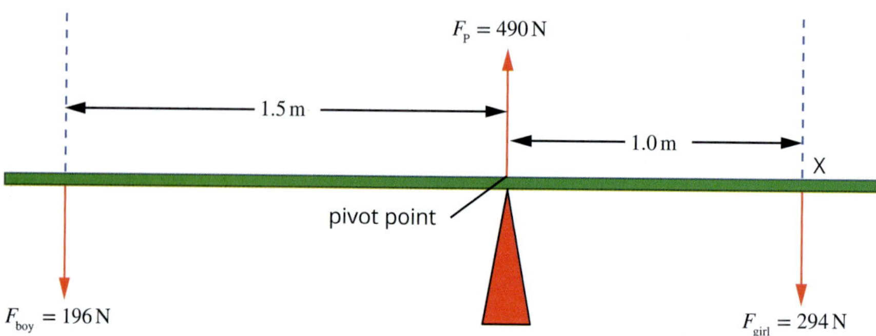

FIGURE 14.3.10 A force diagram for the see-saw problem from Worked example 14.3.2 in which the point at which the girl sits (labelled X) has been chosen as the reference point.

The boy will create an anticlockwise torque around the girl, and the normal force at the pivot for the see-saw, F_P, creates an equal clockwise torque around the girl. In Worked example 14.3.3 the see-saw is in rotational equilibrium. This is verified by calculating the torques around the position of the girl. The plank will be in rotational equilibrium if the clockwise torque equals the anticlockwise torque.

Worked example 14.3.3

CALCULATING STATIC EQUILIBRIUM USING A DIFFERENT REFERENCE POINT

Verify that the see-saw plank in the image below is in rotational equilibrium about the reference point X, where the girl is sitting. The forces due to gravity of the boy and girl are 196 N and 294 N respectively, and the force of the pivot on the plank is 490 N upwards. Assume that the mass of the plank is negligible.

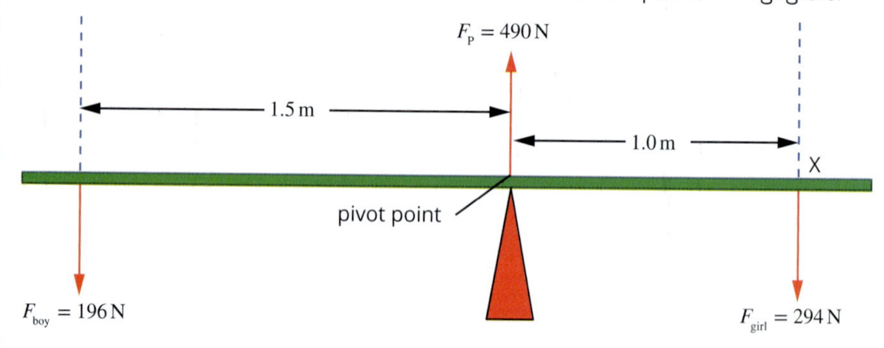

Thinking	Working
Identify the object that is in rotational equilibrium. This is the object on which all the torques are acting.	The object experiencing rotational equilibrium is the see-saw.
Decide the reference point about which the torques will be calculated.	The reference point is the position of the girl, at X.
Identify the variables involved and state them in their standard form.	$F_P = 490\,N$ $F_g = 294\,N$ $F_b = 196\,N$ $r_{\perp b} = 2.50\,m$ $r_{\perp P} = 1.00\,m$
Decide which force causes the clockwise torque, and which force causes the anticlockwise torque around the chosen reference point.	The force of the pivot on the plank provides the clockwise torque. The force of the boy on the plank provides the anticlockwise torque.
Apply the equation for rotational equilibrium.	$\Sigma\tau_{clockwise} = \Sigma\tau_{anticlockwise}$

Expand the equation to include each of the torques acting on the see-saw. Note that the girl's torque is not included here as her torque is zero.	$r_{\perp P}F_P = r_{\perp b}F_b$
Substitute the data into the equation and solve for the unknown.	$r_{\perp P}F_P = r_{\perp b}F_b$ $1.00 \times 490 = 2.5 \times 196$ $490 = 490$
Describe the magnitude of the clockwise torque compared to the magnitude of the anticlockwise torque.	Around reference point X (the position of the girl), the clockwise torque due to the pivot point on the plank is equal to the anticlockwise torque due to the boy on the plank. So the plank is in rotational equilibrium.

Worked example: Try yourself 14.3.3

CALCULATING STATIC EQUILIBRIUM USING A DIFFERENT REFERENCE POINT

Verify that the see-saw plank in the figure below is in rotational equilibrium about the reference point at point Y, where the boy is sitting. The forces due to gravity of the boy and girl are 196 N and 294 N respectively, and the force of the pivot on the plank is 490 N upwards. Assume that the plank's mass is negligible.

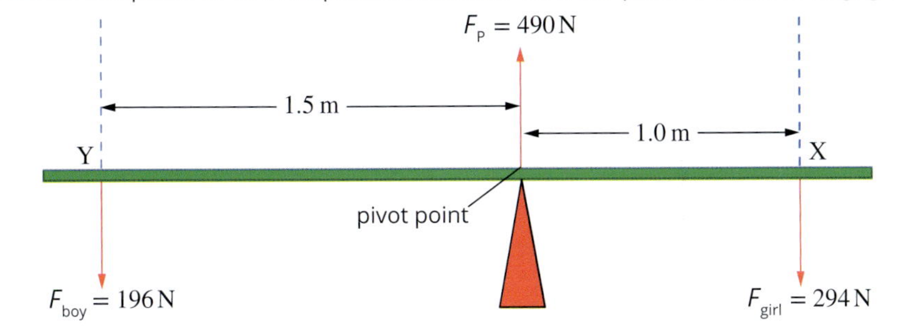

Static equilibrium with two unknown forces

The see-saw problem is relatively straightforward since, in each situation, there is only one unknown force. If there are two unknown forces, the reference point can be chosen to coincide with one of the forces. By using this strategy it means that the force acting at the reference point contributes no torque (since $r = 0$). The remaining unknown force can be found using the relationship $\sum\tau_{clockwise} = \sum\tau_{anticlockwise}$. Worked example 14.3.4 below employs this strategy.

Worked example 14.3.4

CALCULATING STATIC EQUILIBRIUM WITH TWO UNKNOWNS

While painting a tall building, a 70 kg painter stands 4.00 m from the end of a 6.00 m long plank that is supported by a rope at each end. The plank has a mass of 20 kg. Use g = 9.8 N kg^{-1} when answering this question.

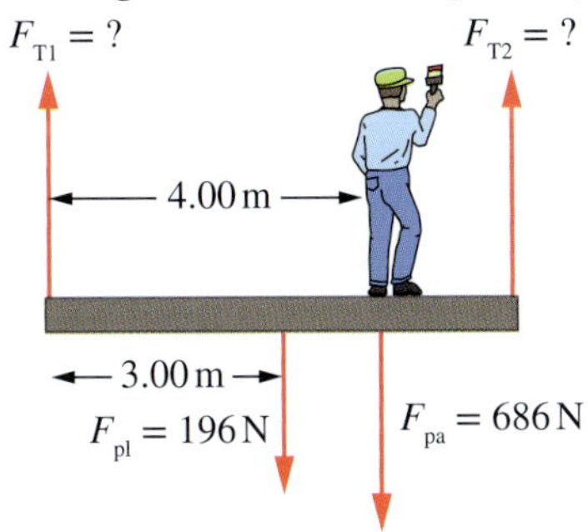

Determine the tension on the left-hand rope (F_{T1}).

Thinking	Working
Decide the reference point about which the torques will be calculated. Note that this must be the point at which the other unknown force acts.	The reference point is the point at which the rope providing the other tension force, F_{T2}, is attached.
Identify the variables involved and state them in their standard form. Remember to quote all $r_\perp$ lengths from the reference point, F_{T2}. Take the centre of the plank to be the point at which its force due to gravity acts.	$m_{pl} = 20\,kg$ $m_{pa} = 70\,kg$ $r_{\perp FT1} = 6.00\,m$ $r_{\perp pl} = 3.00\,m$ $r_{\perp pa} = 2.00\,m$ $g = 9.8\,N\,kg^{-1}$
Identify the object that is in rotational equilibrium. This is the object on which all the torques are acting.	The object in rotational equilibrium is the plank.
Decide which force causes the clockwise torques and which forces cause the anticlockwise torques around the chosen reference point.	The tension force of the left-hand rope on the plank provides the clockwise torque. The force of the painter on the plank provides an anticlockwise torque. The force of gravity on the plank provides another anticlockwise torque.
Apply the equation for rotational equilibrium.	$\Sigma\tau_{clockwise} = \Sigma\tau_{anticlockwise}$
Expand the equation to include each of the torques acting on the plank. Note the torque of the right-hand rope is not included as it acts through the reference point.	$r_{\perp FT1}F_{T1} = r_{\perp pl}F_{pl} + r_{\perp pa}F_{pa}$
Substitute the data into the equation and solve for the unknown, ensuring to give a direction.	$6.00 \times F_{T1} = 3.00 \times 20 \times 9.8 + 2.00 \times 70 \times 9.8$ $F_{T1} = \frac{(3.00 \times 20 \times 9.8) + (2.00 \times 70 \times 9.8)}{6.00}$ $= \frac{588 + 1372}{6.00}$ $= 326.67$ $= 3.3 \times 10^2\,N$ upwards

Worked example: Try yourself 14.3.4

CALCULATING STATIC EQUILIBRIUM WITH TWO UNKNOWNS

For the painter on the plank scenario in Worked example 14.3.4, determine the tension on the right-hand rope (F_{T2}).

Another way to determine the second unknown force is to apply the conditions for translational equilibrium. You can check these values using $F_{net} = 0$: the sum of the two upwards forces (tensions), 555 N + 327 N = 882 N. This balances the sum of the two downwards forces, 196 + 686 = 882 N.

OTHER STATIC EQUILIBRIUM SCENARIOS

The see-saw scenario and the supported plank are just two of many situations in which static equilibrium can be used to solve the forces involved. The conditions of static equilibrium will be applied to the following scenarios.

Cantilevers

A beam that extends beyond its support structure is called a **cantilever**. Cantilevers are common structural elements. For example, the diving board at the local pool is a cantilever. A cantilever bridge might be used to span a river or valley, as shown in Figure 14.3.11. Pillars are built at regular intervals across a river in order to support a number of beams. The cantilever beams are then joined at the centre of each span. The forces on the pillars are not affected by joining the beams; these are the same as if the beams were not connected. All the support for the cantilever is supplied by the pillars. Other structures that can involve cantilevers include shelving, awnings over the footpath outside some shops, and the wings of a plane.

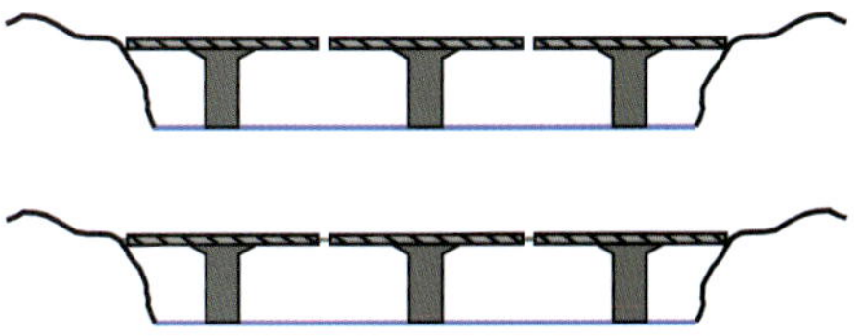

FIGURE 14.3.11 Each beam in the cantilever bridge is fully supported by the pillar below the beam. No added support is provided by connecting the beams.

When the centre of mass of the beam is not directly above a support, then the force due to gravity that acts on the centre of mass will provide a torque. This must be factored into the conditions of rotational equilibrium. In this case two supports are usually required, with one support providing an upwards force on the beam and the other support providing a downwards force on the beam.

FIGURE 14.3.12 A diving board showing that a metal strap is needed to provide the downwards force on the fixed end of the board

Figure 14.3.12 shows a swimming-pool diving board, with one support providing an upwards force on the board and the other support providing a downwards force on the board. A diving board must have a downwards force on the fixed end of the board to provide an opposing torque to the torque provided by the force due to gravity on the board.

Worked example 14.3.5

USING STATIC EQUILIBRIUM TO CALCULATE THE FORCE ON A CANTILEVER

A uniform cantilever beam 18.0 m long is used as a viewing platform. It extends 10.0 m beyond two supports that are 8.0 m apart. The beam has a mass of 300 kg.

(a)

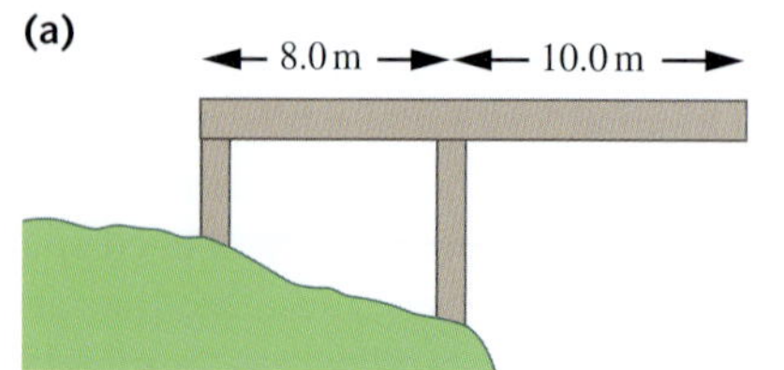

(b)

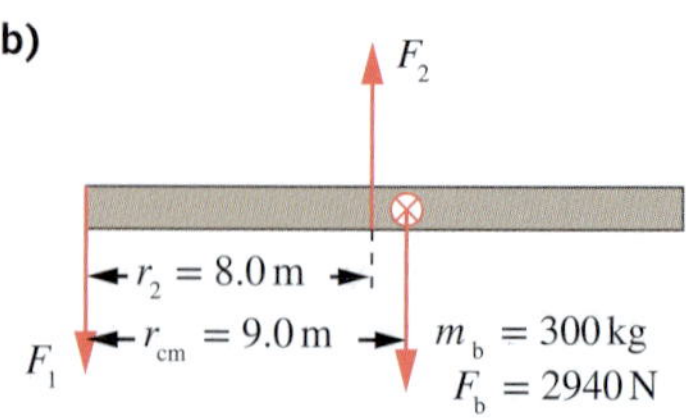

Determine the magnitude and direction of the force that the right-hand support must supply so that the beam is in static equilibrium (F_2).

Thinking	Working
Decide the reference point about which the torques will be calculated. Note that this must be the point at which the other unknown forces act.	The reference point is the point at which the left-hand support providing the force F_1 is attached to the beam.
Identify the variables involved and state them in their standard form.	$m_b = 300$ kg $r_{\perp F2} = 8.0$ m $r_{\perp cm} = 9.0$ m $g = 9.8$ N kg^{-1}
Identify the object that is in rotational equilibrium. This is the object upon which all the torques are acting.	The object in rotational equilibrium is the beam.
Decide which force causes the clockwise torque and which force causes the anticlockwise torque around the chosen reference point.	The force of the right-hand support on the beam provides the anticlockwise torque. The force of gravity on the beam provides the clockwise torque.
Apply the equation for rotational equilibrium.	$\Sigma\tau_{clockwise} = \Sigma\tau_{anticlockwise}$
Expand the equation to include each of the torques acting on the beam. Note the torque of the left-hand support is not included, as it acts through the reference point.	$r_{\perp cm}F_b = r_{\perp F2}F_2$
Substitute the data into the equation and solve for the unknown.	$9.0 \times 300 \times 9.8 = 8.0 \times F_2$ $F_2 = \frac{9.0 \times 300 \times 9.8}{8.0}$ $= \frac{24\,460}{8.0} = 3307.5$ $= 3.3 \times 10^3$ N
State the direction of the force acting on the object in equilibrium.	The force is upwards on the beam.

Worked example: Try yourself 14.3.5

USING STATIC EQUILIBRIUM TO CALCULATE THE FORCE ON A CANTILEVER

For the cantilever shown in Worked example 14.3.5, determine the magnitude and direction of the force that the left-hand support must supply so that the beam is in static equilibrium (F_1).

Another way to determine the second unknown force is to apply the conditions for translational equilibrium. We can check these values using $\Sigma F = 0$: the sum of the upwards force (F_2) is 3308 N. This balances the sum of the two downwards forces (F_b and F_1): 2940 + 368 = 3308 N.

CASE STUDY **ANALYSIS**

The Bolte Bridge

A cantilever bridge you may be very familiar with is the Bolte Bridge in Melbourne (Figure 14.3.13), a 490 m long bridge spanning the Yarra River and Victoria Harbour. The Bolte Bridge is known as a twin cantilever bridge because it comprises two independent, balanced structures supported together in the middle by a main pier—hidden by the central towers (these towers are purely decorative and serve no structural purpose). One significant advantage of cantilever bridges is that they require minimal temporary structures during their construction. For example, in the case of the Bolte Bridge, the central pier was first supported with 2800 m^3 of concrete, with the first segments on either side attached to the pier column and temporarily supported. Segments were then cast in place, incrementally adding to the cantilever by about 4.3 metres at a time. At the same time, the left and right piers were constructed and their cantilever segments continually added until the bridge formed one continuous structure. This took three years in total and was the largest urban road project ever undertaken in Australia.

FIGURE 14.3.13 The Bolte Bridge in Melbourne is a large twin cantilever bridge. The piers are used to support different sections of road that project out horizontally.

Analysis

Imagine you are an engineer responsible for constructing the Bolte Bridge. You need to perform some calculations to accurately predict the different stresses to which the bridge will be subjected during both its construction and when it is operational.

1. The bridge is large enough to support approximately 850 cars. Assuming that the average mass of a car is 1800 kg, what is the minimum force due to gravity that each of the ten main pillars should be able to support?
2. Cantilever bridges are built by starting from a central pier and moving outwards incrementally with pairs of 'form travellers' that become the bridge. In the Bolte Bridge, these form travellers were 110 tonnes each and extended 4.325 metres horizontally.
   - **a** Explain why cantilevered construction would require a pair of two form travellers to be moved away from the central pier in tandem and why this minimises the need for falsework (temporary structures to support the bridge during construction).
   - **b** Suppose that it was estimated that 50 tonnes of workers and machinery would need to be at the edge of the right-most form traveller. In this event, it is necessary to place a temporary foundation halfway across the right form traveller. Calculate the magnitude and direction of the force that the support must supply so that the structure is in static equilibrium.

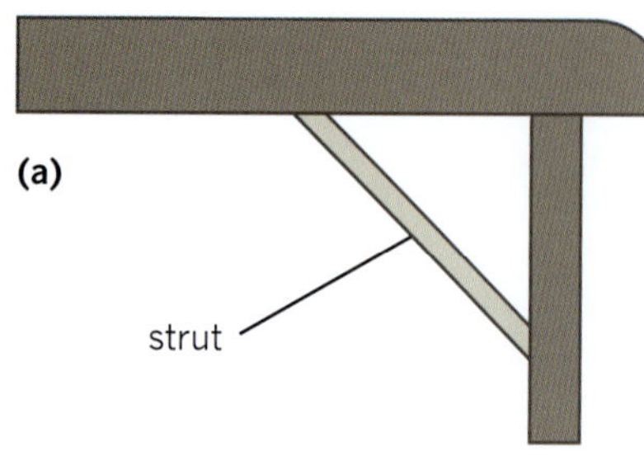

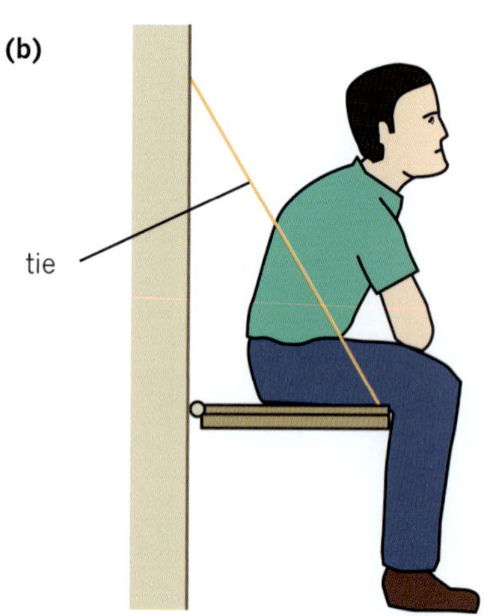

FIGURE 14.3.14 A strut (a) helps to support a cantilevered beam and is under compression. (b) A tie helps to support a fold-out bench and is under tension.

Struts and ties

As well as their main beams and pillars, many structures have additional ways to strengthen them. A structure, such as a cantilever, may be supported by struts and ties. A strut will be under compression (squeezed) and must be rigid. A tie will be under tension (stretched) and may be rigid or flexible. Struts and ties are shown in Figure 14.3.14.

Worked example 14.3.6

USING STATIC EQUILIBRIUM TO CALCULATE THE TENSION IN A TIE THAT IS SUPPORTING A BEAM

A sign of mass 10 kg is suspended from the end of a uniform 2.0 m long cantilevered beam. The other end of the beam is attached to the wall by a hinge labelled *h*. The beam has a mass of 25 kg and is further supported by a wire tie that makes an angle of 30° to the beam. The wire is attached to the beam at a point 1.5 m from the wall. Use $g = 9.8\,\text{N kg}^{-1}$ and ignore the mass of the wire for these calculations.

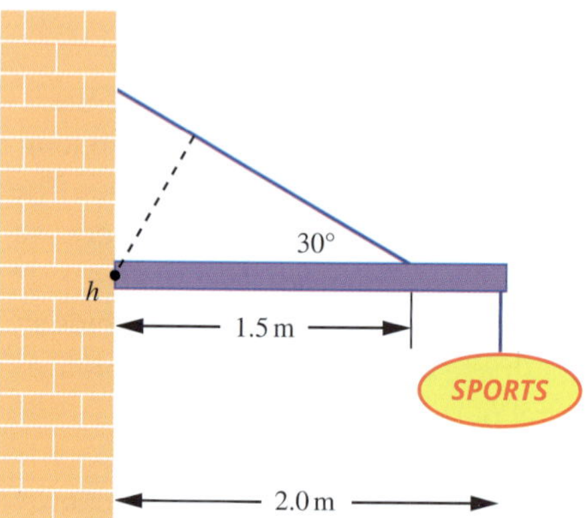

Calculate the magnitude of the tension (F_T) in the wire that is supporting the beam.

Thinking	Working
Decide the reference point about which the torques will be calculated. Note that this must be the point at which the other unknown force acts.	The reference point is the hinge (*h*) at which the beam is connected to the wall.
Identify the variables involved and state them in their standard form. Here the subscripts used are: b = beam, cm = centre of mass of the beam, s = sign and w = wire.	$m_b = 25\,\text{kg}$ $m_s = 10\,\text{kg}$ $r_{\perp cm} = 1.0\,\text{m}$ $r_{\perp s} = 2.0\,\text{m}$ $r_w = 1.5\,\text{m}$ $g = 9.8\,\text{N kg}^{-1}$
Identify the object that is in rotational equilibrium. This is the object on which all the torques are acting.	The object in rotational equilibrium is the beam.
Decide which force causes the anticlockwise torque and which forces cause the clockwise torque around the chosen reference point.	The force of the wire tie on the beam provides the anticlockwise torque. The force of gravity on the beam provides one clockwise torque. The force of gravity on the sign provides another clockwise torque.

Apply the equation for rotational equilibrium.	$\Sigma\tau_{clockwise} = \Sigma\tau_{anticlockwise}$
Expand the equation to include each of the torques acting on the beam.	$r_{\perp cm}F_b + r_{\perp s}F_s = r_{\perp w}F_w$
Solve for the perpendicular distances from the force arm to the line of action of the force.	$r_{\perp w} = r_w \sin 30°$ $= 1.5 \times \sin 30°$ $= 0.75\,m$
Substitute the data into the equation and solve for the unknown force.	$(1.00 \times 25 \times 9.8) + (2.00 \times 10 \times 9.8) = 0.75 \times F_T$ $F_T = \frac{245+196}{0.75}$ $= \frac{441}{0.75}$ $= 5.9 \times 10^2\,N$ Note that by finding the perpendicular distance, the tension force can be calculated directly by the equation. If the vertical component of the tension force was found then the tension force would need to be calculated.

Worked example: Try yourself 14.3.6

USING STATIC EQUILIBRIUM TO CALCULATE THE TENSION IN A TIE THAT IS SUPPORTING A BEAM

A uniform 5.0 kg beam, 1.8 m long, extends from the side of a building and is supported by a wire tie that is attached to the beam 1.2 m from a hinge (h) at an angle of 45°. Calculate the magnitude of the tension (F_T) in the wire that is supporting the beam.

Use $g = 9.8\,N\,kg^{-1}$ and ignore the mass of the wire for these calculations.

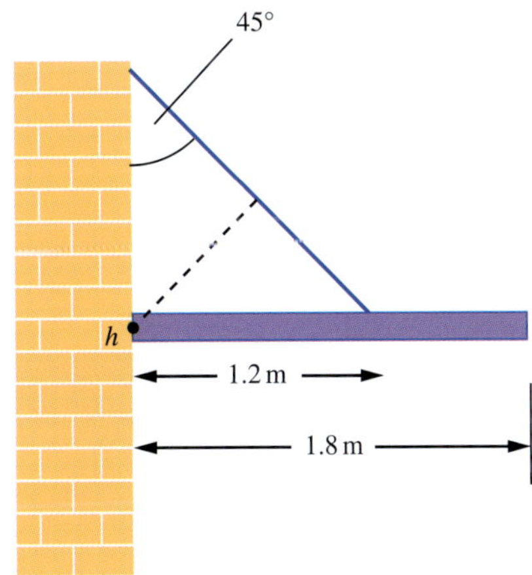

14.3 Review

SUMMARY

- The centre of mass is a single point in an object at which the mass can be considered to be 'concentrated' for the purposes of analysing motion.
- The net external force acting on a body can be considered to act on its centre of mass.
- The centre of mass of a system of masses m_i located at positions x_i may be calculated by:

$$x_{cm} = \frac{m_1x_1 + m_2x_2 \ldots + m_nx_n}{m_1 + m_2 + \ldots m_n}$$

- For stability, the centre of mass must lie above the base of support.
- Static equilibrium occurs when an object experiences translational equilibrium and rotational equilibrium.
- Static equilibrium can be represented mathematically as $F_{net} = 0$ (translational equilibrium) and $\tau_{net} = 0$ (rotational equilibrium).
- Rotational equilibrium can be represented mathematically as $\Sigma\tau_{clockwise} = \Sigma\tau_{anticlockwise}$.
- In calculations of static equilibrium with only one unknown force, the reference point is the point about which the torques act.
- In calculations of static equilibrium with two unknown forces, the reference point can be placed at the point at which one of the unknown forces acts. This eliminates any torque due to this force as the distance from the force to the reference point is zero.

KEY QUESTIONS

Assume $g = 9.8\,\text{N kg}^{-1}$ when answering these questions.

Knowledge and understanding

1 Which one of the following describes the torques acting on an object in rotational equilibrium?

A It must have only clockwise torques.

B It must have only anticlockwise torques.

C It must have unequal clockwise and anticlockwise torques.

D It must have equal clockwise and anticlockwise torques.

2 A cyclist is coasting down a road at a constant velocity while standing on the pedals. Which object in the list below is in static equilibrium?

A the rear wheel

B the front and rear spokes

C the front cog connected to the pedals

D the front wheel

3 An adult of mass 68.0 kg sits on a see-saw at a playground with a 32.0 kg child sitting 3.15 m from the pivot point on the other side from the adult. Calculate where the adult must sit for the see-saw to remain balanced and horizontal.

4 Where, approximately, is the centre of mass for the human body when standing upright?

Analysis

5 A makeshift shelf is used in a bakery to store sacks of flour. The shelf is constructed using an 8.0 m beam with a mass of 40 kg, with a support positioned at each end. The shelf holds sacks of mass 75 kg, 120 kg and 160 kg at the positions shown in the figure below. Calculate the forces on the beam due to the left and right supports.

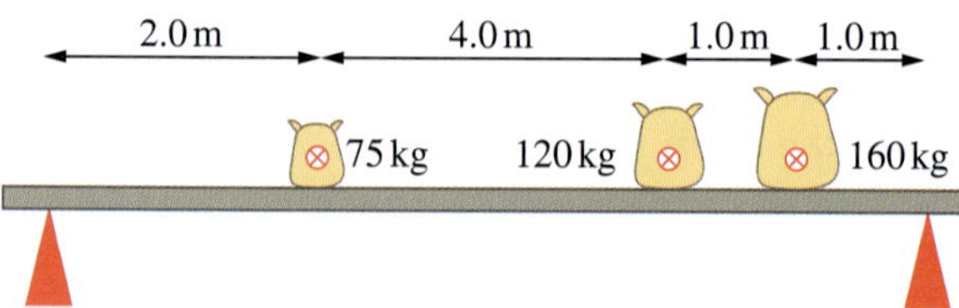

6 A 6.0 m cantilever-type awning is constructed on the roof of a shop so that it shades the front. The awning has a mass of 1200 kg and is supported by two columns, X and Y, which produce forces on the awning of F_X and F_Y respectively.

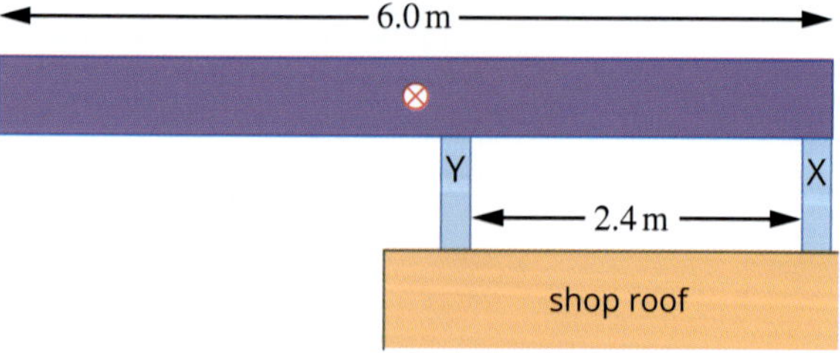

a Determine the direction in which F_X acts on the awning.

b Determine the force supplied by column Y on the awning (F_Y).

c Determine the force supplied by column X on the awning (F_X).

7 An adult of mass 86.0 kg sits on a see-saw at a playground with two 23.0 kg children. The adult sits on the opposite side from the children. One child is sitting 1.35 m from the pivot point and the other sitting 2.70 m from the pivot point. Calculate where the adult must sit for the see-saw to remain balanced and horizontal.

8 The end-post of a fence is held in position by a backstay wire that is under a tension of 620 N at an angle of 45° to the horizontal. A horizontal fence wire connects to the other posts in the fence. A diagram of the fence is below.

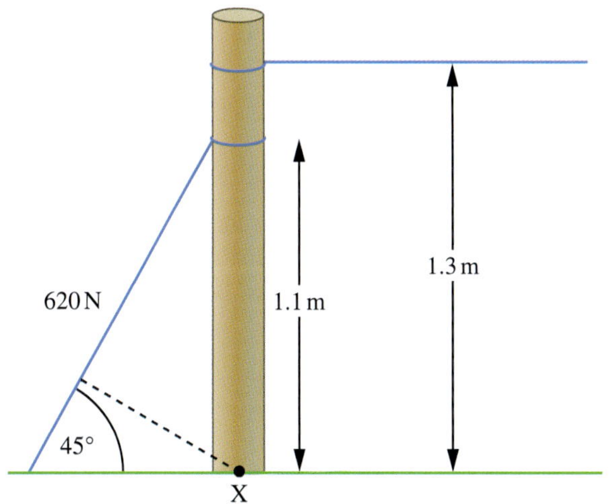

a Determine the values for the horizontal and vertical components of the tension in the backstay wire.

b By considering the base of the post to be a pivot point, determine the size of the tension in the fence wire, F_T.

9 Consider a person with a height of 180 cm and a mass of 80 kg. Assume that you can simplify the person so they are made up of five symmetrical parts and each part has a uniform mass distribution. The parts are feet, with height 8.0 cm and mass 2.0 kg; lower legs, with height 42 cm and combined mass 6.0 kg; upper legs, with height 40 cm and combined mass 14 kg; torso, with height 60 cm and mass 50 kg; and head, with height 30 cm and mass 8.0 kg. The diagram to the right is a schematic representation of this person. Use this information to calculate the height of the person's centre of mass when they are standing upright.

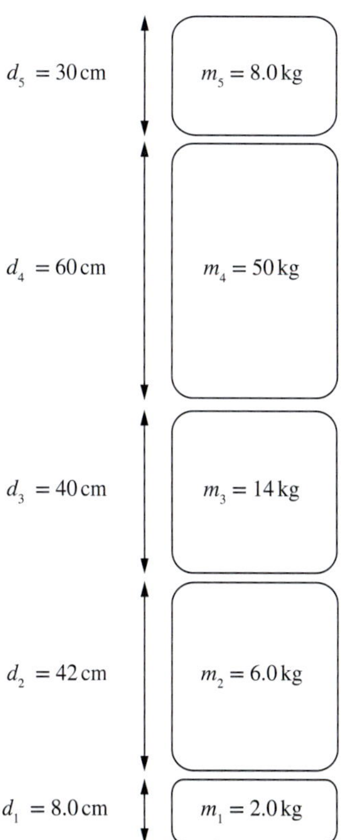

10 Explain in your own words why the centre of mass of the person shown lies outside their body, and why they are unstable as they are currently standing.

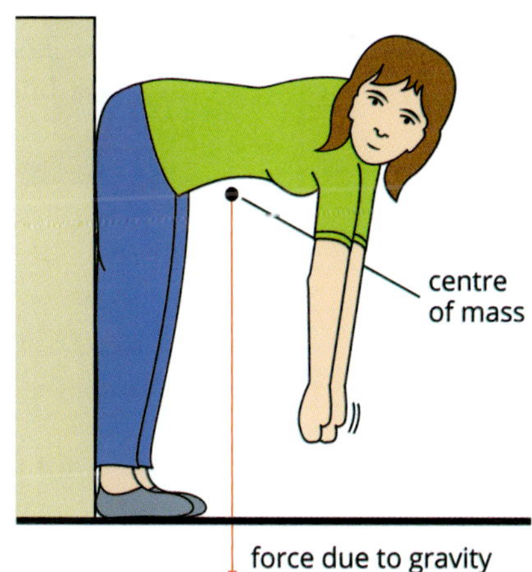

continued over page

14.3 Review *continued*

11 Fred has a plain wall in his backyard on which he decides to build a fold-out bench. He would like the bench to be able to support two people with a total mass of 150 kg and with their force due to gravity evenly distributed. He opts for the design shown in the figure below: a 6.0 kg bench, which is 0.80 m deep, and attached to a hinge. He purchases some wires to act as a tie for the bench. One end will be attached to the edge of the seat, and one end attached to a point on the wall. There is one wire to support each end of the bench. The wires are rated at being able to support 467 N of tension before they will snap, and Fred is able to cut them to any length to suit the design.

a Is Fred safer to attach the wire higher up or lower down on the wall? Explain your reasoning.

b Calculate the exact angle at which the tension in the wire matches its rating with the two people sitting on it.

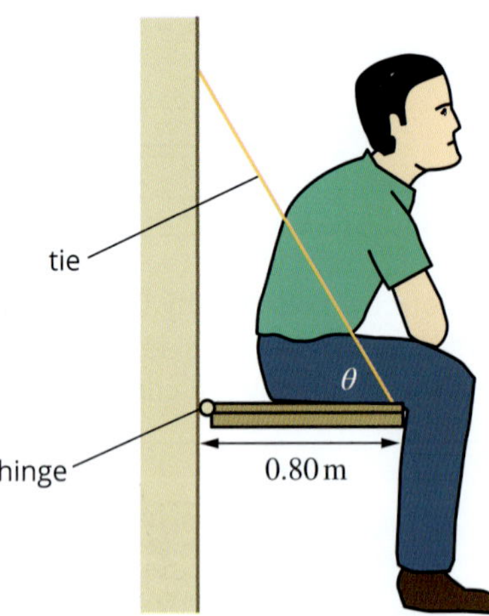

Chapter review

KEY TERMS

axis of rotation
base
cantilever
centre of gravity
components
force arm
line of action of the force
neutral equilibrium
pivot point
principle of moments
reference point
rotational equilibrium
stable equilibrium
static equilibrium
torque
translational equilibrium
unstable equilibrium

REVIEW QUESTIONS

Assume $g = 9.8\,\text{N kg}^{-1}$ when answering these questions.

Knowledge and understanding

1 Which option below explains the factors affecting the torque on an object?
- **A** Torque only increases if the force applied to the object is increased.
- **B** Torque increases when either the force or the force arm increases, or both increase.
- **C** Torque only increases when the force arm decreases.
- **D** Torque only increases when the force arm increases.

2 Calculate the torque applied to a bolt by a 26.0 cm long spanner that has a force of 35.0 N acting at 90° to its length and at the end of the spanner.

3 Calculate the radius of the wheel on a pressure valve that supplies a torque on the valve of 4.35 N m when a force of 15.0 N is applied.

4 Which of the following options would provide you with the greatest torque when opening a 1 metre–wide door?
- **A** pushing with a force of 40 N at right angles to the door, in the middle of the door
- **B** pushing with a force of 32 N at right angles to the door, 32 cm from the handle edge of the door
- **C** pushing with a force of 60 N at right angles to the door 15 cm from the hinges of the door
- **D** pushing with a force of 19 N at right angles to the door, at the handle edge of the door

5 Demolishers wish to knock over a concrete wall. They plan to use a wrecking ball that exerts a 7.5 kN force as it hits the wall. If the ball hits at a point that is 4.2 m above the ground, calculate the torque that is developed on the wall if it pivots at its base.

6 Calculate the length of a spanner that is used to tighten a nut to a torque of 28.0 N m when a force of 14.0 N is applied at right angles to the spanner, at the end of the spanner.

7 Which of the following is the correct description of the maximum torque on a door?
- **A** The maximum torque will be achieved when the force is at a 60° angle to the door.
- **B** The maximum torque will be achieved when the force is parallel to the door.
- **C** The maximum torque will be achieved when the force is at a 90° angle to the door.
- **D** The maximum torque will be achieved when the force is at a 120° angle to the door.

8 A camper ties a rope from the top of a 1.90 m tent pole to a peg on the ground. The rope is tightened so that the rope applies a 45.0 N force at an angle of 60.0° to the pole. Calculate the torque that is applied on the tent pole due to the rope if the pole pivots at its base.

9 Calculate the torque on a 65.0 cm spanner that is used to tighten a nut when a force of 45.0 N is applied at 50.0° to the spanner, at the end of the spanner.

10 A rope is attached at 40.0° to a freshly planted tree. The line of action of the force is along the same line as the rope, and the rope is attached to the tree 1.50 m up from the bottom of the tree. Assume the base of the tree is the pivot point.
- **a** Calculate the length of the perpendicular force arm of the rope.
- **b** Calculate the torque on the tree if the force applied to the tree by the rope is 16.5 N.

continued over page

11 A piece of flat-pack furniture is being assembled using an allen key. Calculate the torque applied on a screw by a 12.0 cm long allen key that has a force of 14.5 N acting at 90° to its length and the end of the allen key.

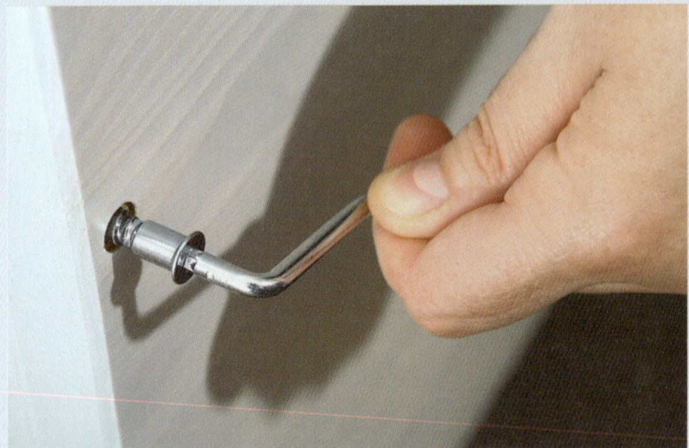

12 A mechanic uses a trolley jack to jack up a car. The mechanic pushes vertically down on the end of a 120 cm lever with a force of 76.0 N, as shown in the diagram below. The lever is at an angle of 35.0° upwards from the horizontal. Calculate the torque acting on the pivot.

13 A street performer stands on a rope tied between two bollards. When the performer is standing at the centre, the rope makes an angle of 25.0° to the horizontal. Assuming the mass of the rope is negligible and that the mass of the performer is 85.0 kg, calculate the tension in the rope.

14 Calculate the mass of a pendulum bob that hangs stationary from a chain if the tension in the chain is 12.50 N.

15 The manager of a bowling alley wants to promote the business by suspending a 125 kg fibreglass bowling ball on a frame outside the alley. The bowling ball is supported by two steel cables capable of withstanding up to 2250 N of tension before breaking. Cable A is at an angle of 45.0° to the horizontal frame member and cable B is perpendicular to the vertical frame member. Assuming the mass of the cables is negligible, calculate the tension in cable A and in cable B.

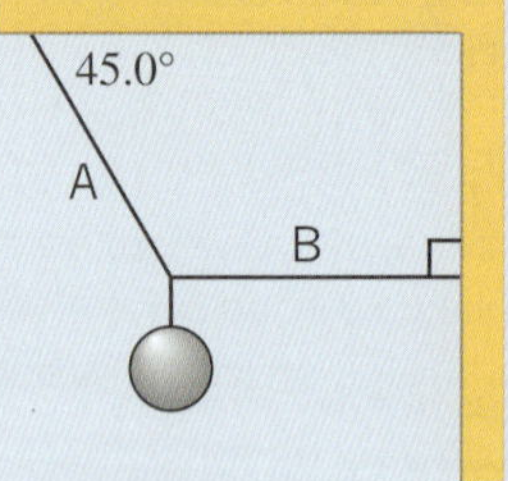

16 Which of the following correctly describes rotational equilibrium?

A A net torque acts perpendicular to a reference point and rotation does not occur.

B A net torque acts about the reference point and rotation occurs.

C No net torque acts about the reference point and rotation does not occur.

D No net torque acts about the reference point and rotation occurs.

17 When an object is in static equilibrium, it experiences

A rotational equilibrium, but not translational equilibrium.

B rotational equilibrium and translational equilibrium.

C neither rotational equilibrium nor translational equilibrium.

D constant translational velocity and constant rotational motion.

18 Tom is riding his skateboard and is in a state of translational equilibrium. Which one or more of the following statements could be true of Tom's motion?

A He is maintaining a constant velocity.

B He is slowing down.

C He is experiencing a translational acceleration.

D He is experiencing a rotational acceleration.

Application and analysis

19 Consider the situation shown below.

a Give the magnitude and direction of the torque, τ_1, on the see-saw caused by the force due to gravity of the child on the left.

b Give the magnitude and direction of the torque, τ_2, on the see-saw caused by the force due to gravity of the child on the right.

c Give the magnitude and direction of the net torque acting on the see-saw.

20 For the situation shown below, determine the magnitude and direction of the resultant torque acting on the see-saw.

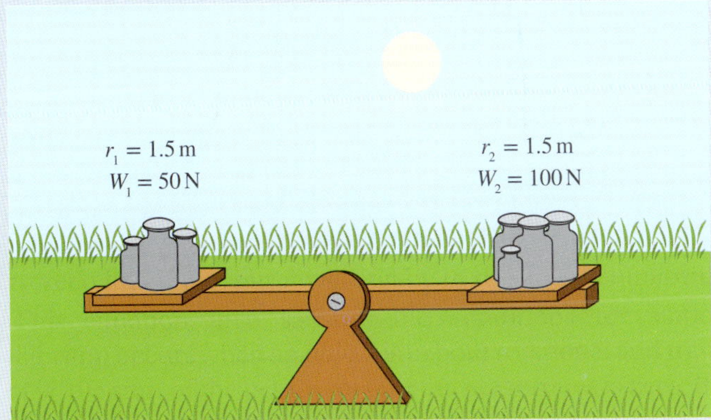

21 A child pulls down on a lever-type door handle with a force of 45.0 N. If the length of the handle is 7.5 cm then calculate the torque acting on the handle's pivot. Assume the force is applied at right angles to the handle.

22 A signalman is responsible for the safe routing of trains through their section via the use of levers that control points and signals, as shown in the figure below. Calculate the torque on the axle at the bottom of a 1.42 m lever if an 80.0 N force acts at an angle of 30.0° to the lever's perpendicular radius.

23 Hand-pumps are capable of pumping water from wells instead of using a bucket and rope. A man pushes horizontally on the end of a 125 cm lever with a force of 70.0 N, as shown in the diagram below. The lever is shown at an angle of 65.0° down from the horizontal. Calculate the torque acting on the pivot.

24 A climber is hanging from a rope halfway up a rock wall. If the climber has a mass of 95.5 kg, calculate the tension force provided by the rope in order for the climber to be in translational equilibrium.

continued over page

25 A dining table top of mass 75.0 kg is supported by four legs. On the table there is 3.25 kg of food. The table is stationary, and each leg provides the same support as each of the other legs. Calculate the force of one of the legs on the table.

26 A 15.0 kg sign is hung between two poles by two cables of negligible mass, as shown in the diagram below. The sign is in translational equilibrium. Calculate the tension in the shorter left-hand cable and the longer right-hand cable.

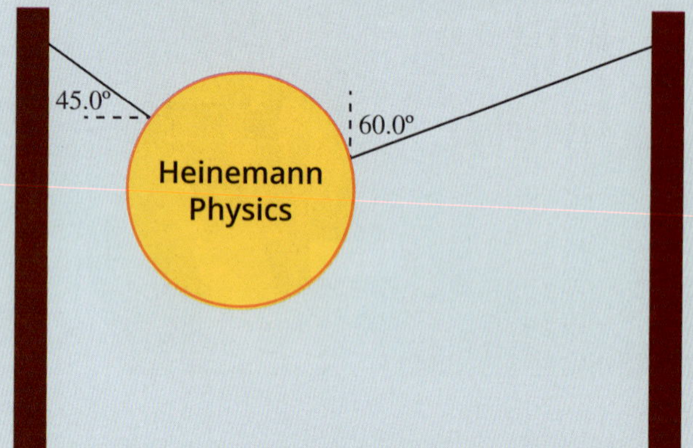

27 A child's toy is suspended above her bed from a string attached to the ceiling. The toy is made from a 0.80 m long aluminium rod of mass 12.5 g, with a 225 g Sun hanging at one end of the rod and a 30 g Earth hanging at the other end. Calculate how far from the Sun, in centimetres, the string needs to be tied to the bar so that the whole toy is in static equilibrium.

28 A pedestrian bridge over a small creek is made from two identical 3.0 m long cantilevers, each of mass 400 kg.

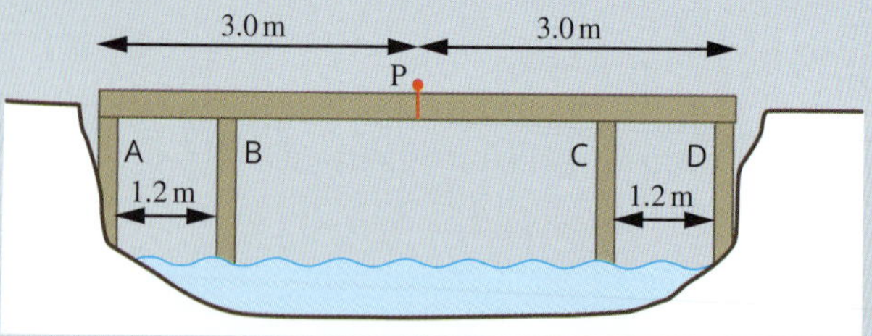

a Calculate the reaction forces produced by pillars A, B, C and D when:

i there are no pedestrians on the bridge

ii a 70 kg woman stands at position P with half her force due to gravity on each cantilever.

b What happens to the values of the forces in A and B as the woman walks from A past B to P?

OA ✓✓

UNIT 2 • Area of Study 1

REVIEW QUESTIONS

WS 37

How is motion understood?

Multiple-choice questions

1 Select the list below that contains only vector quantities.

A displacement, velocity, acceleration, force
B displacement, speed, acceleration, force due to gravity
C distance travelled, velocity, acceleration, force
D displacement, velocity, acceleration, mass

The following information relates to questions 2 and 3.

A car accelerates in a straight line at a rate of 5.5 m s^{-2} from rest.

2 What distance has the car travelled at the end of three seconds?

A 8.3 m
B 11 m
C 16 m
D 25 m

3 What distance does the car travel in the third second of its motion?

A 8.3 m
B 14 m
C 19 m
D 25 m

4 Two boats are tied together with an inextensible rope as shown in the diagram below. Boat A has twice the mass of boat B.

Boat A exerts a force of +*F* newtons on boat B.

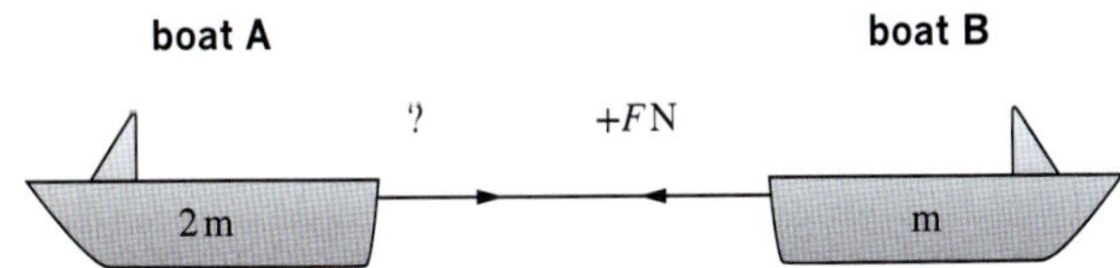

What is the magnitude and direction of the force on boat A by boat B?

A $-\frac{1}{2}F$N
B +*F*N
C −*F*N
D −2*F*N

The following information relates to questions 5 and 6.

A graph depicting the velocity of a small toy train versus time is shown below. The train is moving on a straight section of track, and is initially moving in the easterly direction.

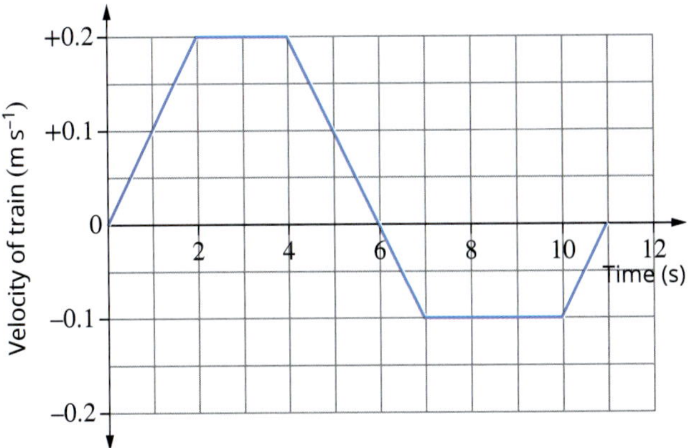

5 What distance does the train travel in the first 6 seconds of its motion?

A 0 m
B 0.4 m
C 0.8 m
D 1.2 m

6 What is the displacement of the train after the first 11 seconds of its motion?

A 0 m
B 0.4 m east
C 0.8 m east
D 1.2 m east

7 A ball is dropped, falls vertically and strikes the ground with a velocity of +5 m s^{-1}. It rebounds and leaves the ground with a velocity of −3 m s^{-1}. What is the change in velocity that the ball experiences?

A −8 m s^{-1}
B +8 m s^{-1}
C −2 m s^{-1}
D +2 m s^{-1}

The following information relates to questions 8 and 9.

A 60 kg student stands on a set of digital scales in an elevator, as shown in the diagram below.

8 In which of the following situation/s will the digital scales show a reading of 60 kg?

A when the lift is travelling upwards at constant velocity

B when the lift is travelling downwards at constant velocity

C when the lift is stationary

D all of the above

9 The lift travels upwards with an acceleration of *a*. Taking up as positive (+), which of the following is the correct expression for the normal force (F_N) on the student? m = mass of the student, g = acceleration due to gravity

A $F_N = mg + ma$

B $F_N = mg - ma$

C $F_N = ma - mg$

D $-F_N = mg + ma$

10 The orbit of the Moon around Earth can be modelled as circular motion in which the Moon can be considered to orbit at a fixed height above the surface of Earth (and so at a fixed radius from the centre of Earth) and at a constant speed. Earth exerts a gravitational force on the Moon that acts at right angles to the velocity of the Moon, as shown in the diagram below. The diagram is not to scale, as the distance between Earth and the Moon is much larger.

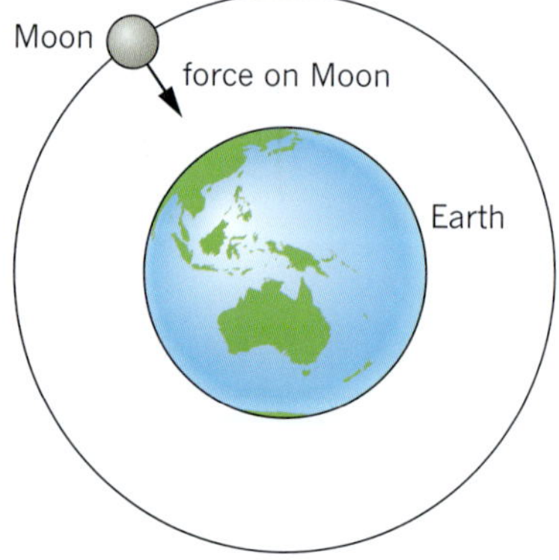

Which of the following statements about the Moon is correct?

A The Moon experiences no change in kinetic energy during an orbit.

B The Moon experiences no change in gravitational potential energy during an orbit.

C Earth's gravitational force does no work on the Moon.

D All of the above statements are correct.

11 Two dynamics carts are being used in the physics laboratory to study momentum. The two carts have the same mass, *m* kg, and are travelling towards each other with a speed v m s^{-1}, as shown in the diagram. Take the right-hand direction to be positive.

What is the total momentum of this system?

A 0 kg m s^{-1}

B $+mv$ kg m s^{-1}

C $+2mv$ kg m s^{-1}

D $-2mv$ kg m s^{-1}

12 A single spring has a spring constant of 10 N m^{-1}. What mass must be suspended from the spring to cause the spring to stretch (extend) by 20 cm? Choose the closest answer.

A 2.0 g

B 20 g

C 0.20 kg

D 2.0 kg

Short-answer questions

13 A car with good brakes, but smooth tyres, can slow down at a rate of 4.0 m s^{-2} on a wet road. The driver has a reaction time of 0.50 s. The car is travelling at 72 km h^{-1} when the driver sees a danger and reacts by braking.

a Calculate the distance travelled by the car during the reaction time.

b Calculate the braking time.

c Determine the total distance travelled by the car from the time the driver realises the danger to the time the car finally stops.

14 Yindi and Emily conduct the following experiment from a skyscraper. Yindi drops a platinum sphere from a vertical height of 122 m while at exactly the same time Emily throws a lead sphere with an initial downwards vertical velocity of 10.0 m s^{-1} from a vertical height of 140 m. Assume $g = -9.8$ m s^{-2} and ignore friction and air resistance.

a Determine the time taken by the platinum sphere to strike the ground.

b Calculate the time taken by the lead sphere to strike the ground.

c Using your values from parts a and b, determine the average velocity of each sphere over their respective distances.

15 An Olympic archery competitor tests a bow by firing an arrow of mass 25 g vertically into the air. The arrow leaves the bow with an initial vertical velocity of $98\,m\,s^{-1}$. The acceleration due to gravity may be taken as $g = 9.8\,m\,s^{-2}$ and the effects of air resistance can be ignored.

a Determine the time the arrow reaches its maximum height.

b Calculate the maximum vertical distance that this arrow reaches.

c Explain what value of acceleration the arrow has when it reaches its maximum height.

d If the arrow lands at the same height from which it was vertically fired, and the archer has moved safely out of the way, calculate the average speed and average velocity of the arrow.

e Describe the energy transformations occurring during the flight, from after the arrow leaves the bow to when it comes back down.

16 a Describe the acceleration depicted by each of the following graphs.

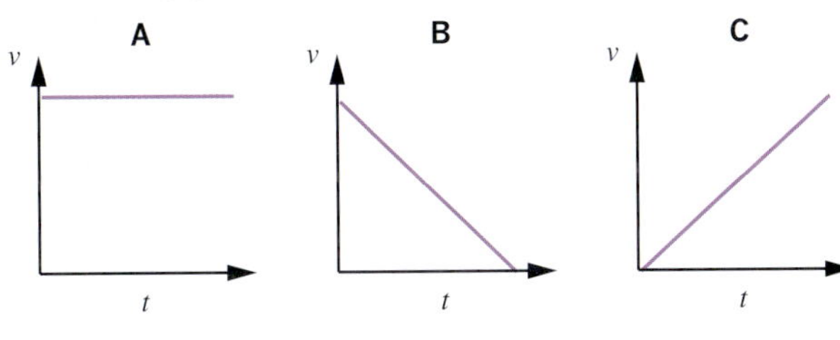

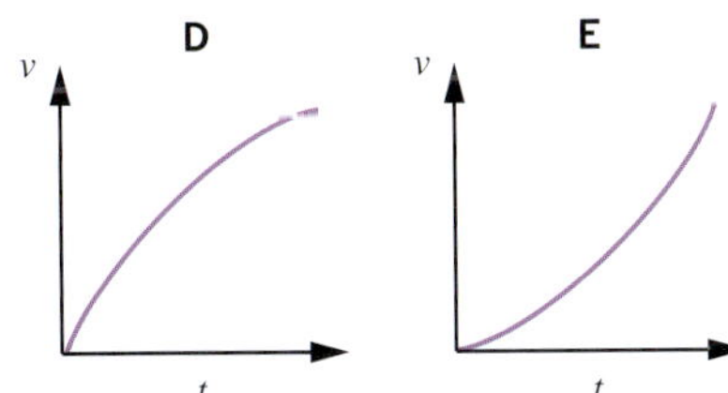

b If the graphs are taken to depict the same time interval, state and explain which one represents the greatest displacement.

c State which graph could describe the following motion.

i a ball being dropped from a height with air resistance neglected

ii a skydiver jumping out of a plane, allowing for air resistance

iii an ice skater gliding on the ice with negligible friction

17 During a physics experiment, a student sets a multiflash timer at a frequency of 10 Hz, so that the time between photos is 0.1 s. A force is applied to mass that slides across a frictionless horizontal table. The diagram shows the position, A, B, C and D, of the mass for the first four flashes.

Assume that when flash A occurred $t = 0$, at which time the mass was at rest.

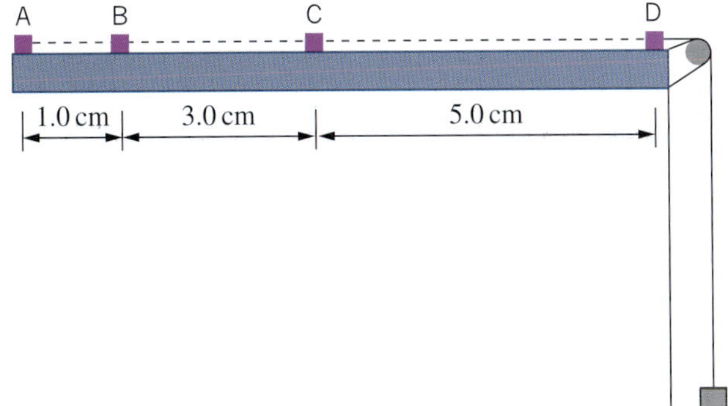

a Determine the average speed of the mass for these distance intervals:

i A to B

ii B to C

iii C to D

b Determine the instantaneous speed of the mass for these times:

i $t = 0.05\,s$

ii $t = 0.15\,s$

iii $t = 0.25\,s$

c Describe the motion of the mass.

18 A 100 kg man is standing at rest on the ground. Use $g = 9.8\,m\,s^{-2}$.

a Name the forces acting on the man, using the convention $F_{\text{on A by B}}$ to describe the force on A by B.

b Compare the relative magnitudes of these forces.

c Describe the forces that form Newton's third law pairs with the forces acting on the man.

19 A tow truck, pulling a car of mass 1000 kg along a straight road, causes the velocity of the car to increase from 5.00 m s^{-1} west to 10.0 m s^{-1} west over a distance of 100 m. A constant frictional force of 200 N acts on the car.

a Calculate the acceleration of the car.

b Calculate the net force acting on the car during this 100 m.

c Calculate the force exerted on the car by the tow truck.

d What force does the car exert on the tow truck?

20 A car that is initially at rest begins to roll down a steep road that makes an angle of 11.3° with the horizontal. Ignoring friction, determine the speed of the car in km h^{-1} after it has travelled a distance of 100 m. (Assume g = 9.8 m s^{-2}.)

21 A 100 kg trolley is being pushed up a rough 30° incline by a constant force F. The frictional force F_f between the incline and the trolley is 110 N.

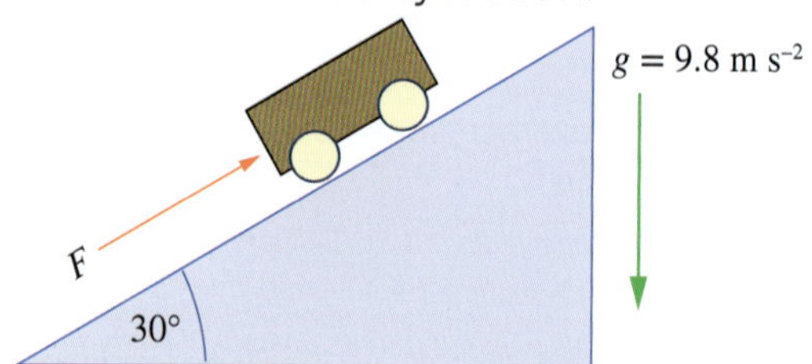

a Determine the value of F that will move the trolley up the incline at a constant velocity of 5.0 m s^{-1}.

b Determine the value of F that will accelerate the trolley up the incline at 2.0 m s^{-2}.

c Calculate the acceleration of the trolley when F = 1000 N.

22 Two masses, 10 kg and 20 kg, are attached via an inextensible steel cable to a frictionless pulley, as shown in the following diagram.

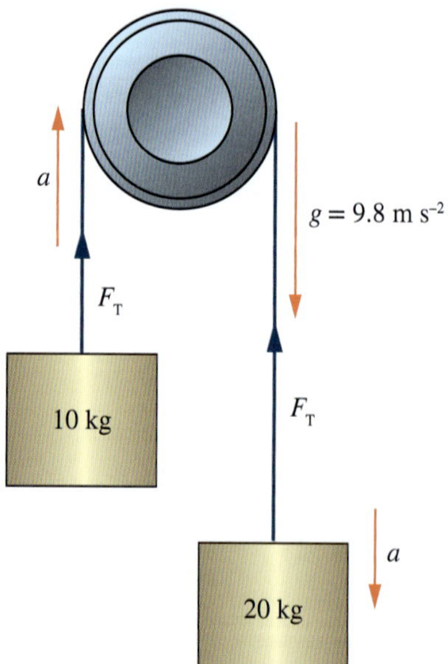

a Determine the acceleration for each mass.

b Calculate the magnitude of the tension in the cable.

23 An 800 N force is applied, as shown, to a 20.0 kg mass that is initially at rest on a horizontal surface. During its subsequent motion, the mass encounters a constant frictional force of 100 N while moving along a horizontal distance of 10 m.

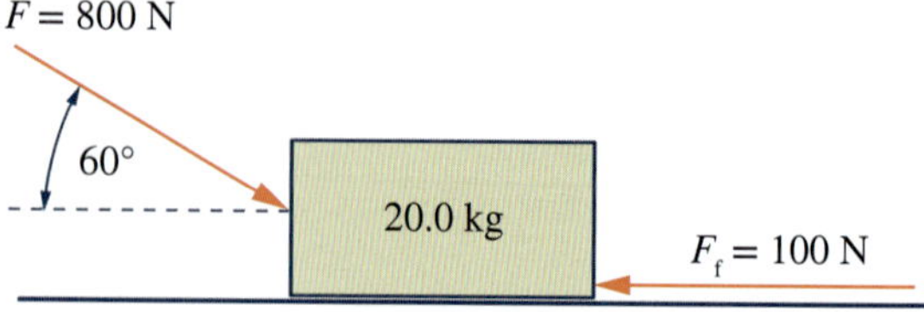

a Determine the resultant horizontal force acting on the 20.0 kg mass.

b Calculate the work done by the frictional force in kJ.

c Calculate the work done by the resultant horizontal force in kJ.

d Determine the change in kinetic energy of the mass in kJ.

e Calculate the final speed of the mass.

24 Three children are playing on a see-saw. Farah has a mass of 35 kg and is sitting 0.7 m from the pivot point. Her younger brother Karim has mass 25 kg and is sitting 1.0 m from the pivot on the same side as Farah.

a Calculate and explain where their older brother Amir (mass 42 kg) should position himself in order to balance his two younger siblings.

b The siblings have a very advanced physics discussion and disagree on what kind of equilibrium the see-saw is in when they are all balanced.

Karim asserts that the see-saw is in translational equilibrium, but not in rotational equilibrium.

Farah thinks that the see-saw is in rotational equilibrium, but not in translational equilibrium.

Amir thinks that his siblings are both incorrect.

Explain why Amir is correct, by evaluating the statements of his siblings.

c Describe a scenario in which Karim is correct. You may wish to think about how to change the see-saw example to demonstrate this.

d Describe a scenario in which Farah is correct.

25 Calculate the upwards forces applied by the ground and the supporting column on the cantilever beam shown below given that the mass of the beam is 300 kg.

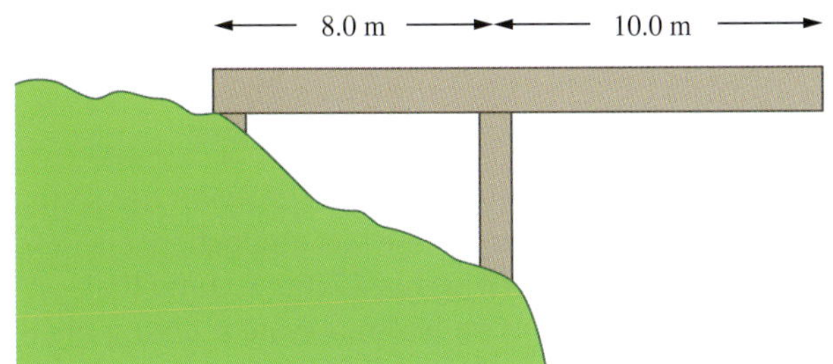

26 The figure below shows the velocity–time graph for a car of mass 2000 kg. The engine of the car is providing a constant driving force. During the 5.00 s interval, the car encounters a constant frictional force of 400 N. At t = 5.00 s, v = 40.0 m s^{-1}.

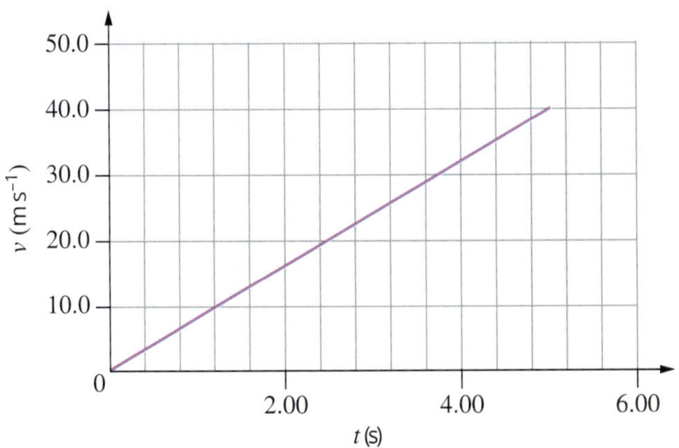

a Calculate the kinetic energy (in MJ) of the car at t = 5.00 s.

b Calculate the resultant force acting on the car.

c Determine the force provided by the car's engine (in kN) during the 5.00 s interval.

d Calculate how much work is done (in MJ) on the car during the 5.00 s interval.

e Determine the power output of the car's engine during the 5.00 s interval.

f How much heat energy (in kJ) is produced due to friction during the 5.00 s interval?

g Calculate the efficiency with which the energy provided by the engine is transformed into kinetic energy.

27 The following diagram shows the trajectory of a 2.0 kg shot-put recorded by a physics student during a practical investigation. The sphere is projected at a vertical height of 2.0 m above the ground with an initial speed of v = 10 m s^{-1}. The maximum vertical height of the shot-put is 5.0 m. (Ignore friction and assume g = 9.8 N kg^{-1}.)

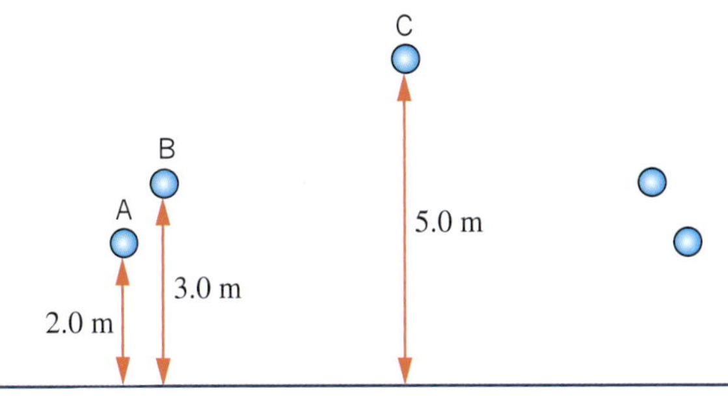

a Calculate the total energy of the shot-put just after it is released at point A.

b Determine the kinetic energy of the shot-put at point B.

c Calculate the minimum speed of the shot-put during its flight.

d State and explain the total energy of the shot-put at point C.

28 A 5.0 kg trolley approaches a spring that is fixed to a wall. During the collision, the spring undergoes a compression Δx and the trolley is momentarily brought to rest, before bouncing back at 10 m s^{-1}. The force–compression graph for the spring is shown below. (Ignore friction.)

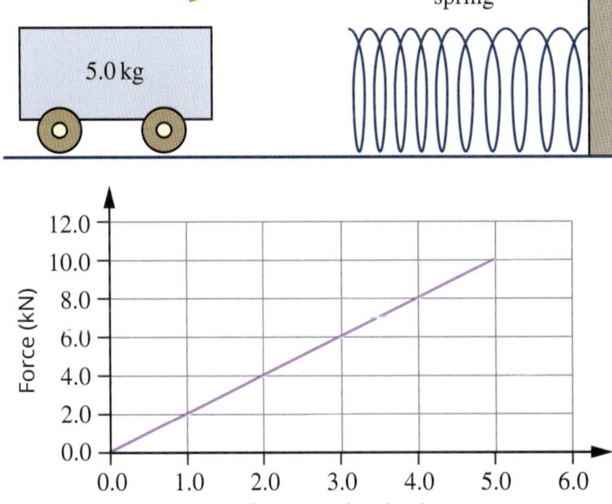

a Calculate the value of the spring constant.

b Calculate the elastic potential energy stored in the spring when its compression is equal to 2.0 cm.

c Calculate the elastic potential energy stored in the spring when the trolley momentarily comes to rest.

d Determine at what compression (in cm) the trolley will come to rest.

e Explain why the trolley starts moving again.

f State the property of a spring that accounts for the situation described above.

29 A child of mass 34 kg drops from a height of 3.50 m above the surface of a trampoline. When the child lands on the trampoline, it stretches so that she is 50 cm below the initial trampoline position. (Use $g = 9.8\,m\,s^{-1}$.)

- **a** What is the spring constant of the trampoline?
- **b** The child rebounds to a height of 0.75 m below her original position. Calculate the efficiency of the energy transfer of the trampoline.
- **c** Identify the point at which the child has maximum kinetic energy.

30 A nickel cube of mass 200 g is sliding across a horizontal surface. One section of the surface is frictionless while the other is rough. The graph shows the kinetic energy, E_k, of the cube versus distance, x cm, along the surface.

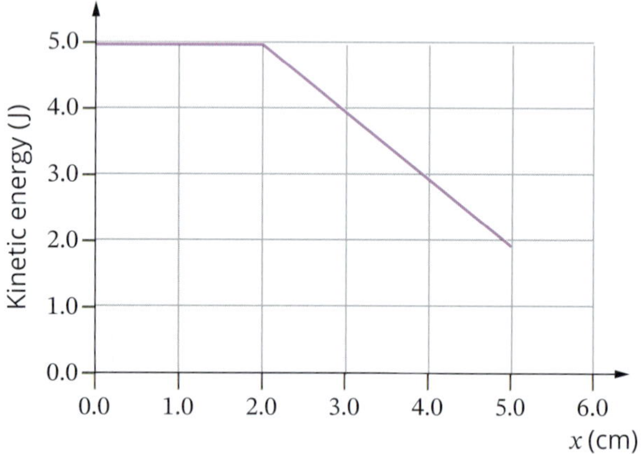

- **a** Which section of the surface is rough? Justify your answer.
- **b** Determine the speed of the cube during the first 2.0 cm.
- **c** How much kinetic energy is lost by the cube between $x = 2.0$ cm and $x = 5.0$ cm?
- **d** What has happened to the kinetic energy that has been lost by the cube?
- **e** Calculate the value of the average frictional force acting on the cube as it travels over the rough surface.

31 A football player of mass 86.0 kg and travelling at $7.50\,m\,s^{-1}$ collides with a goalpost.

- **a** Calculate the impulse of the post on the football player if the player comes to rest at the goalpost and does not bounce back.
- **b** Is momentum conserved in this collision? Explain.
- **c** The player hits his head in the collision. Explain, using Newton's laws, why he might sustain a concussion.

32 A goods train wagon of mass 4.0×10^4 kg travelling at $3.0\,m\,s^{-1}$ in a shunting yard couples with a stationary wagon of mass 1.5×10^4 kg and the two wagons move on at a reduced speed.

- **a** Calculate the speed of the coupled wagons.
- **b** Check to see if kinetic energy is conserved in this collision.
- **c** Explain your finding.

33 Jordy is playing softball and hits a ball with her softball bat. The force–time graph for this interaction is shown below. The ball has a mass of 170 g. Assume that the bat and ball form an isolated system during the interaction.

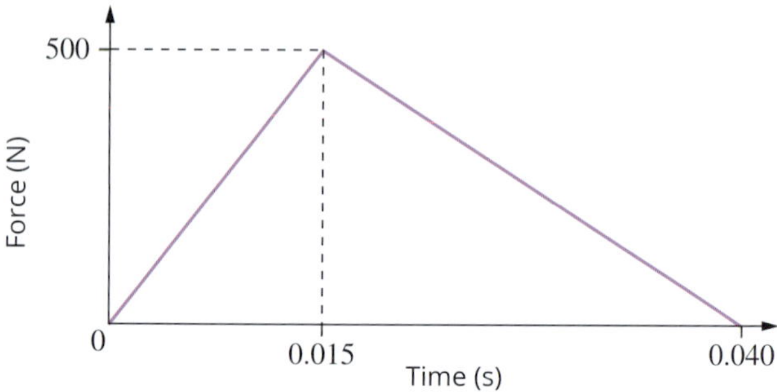

- **a** Determine the impulse experienced by the ball.
- **b** Calculate the change in momentum of the bat.
- **c** Determine the magnitude of the change in velocity of the ball.

34 A naval cannon with mass 1.08×10^5 kg fires projectiles of mass 5.5×10^2 kg with a muzzle speed of $8.0 \times 10^2\,m\,s^{-1}$. The barrel of the cannon is 20 m long, and it can be assumed that the propellant acts on the projectile for the time that it is in the barrel.

- **a** Calculate the magnitude of the average acceleration of the projectile down the barrel.
- **b** Using Newton's second law, calculate the average force exerted by the propellant as the projectile travels down the barrel.
- **c** Calculate the momentum of the projectile as it leaves the barrel.
- **d** Calculate the recoil velocity of the gun.
- **e** Calculate the average force of the propellant from the change in momentum of the projectile.
- **f** Calculate the average work done by the propellant on the projectile and compare this with the kinetic energy gained by the projectile.

35 Two students have designed a practical investigation to determine the relationship between the initial velocity of a projectile and the maximum height reached. They obtained the following results.

Initial velocity ($m\,s^{-1}$)	Final height (m)
0.25	0.3
0.50	1.3
0.75	2.8
1.00	4.9
1.25	7.5
1.50	11.8
1.75	15.4
2.00	19.4
2.25	24.9
2.50	31.1

a Identify the independent and dependent variables.

b Suggest a variable that would need to be controlled in this experiment.

c Draw a graph of the results of this practical investigation.

d Justify why the relationship between the initial velocity and the final height is not linear.

e The teacher suggests that if they graph the initial velocity squared against the final height, the students would be able to linearise the data. Complete the table below and graph the data again. Round the initial velocity squared values to two decimal places.

(Initial velocity)^2 ($m^2\,s^{-2}$)	Final height (m)
	0.3
	1.3
	2.8
	4.9
	7.5
	11.8
	15.4
	19.4
	24.9
	31.1

WS 38

EQ U201

Answers

The answers to questions that involve calculations are given to the least number of significant figures as given in the question.

See page 25 in Chapter 1 for more details.

Chapter 1 Scientific investigation

1.1 Planning scientific investigations

1 The aim of an experiment is a statement describing in detail the purpose of the experiment.

2 independent variable: the variable controlled by the researcher; dependent variable: the variable that may change in response to a change in the independent variable; controlled variables: all the variables that must be kept constant during the investigation

3 (I) state the research question to be investigated, (II) form a hypothesis, (III) plan the experiment and equipment, (IV) perform the experiment, (V) collect the results, (VI) question whether the results support the hypothesis, (VII) draw conclusions

4 slope of the ramp

5 dependent variable: current through the resistor; independent variable: voltage; controlled variable: the resistor

6 **a** B **b** A **c** C

7 **a** independent: distance travelled; dependent: speed
b independent: angle of inclined plane; dependent: horizontal distance
c independent: voltage; dependent: current
d independent: distance from source; dependent: intensity of light

8 **a** C **b** B **c** A

9 B, as it specifies what is being measured, how it is measured and how the dependent variable is expected to relate to the independent variable.

1.2 Conducting investigations

1 Accuracy refers to the ability to obtain the correct measurement, including during repeated trials of the experiment. Precision refers to how close two or more measurements are to each other. A set of precise measurements will have values very close to the mean value of the measurements.

2 **a** systematic error **b** random error
c mistake **d** mistake
e random error **f** systematic error

3 It will influence the repeatability of the experiment.

4 taking repeat readings/measurements and running the experiment or trial more than once

5 **I** Obtain a video recording device (e.g. mobile phone camera with video recording), ball and measuring tape.
II Mark off different heights on a wall: 100, 120, 140, 160, 180 and 200 cm.
III Drop the ball from 100 cm high and record the video of the ball falling. From the recorded video, determine the time it takes for the ball to fall from this height.
IV In a table, record the height and the time the ball took to fall.
V Repeat three times for each height.

1.3 Data collection and quality

1 to be able to substantiate whether or not a relationship exists between the variables being investigated, and to ensure repeatability of the experiment

2 Raw data is the data you record in your logbook (the actual, unmodified and unprocessed data). Processed data is raw data that has been organised, altered or analysed to produce meaningful information.

3 to communicate to others the degree of precision of the instruments used to measure the data collected

4 **a** 3 **b** 3 **c** 4 **d** 4

5 **a** 6.2×10^{-4} **b** 3.5×10^{4} **c** 4.6×10^{2} **d** 6.3×10^{0}

6 **a** 1.32×10^{5} V **b** 1.013×10^{5} Pa **c** 6×10^{6} J **d** 1.00×10^{-4} A

1.4 Data analysis and presentation

TY 1.4.1 9.9 m s^{-2} **TY 1.4.2** 0.04 W

1 discrete data: bar graph or pie chart; continuous data: line graphs and scatter plots

2 title, x- and y-axis labels, x- and y-axis units

3 a form of systematic error resulting from the researcher's personal preferences or motivations; e.g. poor definitions of concepts or variables, incorrect assumptions, and errors in the design or methodology of the investigation

4

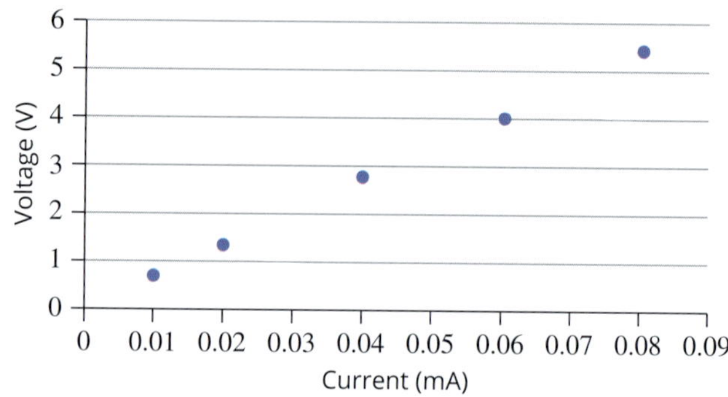

5 outlier: a data point that significantly does not fit the trend; should be included on the graph and in the discussion, but excluded from any trend lines or data analysis

6 Two of the following: mean, mode and median. The mean is the average of the data, and can be obtained by adding all the measurements and dividing by the total number of data points. The mode is the value that appears most often in a data set. The median is the middle value of an ordered list of values.

7 $y = 67.732x + 0.0093$; $m = 67.7$ V A^{-1} (or 67.7 Ω), $c = 0.0093$ V (9.3 mV)

8 **a**

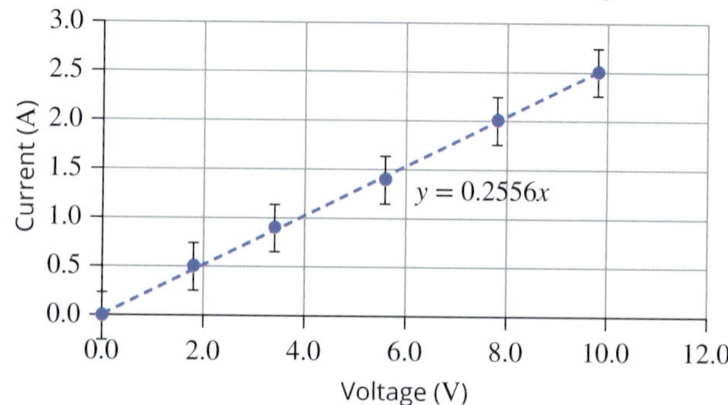

b 0.2556 Ω **c** $R = 3.9$ Ω

1.5 Conclusion and evaluation

1 to explain and evaluate the methods and interpret results, the relevance of the experiment and results, and whether the hypothesis was proven or disproven

2 At the conclusion of the discussion, the audience must have a clear idea of the results of the investigation, whether the methods were appropriate, and what are the overall context and implications of the investigation.

3 an evaluation of the investigative method, an analysis and evaluation of the data, an explanation of the link between the findings and relevant concepts in physics

4 The conclusion links the evidence and results of the investigation to the hypothesis and provides a justified response to the research question. The conclusion should not provide any new information that has not already been presented in the report or findings.

5 In your conclusion, you should mention possible future experiments. In this example, further investigation could be made of the effect of air resistance (i.e. drag) on the drop time. References to other, similar experiments that were conducted in a vacuum could also be made.

1.6 Reporting investigations

1 poster or oral presentation: Poster allows researcher to summarise the research topic, key results and conclusions, and should be presented in a manner that readers can easily disseminate the information from. Oral presentations are suited to a large audience, when the audience is listening at the same time.

2 A and C

3 written report

4 using other people's work without acknowledging them as the author or creator

5 They allow the reader to quickly ascertain whether the research report, poster or presentation is of interest to them or relevant to the work they are doing.

6 **a** accuracy **b** precision

Chapter 2 Waves and electromagnetic radiation

2.1 Longitudinal and transverse waves

1 The particles oscillate back and forth or up and down around a central or average position, but do not move along with the wave. Energy is carried along by the wave.

2 sound, ripples on a pond, vibrations in a rope

3 **a** longitudinal **b** transverse **c** transverse **d** longitudinal

4 The energy travels towards the right; that is, the energy is transferred away from the tuning fork (the source) towards point X. The compressions become more dispersed as the energy moves further from the source.

5 Particle A moves down the page, until it reaches the minimum point or trough, then moves up the page through the rest point (dashed line) and back to its position shown. Particle B initially moves upwards to its maximum point (crest) then moves down past the rest position (dashed line) to a maximum position below the dashed line (trough) and then proceeds upwards past the dashed line to its original position.

6 Particle A has moved right and particle B has moved left. As the sound wave moves to the right, particles ahead of the compression must move to the left initially to meet the compression and then move forwards to carry the compression to the right.

7 Both waves transfer energy throughout the medium; the particles oscillate (vibrate) about a mean position, but do not travel.

longitudinal wave: motion of particles is in the direction of propagation of the wave; e.g. sound, seismic P-waves from earthquakes

transverse wave: motion of particles is at right angles to the direction of travel of wave; e.g. light and all electromagnetic radiation, vibrations on a guitar, seismic S-waves, ripples in water, tsunamis

8 Sound waves are mechanical waves and require a medium to be produced and to transfer energy. In the vacuum of space, there is no medium, so there is no energy transfer.

9 Similarities: Both are transverse waves.

Differences: The guitar string wave is a mechanical wave and uses the medium of the string. Light is electromagnetic radiation and does not require a medium.

2.2 Measuring waves

TY 2.2.1 0.4 m

TY 2.2.2 2.0 Hz

TY 2.2.3 7.5×10^{14} Hz

TY 2.2.4 1.3×10^{-15} s

CSA: Seismic waves and the composition of Earth

1 *P*-waves are longitudinal waves and can travel through both solid and molten substances. *S*-waves are transverse waves and can only travel through solids. Therefore, because they cannot travel through the centre of Earth, the core (or part of it) must be molten rock.

2 $\lambda = 1200$ m, $f = 5.0$ Hz

3 $\Delta t = d\left(\frac{1}{v_S} - \frac{1}{v_P}\right)$

4 54.6 km

Key questions

1 **a** C, F **b** wavelength **c** B and D **d** amplitude

2 $\lambda = 1.6$ m, amplitude = 0.2 m

3 **a** 0.4 s **b** 2.5 Hz

4 5×10^{-6} s

5 6.5 m s^{-1}

6 **a** $\lambda = 4$ cm; amplitude = 0.5 cm **b** 0.02 m s^{-1} **c** A

7 0.75 m s^{-1}

2.3 The electromagnetic spectrum

TY 2.3.1 7.0×10^{-7} m

1 A mechanical wave requires a medium, light does not.

2 X-rays, visible light, infrared radiation, microwaves, FM radio waves

3 D

4 **a** infrared **b** microwave **c** radio waves **d** visible

e Astronomy uses a variety of EM radiation, from gamma rays to radio waves.

f X-rays

5 **a** 4.57×10^{14} Hz **b** 5.09×10^{14} Hz

c 6.17×10^{14} Hz **d** 7.56×10^{14} Hz

6 500 nm

7 4.3 m (to 2 s.f.)

8 1.5×10^{18} Hz

Chapter 2 Review

1 The particles move up and down as the waves radiate outwards carrying energy away from the point on the surface of the water where the stone entered the water.

2 **a** false: Longitudinal waves occur when particles of the medium vibrate in the *same* direction or *parallel* to the direction of the wave.

b true **c** true **d** true

3 C, D. Only energy is transferred by a wave.

4 a true
   b false: The period of a wave is *proportional* to its wavelength.
   c true
   d false: The wavelength *and frequency* of a wave determine its speed.

5 Frequency would increase.

6 a infrared waves b X-rays

7 Sound waves are longitudinal mechanical waves in which the particles only move back and forth around an equilibrium position, parallel to the direction of travel of the wave. When these particles move in the direction of the wave, they collide with adjacent particles and transfer energy to the particles in front of them. This means that kinetic energy is transferred between particles in the direction of the wave through collisions. Therefore, the particles cannot move along with the wave from the source as they lose their kinetic energy to the particles in front of them during the collisions.

8 U is moving down; V is momentarily stationary (and will then move downwards).

9 amplitude and period: amplitude = 10 cm, period = 2 s;
$f = \frac{1}{T} = 0.5\,\text{Hz}$

10 amplitude and wavelength: amplitude = 5 cm, λ = 40 cm

11 $0.300\,\text{m s}^{-1}$ 12 0.044 m 13 5.00 m

14 $2.25 \times 10^{8}\,\text{m s}^{-1}$ 15 490 m

16 2.4 GHz: 0.125 m to 0.120 m; 5.180 GHz: 0.052 m to 0.058 m

17 The microwave frequency is 2.45 GHz, which is within the 2.4 GHz to 2.5 GHz range. Microwave ovens are shielded to protect the user. If there is microwave leakage, interference depends on the proximity of the router to the microwave oven.

18 X-rays and ultraviolet waves are a type of ionising radiation and are known to damage cells and cause cancer. From Table 2.3.1, these waves have wavelengths of 10^{-10} and 10^{-8} m respectively, which are much smaller than the human cell size. From question 16, the wavelengths for the 2.4 GHz microwave band vary from 0.125 m to 0.120 m, which is much larger than the human cell size of 100 µm, so they would not be expected to ionise and damage cells in the same way as X-rays and UV waves.

Chapter 3 Light

3.1 Reflection and refraction

TY 3.1.1

TY 3.1.2 1.52

TY 3.1.3 $1.62 \times 10^{8}\,\text{m s}^{-1}$

TY 3.1.4 28.2°

TY 3.1.5 24.4°

1 The new wavefront should be a straight line across the front of the secondary wavelets.

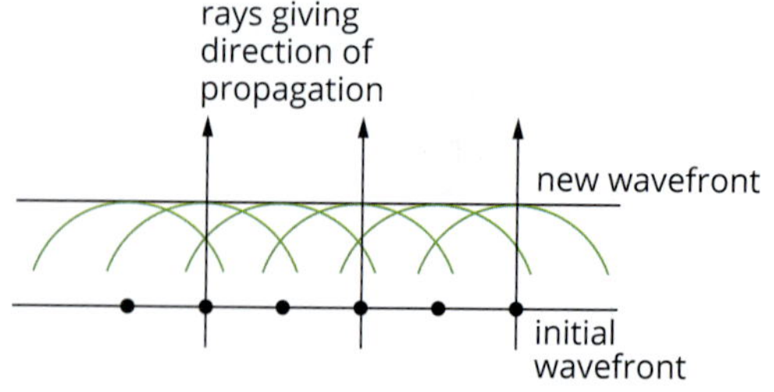

2 slower than 3 $2.17 \times 10^{8}\,\text{m s}^{-1}$ 4 1.31

5 35.3° 6 A

7 a true
   b False. The lower part of the Sun is closer to the horizon and its light is refracted more than that from the higher part of the Sun further from the horizon. The observed effect is that the bottom of the Sun is lifted up more than the top of the Sun.
   c False. An object in water appears higher than is actually is.

8 a lower
   b The lower refractive index of the cladding means that total internal reflection will occur at the core–cladding interface, keeping the light in the core.

9 a 1.522 b $1.97 \times 10^{8}\,\text{m s}^{-1}$ c 0° to 27.7°

10 Total internal reflection occurs when light passes from a more-dense medium into a less-dense medium and refracts away from the normal. This corresponds to a change in refractive index from high refractive index to a lower refractive index
   a no b yes c yes d no

3.2 Dispersion and polarisation

TY 3.2.1 38.5° to 39.2°

1 a False. Light travels through a water droplet at a *slower* velocity than in air, which leads to refraction effects in water.
   b False. Red light travels at *faster* speeds than blue light, which leads to a greater angle of refraction of red light.

2 In lenses, coloured fringes can occur on images. This is because the refractive index varies with wavelength and therefore the image is not formed at exactly the same point for all wavelengths.

3 Polarising filters need to be orientated perpendicular to the surface.

4 22.2°

5 a Yes, dispersion will occur because the speed of the different wavelengths of light will vary in the acrylic prism.
   b red: 35.6°; blue: 35.0°

Chapter 3 Review

1 A: incident wave B: normal C: reflected wave
   D: boundary between media E: refracted wave

2 Chromatic aberration occurs because different wavelengths of light travel at different speeds in glass, and so have different refractive indices. Therefore the different colours do not form an image at exactly the same point.

3 The observation of water on the road is a mirage and is due to variations in the refractive indices of layers in the atmosphere on a hot day. The path of light from the sky is refracted in a curve, leading to the illusion there is water on the road.

4 The light reflected from water and snow is partially polarised parallel to the surface. Polarising sunglasses will block this polarised light.

5 increases, away from 6 $2.25 \times 10^{8}\,\text{m s}^{-1}$

7 a The light wave does not refract when striking the glass boundary at 90°, which corresponds to an angle of incidence of zero degrees. The wavefronts do not travel as far in the glass medium in the same time interval (period). As the wavefronts are parallel with the glass boundary, there is no change in direction but a bunching up (shortening) of wavefronts in the glass.

b The wave slows down in glass and its wavelength also decreases, but the frequency does not change.

8 C 9 2.5×10^8 m s^{-1} 10 27.6°

11 2.10×10^8 m s^{-1} 12 $a = b = 25.4°$, $c = 28.9°$

13 a 32.0° b 53.7°
c 21.7° d 1.97×10^8 m s^{-1}

14 a 49.8° b 40.5° c 27.6°

15 a 19.5° b 19.1°
c 0.4° d 1.96×10^8 m s^{-1}

16 B, D, A, C

17 The angle results, θ_1 and θ_2, need to be converted to $\sin\theta_1$ and $\sin\theta_2$ and then plotted:

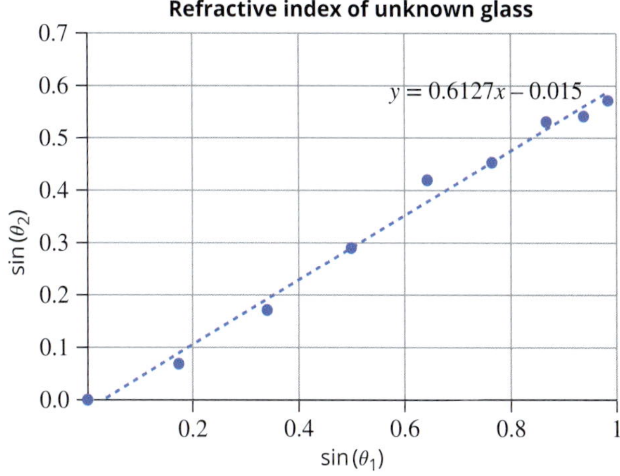

A value for the refractive index of between 1.60 and 1.70 is reasonable.

18 a core: 1.92×10^8 m s^{-1}; cladding: 2.23×10^8 m s^{-1}
b critical angle: 59.6° (relative to the normal)
c 0° to 30.4°
d Total internal reflection will occur.
e Some of the signal will undergo refraction and not be reflected, so signal loses intensity.

Chapter 4 Thermal energy

4.1 Heat and temperature

TY 4.1.1 $\Delta U = 2770$ J

1 C

2 a C, D b B

3 C, D

4 a 303.15 K b 101.85°C

5 300 K is 26.75°C. Higher temperatures mean molecules have greater average kinetic energy. So the average kinetic energy of the hydrogen particles in tank B is greater than the average kinetic energy of the hydrogen particles in tank A.

6 $\Delta U = -70$ kJ

7 $\Delta U = 230$ J

8 $W = -600$ J, so the scientist did 600 J of work on the sodium.

4.2 Specific heat capacity

TY 4.2.1 $Q = -6.30 \times 10^6$ J (i.e. energy was transferred from the water)

TY 4.2.2 ratio = 2.1

1 B 2 aluminium 3 aluminium

4 $Q = 7.88 \times 10^4$ J 5 $Q = 2x$ J 6 ratio = 4.67

7 30°C 8 1.0 kg

4.3 Latent heat

TY 4.3.1 $Q = 1.27 \times 10^5$ J

TY 4.3.2 $Q = 7.96 \times 10^6$ J

CSA: Extinguishing fire

1 $Q = 3.78 \times 10^5$ J

2 The heat will flow from the hotter gases to the cooler water.

3 $Q = 2.25 \times 10^6$ J 4 $Q = 4.00 \times 10^5$ J

5 $= 3.03 \times 10^6$ J

Key questions

1 a The mercury is changing state from solid to liquid. It is melting; temperature does not change during phase transitions as average kinetic energy does not change.
b −39.0°C
c 357°C
d $L = 1.26 \times 10^4$ J kg^{-1}
e $L = 2.85 \times 10^5$ J kg^{-1}

2 B

3 A hot, dry, windy day will dry clothes faster. This is because the heat, low humidity and moving air will result in faster evaporation of water from the surface of the clothes.

4 $Q = 2.25 \times 10^5$ J

5 $E_T = 3.42 \times 10^4$ J

6 Because ice, water and water vapour are different phases of the same substance, the water vapour (gas) will have more kinetic and potential energy than the liquid water, which will have more potential energy than the solid ice. The particles of ice are closely packed in rigid shapes, whereas in liquid water they are close but free to slide past each other. In water vapour the particles are a significant distance apart compared to their size.

4.4 Conduction, convection and radiation

1 a Metals conduct heat by free-moving electrons as well as by molecular collisions. Wood does not have any free moving electrons, so it is a poor conductor of heat.
b thickness, surface area, nature of the material and the temperature difference between it and the second material
c Copper is a better conductor of heat than stainless steel.

2 a liquids and gases
b The source of heat, the Sun, is above the water. It takes much longer to heat a liquid when the source is at the top, as the convection currents will also remain near the top. The warm water is less dense than the cool water and will not allow convection currents to form throughout the water.

3 a partially reflected, partially transmitted and partially absorbed
b absorption

4 Conduction and convection require the presence of particles to transfer heat. Heat transfer by radiation can occur in a vacuum as the movement of particles is not required.

5 Plastic and rubber have low conductivity, so they do not allow the heat from your hand to transfer very easily. Metal has high conductivity, so heat transfers easily from your hand and your hand feels cold.

6 No, because solids do not contain the free molecules that are required to establish convection currents.

Chapter 4 Review

1 Heat refers to the energy that is transferred between objects, whereas temperature is a measure of the average kinetic energy of the particles within a substance.

2 a 278.15 K b −73.15°C

3 The temperature of the gas is just above absolute zero so the particles have very little energy.

4 absolute zero, 10 K, −180°C, 100 K, freezing point of water

5 a The copper ball will have three times the thermal energy of the lead ball, as its specific heat capacity is three times that of lead.
  b Thermal equilibrium is reached, so the balls must be at the same temperature.

6 No, he is not correct. The metals have different specific heat capacities, so will have had different amounts of heat energy removed according to $Q = mc\Delta T$.

7 a convection b convection c radiation

8 This situation describes a change of state, in this case, melting. It occurs because the heat energy is used to increase the potential energy of the particles in the solid instead of increasing their kinetic energy. The energy needed to change from solid to liquid is the latent heat of fusion.

9 Steam. Both have the same kinetic energy as their temperatures are the same; however the steam has more potential energy due to its change in state, and so has greater internal energy.

10 10 kJ

11 The stopper reduces heat loss by convection and conduction. The air between the walls reduces heat transfer by conduction, and the space between the walls is small enough that convection currents will not form. The flask's shiny surface reduces heat transfer by radiation.

12 a Solar energy is transferred to the pipes primarily through radiation. A small amount of energy is also transferred to the pipes via conduction and convection from the air on hot days, but this would be insignificant compared to the radiation from the Sun. The energy is then transferred to the water through conduction from the hot pipe.
  b The warm water moves up to the top of the solar panel because as the water temperature increases, the density decreases due to the molecules moving further apart. The less dense, warmer water rises to the top through natural convection.

13 $Q = 3.28 \times 10^4$ J 14 $Q = 2.52 \times 10^4$ J

15 $Q = 340$ J kg^{-1} K^{-1} 16 $E_T = 1.10 \times 10^5$ J

17 $T = 59.9$°C

18 Students' answers will vary. One example is as follows.

Gather a long, thin strip of two different solids. Ensure each strip is the same dimension and at the same temperature by placing them in a thermally controlled area to reach equilibrium. Measure the temperature of that area. Prepare a water bath set at a high temperature. Measure the temperature of the water. Place both strips with one end in the water bath. Measure the temperature at different intervals along the strips every 10 seconds. This can be done using an infrared camera, or thermocouples placed at intervals with good thermal conductivity to the solid strips.

Chapter 5 Thermal energy, electromagnetic radiation and Earth's climate

5.1 Wien's law and black-body radiation

TY 5.1.1 $T = 3.2 \times 10^4$ K

1 Wavelength decreases and frequency increases.

2 C

3 a matte black beaker cools fastest; glossy white surface will cool slowest
  b matte black beaker will warm fastest; glossy white beaker will warm slowest

4 Heat sinks are made of dark-coloured metals that radiate heat energy strongly and keep the computer cool.

5 $\lambda_{max} = 322.0$ nm 6 $T = 3.62 \times 10^3$ K 7 $T = 4.14 \times 10^3$ K

5.2 Radiation and the enhanced greenhouse effect

TY 5.2.1 $\lambda_{max} = 9.66\,\mu$m

1 radiant energy from the Sun

2 Radiant energy from the Sun reaches Earth largely as electromagnetic radiation in or near the visible wavelengths. Earth re-emits energy as longer wavelength, infrared radiation.

3 a C
  b i combustion of fossil fuels
    ii agriculture
    iii air-conditioners
    iv artificial fertilisers

4 a Thermal energy moves through Earth's mantle by convection currents that rise to the upper layer of the mantle and fan out before sinking as cooler magma.
  b While some energy is transferred by conduction and radiation in the atmosphere, most of it is transferred is by convection. Locally, sea breezes and land breezes move heat energy around. On a global scale, radiant energy from the Sun at the equator heats the air. Cool, dense air moves in, forcing the less dense warm air upwards where it spreads out towards the poles. There it cools and sinks.

5 Earth stays warm because the greenhouse gases in the atmosphere absorb some of the energy re-radiated by Earth and re-radiate it back down towards Earth's surface. This is the greenhouse effect. This has led to relatively stable temperatures, allowing life to evolve. Human activity since the Industrial Revolution has resulted in increased amounts of greenhouse gases in the atmosphere, which are absorbing and retaining more of the long-wavelength infrared radiation emitted from Earth's surface and causing the temperature on Earth to increase. This is called the enhanced greenhouse effect.

6 Students' answers will vary. While answers may describe effects of a changing climate such as higher average global temperatures, warming oceans, shrinking ice sheets and glaciers, lower amounts of sea ice, and extreme weather, they must link these to human activity. This could be the increased concentration of greenhouse gases since the Industrial Revolution from historic levels found in Antarctic ice cores, or that the rate of current climate change is greater than previous changes in climate.

Chapter 5 Review

1 It is the increase in the retention of radiated thermal energy by Earth's atmosphere due to the increase of greenhouse gases.

2 The rate at which energy is re-radiated by Earth's surface depends on the surface material. Materials with a high emissivity will not only readily absorb thermal energy, but will also re-radiate the energy very well. This contributes to the greenhouse effect. Building cities and land clearing on a large scale changes the amount of thermal energy retained or reflected by changing the surface characteristics by increasing the emissivity.

3 A

4 visible, longer, infrared

5 Carbon dioxide, like other greenhouse gases, absorbs and emits long-wavelength infrared radiation, and re-radiates it back to the surface of Earth rather than reflecting it back out to space. Carbon dioxide is present in the largest proportion relative to other greenhouse gases and so has a greater impact than other greenhouse gases.

6 a convection
  b convection
  c radiation
  d conduction (through the crust, at the surface) and convection (through the mantle, deep inside Earth)

7 On a hot day, the surface of a land mass heats up more quickly than the ocean. Hot air rises over the land and cooler, denser air moves in from over the ocean, creating a breeze.

8 The person, due to their temperature, emits stronger infrared radiation than their surroundings.

9 B

10 a It is a very good absorber of radiant thermal energy.
   b It is also a good emitter of radiant energy.

11 $\lambda_{max} = 414.0$ nm 12 $\lambda_{max} = 9600$ nm 13 $T = 4.46 \times 10^3$ K

14 $\lambda_{max} = 311.6$ nm, which corresponds to UV light

15 a Greenhouse gases in Earth's atmosphere absorb and re-radiate most of the infrared radiation.
   b The radiation from objects viewed by the James Webb Space Telescope is longer than radiation from objects viewed by the Hubble telescope, indicating that they are colder according to Wien's law.
   c hotter

Unit 1 Area of Study 1 review

1	C	2	A	3	D	4	B	5	A
6	B	7	C	8	C	9	D	10	B and C
11	C	12	B						

13 A mechanical wave involves energy being transferred from one location to another, without any net transfer of matter. Mechanical waves need a medium through which to travel and cannot transfer energy through a vacuum.

14 As a transverse wave passes through a medium, the particles of the medium move up and down perpendicular to the direction in which the wave travels, whereas in a longitudinal wave the particles travel back and forth parallel to the direction of travel of the wave.

Mechanical waves can be transverse waves or longitudinal waves, whereas electromagnetic waves are transverse waves.

15 wavelength = 4 cm, amplitude = 2 cm

16 a 3×10^8 m s^{-1} b $\lambda = 3 \times 10^{-8}$ m
   c ultraviolet waves
   d Any one of:
   - UV lamps are used to sterilise surgical equipment in hospitals.
   - UV lamps are used to sterilise food and drugs.
   - UV rays help the body to produce vitamin D.
   - Any other suitable use of UV.

17 $\theta_2 = 12°$ 18 $v_2 = 1.3 \times 10^8$ m s^{-1} 19 $\theta_c = 25°$

20 The students would need to pass white light through a triangular glass prism.

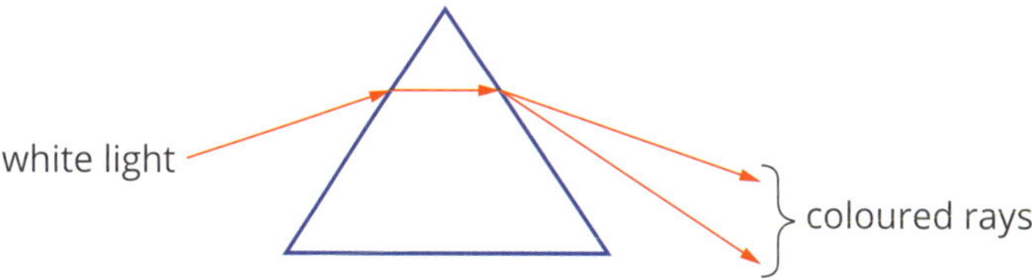

21 The white light will separate into the component colours which, from top to bottom, will be red, orange, yellow, green, blue, indigo, violet.

22 Energy flows from an object at a higher temperature to one at a lower temperature until both furnace and copper are at the same temperature.

23 When energy is added to the rod, the atoms vibrate faster about their fixed position, increasing their average kinetic energy. Temperature is a measure of the average kinetic energy of the particles, and so the temperature increases.

24 The rod will lose heat by conduction to the steel plate and the nitrogen gas, and by radiation. There is no convection in the steel or the copper solids, but convection currents in the gas could assist with cooling.

25 The temperature difference between the hot water pipe and the surrounding air will be quite high, so the rate at which heat is transferred to its surroundings will be high and the water may cool before reaching the tap inside the house. The hot water pipe will lose heat mainly through radiation. There will be some heat lost by conduction to the air surrounding the pipe, and as this air heats it will move away due to convection currents to be replaced by more cool air. The foam insulates the pipe to slow this heat transfer. Heat exchange between the pipe carrying cool water and the surrounding air will be small because they will be closer in temperature, so insulation is unnecessary.

26 Specific heat capacity of unknown metal = 867 J kg^{-1} K^{-1}, and this would most closely match that of aluminium.

27 199 kJ 28 263 kJ 29 52.8 kJ

30 0.114 kg 31 $c_c = 462$ J kg^{-1} K^{-1}

32 a $\lambda_{max} = 502$ nm; this is in the visible range of the electromagnetic spectrum
   b 700 K

33 a Because Mercury is close to the Sun it receives a high level of solar radiation, which is absorbed by the surface causing it to heat up. When the surface is facing away from the Sun, the heat absorbed is radiated back into space and, because Mercury has no atmosphere, none of the radiated heat is absorbed or trapped by gases, so the surface cools rapidly. (Additionally, without an atmosphere there is no convection of heat from warmer areas to cooler areas.)
   b Although further from the Sun, Venus has an atmosphere with a high concentration of the greenhouse gas carbon dioxide, so the heat absorbed from solar radiation is trapped in the atmosphere. The atmosphere also allows convection currents to develop and so transfers the heat around the planet, including the side not facing the Sun.

Chapter 6 Radiation from the nucleus

6.1 Atoms, isotopes and radioisotopes

TY 6.1.1 90 protons, 230 nucleons, 140 neutrons

1 nucleons

2 It is the same as the number of protons, which is given by the atomic number.

3 Isotopes are atoms with the same number of protons but different numbers of neutrons.

4 A radioisotope is an unstable isotope. At some time, it will spontaneously decay by emitting a particle from the nucleus.

5 Yes, e.g. uranium is naturally occurring and every isotope of uranium is radioactive.

6.2 Radioactivity

TY 6.2.1 Element 82 is lead.

1 beta-plus particle (β^+) decay

2 An alpha particle is a helium nucleus. A beta particle is a positively or negatively charged electron. Gamma radiation is electromagnetic radiation.

3 **a** X: atomic number = 90, mass number = 231, X is thorium
b Y: atomic number = 89, mass number = 228, Y is actinium

4 **a** 7 protons and 7 neutrons
b A neutron has changed into a proton, an electron and an antineutrino.

5 **a** nucleus **b** nucleus **c** nucleus

6 The strong nuclear force is a force of attraction that acts between every nucleon but only over relatively short distances. This force acts like a nuclear cement to stop protons repelling each other.

7 Alpha particles travel through air at a relatively low speed and have a double positive charge. Their charge and their relatively slow speed make them very easy to stop and so have a very poor penetrating ability.

8 The wire should be primarily a beta emitter, since the irradiation needs to be confined to a relatively small area. Alpha radiation does not have sufficient penetrating power, while gamma radiation would also irradiate adjacent healthy cells.

6.3 Half-life and decay series

TY 6.3.1 3.9×10^7 nuclei

CSA: Radiocarbon dating

1 about 11 500 years old

2 about 17 190 years old

3 1.25 g

Key questions

1 The 'activity' is the count rate or the number of decays each second.

2 15 minutes

3 **a** time to drop from 800 → 400 = 10 minutes or from 400 → 200 = 10 minutes
b $A = 50$ Bq

4 192 μg

5 beta decay; half-life is 20 years

6.4 Radiation and the human body

TY 6.4.1 4.0 Sv

TY 6.4.2 3.0 mSv

1 Somatic effects arise when ordinary body cells are damaged. Examples include skin rash, hair loss and nausea.

2 Beta-plus radiation (i.e. positrons). Positrons are used because they annihilate when they interact with electrons in the tissue and produce gamma rays, which can be detected by a camera.

3 2.0 Gy

4 Any two of:
- have a short half-life (hours or days) that is appropriate for the time taken for the diagnostic procedure
- emit only gamma radiation of an energy that can be detected by a γ camera
- do not emit alpha or beta radiation because these particles would be trapped in the patient's tissues and they would not be detected externally
- be available in the highest possible activity but not be toxic to the patient or react with drugs used at the same time.

5 ED = AD × QF
$E_\gamma D = 200 \times 1 = 200$ μSv
$E_\alpha D = 20 \times 20 = 400$ μSv
$E_\beta D = 50 \times 1 = 50$ μSv
$E_{neutron} D = 300 \times 10 = 300$ μSv
The 20 μGy of alpha radiation has the highest ED of 400 μSv and would be the most damaging.

6 The dose of radiation is not being delivered in one treatment session but over multiple treatment sessions.

7 15 mSv

8 Somatic. The radiation has affected her bone marrow, where red blood cells are made. Genetic damage affects the DNA and is passed on to subsequent generations.

9 The small amount of harm caused by the radiation is outweighed by the benefits of the diagnostic procedure.

Chapter 6 Review

1 20 protons and 25 neutrons

2 Cobalt-60 has 27 protons, 33 neutrons and 60 nucleons.

3 It emits a beta-minus particle.

4 Atomic number = 3, mass number = 7, so X is lithium.

5 **a** Atomic number = 1, mass number = 1, so X is a proton.
b Atomic number = 0, mass number = 1, so Y is a neutron.
c $a = 4$, $b = 2$, so X is helium, He.

6 Half-life is an average time calculated from a very large number of nuclei. An individual nucleus may decay at any time, but half of a large number of nuclei will decay by one half-life. Therefore, for uranium-236, the probability that a single nucleus will decay is very small, but over a large number of nuclei, half of them will be expected to have decayed in 23.4 million years.

7 **a** a radioactive tracer that emits positrons (e.g. ^{18}F, ^{15}O, ^{82}Rb or ^{64}Cu)
b Positrons interact with electrons (annihilate) in the body tissue and create gamma rays that are detected by a gamma camera.

8 1 Gy of alpha radiation is the most damaging as it has a quality factor of 20, so 1 Gy of alpha radiation is 20 times more damaging than 1 Gy of the beta or gamma radiation.

9 They are all equally damaging as they are already in the units of dose equivalent, sieverts (Sv).

10 300 mSv

11 Technetium. It is produced in small nuclear generators by the beta decay of Mo-99.

12 Gamma radiation. Alpha and beta radiation would be stopped by skin and tissue, whereas gamma radiation can penetrate through these layers to the internal organs, so lead shielding is needed to stop it.

13 3×10^{11} nuclei **14** alpha decay; half-life is 3 min

15 Electrostatic forces of repulsion between the positively charged protons in the nucleus are balanced by the strong nuclear force acting between all nucleons (protons and neutrons) in close proximity.

16 **a** 3.8×10^2 J **b** 5.0 Sv

17 **a** about 3.5 μSv per X-ray
b about 4 times background dose

18 Tracers need to able to pass through body tissue and be detected externally. Gamma rays can pass through the body, whereas alpha and beta particles cannot. Alpha and beta radiation would also cause too much local damage to tissues.

19 a

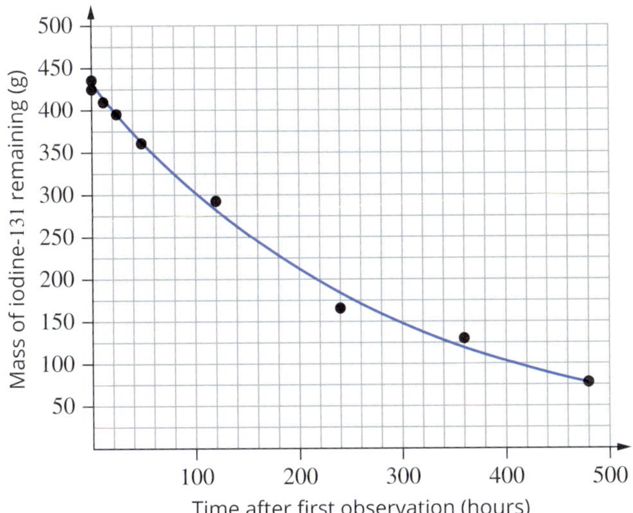

b 7.9 days

c He measured the mass remaining at a small time interval after the initial measurement and found there was no change in mass, therefore the half-life was longer than a few minutes. He then increased the time intervals between measurements in order to obtain a large enough change in mass to indicate the half-life is measured in days, but not too large to lose accuracy in his mass measurements.

d Max would need to keep the sample in a lead-lined container and wear protective clothing (also probably lead-lined) and glasses to prevent any electrons from entering his body or eyes.

e The daughter nuclide of iodine-131 is xenon-131, a gas. Any xenon-131 produced in the outer layers of the crystal will escape to the air and be lost. Any xenon-131 produced inside the crystal will probably be trapped between the iodine-131 nuclei and therefore keep contributing to the recorded mass. Max could grind the iodine crystal to a powder to increase the surface area and minimise the xenon-131 gas trapped inside the crystals. In this way, the measured mass would reflect the remaining mass of the iodine-131 only.

Chapter 7 Nuclear energy

7.1 Energy from mass

TY 7.1.1 140 keV, 0.14 MeV

TY 7.1.2 2.917×10^{-10} J, 1823 MeV

1 8.0×10^{-13} J **2** 3.8×10^{4} eV = 0.038 MeV

3 The mass defect is the difference between the total mass of the individual nucleons that make up a nucleus and the actual mass of the nucleus.

4 The binding energy is equivalent to the mass defect according to $\Delta E = \Delta mc^2$. The binding energy is a measure of how strongly a nucleus is held together

5 ΔE = 1013 MeV

7.2 Fission, fusion and the future of nuclear energy in Australia

TY 7.2.1 3 neutrons

1 a fissile: can undergo nuclear fission, i.e. can be made to split into two smaller nuclei

b A chain reaction is one that occurs when a single fission reaction causes a subsequently larger number of fission reactions, which then cause even more fission reactions.

2 Fusion is the joining together of two small nuclei to form a larger nucleus. Fission is the splitting apart of one large nucleus into smaller fragments.

3 In both reactions, the mass of the products is less than the mass of the reactants. This mass difference is related to the energy released via $\Delta E = \Delta mc^2$.

4 uranium-235, plutonium-239, and several others

5 hydrogen-1, hydrogen-2, hydrogen-3, helium-3, helium-4 and most other nuclei lighter than nickel-62

6 The decay products of the nuclear fission process comprise many different, often highly radioactive isotopes. This is what makes up the waste.

7 Student responses may vary. Answers will cover the possible points:

- In efforts to improve Australia's impact on climate change, utilising nuclear energy is a cleaner option than fossil fuels and avoids peaks and troughs in supply that can occur with renewable energy sources.
- Fission reactors produce large amounts of energy at relatively low cost.
- Current fission reactors create radioactive waste for which we do not have good management plans and which is expensive and difficult to contain.
- There is a risk of accident or terrorist attack, which would impact people and the environment.

8 News article responses may address the following:

- Australia does not have a nuclear energy plan. A royal commission in 2019 concluded nuclear power was currently unattainable.
- Australia's OPAL reactor conducts research into fission capabilities.
- Australia has a history in fusion research with the construction of tokamak reactors at Australian National University (ANU).
- ANSTO and ANU have a partnership with ITER such that the Australian research community is able to contribute to the ITER project.

Chapter 7 Review

1 fissile: uranium-235 and plutonium-239

non-fissile: uranium-238 and cobalt-60

2 The amount of energy per nucleon released with fusion is greater than with fission and there is no radioactive waste produced.

3 The binding energy per nucleon increases. The new nuclei are more stable than the original uranium nucleus.

4 The higher the binding energy, the more stable the nucleus. This is because higher binding energy means that it takes more energy to completely separate particles in the nucleus. Iron therefore has the most stable nuclei of all the elements.

5 Fissile nuclei such as uranium are much larger than nuclei that can fuse, such as hydrogen. Although the energy released per nucleon is greater for fusion than for fission, because of the large number of nucleons in a fissile nucleus, the energy released per single fission is greater.

6 No, only a few nuclides (e.g. uranium-235 and plutonium-239) are fissile. Some nuclei can be made to split when bombarded by a slow-moving neutron and so are considered to be fissile. Others can be induced to undergo fission by bombardment with a very high-energy neutron, so they are regarded as fissionable, but are considered to be non-fissile.

7 Electrostatic forces of repulsion act on the protons. If the protons are moving slowly they will not have enough energy to overcome the repulsive forces and they will not fuse together.

8 Initially, electrostatic forces of repulsion act on the protons, but they are travelling fast enough to overcome these forces. The protons will get close enough for the strong nuclear force to take effect and they will fuse together.

9 a Technetium-99 is the most stable as it has the longest half-life.

b Technetium-99m is the least stable as it has the shortest half-life.

10 $x = 5$

11 The nuclei are all positively charged and so repel each other. They need the very large amount of energy (e.g. provided by100 million degrees) to overcome these forces and get close enough for the strong nuclear force to take effect.

12 **a** The combined mass of the hydrogen and helium-3 nuclei is greater than the combined mass of the helium-4 nucleus, positron and neutrino.

b The energy has come from the lost mass (or mass defect) via $E = mc^2$.

c 3.4×10^{-12} J

13 The binding energy per nucleon increases and the nucleus becomes more stable.

14 8.28 MeV per nucleon

Unit 1 Area of Study 2 review

1 D **2** A **3** C **4** D **5** B **6** C **7** B
8 A **9** C **10** D **11** B **12** D

13 caesium-137: 55 protons, 82 neutrons, 137 nucleons; iodine-131: 53 protons, 78 neutrons, 131 nucleons

14

Time ($\times 10^9$ years)	No. of K nuclei	No. of Ar nuclei	Ratio K:Ar
0	1000	0	–
1.3	500	500	1:1
2.6	250	750	1:3
3.9	125	875	1:7

15 **a** A C-14:C-12 ratio of 1:8 means approximately 3 half-lives have occurred.

b Wasps nests, and therefore the painting, are approximately 17 190 years old.

c Radiocarbon dating is used to date organic objects. The ratio of carbon-14 to carbon-12 remains the same in living organisms because carbon is continually taken in from the environment. However, when the organism dies, carbon is no longer taken in from the environment. The ratio of carbon-14 to carbon-12 changes because the radioactive carbon-14 decays. The organism's age can be found from the proportion of carbon-14 to carbon-12 in the sample. However, it cannot be used for inorganic substances such as rocks (ochre).

16 ${}^{185}_{79}\text{Au} \rightarrow {}^{181}_{77}\text{Ir} + {}^{4}_{2}\text{He}$

17 ${}^{26}_{11}\text{Na} \rightarrow {}^{0}_{-1}\beta + {}^{26}_{12}\text{Mg} + \text{energy}$

18 half-life is 1 minute

19 remaining mass = 4.7 g

20 The beams from linear accelerators need to have high energy so they are able to penetrate human tissue. Although they have high energy they have low ionising power, which means they are less likely to damage healthy tissue. As the exact location of tumours can be found using a CT scan, high-energy radiation can be used to treat these areas while minimising damage to healthy tissue.

21 equivalent dose = 0.20 Sv

22 effective dose = 4.0 mSv

23 Comparing these effective doses (9.6 : 4.0 = 2.4), cancer is 2.4 times more likely to develop in the lungs than the esophagus, due to the lungs having a greater weighting factor.

24 Th, Pa and U have different numbers of protons and this is what makes them distinct elements.

25 **a** The concentration of ^{222}Rn gas can be quite high in uranium mines and, as it is a gas, the ^{222}Rn can be breathed in by workers. As it is an alpha emitter, it can be potentially very damaging to the lung tissue of workers.

b Alpha particles have low penetrative properties. Appropriate safety precautions include anything that will reduce the inhalation of ^{222}Rn, such as PPE (i.e. masks) to protect the lungs, radon monitoring systems, ventilation to prevent the buildup of radioactive gas, water sprays for dust suppression, etc.

26 Fe is at the peak of the binding energy curve with a binding energy of approximately 8.8 MeV per nucleon. As Fe requires the most energy per nucleon to break up the nucleus, it is the most stable nucleus.

27 The binding energy per nucleon for uranium is about 7.5 MeV and the binding energy per nucleon for fragments of mass number 118 is 8.5 MeV, which means that when the smaller fragments are formed, the nucleons are more tightly bound and the difference in energy is released in the fission reaction. This is about 1.0 MeV for each nucleon.

28 Energy yield is approximately 235 MeV for each uranium nucleus.

29
- Fusion joins lighter elements to form a heavier element and fission splits a heavy element to form lighter fragments.
- Fusion usually requires the reacting nucleons to have very high energy to overcome the electrostatic repulsion. Fission can be achieved with the capture of relatively slow-moving neutrons.
- Fusion does not create harmful and radioactive waste products, but fission wastes are often highly radioactive.
- At this point fission has been achieved commercially, but fusion reactors are still under development.
- Energy released per nucleon is higher for fusion than for fission.

30 Fission reactions occur when a neutron collides with a fissile isotope. This fission reaction produces neutrons that can contribute to further fission reactions, causing a chain reaction. Neutrons that are absorbed cannot contribute to further fission reactions, which controls the chain reaction in the reactor.

31 Answers will vary, but could include any of the following:

Reduced greenhouse emissions: Nuclear fission produces almost no carbon dioxide.

Availability of resources: Australia is one of the largest producers of uranium and has approximately one third of the world's uranium deposits.

Efficiency of energy production: By weight, uranium is able to produce millions of times more energy.

32 Possible safety concerns of using nuclear power may include:

Produces radioactive wastes that remain radioactive for hundreds of thousands of years, requiring safe storage.

Accidents or attacks on nuclear power plants can release dangerous radioactive isotopes.

Mining, processing and transporting of uranium can have detrimental health outcomes for workers.

Chapter 8 Electrical physics

8.1 Behaviour of charged particles

TY 8.1.1 -6.4×10^{-13} C

TY 8.1.2 3.0×10^{13} electrons

1 They will attract as they will be oppositely charged.

2 Copper is a good conductor but plastic is not. Charge can move freely in the copper wire but cannot escape the circuit through the plastic coating.

3 Metals have valence electrons that are not tightly bound to the nuclei. These electrons are free to move when a potential difference is applied between the ends of the conductor. The electrons in poor conductors such as plastic are tightly bound to their nuclei and are not free to move when a potential difference is applied.

4 3.1×10^{19} electrons　　5 +6.7C

8.2 Electric current and circuits

TY 8.2.1 4.7×10^{18} electrons

1 A continuous conducting loop (closed circuit) must be created from one terminal of a power supply to the other terminal.

2 **a** cell **b** light bulb **c** open switch **d** resistor **e** ammeter

3 C　　4 B

5 **a** 3A **b** 0.5A **c** 0.008A

6 **a** 5C **b** 300C **c** 18000C

7 **a** 9A to the right **b** 2A to the left

8 **a** 3C **b** 1000C **c** 1440C

9 **a** 16C **b** 4A

10 **a** 2×10^{19} electrons **b** 0.32A

8.3 Energy in electric circuits

TY 8.3.1 4.32×10^4 J

TY 8.3.2 9.45×10^3 J

TY 8.3.3 720W

CSA: Lightning

1 Static charge is charge that is not moving. It can only build up on an insulator and is usually created by friction between two insulating materials (e.g. glass and wool).

2 The passing contact between two non-conductors will cause electrons to be transferred from one to the other. In this case, the electrons will be transferred from the moving ice crystals to the grauphel.

3 The plasma is formed due to the strong electrical force fields ripping electrons away from the gas atoms (air molecules) and ionising the molecules. This high velocity stream of electrons and ions is at a very high temperature and travelling so fast that recombination cannot occur.

4 **a** 4.8×10^{-11} J **b** 1.7×10^5 A **c** 5.0×10^{13} W

Key questions

1 A

2 the gravitational potential energy of the water

3 **a** The voltmeter must always be in parallel with the light bulb, so M2 or M3.
  b The ammeter must always be in series with the light bulb, so M1 or M4.

4 **a** 1.38×10^5 J (or 138kJ) **b** 2.00A

5 **a** 4V **b** 1A

6 20V　　7 1.7×10^2 C

8 **a** heat and light **b** 60W **c** 0.25A

8.4 Resistance

TY 8.4.1 24Ω

TY 8.4.2 $I_1 = 0.60$ A, $V_2 = 12$ V

TY 8.4.3 778Ω

TY 8.4.4 9600mA

TY 8.4.5 4.0V

1 **a** A, B, C **b** C, B, A

2 $I_1 = 0.375$ A, $V_2 = 4.8$ V

3 **a** The wire is ohmic. This is because there is a proportional relationship between the voltage and the current, as shown by the linear nature of the I–V graph, which means that the resistance is a constant.
  b 3A
  c 2.5Ω

4 **a** 0.71Ω
  b The resistor is ohmic, as its resistance is constant.

5 They are both right. The resistance of the device is different for different voltages. Therefore, the device is non-ohmic.

6 72mA

7 **a** 2Ω **b** 5A

8 **a** It is non-ohmic, as the I–V relationship is non-linear.
  b 0.5A
  c 15V
  d **i** 20Ω
    ii 13Ω

Chapter 8 Review

1 1.9×10^{19} electrons　　2 6.7C

3 A

4 The potential difference between the ends of the conductor is causing the electrons to all move simultaneously in the same direction in the same way that links in a bicycle chain move together.

5 C

6 Conventional current represents the flow of charge around a circuit as if the moving charges were positive, which means the direction is from the positive terminal to the negative terminal. In reality, the moving particles in a metal wire are negatively charged electrons. Electron flow describes the movement of these electrons from the negative terminal to the positive terminal.

7 As electrons travel through a piece of copper wire, they constantly bump into copper ions, which slows them down. Resistance is a measure of how much energy electrons need to be given to maintain a constant flow of charge through the wire.

8 3.8×10^{-3} A　　9 3.2×10^{-19} C

10 **a** 160C **b** 10^{21} electrons

11 **a** 0.8C **b** 20s

12 7.6J　　13 4V　　14 1.4W

15 8.8V　　16 8.7A　　17 0.5Ω

18 **a** 0.25A **b** 240 LED bulbs

19 430Ω　　20 20V　　21 30Ω

22 **a** 1Ω **b** 2Ω **c** 3Ω

23 2.6MJ

24 In general, as the voltage increases the resistance of the conductor increases. Supporting data might involve calculating R for three voltages and current values. For example. For $V = 1.0$ V, $I = 2.5$ A and $R = 0.4$ Ω; for $V = 2.0$ V, $I = 3.5$ A and $R = 0.57$ Ω; and $V = 5.0$ V, $I = 5.0$ A, $R = 1.0$ Ω.

25 18.6W　　26 36Ω

27 $R_{cold} = 32$ Ω, $R_{hot} = 400$ Ω

28 3.75×10^{18} electrons

29 **a** 48C
  b 216J
  c The energy is provided by chemical reactions in the batteries.

Chapter 9 Electric circuits

9.1 Series and parallel circuits

TY 9.1.1 $R_{equivalent} = 160$ Ω

TY 9.1.2 $I = 0.011$ A; $V_{100} = 1.1$ V, $V_{690} = 7.4$ V, $V_{330} = 3.5$ V

TY 9.1.3 $R_{equivalent} = 14.3$ Ω

TY 9.1.4 $I_{circuit} = 0.53$ A; $I_{30} = 0.33$ A, $I_{50} = 0.20$ A

TY 9.1.5 $\Delta V_1 = 29.6$ V, $\Delta V_{2\text{–}4} = 21.2$ V, $\Delta V_{5\text{–}6} = 4.93$ V, $\Delta V_7 = 44.4$ V; $I_1 = 1.48$ A, $I_2 = 0.424$ A, $I_3 = 0.848$ A, $I_4 = 0.212$ A, $I_5 = 0.986$ A, $I_6 = 0.493$ A, $I_7 = 1.48$ A

TY 9.1.6 $P_{series} = 0.9$ W, $P_{parallel} = 0.144$ W; parallel circuit draws 6.25 times as much power as the series circuit

CSA: High power–low power

1 a $R_{equivalent} = R_1 + R_2$
  b $R_{equivalent} = R_1$
  c $\frac{1}{R_{equivalent}} = \frac{1}{R_1} + \frac{1}{R_2}$

2 a $240\,\Omega$ b $240\,W$ c $1\,A$

3 $P_{highest} : P_{lowest} = 4:1$

Key questions

1 B

2 a $I_T = 0.0075\,A$ or $7.5\,mA$ b $V_{100} = 0.75\,V$

3 $R_1 = R_2 = 136\,\Omega$

4 a $I_T = 0.75\,A$ b $I_{20} = 0.25\,A$ c $I_{10} = 0.5\,A$

5 a $V = 12\,V$ b $I_{60} = 0.2\,A$ (or 200 mA)

6 $V_1 = 6\,V, V_2 = 4.5\,V, V_3 = V_4 = 1.5\,V; I_1 = I_2 = 0.3\,A, I_3 = I_4 = 0.15\,A$

7 $4.3\,\Omega$

8 a $P = 1.3\,W$ b $20\,W$

9.2 Using electricity

TY 9.2.1 $V_{out} = 4\,V$

TY 9.2.2 a $20\,k\Omega$ b $0.20\,mA$ c $1\,V$

TY 9.2.3 a $9.8\,k\Omega$ b $0.92\,mA$ c $6.3\,V$

TY 9.2.4 a $0.6\,V$ b $R_1 = 1700\,\Omega$

TY 9.2.5 $140\,\Omega$

1

Input transducer	Signal-processing component	Output transducer
LDR microphone thermistor	diode potentiometer	LED light bulb speaker

2 A potentiometer. It divides the voltage depending on the position of the wiper. When the least resistance is across the bulb it will glow brightest.

3 The bulb will light up when the LDR has its highest resistance. This will occur when there no light falling on the LDR.

4 LEDs. They operate at much lower voltages and currents than incandescent and fluorescent bulbs, are more efficient and have a much longer life.

5 a $7.5\,V$ b $2.5\,V$

6 $V_{2000} = 16\,V$

7 $R_L = 350\,\Omega$

8 a $R_{20°} = 500\,\Omega$ b $R_R = 625\,\Omega$ c $T = 10°C$

9.3 Electrical safety

TY 9.3.1 $1.63

1 In the event of an electrical fault, the current will rapidly increase through the zero resistance path offered by the earth connection. Once this current exceeds the rating of the fuse or circuit breaker, it will blow, shutting off power to the appliance.

2 Double-insulated appliances have two barriers of insulation in their construction, with the outer case often made of plastic. Should one of these barriers become damaged and expose a conducting wire, the second barrier insulates this wire from the person using the appliance. Such appliances usually do not need an earth connection.

3 The earth stake ensures that the neutral and earth conductors are at zero potential.

4 The toaster will work normally, but the connection is very unsafe because it will remain live even when switched off. Peter has connected the active wire of the toaster to the neutral wire of the cable and the neutral wire of the toaster to the active wire of the cable.

5 The outer casing of the appliance could become live.

6 $3.6 \times 10^7\,J$

7 Air conditioner would cost approx. $1 to run for 5 hours, not $10 as in the statement.

8 $I = 2.4\,mA$

Chapter 9 Review

1 A

2 The circuit breaker is connected in series in the active wire. If a current greater than its rated value occurs, the circuit breaker will act like a switch and shut the circuit down. This prevents a current greater than is safe for the circuit, thereby preventing overheating and a possible fire.

3 The circuit will need to have two resistors connected in parallel and this combination connected in series with the third resistor.

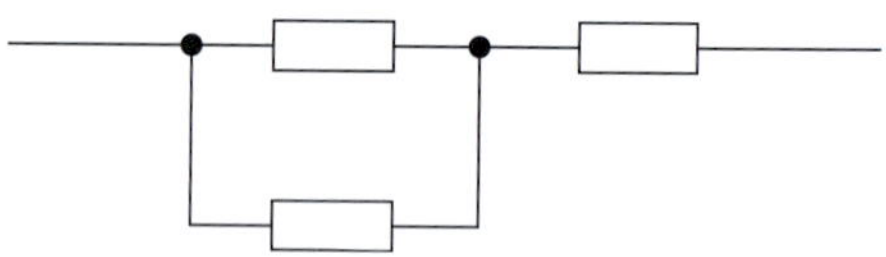

4 An LDR is a type of input transducer. It provides a variable resistance that depends on the amount of light falling on it.

5 D. The effects of an electric shock depend on the amount of current through the body and the duration of the shock. These are highest in D.

6 Power is the rate at which energy is consumed, supplied or transformed in or to that component (measured in watts or joules per second). The parallel circuit will consume more energy over the same time period because it has the smaller resistance and will therefore draw a larger current:
$P = VI = \frac{V^2}{R}$.

7 The finger provides less contact with the live wire and hence has more resistance. This means a smaller current.

8 A fuse will melt when the current exceeds its rating. Without the fuse the heat generated from a high current could be enough to start a fire and burn the house down. An RCD switches off a circuit when the currents in the active and neutral wires are not equal, thus preventing possible electrocution.

9 Parallel wiring allows each appliance to be switched on and off independently, with all receiving the mains voltage supply. See Figures 9.1.16 and 9.1.17 for suitable sample diagrams.

10 The resistor acts as a voltage divider and ensures the forward biased current in the LED is within the maximum allowed rating.

11 A thermistor could be used as an input transducer in a circuit to control the temperature inside a refrigerator because its resistance varies with temperature.

12 Only the neutral and active enter the house from the street. The earth and neutral are common with the earth connected from the neutral bar in the power board to the copper earth rod on the property.

13 It is much safer to place the fuse in the active circuit because then it cuts off the supply to the circuit.

14 a $6.0\,\Omega$ b $0.56\,A$ c $2.7\,V$
  d $0.33\,A$ e $0.22\,A$ f $12\,\Omega$

15 a Ammeter. The meter is connected in series so it must be an ammeter. It measures the total current in the circuit.
  b $7.71\,\Omega$

16 $4.2\,V$ 17 $1200\,\Omega$

18 a $1.2\,W$ b $0.60\,W$ c $2.4\,W$
  d The bulbs in **c** would be brighter but they consume more power per bulb. The two sets should be connected in parallel if the householder wants the bulbs to be as bright as possible, and in series if they want to use less power.

19 $3.60

20 **a** 60 Ω **b** 2.0 A

c I_1 = 1.20 A, I_2 = 0.60 A, I_3 = 0.20 A

d 240 W **e** 240 W

21 **a** 4 V

b Above, because as light increases, the resistance of the LDR decreases, hence V_{out} rises.

c V_{out} approaches zero, as the LDR has increased resistance and therefore the voltage drop across the LDR approaches 12 V.

22 **a** A thermistor is a temperature-sensitive resistor whose resistance decreases as its temperature increases.

b It is not linear.

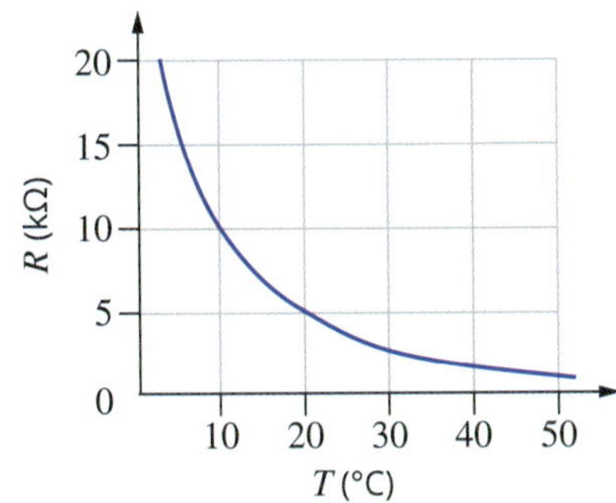

23 20°C

24 **a** 75 V **b** 38 W

25 **a** **i** R_L = 117 Ω **ii** R_L = 150 Ω

b Both circuits will emit the same light—they are operating at the same *V* and *I*.

c Circuit **ii** requires only 20 mA, so will run longer.

26 **a** 3 **b** 1 **c** 2

Unit 1 Area of Study 3 review

1 C **2** B **3** A **4** B **5** C **6** D **7** B
8 B **9** C **10** A **11** D **12** C **13** A **14** A
15 D **16** A

17 There will only be a current when there is a potential difference. If the large bird touches only one wire, the potential difference between its feet is negligible. If it touches two different wires, they will be at different voltages, and so charges will flow, meaning a current (sadly!) occurs in the bird.

18 **a** 6.7 kΩ **b** 2.5 kΩ

19 **a** 12 V **b** 8.0 V **c** 16 Ω

20 **a** 1.3×10^{19} electrons s^{-1}

b 1.6×10^{-17} J

c The electrical energy is converted into thermal (heat) energy in the wire.

d 2.0×10^2 W

e 2.0×10^2 J

f 2.0×10^2 W

g These answers are the same because of conservation of energy. The power provided by the battery (i.e. the energy given to each unit of charge (volt) per unit time) is the same as the power dissipated by the resistance wire.

21 **a** 50 Ω **b** 1.4×10^{-18} J

c 1.6 W **d** 1.1×10^{19} electrons

22 **a** 60.0 Ω

b 2.00 A

c I_1 = 1.20 A, I_2 = 0.600 A, I_3 = 0.200 A

d 240 W

e 240 W, as the power supplied to the circuit = the power consumed in the circuit

23 You might find it helpful to redraw the circuit to a more recognisable shape to find the equivalent resistance:

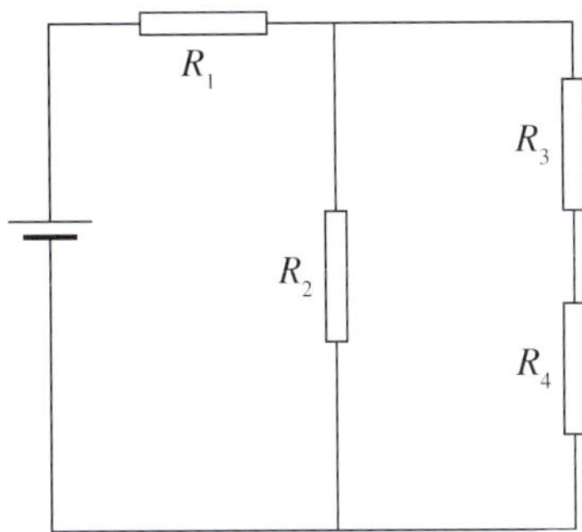

I_3 = 0.012 A, P_{R2} = 0.12 W

24 **a** The device is non-ohmic. The purpose of this device is to limit the current through a particular section of the circuit to a constant value regardless of the voltage across that part of the circuit.

b The resistance of the device increases with voltage.

c V_1 = 150 V, V_2 = 100 V

d 0.30 W

e 0.20 W

f 0.50 W

25 **a** 60 V, 60 W each

b The total current in the parallel circuit is the same as the total current in the series circuit = 1.0 A.

c The power bill is calculated on the total power consumption, which is the same for both arrangements. Mary is correct.

d Mary is correct, in that in a parallel arrangement, if one light bulb blows, the rest of the bulbs will not be affected, as the circuit would remain complete with the other bulbs. Conversely, in a series circuit, if one light bulb blows, none of the lights would be able to operate as there would be no current. (Furthermore, changing the arrangement would incur costs, which are not worth it if no money is saved due to the lack of change of power consumption!)

26 A fuse protects against overload current—too much current poses a fire hazard. A RCD detects an imbalance between current entering and leaving a device, which suggests that current is going to earth. Both will throw a circuit breaker.

27 **a** Household circuits are connected in parallel, so that each device is supplied with 240 V and can be turned on and off individually. Within a single circuit all the current to the devices in parallel passes through the one circuit breaker.

b A circuit breaker is used to protect against drawing too much current and exceeding the rating, and thus posing a fire risk.

c 3.33 V

d 28.8 Ω

e Total power = 4000 W. Current drawn by circuit = 16.6 A. Circuit has a 15 A circuit breaker, so the circuit breaker will trip.

f 936 kWh

28 20°C

29 Resistors R_1 and R_2 in the potentiometer form a voltage divider, which can set a variable voltage across the load. This, in turn, will control the current in the load resistor.

30 Transducers convert one form of energy to another. A microphone picks up the kinetic energy of air molecules and converts this to an electrical signal, and a speaker takes an electrical signal and converts it into the motion of air molecules.

31 A short circuit effectively bypasses the load in the circuit and connects the active and the neutral wires. This results in a greatly reduced resistance, and a high current. This condition will trigger the circuit breaker.

32 Plugs with three prongs have an earth. This is required when there is any possibility that the active lead could contact a metal and risk electrocuting the user. Some smaller devices are double insulated and so the active wire cannot deliver charge to any part of the device that a user can touch. In this case the earth is not needed and the plug can safely have only two prongs.

33 **a** **i** Elements A and B are connected in parallel and therefore have the same voltage drop of 20 V across each of them.
ii element A: 2.0 A, element B: 0.8 A
b **i** 3.2 A. Elements connected in series have the same current.
ii 72 V
c **i** 10 A **ii** = 5.0 Ω
d Neither student is correct in their entire statement. Evie's statement is partially correct, in that element A has a constant ratio of the voltage and current, which is the definition of being ohmic. This is not true in the case of element B, as the ratio of voltage and current changes, meaning it is non-ohmic. Nick is incorrect, he seems to be confusing the constant value of the resistance to mean that both elements need to have the same value of the resistance (which is the case where they intercept).

34 **a** The copper wire has a low resistance, which means that the supplied voltage will cause a large current. The current is a measure of the number of electrons passing a point per second. If large numbers of electrons are moving, they will be bumping into the atoms in the wire, transferring some of their kinetic energy. Temperature is a measure of the average kinetic energy of the molecules, so as the kinetic energy of the molecules increases the temperature also increases. Plastic is an insulator. It does not heat up as no current is passing through it.
b Electrical resistance is a measure of how difficult it is for electric charges to flow through the material. The copper wire has delocalised electrons that are able to move fairly easily, so it has a low resistance. The plastic has tightly held electrons that are unable to move, so it has a very high resistance.
c Various answers are possible. Ensure method allows for repeatable measurements, with at least five different data points collected. Electrical safety should be considered, including connection of components and choice of voltage supplied and resistors to have suitable currents. Check with your teacher if you are unsure of your answer.
d Three data points are not enough to determine the resistance of the component. More measurements are needed. The voltages being tested (0.5–1.5 V) are much lower than the voltage that needs to be predicted (24 V). This is extrapolation and is unreliable. Voltages above and below 24 V need to be included in the practical.

35 **a** 6.0 A **b** 2.8 Ω
c 17 V **d**

2.0 A
A
4.0 Ω
2.0 Ω
V
100 V
3.0 Ω
3.0 Ω

Answer diagrams will vary, but should include:
- correct circuit symbols
- ammeter placed in series with the resistor
- voltmeter placed in parallel with the resistor
- all connections closed.

e A multimeter can act as both components and has a larger range of measurements it can make.
f Various answers are possible. One solution includes placing the 2.0 Ω resistor and 4.0 Ω resistor in series, and placing the other two in a parallel arm.

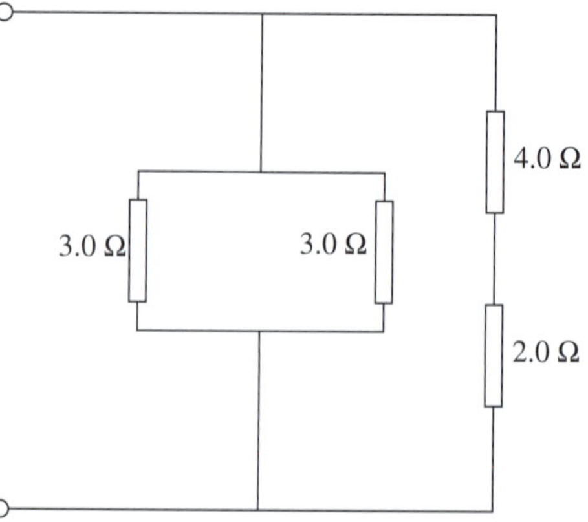

g Place both 3.0 Ω resistors in series in one branch, and place this branch in parallel with the 4.0 Ω and 2.0 Ω resistors, as shown:
$R_{equivalent} = 3.0\ \Omega$

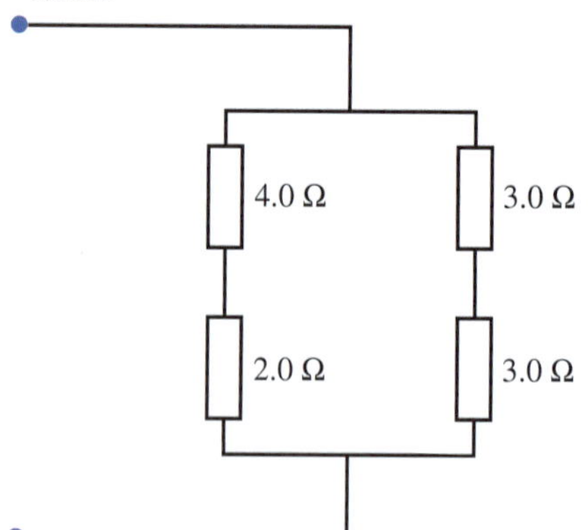

Chapter 10 Scalars and vectors

10.1 Scalars and vectors

TY 10.1.1 **a** 50 N west **b** −50 N
TY 10.1.2 This vector is 50.0° clockwise from the right direction.

1 Scalar quantities require a magnitude (size) and a unit.
2 Vectors require a magnitude, a unit and a direction.
3

Scalar	Vector
time distance volume speed temperature	force acceleration position displacement momentum velocity

4 **a** downwards **b** south **c** forwards
d upwards **e** east **f** positive
5 **a** 5.4 N **b** 2.7 N **c** 8.1 N
6 **a** 10.8 N **b** −5.4 N **c** 16.2 N
7 −35 N

8 a i 225°T
ii S45°W (or W45°S)
b i 120°T
ii S60°E (or E30°S)

10.2 Adding and subtracting vectors in one and two dimensions

TY 10.2.1 19.0 N downwards

TY 10.2.2 622 m s^{-1} upwards

TY 10.2.3 5.83 N, N59.0°E

TY 10.2.4 9.22 m s^{-1} N40.6°E

CSA: Surveying

1 19° up from the horizontal

2 A = 26 m at 160°, B = 40 m at 127°, C = 36 m at 66°, D = 17 m at 10.0°

3

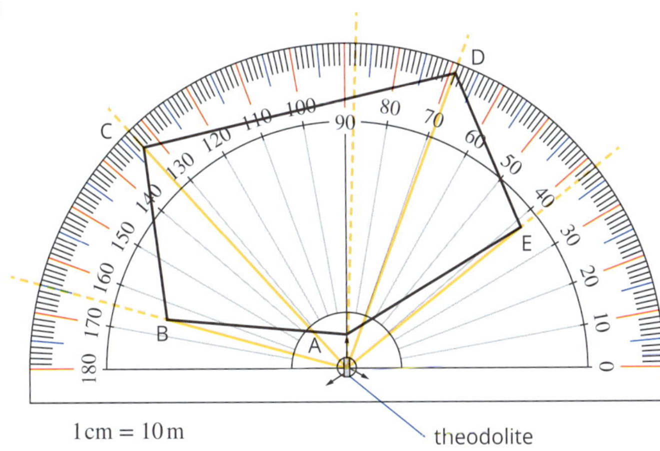

Key questions

1 Draw vector A first, then draw vector B with its tail at the head of A. The resultant is drawn from the tail of the first vector (A) to the head of the last vector (B).

2 a 16.0 N right b 6.00 N north c 14.4 N N56.3°E

3 a 5.00 m s^{-1} right b 13.00 m s^{-1} west
c 12.00 m s^{-1} north d 62.00 m s^{-1} east

4 a 11.0 m s^{-1} left b 4.0 m s^{-1} east
c 17.0 m s^{-1} S45.0°W d 12.8 m s^{-1} N38.7°E

5 4.00 m west 6 Δv = 14.0 m s^{-1} backwards

7 50.0 m 8 Δv = 8.79 m s^{-1} N36.7°W

10.3 Vector components

TY 10.3.1 F_D = 1580 N downwards

1 a

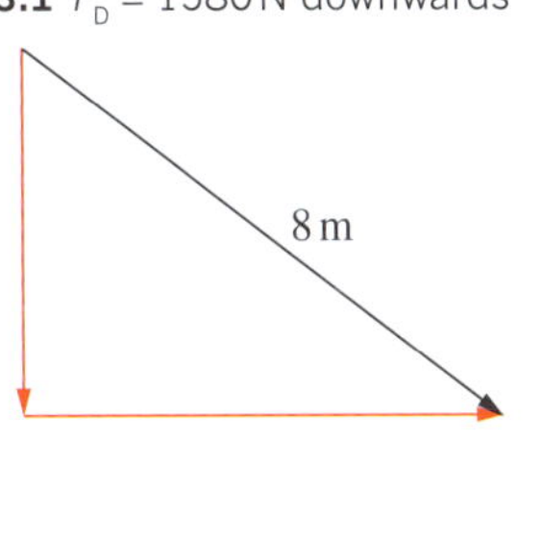

b

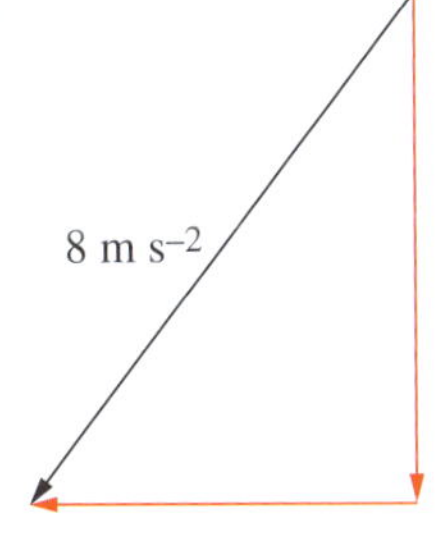

c

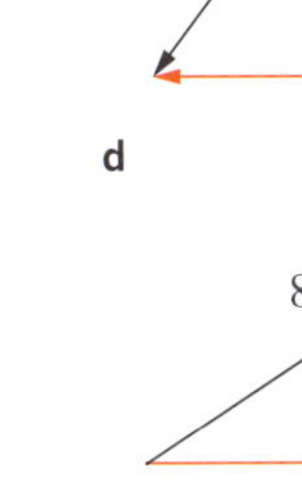

d

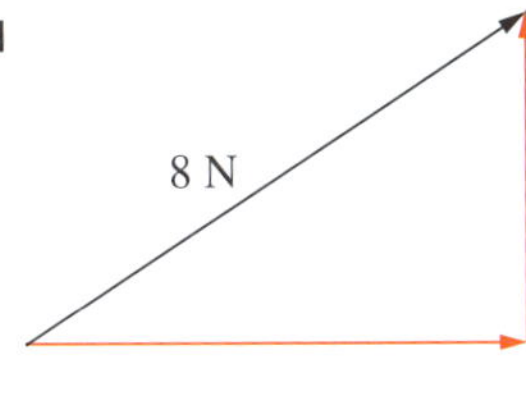

2 a 265 N downwards b 378 N right

3 19.8 N south and 16.6 N east

4 4.55 m s^{-1} south and 17.7 m s^{-1} east

5 18.9 m south and 43 m east of the starting point

6 $F_N = 1.09 \times 10^5$ N north, $F_W = 2.08 \times 10^5$ N west

7 a F_S = 50 N south, F_E = 87 N east
b F_N = 60.0 N north
c $F_S = 2.8 \times 10^2$ N south, $F_E = 1.0 \times 10^2$ N east
d $F_V = 1.50 \times 10^5$ N up, $F_H = 2.60 \times 10^5$ N horizontal

8 $F_H = 2.07 \times 10^2$ N, $F_V = 4.05 \times 10^2$ N up

Chapter 10 Review

1 B and D 2 A and D

3 Terms such as north and left cannot be used in a calculation; + and – can be used in calculations with vectors.

4 The vector must be drawn as an arrow with its tail at the point of contact between the hand and the ball. The arrow points in the direction of the 'push' of the hand.

5 Vector A has twice the magnitude of vector B.

6 Signs are useful in mathematical calculations; words such as 'north 'and 'south' cannot be used in an equation.

7 34.0 m s^{-1} north and 12.5 m s^{-1} east

8 –80.0 N

9 70° anticlockwise from the left direction

10 40° clockwise from the left direction

11

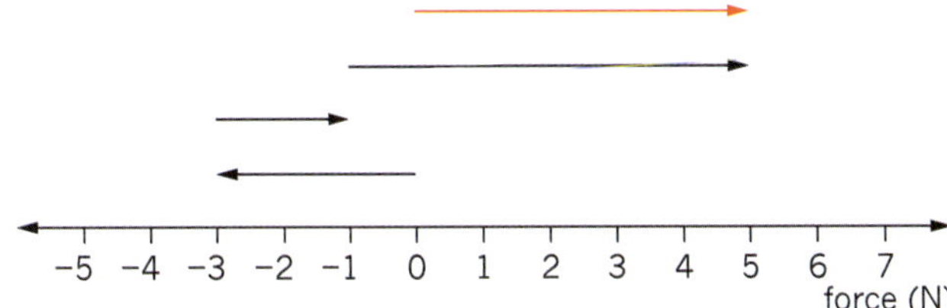

The resultant vector is 5.00 N right.

12 2.00 m downwards 13 21.0 m backwards

14 2.00 m s^{-1} left 15 7.00 m s^{-1} downwards

16 6.00 m s^{-1} left 17 65.7 m S56.8°W

18 813 N, N53.7°E 19 45 m, S63°W

20 6325 N, N71.6°E 21 11.0 m forwards

22 8.00 m s^{-1} east 23 22.8 m s^{-1} N55.2°W

24 67.7 m s^{-1} N35.0°W 25 67.5 m s^{-1} south

26 533 m s^{-1} N49.6°W 27 59.4 m s^{-1} N45.0°W

28 F_E = 39.4 N east, F_S = 22.8 N south

Chapter 11 Linear motion

11.1 Displacement, speed and velocity

CSA: Alternative units for speed and distance

1 a 11.1 m s^{-1} b 16.7 m s^{-1} c 90.0 km h^{-1} d 774 km h^{-1}

2 27.4 m s^{-1}; 98.7 km h^{-1}

3 22.9 Mach 4 2.27×10^{10} km

TY 11.1.1 a 0.923 m s^{-1} east
b 3.32 km h^{-1}
c 4.00 m s^{-1}
d 14.4 km h^{-1}

CSA: Breaking the speed limit

1 An average over two runs in opposite directions accounts for any differences in wind direction or slope of the road.

2 7.2722×10^2 s

3 Superconducting magnets on the train and coils in the track levitate the train, so there is less friction due to the absence of wheels contacting rails. The system is also used to accelerate and decelerate the train so there is no need for brake pads and discs, which would cause heating and wear.

Key questions

1 Distance travelled is 25.0 × 10 = 250 m, but displacement is zero because the swimmer starts and ends at the same place.

2 a displacement = +40 cm, distance travelled = 40 cm
b displacement = −10 cm, distance travelled = 10 cm
c displacement = 20 cm, distance travelled = 20 cm
d displacement = 20 cm, distance covered = 80 cm

3 a d = 80.0 km b s = +20.0 km or 20.0 km north

4 a −10 m or 10 m downwards b +60 m or 60 m upwards
c 70 m d 50 m or 50 m upwards

5 a 33.3 m s^{-1} b 25.0 m

6 a 16.7 km h^{-1} b 4.63 m s^{-1}

7 a 0.900 m s^{-1} b 0.100 m s^{-1} east

8 a 21.0 km b 15.0 km north of the starting point
c 14.0 km h^{-1} d 10.0 km h^{-1}

11.2 Acceleration

TY 11.2.1 a −2.00 m s^{-1}
b 16.0 m s^{-1} upwards

TY 11.2.2 4.57 × 10^2 m s^{-2} upwards

CSA: Human acceleration

1 26.2g 2 maximum gs = 1.76 × average gs

3 −406.3 m s^{-2}, 41.5g

Key questions

1 −7.00 km h^{-1} (negative value indicates a decrease in magnitude, and not a negative direction)

2 +5.00 m s^{-1} or 5.00 m s^{-1} upwards 3 8.00 m s^{-1} upwards

4 5.00 m s^{-2} south 5 42.9 km h^{-1} s^{-1}

6 a −10.0 m s^{-1} b 40.0 m s^{-1} west c 8.00 × 10^2 m s^{-2}

7 a 8.00 m s^{-1} b 8.00 m s^{-1} south c 6.67 m s^{-2}

11.3 Graphing position, velocity and acceleration over time

TY 11.3.1 a −15.0 m s^{-1} or 15.0 m s^{-1} backwards
b Cyclist is stationary, so velocity is 0.0 m s^{-1}.

TY 11.3.2 a 4.0 m west b 2.0 m s^{-1} west

TY 11.3.3 2.0 m s^{-2} west

CSA: Analysing performance in sport

1 ≈ 15 m

2 a ≈ 3.2 m s^{-2} b ≈ 2.6 m s^{-2} c ≈ −1.8 m s^{-2}

3 ≈ 51 m

Key questions

1 The gradient is the displacement over the time taken, hence velocity.

2 a The car initially moves in a positive direction and travels 8.0 m in 2.0 s. It then stops for 2.0 s. The car then reverses direction for 5.0 s, passing back through its starting point after 8.0 s. It travels a further 2.0 m in a negative direction before stopping after 9.0 s.
b i +8.0 m ii +8.0 m iii +4.0 m iv −2.0 m
c at t = 8.0 s
d i +4.0 m s^{-1} ii velocity is zero iii −2.0 m s^{-1}
iv −2.0 m s^{-1} v −2.0 m s^{-1}
e i 18 m ii −2.0 m

3 a 5.00 m s^{-1} b 20.0 m s^{-1} north
c 10.0 m s^{-1} north

4 a 0.0 m s^{-2} b 1.0 m s^{-2} c 10.5 m d 1.5 m s^{-1}

5 a 20.0 m s^{-1} north b −40.0 or 40.0 m s^{-1} south

6 a after 80.0 s b 1.27 or 1.3 m s^{-2} (answers may vary slightly)
c 0.41 m s^{-2} (answers may vary slightly)
d 4900 m or 4.9 km

7 a 2.0 m s^{-2} b t = 10 s c 80 m d 7.0 m s^{-1}

8 a

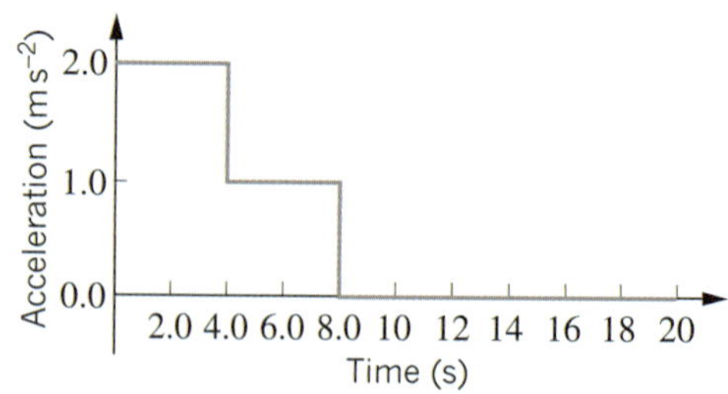

b 12 m s^{-1}

11.4 Equations for uniform acceleration

TY 11.4.1 a 3.75 m s^{-2} west
b 4.00 s
c 7.50 m s^{-1} east

1 $v^2 = u^2 + 2as$

2 The stone is travelling downwards, so the velocity is downwards. As the stone strikes the water, it quickly decelerates, so the acceleration is upwards.

3 a 3.13 m s^{-2} b 50.0 m s^{-1} c 180 km h^{-1}

4 a 2.00 m s^{-2} b 8.00 m s^{-1} c 64.0 m

5 a 40.0 m s^{-2} upwards b 1.12 km
c 576 km h^{-1} d 80.0 m s^{-1} e 124 m s^{-1}

6 a 20.8 m s^{-1} b 5.21 m c 36.2 m d 41.4 m

7 a 4.00 m s^{-1} b 5.70 m s^{-1} c 2.00 s d 0.83 s

8 a at 8.00 s b 16.0 s c 192 m

11.5 Vertical motion

CSA: Theories of motion—Aristotle and Galileo

1 Birds seem to fly in the air with little effort, so perhaps the feathers contain a small amount of earth substances but are mostly made of air substances. A rock, on the other hand, is made entirely of earth substances so the feather will not be attracted to the earth as strongly as the rock is.

2 Galileo observed both large and small hailstones striking the ground at the same time. According to Aristotle's theory, the larger hailstones should have reached the ground before the small ones.

3 Proof; see fully worked solutions.

TY 11.5.1 a 2.5 s
b 3.5 s
c −34 m s^{-1} or 34 m s^{-1} downwards

TY 11.5.2 a 12 m
b 1.5 s

Key questions

1 The force due to gravity of any object will be less on the Moon than on Earth because gravity is weaker on the Moon, due to the Moon's smaller mass.

2 The acceleration of a falling object is due to gravity, so it is constant, while the velocity increases at a uniform rate.

3 Acceleration due to gravity is constant (downwards) throughout the bounce; however, velocity decreases until it is zero at the top of the flight.

4 735 N downwards

5 a $15\,m\,s^{-1}$ b 11 m
6 a $-3.9\,m\,s^{-1}$ b 0.78 m c 0.20 m d 0.59 m
7 a 2.00 s b $20\,m\,s^{-1}$ c 20 m
d $20\,m\,s^{-1}$ downwards
8 a 3.5 s b 2.9 s

Chapter 11 Review

1 $26.4\,m\,s^{-1}$ 2 $54.0\,km\,h^{-1}$ 3 $-2.00\,m\,s^{-1}$

4 The car is moving in a positive direction so its velocity is positive. The car is slowing down so its acceleration is negative.

5 a The acceleration of an object in vertical motion is due to gravity, so it is constant no matter the direction of vertical travel (upwards or downwards).
b The flight is symmetrical, so the starting and landing speeds are the same, but in opposite directions.

6 a $15.0\,km\,h^{-1}$
b i $10.0\,km\,h^{-1}$ north
ii $2.78\,m\,s^{-1}$ north

7 a $10.0\,km\,h^{-1}$
b $2.80\,m\,s^{-1}$ south

8 $-6.00\,m\,s^{-2}$

9 a from 10 to 25 s
b from 30 to 45 s
c from 0.0 to 10 s, from 25 to 30 s and from 45 to 60 s
d 42.5 s or 43 s

10 The car was not speeding and was slowing down.

11 a B b A c C

12 a 114 m north
b $10.4\,m\,s^{-1}$
c $0.0\,m\,s^{-2}$
d $-7.0\,m\,s^{-2}$ or $7.0\,m\,s^{-2}$ south
e A

13 $15.8\,m\,s^{-1}$ 14 98 N

15 $m = 1.5$ kg, $F_g = 5.4$ N

16 a $4.00\,m\,s^{-2}$ away from the beach
b $4.00\,m\,s^{-1}$
c 8.00 m

17 a $-5.00\,m\,s^{-2}$ b 2.00 s

18 a +4.0 m b A and C c B; $v = +0.80\,m\,s^{-1}$
d D; $v = -2.4\,m\,s^{-1}$ e $0.800\,m\,s^{-1}$

19 The marble slows down by $9.8\,m\,s^{-1}$ each second, so it will take 4.00 s to stop momentarily at the top of its motion. It has a positive velocity that decreases to zero on the way up. Its acceleration is constant at $-9.8\,m\,s^{-2}$ due to gravity.

20 a $5.00\,m\,s^{-2}$ b 2.73 m c $5.45\,m\,s^{-1}$

21 a $98.0\,m\,s^{-2}$ upwards b 0.290 s
c $19.8\,m\,s^{-1}$ downwards

22 D

23 upwards, downwards, zero, downwards; downwards, downwards

24 a 45 m b 6.0 s c $20\,m\,s^{-1}$ downwards
d $10\,m\,s^{-2}$ downwards

25 a $29\,m\,s^{-1}$ b $24\,m\,s^{-1}$ c $12\,m\,s^{-1}$ downwards

26 2.1 kg

27 a 85 kg b 85 kg c 3.1×10^2 N downwards

28 Earth, Mars, Moon

29 a 10.0 s b $40\,m\,s^{-1}$ c 6.7 s

30 a $15\,m\,s^{-1}$ upwards b 11 m

31 a 1.7 s b 3.25 s

Chapter 12 Momentum and force

12.1 Newton's first law

CSA: Terminal velocity of raindrops

1 It is a direct relationship.

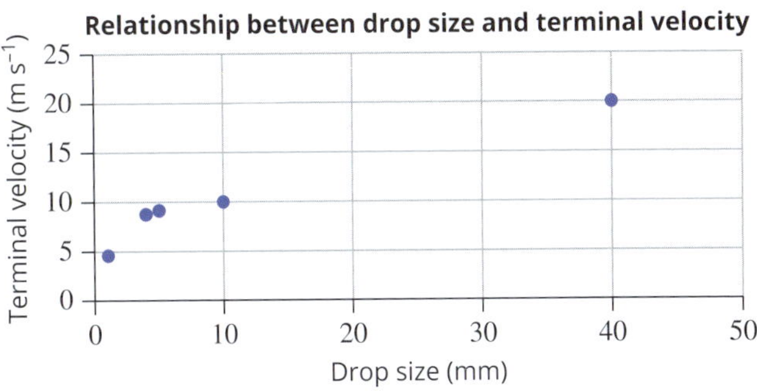

2 $7–8\,m\,s^{-1}$ 3 $125\,m\,s^{-1}$

Key questions

1 The box has changed its velocity, so the student can use Newton's first law to conclude that an unbalanced force must have acted on the box to slow it down.

2 Even though the car has maintained its speed, the direction has changed, which means the velocity has changed. Using Newton's first law, it can be concluded that an unbalanced force has acted on the car to change its direction.

3 The plane slows down as it travels along the runway because of the large retarding forces acting on it. The passengers wearing seatbelts would have retarding forces provided by the seatbelt and would slow down at the same rate as the plane. A passenger standing in the aisle, if they were not hanging on to anything, would have no retarding forces acting and so would tend to maintain their original velocity and move towards the front of the plane.

4 a gravitational force of attraction between the two masses
b electrical force of attraction between the negative electron and the positive nucleus
c friction between the tyres and the road
d tension in the wire

5 a 25 N forwards b 25 N forwards
c 25 N
$F = 29$ N forwards at an angle of 30° to the horizontal

6 Using a full glass makes the trick easier, because the force will have even less effect on the glass with a greater mass because the inertia of the full glass is greater than that of an empty glass.

7 The fully laden semitrailer. Its large mass means that more force is required to bring it to a stop.

8 lift = 50 kN upwards, drag = 12 kN west

12.2 Newton's second law

TY 12.2.1 307 N south
TY 12.2.2 $4.23\,m\,s^{-1}$ left
TY 12.2.3 $2.50\,m\,s^{-2}$ forwards
TY 12.2.4 a $7.0\,m\,s^{-2}$ forwards
b $5.0\,m\,s^{-2}$ forwards

1 In this experiment, a hammer and a feather are dropped at the same time to show that objects of different mass accelerate at the same rate due to gravity. On Earth, air resistance causes the feather to fall more slowly than the hammer, whereas on the Moon, where there is no air resistance, they fall at the same rate.

2 $3.56\,m\,s^{-1}$ north 3 $9.80\,m\,s^{-2}$ downwards

4 $0.547\,m\,s^{-1}$ east

5 a 490 N b 20 N south c $0.31\,m\,s^{-2}$

6 a $1.6\,m\,s^{-2}$ b $0.80\,m\,s^{-1}$ c $0.2\,m\,s^{-2}$

7 4 boxes 8 $10.2\,m\,s^{-2}$

12.3 Newton's third law

TY 12.3.1 **a** $F_{\text{on floor by bowling ball}}$
b $F_{\text{on bowling ball by floor}}$
c

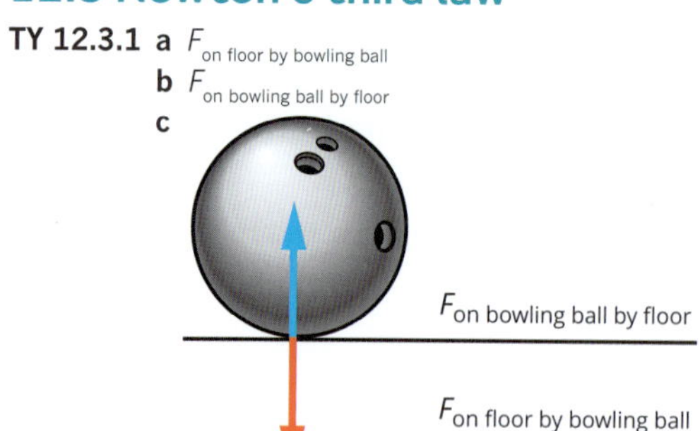

TY 12.3.2 **a** Action: The force from the beach volleyball player on the ground. Reaction: The force from the ground on the beach volleyball player.
b Action: The force from the top of the ladder pushing against the wall. Reaction: The force from the wall pushing against the top of the ladder.

1 There is a force on the hammer by the nail, and a force on the nail by the hammer. These two forces are equal in magnitude and opposite in direction.
2 **a** $F_{\text{on the astronaut by Earth}}$
b $F_{\text{on Earth by the astronaut}}$
3 the reaction force on the hand by the water
4 the reaction force on the balloon by the escaping air
5 5 N downwards
6 **a** 80 N towards the spacecraft
b $16\,\text{m s}^{-2}$ away from the spacecraft
c speed of the astronaut: $0.32\,\text{m s}^{-1}$; speed of the toolkit: $6.4\,\text{m s}^{-1}$
7 Jessie is correct. Action–reaction pairs of forces occur with contact forces and also with forces at a distance (such as gravity).

12.4 Momentum and conservation of momentum

TY 12.4.1 $2.05 \times 10^4\,\text{kg m s}^{-1}$ north
TY 12.4.2 $0.80\,\text{m s}^{-1}$ north
TY 12.4.3 $1.00\,\text{m s}^{-1}$ south
TY 12.4.4 $1630\,\text{m s}^{-1}$ south

1 $m_Au_A + m_Bu_B = m_Cv_C$
2 $m_Au_A = m_Bv_B + m_Cv_C + m_Dv_D$
3 $p = 8.75\,\text{kg m s}^{-1}$ south
4 $p = 9.61 \times 10^3\,\text{kg m s}^{-1}$ west
5 $3.0\,\text{m s}^{-1}$ in the direction opposite to that of the exhaust gases
6 An external force (gravity) is acting so momentum is not conserved.

12.5 Momentum transfer

TY 12.5.1 $\Delta p = 803\,\text{kg m s}^{-1}$ south
TY 12.5.2 $\Delta p = 0.0208\,\text{kg m s}^{-1}$ N38.7°E

1 kg m s^{-1}
2 Yes. Momentum is a vector quantity, so a change in direction (i.e. rebounding) means that momentum has changed.
3 $\Delta p = 83.1\,\text{kg m s}^{-1}$ south
4 $\Delta p = 2.35 \times 10^5\,\text{kg m s}^{-1}$ east
5 $v = 2.4\,\text{m s}^{-1}$ north
6 $\Delta p = 377\,\text{kg m s}^{-1}$ S42°W

12.6 Momentum and net force

TY 12.6.1 **a** $\Delta p = 0.192\,\text{kg m s}^{-1}$ upwards
b $F_{net} = 54.1\,\text{N}$ upwards
TY 12.6.2 **a** $\Delta p = 0.192\,\text{kg m s}^{-1}$ upwards
b $F_{net} = 0.591\,\text{N}$ upwards
TY 12.6.3 **a** $F = 32\,\text{N}$ upwards
b $\Delta p = 0.4\,\text{kg m s}^{-1}$ upwards

1 The impulse is the area below the force–time graph.
2 Airbags are designed to increase the duration of the collision, which changes the momentum of a person's head during a car accident. Increasing the duration of the collision decreases the force, which reduces the severity of injury.
3 The impulse due to the collision of the egg with the cardboard can't be changed, but the corrugated cardboard has more ability to compress, making the time of collision longer and the force on the egg smaller. The egg falling onto the corrugated cardboard is less likely to break.
4 **a** $450\,\text{kg m s}^{-1}$ east **b** $450\,\text{kg m s}^{-1}$ east
c 129 N east
5 **a** $9.0\,\text{kg m}^{-1}$ **b** 180 N in the direction of travel of the ball
c 180 N in the opposite direction to the ball's travel
6 **a** 1200 N **b** 62 N s
7 **a** $1.25\,\text{kg m s}^{-1}$ opposite in direction to its initial velocity
b $1.25\,\text{kg m s}^{-1}$ opposite in direction to its initial velocity
c $1.6 \times 10^3\,\text{N}$ in the opposite direction to the initial velocity of the arrow
8 **a** The crash helmet is designed so that the stopping time is increased by the collapsing shell during impact. This will reduce the force, since impulse $= F\Delta t = \Delta p$.
b No; a rigid shell would reduce the stopping time, therefore increasing the force.

Chapter 12 Review

1 According to Newton's first law, the passengers' bodies have inertia. As the bus turns to the left, the passengers' bodies maintain their original state of moving forwards in a straight line at a constant speed. The handrail provides an unbalanced force that acts to accelerate them to the left.
2 B
3 D
4 $9.80\,\text{m s}^{-2}$ downwards
5 50 N
6 38.3 kg
7 $6.90\,\text{m s}^{-2}$ north
8 27 N
9 $3.3\,\text{m s}^{-1}$ south
10 150 N backwards at an angle of 45° downwards into the ground
11 $51\,\text{kg m s}^{-1}$ forwards
12 $2.92 \times 10^4\,\text{kg m s}^{-1}$ east
13 $7.8 \times 10^2\,\text{N}$ forwards
14 $5.75\,\text{kg m s}^{-1}$ backwards
15 $-3.37\,\text{m s}^{-1}$ or $3.37\,\text{m s}^{-1}$ backwards
16 $32.5\,\text{m s}^{-1}$ in the direction of the ball
17 **a** $1.5\,\text{m s}^{-2}$ forwards **b** $1.05\,\text{m s}^{-2}$ forwards
c $F_{fr} = 90\,\text{N}$ backwards **d** 130 N
18 $\Delta p = 280\,\text{kg m s}^{-1}$ N53.1°E
19 **a** $8.5 \times 10^2\,\text{N}$ east
b If the deceleration time was much smaller, the force on the person's head would be much larger.
20 When a car is in an accident, the bodies of the passengers undergo a change in momentum. The concept of impulse tells us that, for a given change in a momentum, force is inversely proportional to time. This means that the more time it takes for the change in momentum to occur, the smaller the force needed. Crumple zones increase the time of a motor vehicle collision and therefore decrease the amount of force experienced by the bodies of the passengers of the cars involved.
21 **a** $\Delta p = 10\,\text{kg m s}^{-1}$ **b** $\Delta p_{bat} = 10\,\text{kg m s}^{-1}$
c $\Delta v_{ball} = 59\,\text{m s}^{-1}$

Chapter 13 Energy and motion

13.1 Work

TY 13.1.1 $W = 250\,\text{J}$
TY 13.1.2 $W = 1.1 \times 10^3\,\text{J}$
TY 13.1.3 $W = 0.05\,\text{J}$

1 $9.6 \times 10^3\,\text{J}$
2 The person exerts a force on the wall but there is no displacement of the wall ($s = 0$), so no work is done.

3 0.90 J 4 17.1 N 5 3.0×10^2 J

6 Since the box does not move, no work is done.

7 6.4 J

8 W_A is approximately 9 squares = 1.8 J, W_B is approximately 7 squares = 1.4 J, W_C is approximately 3 squares = 0.6 J

9 a 12 J b 9.0 J
c As the basketball bounces, some energy is lost as heat and sound, so the work done when the ball rebounds is less than the work done when the ball compresses.

13.2 Mechanical energy

TY 13.2.1 78 J

TY 13.2.2 **a** -8.40×10^2 kJ **b** 2.1×10^4 N

TY 13.2.3 114 km h^{-1}

TY 13.2.4 15 J

TY 13.2.5 32 J

TY 13.2.6 8.30 J

1 31 kJ 2 2.0×10^2 kJ 3 12 m s^{-1} or 43 km h^{-1}

4 $E_k \propto m$, so tripling the mass causes E_k to also increase by a factor of 3.

5 a 7.6 J b 3.8 J

6 0.323 m 7 3.04 J 8 2.5×10^6 J

9 spring A: E_s = 6.90 J, spring B: E_s = 26.9 J, spring C: E_s = 37.2 J

10 Spring A, as it is a stiffer spring and so requires more force to extend the same amount.

13.3 Using energy: power and efficiency

TY 13.3.1 50 J

TY 13.3.2 3.8 m s^{-1}

TY 13.3.3 76 m s^{-1}

CSA: Elastic potential energy

1 k = 5150 N m^{-1}

2 The teacher would have had to measure the relationship between energy supplied and energy lost across a variety of different values. One way in which she might have done this would be to drop the ball from many different heights and then measure the return height. From the initial drop, she can calculate the initial gravitational potential energy. This amount of energy will be stored as potential energy in the tennis ball. Using Hooke's law, she can determine by how much the ball would have compressed with that amount of energy. Finally, when the ball bounces back up, its final height can be used to determine the amount of final energy. The difference between this and the initial energy will be the total energy lost.

3 group 1: h_f = 11.4 m, group 2: h_f = 6.46 m, group 3: h_f = 2.85 m

4 group 1: h_f = 1.4 m, group 2: h_f = 2.5 m, group 3: h_f = 1.8 m; group 2's ball will bounce the highest

TY 13.3.4 output = 4.19 kJ

TY 13.3.5 7.0×10^2 W

TY 13.3.6 2.64×10^4 W

Key questions

1 a 4.8×10^4 J b 3.4×10^4 J

2 a 19 m s^{-1} b 16 m s^{-1}

3 2.0 m

4 a 318 J b 32.0 m s^{-1}

5 3.3×10^3 J 6 1.75 m

7 75414 W or 75 kW 8 198 kW

9 2.0×10^2 N 10 13.6 m s^{-1}

Chapter 13 Review

1 $\Delta E_p = mg\Delta h$, where ΔE_p is the gravitational potential energy and g is the acceleration due to gravity (9.8 m s^{-2}).

2 It is mechanical energy.

3 $E_k \propto v^2$, so triple the velocity causes E_k to change by a factor of 3^2 or 9.

4 When a pendulum is let go from its highest point, it will start with a large amount of gravitational potential energy, and no kinetic energy. As it travels to its lowest point, it accelerates, and energy is transformed from gravitational potential energy into kinetic energy. At the lowest point, it will reach its maximum kinetic energy. Assuming no energy is lost from the system, conservation of mechanical energy says that the sum of kinetic and gravitational potential energy is constant. Therefore, the exact amount of energy that is lost from one type will be converted into the second type.

5 2.22×10^5 J

6 approximately 40 squares, so 100 J

7 40572 J or 41 kJ 8 2.7×10^3 J

9 8.2 m 10 F = 73.3 N 11 160 J

12 10.3 m s^{-1} 13 2.2×10^2 J

14 k_A = 7500 N m^{-1}, k_B = 2500 N m^{-1}, k_C = 1071 N m^{-1}

15 20.9 m s^{-1}

16 a 8.6 J
b The gain in gravitational potential energy of the pendulum (8.6 J) is equal to the kinetic energy of the pendulum as it starts to swing upwards, so the pendulum had 8.6 J of kinetic energy.
c 2.2 m s^{-1}

17 238875 W or 2.4×10^2 kW

18 42 kW 19 2300 N

20 a 2900 J b 160 N

21 108 J 22 2600 J

Chapter 14 Forces and equilibrium

14.1 Torque

TY 14.1.1 τ = 39.5 N m anticlockwise

TY 14.1.2 $r_\perp$ = 62.1 cm, so 80 cm spanner is long enough

TY 14.1.3 τ = 17.1 N m

TY 14.1.4 τ = 17.1 N m

1 a The magnitude of the torque produced by a given force is proportional to the length of the force arm. By pushing the door at the handle, rather than the middle, the length to the force arm is increased.
b A crowbar can be used to generate a large torque because the force can be applied at a large distance from the pivot.

2 A torque is generated when the acting force is perpendicular to the lever arm of the object. If you were to push a door handle to the left, the entire force vector would be parallel to the lever arm of the door handle. Hence, no torque will be generated, and the door handle will not be able to rotate in order to open.

3 a 2×10^2 N m anticlockwise
b The torque is zero because the line of action passes through the pivot point.

4 0.50 m 5 18 N 6 90 N m 7 45 N m

8 a 4.9 N m b 9.8 N m c 4.9 N m

9 a 3.4×10^4 N
b The effective force arm remains at 15 m throughout, so the torque does not change.
c 5.2×10^5 N m clockwise about the pivot

14.2 Translational equilibrium

TY 14.2.1 $F_Y = 2.9 \times 10^4$ N upwards

TY 14.2.2 $F_{T2} = F_{T1} = 8.5 \times 10^3$ N

1 D 2 A, B and D

3 a 150 N upwards b 40 N west
c zero d 14 N north-west

4 2.4 N upwards

5 $F_T = 1.2 \times 10^3$ N upwards

6 a $F_{net} = 613$ N to the right b $F_{fr} = 613$ N to the left

7 $F_T = 32$ N

14.3 Static equilibrium

TY 14.3.1 The object is best supported at the centre of mass, 0.94 m from the origin.

TY 14.3.2 a 2.0 N upwards b 0.30 m

TY 14.3.3 Around reference point Y (the position of the boy), the clockwise torque due to the girl on the plank is equal to the anticlockwise torque due to the pivot point on the plank, so the plank is in rotational equilibrium.

TY 14.3.4 5.6×10^2 N upwards

TY 14.3.5 3.7×10^2 N downwards

CSA: The Bolte Bridge

1 $F_{min} = 1.5 \times 10^6$ N

2 a By moving the form travellers in pairs outwards from the central pillars, the force due to gravity is always balanced across the bridge and the static forces on the pier maintain static equilibrium. Since these components of the bridge are already balanced even in the temporary structure, there are minimal unbalanced forces that any temporary falsework would need to support while the bridge is being finished.

b $F_s = 9.8 \times 10^5$ N upwards

TY 14.3.6 $F_T = 52$ N

Key questions

1 D 2 C 3 1.5 m from the pivot

4 Depending on body position, the centre of mass can vary significantly and may fall outside the body. The centre of mass is only roughly above the navel when the person is standing upright with their arms by their sides.

5 $F_L = 1.2 \times 10^3$ N upwards, $F_R = 2.6 \times 10^3$ N upwards

6 a F_X acts downwards so that torques balance.

b $F_Y = 1.5 \times 10^4$ N c $F_X = 2.9 \times 10^3$ N

7 1.1 m from pivot

8 a $F_H = 438$ N, $F_V = 438$ N b $F_T = 371$ N

9 $x_{cm} = 1.1 \times 10^2$ cm

10 The centre of mass of each of the body parts—head, torso, arms, and legs—would lie within that body part. Since there is considerable mass from legs and arms below the torso, this moves the centre of mass lower. The overall centre of mass is outside their body. The force due to gravity can be considered to act downwards from the centre of mass. It creates a torque causing them to fall forwards. The person's feet cannot provide a balancing force because the force due to gravity acts beyond the base of support.

11 a Fred should attach the wires as high up on the wall as possible. This is because the wire needs to provide a torque that counterbalances the force due to gravity of the people sitting on the bench. The force that provides the torque comes from the vertical component of the tension in the wire. When the angle to the horizontal is increased, more of the tension is used for counterbalancing the force due to gravity. Therefore, a larger angle is better because it results in less tension.

b $\theta = 55°$, but really Fred should increase this angle to make the bench as safe as possible.

Chapter 14 Review

1 B 2 9.10 N m 3 0.290 m 4 B

5 31.5×10^4 N m 6 2.00 m 7 C

8 74.0 N m 9 22.4 N m

10 a 0.964 m b 15.9 N m

11 1.74 N m 12 74.7 N m

13 $F_{TR} = F_{TL} = 9.9 \times 10^2$ N 14 1.28 kg

15 $F_{TA} = 1732$ N, $F_{TB} = -1.2 \times 10^3$ N (1.2×10^3 N opposing A)

16 C 17 B 18 A

19 a 1.5×10^2 N m anticlockwise b 2.0×10^2 N m clockwise

c 49 N m clockwise

20 75 N m clockwise 21 3.40 N m 22 56.8 N m

23 79.3 N m 24 9.4×10^2 N upwards

25 $F_L = 1.9 \times 10^2$ N 26 $F_L = 131.8$ N, $F_R = 1.2 \times 10^2$ N

27 11 cm from the Sun

28 a i $F_A = 980$ N downwards, $F_B = 4.9$ kN upwards, $F_C = 4.9$ kN upwards, $F_D = 980$ N downwards

ii $F_A = 1.5$ kN downwards, $F_B = 5.8$ kN upwards, $F_C = 5.8$ kN upwards, $F_D = 1.5$ kN downwards

b As the woman walks from A to B, the force acting in pillar A decreases and the force acting in B increases. When the woman passes point B and continues on to point P, the forces in both A and B increase in order to produce a greater torque to counterbalance the increase in torque as she moves to point P.

Unit 2 Area of Study 1 review

1 A 2 D 3 B 4 C 5 C 6 B 7 A

8 D 9 A 10 D 11 A 12 C

13 a 10 m b 5.0 s c 60 m

14 a 5.0 s

b 4.4 s

c platinum sphere: $v = 24$ m s^{-1} downwards

lead sphere: $v = 32$ m s^{-1} downwards

15 a $t = 10$ s

b 4.9×10^2 m

c The acceleration of the arrow is constant during its flight and is equal to 9.8 m s^{-2} downwards; so, at the top of its flight, it is still 9.8 m s^{-2} downwards.

d average speed = 49 m s^{-1}, average velocity = 0 m s^{-1}

e When the arrow has just left the bow, it has maximum kinetic energy and minimum gravitational potential energy. As it rises, the kinetic energy is converted to gravitational potential energy, until the arrow reaches the top of its motion, where the kinetic energy is momentarily zero and the gravitational potential energy is at a maximum. Then, the process reverses as the arrow returns to the ground: the gravitational potential energy decreases to a minimum again, and the kinetic energy increases to a maximum again.

16 a Acceleration is given by the gradient of a velocity–time graph. Graph A, being horizontal and therefore having a gradient of zero, has an acceleration of zero. Graphs B and C both have constant gradients, meaning both depict constant acceleration (negative and positive respectively). D and E both have varying gradients, so represent changing accelerations (non-uniform acceleration). Graph D starts with a large gradient, so large acceleration, which slowly reduces. Graph E starts with a smaller gradient, therefore smaller acceleration, which gradually increases.

b Displacement is the area under the curve, so A has the greatest displacement.

c i C ii D iii A

17 a i 0.10 m s^{-1} ii 0.30 m s^{-1} iii 0.50 m s^{-1}

b i $v = 0.10$ m s^{-1} ii $v = 0.30$ m s^{-1} iii $v = 0.50$ m s^{-1}

c The mass is moving with constant acceleration to the right (due to the constant net force).

18 a $F_{\text{on man by Earth}}$ is the gravitational force on the man by Earth and $F_N = F_{\text{on man by ground}}$, the normal reaction force exerted on the man by the ground.

b $F_{\text{on man by Earth}} = mg = 980$ N downwards. As the man is in equilibrium, $F_N = F_{\text{on man by ground}}$ must balance this, so it is 980 N upwards.

c There are two separate Newton's third law pairs:

$F_{\text{on Earth by man}}$ is the gravitational attraction that the man exerts on the Earth, which is the Newton's third law pair to $F_{\text{on man by Earth}}$.

$F_{\text{on ground by man}}$ is the force that the man exerts on the ground, which is the Newton's third law pair to $F_{\text{on man by ground}}$.

19 a $0.375\,\text{m s}^{-2}$ west b $\Sigma F = 375\,\text{N}$ west c 575 N west d 575 N east

20 $v = 71\,\text{km h}^{-1}$

21 a $6.0 \times 10^2\,\text{N}$ b $8.0 \times 10^2\,\text{N}$ c $a = 4.0\,\text{m s}^{-2}$ up the incline

22 a $a = 3.3\,\text{m s}^{-2}$ upwards for the 10 kg mass and $3.3\,\text{m s}^{-2}$ downwards for the 20 kg mass

b $1.3 \times 10^2\,\text{N}$

23 a $3.0 \times 10^2\,\text{N}$ to the right b 1.0 kJ c 3.0 kJ d $\Delta E_k = 3.0\,\text{kJ}$ e $17\,\text{m s}^{-1}$

24 a Amir should sit 4.6 m from the pivot opposite his siblings.

b Neither Karim nor Farah are wholly correct, as the see-saw is in both translational and rotational equilibrium. The forces are equal in magnitude and opposite in direction, hence causing no translation because there is no net force. The object is also in rotational equilibrium as the sum of the clockwise torques equals the sum of the anticlockwise torques.

c Various answers are possible.

d Various answers are possible.

25 $F_{\text{ground}} = 367.5\,\text{N} = 370\,\text{N}$, $F_{\text{column}} = 3307.5\,\text{N}$

26 a 1.60 MJ b $\Sigma F = 1.60 \times 10^4\,\text{N}$ c $F = 16.4\,\text{kN}$ d $W = 1.64 \times 10^6\,\text{J} = 1.64\,\text{MJ}$ e $P = 3.28 \times 10^5\,\text{W}$ f 40.0 kJ g 97.6%

27 a $1.4 \times 10^2\,\text{J}$

b 80 J

c $6.4\,\text{m s}^{-1}$

d As we assume no losses due to friction, E_{total} remains constant $= 1.4 \times 10^2\,\text{J}$.

28 a $k = 2.0 \times 10^5\,\text{N m}^{-1}$

b 40 J

c $2.5 \times 10^2\,\text{J}$

d $\Delta x = 0.050\,\text{m} = 5.0\,\text{cm}$

e The elastic potential energy stored in the spring is transferred back to the trolley as kinetic energy when the spring starts to regain its original shape by re-expanding.

f The spring is elastic. This means it can return to its original shape after the compression force has been removed.

29 a $k = 1.1 \times 10^4\,\text{N m}^{-1}$

b 81%

c Child has maximum kinetic energy just before striking the trampoline on first descent.

30 a The section from 2.0 cm to 5.0 cm is rough because the cube loses kinetic energy (E_k) in this section.

b $7.1\,\text{m s}^{-1}$

c $\Delta E_k = 3.0\,\text{J}$

d The kinetic energy has been converted into heat and sound.

e $F_{fr} = 1.0 \times 10^2\,\text{N}$

31 a $645\,\text{kg m s}^{-1}$

b Momentum is always conserved. The motion of the pole is minute because it is anchored in the ground and very heavy.

c As his skull comes to rest against the pole, according to Newton's first law, his brain would continue in its motion at $7.50\,\text{m s}^{-1}$ until it collides with the skull, incurring potential damage in the collision. If he was wearing a helmet, he would increase his time of collision, and therefore decrease the force for the same change in momentum, reducing the risk of a concussion.

32 a $v_{\text{final}} = 2.2\,\text{m s}^{-1}$

b E_k before $= 1.8 \times 10^5\,\text{J}$, E_k after $= 1.3 \times 10^5\,\text{J}$; kinetic energy is not conserved.

c Total energy is always conserved, but kinetic energy is only conserved for perfectly elastic collisions. In this case, there is considerable loss to heat and sound in the collision.

33 a $10\,\text{kg m s}^{-1}$ b $10\,\text{kg m s}^{-1}$ c $59\,\text{m s}^{-1}$

34 a $1.6 \times 10^4\,\text{m s}^{-2}$ b $8.8 \times 10^6\,\text{N}$ c $4.4 \times 10^5\,\text{kg m s}^{-1}$ d $4.1\,\text{m s}^{-1}$ e $8.8 \times 10^6\,\text{N}$

f $W = 1.8 \times 10^8\,\text{J}$, E_k of projectile $= 1.8 \times 10^8\,\text{J}$

This represents an ideal situation; realistically there would be significant energy losses to heat and sound.

35 a independent variable: initial velocity; dependent variable: final height

b Various suggestions are possible. The students would need to control the type of projectile, they would want to complete the practical under the same conditions—ideally on a non-windy day.

c

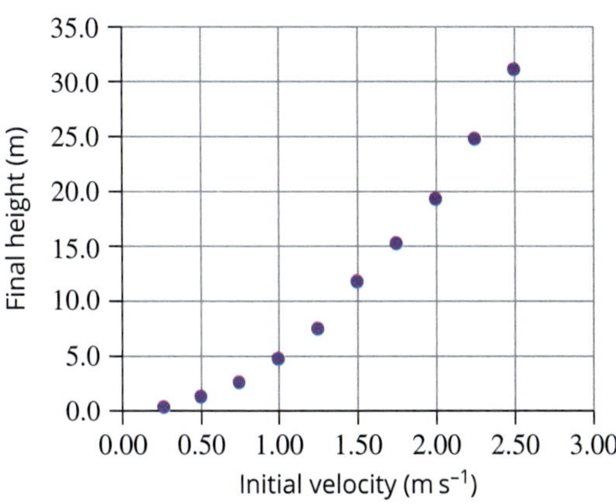

d A straight line of best fit/trendline does not fit with the data, as there appears to be an upwards curve in the data.

e

(Initial velocity)^2 ($\text{m}^2\,\text{s}^{-2}$)	Final height (m)
0.06	0.3
0.25	1.3
0.56	2.8
1.00	4.9
1.56	7.5
2.25	11.8
3.06	15.4
4.00	19.4
5.06	24.9
6.25	31.1

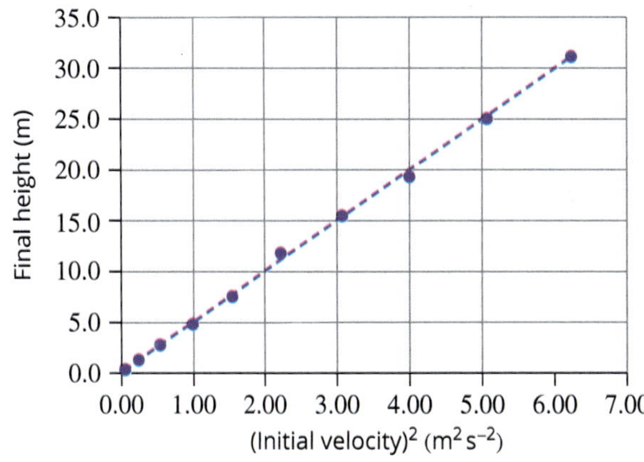

Glossary

RED ENTRIES VCE Physics Study Design extracts © VCAA (2022); reproduced by permission

A

absolute zero A temperature of −273°C or 0 K. Molecules and atoms have minimum kinetic energy at this temperature.

absorb To take up and store energy, such as radiation, light or sound, without it being reflected or transmitted. During absorption the energy may change from one form into another. When radiation strikes the electrons in an atom, the electrons move to a higher orbit or state of excitement by absorption of the energy of the radiation.

absorbed dose The amount of ionising radiation absorbed per kilogram of irradiated material, measured in grays (Gy).

acceleration The rate of change of velocity. Acceleration is a vector quantity. The SI unit for acceleration is $m\,s^{-2}$.

accuracy A measurement value is considered to be accurate if it is judged to be close to the true value of the quantity being measured.

activity The number of nuclei of a radioactive substance that decay each second, measured in becquerels (Bq).

aim A statement describing in detail what will be investigated in an experiment.

air resistance The retarding force (drag) caused by collisions between air and moving objects.

alpha particle A particle consisting of two protons and two neutrons ejected from the nucleus of a radioactive nuclide.

alternating current In an alternating current (AC), electrons oscillate backwards and forwards around a mean position, as opposed to direct current (DC). Household power supplies usually operate at 240 V AC.

ammeter An instrument used to measure the electric current in a circuit. Electric current is measured in amperes (A), which is why it is called an ammeter.

amplitude The absolute value of the maximum displacement from a zero value during one period of an oscillation.

angle of incidence The angle measured between the incident ray and the normal to the surface that it strikes.

angle of reflection The angle measured between the reflected ray and the normal to the reflective surface.

antineutrino A neutral subatomic particle that interacts very weakly with other matter; the antimatter particle of the neutrino.

apparent depth The apparent distance of the virtual image from the surface of the water.

atomic number The number of protons in a nucleus.

axis of rotation An imaginary line through the centre of mass or pivot point of an object. The axis is perpendicular to the plane of rotation of the object.

B

base The support for a structure. For example, the base of a car is a rectangle with each of the four tyres at the corners of the rectangle.

beta particle An electron or positron ejected from the nucleus of a radioactive nuclide.

bias A form of systematic error resulting from the researcher's personal preferences or motivations.

binding energy Energy required to split a nucleus into its separate nucleons.

binding energy per nucleon The binding energy of a nucleus divided by the number of nucleons in that nucleus.

black body A black body is a perfect absorber and emitter of electromagnetic radiation; i.e. it does not reflect any radiation. It does not necessarily have to be black; for instance, the Sun can be modelled as a black body.

black-body radiation Radiation emitted by a black body in a continuous spectrum.

C

cantilever A beam that extends out horizontally beyond its supporting structure.

centre of gravity The position from which the entire weight of the body or system is considered to act; the position at which the body will balance.

centre of mass A single point in an object where the mass can be considered to be 'concentrated' for the purposes of analysing motion.

chain reaction A reaction that can sustain itself; in nuclear fission, it is the exponential cascade of more neutrons initiating more fission reactions.

charge A property of matter that causes electric effects. Protons have positive charge, electrons have negative charge and neutrons have no charge.

chemotherapy The use of drugs to destroy or slow the rapid growth of cancer cells in the body. These drugs are highly toxic to the cells and therefore may cause secondary side effects due to damage of surrounding healthy cells. Chemotherapy can use one drug or a combination of drugs. It can also use drugs that emit radiation (radiopharmaceuticals).

circuit breaker A device that automatically switches off an excessive current by detecting the magnetic field associated with it.

collinear Lying on the same straight line.

components The components of a force are two vectors at right angles to each other that when added together will be equivalent to the original force.

compression A material being squashed or squeezed is said to be under compression.

conclusion Evidence-based statements that are developed from the analysis of experimental results.

conduction The movement of energy (such as heat) from one object to another without the net movement of particles (atoms or molecules).

conductor A substance, body or system that readily conducts heat, electricity, sound or light.

conservation of mechanical energy The sum of all energy in a system (potential and kinetic) will remain constant.

conserved When a quantity that exists before an interaction is exactly equal to the quantity that exists after the interaction.

contact forces Forces that exist when one object or material is touching another. Friction, drag and the normal force are contact forces.

continuous variable A variable that can have any number value within a given range.

controlled variable A variable that is kept constant in order to reliably find the effect of changing the dependent variable.

convection A process of heat transfer through a gas or liquid by bulk motion of hotter material into a cooler region.

conventional current The same as electric current, but in the opposite direction to electron flow.

coulomb The SI unit of charge; 1 C is equivalent to the combined charge of 6.2×10^{18} protons.

crest The highest part or top of a transverse wave.

critical angle The angle of incidence that produces an angle of refraction of 90°. The largest angle for which refraction will occur; at angles larger than the critical angle, light undergoes total internal reflection.

critical mass The amount of material needed to sustain a nuclear fission chain reaction.

current The net flow of electric charge. Current is measured in amperes (A) where $1\,A = 1\,C\,s^{-1}$. By convention, electric current is assumed to flow from positive to negative.

D

daughter nucleus The resulting nuclide formed after the parent nucleus has decayed.

decay series A sequence of radioactive decays that results in the formation of a stable isotope.

dependent variable The variable that is to be measured.

diffuse Spread out. Diffuse reflection occurs when light is reflected from a rough or uneven surface and is scattered in all directions.

dimension Space can be considered to consist of three length dimensions. These length dimensions are arranged at 90 degrees to each other with their point of intersection being the origin. The position of an object can be defined in relation to its position along each of the three dimensions. Typically, these three dimensions are labelled x, y and z. However, left–right, upwards–downwards and backwards-forwards are also appropriate.

dimensional analysis Using the units in a graph or formula to check that the derived term is correct.

diode A semiconductor device that has the special property that it will only allow electric current to flow through it in one direction.

direct current In a direct current (DC), electrons travel in one direction only, as opposed to alternating current (AC). Batteries and electric cells provide direct current.

direction convention Standardised system for describing the direction in which an object is travelling. The use of cardinal points of a compass (N, S, E and W) is an example of a direction convention.

discrete variable A variable that can be counted or measured but which can only have certain values, for example, the number of times a lever is pressed.

dispersion The spreading out of white light into its component colours; occurrs when different colours of light take slightly different paths when refracted.

displacement The change in the position of an object relative to its starting position and final position. Displacement does not consider the route the object took to change position, only where it started and where it ended. Displacement is a vector quantity. It is measured in metres (m) and given the symbol *s*.

distance travelled How far an object travels during a particular motion or journey. Distance is a scalar value. Direction is not required when expressing magnitude. It is measured in metres (m) and given the symbol *d*.

E

earth The third wire (usually green or green and yellow) in electrical devices that acts as an important safety feature by carrying excess current due to a device malfunction directly into the earth.

effective dose Measurement of the amount of radiation a tissue or organ has been exposed to which takes into account the sensitivity of the tissue towards the particular type of radiation. It is calculated by multiplying the value of equivalent dose by the 'weighting factor' (W). If more than one organ has been exposed to a radiation source, then the total value of the effective dose is the sum of all the effective doses for all of the organs/tissues involved.

efficiency The proportion of energy transformed by a device. It is the energy output of the device divided by the energy input, expressed as a decimal or a percentage.

elastic Material that returns to its original shape after being deformed.

elastic potential energy Stored energy in a stretched or compressed material, measured in joules (J).

electric circuit A continuous conducting loop that allows electric current to flow.

electric shock Also known as electrocution, in which excess electricity flows into the human body due to a device malfunction or electrical accident.

electrical potential energy Potential energy due to the concentration of charge in part of an electric circuit.

electricity A form of energy resulting from the existence of charged particles (electrons or protons). Electricity is fuelled by the attraction of particles with opposite charges and the repulsion of particles with the same charge.

electromagnetic spectrum The entire range of electromagnetic radiation. Consists of radio waves, microwaves, infrared radiation, visible light, ultraviolet light, X-rays and gamma rays. In a vacuum, all electromagnetic radiation travels at $3.0 \times 10^8 \, m \, s^{-1}$.

electromagnetic wave A wave with two transverse, mutually perpendicular components: a varying magnetic field and a varying electric field.

electron A negatively charged particle in the outer region of an atom; it can move from one object to another, creating an electrostatic charge. When electrons move in a conductor, they constitute an electric current.

electron flow The net flow of electrons. Although electric current is assumed to flow from positive to negative, electrons physically move from negative to positive.

electron volt (eV) A small unit of energy. One electronvolt (1 eV) is the energy an electron would gain when accelerated across a potential difference of one volt: $1 \, eV = 1.6 \times 10^{-19} \, J$.

electrostatic force The attractive (for particles of opposite charge) or repulsive (for particles of the same charge) force that exists between charged particles over a distance.

elementary charge The magnitude of the charge on an electron or proton: $e = 1.6 \times 10^{-19} \, C$.

emissivity An object's effectiveness at emitting energy. It is defined as the fraction of energy that is emitted relative to the energy emitted by a black body.

enhanced greenhouse effect Also known as climate change or global warming, it is the impact on the climate from the additional heat retained due to the increased amounts of carbon dioxide and other greenhouse gases in the lower atmosphere.

equivalent dose A measure of the biological damage inflicted on a tissue due to absorption of a defined quantity of radiation. Equivalent dose measurements take into account the nature of the radiation applied. It is measured in sieverts (Sv).

equivalent resistance A single resistance that could be used to replace a number of individual resistors for the purpose of circuit analysis.

evaporation The changing of a liquid into a gas, often under the influence of heat (as in the boiling of water).

F

first law of thermodynamics Energy within an isolated system can change form but the total internal energy is constant. If heating or cooling is applied to the system, or if work is done on or by the system, the system is no longer isolated and the total internal energy changes.

fissile Capable of undergoing nuclear fission after capturing low-energy neutrons.

fission The splitting of a nucleus into two or more pieces, usually after bombardment by neutrons.

fission fragments Nuclides formed during nuclear fission; these are usually radioactive.

force A vector quantity that measures the magnitude and direction of a push or a pull. It is measured in newtons (N).

force arm The perpendicular distance between the axis of rotation and the line of action of the force.

fossil fuel A natural fuel such as coal or gas, formed in the geological past from the remains of living organisms.

free fall The motion of a falling body under the effect of gravity only. No air resistance or propulsive forces are acting.

frequency A measure of the rate at which something occurs; for example, the number of vibrations or cycles that are completed per second or the number of complete waves that pass a given point per second. Measured in hertz (Hz).

fuse A circuit device that melts when too much current flows through it, breaking the circuit in the process and protecting the other circuit components.

fusion A process taking place inside stars in which small nuclei are forced together to make larger nuclei. Energy is released in the process.

G

gamma ray High-energy electromagnetic radiation ejected from the nucleus of a radioactive nuclide.

Geiger counter A device for measuring radioactive emissions.

genetic Refers to the characteristics or modifications introduced to the DNA in the cell and are passed on to the offspring by sexual reproduction.

geothermal energy Energy produced by the internal heat of Earth.

gravitational potential energy Energy available to an object due to its position in a gravitational field. Measured in joules (J).

greenhouse effect The trapping of the Sun's energy in a planet's atmosphere, which warms the planet. Thermal radiation from a planetary surface is absorbed by atmospheric greenhouse gases, and re-radiated in all directions.

greenhouse gas A gas that contributes to the greenhouse effect by absorbing infrared radiation. Carbon dioxide and methane are examples of greenhouse gases.

H

half-life The time taken for half of the nuclei of a radioactive isotope to decay.

heat The energy transferred from a hotter object to a cooler one that increases the kinetic and/or potential energy of the particles in the cooler object.

Hooke's law Elastic materials for which there is a direct relationship between the force acting on them and the extension or compression that they undergo are said to obey Hooke's law.

hypothesis A proposed explanation for an observed phenomenon. A prediction based on scientific reasoning that can be tested experimentally.

I

impulse The change in momentum of an object is also called the impulse of the object. The impulse is calculated by the final momentum minus the initial momentum.

independent variable The variable that the experimenter is manipulating.

inertia A property of an object, related to its mass, that opposes a change in motion.

insulator A material or an object that does not easily allow heat, electricity, light or sound to pass through it. Air, cloth and rubber are good electrical insulators; feathers and wool are good thermal insulators.

internal energy The total kinetic and potential energy of the particles within a substance.

ion Atom of a chemical element in which the number of electrons and protons is not equal and therefore the atom is electrically charged. If extra electrons are present, the ion has a negative charge. If electrons are missing, the ion has a positive charge.

ionised An atom or molecule that has become electrically charged.

isotope Atoms with the same number of protons but with different numbers of neutrons.

J

junction A point in an electric circuit from which current can flow into or out of from more than one direction.

K

kelvin An absolute temperature scale based on the triple point of water.

kilowatt hour (kWh) Unit of energy equivalent to 3.6 megajoules. The equivalent amount of energy as a 1000 W device turned on for one hour. It is the unit of measure of electricity usage that is measured by electricity meters and appears on electricity bills.

kinetic energy The energy of a moving body, measured in joules (J).

kinetic particle model A model that states that the small particles (atoms or molecules) that make up all matter have kinetic energy, which means that all particles are in constant motion, even in solids.

L

latent heat The 'hidden' energy used to change the state of a substance at the same temperature; i.e. the energy is not seen as a change in temperature.

latent heat of fusion The energy required to change 1 kg of solid to a liquid at its melting point.

latent heat of vaporisation The energy required to change 1 kg of liquid to a gas at its boiling point.

law of conservation of momentum In any collision or interaction between two or more objects in an isolated system, the total momentum of the system will be conserved. That is, the total initial momentum is equal to the total final momentum.

light dependent resistor (LDR) A non-ohmic device whose resistance varies with the light intensity that falls upon it.

light-emitting diode (LED) A type of diode that emits light as current passes through it.

line of action of the force The line along which a force is acting. The line of action extends forwards and backwards from the force vector.

line of best fit A straight line that best represents scatter-plot data but does not necessarily pass through all points.

linear relationship Variables that produce a straight trend line.

longitudinal Extending in the direction of the length rather than across something. The vibrations of a longitudinal wave are in the same direction as, or parallel to, the direction of travel of the wave.

M

magnitude The size or extent of something, with no need for direction. In physics, this is usually a quantitative measure expressed as a number of a standard unit.

mantle The interior section of Earth, found between the crust and the outer core.

mass An amount of matter. One kilogram can be defined as the amount of matter that would result in an acceleration of $1\ \mathrm{m\,s^{-2}}$ when a force of 1 N is applied in a frictionless environment.

mass defect The difference between the mass of the nucleus as a whole and the sum of the individual nucleons.

mass number The number of nucleons (protons and neutrons) in a nucleus.

mean Equal to the average of a set of data.

mechanical energy The energy that a body possesses due to its position or motion. Kinetic energy, gravitational energy and elastic potential energy are all forms of mechanical energy.

mechanical wave A mechanical wave is a wave that propagates as an oscillation of matter, and therefore transfers energy through a medium.

median The middle number for a set of data.

medium A physical substance, such as air or water, through which a mechanical wave is propagated.

metal Material in which some of the electrons are only loosely attracted to their atomic nuclei.

method The specific steps that are taken to collect data during the investigation.

methodology The approach taken to investigate the research question or hypothesis. Examples include controlled experiments, fieldwork, literature reviews, modelling and simulation.

mistake An error made by an experimenter that could have been avoided.

mode A value that appears the most number of times within a data set.

momentum The product of the mass and velocity of an object. Objects with larger momentum require a larger force to stop them in the same time that an object with smaller momentum takes to stop. It is given by the equation $p = mv$.

mutation Any change in the structure or composition of DNA, which in turn alters the genetic information stored in a cell.

N

net charge When the number of positive and negative charges in an object are not balanced.

net force The vector sum of all the individual forces acting on a body.

neutral No electric charge, or a situation in which positive and negative charges are balanced.

neutral equilibrium A situation in which an object will remain stationary no matter where it is; for example, a ball on a horizontal table.

neutron An uncharged subatomic particle.

newton SI unit of force. One newton (1 N) is defined as the force required to make a mass of 1 kg accelerate at $1\ \mathrm{m\,s^{-2}}$.

Newton's first law States that an object will maintain a constant velocity unless an unbalanced, external force acts on it.

Newton's second law States that force is equal to the rate of change of momentum. This can be processed mathematically to: the acceleration of an object is directly proportional to the force on the object and inversely proportional to the mass of the object.

Newton's third law States that for every action (force), there is an equal and opposite reaction (force).

non-contact forces Forces that act at a distance and do not require the bodies to actually touch each other. Gravitational, magnetic and electric forces are non-contact forces.

non-metal Material in which all of the electrons are strongly attracted to their atomic nuclei.

non-ohmic Not behaving according to Ohm's law; resistance changes depending on the potential difference.

normal A line constructed at 90° to the surface at the point that a wave strikes the surface.

nuclear transmutation The changing of one element into another.

nucleon A particle located in the nucleus of an atom.

nucleus The central part of an atom.

nuclide The range of atomic nuclei associated with a particular atom, which is defined by its atomic number, and the various isotopes of that atom as identified by the mass number.

O

observation Gathering information using all your senses and a variety of instruments and laboratory techniques.

ohmic A conductor that follows Ohm's law; the relationship between voltage and current is linear.

oscillate To move about an average position in a regular, repetitive or periodic pattern.

outliers Data points or observations that differ significantly from other data points or observations.

overload When an unsafe amount of current flows through a wire; for example, when too many electrical appliances are connected to the same power point.

P

parallel circuit A circuit that contains junctions; the current drawn from the battery, cell or electricity supply splits before it reaches the components and rejoins afterwards.

parent nucleus A nucleus on the reactant side of a nuclear equation that when struck by a neutron undergoes fission or simply decays by natural means.

peak wavelength The wavelength at which an object emits the maximum intensity of radiation. It is dependent on the object's surface temperature.

penetrating ability A measure of how easily radiation passes through matter.

period The time interval for one vibration or cycle to be completed.

personal protective equipment (PPE) Clothing that is worn to minimise personal risk in an investigation. Examples include safety glasses, protective gloves, ear muffs and a laboratory coat.

pivot point A point about which an object can rotate.

plane wave A constant frequency wave with wavefronts that are infinite parallel lines or planes.

polarisation The phenomenon in which transverse waves are restricted in their direction of vibration.

position The location of an object with respect to a reference point. Position is a vector quantity.

positron The antimatter pair of the electron. This means it has the same mass as an electron but has opposite properties such as electromagnetic charge and spin.

potential difference The difference in electric potential between two points in a circuit; measured by a voltmeter when placed across a circuit. A battery creates the potential difference across a circuit, which drives the current.

potential energy Energy that can be considered to be 'stored' within a body due to its position, composition or molecular arrangement.

potentiometer A circuit device consisting of a three-terminal resistor with a sliding or rotating contact called the wiper. It can also be connected at one end and at the wiper to create a variable resistor.

power The rate at which work is done. It is a scalar quantity measured in watts (W).

precision A measure of the repeatability or reproducibility of scientific measurements and refers to how close two or more measurements are to each other. A set of precise measurements will have values very close to the mean value of the measurements.

primary data Data you have collected yourself.

primary source Information created by a person directly involved in an investigation.

principle of moments In order for an object to be in rotational equilibrium, the sum of the moments in a clockwise direction must balance the sum of the moments in an anticlockwise direction.

processed data Raw data that has been organised, altered or analysed to produce meaningful information.

proton A positively charged subatomic particle.

pulse A single disturbance moving through a medium from one point to the next.

Q

qualitative data Non-numeric data from categorical variables that can be counted but not measured. Examples include types of star and states such as on/off.

qualitative variable Variables that can be observed but not measured. Examples include types of animals and brightness.

quality factor (QF) The number used to indicate the weighting of the biological impact of radiation.

quantitative data Numeric data from variables that can be measured. Examples include wavelength and temperature.

quantitative variable Variables that can be measured. Examples include wavelength and temperature.

R

radiation Rays or particles that carry energy. Also, the process by which energy is emitted by an object or system, transmitted through an intervening medium or space, and absorbed by another object or system.

radioactive Something that spontaneously emits radiation in the form of alpha particles, beta particles and gamma rays.

radioactive decay The process by which an unstable nucleus emits alpha or beta particles and electromagnetic radiation.

radioactive tracer A chemical compound or atom that emits radioactivity that can be used to trace the position and localisation of a molecule within an organ or tissue in the body.

radioisotope An isotope of a chemical element that emits radioactivity due to its unstable combination of neutrons and protons in the nucleus.

radiopharmaceutical One of a group of drugs that contains a radioactive tracer attached to it.

random errors Affect the precision of a measurement and may be present in all measurements. Random errors are unpredictable variations in the measurement process and result in a spread of readings.

range The difference between the highest and lowest values in a data set.

rarefaction An area of decreased pressure within a longitudinal wave.

raw data Data that has not been processed or analysed.

ray An artificially constructed line, drawn perpendicular to the wavefront, in the direction of energy transmission or transfer.

real depth The actual distance of an object beneath the surface of the water, as would be measured by submerging a ruler along with it.

reference point A point about which a rotational equilibrium can be calculated for an object that is in static equilibrium. This point can be anywhere, but is best selected to cancel the torque of an unknown force in a problem.

reflect To cause light, other electromagnetic radiation, sound, particles or waves to bounce back after reaching a boundary or surface.

refraction The bending of the direction of travel of a ray of light, sound or other wave as it enters a medium of a different density.

refractive index An index or number that is allocated to a medium and indicates its refracting properties; ratio of the speed of light in a vacuum (c) to the speed of light in the medium (v), i.e. $n = c/v$.

repeatability The closeness of the agreement between the results of successive measurements of the same quantity being measured in an experiment, carried out under the same conditions of measurement.

reproducibility The closeness of the agreement between the results of measurements of the same quantity being measured, carried out under changed conditions of measurement.

research question A question that defines the focus of an investigation.

residual current device A device that can detect a difference in the active and neutral wires and switch off current in dangerous situations to help prevent electrocution.

resistance A measure of how much an object or material resists the flow of current; the ratio of the potential difference across a circuit component and the current flowing through it: $R = V/I$. Resistance is measured in ohms (Ω).

resistor A circuit component, often used to control the amount of current in a circuit by providing a constant resistance. Resistors are ohmic conductors, i.e. they obey Ohm's law.

resultant One vector that is the sum of two or more vectors.

risk assessment Performed to identify, assess and control hazards.

rotational equilibrium A situation in which the sum of all the clockwise torques is equal to the sum of all the anticlockwise torques.

S

safety data sheet (SDS) Important information about the possible hazards in using the substance and how it should be handled and stored.

scalar A physical quantity that is represented by magnitude and units only. Mass, time and speed are examples of scalar quantities.

scientific method The process scientists use to construct theories that explain practical observations.

secondary data Data you have not collected yourself.

secondary source A synthesis, review or interpretation of primary sources. For example, textbooks or news articles.

series circuit A circuit in which components are connected one after another in a continuous loop so that the same current passes through each component.

short circuit The situation in which a good conductor is inadvertently placed across a battery. This causes an excessive current to flow, which may cause damage.

significant figures The number of digits used. For example, 5.1 has two significant figures, whereas 5 has just one significant figure.

Snell's law When light passing from one medium to another is bent, the geometry is described by the relationship $n_1 \sin\theta_1 = n_2 \sin\theta_2$ where n_1 and n_2 are the refractive indices of the media, and θ_1 and θ_2 are the angles measured from the normal.

somatic Refers to all cells in the human body, except those present in the reproductive organs (i.e. the ovaries and testes). Mutations affecting somatic cells are not passed on to offspring.

specific heat capacity The amount of energy that must be transferred to change the temperature of 1 kg of material by 1°C or 1 K.

speed The ratio of distance travelled to time taken. Speed is a scalar quantity. The SI unit for speed is m s^{-1}.

spring constant A measure of the stiffness of a spring or material. It is the gradient of a force–extension graph.

stable equilibrium A situation in which an object will return to its equilibrium position, even when it is displaced by a force. For example, a ball placed in a large bowl will always return to the bottom of the bowl.

static equilibrium A situation in which an object is in both translational equilibrium and rotational equilibrium.

strong nuclear force A short-range but powerful force of attraction that acts between all nucleons in a nucleus and holds the nucleus together.

systematic errors Cause readings to differ from the true value in a systematic manner so when a particular value is measured repeatedly, the error is the same. Systematic errors result from limitations in the instrument itself or incorrect calibration, or inappropriate methods (including parallax).

T

temperature A measure of the average kinetic energy of the particles in a substance. Temperature can be measured in degrees Celsius (°C) or in kelvin (K).

thermal contact When two objects are in contact so that energy exchange via heat transfer is possible.

thermal energy A form of energy transferred as a result of a difference in temperature or average kinetic energy within a system.

thermal equilibrium For two bodies in thermal contact, the point at which the two reach the same temperature and there is no further net transfer of thermal energy.

thermistor A non-ohmic device whose resistance varies with temperature.

torque A turning effect caused by a force acting along a line that is not directed through a pivot point or a centre of mass.

total internal reflection Occurs when the angle of incidence exceeds the critical angle for refraction. Light or waves are reflected back into the medium; there is no transmission of light.

transducer A device that receives a signal in the form of one type of energy and converts it into another form of energy.

transfer The conversion of energy from one system to another.

transform To change from one thing to another; for example, to change energy from electrical potential energy to kinetic energy.

translational equilibrium A situation in which the sum of the forces acting on an object are equal to zero.

transmit To cause light, heat, sound, etc. to pass through into a medium.

transverse Lying or extending across something. The vibrations of a transverse wave are at right angles to the direction of travel of the wave.

trough The lowest part or bottom of a transverse wave.

true value The value that would be found if the quantity could be measured perfectly.

U

uncertainty The true value is expected to lie in the interval expressed as mean ± uncertainty. For example, for the measurement result 20 ± 1 mm, the 'uncertainty' is 1 mm and the 'interval' is between 19 to 21 mm.

uncertainty bars Graphical representations on graphs to show the variability of data and thus indicate the uncertainty in a measurement.

units Properties related to physical measurements. Units can be fundamental, such as metres (m), seconds (s) or kilograms (kg), or they can be derived by combining fundamental units, such as metres per second ($m\,s^{-1}$).

unstable equilibrium A situation in which an object will accelerate and will not return to its equilibrium position when it is displaced by a force; for example, a sphere lying on top of a dome.

V

validity A valid experiment investigates what it sets out and/or claims to investigate.

variable A factor or condition that can change.

vector A physical quantity that requires magnitude, units and a direction in order to be fully defined. Velocity, acceleration and force are examples of vector quantities.

vector diagram A system of adding vectors in which each vector is drawn head-to-tail, with the resultant vector drawn from the tail of the first vector to the head of the last vector.

velocity The ratio of displacement to time taken. Velocity is a vector quantity. The SI unit for velocity is $m\,s^{-1}$.

volatile Liquids with weak surface bonds that evaporate rapidly.

volt The unit of electrical potential. One volt is equal to one joule of potential energy given to one coulomb of charge in a source of potential difference. Voltage (or the number of volts) is another name for the potential difference.

voltage divider A series circuit with two or more components and in which the voltage supplied to the circuit is shared (or divided) between the components in the circuit.

voltmeter A device used to measure the voltage change between two points in a circuit.

W

wavefront A continuous line or surface that includes all the points reached by a wave or vibration at the same instant as it travels through a medium.

wavelength The distance between one peak or crest of a wave of light, heat or other energy and the next corresponding peak or crest (symbol λ).

weak nuclear force A nuclear force that causes radioactive decay. It acts to make the nucleus more stable by rearranging the number of protons and neutrons into a more energy-favourable ratio via beta decay.

weighting factor The number used to indicate the different sensitivities towards radiation by each organ in the body (W).

Wien's law The peak wavelength of radiation emitted by an object is inversely proportional to the surface temperature of the object.

work The transfer of energy as a result of the application of a force; measured by multiplying the force and the displacement of its point of application along the line of action. Measured in joules (J).

work–energy theorem The work done on an object by a net force is equal to the change in the kinetic energy of the object.

Z

zeroth law of thermodynamics If objects A and B are each in thermal equilibrium with object C, then objects A and B are in thermal equilibrium with each other.

Index

D

E

Q

R

S

W

X

Z

Attributions

Front and back covers: Pasieka, Alfred/Science Photo Library.

123RF.com: Andrey, Pashkov, p. 443t; arenaphotouk, p. 475tr; claudiodivizia, p. 96b; Dario Lo Presti, p. 283–84; Gaca, Pawel, p. 443c; grigvovan, p. 474tl; lianem, p. 437l; maigi, p. 241; michalex, p. 474cl; NATALE, matteo, p. 257c; Sean Nel, p. 363b; verdateo, p. 434bl; xie, yuanyuan, p. 432–33.

Aaron Gold: Aaron Gold/About.com:Cars, p. 458b.

Air Liquide: p. 13.

Alamy Stock Photo: Archive Pics, p. 129r; CBW, p. 197; Chillmaid, Martyn F./Science Photo Library, p. 17b; Dunkley, Dan/Science Photo Library, p. 46; frantic, p. 4; Indiapicture, p. 251c; koene, ton, p. 10; Mittleman, Richard/Gon2Foto, p. 90t; NASA Goddard Space Flight Center/Science Photo Library, p. 58–9; Photo Researchers, p. 198r; Posnov, Viktor p. 214t; Science History Images, p. 1–3; Scientific Picture Research/ Science Photo Library, p. 12t; US National Archives and Records Administration/Science Photo Library, p. 203b; World History Archive, p. 203t; YAY Media AS, p. 355.

Australian Nuclear Science and Technology Organisation (ANSTO): OPAL Research Reactor — ANSTO, p. 187–88.

Australian Government's Department of the Environment, Water, Heritage and the Arts: © Commonwealth of Australia, p. 138.

Burrows, Keith: pp. 266b, 270t.

European Organization for Nuclear Research (CERN): "View of an open LHC interconnection" by 2005–2022 CERN is licensed under CC BY 4.0. Link to licence: https://creativecommons.org/licenses/by/4.0/.

City of Melbourne: p. 137.

Corbis: p. 356bl, Escobar, Alfredo, p. 317; ITAR-TASS, p. 171; Visuals Unlimited, p. 356tl.

DK Images: Ridley, Tim, p. 261t.

Fotolia: Akinshin, Ilya, p. 403cl; Am, p. 407rt; Amphotolt, p. 221b; Apops, p. 301tl; Bizgaimer, Yuri, p. 398t; Chestnut, Corey, p. 457b; ChrisVanLennepPhoto, p. 62bl; Coprid, p. 262bl; Cunliffe, Les, p. 383; Cutimage, p. 221t; demarco, p. 212l; destina, p. 407rb; Dolgikh, George, p. 225; Dreaming Andy, p. 267br; Ezume Images, p. 8; Flippo, Michael, p. 104l; Folsom, Bert, p. 230; Generalfmv, p. 152t; Gil, Francisco Javier p. 234b; itestro, p. 64tl; JM Fotografie, p. 223; Kadmy, p. 301bl; Koslov, Oleg, p. 384; manfredxy, p. 162b; Monasevich, Elisheva, p. 285t (hand and toy car); nobeastsofierce, p. 262br; Sahua D, p. 412r; Science Photo, p. 82tr; Scott, Ian, p. 267bl; Siegmar, p. 465b; Silver, p. 119t; Stuporter, p. 407rc; Tagstock, p. 121br; William87, p. 403bl.

Getty Images: p. 218t; AFP/Stringer, pp. 173t, 342cr; BBC Motion Gallery, p. 345t; Dr. Settles, Gary, p. 423b; Finney, Julian, p. 390l; Lennon, Bryn, p. 312; Loomis, Dean, p. 323l;

Mead, Ted, p. 126–27; NurPhoto, p. 333t; Parker, David, p. 150–51; Pasieka, p. 209; Print Collector, p. 247b; Science Photo Library, pp. 157b, 174, 198l, 412l; The Asahi Shimbun, p. 178; Traynor, Darrian/Stringer, p. 387.

Google Earth: Map data ©2022 Google, p. 285cl.

Mann, Dale: p. 62c.

NASA: Nasa Images, p. 367.

National Institute of Standards and Technology (NIST): J. L. Lee, p. 342cl.

Newspix: Hargest, Jon, p. 316c.

Pasco Media: p. 23.

Pearson Education Ltd.: pp. 235, 418.

Rice University: Linda Welzenbach, Rice University/Permut number FCO UK No. 29/2018. Seal tagging was part of ITGC collaboration with NSF science cruise NBP19-02), p. 111.

Science Photo Library: Andrew Lambert Photography, p. 94; Bartel, Alex, p. 270b; Chumack, John, p. 361b; Dlugokencky, Ed, NOAA/GML, p. 28; Dr. Marzzi, P., p. 182; Dumas, Patrick, p. 269cr; Fraser, Simon, p. 179r; Giphotostock, pp. 84bl; GustoImages, p. 222tl; Hardy, David A., p. 134; Hincks, Gary, p. 139t; Lawrence Berkeley Laboratory, pp. 56–7, 147–49, 206–08, 275–79; Lawrence Livermore National Laboratory, pp. 203c; LMSAL/Stanford University/NASA, p. 77t; Loren Winters/Visuals Unlimited, p. 344l; Macrovector, p. 76t; NASA, p. 323r; Public Health England, p. 163br; Ria Novosti, p. 163tr; Science Photo Library, p. 346; Sirulnikoff, Alan, p. 11t; Trevor Clifford Photography, pp. 224, 255b; U.S. Department of Energy, p. 193t; Van Ravenswaay, Detlev, p. 60l; Van Sant, Tom, p. 142t; Victor Habbick Visions, pp. 280–81, 477–83.

Shutterstock.com: 3Dsculptor, p. 397; A.RICARDO, pp. 403tl, 403tc; AGorohov, pp. 434br, 435t, 435c, 435b, 436t, 436b; Allison, Shelby, p. 399; Andrushko, Galyna, p. 356tc; ArliftAtoz2205, p. 453; bikeriderlondon, p. 293rt; Brave, Greg, p. 132t; Cecconi, Michel, p. 120b; Chairak, Jaruek, p. 419; Cholakov, Ivan, p. 313t; chotirat, ommaphat, p. 459; Cousland, Neale, p. 377; Crone, Benjamin, p. 276t; CRStocker, p. 93t; Cuson, p. 81; Frentz, Hank, p. 181; gilardelli, angelo, p. 370l; Gojda, Lukas, p. 100–01; Greg Brave, p. 467; Hodgkinson, Timothy, p. 259t; itsjustme, p. 77b; jdwfoto, p. 474bl; Kalinovsky, Dmitry, p. 440b; Kalinovsky, Dmitry, p. 442b; kamilpetran, p. 122l; kiya, saied shahin, p. 76b; Kneschke, Robert, p. 60r; Kostsov, p. 474tr;liu, fuyu, p. 308–9; Lund, Jacob, p. 285cr; M.Aurelius, p. 285t (astronaut); Maxisport, p. 454 (high jumper); mimagephotography, pp. 401t, 401b; MXW Stock, p. 91l; Neliubov, Iaroslav, p. 356br; nuwatphoto, p. 475br; Perugini, William, pp. 446t, 446b; PeterPhoto123, p. 89tl; Photobac, p. 415t; Plumridge, Claire, p. 434t; Postnikova, Kristina, p. 95rt; Sangasaeng, Atiketta, p. 89br; SeventyFour, p. 122r;

Speed, Beverly, p. 74; Surakit, pp. 437r, 440t, 442t; Thieury, p. 393bl; Thirteen, p. 356tr; violetkaipa, p. 226; YMZK-Photo, p. 251t.

The Library of Congress: Prints & Photographs Division, LC-DIG-ggbain-36570, p. 157t.

Thermal Image UK: Tom Barbour, www.thermalimageuk.com, p. 104r.

United Nations Economic Commission for Europe (UNECE): p. 12b.

University of California: Drs. Landau, Drs Susan M and Jagust, William, University of California, Berkeley, p. 180.

Victorian Curriculum and Assessment Authority (VCAA): Selected extracts from the VCE Physics Study Design (2023–2027) are copyright Victorian Curriculum and Assessment Authority (VCAA), reproduced by permission. VCE® is a registered trademark of the VCAA. The VCAA does not endorse this product and makes no warranties regarding the correctness or accuracy of its content. To the extent permitted by law, the VCAA excludes all liability for any loss or damage suffered or incurred as a result of accessing, using or relying on the content. Current VCE Study Designs and related content can be accessed directly at www.vcaa.vic.edu.au, pp. 1–3, 57, 59, 81, 101, 127, 151, 188, 209, 241, 281, 284, 309, 397, 433.

Wikicommons: Bielasko, p.443b.

Wikipedia: United States Department of Energy, p. 199bc.